REAGENTS FOR TRANSITION METAL COMPLEX AND ORGANOMETALLIC SYNTHESES

INORGANIC
SYNTHESES

Volume 28

Board of Directors

JOHN P. FACKLER, JR.   *Texas A & M University*
BODIE E. DOUGLAS   *University of Pittsburgh*
DARYLE H. BUSCH   *University of Kansas*
JAY H. WORRELL   *University of South Florida*
HERBERT D. KAESZ   *University of California, Los Angeles*
ALVIN P. GINSBERG   *AT&T Bell Laboratories*

Future Volumes

29 RUSSELL N. GRIMES   *University of Virginia*
30 LEONARD V. INTERRANTE   *Rensselaer Polytechnic Institute*
    and DONALD MURPHY   *AT&T Bell Laboratories*
31 ALAN H. COWLEY   *University of Texas*
32 MARCETTA Y. DARENSBOURG   *Texas A & M University*

International Associates

MARTIN A. BENNETT   *Australian National University, Canberra*
FAUSTO CALDERAZZO   *University of Pisa*
E. O. FISCHER   *Technical University, Munich*
JACK LEWIS   *Cambridge University*
LAMBERTO MALATESTA   *University of Milan*
RENE POILBLANC   *University of Toulouse*
HERBERT W. ROESKY   *University of Göttingen*
F. G. A. STONE   *University of Bristol*
AKIO YAMAMOTO   *Tokyo Institute of Technology, Yokohama*

Editor-in-Chief
ROBERT J. ANGELICI
Department of Chemistry
Iowa State University

REAGENTS FOR TRANSITION METAL COMPLEX AND ORGANOMETALLIC SYNTHESES

INORGANIC SYNTHESES

Volume 28

A WILEY-INTERSCIENCE PUBLICATION
John Wiley & Sons, Inc.
NEW YORK / CHICHESTER / BRISBANE / TORONTO / SINGAPORE

In recognition of the importance of preserving what has been written, it is a policy of John Wiley & Sons, Inc. to have books of enduring value published in the United States printed on acid-free paper, and we exert our best efforts to that end.

Published by John Wiley & Sons, Inc.

Copyright © 1990 Inorganic Syntheses, Inc.

All rights reserved. Published simultaneously in Canada.

Reproduction or translation of any part of this work beyond that permitted by Section 107 or 108 of the 1976 United States Copyright Act without the permission of the copyright owner is unlawful. Requests for permission or further information should be addressed to the Permissions Department, John Wiley & Sons, Inc.

Library of Congress Catalog Number: 39-23015
ISBN 0-471-52619-3

Printed in the United States of America

10 9 8 7 6 5 4 3 2 1

This volume is dedicated to Professor Fred Basolo on the occasion of his 70th birthday. We are all indebted to him for his many contributions to the field of inorganic chemistry.

PREFACE

This is a special volume of *Inorganic Syntheses* that focuses on complexes that are likely to be useful as starting materials for the preparations of new transition metal coordination and organometallic compounds. There are chapters on complexes with weakly coordinated and therefore easily displaced ligands, low-valent complexes that undergo oxidative–addition reactions, substituted metal carbonyl complexes, nucleophilic metal carbonyl anions, transition metal clusters, a variety of cyclopentadienyl complexes, lanthanide and actinide complexes, and a range of other useful ligands and complexes.

Most of the syntheses are taken from previous volumes of *Inorganic Syntheses*; however, in each case the original author(s), or in one case someone very familiar with the synthesis, was asked to correct errors, add safety notes, or make changes in the references. In several instances, experimental procedures were modified; where these modifications were substantive, the new procedures were repeated by checkers. There are nine totally new syntheses that I solicited, and these were checked in the usual manner.

This project began in 1983 when Duward F. Shriver, who was President of Inorganic Syntheses, Inc. at the time, asked me to chair a committee with Russell N. Grimes and Herbert W. Roesky as members to explore the advisability of publishing a collective topical volume of *Inorganic Syntheses* on "basic starting materials." The committee recommended that a volume on starting materials for the synthesis of transition metal complexes be published. I agreed to serve as Editor-in-Chief of the volume and solicited opinions of all members of *Inorganic Syntheses* for suggestions of syntheses that should be included. From these many very helpful comments, I compiled a list of preparations, most of which are in this volume. The compounds were

---

Previous volumes of *Inorganic Syntheses* are available. Many of the volumes originally published by McGraw Hill, Inc. are available from R. E. Krieger Publishing Co., Inc., P.O. Box 9542, Melbourne, FL 32901. Please write this publisher for a current list. Volumes out of print with John Wiley & Sons, Inc. are also available from Krieger Publishing. Recent back volumes can be obtained from John Wiley & Sons, Inc., 605 Third Avenue, New York, NY 10158. Please write the publisher for a current list of available volumes.

chosen with the thought in mind that this volume would be a primary source of synthetic procedures for those who have need to prepare transition metal complexes.

The success of a volume like this depends upon a large number of people—those whose manuscripts are included, those who made suggestions, those who prepared new manuscripts, and especially those who verified the procedures by repeating them. Their cooperation made my task easier than it might have been. I am grateful to Thomas Sloan for compiling the indices. Also, many thanks to Carla Holbrook who provided cheerful, accurate, and efficient secretarial support of this project.

ROBERT J. ANGELICI

Ames, Iowa
September 1989

TOXIC SUBSTANCES AND LABORATORY HAZARDS

Chemicals and chemistry are by their very nature hazardous. Chemical reactivity implies that reagents have the ability to combine. This process can be sufficiently vigorous as to cause flame, an explosion, or, often less immediately obvious, a toxic reaction.

The obvious hazards in the syntheses reported in this volume are delineated, where appropriate, in the experimental procedure. It is impossible, however, to foresee every eventuality, such as a new biological effect of a common laboratory reagent. As a consequence, *all* chemicals used and *all* reactions described in this volume should be viewed as potentially hazardous. Care should be taken to avoid inhalation or other physical contact with all reagents and solvents used in procedures described in this volume. In addition, particular attention should be paid to avoiding sparks, open flames, or other potential sources that could set fire to combustible vapors or gases.

A list of 400 toxic substances may be found in the *Federal Register*, Vol. 40, No. 23072, May 28, 1975. An abbreviated list may be obtained from *Inorganic Syntheses*, Volume 18, p. xv, 1978. A current assessment of the hazards associated with a particular chemical is available in the most recent edition of *Threshold Limit Values for Chemical Substances and Physical Agents in the Workroom Environment* published by the American Conference of Governmental Industrial Hygienists.

The drying of impure ethers can produce a violent explosion. Further information about this hazard may be found in *Inorganic Syntheses*, Volume 12, p. 317. A hazard associated with the synthesis of tetramethyldiphosphine disulfide [*Inorg. Synth.*, **15**, 186 (1974)] is cited in *Inorganic Syntheses*, Volume 23, p. 199.

NOTICE TO CONTRIBUTORS AND CHECKERS

The *Inorganic Syntheses* series is published to provide all users of inorganic substances with detailed and foolproof procedures for the preparation of important and timely compounds. Thus the series is the concern of the entire scientific community. The Editorial Board hopes that all chemists will share in the responsibility of producing *Inorganic Syntheses* by offering their advice and assistance in both the formulation of and the laboratory evaluation of outstanding syntheses. Help of this kind will be invaluable in achieving excellence and pertinence to current scientific interests.

There is no rigid definition of what constitutes a suitable synthesis. The major criterion by which syntheses are judged is the potential value to the scientific community. For example, starting materials or intermediates that are useful for synthetic chemistry are appropriate. The synthesis also should represent the best available procedure, and new or improved syntheses are particularly appropriate. Syntheses of compounds that are available commercially at reasonable prices are not acceptable. We do not encourage the submission of compounds that are unreasonably hazardous, and in this connection, less dangerous anions generally should be employed in place of perchlorate.

The Editorial Board lists the following criteria of content for submitted manuscripts. Style should conform with that of previous volumes of *Inorganic Syntheses*. The introductory section should include a concise and critical summary of the available procedures for synthesis of the product in question. It should also include an estimate of the time required for the synthesis, an indication of the importance and utility of the product, and an admonition if any potential hazards are associated with the procedure. The Procedure should present detailed and unambiguous laboratory directions and be written so that it anticipates possible mistakes and misunderstandings on the part of the person who attempts to duplicate the procedure. Any unusual equipment or procedure should be clearly described. Line drawings should be included when they can be helpful. All safety measures should be stated clearly. Sources of unusual starting materials must be given, and, if possible, minimal standards of purity of reagents and solvents should be stated. The scale should be reasonable for normal laboratory operation, and any problems involved in scaling the procedure either up or down should be discussed. The criteria for judging the purity of the final product should be delineated clearly. The section on Properties should supply and discuss those physical and chemical characteristics that are relevant to judging the purity of the product and to permitting its handling and use in an intelligent manner. Under References, all pertinent literature citations should be listed in order. A style sheet is available from the Secretary of the Editorial Board. Authors are requested to avoid procedures involving perchlorate salts due to the high risk of explosion in combination with organic or organometallic substances. Authors are also requested to avoid the use of solvents known to be carcinogenic. As stated above, adequate warning must be given when potentially hazardous materials are associated with the synthesis.

The Editorial Board determines whether submitted syntheses meet the general specifications outlined above, and the Editor-in-Chief sends the manuscripts to an independent laboratory where the procedure must be satisfactorily reproduced.

Each manuscript should be submitted in duplicate to the Secretary of the Editorial Board, Professor Jay H. Worrell, Department of Chemistry, University of South Florida, Tampa, FL 33620. The manuscript should be typewritten in English. Nomenclature should be consistent and should follow the recommendations presented in *Nomenclature of Inorganic Chemistry*, 2nd ed., Butterworths & Co., London, 1970, and in *Pure and Applied Chemistry*, Volume 28, No. 1 (1971). Abbreviations should conform to those used in publications of the American Chemical Society, particularly *Inorganic Chemistry*.

Chemists willing to check syntheses should contact the editor of a future volume or make this information known to Professor Worrell.

CONTENTS

Chapter One COMPLEXES WITH WEAKLY COORDINATED LIGANDS

1. Metal Complexes with Weakly Bonded Anions: General Considerations.. 1
2. Carbonyl(η^5-cyclopentadienyl)(tetrafluoroborato)molybdenum and -tungsten complexes: CpM(CO)$_2$(L)(BF$_4$) (M = Mo, W; L = CO, PPh$_3$).. 5
   - A. Tricarbonyl(η^5-cyclopentadienyl)(tetrafluoroborato) molybdenum and -tungsten........................... 5
   - B. Dicarbonyl(η^5-cyclopentadienyl)(tetrafluoroborato) (triphenylphosphine)molybdenum and -tungsten, MCp(CO)$_2$(PPh$_3$)(FBF$_3$) (M = Mo, W)................ 7
   - C. Tricarbonyl(η^5-cyclopentadienyl)(η^2-ethene)molybdenum (1 +) Tetrafluoroborate(1 −)........................... 11
   - D. Carbonyl(η^5-cyclopentadienyl)bis(diphenylacetylene) molybdenum (1 +) Tetrafluoroborate(1 −)............. 11
   - E. Carbonyl(η^5-cyclopentadienyl)(diphenylacetylene) (triphenylphosphine) molybdenum(1 +) Tetrafluoroborate(1 −)................................ 13
   - F. (Acetone)(tricarbonyl)(η^5-cyclopentadienyl)molybdenum- (1 +) and -tungsten(1 +) Tetrafluoroborate(1 −)....... 14
3. Pentacarbonyl(tetrafluoroborato)rhenium and -manganese and Reactions Thereof: M(CO)$_5$(BF$_4$) (M = Mn, Re)............. 15
   - A. Pentacarbonylmethylrhenium: Re(CO)$_5$Me.............. 16
   - B. Pentacarbonyl(tetrafluoroborato)rhenium(I)............. 17
   - C. Pentacarbonyl(η^2-ethene)rhenium(1 +) Tetrafluoroborate(1 −): Re(CO)$_5$(η^2-C$_2$H$_4$)$^+$...................... 19
   - D. Octadecacarbonyl-bis(μ_3-carbon dioxide)-tetrarhenium: Re$_4$(CO)$_{18}$(CO$_2$)$_2$...................................... 20
4. Iridium(III) Complexes with the Weakly Bonded Anions [BF$_4^-$] and [OSO$_2$CF$_3^-$].. 22
   - A. Carbonylchlorohydrido(tetrafluoroborato)bis(triphenylphosphine)iridium(III): Ir(Cl)(CO)(PPh$_3$)$_2$(H)(BF$_4$)   23

B. Carbonylchloromethyl(tetrafluoroborato)bis(triphenylphosphine)iridium(III): Ir(Cl)(CO)(PPh$_3$)$_2$(CH$_3$)(BF$_4$) 24
C. Chloro(dinitrogen)hydrido(tetrafluoroborato)bis(triphenylphosphine)iridium(III): Ir(Cl)(N$_2$)(PPh$_3$)$_2$(H)(BF$_4$). 25
D. Carbonylhydridobis(trifluoromethanesulfonato)bis(triphenylphosphine)iridium(III): Ir(CO)(PPh$_3$)$_2$(H)(O$_3$SCF$_3$)$_2$ 26
5. cis-Chlorobis(triethylphosphine)(trifluoromethanesulfonato)platinum(II) 27
6. Tricarbonyltris(nitrile) Complexes of Cr, Mo, and W: M(CO)$_3$(NCEt)$_3$ (M = Cr, Mo, W) 29
   A. Tricarbonyltris(propionitrile)tungsten(0) 30
   B. Tricarbonyltris(propionitrile)molybdenum(0). 31
   C. Tricarbonyltris(propionitrile)chromium(0) 32
7. The Synthesis of Molybdenum and Tungsten Dinitrogen Complexes: MoCl$_4$(L)$_2$, MoCl$_3$(L)$_3$ (L = NCMe, THF), Mo(N$_2$)$_2$(dppe)$_2$, W(N$_2$)$_2$(dppe)$_2$ 33
   A. Bis(acetonitrile)tetrachloromolybdenum(IV), [MoCl$_4$(CH$_3$CN)$_2$] 34
   B. Tetrachlorobis(tetrahydrofuran)molybdenum(IV), [MoCl$_4$(THF)$_2$] 35
   C. Trichlorotris(tetrahydrofuran)molybdenum(III), [MoCl$_3$(THF)$_3$] 36
   D. Trichlorotris(acetonitrile)molybdenum(III), [MoCl$_3$(MeCN)$_3$] 37
   E. trans-Bis(dinitrogen)bis[1,2-ethanediylbis(diphenylphosphine)]molybdenum(0), trans-[Mo(N$_2$)$_2$(Ph$_2$PCH$_2$CH$_2$PPh$_2$)$_2$] 38
   F. Tetrachlorobis(triphenylphosphine)tungsten(IV), [WCl$_4$(PPh$_3$)$_2$]. 40
   G. Tetrachloro[1,2-ethanediylbis(diphenylphosphine)]tungsten(IV), [WCl$_4$(Ph$_2$PCH$_2$CH$_2$PPh$_2$)] ... 41
   H. trans-Bis(dinitrogen)bis[1,2-ethanediylbis(diphenylphosphine)]tungsten(0), trans-[W(N$_2$)$_2$(Ph$_2$PCH$_2$CH$_2$PPh$_2$)$_2$] 41
8. trans-Bis[1,2-ethanediylbis(diphenylphosphine)]bis(isocyanomethane)tungsten(0): trans-[W(CNR)$_2$(Ph$_2$PCH$_2$CH$_2$PPh$_2$)$_2$] 43
9. Tricarbonyl(cycloheptatriene)molybdenum(0): Mo(CO)$_3$(C$_7$H$_8$) 45
10. Tetracarbonyl(η^2-Methyl Acrylate)Ruthenium: Ru(CO)$_4$(η^2-CH$_2$=C(H)(CO$_2$Me)). 47
11. Reagents for the Synthesis of η-Diene Complexes of Tricarbonyliron and Tricarbonylruthenium 52

	A.	(Benzylideneacetone)tricarbonyliron(0)	52
	B.	(1,5-Cyclooctadiene)tricarbonylruthenium(0)..............	54
12.	Dihydridobis(solvent)bis(triphenylphosphine)iridium(III) Tetrafluoroborates: Ir(COD)(PPh$_3$)$_2^+$, Ir(H)$_2$(PPh$_3$)$_2$(L)$_2^+$ (L = acetone, H$_2$O)	56
	A.	(η^4-1,5-Cyclooctadiene)bis(triphenylphosphine)iridium(I) Tetrafluoroborate(1 −)	56
	B.	Bis(acetone)dihydridobis(triphenylphosphine)iridium(III) Tetrafluoroborate(1 −)	57
	C.	Diaquadihydridobis(triphenylphosphine)iridium(III) Tetrafluoroborate(1 −)	58
	D.	(1,2-Diiodobenzene)dihydridobis(triphenylphosphine)- iridium(III) Tetrafluoroborate(1 −)	59
13.	Bis(benzonitrile)dichloro Complexes of Palladium and Platinum: MCl$_2$(NCPh)$_2$ (M = Pd, Pt)......................	60
	A.	Bis(benzonitrile)dichloropalladium.......................	61
	B.	Bis(benzonitrile)dichloroplatinum........................	62
14.	Acetonitrile Complexes of Selected Transition Metal Cations: Pd(NCMe)$_4^{2+}$, M(NO)$_2$(NCMe)$_4^{2+}$ (M = Mo, W).............	63
	A.	Tetrakis(acetonitrile)palladium(2 +) Bis[tetrafluoroborate- (1 −)]...	63
	B.	cis-Tetrakis(acetonitrile)dinitrosylmolybdenum(2 +) Bis[tetrafluoroborate(1 −)].............................	65
	C.	cis-Tetrakis(acetonitrile)dinitrosyltungsten(2 +) Bis[tetrafluoroborate(1 −)].............................	66
15.	Tetrakis(acetonitrile)copper(1 +) Hexafluorophosphate(1 −): [Cu(NCMe)$_4$]$^+$	68
16.	Trifluoromethanesulfonates and Trifluoromethanesulfonato-O Complexes	70
	A.	Triflate Salts from Chloride Salts.......................	72
	B.	Triflate Salts from Sulphate Salts.......................	73
	C.	Triflate Salts Using Silver Triflate	73
	D.	Triflato Complexes from Chloro Complexes.............	74
	E.	Triflato Complexes from Solid State Reactions of Triflato Salts	75
	F.	Regeneration of Triflato Complexes	76

**Chapter Two   LOW-VALENT COMPLEXES OF
Rh, Ir, Ni, Pd, and Pt**

17.	Chlorotris(triphenylphosphine)rhodium(I) (Wilkinson's Catalyst): Rh(Cl)(PPh$_3$)$_3$..................................	77

18. *trans*-Carbonylchlorobis(triphenylphosphine)rhodium and
    Related Complexes: Rh(Cl)(CO)(PPh$_3$)$_2$ 79
19. Hydrido Phosphine Complexes of Rhodium(I): Rh(H)(PPh$_3$)$_4$,
    Rh(H)(CO)(PPh$_3$)$_3$ 81
    A. Hydridotetrakis(triphenylphosphine)rhodium(I) 81
    B. Carbonylhydridotris(triphenylphosphine)rhodium(I) 82
20. Tetracarbonyldichlorodirhodium: [Rh(Cl)(CO)$_2$]$_2$ 84
21. Di-μ-chlorotetrakis(ethene)dirhodium(I): [Rh(Cl)(C$_2$H$_4$)$_2$]$_2$ 86
22. Di-μ-chloro-bis(η^4-1,5-cyclooctadiene)dirhodium(I):
    [Rh(Cl)(COD)]$_2$ 88
23. Chlorobis(cyclooctene)rhodium(I) and -iridium(I) Complexes:
    [M(Cl)(C$_8$H$_{14}$)$_2$]$_2$(M = Rh, Ir) 90
    A. Chlorobis(cyclooctene)rhodium(I) 90
    B. Di-μ-chlorobis[bis(cyclooctene)iridium] 91
24. *trans*-Carbonylchlorobis(triphenylphosphine)iridium:
    Ir(Cl)(CO)(PPh$_3$)$_2$ 92
25. Bis(1,5-cyclooctadiene)nickel(0): Ni(COD)$_2$ 94
26. Complexes of Nickel(0): Ni(CN-*t*-Bu)$_4$, Ni(PPh$_3$)$_4$ 98
    A. Tetrakis(*tert*-butyl isocyanide)nickel(0) 99
    B. Tetrakis(triphenylphosphine)nickel(0). 102
27. Tetrakis(triethyl phosphite)nickel(0), palladium(0), and
    platinum(0) Complexes: M[P(OEt)$_3$]$_4$ 104
    A. Tetrakis(triethyl phosphite)nickel(0). 104
    B. Tetrakis(triethyl phosphite)palladium(0) 105
    C. Tetrakis(triethyl phosphite)platinum(0) 106
28. Tetrakis(triphenylphosphine)palladium(0): Pd(PPh$_3$)$_4$ 107
29. Tetrakis(*tert*-butyl isocyanide)di-μ-chlorodipalladium(I):
    Pd(dba)$_2$, Pd$_2$Cl$_2$(CN-*t*-Bu)$_4$ 110
30. Two-Coordinate Phosphine Complexes of Palladium(0) and
    Platinum(0): M(PR$_3$)$_2$(M = Pd, Pt) 113
    A. Bis(di-*tert*-butylphenylphosphine)palladium(0) 114
    B. Bis(tricyclohexylphosphine)palladium(0). 114
    C. Bis(tri-*tert*-butylphosphine)palladium(0). 115
    D. Bis(di-*tert*-butylphenylphosphine)platinum(0) 116
    E. Bis(tricyclohexylphosphine)platinum(0). 116
31. Three-Coordinate Phosphine Complexes of Platinum(0):
    Pt(PR$_3$)$_3$ 119
    A. Tris(triethylphosphine)platinum(0) 120
    B. Tris(triisopropylphosphine)platinum(0) 120
32. Tetrakis(triethylphosphine)platinum(0): Pt(PEt$_3$)$_4$ 122
33. Tris- and tetrakis(triphenylphosphine)platinum(0): Pt(PPh$_3$)$_{3,4}$ 123
    A. Tetrakis(triphenylphosphine)platinum(0). 124
    B. Tris(triphenylphosphine)platinum(0) 125

34. Olefin Complexes of Platinum: $Pt(COD)_2$, $Pt(C_2H_4)_3$, $Pt(C_2H_4)_2(PR_3)$ 126
   A. Bis(1,5-cyclooctadiene)platinum(0) 126
      1. (1,3,5,7-Cyclooctatetraene)dilithium ($Li_2C_8H_8$) 127
      2. Tris(bicyclo[2.2.1]heptene)platinum(0) 127
      3. Bis(1,5-cyclooctadiene)platinum(0). 128
   B. Tris(ethene)platinum(0) 129
   C. Bis(ethene)(tricyclohexylphosphine)platinum(0). 130
35. Ethene Complexes of Bis(trialkylphosphine)platinum(0): $Pt(C_2H_4)(PR_3)_2$ 132
   A. (Ethene)bis(triethylphosphine)platinum(0). 133
   B. (Ethene)bis(phosphine)platinum(0) 135

Chapter Three SUBSTITUTED METAL CARBONYL COMPLEXES

36. (η^6-Arene)tricarbonylchromium Complexes: $Cr(\eta^6$-Arene)$(CO)_3$ 136
   A. (η^6-Anisole)tricarbonylchromium 137
   B. Other Arenetricarbonylchromium Complexes. 138
37. Zero-Valent Isocyanide Complexes of Chromium, Molybdenum, and Tungsten: $M(CO)_x(CNR)_{6-x}$ (M = Cr, Mo, W)... 140
   A. $[M(CO)_{6-n}(CNBu^t)_n]$ (M = Cr, Mo, W; n = 1–3) Using Cobalt(II) Chloride as Catalyst 142
   B. $[M(CO)_{6-n}(CNBu^t)_n]$ (M = Cr, Mo, W; n = 1–3) Using Palladium(II) Oxide as Catalyst...................... 144
   C. Other Zero-Valent Isocyanide Complexes of Chromium, Molybdenum and Tungsten 144
38. cis-Bis(diethyldithiocarbamato)dinitrosylmolybdenum, Dibromotetracarbonylmolybdenum(II), and Di- and Tricarbonylbis(diethyldithiocarbamato)molybdenum(II): $Mo(Br)_2(CO)_4$, $Mo(CO)_{2,3}[S_2CNEt_2]_2$, $Mo(NO)_2[S_2CNEt_2]_2$ 145
39. Hexacarbonylbis(η^5-cyclopentadienyl)dichromium, molybdenum, and tungsten and Their Analogs: $M_2(\eta^5$-$C_5H_4R)_2(CO)_6$ (M = Cr, Mo, and W; R = H, Me, or $PhCH_2$). 148
40. Cyclopentadienyl Metal Carbonyl Dimers of Molybdenum and Tungsten: $[CpM(CO)_3]_2$, $[CpM(CO)_2]_2$ (M = Mo, W) 150
   A. Hexacarbonylbis(η^5-cyclopentadienyl)dimolybdenum(Mo–Mo)....................................... 151
   B. Tetracarbonylbis(η^5-cyclopentadienyl)dimolybdenum-(Mo≡Mo) 152
41. Pentacarbonylmanganese Halides: $Mn(CO)_5X$ (X = Cl, Br, I) . 154
   A. Pentacarbonylchloromanganese, $MnCl(CO)_5$ 155

	B. Bromopentacarbonylmanganese, $MnBr(CO)_5$	156
	C. Pentacarbonyliodomanganese, $MnI(CO)_5$	157
	D. Pentacarbonyliodomanganese, Alternate Procedure	158
42.	Pentacarbonylrhenium Halides: $Re(CO)_5X$ (X = Cl, Br, I)	160
	A. Pentacarbonylchlororhenium, $[ReCl(CO)_5]$	161
	B. Pentacarbonylchlororhenium (Alternate Procedure)	161
	C. Bromopentacarbonylrhenium, $[ReBr(CO)_5]$	162
	D. Pentacarbonyliodorhenium, $[ReI(CO)_5]$	163
43.	Pentacarbonylhydridorhenium: $Re(CO)_5H$	165
44.	Tetracarbonyliron(0) Complexes Containing Group V Donor Ligands: $Fe(CO)_4(PR_3)$	168
	A. Tetracarbonyl(triphenylphosphine)iron(0)	170
	B. Other Tetracarbonyl(Group V Donor Ligand)iron(0) Complexes	170
45.	Bis(phosphine) Derivatives of Iron Pentacarbonyl and Tetracarbonyl(tri-*tert*-butylphosphine)iron(0): $Fe(CO)_3(PR_3)_2$	173
	A. Tricarbonylbis(triphenylphosphine)iron(0)	176
	B. Tricarbonylbis(tricyclohexylphosphine)iron(0)	176
	C. Tricarbonylbis(tri-*n*-butylphosphine)iron(0)	177
	D. Tricarbonylbis(trimethylphosphine)iron(0)	177
	E. Tetracarbonyl(tri-*tert*-butylphosphine)iron(0)	177
46.	Zero-Valent Iron Isocyanide Complexes: $Fe(CO)_x(CNR)_{5-x}$	179
	A. Tetracarbonyl(2-isocyano-1,3-dimethylbenzene)iron(0)	180
	B. Tricarbonylbis(2-isocyano-1,3-dimethylbenzene)iron(0)	181
	C. Dicarbonyltris(2-isocyano-1,3-dimethylbenzene)iron(0)	182
	D. Carbonyltetrakis(2-isocyano-1,3-dimethylbenzene)iron(0)	183
	E. Pentakis(2-isocyano-1,3-dimethylbenzene)iron(0)	184
47.	Dicarbonyl(η^5-cyclopentadienyl)(thiocarbonyl)iron(1+) Trifluoromethanesulfonate (1−) and Dicarbonyl(η^5-cyclopentadienyl)[methylthio)thiocarbonyl]iron: $CpFe(CO)_2(CS)^+$, $CpFe(CO)_2[C(=S)SMe]$	186
48.	Tetracarbonylbis(η^5-cyclopentadienyl)diruthenium: $[CpRu(CO)_2]_2$	189

Chapter Four   METAL CARBONYL ANION COMPLEXES

49.	Sodium Hexacarbonylniobate(1−): $Nb(CO)_6^-$	192
50.	Dicarbonyl(η^5-cyclopentadienyl)nitrosyl Complexes of Chromium, Molybdenum, and Tungsten: $CpM(CO)_3^-$, $CpM(CO)_2(NO)$ (M = Cr, Mo, W)	196

51. Acetylpentacarbonylmanganese and Acetylpentacarbonylrhenium: $M(CO)_5^-$, $M(CO)_5[C(=O)Me]$ (M = Mn, Re) 199
   A. Acetylpentacarbonylmanganese 199
   B. Acetylpentacarbonylrhenium 201
52. Sodium Carbonyl Ferrates, $Na_2[Fe(CO)_4]$, $Na_2[Fe_2(CO)_8]$, and $Na_2[Fe_3(CO)_{11}]$. Bis[μ-nitrido-bis(triphenylphosphorus)(1 +)] Undecacarbonyltriferrate(2 −), $[(Ph_3P)_2N]_2[Fe_3(CO)_{11}]$ 203
53. Dicarbonyl(η^5-cyclopentadienyl)(2-methyl-1-propenyl-κC^1)iron and Dicarbonyl(η^5-cyclopentadienyl)(η^2-2-methyl-1-propene)iron(1 +) Tetrafluoroborate(1 −): $CpFe(CO)_2^-$, $CpFe(CO)_2[CH_2{=}CMe_2]^+$ 207
   A. Dicarbonyl(η^5-cyclopentadienyl)(2-methyl-1-propenyl-κC^1)iron 208
   B. Dicarbonyl(η^5-cyclopentadienyl)(η^2-2-methyl-1-propene)iron(1 +) Tetrafluoroborate 210
54. μ-Nitrido-bis(triphenylphosphorus)(1 +) Tetracarbonylrhodate(1 −) and μ-Nitrido-bis(triphenylphosphorus)(1 +) Tetracarbonyliridate(1 −): $[(Ph_3P)_2N][M(CO)_4]$ (M = Rh, Ir) 211
   A. μ-Nitrido-bis(triphenylphosphorus)(1 +) Tetracarbonylrhodate(1 −) 213
   B. μ-Nitrido-bis(triphenylphosphorus)(1 +) Tetracarbonyliridate(1 −) 214

Chapter Five METAL CLUSTER COMPLEXES

55. Polynuclear Ruthenium Complexes: $Ru_3(CO)_{12}$, $H_4Ru_4(CO)_{12}$ 216
   A. Dodecacarbonyltriruthenium, $Ru_3(CO)_{12}$ 216
   B. Dodecacarbonyltetra(μ-hydrido)tetraruthenium, $Ru_4(\mu\text{-H})_4(CO)_{12}$ 219
56. Tri- and Tetranuclear Carbonyl-Ruthenium Clusters Containing Isocyanide, Tertiary Phosphine and Phosphite Ligands. Radical Ion-Initiated Substitution of Metal Cluster Carbonyl Complexes Under Mild Conditions: $Ru_3(CO)_{12-x}(L)_x$, $H_4Ru_4(CO)_{12-x}(L)_x$ 221
   A. Sodium Benzophenone Ketyl Solution 222
   B. Undecacarbonyl(dimethylphenylphosphine)triruthenium, $Ru_3(CO)_{11}(PMe_2Ph)$ 223
   C. Decacarbonyl(dimethylphenylphosphine)(2-isocyano-2-methylpropane)triruthenium, $Ru_3(CO)_{10}(CNBu^t)(PMe_2Ph)$ 224

D. Decacarbonyl[methylenebis(diphenylphosphine)]triruthenium, $Ru_3(CO)_{10}[(Ph_2P)_2CH_2]$ 225
E. [μ-Ethynediyl-bis(diphenylphosphine)]bis[undecacarbonyltriruthenium], $[Ru_3(CO)_{11}]_2[\mu-C_2(PPh_2)_2]$ 226
F. Undecacarbonyltetrahydrido[tris(4-methylphenyl)phosphite]tetraruthenium, $Ru_4H_4(CO)_{11}[P(OC_6H_4Me-p)_3]$ 227
G. Decacarbonyl(dimethylphenylphosphine)tetrahydrido[tris-(4-methylphenyl)phosphite]tetraruthenium, $Ru_4H_4(CO)_{10}(PMe_2Ph)[P(OC_6H_4Me-p)_3]$ 228
57. Dodecacarbonyltriosmium: $Os_3(CO)_{12}$ 230
58. Some Useful Derivatives of Dodecacarbonyltriosmium: $Os_3(CO)_{11}(NCMe)$, $Os_3(CO)_{10}(NCMe)_2$ 232
  A. (Acetonitrile)undecacarbonyltriosmium................... 232
  B. Undecacarbonyl(pyridine)triosmium 234
  C. Bis(acetonitrile)decacarbonyltriosmium.................. 234
59. μ-Nitrido-bis(triphenylphosphorus)(1 +)μ-carbonyl-decacarbonyl-μ-hydridotriosmate(1 −): $HOs_3(CO)_{11}^-$ 236
60. Decacarbonyldi-μ-hydridotriosmium: $Os_3(\mu-H)_2(CO)_{10}$ 238
61. Dodecacarbonyltetra-μ-hydrido-*tetrahedro*-tetraosmium: $H_4Os_4(CO)_{12}$... 240
62. Tri-μ-carbonyl-nonacarbonyltetrarhodium: $Rh_4(CO)_{12}$........ 242
63. Dodecacarbonyltetrairidium: $Ir_4(CO)_{12}$ 245

Chapter Six CYCLOPENTADIENYL COMPLEXES

64. Dicarbonylbis(η^5-cyclopentadienyl) Complexes of Titanium, Zirconium, and Hafnium: $Cp'_2M(CO)_2$ (Cp' = C_5H_5, C_5Me_5; M = Ti, Zr, Hf).. 248
  A. Dicarbonylbis(η^5-cyclopentadienyl)titanium 250
  B. Dicarbonylbis(η^5-cyclopentadienyl)zirconium............ 251
  C. Dicarbonylbis(η^5-cyclopentadienyl)hafnium............. 252
  D. Dicarbonylbis(η^5-pentamethylcyclopentadienyl)titanium .. 253
  E. Dicarbonylbis(η^5-pentamethylcyclopentadienyl)zirconium. 254
  F. Dicarbonylbis(η^5-pentamethylcyclopentadienyl)hafnium .. 255
65. (η^5-Cyclopentadienyl)hydridozirconium Complexes: Cp_2ZrH_2, $Cp_2Zr(H)(Cl)$... 257
  A. Bis(η^5-cyclopentadienyl)dihydridozirconium 257
    1. μ-Oxo-bis[chlorobis(η^5-cyclopentadienyl)zirconium].. 257
    2. Bis(η^5-cyclopentadienyl)dihydridozirconium 258
  B. Chorobis(η^5-cyclopentadienyl)hydridozirconium 259

66. Cyclopentadienyl Complexes of Titanium(III) and
    Vanadium(III): Cp_2MCl (M = Ti, V)...................... 260
    A. Chlorobis(η^5-cyclopentadienyl)titanium(III).............. 261
    B. Chlorobis(η^5-cyclopentadienyl)vanadium(III) 262
67. Vanadocene, Bis(η^5-cyclopentadienyl)vanadium: Cp_2V,
    $V_2(\mu\text{-Cl})_3(THF)_6^{2+}$ 263
68. Dichlorobis(η^5-cyclopentadienyl)niobium(IV): Cp_2NbCl_2...... 267
69. Chloro(η^5-cyclopentadienyl)bis(triphenylphosphine)-
    ruthenium(II): $RuCl(PPh_3)_2(\eta^5\text{-}C_5H_5)$...................... 270
70. (η^5-Pentamethylcyclopentadienyl)cobalt Complexes:
    $Cp^*Co(CO)_2$, $Cp^*Co(C_2H_4)_2$............................... 273
    A. Dicarbonyl(η^5-pentamethylcyclopentadienyl)cobalt(I)..... 273
    B. Carbonyldiiodo(η^5-pentamethylcyclopenta-
       dienyl)cobalt(III)...................................... 275
    C. Di-μ-iodo-bis[iodo(η^5-pentamethylcyclopenta-
       dienyl)cobalt(III)]..................................... 276
    D. Bis(η^2-ethene)(η^5-pentamethylcyclopentadienyl)cobalt(I) .. 278
71. Cyclopentadienylbis(trimethylphosphine) and
    Cyclopentadienylbis(trimethylphosphite) Complexes of Co and
    Rh: $CpM(PMe_3)_2$, $CpM[P(OMe)_3]_2$ (M = Co, Rh).......... 280
    A. (η^5-Cyclopentadienyl)bis(trimethylphosphine)rhodium(I).. 280
    B. (η^5-Cyclopentadienyl)bis(trimethylphosphine)cobalt(I) 281
    C. (η^5-Cyclopentadienyl)bis(trimethylphosphite)cobalt(I)..... 283
    D. (η^5-Cyclopentadienyl)bis(trimethylphosphite)rhodium(I)... 284

Chapter Seven LANTHANIDE AND ACTINIDE COMPLEXES

72. Lanthanide Trichlorides by Reaction of Lanthanide Metals with
    Mercury(II) Chloride in Tetrahydrofuran: $LnCl_3(THF)_x$ 286
    A. Ytterbium Trichloride–Tetrahydrofuran (1/3)............ 289
    B. Erbium Trichloride–Tetrahydrofuran (2/7) 290
    C. Samarium Trichloride–Tetrahydrofuran (1/2)............ 290
    D. Neodymium Trichloride–Tetrahydrofuran (2/3).......... 290
73. (η^5-Cyclopentadienyl)lanthanide Complexes from the Metallic
    Elements: Cp_3Ln (Ln = Nd, Sm), $Cp_2Y(DME)$ 291
    A. Tris(η^5-cyclopentadienyl)(tetrahydrofuran)neodymium(III) 293
    B. Tris(η^5-cyclopentadienyl)(tetrahydrofuran)samarium(III) .. 294
    C. Bis(η^5-cyclopentadienyl)(1,2-dimethoxyethane)-
       ytterbium(II).. 295
74. Bis(η^5-pentamethylcyclopentadienyl)bis(tetrahydrofuran)-
    samarium(II): $Cp_2^*Sm(THF)_2$ 297

75. Chlorotris(η^5-cyclopentadienyl) Complexes of Uranium(IV) and Thorium(IV): Cp$_3$MCl (M = U, Th) 300
   A. Chlorotris(η^5-cyclopentadienyl)uranium(IV).............. 301
   B. Chlorotris(η^5-cyclopentadienyl)thorium(IV).............. 302

Chapter Eight LIGANDS AND OTHER TRANSITION METAL COMPLEXES

76. Trimethylphosphine: PMe$_3$ 305
77. Phosphorus Trifluoride: PF$_3$ 311
78. Cyclopentadienylthallium: CpTl 315
79. 1,2,3,4,5-Pentamethylcyclopentadiene: C$_5$Me$_5$H 317
80. Anhydrous Metal Chlorides................................. 321
81. Tungsten and Molybdenum Tetrachloride Oxides: MOCl$_4$ (M = Mo, W)...................................... 323
   A. Tungsten Tetrachloride Oxide......................... 324
   B. Molybdenum Tetrachloride Oxide...................... 325
82. Tungsten Chloro Phosphine Complexes: WCl$_4$(PR$_3$)$_{2, 3}$, WCl$_2$(PR$_3$)$_4$ 326
   A. Tetrachlorotris(trimethylphosphine)tungsten(IV) 327
   B. Tetrachlorobis(diphenylmethylphosphine)tungsten(IV) 328
   C. Dichlorotetrakis(trimethylphosphine)tungsten(II)......... 329
   D. Dichlorotetrakis(dimethylphenylphosphine)tungsten(II)... 330
   E. Dichlorotetrakis(methyldiphenylphosphine)tungsten(II) ... 331
83. Tetrabutylammonium Octachlorodirhenate(III): Re$_2$Cl$_8^{2-}$ 332
84. Di-μ-chloro-bis[tricarbonylchlororuthenium(II)]: [Ru(CO)$_3$Cl$_2$]$_2$ 334
85. Dihydridotetrakis(triphenylphosphine)ruthenium(II): H$_2$Ru(PPh$_3$)$_4$.. 337
86. Tris(2,2'-bipyridine)ruthenium(II) Dichloride Hexahydrate: Ru(bipy)$_3^{2+}$ 338
87. Dichlorobis[μ-methylenebis(diphenylphosphine)]-dipalladium(I)(Pd—Pd): Pd$_2$(dppm)$_2$Cl$_2$................... 340
88. (η^3-Allyl)palladium(II) Complexes: (η^3-C$_3$H$_5$)$_2$Pd$_2$Cl$_2$, (η^3-C$_3$H$_5$)(η^5-C$_5$H$_5$)Pd.................................... 342
   A. Bis(η^3-allyl)di-μ-chloro-dipalladium(II) 342
   B. (η^3-Allyl)(η^5-cyclopentadienyl)palladium(II).............. 343
89. Cyclic Diolefin Complexes of Platinum and Palladium: MCl$_2$(COD) (M = Pd, Pt) 346
   A. Dichloro(η^4-1,5-cyclooctadiene)platinum(II).............. 346
   B. Dichloro(η^4-1,5-cyclooctadiene)palladium(II)............. 348

90. Potassium Trichloro(ethene)platinate(II) (Zeise's Salt): K[PtCl$_3$(C$_2$H$_4$)] 349

Index of Contributors... 353
Subject Index... 361
Formula Index... 387

# REAGENTS FOR TRANSITION METAL COMPLEX AND ORGANOMETALLIC SYNTHESES

INORGANIC SYNTHESES

Volume 28

Chapter One

COMPLEXES WITH WEAKLY COORDINATED LIGANDS

1. METAL COMPLEXES WITH WEAKLY BONDED ANIONS: GENERAL CONSIDERATIONS

Preface W. Beck*

In the following procedures, the preparation and reactions are given for complexes that behave as strong electrophilic Lewis acids. Such complexes,

L_nM-X

$L = CO, \pi\text{-}C_5H_5$

$X = [BF_4]^-, [ClO_4]^-, [PF_6]^-, [PO_2F_2]^-, [AsF_6]^-, [SbF_6]^-, [SO_3CF_3]^-$

contain a "soft" metal in a low oxidation state with strongly bonded π-acceptor ligands and a "hard" labile anionic or neutral ligand[1] that can be easily substituted by other even weakly nucleophilic ligands, usually under very mild conditions. These complexes are precursors of coordinatively and electronically unsaturated compounds and have proved to be excellent starting materials in preparative organometallic chemistry. Complexes with easily dissociating ligands are also possible precursors of catalysts.

Although compounds with weakly coordinated anions have been isolated with some main group elements and with the more electron rich transition

---

*Institut für Anorganische Chemie Universität München, Meiserstr. 1, 8000 München, Federal Republic of Germany.

metals such as Cu, Zn, or Ni^2 only a few compounds with carbonyl ligands were reported prior to 1978.

A series of organometallic compounds with good leaving groups has been isolated and fully characterized by physical methods (for further examples, see the following procedures):

$Cr(CO)_3(PMe_3)(CMe)(FBF_3)$ (Ref. 3), $W(CO)_4(CNEt_2)(FBF_3)$ (Ref. 3), $M(CO)_5X$ (Ref. 4) (M = Mn, Re; X = FAsF$_5$, OClO$_3$, O$_2$PF$_2$, FBF$_3$), $M(CO)_3(\pi\text{-}C_5H_5)X$ (Ref. 5) (X = FBF$_3$, FPF$_5$, FAsF$_5$, FSbF$_5$), $Cr(NO)_2(\pi\text{-}C_5H_5)X$ (Ref. 6) (X = FBF$_3$, FPF$_5$, FAsF$_5$), $Fe(CO)_2(\pi\text{-}C_5H_5)FBF_3$ (Ref. 7), $Ir(Cl)(CO)(PPh_3)_2$ (H)FBF$_3$ (Ref. 8).

However, it is not always necessary to isolate these complexes. Often the complexes generated *in situ* or the solvent containing compounds, for example, $[(M(CO)_2(\pi\text{-}C_5H_5)(\text{solvent})][BF_4]$ (Ref. 9) (M = Fe, Ru, Os), $W(NO)_2(\pi\text{-}C_5H_5)(BF_4)$ (Ref. 10), and $Mo(CO)_3(\pi\text{-}C_5H_5)(BF_4)$ (Ref. 11) can be used in synthesis.

Various routes have been used by many groups for the preparation of metal complexes with weakly bonded anions. These are

(a) Abstraction of an anionic ligand Y by a cationic Lewis acid A$^+$X$^-$ to give a very stable compound AY and the desired complex. Hereby the Lewis acidity of A$^+$ is transferred to the metal complex

$$L_nM - Y + A^+X^- \rightarrow L_nM - X + AY$$

$Y^- = H^-$, CH_3^-, halide, $[N_3]^-$

$A^+ = H^+$, Ag^+, Me_3O^+, CPh_3^+, NO^+

$X^- = [FBF_3]^-$, $[FPF_5]^-$, $[FAsF_5]^-$, $[FSbF_5]^-$, $[OClO_3]^-$, $[OSO_2CF_3]^-$

$AY = H_2$, CH_4, Ag halide, Ph_3CH, Ph_3CCH_3, N_2O, respectively

(b) Addition of main group Lewis acids to a coordinated halide

$$L_nM - Y + EX_n \rightarrow L_nM - YEX_n$$

Y = halide
EX_n = for example, BF$_3$, AlCl$_3$, AsF$_5$

(c) Oxidation of dimeric complexes

$$[L_nM]_2 \xrightarrow[\text{solvent}]{-2e} 2[L_nM(\text{solvent})]^+$$

(d) Oxidative addition of AX to low-valent coordinatively unsaturated complexes

$$L_nM + AX \rightarrow L_nM{<}^A_X$$

A comprehensive survey on organometallic Lewis acids is given in Ref. 12.

General Remarks

The user should be familiar with the Schlenk technique for handling air- and moisture-sensitive compounds.[13] Prepurified argon or nitrogen are recommended as inert gas. The argon is dried by use of a column (100-cm length, 5-cm diameter) filled with molecular sieves 4 and 5 Å. Traces of oxygen are removed by another column filled with chromium(II) oxide on silica gel, which was prepared as reported[14] by reduction of CrO_3 on silica gel with carbon monoxide.

■ **Caution.** *Chromium(VI) oxide, CrO_3, is carcinogenic. Especially after the drying procedure, inhalation of the fine powder must be avoided by working in a well-ventilated hood. Also the use of the toxic CO gas makes an efficient hood absolutely necessary. The final Cr(II) catalyst is extremely pyrophoric and should never be allowed to get into contact with air.*

Solvents are purified as follows: Dichloromethane is passed through a column (200-cm length, 3-cm diameter) with molecular sieve 4 Å, then heated to reflux over P_2O_5 for 1 day and subsequently distilled under argon; it is stored in a (100-cm length, 2-cm diameter) column, filled with molecular sieves 4 Å; pentane is degassed by evacuating and filling with argon for three times; for very sensitive compounds [e.g., $Mo(CO)_3Cp(FBF_3)$] it should be stored over Na–K alloy (prepared by mixing 3 g Na and 9 g K in 250 mL of xylene and refluxing for 1 day) and refluxed and distilled prior to use. The same procedure is used for hexane (Na–K alloy might here be replaced by potassium alone).

■ **Caution.** *Potassium and especially Na–K alloy are extremely flammable in moist air and explode with liquid water. Potassium should be cut under paraffin oil and the alloy should be stored under xylene. The alloy can be transferred via a syringe that has been thoroughly dried. t-Butanol is recommended for destroying unused portions of both K and Na–K alloy.*

Easy separation of solids from the solution can be achieved by using a centrifuge (Macrofuge C-4, Heraeus-Christ). The centrifuge can be provided with polyethylene blocks each having a hole just fitting the Schlenk tube.

These blocks are usually stored in Dry Ice so that for short periods of time, sufficiently low temperatures during centrifugation can be maintained.

Before the reaction, the Schlenk tube, provided with the magnetic stirring bar, is flamed in a high vacuum for at least 10 min with a Bunsen burner, and cooled in a stream of argon.

■ **Caution.** *Since metal carbonyls and their derivatives are toxic volatile compounds, all operations must be performed in an efficient hood.*

References

1. J. A. Davies and F. R. Hartley, *Chem. Rev.*, **81**, 79 (1981).
2. M. R. Rosenthal, *J. Chem. Educ.*, **50**, 331 (1973); A. P. Gaughan, Jr., Z. Dori, and J. A. Ibers, *Inorg. Chem.*, **13**, 1657 (1974).
3. K. Richter, E. O. Fischer, and C. G. Kreiter, *J. Organomet. Chem.*, **122**, 187 (1976); E. O. Fischer, D. Wittmann, D. Himmelreich, U. Schubert, and K. Ackermann, *Chem. Ber.*, **115**, 3141 (1982).
4. R. Mews, *Angew. Chem. Int. Ed. Engl.*, **14**, 640 (1975); F. L. Wimmer and M. R. Snow, **31**, 267 (1978); R. Uson, V. Riera, J. Gimeno, M. Laguna, and M. P. Gamasa, *J. Chem. Soc. Dalton Trans.*, **1979**, 966; K. Raab, U. Nagel, and W. Beck, *Z. Naturforsch. Teil B*, **38**, 1466 (1983) and references cited therein.
5. W. Beck and K. Schloter, *Z. Naturforsch. Teil B*, **33**, 1214 (1978); K. Sünkel, U. Nagel, and W. Beck, *J. Organometal. Chem.*, **251**, 227 (1983) and references cited therein.
6. F. J. Regina and A. Wojcicki, *Inorg. Chem.*, **19**, 3803 (1980); G. Hartmann, R. Froböse, R. Mews, and G. M. Sheldrick, *Z. Naturforsch. Teil B*, **37**, 1234 (1982); P. Legzdins, D. T. Martin, Ch. R. Nurse, and B. Wassink, *Organometallics*, **2**, 1238 (1983).
7. B. M. Mattson and W. A. G. Graham, *Inorg. Chem.*, **20**, 3186 (1981).
8. B. Olgemöller, H. Bauer, H. Löbermann, U. Nagel, and W. Beck, *Chem. Ber.*, **115**, 2271 (1982).
9. D. L. Reger, C. J. Coleman, and P. J. Mc Elligott, *J. Organometal. Chem.*, **171**, 73 (1979); E. K. G. Schmidt and C. H. Thiel, *J. Organometal. Chem.*, **209**, 373 (1981); J. K. Hoyano, C. J. May, and W. A. G. Graham, *Inorg. Chem.*, **21**, 3095 (1982); and references cited therein.
10. P. Legzdins and D. T. Martin, *Organometallics*, **2**, 1785 (1983).
11. S. J. La Croce and A. R. Cutler, *J. Am. Chem. Soc.*, **104**, 2312 (1982); T. C. Forschner and A. R. Cutler, *Inorg. Synth.*, **26**, 000 (1989).
12. W. Beck and K. Sünkel, *Chem. Rev.*, **88**, 1405–1421 (1988).
13. D. F. Shriver and M. A. Drezdzon, *The Manipulation of Air Sensitive Compounds*, 2nd ed., McGraw-Hill, New York, 1986.
14. H. L. Krauss and H. Stach, *Z. Anorg. Allgem. Chem.*, **366**, 34 (1969).

2. CARBONYL(η^5-CYCLOPENTADIENYL)-(TETRAFLUOROBORATO)MOLYBDENUM AND -TUNGSTEN COMPLEXES

Submitted by WOLFGANG BECK, KLAUS SCHLOTER,
KARLHEINZ SÜNKEL, and GÜNTER URBAN*
Checked by THOMAS FORSCHNER, ALICIA TODARO, and ALAN CUTLER†

A. TRICARBONYL(η^5-CYCLOPENTADIENYL) (TETRAFLUOROBORATO)MOLYBDENUM AND -TUNGSTEN[1a]

An efficient method for the preparation of tetrafluoroborato complexes is hydride abstraction from metal hydrides using triphenylmethylium‡ tetrafluoroborate.[1a] This method has been first reported by Sanders for hydridoruthenium complexes.[2]

$$[Ph_3C][BF_4] + MCp(CO)_3H \rightarrow MCp(CO)_3(FBF_3) + Ph_3CH$$

$$M = Mo, W$$

In a similar way, the hexafluoroarsenato and hexafluoroantimonato complexes $MCp(CO)_3FEF_5$ (M = Mo, W; E = As, Sb) have been prepared from $MCp(CO)_3H$ and $[CPh_3][EF_6]$.[1b]

An alternative method for the preparation of $MCp(CO)_3(FBF_3)$ (M = Mo, W) is protonation of $MCp(CO)_3CH_3$ by $HBF_4 \cdot Et_2O$.[1c]

Procedure

Tritylium tetrafluoroborate is commercially available (Fluka AG) and should be freshly recrystallized from dichloromethane or dichloromethane–ethyl acetate prior to use. The hydrido complexes, $MCp(CO)_3H^3$ should be purified by sublimation or by chromatography (neutral alumina, activity 3, pentane eluant) prior to use. All solvents must be rigorously dried and handled under an inert atmosphere, see the preceding general comments.

A quantity of $Ph_3C[BF_4]$ (0.33 g, 1.0 mmol) is dissolved in 10 mL of CH_2Cl_2 in a 50-mL Schlenk tube, under an inert atmosphere. The solution is

---

*Institut für Anorganische Chemie der Universität München, Meiserstr. 1, 8000 München 2, Federal Republic of Germany.
†Department of Chemistry, Rensselaer Polytechnic Institute, Troy, 12180-3590.
‡ triphenylmethylium = tritylium

cooled to $-40°C$ (using Dry Ice–acetone and a low-temperature thermometer). To this is added MoCp(CO)$_3$H (Ref. 3) (0.22 g, 0.89 mmol) or WCp(CO)$_3$H (Ref. 3) (0.30 g, 0.90 mmol). An immediate color change from yellow to purple-red is observed. After stirring for 10 min, 0.2 mL of the solution is syringed into an infrared (IR) solution cell and a spectrum is taken. If a more or less intense band is observed at ~ 1355 cm^{-1}, indicating the presence of unreacted tritylium salt, small amounts of the corresponding hydrides are then added via a spatula. After stirring for 5 min, the IR spectrum is recorded for another solution aliquot. The addition of hydride is repeated until the IR spectrum of the solution shows no band at 1355 cm^{-1}. As soon as this equivalence point is reached, a sudden color change from dark red to lilac or violet is observed (see the solution in Section A). If this color change does not occur, the presence of moisture can be suspected. In this case the solution may be used for a reaction with stronger ligands than water, otherwise the preparation has to be tried again.

Two procedures are given for the treatment of the solution in Section A. In the first the solution is cooled down to $-60°C$, and 20 mL of hexane is added. Careful evaporation under vacuum to ~ 20 mL removes most of the CH$_2$Cl$_2$. The lilac precipitate is isolated by centrifugation (~ 2 min at 1500 rpm) and decanting off the solution. Hexane (20 mL) is added at $-60°C$ and the suspension is stirred for 10 min. Centrifugation, decanting, and washing are repeated three times. Then the product is dried at $-20°C$ for 8 h on a high-vacuum line (10^{-3} torr).

Alternate Procedure for Treatment of the Solution in Section A

The lilac-colored reaction mixture is transferred into a second Schlenk flask (100 mL) using a double-ended stainless steel cannula. The second flask contains hexane previously cooled to $-78°C$ (Dry Ice–acetone bath). A lilac colored solid precipitates. The solvent is siphoned off and the solid is washed three times with hexane (20 mL) previously cooled to $-78°C$ and transferred into the flask using the double ended cannula technique. The wash solvent is siphoned off and remaining solid is dried under vacuum (10^{-3} torr, oil pump) at $-40°C$ for 8 h. Yields: for MoCp(CO)$_3$(FBF$_3$): 282–319 mg (85–96%), for WCp(CO)$_3$(FBF$_3$): 357–378 mg (85–90%).

Anal. Calcd. for C$_8$H$_5$BF$_4$MoO$_3$: C, 28.95; H, 1.52. Found: C, 28.27; H, 1.64. Calcd. for C$_8$H$_5$BF$_4$O$_3$W: C, 22.28; H, 1.20. Found: C, 23.63; H, 1.39.

Properties*

*See Section B.

B. DICARBONYL(η^5-CYCLOPENTADIENYL)(TETRAFLUORO-BORATO)(TRIPHENYLPHOSPHINE)MOLYBDENUM AND -TUNGSTEN, MCp(CO)$_2$(PPh$_3$)(FBF$_3$)(M = Mo, W)[4]

Substitution of a CO group by a phosphine ligand makes the metal center electron-richer and therefore less Lewis acidic. This weakens the coordination of the [BF$_4$]$^-$ ion. In addition, steric interactions with the phosphine ligands, the possibility of cis–trans isomerism in the complexes with "four-legged piano stool" geometry,[5] and the introduction of the ^{31}P nucleus as another sensitive NMR probe make this variation of the synthesis described in Section A, an interesting field of further investigation. The preparation described here for the PPh$_3$ compounds, can also be used with other PR$_3$ ligands such as PMe$_3$, PEt$_3$, P(OPh)$_3$, or $\frac{1}{2}$(dppe)[dppe = 1,2-ethanediyl-bis(diphenyl-phosphine)].†

Dicarbonyl(η^5-cyclopentadienyl)hydrido(triphenylphosphine) molybdenum and -tungsten

$$MCp(CO)_3H + PPh_3 \rightarrow MCp(CO)_2(PPh_3)H + CO$$

$$M = Mo, W$$

Monophosphine substituted carbonylcyclopentadienylhydrido complexes of molybdenum and tungsten have been obtained by protonation of the anions [MCp(CO)$_2$(PR$_3$)]$^-$,[6] or by substitution of CO with phosphines in the hydrides MH(CO)$_3$Cp.[7] The straightforward synthesis of the hydrides MH(CO)$_3$Cp(M = Mo, W)[3] makes the latter procedure preferable, at least for PPh$_3$, P(OPh)$_3$, PMe$_3$, and PEt$_3$, where fast reactions and good yields can always be obtained. For the analogous syntheses of PMe$_3$ or PEt$_3$ substituted hydrides, special precautions for handling these highly toxic, malodorous, and highly inflammable phosphines must be taken.[8]

Procedure

A quantity of freshly sublimed MCp(CO)$_3$H (0.49 g, M = Mo or 0.66 g, M = W, each 2.0 mmol) is dissolved in 15 mL of hexane at room temperature in a 100-mL Schlenk flask. To this is added PPh$_3$ (0.58 g, 2.1 mmol) under vigorous stirring. The Schlenk tube is then connected to a mercury bubbler and a stream of argon (0.5 L min^{-1}) is passed over the solution for 2 h (it is not necessary to bubble the argon through the solution). Soon a white precipitate

---

†Commonly known as 1,2-bis(diphenylphosphino)ethane.

forms, which is isolated by filtration under argon and washed twice with 5 mL of hexane. The product is dried at room temperature for 6 h *in vacuo*. It may be recrystallized from CH_2Cl_2–hexane. Yields: $MoCp(CO)_2(PPh_3)H$, 625 mg, 65%; $WCp(CO)_2(PPh_3)H$, 636 mg (56%).

Properties

The hydrides $MCp(CO)_2(PPh_3)H$ are yellowish-white powders. They are air stable for several minutes exposure as solids, however, for extended storage they should be kept under argon. IR spectra (in CH_2Cl_2): $v_{CO} = 1936$, 1856 cm^{-1} (Mo); 1923, 1835 cm^{-1} (W); ^{1}H NMR [in CD_2Cl_2 (Mo), $CDCl_3$ (W)]: $\delta_{C_5H_5} = 5.08$ (Mo); 5.10 ppm (W); $\delta_{M-H} = -5.56$ (Mo), -7.06 ppm (W), "doublets" $^2J_{(^{31}P^1H)_{av}} = 47$ Hz (Mo), 55 Hz (W); ^{31}P NMR (in CD_2Cl_2): $\delta_{PPh_3} = 74.3$ (Mo), 40.9 ppm (W) (relative to H_3PO_4). A fast equilibrium between the cis and trans isomers[5] leads to averaging of the signals and coupling constants at room temperature. Both isomers can be distinguished by low-temperature ^{1}H NMR [δ_{Mo-H}: -5.33 d and -6.14 d; $^2J_{^{31}P^1H} = 64$ and 21.4 Hz; δ_{W-H}: -6.90d, -7.36d, $^2J_{^{31}P^1H} = 65$ and 22 Hz].

The analogous PMe_3 and PEt_3 containing hydrides tend to form oils and decompose quickly on contact with air; the tungsten compounds are more stable than the molybdenum analogs. Their spectral properties are similar to those of the PPh_3 compounds. The best yields are obtained with the $P(OPh)_3$ ligand, which leads exclusively to the stable cis isomers.[5]

Dicarbonyl(η^5-cyclopentadienyl)(tetrafluoroborato)(triphenylphosphine)-molybdenum and -tungsten[4]

$$[Ph_3C][BF_4] + MCp(CO)_2(PPh_3)H \rightarrow MCp(CO)_2(PPh_3)(FBF_3)$$
$$+ Ph_3CH$$
$$M = Mo, W$$

Procedure

Generally, the same guidelines as described in Section A have to be followed. A quantity of $MoCp(CO)_2(PPh_3)H$ (0.45 g, 0.94 mmol) or $WCp(CO)_2(PPh_3)H$ (0.55 g, 0.97 mmol) is added to a solution of $[Ph_3C][BF_4]$ (0.33 g, 1.00 mmol) in 10 mL CH_2Cl_2 at $-40°C$ contained in a Schlenk flask (100 mL) equipped with a magnetic stirring bar. The mixture is stirred for 20 min, after which an IR spectrum is recorded of an aliquot (0.2 mL) to inspect the intensity of the band at 1355 cm^{-1}. Small amounts of the hydride are added until the IR spectrum, recorded at 5-min intervals shows no band at 1355 cm^{-1}. Usually, a lilac precipitate forms before the equivalence point is reached. The equivalence point is again indicated by a lilac color of the solution. Complete precipitation of the product is obtained by addition of 20 mL of hexane at $-60°C$, or by transfer of the complete reaction mixture to another Schlenk flask containing the hexane cooled to $-60°C$ (see the procedure in Section A). Isolation of the product is the same as described in Section A. Yields: $MoCp(CO)_2(PPh_3)(FBF_3) \cdot 2CH_2Cl_2$ 632 mg. (86%); $WCp(CO)_2(PPh_3)(FBF_3)$ 556 mg (85%).

Anal. Calcd. for $C_{25}H_{20}BF_4MoO_2P \cdot 2CH_2Cl_2$: C, 44.1; H, 3.29. Found: C, 45.0; H, 3.32.

Properties

All tetrafluoroborato complexes are very sensitive to moisture. Schlenk tubes used for storage therefore have to be heated to 400°C or more under vacuum for several hours; O-ring stopcocks or similar grease-free stopcocks are superior to the usual ground-glass stopcocks. Although the phosphine containing BF_4 complexes are thermally more stable than the unsubstituted compounds, storage at temperatures below $-25°C$ under Ar is recommended for all these compounds. They dissolve in CH_2Cl_2 and $CHCl_3$ below $-40°C$ without decomposition, while solvents with donor properties like acetone or acetonitrile dissolve these complexes under substitution of tetrafluoroborato ligands by the solvent to give ionic complexes, for example, $[MoCp(CO)_3(acetone)][BF_4]$. They can be characterized by their IR spectra in the region from 1200 to 700 cm^{-1} and by their low-temperature ^{19}F and, where appropriate, ^{31}P NMR spectra.[9]

TABLE I. Spectroscopic Data of $MoCp(CO)_2L(FBF_3)$

L=CO
IR: $v_{CO}=2071, 1988$ cm^{-1} (in CH_2Cl_2)
$v_{11BF}=1130, 884, 722$ cm^{-1} (in Nujol)
^{19}F NMR:[a] -155d (MoFBF$_3$), -370q (MoFBF_3), 95 Hz ($^2J_{F-F}$)
L=PPh$_3$
IR: $v_{CO}=1991, 1903$ cm^{-1} (in CH_2Cl_2)
$v_{11BF}=1119, 901, 732$ cm^{-1} (in Nujol)
^{19}F NMR:[a] -155d (MoFBF$_3$), -344q, -391q (MoFBF_3), 90 Hz ($^2J_{F-F}$)

[a] δ in ppm, relative to CFCl$_3$, in CD$_2$Cl$_2$, $-80°C$.

TABLE II. Spectroscopic Data of WCp(CO)$_2$L(FBF$_3$)

L = CO
IR: v_{CO} = 2067, 1975 cm^{-1} (in CH$_2$Cl$_2$)
$v_{^{11}BF}$ = 1149, 874, 704 cm^{-1} (in Nujol)
^{19}F NMR:[a] −153d (WFBF$_3$), −394q (WFBF$_3$), 99 Hz ($^2J_{F-F}$)

L = PPh$_3$
IR: v_{CO} = 1988, 1963, 1877 cm^{-1} (in Nujol)
$v_{^{11}BF}$ = 1148, 887, 720 cm^{-1} (in Nujol)
^{19}F NMR:[b] −156d (WFBF$_3$), −371q (WFBF$_3$), 98 Hz ($^2J_{F-F}$)

[a] δ in ppm, relative to CFCl$_3$, in CD$_2$Cl$_2$, −52°C.
[b] δ in ppm, relative to CFCl$_3$, in CD$_2$Cl$_2$, −80°C.

Coordination of the BF$_4$ ion lowers the T_d symmetry of [BF$_4$]$^-$ and makes the fluorine atoms nonequivalent. Therefore the IR spectra show three instead of one $v_{^{11}B-F}$ absorptions[1,4] (Tables I and II); the low-temperature ^{19}F NMR spectra[9] show two distinct fluorine resonances, a high-field quartet (which may be split by coupling to the phosphorus in the PR$_3$ substituted compounds and a doublet at lower field, close to the resonance of free [BF$_4$]$^-$;[13] the ^{31}P NMR spectrum shows at low temperature a pseudodoublet, produced by coupling with the coordinated fluorine (Tables I and II). Compounds MoCp(CO)$_2$(PR$_3$)(FBF$_3$) are obtained as cis and trans isomers. In WCp(CO)$_2$[P(OPh)$_3$](FBF$_3$) total isomerization from the pure cis hydride to the pure *trans*-BF$_4$ compound could be followed via NMR.[9]

Reactions of Tetrafluoroborato Complexes with Ethylene, Diphenylacetylene, and Acetone

General Remarks

The tetrafluoroborate ligand of these highly reactive complexes can be easily substituted by a series of N, O, P, and S σ donors[1a,4,10,11] and π donors (see Sections C–F).

As described in Sections A and B, a lilac solution of the corresponding tetrafluoroborato complex is prepared at −30°C in 10 mL of CH$_2$Cl$_2$. Complete reaction of the tritylium salt is verified by checking for the disappearance of the 1355 cm^{-1} absorption in the IR spectrum of the solution. This solution is used for the following reactions without isolation of the tetrafluoroborato complex.

C. TRICARBONYL(η^5-CYCLOPENTADIENYL)(η^2-ETHENE)-MOLYBDENUM(1+) TETRAFLUOROBORATE(1−)[1a]

$$MoCp(CO)_3(FBF_3) + C_2H_4 \rightarrow [MoCp(CO)_3(\eta^2\text{-}C_2H_4)][BF_4]$$

The title compound can be obtained in three ways. One method starts from $Mo(C_5H_5)(CO)_3Cl$, which is reacted with C_2H_4 at a pressure of 70 bar in the presence of $AlCl_3$ and consecutive precipitation with ammonium salt.[12] A second method involves β-hydride abstraction from the ethyl group in $MoCp(CO)_3(C_2H_5)$ by $[Ph_3C][BF_4]$.[13] The third method, described here, has the advantage of mild reaction conditions and a good overall yield. Analogous complexes with other olefins have been prepared similarly.[4]

Procedure

Ethylene (1 bar), dried over P_2O_5, is bubbled through a vigorously stirred lilac solution of $MoCp(CO)_3(FBF_3)$ (1 mmol) in CH_2Cl_2 in a Schlenk flask (50 mL) cooled to $-30°C$. With continuous ethylene bubbling, the cooling bath is removed and the flask is permitted to warm up to $+20°C$ over a 4-h period. The flow of ethylene is then stopped, and the reaction mixture stirred under argon for another 30 min. The yellow precipitate is isolated by centrifugation or filtration under Ar. After washing four times with 5 mL of CH_2Cl_2, the product is dried 1 h at $+40°C$ on a high-vacuum line. It may be recrystallized from acetone–diethyl ether. Yield: 282 mg (79%).

Properties

The compound decomposes on heating at 102–108°C. Infrared spectra (in Nujol): $v_{CO} = 2104$, 2053, 2001 cm^{-1}; ^{1}H NMR (acetone-d_6) $\delta = 6.35$ ppm (C_5H_5).

D. CARBONYL(η^5-CYCLOPENTADIENYL)BIS(DIPHENYL-ACETYLENE)MOLYBDENUM(1+) TETRAFLUORO-BORATE(1−)[1a]

$$MoCp(CO)_3(FBF_3) + 2PhCCPh \rightarrow [MoCp(CO)(PhCCPh)_2][BF_4] + 2CO$$

Other syntheses of cationic bis(alkyne) complexes of molybdenum and tungsten of the same type include $AgBF_4$ oxidation of the dimer $[MoCp(CO)_3]_2$ in CH_2Cl_2 in the presence of diphenylacetylene or several other alkynes.[14] Protonation of MoCp(CO)$_3$CH$_3$ with CF$_3$COOH and consecutive addition of 2-butyne in acetonitrile, followed by precipitation with a methanolic solution of [NH$_4$][PF$_6$] gives the corresponding 2-butyne complex.[15] Refluxing a solution of MoCp(CO)$_3$Cl with (HOCH$_2$)CC(CH$_2$OH) leads to an analogous compound.[16] The method described here uses very mild conditions and can be applied also for other alkynes, like 2-butyne or acetylene.[17]

Procedure

Diphenylacetylene (535 mg, 3.0 mmol) is added to a lilac solution of MoCp(CO)$_3$(FBF$_3$) (1 mmol in 10 mL CH$_2$Cl$_2$, prepared as described above) at $-30°$C under vigorous stirring in a Schlenk flask (50 mL) equipped with a magnetic stirring bar. The flask is connected to a mercury bubbler and flushed by a constant flow of argon or nitrogen gas. After 30 min, the gas flow is stopped and the cooling bath removed. Stirring is continued for 4 days at room temperature, over which time a yellow precipitate is formed. Diethyl ether (20 mL) is added and the yellowish-red suspension is filtered under argon. The residue on the frit is extracted three times with 10 mL of CH$_2$Cl$_2$. The combined extracts are evaporated to 5 mL, to which is added diethyl ether (20 mL). The orange-yellow precipitate is isolated by centrifugation or filtration under argon, washed three times with 10-mL aliquots of diethyl ether, and then dried for 1 h *in vacuo* at 40°C. The product may be recrystallized from CH$_2$Cl$_2$–pentane. Yield: 235 mg (37%).

Anal. Calcd. for C$_{34}$H$_{25}$BF$_4$MoO: C, 64.58; H, 3.99. Found: C, 63.98; H, 4.09.

Properties

The yellow compound is soluble in polar solvents such as CH$_2$Cl$_2$, acetone, or acetonitrile. Although prolonged exposure to air leads to decomposition, the compound can be handled in air for short periods of time. Its IR spectrum in Nujol shows one $\nu_{^{12}CO}$ vibration at 2088 cm^{-1} and a weak $\nu_{^{13}CO}$ band at 2040 cm^{-1}. Also a weak absorption at 1741 cm^{-1} occurs, which may be due to the $\nu_{C=C}$ band of the coordinated alkyne. The ^{1}H NMR spectrum in CH$_2$Cl$_2$ has a sharp singlet for the C$_2$H$_5$ protons at $\delta = 6.20$ ppm, besides the broad resonance of the phenyl protons of the diphenylacetylene. Interestingly, KBr pellets of the compound several hours after initially formed show a bathochromic shift of the ν_{CO} band, which is also observed with other cationic alkyne complexes.[17]

E. CARBONYL(η^5-CYCLOPENTADIENYL)(DIPHENYL-ACETYLENE)(TRIPHENYLPHOSPHINE)MOLYBDENUM(1+) TETRAFLUOROBORATE(1−)[18]

$$\text{MoCp(CO)}_2(\text{PPh}_3)(\text{FBF}_3) + \text{PhCCPh} \rightarrow$$
$$[\text{MoCp(CO)}(\text{PPh}_3)(\text{PhCCPh})][\text{BF}_4] + \text{CO}$$

Green and coworkers[19] prepared the title compound and other related monoalkyne complexes by reaction of the corresponding bis(alkyne) complex with triphenylphosphine (or other phosphines) in good yields. The method described here works for several alkynes, for example, 2-butyne or phenylacetylene, and also for phosphines, for example, PEt_3 or P(OPh)_3.

Procedure

Diphenylacetylene (1.78 g, 10.0 mmol) is added to a magnetically stirred lilac suspension of $\text{MoCp(CO)}_2(\text{PPh}_3)(\text{FBF}_3)$ (1.0 mmol in 10 mL of CH_2Cl_2, as described previously) in a Schlenk flask (50 mL) cooled to $-30°\text{C}$. The flask is connected to a mercury bubbler and purged with argon for 15 min. The gas flow is stopped and the cooling bath is allowed to warm up to room temperature. Stirring is continued for 2 days, during which time the flask is purged several times with argon to remove the carbon monoxide evolved in the reaction. Then hexane (20 mL) is added. Stirring is continued for another day at ambient temperature, after which the dark green suspension is filtered under argon. The residue on the filter is washed four times with 15-mL aliquots of hexane and then dried 3 h under vacuum at 25°C. Yield: 408 mg (57%).

Anal. Calcd. for $\text{C}_{38}\text{H}_{30}\text{BF}_4\text{MoOP}$: C, 63.7; H, 4.22. Found: C, 62.8; H, 4.15.

Properties

The title compound is soluble in polar organic solvents, for example, acetone, acetonitrile, or dichloromethane. Although storing under inert gas is recommended, no decomposition can be observed when handled as a solid in air for short periods of time. IR (in CH_2Cl_2): $\nu_{^{12}\text{CO}} = 1987 \text{ cm}^{-1}$; ^1H NMR (CD_2Cl_2): $\delta_{\text{C}_5\text{H}_5} = 5.77$ ppm, $\delta_{\text{C-C}_6\text{H}_5,\text{P(C}_6\text{H}_5)_3}$ 8–7 ppm; ^{31}P NMR (in CD_2Cl_2) $\delta_{\text{PPh}_3} = 54.8$ ppm.

The crystal structure of this compound shows a slightly elongated C≡C bond of the alkyne and the usual deviation from linearity at the two carbon atoms of the triple bond.[18,19]

F. (ACETONE)(TRICARBONYL)(η^5-CYCLOPENTADIENYL)-MOLYBDENUM(1+) AND -TUNGSTEN(1+) TETRAFLUOROBORATE(1−)[1a,10]

$$MCp(CO)_3(FBF_3) + (CH_3)_2CO \rightarrow [MCp(CO)_3OC(CH_3)_2][BF_4]$$

$$M = Mo, W$$

Procedure

A lilac solution of the tetrafluoroborato complex $MCp(CO)_3(FBF_3)$, M = Mo or W (1 mmol in 10 mL of CH_2Cl_2) is prepared as indicated above in a Schlenk flask (50 mL) and cooled to −30°C. To this is added acetone (0.1 mL, 1.38 mmol). An immediate color change to red occurs, and stirring is continued for 3 h. Hexane (15 mL) is then added. The solution is cooled to −78°C (Dry Ice) and stored overnight, giving a dark red precipitate. This is isolated by centrifugation and washed twice with 19 mL of hexane at 0°C. Alternatively, the supernatant solution may be removed by a stainless steel cannula fitted with a sintered-glass frit. The solids are washed with two aliquots of cold (0°C) hexane (19 mL), each removed by use of the stainless steel cannula fitted with the glass frit. The product is then dried for 6 h at 0°C on a high-vacuum line. Yield: $MoCp(CO)_3(OC(CH_3)_2)(BF_4)$ 350 mg (90%); $WCp(CO)_3[OC(CH_3)_2](BF_4)$ 420 mg (88%).

Anal. Calcd. for $C_{11}H_{11}BF_4MoO_4$: C, 33.88; H, 2.84. Found: C, 33.87; H, 2.85. Calcd. for $C_{11}H_{11}BF_4O_4W$: C, 27.65; H, 2.32. Found: C, 26.90; H, 2.45.

Similar acetone complexes can also be prepared from the PPh_3 containing tetrafluoroborate complexes.

Properties

Solutions of the compounds in CH_2Cl_2 or acetone decompose at 20°C within a short time, especially when traces of water are present. The solid compounds can be stored under argon at −30°C for several weeks without decomposition.

IR(CH_2Cl_2): v_{CO} = 2072, 1987 cm^{-1} (Mo) IR (in Nujol): v_{CO} = 2050, 1930 cm^{-1} (W); $v_{M-O=C}$ = 1660 cm^{-1} (Mo), 1640 cm^{-1} (W). ^{1}H NMR (in CD_2Cl_2): $\delta_{C_5H_5}$ = 6.11 (Mo), 6.19 ppm (W); δ_{CH_3} = 2.39 (Mo), 2.43 ppm (W).

References

1. (a) W. Beck and K. Schloter, Z. *Naturforsch. Teil B*, **33**, 1214 (1978); (b) K. Sünkel, U. Nagel, and W. Beck, *J. Organomet. Chem.*, **251**, 227 (1983); (c) M. Appel, K. Schloter, J. Heidrich, and W. Beck, *J. Organomet. Chem.* **322**, 77 (1987).
2. J. R. Sanders, *J. Chem. Soc. Dalton Trans.*, **1972**, 1333.
3. (a) R. B. King and F. G. A. Stone, *Inorg. Synth.*, **7**, 107 (1963). E. O. Fischer, *Inorg. Synth.* **7**, 136 (1963); (b) R. B. King, *Organometallic Syntheses*, Vol. 1, Academic Press, New York, 1965, p. 156; (c) W. P. Fehlhammer, W. A. Herrmann, and G. K. Öfele, in *Handbuch der Präparativen Anorganischen Chemie*, Vol. 2, 3th ed., F. Enke Verlag, Stuttgart, 1981.
4. K. Sünkel, H. Ernst, and W. Beck, *Z. Naturforsch. Teil B*, **36**, 474 (1980).
5. J. W. Faller and A. S. Anderson, *J. Am. Chem. Soc.*, **92**, 5852 (1970).
6. (a) A. R. Manning, *J. Chem. Soc. A*, **1968**, 651; (b) M. J. Mays and S. M. Pearson, *J. Chem. Soc. A*, **1968**, 2291.
7. (a) A. Bainbridge, P. J. Craig, and M. Green, *J. Chem. Soc. A*, **1968**, 2715; (b) P. Kalck, R. Pince, R. Poilblanc, and J. Roussel, *J. Organomet. Chem.*, **24**, 445 (1970).
8. R. T. Markham, E. A. Dietz, Jr., and D. R. Martin, *Inorg. Synth.*, **16**, 153 (1976).
9. K. Sünkel, G. Urban, and W. Beck, *J. Organomet. Chem.*, **252**, 187 (1983); M. Appel and W. Beck, *J. Organomet. Chem.*, **319**, C1 (1987).
10. K. Sünkel, G. Urban, and W. Beck, *J. Organomet. Chem.*, **290**, 231 (1985).
11. K. Schloter and W. Beck, *Z. Naturforsch. Teil B*, **35**, 985 (1980); G. Urban, K. Sünkel, and W. Beck, *J. Organomet. Chem.*, **290**, 329 (1985).
12. E. O. Fischer and K. Fichtel, *Chem. Ber.*, **94**, 1200 (1961).
13. M. Cousins and M. L. H. Green, *J. Chem. Soc.*, **1963**, 889.
14. M. Bottrill and M. Green, *J. Chem. Soc. Dalton Trans.*, **1977**, 2365.
15. P. L. Watson and R. G. Bergman, *J. Am. Chem. Soc.*, **102**, 2698 (1980).
16. J. W. Faller and H. H. Murray, *J. Organomet. Chem.*, **172**, 171 (1979).
17. K. Schloter, K. Sünkel and W. Beck, unpublished results.
18. K. Sünkel, U. Nagel, and W. Beck, *J. Organomet. Chem.*, **222**, 251 (1981).
19. S. R. Allen, P. K. Baker, S. G. Barnes, M. Green, L. Trollope, L. Manojlovic-Muir, and K. W. Muir, *J. Chem. Soc. Dalton Trans.*, **1981**, 873.

3. PENTACARBONYL(TETRAFLUOROBORATO)RHENIUM AND -MANGANESE AND REACTIONS THEREOF

Submitted by WOLFGANG BECK and KLAUS RAAB*
Checked by J. R. SHAPLEY and B. R. WHITTLESEY†

Highly reactive pentacarbonylmanganese and rhenium complexes with weakly coordinated anions include $M(CO)_5(FAsF_5)$ (Ref. 1), $M(CO)_5(OClO_3)$ (Ref. 2), $M(CO)_5(OPOF_2)$ (Ref. 2), $M(CO)_5(OSO_2CF_3)$ (Ref. 3), $M(CO)_5OTeF_5$ (Ref. 4) (M = Mn, Re), and $Mn(CO)_5(O_2CCF_3)$

---

*Institut für Anorganische Chemie, Universität München, Meiserstr.1, 8000 München 2, Federal Republic of Germany.
†452 Noyes Laboratory, Box 20, 505S. Mathews, University of Illinois Urbana, IL 61801.

(Ref. 5), which are usually prepared from pentacarbonyl halides and the silver salt of the corresponding anion. In the following procedures, the corresponding tetrafluoroborates and some of their reactions are described. These tetrafluoroborato complexes $M(CO)_5(FBF_3)$ (M = Mn, Re) are accessible from the corresponding methyl complexes and triphenylmethylium tetrafluoroborate[6] or tetrafluoroboric acid, respectively.[7] Interestingly, methyl metal compounds may react with the triphenylmethylium ion by abstraction of the methyl group[6,8] or by abstraction of hydride to give methylene carbene complexes.[9]

A. PENTACARBONYLMETHYLRHENIUM[10]

$$Re_2(CO)_{10} + 2Na(Hg) \rightarrow 2Na[Re(CO)_5]$$

$$Na[Re(CO)_5] + CH_3I \rightarrow Re(CH_3)(CO)_5 + NaI$$

Procedure

■ **Caution.** *Pentacarbonylmethylrhenium is a volatile metal carbonyl derivative. Metal carbonyls usually are very toxic and must be handled in a well-ventilated hood.*

The starting material $Re_2(CO)_{10}$ may be purchased either from Strem or from Pressure Chemicals. The reactions are conducted in Schlenk tubes under a dry argon or nitrogen atmosphere. Sodium amalgam is prepared by addition of sodium metal (0.3 g) to 3 mL of mercury under nitrogen.

■ **Caution.** *The dissolution of sodium metal in mercury is an exothermic reaction; therefore sodium must be added in small pieces.*

A quantity of $Re_2(CO)_{10}$ (3.00 g, 4.6 mmol) is dissolved in 10–12 mL of dry tetrahydrofuran (THF) previously saturated with argon or nitrogen in a Schlenk flask (100 mL). After all the solid has dissolved, the flask is cooled to 0°C, and the solution is transferred to the sodium amalgam, also in a Schlenk flask (100 mL) equipped with a stirring bar and cooled to 0°C. The mixture is stirred for 60 min at 0°C and for another hour at room temperature. A third Schlenk flask (100 mL) is flushed with argon or nitrogen. The red, air sensitive solution obtained in the previous step is transferred away from the excess sodium amalgam into the third Schlenk flask, using Teflon tubing passing from one flask to the next through rubber septa. A light over pressure is applied over the solution to be transferred. The transferred solution is cooled to ~ −25°C (a Dry Ice–isopropyl alcohol bath) (no precipitate of $Na[Re(CO)_5]$ should be formed) and iodine-free iodomethane (0.6 mL, 9.7 mmol) is added dropwise.

■ **Caution.** *Iodomethane is volatile and carcinogenic.*

After stirring for 10 min the solution is warmed up to room temperature. After stirring for another 75 min the solvent is evaporated at $-20°C$ under vacuum. A trap cooled with liquid nitrogen is placed between the Schlenk tube and the pump. To remove the last traces of THF the yellow residue is dried for a short period in an oil pump vacuum. Finally, the yellow residue is sublimed *in vacuo* at 30–40°C for 2–3 days. Since $Re(CO)_5(CH_3)$ is volatile, the stopcock between the sublimation apparatus and the pump is opened only briefly several times. Yield: 2.3–2.9 g (60–76%).

Another 1–5% of $Re(CO)_5(CH_3)$ can be isolated by adding 300 mL of water to the THF in the trap at room temperature. The formed precipitate is washed with water and dried over P_4O_{10} in a small evacuated desiccator.

Properties

Pentacarbonylmethylrhenium is a colorless, volatile solid. It is air and moisture stable and soluble in most organic solvents. The IR shows CO bands at 2129 (w), 2012 (s), 1975 cm^{-1} (in CH_2Cl_2). The structure of $Re(CO)_5(CH_3)$ has been determined by electron diffraction.[11]

B. PENTACARBONYL(TETRAFLUOROBORATO)RHENIUM(I)[6,7]

Method a

$$Re(CO)_5(CH_3) + H[BF_4]Et_2O \rightarrow Re(CO)_5(FBF_3) + CH_4 + Et_2O$$

Method b

$$Re(CO)_5(CH_3) + Ph_3C[BF_4] \rightarrow Re(CO)_5(FBF_3) + H_3CCPh_3$$

Procedure

See General Remarks for this chapter for the preparation and handling of tetrafluoroborato complexes.

■ **Caution.** *Pentacarbonylmethylrhenium is a volatile metal carbonyl derivative. Metal carbonyls usually are very toxic and must be handled in a well-ventilated hood. Tetrafluoroboric acid is a very corrosive chemical. Contact with the skin has to be avoided. Gloves should be worn.*

The reagents may be purchased as indicated: Tetrafluoroboric acid from Merck; $[Ph_3C][BF_4]$ from Fluka.

Method a. A quantity of $Re(CO)_5(CH_3)$ (342 mg, 1.0 mmol) is dissolved in 3 mL of dichloromethane in a dried Schlenk flask (50 mL). To the stirred solution tetrafluoroboric acid (54% in diethyl ether, $d=1.18$ g cm^{-3}; 138, 5 µL, 1.0 mmol) is added at room temperature using a plastic micropipette. The tetrafluoroboric acid solution in diethyl ether is transferred from the original bottle under an atmosphere of argon. Excess tetrafluoroboric acid should be avoided. After addition of H[BF$_4$]·Et$_2$O to the solution a vigorous evolution of methane occurs and a colorless precipitate is formed, which—after 20 min—is centrifugated off or collected on a glass frit, washed several times each with 5 mL of dichloromethane, and dried in a high vacuum (2 h). Yield: 380–401 mg (92–97%).

Anal. Calcd. for C$_5$BF$_4$O$_5$Re(MW 413.0): C, 14.54. Found: C, 14.17.

Properties

The colorless compound Re(CO)$_5$(FBF$_3$) is very sensitive to moisture. On exposure to moist air pentacarbonyl(trifluorohydroxoborato)rhenium, Re(CO)$_5$(OHBF$_3$) is formed. The complex Re(CO)$_5$(FBF$_3$) is only sparingly soluble in dichloromethane. The coordinated [BF$_4$]$^-$ ligand shows the following v_{B-F} stretching bands in the IR spectrum ($v_{^{10}B-F}$): 1203 (m), 1172 (sh), 930 (m), 757 (m); ($v_{^{11}B-F}$): 1162 (s), 1128 (s), 902 (s), 738 (s) cm^{-1} (in Nujol). The three v_{CO} bands (in Nujol) at 2165(w, A$_1$), 2055(vs, br, E), 2014(s, A$_1$) cm^{-1} are characteristic for the Re(CO)$_5$ group. The complexes Re(CO)$_5$(FBF$_3$) and Re(CO)$_5$(FAsF$_5$)[1] undergo many reactions with anionic and neutral donor ligands, usually under substitution of tetrafluoroborate.[6,12,13]

Ionic complexes are formed under mild conditions with various soft and hard neutral donor molecules:

$$Re(CO)_5(FBF_3) + L \rightarrow [(OC)_5ReL][BF_4]$$

(L = CO, ethylene, propene, 1-pentene, butadiene, H$_2$S, THF,* acetone, acetonitrile, nitromethane). The moiety [Re(CO)$_5$]$^+$ can also be added to a nucleophilic atom of a coordinated ligand, which provides a systematic way to prepare ligand-bridged complexes,[12] for example,

$$(OC)_5Re\text{—}O(H)C=O + Re(CO)_5(FBF_3) \rightarrow$$
$$[(OC)_5ReOC(H)ORe(CO)_5][BF_4]$$

$$[Au(CN)_2]^- + 2Re(CO)_5(FBF_3) \rightarrow$$
$$[(OC)_5Re\text{—}NCAuCN\text{—}Re(CO)_5]^+[BF_4]^- + [BF_4]^-$$

*THF = tetrahydrofuran

The complex $Re(CO)_5(FBF_3)$ also reacts with 2-butyne or 2-pentyne via cyclodimerization of the alkyne to give complexes with a coordinated methylenecyclobutene derivative.[7]

The compound $Re(CO)_5(FBF_3)$ is soluble in water to give a solution of $[Re(CO)_5(OH_2)]^+$. A series of water insoluble neutral pentacarbonylrhenium derivatives has been obtained from aqueous solutions of pentacarbonylrhenium(tetrafluoroborate) and various salts:[12]

$$[Re(CO)_5(OH_2)]^+ + X^- \xrightarrow[-H_2O]{} Re(CO)_5X$$

$X^- = Cl^-, Br^-, I^-, NO_2^-, NO_3^-, OOCH^-, NCO^-, SCN^-, SeCN^-, RS^-,$
$Au(CN)_2^-, \frac{1}{2}Pt(CN)_4^{2-}, \frac{1}{2}C_4O_4^{2-}, \frac{1}{2}C_4S_4^{2-}$

C. PENTACARBONYL(η^2-ETHENE)RHENIUM(1+) TETRAFLUOROBORATE(1−)[6]

The complexes $[M(CO)_5(\eta^2\text{-}C_2H_4)][AlCl_4]$ (M = Mn, Re) were first prepared by abstraction of the chloride ligand in $M(CO)_5Cl$ using aluminumtrichloride under ethylene pressure.[14] The preparation of $[BF_4]^-$ salts of these cationic pentacarbonylethene complexes of manganese and rhenium proceeds under very mild conditions (1 bar) and gives high yields.[6]

$$Re(CO)_5(FBF_3) + C_2H_4 \rightarrow [Re(CO)_5(\eta^2\text{-}C_2H_4)][BF_4]$$

Procedure

Moisture has to be carefully excluded. A quantity of $Re(CO)_5(FBF_3)$ (1.21 g, 2.93 mmol) is weighed into a Schlenk tube (100 mL) under a dry argon atmosphere. A magnetic stirring bar is added. The Schlenk tube is connected with a mercury bubbler and evacuated and flushed with ethene several times. (Drying of ethene with molecular sieve 4 Å is not necessary since it is not used in large excess in this procedure.) Dried dichloromethane (10 mL) is added with a pipette under a flush of ethene. The suspension is stirred magnetically for 1–2 days. If the mercury bubbler shows reduced pressure the tube is again filled with ethene (1 bar). The complex is centrifugated off or collected on a glass frit, washed with dichloromethane and dried in a high vacuum. Yield: 1.23–1.29 g (95–100%).

The complex can be dissolved in acetone at $-20°C$ and precipitated by addition of dichloromethane or diethyl ether.

Anal. Calcd. for $C_7H_4BF_4O_5Re$ (MW 441.1): C, 19.06; H, 0.91. Found: C, 19.21; H, 0.78.

The complex [Re(CO)$_5$(η^2-C$_2$H$_4$)][BF$_4$] may also be obtained directly from the reaction of [Ph$_3$C][BF$_4$] with Re(CO)$_5$(CH$_3$) in an ethene atmosphere. Reaction time: 3–4 days. Yield: (95–98%).

Properties

The colorless complex is nearly insoluble in dichloromethane and readily soluble in acetone, acetonitrile, and nitromethane. In these solvents ethene is very slowly substituted to give [Re(CO)$_5$(solvent)][BF$_4$]. IR (in Nujol): 2174 (m), 2055 (s, br) (ν_{CO}); 1055 (vs) (ν_{BF_4}); 1538 (vw) ($\nu_{C=C}$) cm^{-1}. IR (in CH$_3$NO$_2$): 2172 (m), 2071 (s) cm^{-1} (ν_{CO}). ^{1}H NMR (acetone-d_6, i-TMS): $\delta = 5.12$ ppm (singlet). The complex [Re(CO)$_5$(η^2-C$_2$H$_4$)][BF$_4$] has been used for the preparation of the ethene bridged complex (OC)$_5$ReCH$_2$CH$_2$Re(CO)$_5$ via nucleophilic attack of pentacarbonylrhenate (1−) at the coordinated ethene of [Re(CO)$_5$(C$_2$H$_4$)]$^+$.[6]

Other alkene complexes [Re(CO)$_5$(alkene)][BF$_4$] (alkene = propene, 1-pentene, 1,3-butadiene) can also be obtained from Re(CO)$_5$(FBF$_3$) and alkene.[13]

D. OCTADECACARBONYL-BIS(μ_3-CARBON DIOXIDE)-TETRARHENIUM[15]

The classic "Hieber-base reaction"[16] is that of a hydroxide with metal carbonyls, which proceeds by nucleophilic attack of the hydroxide at a carbon atom of a carbonyl ligand to give a carboxy group or consequently carbon dioxide and a metal hydride.[17] Metal carbonyls are catalysts for the water–gas shift reaction.[18] Pentacarbonyl tetrafluoroboratorhenium reacts with alkali hydroxide in a similar way; however, due to the coordinatively unsaturated nature of the [Re(CO)$_5$]$^+$ group polynuclear compounds are formed.[15]

$$Re(CO)_5(FBF_3) + OH^- \xrightarrow{H_2O} [Re(CO)_4(COOH)]_n$$

$$\xrightarrow[-H_2O]{acetone} (OC)_4Re\underset{\underset{ORe(CO)_5}{O=C}}{\overset{\overset{(OC)_5ReO}{C=O}}{}}Re(CO)_4$$

Tetracarbonyl(hydroxycarbonyl)rhenium

Procedure

The preparation can be carried out in air. A quantity of $Re(CO)_5(FBF_3)$ (0.85 g, 2.06 mmol) is dissolved in water (15 mL) in a Schlenk flask (25 mL). The solution is filtered into a second flask (25 mL) equipped with a magnetic stirring bar. Aqueous NaOH (2.15 mL of a 1 M solution) is added to the filtrate under stirring. After a few minutes the colorless precipitate is collected on a glass frit and washed several times with water. The solid is dried over P_4O_{10} for 2 days and for 20 h in a high vacuum. Yield: 643–693 mg (91–98%).

Anal. Calcd. for C_5HO_6Re: C, 17.50; H, 0.29. Found: C, 17.33; H, 0.36.

Properties

IR (KBr): 2145 (m), 2098 (m), 2073 (m), 2050 (sh), 2030 (vs), 1981 (vs), 1963 (vs), 1900 (vs), 1870 (vs), (v_{CO}); 1458 (s), 1180 (s), 1165 (sh) (v_{CO_2}); 3270 (m, br), 2900–2850 cm^{-1} (v_{OH}, fluorinated Nujol).

Octadecacarbonyl-bis(μ_3-carbon dioxide)tetrarhenium

A quantity of $[Re(CO)_4(COOH)]_n$ (0.63 g) is dissolved in acetone (40 mL) in a Schlenk flask (100 mL). The pale yellow solution is filtered quickly, if necessary. After a few minutes a colorless precipitate is formed. The suspension is stirred for 30–40 min and the solvent is evaporated *in vacuo* to ~4 mL. The solid is centrifuged off or collected on a glass frit, washed two times each with 2 mL of acetone, and dried in a high vacuum. Yield: 344 mg (56%).

Anal. Calcd. for $C_{20}O_{22}Re_4$ (MW 1337.0): C, 17.97. Found: C, 18.15.

Properties

The colorless CO_2 bridged complex is stable in air and soluble in THF. It is only slightly soluble in acetone and dichloromethane. IR (KBr): 2147 (m), 2088 (m), 2084 (sh), 2054 (s), 2033 (s), 1987 (s), 1968 (s), 1921 (s), 1898 (w) (v_{CO}); 1379 (s), 1294 (m), 1259 (w) cm^{-1} (v_{CO_2}). The X-ray structure of the complex has been determined.[15]

References

1. R. Mews, *Angew. Chem.*, **87**, 669 (1975); *Angew. Chem. Int. Ed. Engl.*, **14**, 640 (1975); M. Oltmanns and R. Mews, *Z. Naturforsch. Teil B*, **35**, 1324 (1980); G. Hartmann, R. Froböse, R. Mews, and G. M. Sheldrick, *Z. Naturforsch. Teil B*, **37**, 1234 (1982).
2. F. L. Wimmer and M. R. Snow, *Aust. J. Chem.*, **31**, 267 (1978); E. Horn and M. R. Snow, *Aust. J. Chem.*, **33**, 2369 (1980); R. Uson, V. Riera, J. Gimeno, M. Laguna, and M. P. Gamasa, *J. Chem. Soc. Dalton Trans.*, **1979**, 996.
3. W. C. Trogler, *J. Am. Chem. Soc.*, **101**, 6459 (1979); St. P. Schmidt, J. Nitschze, and W. C. Trogler, *Inorg. Synth.*, **26**, 113 (1989).
4. K. D. Abney, K. M. Long, O. P. Anderson, and S. H. Strauss, *Inorg. Chem.*, **26**, 2638 (1987).
5. F. A. Cotton, D. J. Darensbourg, and B. W. S. Kolthammer, *Inorg. Chem.*, **20**, 1287 (1981).
6. K. Raab, B. Olgemöller, K. Schloter, and W. Beck, *J. Organomet. Chem.*, **214**, 81 (1981); K. Raab, U. Nagel, and W. Beck, *Z. Naturforsch. Teil B*, **38**, 1466 (1983).
7. K. Raab and W. Beck, *Chem. Ber.*, **117**, 3169 (1984).
8. P. J. Harris, S. A. R. Knox, R. J. McKinney, and F. G. A. Stone, *J. Chem. Soc. Dalton Trans.*, **1978**, 1009.
9. A. T. Patton, C. E. Strouse, C. B. Knobler, and J. A. Gladysz, *J. Am. Chem. Soc.*, **105**, 5804 (1983).
10. W. Hieber and G. Braun, *Z. Naturforsch. Teil B*, **14**, 132 (1959); W. Hieber, G. Braun, and W. Beck, *Chem. Ber.*, **93**, 901 (1960).
11. D. W. H. Rankin and A. Robertson, *J. Organomet. Chem.*, **105**, 331 (1976).
12. K. Raab and W. Beck, *Chem. Ber.*, **118**, 3830 (1985).
13. W. Beck, K. Raab, U. Nagel, and W. Sacher, *Angew. Chem. Int. Ed. Engl.*, **24**, 505 (1985).
14. E. O. Fischer and K. Fichtel, *Chem. Ber.*, **94**, 1200 (1961); E. O. Fischer, K. Fichtel, and K. Öfele, *Chem. Ber.*, **95**, 249 (1962); E. O. Fischer und K. Öfele, *Angew. Chem.*, **73**, 581 (1961); **74**, 76 (1962); A. M. Brodie, B. F. G. Hulley, B. F. G. Johnson, and J. Lewis, *J. Organomet. Chem.*, **24**, 201 (1970).
15. W. Beck, K. Raab, U. Nagel, and M. Steimann, *Angew. Chem.*, **94**, 556 (1982); *Angew. Chem. Int. Ed. Engl.*, **21**, 526 (1982).
16. W. Hieber and F. Leutert, *Z. Anorg. Allg. Chem.*, **204**, 145 (1932).
17. Th. Kruck, M. Höfler, and M. Noack, *Chem. Ber.*, **99**, 1153 (1966).
18. D. J. Darensbourg and A. Rokiki, *Organometallics*, **1**, 1685 (1982); R. M. Laine and E. J. Crawford, *J. Mol. Catal.*, **44**, 357 (1988).

4. IRIDIUM(III) COMPLEXES WITH THE WEAKLY BONDED ANIONS [BF$_4$]$^-$ AND [OSO$_2$CF$_3$]$^-$

Submitted by WOLFGANG BECK,* HERBERT BAUER,* and BERNHARD OLGEMÖLLER*
Checked by DEVEREAUX A. CLIFFORD† and ROBERT H. CRABTREE†

trans-Carbonylchlorobis(triphenylphosphine)iridium(I) (Vaska's compound) forms adducts with many substrates such as acids and alkyl halides.[1] The

---
*Institut für Anorganische Chemie, Universität München, Meiserstr. 1, 8000 München 2, Federal Republic of Germany.
†Department of Chemistry, Yale University, New Haven CT 06511.

oxidative addition of CH_3SO_3F or $CH_3OSO_2CF_3$ to trans-$Ir(Cl)(CO)(PPh_3)_2$ gives iridium(III) complexes $Ir(Cl)(CO)(PPh_3)_2(CH_3)X$ (X = SO_3F, SO_3CF_3) in which the weakly bonded anion X can be easily substituted by other ligands.[2] Highly reactive iridium(III) complexes are also obtained by oxidative addition of tetrafluoroboric acid and trialkyloxonium tetrafluoroborate to Vaska's compound.[3,4] The oxidative addition of $H_3COSO_2CF_3$ and tetrafluoroboric acid to the nitrogen complex[5] trans-$Ir(Cl)(N_2)(PPh_3)_2$ gives iridium(III) complexes $Ir(Cl)(N_2)(PPh_3)(R)(X)$ (R = CH_3, H; X = SO_3CF_3, BF_4) with two excellent leaving groups (N_2 and X).[6,7] They are precursors for 14 electron iridium(III) complexes.

A. CARBONYLCHLOROHYDRIDO(TETRAFLUOROBORATO)-BIS(TRIPHENYLPHOSPHINE)IRIDIUM(III)[3]

trans-$Ir(Cl)(CO)(PPh_3)_2$ + HBF_4 ⟶ [octahedral Ir complex with Cl, H, two PPh$_3$, CO, and FBF$_3$ ligands]

Procedure

■ **Caution.** *Tetrafluoroboric acid is a very corrosive chemical. Contact with skin has to be avoided. Gloves should be worn. The procedure should be carried out in a well-ventilated hood.*

A quantity of trans-$Ir(Cl)(CO)(PPh_3)_2$ (780 mg, 1.00 mmol)[8] is suspended in 20 mL of dry dichloromethane in a Schlenk tube (100 mL) under dry argon, and 165 mg (0.139 mL, 1.00 mmol) of tetrafluoroboric acid ($d = 1.18$ g cm^{-3}, 54% in diethyl ether, Merck), is added by means of a syringe at room temperature. The mixture is stirred magnetically for 15 min. The solid dissolves within a few seconds and the product crystallizes after several minutes. The precipitation is completed by addition of 60 mL of pentane. The mixture is centrifuged and the solvent is decanted. The solid is washed with five portions of pentane, 20 mL each, and dried *in vacuo*. Yield: 860 mg (99%).

Anal. Calcd. for $C_{37}H_{31}BClF_4IrOP_2$ (MW 868.1): C, 51.19; H, 3.60. Found: C, 50.04; H, 3.53.

Properties

The hydrido(tetrafluoroborato)iridium(III) complex is a white crystalline solid, highly sensitive to air and moisture. It is soluble in dichloromethane. The IR spectrum (Nujol) shows absorptions at 2061 (v_{CO}), 2333 (v_{IrH}), 322

(ν_{IrCl}), and at 1137, 910, 730 cm^{-1} (ν_{11BF_4}) for the coordinated BF$_4$ ligand. The ^{1}H NMR spectrum (in CH$_2$Cl$_2$) shows a multiplet at δ −26.5 ppm. The hydride and the BF$_4$ group are in trans position, and the [BF$_4$]$^-$ ligand is coordinated via a fluorine atom as shown by an X-ray structural determination.[3] The BF$_4$ ligand can be easily substituted by neutral σ or π donors to give cationic complexes [Ir(Cl)(CO)(PPh$_3$)$_2$(H)L][BF$_4$] (L = PPh$_3$, CH$_3$CN, H$_2$O, tetrahydrofuran (THF), C$_2$H$_4$).[3] Substitution of the [BF$_4$]$^-$ ligand by various anions gives neutral iridium(III) complexes.[9] The complex is deprotonated by strong bases to give *trans*-Ir(Cl)(CO)(PPh$_3$)$_2$.

B. CARBONYLCHLOROMETHYL(TETRAFLUOROBORATO)BIS-(TRIPHENYLPHOSPHINE)IRIDIUM(III)[3]

trans-Ir(Cl)(CO)(PPh$_3$)$_2$ + [(H$_3$C)$_3$O][BF$_4$]

⟶

$$\text{Ph}_3\text{P} \underset{\underset{\text{Cl}}{|}}{\overset{\overset{\text{CH}_3}{|}}{\underset{\text{Ir}}{}}} \begin{matrix} \text{FBF}_3 \\ \text{PPh}_3 \end{matrix} + (\text{H}_3\text{C})_2\text{O}$$

OC

Procedure

■ **Caution.** *Benzene is a highly toxic solvent. It should be handled in a well-ventilated hood.*

Crystalline and rigorously dry trimethyloxonium tetrafluoroborate (150 mg, 1.02 mmol) is added to a suspension of *trans*-Ir(Cl)(CO)(PPh$_3$)$_2$ (780 mg, 1.00 mmol)[8] in 10 mL of dry benzene in a Schlenk tube (60 mL) and under an argon atmosphere. The mixture is stirred magnetically for ∼1 week until the yellow solid becomes colorless. After addition of 20 mL of pentane, the mixture is centrifuged and the solvent decanted. The crystalline solid is washed with three portions of dry pentane, 10 mL each, and dried *in vacuo*. Yield: 810–836 mg (92–95%).

Anal. Calcd. for C$_{38}$H$_{33}$BClF$_4$IrOP$_2$ (MW 882.1): C, 51.74; H, 3.78. Found: C, 51.28; H, 4.00.

Properties

The methyl(tetrafluoroborato)iridium(III) complex is an air and moisture sensitive white crystalline solid, soluble in dichloromethane. The IR spectrum (in Nujol) shows absorptions at 2070 (ν_{CO}), 310 (ν_{IrCl}) and 1136, 908 cm^{-1} (ν_{11BF_4}).[3]

C. CHLORO(DINITROGEN)HYDRIDO-(TETRAFLUOROBORATO)BIS-(TRIPHENYLPHOSPHINE)IRIDIUM(III)[7]

$$trans\text{-}Ir(Cl)(N_2)(PPh_3)_2 + HBF_4 \longrightarrow \begin{array}{c} H \\ Cl\diagdown | \diagup PPh_3 \\ Ph_3P \diagup \overset{|}{Ir} \diagdown N_2 \\ FBF_3 \end{array}$$

Procedure

■ **Caution.** *See precautionary note in the procedure in Section A.*

A suspension of trans-Ir(Cl)(N$_2$)(PPh$_3$)$_2$ (1.36 g, 1.74 mmol)[5] in 12 mL of dry dichloromethane in a Schlenk tube is cooled to $-25\,°C$, and a solution of tetrafluoroboric acid in diethyl ether (54%, $d = 1.18\,g\,cm^{-3}$) (0.25 mL, 1.89 mmol) is added in one portion by means of a plastic micropipette under stirring and under a flush of dry argon. The mixture is stirred for 1 h at $-25\,°C$. A pale yellow precipitate settles and the brown solution is decanted. The solid is washed three to four times, each with 8 mL of cold ($-25\,°C$) dichloromethane, until the solid becomes colorless. The solid is then washed three times with 10 mL each of cold (-20 to $-25\,°C$) pentane and dried for 8 h at a high vacuum, during which the temperature is raised from -20 to $0\,°C$. Yield: 1.44 g (95%).

Anal. Calcd. for $C_{36}H_{31}BClF_4IrN_2P_2$ (MW 868.1): C, 49.81; H, 3.60; N, 3.23. Found: C, 49.97; H, 4.78; N, 3.12.

Properties

The complex is a colorless air and moisture sensitive solid, which is thermally stable up to $60\,°C$ when all traces of solvent have been removed by careful drying at $0\,°C$ *in vacuo*. Containing traces of dichloromethane, the complex decomposes quickly at room temperature or in moist CH_2Cl_2. The dry complex can be stored in a refrigerator under argon atmosphere for a long period of time without decomposition. In a solution of dichloromethane, nitrogen evolution is observed at temperatures above $0\,°C$. The IR spectrum of the solid (in Nujol) shows absorptions at 2310 (vw, v_{IrH}), 2229 (s, v_{N_2}), 347 (w, v_{IrCl}), and at 1129(s), 908(s), 740(sh) cm^{-1} for the ^{11}B—F stretching vibrations of the coordinated $[BF_4]^-$ ligand. The ^1H NMR spectrum (in CD$_2$Cl$_2$, 240 K) contains multiplets at $\delta = 7.5$ (phenyl) and -30.3 ppm (IrH). The complex reacts at low temperatures with H_2O, CH_3OH, acetone, THF, CH_3CN, CO, or C_2H_4 by substitution of the $[BF_4]^-$ ligand without loss of dinitrogen to give ionic complexes, for example, [Ir(Cl)(N$_2$)(PPh$_3$)$_2$(H)L]$^+$[BF$_4$]$^-$. At higher temperatures or with bidentate ligands [e.g., N,N-diethyldithiocarbamate or valinate, H$_2$NCH(CHMe$_2$)COO$^-$] dinitrogen is also displaced.[7]

D. CARBONYLHYDRIDOBIS(TRIFLUOROMETHANE-SULFONATO) BIS(TRIPHENYLPHOSPHINE)-IRIDIUM(III)

$$\textit{trans-}\text{Ir(Cl)(CO)(PPh}_3)_2 + \text{Ag(OSO}_2\text{CF}_3)$$
$$\longrightarrow \textit{trans-}\text{Ir(OSO}_2\text{CF}_3)(\text{CO})(\text{PPh}_3)_2 + \text{AgCl}$$
$$\textit{trans-}\text{Ir(OSO}_2\text{CF}_3)(\text{CO})(\text{PPh}_3)_2 + \text{HOSO}_2\text{CF}_3$$

$$\longrightarrow \begin{array}{c} \text{H} \\ \text{OC} \diagdown \mid \diagup \text{PPh}_3 \\ \text{Ir} \\ \text{Ph}_3\text{P} \diagup \mid \diagdown \text{OSO}_2\text{CF}_3 \\ \text{OSO}_2\text{CF}_3 \end{array}$$

Procedure

■ **Caution.** *Trifluoromethanesulfonic acid is a corrosive chemical. Contact with skin must be avoided; gloves should be worn.*

A magnetic stirring bar, *trans*-Ir(Cl)(CO)(PPh$_3$)$_2$ (520 mg, 0.67 mmol)[8] and dry AgOSO$_2$CF$_3$ (173 mg, 0.67 mmol) are placed in a Schlenk tube (60 mL) under an argon atmosphere. The mixture is dried for 2 h under high vacuum. After addition of 10 mL of dry dichloromethane, the yellow suspension is stirred magnetically for 3 h. The solution is filtered away from the silver chloride through a Schlenk frit under an argon atmosphere into another dry Schlenk tube. To the yellow solution trifluoromethanesulfonic acid (0.058 mL, 0.66 mmol) (distilled at 43 °C *in vacuo* with an oil pump and stored under argon) is added by means of a micropipette. The mixture is stirred for 1 h. A colorless precipitate forms. Precipitation of the product is completed by addition of 10 mL of pentane. The mixture is centrifuged and the solution decanted. The remaining solid is washed twice with 10 mL of pentane, and dried under high vacuum for 4 h. Yield: 620 mg (89%).

Anal. Calcd. for C$_{39}$H$_{31}$F$_6$IrO$_7$P$_2$S$_2$ (MW 1043.9): C, 44.87; H, 2.99; S, 6.14. Found: C, 44.58; H, 3.12; S, 6.74.

Properties

The iridium(III) complex is an air and moisture sensitive white solid, which is slightly soluble in dichloromethane and melts at 245 °C with decomposition. The IR spectrum of the solid (in Nujol) shows absorptions at 2297 (w, v_{IrH}), 2073 (vs), 2058 (sh) (v_{CO}) and 1347 (vs), 1319 (vs), 1206 (vs), 1005 (vs), 978 (vs) cm^{-1} for the SO-stretching vibrations of the coordinated sulfonate groups. The ^{1}H NMR spectrum (in CD_2Cl_2) contains signals at $\delta = 7.5$ ppm (multiplet, phenyl) and -20.6 ppm [triplet, IrH, $J_{^{31}P-H} = 11.0$ Hz]. In a solution of acetonitrile, the two OSO_2CF_3 ligands are substituted by solvent to give a solution of $[Ir(CO)(H)(PPh_3)_2(CH_3CN)_2]^{2+}(CF_3SO_3^-)_2$. Acetone does not replace the sulfonate ligands.

References

1. L. Vaska and J. W. Diluzio, *J. Am. Chem. Soc.*, **83**, 2784 (1961); L. Vaska, *J. Am. Chem. Soc.*, **88**, 4100, 5325 (1966); H. Singer and G. Wilkinson, *J. Chem. Soc. A*, **1968**, 2516.
2. D. Strope and D. F. Shriver, *Inorg. Chem.*, **13**, 2652 (1974); C. Eaborn, N. Farrell, J. L. Murphy, and A. Pidcock, *J. Chem. Soc. Dalton Trans.*, **1976**, 58.
3. B. Olgemöller, H. Bauer, H. Löbermann, U. Nagel, and W. Beck, *Chem. Ber.*, **115**, 2271 (1982); B. Olgemöller and W. Beck, *Inorg. Chem.*, **22**, 997 (1983).
4. M. Kubota, T. M. McClesky, R. K. Hayashi, and C. G. Webb, *J. Am. Chem. Soc.*, **109**, 7569 (1987).
5. J. P. Collman, N. W. Hoffman, and J. W. Hosking, *Inorg. Synth.*, **12**, 8 (1970); R. J. Fitzgerald and H.-M. Lin, *Inorg. Synth.*, **16**, 42 (1976).
6. L. R. Smith and D. M. Blake, *J. Am. Chem. Soc.*, **99**, 3302 (1977).
7. B. Olgemöller, H. Bauer, and W. Beck, *J. Organomet. Chem.*, **213**, C 57 (1981); H. Bauer and W. Beck, *J. Organomet. Chem.*, **308**, 73 (1986).
8. K. Vrieze, J. P. Collman, G. T. Sears, Jr., and M. Kubota, *Inorg. Synth.*, **11**, 101 (1968).
9. B. Olgemöller and W. Beck, *Chem. Ber.*, **114**, 2360 (1981).

5. cis-CHLOROBIS(TRIETHYLPHOSPHINE)(TRIFLUORO-METHANESULFONATO)PLATINUM(II)

Submitted by WOLFGANG BECK, BERNHARD OLGEMÖLLER, and LUTTGARD OLGEMÖLLER*
Checked by S. CHALOUPKA and L. M. VENANZI†

Palladium(II) and platinum(II) ions are considered as soft Lewis acids, and hard oxygen and nitrogen donors are only weakly bonded to these metals.

*Institut für Anorganische Chemie, Universität München, Meiserstr. 1, 8000 München 2, Federal Republic of Germany.
†Laboratorium für Anorganische Chemie, ETH, Universitätsstrasse 6, 8092 Zürich, Switzerland.

Such complexes are highly reactive and have been implicated in catalytic cycles.[1] A trifluoromethanesulfonato complex of platinum(II) is described in the following sections.[2] The trifluoromethanesulfonato (triflate) anion has been widely used as a leaving group in organic synthesis.[3]

$$\text{cis-Pt(PEt}_3)_2\text{Cl}_2 + \text{HOSO}_2\text{CF}_3 \longrightarrow \begin{array}{c} \text{Et}_3\text{P} \\ \diagdown \\ \text{Pt} \\ \diagup \\ \text{Et}_3\text{P} \end{array} \begin{array}{c} \text{Cl} \\ \diagup \\ \\ \diagdown \\ \text{OSO}_2\text{CF}_3 \end{array} + \text{HCl}$$

Procedure

■ **Caution.** *Trifluoromethanesulfonic acid is a corrosive chemical. Contact with the skin should be avoided and gloves should be worn.*

A quantity of cis-Pt(PEt$_3$)$_2$Cl$_2$ (500 mg, 1.0 mmol)[4] is suspended in 20 mL of dry pentane in a Schlenk tube fitted with a mercury bubbler under a dry inert atmosphere (N$_2$ or Ar) and cooled to -78 °C. To this a quantity of freshly distilled trifluoromethanesulfonic acid (150 mg, 0.088 mL, 1.0 mmol) is added by means of a micropipette. The mixture is stirred magnetically at -20 °C for 8 h. The pentane phase is decanted after centrifugation and the solid is washed three times, each with 20 mL of pentane at ambient temperature and dried *in vacuo*. Yield: 588–616 mg (96–100%).

Anal. Calcd. for $C_{13}H_{30}ClF_3O_3PtS$ (MW 615.9): C, 25.35; H, 4.91. Found: C, 25.60; H, 5.17.

Properties

cis-Chlorobis(triethylphosphine)(trifluoromethanesulfonato)platinum(II) is a white, hygroscopic solid. The infrared (IR) spectrum of the solid in Nujol shows ν_{SO} absorptions for the coordinated sulfonate ligand at 1312 and 1226 cm^{-1}. ^{31}P NMR (CD$_2$Cl$_2$): $\delta = 18.5$ ppm, $^1J_{^{195}Pt-^{31}P} = 3507$ Hz.

The complex reacts with various σ and π donors, L (e.g., L = phosphine, acetonitrile, and ethene) to give the ionic complexes [Pt(PEt$_3$)$_2$(Cl)L]$^+$(O$_3$SCF$_3^-$). With anionic chelate ligands (e.g., L = α-aminoacidate) complexes of the type [Pt(PEt$_3$)$_2$(chelate)]$^+$(O$_3$SCF$_3^-$) are formed. The complex dimerizes in THF or acetone to give [(Et$_3$P)$_2$Pt(μ-Cl)$_2$Pt(PEt$_3$)$_2$]$^{2+}$[O$_3$SCF$_3^-$]$_2$.[2,5] Chloro-bridged complexes of this type have been obtained previously by other routes.[6]

References

1. Review: J. A. Davies and F. R. Hartley, *Chem. Rev.*, **81**, 79 (1981).
2. B. Olgemöller, L. Olgemöller, and W. Beck, *Chem. Ber.*, **114**, 2971 (1981).
3. P. J. Stang, M. Hannack, and L. R. Subramanian, *Synthesis*, **1982**, 85.
4. G. W. Parshall, *Inorg. Synth.*, **13**, 27 (1972).
5. L. Olgemöller and W. Beck, unpublished results.
6. W. P. Fehlhammer, W. A. Hermann, and G. K. Öfele, *Handbuch der Präparativen Anorganischen Chemie*, Vol. 3, Enke, Verlag Stuttgart 1981, p. 2013 and references therein.

6. TRICARBONYLTRIS(NITRILE) COMPLEXES OF Cr, Mo, AND W

Submitted by GREGORY J. KUBAS* and LORI STEPAN VAN DER SLUYS*
Checked by RUTH ANN DOYLE† and ROBERT J. ANGELICI†

Because of the lability of nitrile ligands,[1] the complexes fac-M(CO)$_3$(NCR)$_3$ (M = Cr, Mo, W) are widely utilized as precursors to tricarbonyl Group 6 complexes. Acetonitrile has been most commonly utilized as the displaceable group, and synthesis of M(CO)$_3$(NCMe)$_3$ by refluxing M(CO)$_6$ in acetonitrile was originally reported by Tate et al.[2] However, only minimal experimental detail was given in their short communication, and we have found that formation of the tungsten complex is very slow, especially at laboratories situated at higher altitudes (>2 weeks). [The reaction times for thermally activated reactions can be very sensitive to reflux temperatures, which in turn are influenced by altitude. At the elevation of Los Alamos (7300 ft), atmospheric pressure is 590 torr and boiling points of nitriles are 5–10° lower than at sea level (e.g., 89–90° vs. 97° for EtCN).] Our attempts to accelerate the latter reaction using Me$_3$NO[3] or promoters such as CoCl$_2$[4] failed (the third CO was still difficult to remove), but reasonably good success was achieved by enlisting higher-boiling nitriles, such as propionitrile, as the reaction medium (and reactant).[5] The resultant W(CO)$_3$(EtCN)$_3$ was found to be superior to W(CO)$_3$(MeCN)$_3$ in terms of preparation time, solubility, and facility of nitrile displacement (as judged by uniformly higher yields of substitution product).[5] For the latter two reasons, use of EtCN, despite its less common availability, is likely to be worthwhile even in the case of the more readily formed Cr and Mo complexes. For example, preparation of Cp*M(CO)$_3$H (M = Mo, W) from Cp*H and M(CO)$_3$(EtCN)$_3$ was particularly facile.[6]

*Los Alamos National Laboratory, University of California, Los Alamos, NM 87545.
†Department of Chemistry, Iowa State University, Ames, IA 50011.

Described below are preparations of M(CO)$_3$(NCR)$_3$ for R = Et; the R = Me complexes can be made similarly but at the cost of significantly more synthesis time for M = W and possibly lower reactivity. The IR carbonyl frequencies for the MeCN complexes[7] are similar to those for R = Et and also do not vary greatly with metal.

A. TRICARBONYLTRIS(PROPIONITRILE)TUNGSTEN(0)

$$W(CO)_6 + 3C_2H_5CN \xrightarrow{reflux} W(CO)_3(C_2H_5CN)_3 + 3CO$$

■ **Caution.** *Nitriles are very toxic and the syntheses described below should be carried out in a well-ventilated hood.*

The following manipulations are carried out using inert atmosphere techniques and reagent-grade deoxygenated diethyl ether, propionitrile (Aldrich Chemical Co., Milwaukee, WI), W(CO)$_6$, Mo(CO)$_6$, and Cr(CO)$_6$ (Strem Chemicals, Inc., Newburyport, MA) without further purification or drying.

Procedure

A mixture of 35 g (0.1 mol) of W(CO)$_6$ and 300 mL of propionitrile is placed into a 500-mL flask equipped with a magnetic stirrer and a reflux condenser with an efficient exit for evolved CO. Heating to reflux under nitrogen gradually dissolves the hexacarbonyl and induces stepwise displacement of three CO ligands. Initially the solvent condensation point should be kept at the upper flask level to prevent clogging of the condenser by sublimed W(CO)$_6$. Once all of the latter is consumed, the reflux should then be vigorous. The solution color becomes yellow, deepening to brown-red. After about 4 days, infrared solution spectroscopy of an aliquot (diluted with propionitrile by ~10:1) is used to determine if the intermediate W(CO)$_4$(NCEt)$_2$ (v_{CO} = 2021, 1898, 1840 cm^{-1}) has been converted to W(CO)$_3$(NCEt)$_3$ (v_{CO} = 1909, 1790 cm^{-1}). When the bands due to the tetracarbonyl become weak or nonexistent (~6 days, depending on factors such as altitude and escape of CO from the reaction flask), the solution is cooled to 45–55°C, filtered (if necessary*), and solvent volume is reduced to 100–150 mL *in vacuo*. Fine yellow needles of product begin to crystallize and 200 mL of diethyl ether is added to complete precipitation. Storage of the reaction mixture in a freezer overnight (optional) gives some additional product. The solution is then filtered through a large-volume medium frit

---

*In one case, a small amount of a black precipitate formed during the latter stages of reflux.

(under nitrogen), and the light yellow microcrystalline solid is washed with 3 × 40 mL of diethyl ether and dried briefly *in vacuo* (~20 min; extended pumping gives some decomposition). The yield is 35 g (81%).

Anal. Calcd. for $C_{12}H_{15}N_3O_3W$: C, 33.3; H, 3.5; N, 9.7. Found: C, 32.5; H, 3.3; N, 9.4.

Properties

The solid complex is stable in air for short periods (it can be weighed in air) but decomposes on extended exposure (greenish tinges indicate minor surface oxidation). It is very soluble in dichloromethane, moderately soluble in nitriles and acetone, and slightly soluble in hydrocarbons and diethyl ether. A Nujol mull infrared spectrum shows broad carbonyl bands at 1895 and 1767 cm^{-1}. An even more soluble NCPr analog can be synthesized in an analogous fashion. The higher solubilities (and possibly higher labilities of the larger nitriles) greatly aid substitution reactions that require refluxing alkane solvents as reaction media. For example, vastly improved yields of $W(CO)_3(\eta^6$-cycloheptatriene$)$[5,8] and $W(CO)_2(\eta^4$-1,3-cyclohexadiene$)_2$[5] were obtained when $W(CO)_3(NCEt)_3$ was used as a precursor rather than the NCMe analog.

B. TRICARBONYLTRIS(PROPIONITRILE)MOLYBDENUM(0)

$$Mo(CO)_6 + 3C_2H_5CN \xrightarrow{\text{reflux}} Mo(CO)_3(C_2H_5CN)_3 + 3CO$$

Procedure

The synthesis is similar to that for the tungsten complex, but the reaction is much faster. The formation of $Mo(CO)_3(C_2H_5CN)_3$ can be assumed to be complete within about one day (checking IR spectra of aliquots is unnecessary). A yield of 5.0 g (64%) of light tan fine needles was obtained by refluxing 6.0 g (22.7 mmol) of $Mo(CO)_6$ in 60 mL of EtCN, removal of about one-half of the solvent, and addition of 30 mL of Et_2O. Reduction of filtrate volume to 5 mL gave a second crop (2.09 g, total yield = 91%). In most instances where the complex is used as a precursor, isolation is unnecessary. Complete solvent removal after reflux gives a residue of $Mo(CO)_3(EtCN)_3$ that is usually pure enough to be further used *in situ*.

Anal. Calcd. for $C_{12}H_{15}N_3O_3Mo$: C, 41.7; H, 4.4; N, 12.2. Found: C, 40.7; H, 4.3; N, 11.8.

Properties

The properties of the Mo complex are similar to that of the W analog except that it is somewhat more air sensitive, becoming dark brown in color within an hour. The IR spectrum shows carbonyl bands at 1919 and 1800 cm^{-1}.

C. TRICARBONYLTRIS(PROPIONITRILE)CHROMIUM(0)

$$Cr(CO)_6 + 3C_2H_5CN \xrightarrow{reflux} Cr(CO)_3(C_2H_5CN)_3 + 3CO$$

Procedure

The synthesis of the Cr analog is somewhat more difficult because of the higher volatility of $Cr(CO)_6$, which readily sublimes out of the refluxing reaction mixture. This problem can be alleviated somewhat by increasing the solvent volume, but care must still be taken to periodically return sublimed hexacarbonyl into the solution phase by scraping or swirling the hot solution. A mixture of 3.0 g of $Cr(CO)_6$ in 60 mL of EtCN in a 100-mL 14/20 flask was heated to reflux (mild at first then full when $Cr(CO)_6$ was consumed), and the reaction was monitored by IR as in the synthesis of the W analog. After about 3 days primarily $Cr(CO)_3(EtCN)_3$ was present, and the reflux was halted. On prolonged heating the solution attains a red-brown color indicative of thermal decomposition, and in one case (larger scale) this occurred before substitution was complete. Thus some adjustment of reaction conditions may be necessary (for the analogous reaction in MeCN this decomposition did not appear to occur). After reduction of solvent volume to 25 mL *in vacuo*, a small amount of fine green precipitate resulting from minor air oxidation was removed by filtration. The latter was very slow, but scraping the frit (medium porosity) during filtration was found to reduce clogging. The filtrate volume was reduced to 10 mL and Et_2O (20 mL) was added to precipitate the product as bright yellow microcrystals. After cooling the solution in a refrigerator overnight, the complex was collected on a *course* frit (to allow any fine green impurity to pass through). The solid was washed thoroughly with Et_2O and dried *in vacuo*. The yield was 2.57 g (63%).

Anal. Calcd. for $C_{12}H_{15}N_3O_3Cr$: C, 47.8; H, 5.0; N, 13.9. Found: C, 46.8; H, 5.0; N, 13.4.

Properties

The yellow $Cr(CO)_3(EtCN)_3$ is much more deeply colored and air sensitive than the Mo and W analogs, turning green within seconds. However, it did not appear to be pyrophoric (at 590 torr atmospheric pressure) as reported for $Cr(CO)_3(MeCN)_3$.[2] It is quite soluble in polar solvents and slightly soluble in toluene. The IR carbonyl stretching frequencies are 1919 and 1794 cm^{-1} in EtCN solution and 1914 and 1784 cm^{-1} in Nujol mulls.

Acknowledgement

This work was funded by the U.S. Department of Energy, Division of Chemical Sciences, Office of Basic Energy Sciences.

References

1. B. N. Storhoff and C. H. Lewis, Jr. *Coord. Chem. Rev.*, **23**, 1 (1977).
2. D. P. Tate, W. R. Knipple, and J. M. Augl, *Inorg. Chem.*, **1**, 433 (1962).
3. U. Koelle, *J. Organomet. Chem.*, **133**, 53 (1977).
4. M. O. Albers, N. J. Coville, T. V. Ashworth, E. Singeton, and H. E. Swanepool, *J. Organomet. Chem.*, **199**, 55 (1980).
5. G. J. Kubas, *Inorg. Chem.*, **22**, 692 (1983).
6. G. J. Kubas, H. J. Wasserman, and R. R. Ryan, *Organometallics*, **4**, 2012 (1985); G. J. Kubas, submitted to *Organometallics*.
7. B. L. Ross, J. G. Grasselli, W. R. Ritchey, and H. D. Kaesz, *Inorg. Chem.*, **2**, 1023 (1963).
8. G. J. Kubas, *Inorg. Syn.*, **27** (in press).

7. THE SYNTHESIS OF MOLYBDENUM AND TUNGSTEN DINITROGEN COMPLEXES

Submitted by JONATHAN R. DILWORTH* and RAYMOND L. RICHARDS*
Checked by GRACE J.-J. CHEN† and JOHN W. McDONALD†

The chemistry of bis(dinitrogen) complexes of molybdenum and tungsten is currently of considerable interest because one of the dinitrogen ligands may be induced to form N—H or N—C bonds under mild conditions.[1,2] Where the dinitrogen ligands have monodentate, tertiary phosphine coligands, one dinitrogen may be converted to ammonia (up to 1.98 NH_3 per W atom or 0.7 NH_3 per Mo atom) by treatment of the complexes with sulfuric acid in methanol.[2]

A preparation of *trans*-$[Mo(N_2)_2(dppe)_2]$ (dppe = $Ph_2PCH_2CH_2PPh_2$) that uses triethylaluminum as reductant has already been reported in

---

*A.R.C. Unit of Nitrogen Fixation, University of Sussex, Brighton, Sussex, BN1 9QJ, United Kingdom.
†Charles F. Kettering Research Laboratory, Yellow Springs, OH 45387.

Inorganic Syntheses,[3] but it is less convenient and gives much lower yields than that described below. Other reported methods, using sodium amalgam as reductant, involve a more difficult work-up procedure.[4]

The preparations of the compounds [MoCl$_4$(CH$_3$CN)$_2$], [MoCl$_4$(THF)$_2$], [MoCl$_3$(THF)$_3$], [MoCl$_3$(MeCN)$_3$], [WCl$_4$(PPh$_3$)$_2$], and [WCl$_4$ (dppe)] are also given, since they are critical to the successful synthesis of the dinitrogen complexes, or are useful precursors for further chemistry.

General Procedure

Conventional Schlenk-type glassware[5] is used in the procedures described below and unless otherwise stated, all manipulations are carried out under nitrogen. All solvents are anhydrous and are freshly distilled under nitrogen prior to use. Solutions are stirred with a magnetic stirrer bar unless otherwise stated.

A. BIS(ACETONITRILE)TETRACHLOROMOLYBDENUM(IV), [MoCl$_4$(CH$_3$CN)$_2$]

$$MoCl_5 + \text{excess } CH_3CN \rightarrow [MoCl_4(CH_3CN)_2]$$
$$+ \text{chlorinated organic products}$$

■ **Caution.** *Acetonitrile must be regarded as a toxic material and handled in an efficient hood at all times.*

Procedure

A slight modification of the method reported by Fowles et al.[6] is used. Molybdenum pentachloride (20 g, 0.073 mol) is added slowly over 0.5 h from a Schlenk tube through a glass connecting tube to stirred acetonitrile (100 mL, distilled from CaH$_2$) in a 250-mL Schlenk flask. Addition of acetonitrile to molybdenum pentachloride can produce sufficient local heating in the early stages of addition to cause the acetonitrile to froth from the flask. The reverse addition described here proceeds smoothly, an instantaneous reaction occurring with mild heat evolution and the formation of an orange-brown solid. The resulting suspension is stirred for a further 2 h and is then allowed to stand at room temperature overnight. The bis(acetonitrile) complex is filtered under nitrogen using a 4–6-cm bore Schlenk filter, washed with acetonitrile (20 mL), and dried *in vacuo* (10^{-2} torr) at room temperature. The product is obtained as a fine orange-brown powder (18–20 g, 77–86%).

Anal. Calcd. for $C_4H_6N_2Cl_4Mo$: C, 15.0; H, 1.9; N, 8.8. Found: C, 15.4; H, 1.8; N, 9.0.

Properties

The complex is air sensitive in the solid state and is only sparingly soluble in acetonitrile and nonpolar organic solvents. Its IR spectrum shows medium intensity bands at 2290 and 2310 cm^{-1} assignable to $v(C\equiv N)$. The complex is a convenient precursor for the preparation of a range of molybdenum(IV) complexes by replacement of the relatively labile acetonitrile groups with other ligands.[6]

B. TETRACHLOROBIS(TETRAHYDROFURAN)-MOLYBDENUM(IV), [MoCl$_4$(THF)$_2$][7]

$$[MoCl_4(CH_3CN)_2] + THF(excess) \rightarrow [MoCl_4(THF)_2] + 2CH_3CN$$

Procedure

Tetrahydrofuran (THF) (80 mL, freshly distilled from sodium benzophenone ketyl) is added from a 250-mL Schlenk flask to bis(acetonitrile)tetrachloromolybdenum(IV) (20 g, 0.052 mol) in a 250-mL Schlenk flask and the mixture is stirred rapidly for 2 h to give a yellow suspension of tetrachlorobis-(tetrahydrofuran)molybdenum(IV). The complex is filtered through a Schlenk filter, washed with THF (20 mL), and dried *in vacuo* (10^{-2} torr) at room temperature. The product is obtained as a microcrystalline orange-yellow powder in yields of 15–17 g, 63–71%. The product is not analytically pure.

Anal. Calcd. for $C_8H_{16}O_2Cl_4Mo$: C, 25.1; H, 4.9. Found: C, 26.5; H, 4.2. However, it is sufficiently pure for subsequent reactions.

Properties

The complex is air sensitive in the solid state and almost insoluble in THF, but soluble in dichloromethane. It deteriorates even under nitrogen when stored for a period of weeks and is best used immediately for subsequent syntheses. Its IR spectrum shows an intense band at 820 cm^{-1}, characteristic of coordinated THF. The complex can be used to prepare other molybdenum(IV) complexes by displacement of the THF ligands[7] or nitrido molybdenum(VI) complexes by reaction with trimethylsilyl azide.[8]

C. TRICHLOROTRIS(TETRAHYDROFURAN)MOLYBDENUM(III), [MoCl$_3$(THF)$_3$)][9]

Submitted by JONATHAN R. DILWORTH* and JON ZUBIETA†
Checked by T. ADRIAN GEORGE‡ and JOHN SMITH‡

$$[MoCl_4(THF)_2] \xrightarrow[THF]{Sn\ powder} [MoCl_3(THF)_3]$$

Procedure

All reactions are carried out under nitrogen using dried solvents in conventional Schlenk apparatus. The complex [MoCl$_4$(THF)$_2$] is prepared as above.

Tetrachlorobis(tetrahydrofuran)molybdenum(IV), 5.0 g, is suspended in 60 mL of tetrahydrofuran and stirred with 10 g of coarse tin powder, 20 mesh, at room temperature for 20 min. The solution is filtered, and any [MoCl$_3$(THF)$_3$] product on the sinter is freed from tin by washing through with ~20 mL of dry dichloromethane. The solution is evaporated at 10^{-2} torr to ~30 mL, and the complex is removed by filtration as a pale orange crystalline material. Yield: 3.4 g (62%). The complex is stored under dry argon in a freezer and in the dark. Care should be taken, since the product is extremely moisture-sensitive.

Anal. Calcd. for C$_{12}$H$_{24}$Cl$_3$O$_3$Mo: C, 34.4; H, 5.73. Found: C, 34.1; H, 5.79.

Properties

The complex [MoCl$_3$(THF)$_3$] is crystallized as pale orange needles from dichloromethane/tetrahydrofuran solution. The IR spectrum of the pure complex is free of intense bands in the 900–1000-cm^{-1} region, which is characteristic of molybdenum oxo species. The compound [MoCl$_3$(THF)$_3$] reacts readily with certain 1,1-dithio acids to yield the tris(dithioacid)molybdenum(III) monomers in high yield (~70%).[10] Direct reaction with tertiary phosphines in tetrahydrofuran yields complexes of the type [MoCl$_3$(PR$_3$)$_x$(THF)$_{3-x}$]. Reduction of [MoCl$_3$(THF)$_3$] by sodium amalgam or metallic magnesium in the presence of an excess of the appropriate organophosphine in tetrahydrofuran yields complexes of the type

---

*A.R.C. Unit of Nitrogen Fixation, University of Sussex, Brighton BN1 9RQ, United Kingdom.
†Department of Chemistry, State University of New York at Albany, Albany, NY 12222.
‡Department of Chemistry, University of Nebraska, Lincoln, NE 68588.

[Mo(PRR$'_2$)$_6$] or [Mo(PRR$'_2$)$_4$], depending on the nature of the organo group.[11] Under molecular nitrogen, reaction of [MoCl$_3$(THF)$_3$] with 1,2-ethanediylbis(diphenylphosphine) (diphos) yields trans-[Mo(N$_2$)$_2$(diphos)$_2$] in high yield[2] (see below).

D. TRICHLOROTRIS(ACETONITRILE)MOLYBDENUM(III), [MoCl$_3$(MeCN)$_3$]

Submitted by: P. T. BISHOP and J. R. DILWORTH*
Checked by: R. E. McCARLEY and XIANG ZHENG†

$$[MoCl_4(MeCN)_2] \xrightarrow[MeCN]{Sn} [MoCl_3(MeCN)_3]$$

The complex [MoCl$_4$(MeCN)$_2$] (5.0 g, 0.016 mol) in dry acetonitrile (40 mL) was stirred vigorously under dinitrogen and tin powder (0.9 g, 0.008 mol) added. After stirring for 30 min at room temperature the product formed as a sparingly soluble microcrystalline yellow powder. This was washed twice with small quantities of cold, dry acetonitrile. Yield: ~50%.

Anal. Calcd. for C$_6$H$_9$N$_3$Cl$_3$Mo: C, 22.2; H, 2.8; N, 13.1%. Found: C, 22.1; H, 2.7; N, 12.9.

Properties

The complex is only sparingly soluble in acetonitrile, but dissolves in dichloromethane. Its IR spectrum as a Nujol mull shows medium-intensity bands at 2295 and 2310 cm^{-1} assignable to ν(C≡N). The complex is a convenient alternative to [MoCl$_3$(THF)$_3$] for the synthesis of Mo(III) complexes owing to the greater lability of the MeCN groups.

---

*Department of Chemistry and Biological Chemistry, University of Essex, Colchester, C04 3SQ, United Kingdom.
†Department of Chemistry, Iowa State University, Ames, IA 50011.

E. trans-BIS(DINITROGEN) BIS[1,2-ETHANEDIYLBIS-(DIPHENYLPHOSPHINE)]MOLYBDENUM(0), trans-[Mo(N$_2$)$_2$ (Ph$_2$PCH$_2$CH$_2$PPh$_2$)$_2$]

Submitted by JONATHAN R. DILWORTH* and RAYMOND L. RICHARDS*
Checked by GRACE J.-J. CHEN† and JOHN W. McDONALD†

$$[MoCl_3(THF)_3] + 2Ph_2PCH_2CH_2PPh_2 \xrightarrow[THF]{Mg, N_2}$$

$$trans\text{-}[Mo(N_2)_2(Ph_2PCH_2CH_2PPh_2)_2]$$

Materials

1,2-Ethanediylbis(diphenylphosphine) (1,2-bis(diphenylphosphino)ethane) (dppe) is prepared by the published method[1,2] from triphenylphosphine or obtained commercially from Strem Chemical Co., Andover, MA. The Grignard magnesium turnings are obtained from BDH, Poole, Dorset, United Kingdom and are activated prior to use by heating *in vacuo* (10^{-2} torr) at 150° with iodine.

Procedure

Trichlorotris(tetrahydrofuran)molybdenum(III) (5 g, 0.012 mol), 1,2-bis(diphenylphosphino)ethane (12 g, 0.030 mol), Grignard magnesium turnings (6.0 g, 0.25 mol), and tetrahydrofuran (80 mL, freshly distilled from sodium benzophenone ketyl) in a 250-mL Schlenk flask are stirred as rapidly as possible under nitrogen for 16 h. If initiation of the reduction, as shown by a darkening of the solution from orange to brown, does not occur the solution is warmed to 60° until onset of initiation; thereafter reaction proceeds at 20°. Rapid stirring, with vortexing of the solution is necessary to ensure rapid assimilation of nitrogen from the gas phase. After 16 h the brown solution (some yellow-orange dinitrogen complex may precipitate at this stage) is evaporated to about 40 mL *in vacuo* at 10^{-2} torr and kept at 4° overnight. A mixture of dinitrogen complex and magnesium is then isolated using a Schlenk filter. As the dinitrogen complex can be rather finely divided a wide-bore (~6 cm) Schlenk filter should be used. The mixture of magnesium and dinitrogen complex is transferred to a 500-mL Schlenk flask and sufficient

---

*A.R.C. Unit of Nitrogen Fixation, University of Sussex, Brighton, Sussex, BN1 9QJ, United Kingdom.
†Charles F. Kettering Research Laboratory, Yellow Springs, OH 45387.

warm (50–60°) THF is added to dissolve the dinitrogen complex. Filtration through a Schlenk filter, evaporation *in vacuo* at 10^{-2} torr, and addition of diethyl ether gives the complex as yellow-orange crystals (7–8 g, 62–70%), after drying *in vacuo* (10^{-2} torr) at room temperature.

Alternatively, large quantities of THF can be avoided, and the dinitrogen complex can be separated from the magnesium by extraction at reduced pressure utilizing the apparatus shown in Fig. 1. The mixture of dinitrogen complex and magnesium is filtered using the Schlenk-filter part of the apparatus with stop-cock T_2 closed. The reflux condenser and the 100-mL Schlenk flask containing THF (70 mL) are then attached as shown, stopcock T_2 unopened, and the water bath is warmed to 35–40°. Stopcock T_1 is opened slowly to vacuum until the THF refluxes steadily; then it is closed and

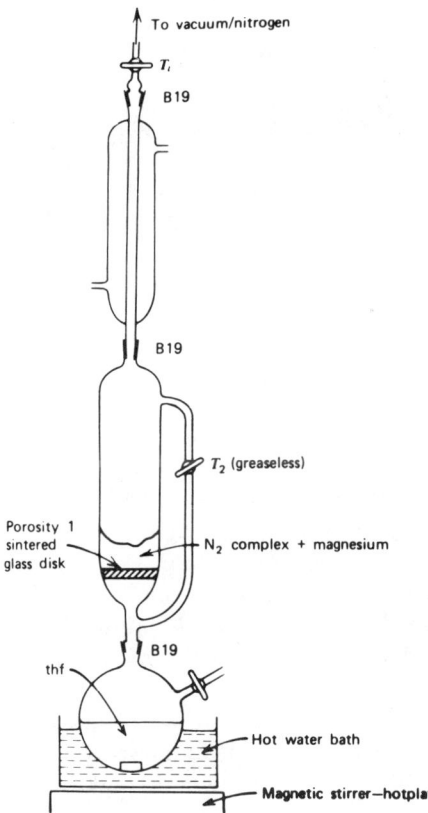

Fig. 1. Apparatus for extraction of the dinitrogen complexes.

extraction is allowed to proceed until all the dinitrogen complex is transferred to the flask. Evaporation of this solution to half volume and addition of diethyl ether then gives the dinitrogen complex as before. The complex can be recrystallized from tetrahydrofuran–diethyl ether.

Anal. Calcd. for $C_{48}H_{52}MoN_4P_4$: C, 65.8; H, 5.1; N, 5.9. Found: C, 65.8; H, 5.0; N, 5.9.

Properties

The complex is only slightly air sensitive in the solid state, but it is rapidly oxidized in solution. It is soluble with very slow decomposition (with loss of dinitrogen) in THF and toluene but reacts with halogenated solvents. Its IR spectrum (Nujol mull) shows an intense absorption at 1970 cm^{-1} and a weak absorption at 2020 cm^{-1}, both assigned to $\nu(N_2)$. The complex reacts with halogen acids to give hydrazido (2—) complexes[2] and with a variety of carbon substrates to give derivatives with N—C bonds.[1]

F. TETRACHLOROBIS(TRIPHENYLPHOSPHINE)TUNGSTEN(IV), $[WCl_4(PPh_3)_2]$[13]

$$WCl_6 + PPh_3(\text{excess}) \xrightarrow[CH_2Cl_2]{Zn} [WCl_4(PPh_3)_2]$$

Procedure

This reaction is carried out in a 500-mL single-necked flask that also has a side arm connected to the nitrogen supply by a stopcock. Dichloromethane (200 mL) is added to tungsten(VI) chloride (25 g, 0.063 mol) and dry granulated zinc (20 g). Then the flask is sealed and shaken for about 1 min, and the pressure is periodically released through the stopcock. Triphenylphosphine (34 g, 0.13 mol) is then slowly added (2–3 min) and the vessel is sealed. It is then shaken until the tungsten(VI) chloride has reacted (about 15 min), while the pressure inside the flask is periodically released. The resulting orange solid is Schlenk filtered and dried *in vacuo* (10^{-2} torr), and any zinc is removed. Alternatively, the reaction mixture is filtered through a Schlenk filter equipped with a perforated rather than sintered-glass disk, which permits the complex to pass through but retains the large lumps of granulated zinc. The product, orange, microcrystalline $[WCl_4(PPh_3)_2]$, is then washed with dichloromethane (30 mL) and dried (10^{-2} torr, 20°) (18 g, 34%). It contains 1 mol of dichloromethane of crystallization and is sufficiently pure for subsequent use.

Anal. Calcd. for $C_{37}H_{32}Cl_6P_2W$: C, 47.4; H, 3.4. Found: C, 47.3; H, 3.5.

Properties

The orange complex is stable in the solid state in the absence of air and is very poorly soluble in organic solvents.

G. TETRACHLORO[1,2-ETHANEDIYLBIS-(DIPHENYLPHOSPHINE)]TUNGSTEN(IV), $[WCl_4(Ph_2PCH_2CH_2PPh_2)]$[13]

$$[WCl_4(PPh_3)_2] + Ph_2PCH_2CH_2PPh_2 \xrightarrow[70°]{\text{toluene}}$$

$$[WCl_4(Ph_2PCH_2CH_2PPh_2)] + 2PPh_3$$

Procedure

Tetrachlorobis(triphenylphosphine)tungsten(IV) (4.5 g, 5.3 mmol) and 1,2-bis(diphenylphosphino)ethane (2.1 g, 5.3 mmol) are heated in toluene (50 mL, distilled from sodium) at 70° for 3 h in a 100-mL flask carrying a reflux condenser. The resulting yellow-green solid is filtered using a Schlenk filter, washed with toluene (30 mL) and diethyl ether (30 mL), and dried (10^{-2} torr, 20°). Yield: 2.4 g, 89%; it is sufficiently pure for subsequent preparations.

Anal. Calcd. for $C_{26}H_{24}Cl_4P_2W$; C, 43.1; H, 3.3. Found: C, 44.5; H, 3.6.

The color of this product is indicative of its efficiency in the subsequent preparation of the dinitrogen complex. Occasionally a pale-yellow or a brown-yellow product has been obtained; both have failed to yield the dinitrogen complex.

Properties

The yellow-green solid is only moderately air sensitive and is poorly soluble in organic solvents.

H. *trans*-BIS(DINITROGEN)BIS[1,2-ETHANEDIYLBIS-(DIPHENYLPHOSPHINE)]TUNGSTEN(0), *trans*-$[W(N_2)_2(Ph_2PCH_2CH_2PPh_2)_2]$[1,2]

$$[WCl_4(Ph_2PCH_2CH_2PPh_2)] + Ph_2PCH_2CH_2PPh_2 \xrightarrow[\text{THF}]{\text{Mg/N}_2}$$

$$\text{trans-}[W(N_2)_2(Ph_2PCH_2CH_2PPh_2)_2]$$

Procedure

Tetrachloro[1,2-ethanediylbis(diphenylphosphine)]tungsten(IV) (3.4 g, 4.7 mmol), magnesium (215 g, 0.1 mol), and 1,2-bis(diphenylphosphino)ethane (2.1 g, 5.3 mmol) are placed in a 250-mL Schlenk flask and pumped under vacuum (10^{-2} torr, 20°) for 2 h. The flask is then filled with nitrogen, tetrahydrofuran (100 mL) is added, and the mixture is stirred for 18 h, after which it is a very dark red-brown. If any yellow-orange precipitate forms, the solution is warmed to about 60° to dissolve it, then filtered, and concentrated in a vacuum to about 40 mL. Orange crystals of *trans*-bis(dinitrogen)-bis[1,2-ethanediylbis(diphenylphosphine)]tungsten(0) deposit on cooling the solution (4°) and are Schlenk filtered, washed with diethyl ether (2 × 30 mL), and dried in a vacuum (10^{-3} torr, 20°). Yield: 3.5 g (72%). The compound may be recrystallized from tetrahydrofuran–diethyl ether.

Anal. Calcd. for $C_{48}H_{52}N_4P_4W$: C, 60.2; H, 4.6; N, 5.4. Found: C, 60.1; H, 4.7; N, 5.4. If the preparation is carried out on a larger scale the dinitrogen complex can be isolated by extraction as described above for *trans*-[Mo(N$_2$)$_2$(dppe)$_2$].

Properties

The complex is slightly air sensitive in the solid state but is quickly oxidized in solution by air. It is soluble in tetrahydrofuran, toluene, and benzene but reacts with chlorinated solvents.[1] It is insoluble in methanol and diethyl ether. Both a strong absorption at 1946 cm^{-1} and a weak one at 2000 cm^{-1} (Nujol) in its IR spectrum are assigned to $\nu(N_2)$.

It reacts with acids to give hydride complexes with or without retention of dinitrogen.[2] The dinitrogen ligands may be replaced by such donor groups as isocyanides.[14] Its reactions, under appropriate conditions, with acids or various organic halides give complexes of —N$_2$R, N$_2$HR, or N—N=CR$_2$ ligands (R=H or organic group).[1,2]

References

1. J. Chatt, A. A. Diamantis, G. A. Heath, N. E. Hooper, and G. J. Leigh, *J. Chem. Soc. Dalton Trans.*, **1977**, 688 and references therein.
2. J. Chatt, A. J. Pearman, and R. L. Richards, *Nature*, **253**, 39 (1975); *J. Chem. Soc. Dalton Trans.*, **1977**, 1852 and references therein.
3. M. Hidai, K. Tominari, Y. Uchida, and A. Misono, *Inorg. Synth.*, **15**, 25 (1974).
4. T. A. George and C. D. Siebold, *Inorg. Chem.*, **12**, 2548 (1973).
5. D. F. Shriver, *The Manipulation of Air-Sensitive Compounds*, McGraw-Hill, New York, 1969.
6. E. A. Allen, B. J. Brisdon, and G. W. A. Fowles, *J. Chem. Soc.*, **1964**, 4531.

7. E. A. Allen, K. Feenan, and G. W. A. Fowles, *J. Chem. Soc.*, **1965**, 1636.
8. J. Chatt and J. R. Dilworth, *J. Indian Chem. Soc.*, **54**, 13 (1977).
9. M. W. Anker, J. Chatt, G. J. Leigh, and A. G. Wedd, *J. Chem. Soc. Dalton Trans.*, **1975**, 2639.
10. J. R. Dilworth and J. Zubieta. *J. Chem. Soc. Dalton Trans.*, **1983**, 397.
11. J. Chatt and A. G. Wedd, *J. Organomet. Chem.*, **27**, C15 (1971).
12. W. Hewertson and H. R. Watson, *J. Chem. Soc.*, **1962**, 1490.
13. A. V. Butcher, J. Chatt, G. J. Leigh, and P. L. Richards, *J. Chem. Soc. Dalton Trans.*, **1972**, 1064.
14. J. Chatt, A. J. L. Pombeiro, R. L. Richards, G. H. D. Royston, K. W. Muir, and R. Walker, *Chem. Commun.*, **1975**, 708.

8. *trans*-BIS[1,2-ETHANEDIYLBIS(DIPHENYL-PHOSPHINE)]BIS(ISOCYANOMETHANE)TUNGSTEN(0)

$$trans\text{-}[W(N_2)_2(dppe)_2] + 2\,CNMe \rightarrow trans\text{-}[W(CNMe)_2(dppe)_2] + 2N_2$$

(dppe = $Ph_2PCH_2CH_2PPh_2$)

Submitted by A. J. L. POMBEIRO* and R. L. RICHARDS†
Checked by B. L. HAYMORE‡

Displacement of ligating dinitrogen by isocyanides constitutes a convenient route for the preparation of isocyanide complexes.[1-4] The first isocyanide complexes prepared by this route[1] involve a molybdenum(0) or tungsten(0) electron-rich metal site, and the general method for their synthesis is given below.

The ligating isocyanide is activated toward electrophilic attack and aminocarbyne-type species are formed by reactions with acids[5-8] or alkylating agents.[9] The nucleophilicity of the isocyanide ligand in those electron-rich complexes[10] is in contrast to its usual electrophilic character[11-13] reported for this ligand when it binds a metal ion in its normal or higher oxidation states.

Procedure

■ **Caution.** *Isocyanides (in particular, isocyanomethane) are extremely toxic and malodorous, and their manipulation should be performed in a well-ventilated hood with great care.*

*Centro de Química Estrutural, Complexo I, Instituto Superior Técnico, 1096 Lisboa 1000 codex. Portugal.
†AFRC-IPSR, Nitrogen Fixation Laboratory. The University of Sussex, Brighton, United Kingdom.
‡Corporate Research Laboratories, Monsanto Co., St. Louis, MO 63167.

The isocyanides can be prepared by the phosgene method of Ugi and co-workers,[14] except for CNMe[15] and CNBu^t,[16] and the dinitrogen trans-[M(N$_2$)$_2$(dppe)$_2$] (M = Mo or W)[17] complexes can also be prepared by published methods.

The trans-[W(N$_2$)$_2$(dppe)$_2$] (2.08 g, 2.0 mmol) is dissolved in dry tetrahydrofuran (THF, 100 mL), under dinitrogen, in a 250 mL single-necked Schlenk flask. The solution is filtered under nitrogen to remove any insoluble impurities. Isocyanomethane (0.28 mL, 5.9 mmol) is added under nitrogen to the filtered solution, which is then refluxed, with stirring, under nitrogen (or argon) for ~6 h. Concentration of the solution almost to dryness, under vacuum and with constant heating at ~50–60°, leads to the precipitation of trans-[W(CNMe)$_2$(dppe)$_2$], as red prisms. The small volume of residual hot solution may be removed by decantation, and the red solid is then thoroughly washed, in air with several (5–10) portions (of ~10–15 mL each) of acetone, to remove a small amount of a yellow, powdery impurity. The pure, red crystals are then dried under vacuum. The yield is about 2.0 g (1.9 mmol, 95%).

Anal. Calcd. for [W(CNMe)$_2$(dppe)$_2$]: C, 63.3; H, 5.1; N, 2.6. Found: C, 63.5; H, 5.3; N, 2.7.

Properties

The product is a red, air-stable, crystalline solid. Its infrared spectrum[1] has a strong, broad band at 1834 cm^{-1} (in a KBr disk), which is assigned to CN stretching. In the mass spectrum the parent ion is observed with a principal peak at m/e = 1062, as required for the predominant ^{184}W isotope of [W(CNMe)$_2$(dppe)$_2$]$^+$.

A series of analogous complexes of Mo(0) or W(0) with a variety of isocyanide ligands may be synthesized similarly:[1] trans-[M(CNR)$_2$(dppe)$_2$] (M = Mo or W; R = Me, Bu^t, Ph, 4-MeC$_6$H$_4$, 4-MeOC$_6$H$_4$, 4-ClC$_6$H$_4$, 2,6-Cl$_2$C$_6$H$_3$. The dinitrogen ligand is more labile when ligating Mo(0) than W(0), and the quantitative formation of the diisocyanide complexes is generally faster for Mo(0) than for W(0) and, in the former, may even occur without heating if it is in the presence of sufficient excess of the appropriate isocyanide.

All the diisocyanide complexes are red (or deep red) in color, and they exhibit a strong broad ν(CN) band in the 1788–1915 cm^{-1} region (in a KBr pellet).[1] Their electronic spectra have been recorded,[18] and their redox properties studied by cyclic voltammetry[1] and by chemical oxidation.[18] The molecular structure has been determined by X-ray diffraction[19] studies for the trans-[Mo(CNMe)$_2$(dppe)$_2$], with the isocyanides showing a CNC bond angle of 156(1)°. The cause of this bending is believed to be electronic in origin, and the electron-rich bent isocyanomethane in the Mo(0) and W(0) complexes is susceptible to attack by alkylating[9] or protonating[5-7] agents, affording aminocarbyne-type complexes.

References

1. J. Chatt, C. M. Elson, A. J. L. Pombeiro, R. L. Richards, and G. H. D. Royston, *J. Chem. Soc. Dalton Trans.*, **1978**, 165.
2. J. Chatt, A. J. L. Pombeiro, and R. L. Richards, *J. Organometal. Chem.*, **190**, 297 (1980).
3. A. J. L. Pombeiro, C. J. Pickett, R. L. Richards, and S. A. Sangokoya, *J. Organometal. Chem.*, **202**, C15 (1980).
4. A. J. L. Pombeiro, C. J. Pickett, and R. L. Richards, *J. Organometal. Chem.*, **224**, 285 (1982).
5. J. Chatt, A. J. L. Pombeiro, and R. L. Richards, *J. Chem. Soc. Dalton Trans.*, **1980**, 492.
6. J. Chatt, A. J. L. Pombeiro, and R. L. Richards, *J. Chem. Soc. Dalton Trans.*, **1979**, 1585.
7. A. J. L. Pombeiro and R. L. Richards, *Transition Met. Chem.*, **5**, 55 (1980).
8. A. J. L. Pombeiro, M. F. N. N. Carvalho, P. B. Hitchcock, and R. L. Richards, *J. Chem. Soc. Dalton Trans.*, **1981**, 1629.
9. J. Chatt, A. J. L. Pombeiro, and R. L. Richards, *J. Organomet. Chem.*, **184**, 357 (1980).
10. A. J. L. Pombeiro, in *New Trends in the Chemistry of Nitrogen Fixation*, J. Chatt, L. M. C. Pina, and R. L. Richards (eds.), Academic Press, New York 1980, Chapter 10.
11. R. L. Richards, *Inorg. Synth.*, **19**, 174 (1979).
12. F. Bonati and G. Minghetti, *Inorg. Chim. Acta*, **9**, 95 (1974).
13. P. M. Treichel, *Adv. Organomet. Chem.*, **11**, 21 (1973).
14. I. Ugi, U. Fetzer, U. Eholzer, H. Knupfer, and K. Offermann, *Angew. Chem. Int. Ed., Engl.*, **4**, 472 (1965).
15. R. E. Schuster, J. E. Scott, and J. Casanova, *Org. Synth.*, **46**, 75 (1966).
16. J. Casanova, N. D. Werner, and A. E. Schuster, *J. Org. Chem.*, **31**, 3473 (1966).
17. J. R. Dilworth and R. L. Richards, *Inorg. Synth.*, **20**, 119 (1980) and section 7, this volume; L. J. Archer, T. A. George, and M. E. Noble, *J. Chem. Ed.*, **58**, 727 (1981).
18. A. J. L. Pombeiro and R. L. Richards, *J. Organomet. Chem.*, **179**, 459 (1979).
19. J. Chatt, K. W. Muir, A. J. L. Pombeiro, R. L. Richards, G. H. D. Royston, and R. Walker, *J. Chem. Soc. Chem. Commun.*, **1975**, 708.

9. TRICARBONYL(CYCLOHEPTATRIENE)-MOLYBDENUM(0)

$$C_7H_8 + Mo(CO)_6 = C_7H_8Mo(CO)_3 + 3CO$$

Submitted by F. A. COTTON,* J. A. McCLEVERTY,* and J. E. WHITE*
Checked by R. B. KING,† A. F. FRONZAGLIA,† and M. B. BISNETTE†

Tricarbonyl(cycloheptatriene)molybdenum has been prepared by refluxing cycloheptatriene with molybdenum hexacarbonyl in benzene.[1] It is a convenient starting material for the preparation of trisubstituted molybdenum carbonyls[2,3] and of tropyliummolybdenum carbonyl salts.[3]

*Massachusetts Institute of Technology, Cambridge, MA.
†Mellon Institute, Pittsburgh, PA.

Procedure

A mixture of 11 g of cycloheptatriene* (this is used in slight excess, without purification), 26 g of molybdenum hexacarbonyl, † and 25 mL of n-octane is refluxed for 8 h. Dibutyl ether is also a good solvent;[4] however, n-octane appears to give slightly better yields. During the refluxing, unreacted molybdenum carbonyl sublimes from the reaction mixture into the condenser and must be poked down into the reaction flask. After cooling, the dark red-brown mixture is filtered, and the brown residue is washed with 20 mL of n-pentane to remove any unreacted cycloheptatriene. Evaporation of the filtrate affords some unreacted molybdenum carbonyl, and this, together with that collected in the condenser, amounts to 3 g. The brown residue is extracted overnight in a Soxhlet apparatus with n-pentane. Red hexagonal prisms precipitate, giving 14 g of the desired product, a yield of 58% based on the Mo(CO)$_6$ consumed in the reaction. (The checkers report 61% yield when using n-octane solvent and only 20% yield using dibutyl ether solvent.) The crystals are separated by filtration and air-dried and are pure enough for most purposes. The compound may be recrystallized, however, from n-hexane using an acetone–Dry Ice cooling bath.

Properties

The product forms as dark red hexagonal prisms or, when finely divided, as orange-red needles. It decomposes at 95° and sublimes rapidly at 85° *in vacuo*. It is very soluble in alcohol, acetone, benzene, chloroform, and dichloromethane, moderately soluble in diethyl ether, and only sparingly soluble in n-hexane. It is scarcely soluble in n-octane and decomposes in carbon tetrachloride. The solutions are sensitive to air and light, and the solid is best stored in the dark.

Normally, replacement of the cycloheptatriene ligand by mono- and tridentate ligands is easily performed, affording the trisubstituted tricarbonyl,[2] but exceptions to this have been noted.[5] The compound may be characterized by its infrared spectrum in the carbonyl region,[1,4] absorptions occurring at 1985, 1919, and 1889 cm^{-1}.

References

1. E. W. Abel, M. A. Bennett, R. Burton, and G. Wilkinson, *J. Chem. Soc.*, **1958**, 4559.
2. E. W. Abel, M. A. Bennett, and G. Wilkinson, *J. Chem. Soc.*, **1959**, 2323.

*Shell Chemical Company.
†Climax Molybdenum Company.

3. H. J. Dauben and H. R. Honnen, *J. Am. Chem. Soc.*, **80**, 5570 (1958).
4. F. A. Cotton and F. Zingales, *Inorg. Chem.*, **1**, 145 (1962).
5. R. B. King, *Inorg. Chem.*, **2**, 936 (1963).

10. TETRACARBONYL(η^2-METHYL ACRYLATE)RUTHENIUM

$$Ru_3(CO)_{12} + 3CH_2=CH-CO_2CH_3 \xrightarrow[t \leq 15°]{h\nu(\lambda \geq 370\,nm)} 3Ru(\eta^2\text{-}CH_2=CHCO_2CH_3)(CO)_4$$

Submitted by F.-W. GREVELS,* J. G. A. REUVERS,* and J. TAKATS†
Checked by B. F. G. JOHNSON‡

Although various synthetic routes have become available for the syntheses of tetracarbonyl(η^2-olefin) complexes of iron,[1] a general high-yield procedure for the preparation of the analogous ruthenium compounds has long been lacking. Photolysis of $Ru_3(CO)_{12}$ in the presence of excess olefin has been reported to yield $Ru(\eta^2\text{-olefin})(CO)_4$ complexes of ethylene,[2] 1-pentene,[3] ethyl acrylate,[4] and diethyl fumarate.[4] However, in no case were analytically pure materials obtained, decomposition often occurring while the excess olefin was being removed from the reaction mixture. A simple method for the photochemical preparation of tetracarbonyl(η^2-methyl acrylate)ruthenium is described here. A similar procedure affords the corresponding (η^2-dimethyl fumarate) and (η^2-dimethyl maleate) complexes in nearly quantitative yield.[5]

Procedure

■ **Caution.** *Ruthenium carbonyl complexes must be handled as toxic compounds in a well-ventilated fume hood. In particular, any kind of bodily contamination, orally or via skin contact, must be strictly avoided. Gloves should be worn.*

All operations are carried out under an atmosphere of argon. Dodecacarbonyltriruthenium can be purchased (Strem Chemicals) or it can be prepared from $RuCl_3 \cdot 3H_2O$.[6,7] Methyl acrylate (synthetic grade, 99%) (Aldrich) is used as received. Hexane (95%) (Aldrich) is distilled under argon before use.

---

*Max-Planck-Institut für Strahlenchemie, Stiftstrasse 34-36, D-4330 Mülheim a.d. Ruhr, Federal Republic of Germany.
†Department of Chemistry, University of Alberta, Edmonton, Alberta, Canada T6G 2G2.
‡University Chemical Laboratory, Lensfield Road, Cambridge, CB2 1EW, United Kingdom.

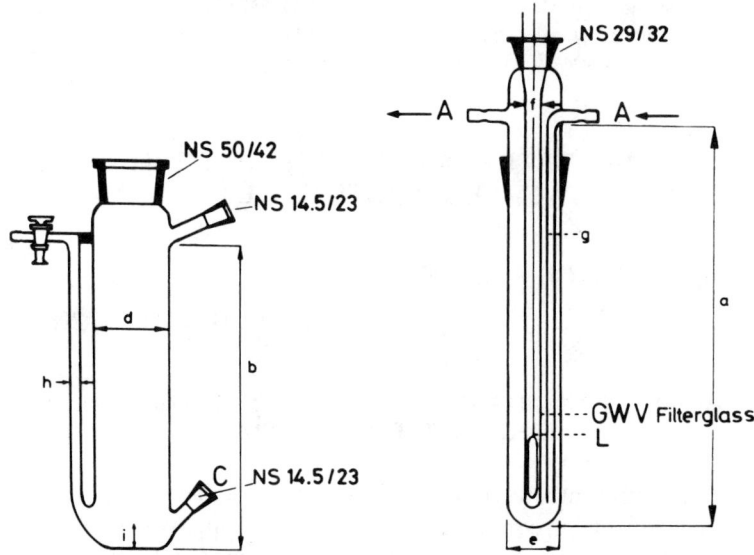

Fig. 1. Irradiation apparatus: A, water cooling; B, Ar inlet; C, septum (rubber) for withdrawing IR samples; L = high-pressure mercury lamp (Philips HPK, 125-W), used in connection with a Philips VG1/HP 125-W power supply converter unit. Dimensions (in millimeters): a = 400, b = 240, d = 70, e = 44, f = 28, g = 6, h = 10, i = 10.

Dodecacarbonyltriruthenium (3.20 g, 5.0 mmol), methyl acrylate (8.6 g, 100 mmol), and hexane (250 mL) are placed in a 300-mL photochemical reaction vessel (Fig. 1). The light source, a high-pressure mercury lamp Philips HPK 125 W, is located inside the reactor and is surrounded by a GWV cutoff filter tube,* $\lambda \geq 370$ nm. Argon is bubbled through the solution, via inlet B, before the light source is turned on. As the reaction proceeds, on irradiation at 10–15°, the solid $Ru_3(CO)_{12}$ gradually dissolves. Irradiation is continued until the orange-yellow color of $Ru_3(CO)_{12}$ has disappeared (5–8 h). The reaction can also be monitored conveniently by means of IR spectroscopy, which shows the exclusive formation of $Ru(\eta^2$-methyl acrylate$)(CO)_4$ [as well as the disappearance of the v_{CO} bands at 2061 (vs), 2031 (s), 2017 (w), and 2011 (m) cm^{-1} due to $Ru_3(CO)_{12}$] with v_{CO} bands at 2121 (w), 2049.5 (s), 2035 (s), and 2008.5 (s) cm^{-1}.

The solution is filtered if necessary, cooled to $-78°$, and allowed to remain at this temperature for several days. The complex precipitates as colorless

---

*Glaswerk Wertheim, Ernst-Abbe-Strasse 1, D-6980 Wertheim/Main, FRG. The Max-Planck-Institut für Strahlenchemie will pass on tube material at cost.

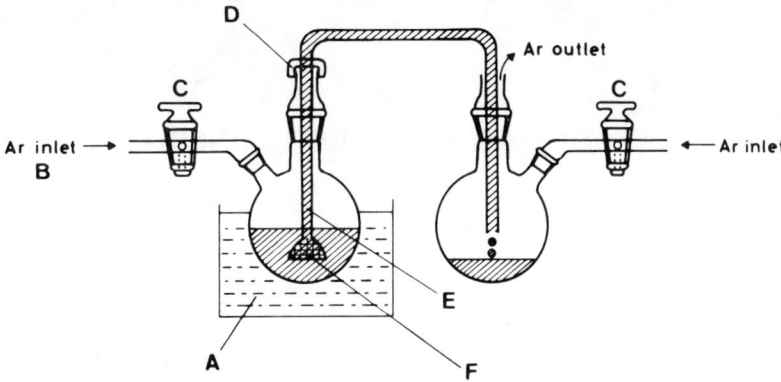

Fig. 2. Inverse filtration at low temperature: A, cooling bath, Dry Ice–acetone; B, argon pressure maintained at 10–20 torr above atmospheric pressure; C, three-way stopcock; D, rubber cap; E, polyethylene tube, 3 mm, widened to 10 mm at one end; F, filter wad, cotton or glass wool.

crystals. The supernatant solution is removed by inverse filtration (Fig. 2) and the crystals are dried under vacuum at $-30°$. Yield: 3.50 g of Ru(η^2-methyl acrylate)(CO)$_4$ (78%).* A second crop can be obtained from the mother liquor by concentrating it to one-fourth of its original volume and cooling to $-78°$ for several days. As described above, the colorless crystals are isolated and dried under vacuum at $-30°$ (0.81 g; 18%). Ru(η^2-methyl acrylate)(CO)$_4$ may be recrystallized from hexane (precooled, $\leq -30°$) to which 0.5% methyl acrylate is added.

Anal. Calcd. for C$_8$H$_6$O$_6$Ru: C, 32.11; H, 2.02. Found: C, 32.16; H, 1.88.

Properties

Tetracarbonyl(η^2-methyl acrylate)ruthenium is obtained as a colorless solid that is stable indefinitely at temperatures below $-30°$ under argon. The complex is soluble in organic solvents but decomposes unless free methyl acrylate is added to the solvent. It appears that an equilibrium is established [eq. (1)], involving the complex, methyl acrylate, and the species [Ru(CO)$_4$]. Excess free methyl acrylate shifts the equilibrium to the left, thereby preventing decomposition of the complex and facilitating the workup of the

---

* The checkers found that isolation of the product is difficult largely because of its instability above $-30°$. However, since its use is that of a precursor to other materials, it can be used in situ without having been isolated.

$$Ru(\eta^2-CH_2=CH-CO_2CH_3)(CO)_4 \rightleftharpoons CH_2=CH-CO_2CH_3 + [Ru(CO)_4] \quad (1)$$
(I)

$$\downarrow h\nu \; (\lambda \geq 280 \text{ nm}) \quad |-CO, +CH_2=CH-CO_2CH_3$$

$$Ru(\eta^2-CH_2=CH-CO_2CH_3)_2(CO)_3$$

$$LRu(CO)_4 \uparrow +L$$

$$\downarrow$$

$$\tfrac{1}{3}Ru_3(CO)_{12}$$
+ other decomposition products

reaction mixture. This moderate stability of Ru(η^2-methyl acrylate)(CO)$_4$ at room temperature establishes it as a useful source of the moiety [Ru(CO)$_4$] under mild conditions. For example, addition of a cooled ($-30°$) hexane solution of a suitable ligand L to a similar solution of **I**, followed by the slow warming up of the reaction mixture to ambient temperature, leads to the mononuclear complex LRu(CO)$_4$ [eq. (1), L = dimethyl fumarate, fumaronitrile, maleic anhydride, trimethyl phosphite, triphenylphosphine].[5,8] In addition to these ligand exchange reactions, compound **I** reacts at ambient temperature (20°) with methyl sorbate and diethyl 2,4-hexadienedioate, respectively, to yield Ru(η^4-diene)(CO)$_3$ complexes, or with diethyl 2,4-hexadienedioate at slightly lower temperature, to a novel triruthenium cluster: undecacarbonyl μ_3-[(1-η^1:2,3-η^2:4-η^1-diethyl 2,4-hexadienedioate)] triruthenium.[8] Dimethyl acetylenedicarboxylate is trimerized by **I** to hexamethyl benzenehexacarboxylate, C$_6$(CO$_2$CH$_3$)$_6$, at $T \leq 25°$.[8]

The use of a cutoff filter with $\lambda \geq 370$ nm is recommended in order to prevent secondary photoreactions, such as the substitution of carbon monoxide for an olefin ligand to give Ru(η^2-methyl acrylate)$_2$(CO)$_3$.[9] The colorless Ru(η^2-methyl acrylate)(CO)$_4$ (**I**) is transparent at $\lambda \geq 370$ nm, that is, in the region of the long-wavelength absorption maximum of the starting material Ru$_3$(CO)$_{12}$ at 390 nm. The absorption curve of **I** increases almost monotonically from about 370 nm to a maximum at 268 nm ($\varepsilon \approx 7000$, in hexane that contains 0.5% methyl acrylate; the same solution is used in the reference cell). The IR spectrum of **I** exhibits four bands in the metal carbonyl region at 2121 (w), 2049.5 (s), 2035 (s), and 2008.5 (s) cm^{-1} and an ester carbonyl band at 1715 cm^{-1}. This is consistent with a trigonal-bipyramidal geometry in which the olefin occupies an equatorial position (C_{2v} local symmetry). ^{1}H NMR data: δ 1.74 (dd, 3 Hz, 8.1 Hz), H^1; 2.47 (dd, 3 Hz, 11.1 Hz), H^2; 2.81 (dd, 8.1 Hz, 11.1 Hz), H^3; 3.29 s, H^4; in toluene-d_8 at $-40°$. ^{13}C NMR data: δ 35.5 [d, 159 Hz, C(1)]; 23.9 [t, 161 Hz, C(2)]; 51.1 [q, 148 Hz, C(4)]; 176.3 [s, C(3)];

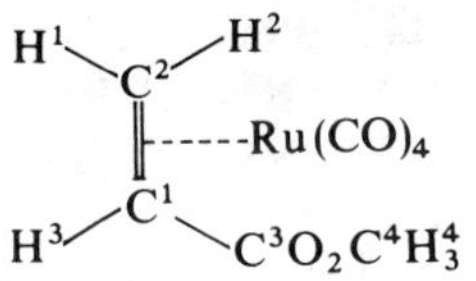

$$\begin{array}{c} H^1 \diagdown \diagup H^2 \\ C^2 \\ \Vert \text{-----Ru(CO)}_4 \\ C^1 \\ H^3 \diagup \diagdown C^3O_2C^4H_3^4 \end{array}$$

Fig. 3. Tetracarbonyl(η^2-methyl acrylate)ruthenium.

and 193.6, 194.8, 195.5, and 197.6 (CO); in toluene-d_8 at $-50°$ (Bruker WH 270; 270 and 67.89 MHz, respectively) (Fig. 3).

The experimental procedure described here for the preparation of Ru(η^2-methyl acrylate)(CO)$_4$ is applicable to a variety of other olefins as manifested by the syntheses of Ru(η^2-olefin)(CO)$_4$ complexes of, for example, dimethyl fumarate, dimethyl maleate, allyl acrylate, methyl vinyl ketone (3-butene-2-one), and acrylonitrile.[5]

Acknowledgment

This work was supported by the Alexander von Humboldt Foundation through a stipend (to J. T.). We thank Mr. K. Schneider for technical assistance.

References

1. R. B. King, in *The Organic Chemistry of Iron*, Vol. 1, E. A. Koerner von Gustorf, F.-W. Grevels, and I. Fischler (eds.), Academic Press, New York, 1978, p. 397.
2. B. F. G. Johnson, J. Lewis, and M. V. Twigg, *J. Organomet. Chem.*, **67**, C75 (1974).
3. R. G. Austin, R. S. Paonessa, P. J. Giordano, and M. S. Wrighton, *Adv. Chem. Ser.*, **168**, 189 (1978).
4. L. Kruczynski, J. L. Martin, and J. Takats, *J. Organomet. Chem.*, **80**, C9 (1974).
5. F.-W. Grevels, J. G. A. Reuvers, and J. Takats, *J. Am. Chem. Soc.*, **103**, 4069 (1981).
6. A. Mantovani and S. Cenini, *Inorg. Synth.*, **17**, 47 (1976).
7. (a) M. I. Bruce, J. G. Matisons, R. C. Wallis, J. M. Patrick, B. W. Skelton, and A. H. White, *J. Chem. Soc. Dalton Trans.*, **1983**, 2365; (b) M. I. Bruce, C. M. Jensen and N. L. Jones, *Inorg. Synth.*, **26**, 259 (1989) and section 55, this volume.
8. F.-W. Grevels and J. G. A. Reuvers, unpublished results; reported in part at the 10th International Conference on Organometallic Chemistry, Toronto, Canada, August 9–15, 1981, Contribution No. 1A-06.
9. F.-W. Grevels, J. G. A. Reuvers, and J. Takats, *Angew. Chem.*, **93**, 475 (1981); *Angew: Chem. Int. Ed. Engl.*, **20**, 452 (1981).

11. REAGENTS FOR THE SYNTHESIS OF η-DIENE COMPLEXES OF TRICARBONYLIRON AND TRICARBONYLRUTHENIUM

Submitted by A. J. P. DOMINGOS,* J. A. S. HOWELL,* B. F. G. JOHNSON,* and J. LEWIS*
Checked by N. GRICE† and R. PETTIT†

Previous syntheses of tricarbonyl(η-diene)iron complexes have relied mainly on the reaction of $Fe(CO)_5$, $Fe_3(CO)_{12}$, or $Fe_2(CO)_9$ with the free diene. The use of the first two carbonyls suffers from the prolonged reflux times and/or ultraviolet irradiation necessary to obtain reaction and the consequent low yields and mixtures of complexes obtained with heat- and ultraviolet-sensitive dienes.[1] The latter reagent, although utilized at lower temperatures, may react with polyenes ($n \geq 3$) to give mixtures containing, in addition to the expected product, binuclear derivatives containing a metal–metal bond.[2]

The only readily available ruthenium carbonyl, $Ru_3(CO)_{12}$, reacts in a complex manner with many polyolefins to give, in addition to mononuclear derivatives, complexes retaining the Ru_3 cluster and products resulting from hydrogen abstraction.[3]

Described here is the preparation of (benzylideneacetone‡)tricarbonyliron(0) and (1,5-cyclooctadiene)tricarbonyl ruthenium(0). Both compounds function as convenient sources of the metal tricarbonyl moiety by displacement of the organic ligand under mild conditions.

A. (BENZYLIDENEACETONE)TRICARBONYLIRON(0)

$$C_6H_5CH=CHCOCH_3 + Fe_2(CO)_9 \xrightarrow[60°]{\text{toluene}}$$

$$(C_6H_5CH=CHCOCH_3)Fe(CO)_3 + Fe(CO)_5 + CO$$

While a great number of tricarbonyl(η-diene)iron complexes have been reported and their reactivity investigated, much less is known of the corresponding heterodiene complexes. In recent years, synthesis of several tricarbonyl(heterodiene)iron systems involving[3] η coordination of the heterodiene unit has been achieved.[4-6] Among the tetracarbonyl(η-olefin)iron complexes prepared by Weiss[7] was tetracarbonyl(cinnamaldehyde)iron,

---

*University Chemical Laboratory, Lensfield Road, Cambridge CB2 1EW, United Kingdom.
†Chemistry Department, University of Texas at Austin, Austin, TX. 78712.
‡4-Phenyl-3-buten-2-one.

which converts on heating to the η-bonded tricarbonyl(cinnamaldehyde)iron.[8] The preparation and synthetic utility of (benzylideneacetone)tricarbonyliron, an analogous complex of an α,β-unsaturated ketone, are reported here.

Procedure

■ **Caution.** *The reaction should be carried out in a well-ventilated hood, since volatile, toxic iron pentacarbonyl is obtained as a by-product.* .

In a 250-mL flask, benzylideneacetone* (10.4 g, 0.07 mol) and $Fe_2(CO)_9$[9] (26 g, 0.07 mol) are heated in toluene† (100 mL) under nitrogen with magnetic stirring for 4–5 h at 60°. After removal of the solvent and iron pentacarbonyl under vacuum, the residue is dissolved in 20 mL of 10% ethyl acetate–toluene, filtered through kieselguhr, and chromatographed on silica gel (2.5 × 40 cm). Elution with the same solvent develops a red band that, on removal of the solvent, gives the product as orange-red crystals (6.07 g, 32%, mp 88–89°). The complex may be recrystallized from hexane.

Anal. Calcd. for $C_{13}H_{10}O_4Fe$: C, 54.5; H, 3.5. Found: C, 54.4; H, 3.6.

Properties

(Benzylideneacetone)tricarbonyliron is stable indefinitely when stored under nitrogen and is soluble in most common organic solvents. The infrared spectrum exhibits three metal carbonyl frequencies at 2065, 2005, and 1985 cm^{-1} (cyclohexane). The NMR spectrum ($H\alpha$, δ 6.02, doublet; $H\beta$, δ 3.10, doublet; $J = 6.0$ Hz; CH_3, δ 2.50, singlet; Ph, δ 7.27, multiplet) is consistent with its η-bonded structure.

The complex functions as a convenient source of the tricarbonyliron moiety by displacement of the α,β-unsaturated ketone. For example, reaction with 1,3-cycloheptadiene results in a 78% yield of (η-1,3-cycloheptadiene)tricarbonyliron. More importantly, it may be used in syntheses of tricarbonyl(diene)iron complexes where the iron carbonyls are not satisfactory. Several complexes of sensitive heptafulvenes have been prepared in this way, and the reagent has been used in the synthesis of tricarbonyliron complexes of several steroids.

These reactions may be carried out under mild conditions (60°, benzene, 4–8 h) and are clean, in that no colloidal iron or other iron carbonyls are produced. The work-up is thus much easier, and the reactions can be

---

*Koch-Light Laboratories, Colnbrook, Bucks., United Kingdom.
†The checkers recommend the use of benzene in place of toluene.

monitored by infrared spectroscopy. In addition, the well-defined stoichiometry of the displacement is in marked contrast to the uncertain stoichiometry of the reactions of the iron carbonyls, which are commonly used in large excess.

B. (1,5-CYCLOOCTADIENE)TRICARBONYLRUTHENIUM(0)

$$1,5\text{-}C_8H_{12} + Ru_3(CO)_{12} \xrightarrow[\text{reflux}]{\text{benzene}} (1,5\text{-}C_8H_{12})Ru(CO)_3$$

In contrast to the very large number of tricarbonyl(η-diene)iron complexes described in the literature,[10–12] the corresponding ruthenium compounds have received very little attention. This may reflect the well-documented tendency of ruthenium to form metal–metal bonds as opposed to iron.[13] In particular, while the metal–metal bonds in $Fe_3(CO)_{12}$ are easily broken, $Ru_3(CO)_{12}$ undergoes a variety of reactions in which the Ru_3 cluster is retained.

Described here is the preparation of (1,5-cyclooctadiene)-tricarbonylruthenium and its use in the synthesis of tricarbonyl(η-diene)ruthenium complexes.

Procedure

In a 100-mL flask, $Ru_3(CO)_{12}$* (960 mg, 1.5 mmol) is reacted for 8 h† with 1,5-cyclooctadiene‡ (15 mL, 0.12 mol) in refluxing benzene (75 mL) under nitrogen and the excess of diene distilled ($\sim 30°$, 7 torr). Unreacted $Ru_3(CO)_{12}$ (307 mg, 32%) is separated by treating the residue with pentane. The pentane solution is concentrated and chromatographed§ with pentane on a silica gel column (1.5 × 30 cm) to separate traces of $(1,2,3,6\text{-}\eta\text{-}C_8H_{12})(RuCO)_3$ and small amounts of $Ru_3(CO)_{12}$. The major band, on concentration, gives a yellowish oil that is sublimed at 0° and 0.005 torr onto a cold finger at $-30°$. A second chromatography and sublimation yield (1,5-cyclooctadiene)tricarbonylruthenium (684 mg, 58%) as white crystals melting at room temperature.

Anal. Calcd. for $C_{11}H_{12}O_3Ru$; C, 44.5; H, 4.1. Found: C, 44.5; H, 4.3.

---

*See synthesis 55 in this volume.
†Refluxing for longer periods results in significant isomerization to $(1,2,3,6\text{-}\eta\text{-}C_8H_{12})Ru(CO)_3$.
‡Koch-Light Laboratories, Colnbrook, Bucks., United Kingdom.
§The chromatography is continuously monitored by passage of the eluant through an infrared solution cell.

Properties

(1,5-Cyclooctadiene)tricarbonylruthenium is unstable even at $-20°$ under nitrogen but may be kept in a frozen benzene solution for a month without decomposition. The infrared spectrum exhibits metal carbonyl absorptions at 2043, 1982 (sh), and 1967 (br) cm^{-1} (heptane). The NMR spectrum shows multiplets at δ 3.72 and δ 2.4–1.8 due to the olefinic and methylenic protons, respectively.

The complex functions as the most suitable source of the tricarbonylruthenium unit in syntheses of tricarbonyl(η-diene)ruthenium complexes. Derivatives of 1,3-cyclohexadiene, 1,3-cycloheptadiene, cycloheptatriene, cyclooctatetraene, 2,4,6-cycloheptatrien-1-one, bicyclo[3.2.1]octa-2,6-diene, bicyclo-[3.2.1]octa-2,4-diene, and butadiene have been prepared by displacement of 1,5-cyclooctadiene.

Although alternative procedures exist for the preparation of some of these compounds, the ligand-displacement method possesses two distinct advantages: (1) the absence of side reactions leads to high yields and facilitates the purifications, and (2) the mild conditions required for the reaction to go to completion (refluxing benzene, 30 min) avoid extensive decomposition of the complex formed; this makes the method invaluable where the complexes are very unstable.

References

1. For example, W. McFarlane, L. Pratt, and G. Wilkinson, *J. Chem. Soc.*, **1963**, 2162.
2. G. F. Emerson, R. Pettit, J. E. Mahler, and R. Collins, *J. Am. Chem. Soc.*, **86**, 3591 (1964).
3. For example, A. J. P. Domingos, B. F. G. Johnson, and J. Lewis, *J. Organomet. Chem.*, **36**, C43 (1972); M. I. Bruce, M. A. Cairns, and M. Green, *J. Chem. Soc., Dalton Trans.*, **1972**, 1293.
4. S. Otsuka, T. Yoshida, and A. Nakamura, *Inorg. Chem.*, **6**, 20 (1967).
5. H. T. Dieck and H. Beck, *Chem. Comm.*, **1968**, 678.
6. M. Dekker and G. R. Knox, *Chem. Comm.*, **1967**, 1243.
7. E. Weiss, K. Stark, J. E. Lancaster, and H. D. Murdoch, *Helv. Chim. Acta*, **46**, 288 (1963).
8. K. Stark, J. E. Lancaster, H. D. Murdoch, and E. Weiss, *Z. Naturforsch.*, **19**, 284 (1964).
9. R. Pettit, G. F. Emerson, and J. Mahler, *J. Chem. Educ.*, **40**, 175 (1963).
10. R. B. King, in *Organometallic Syntheses*, J. J. Eisch and R. B. King (eds.), Vol. I, p. 93, Academic Press, New York, 1965.
11. R. Pettit and G. F. Emerson, *Adv. Organomet. Chem.*, **1**, 1 (1964).
12. H. W. Quinn and J. H. Tsai, *Adv. Inorg. Radiochem.*, **12**, 217 (1969).
13. M. I. Bruce and F. G. A. Stone, *Angew. Chem., Int. Ed.*, **7**, 427 (1968).

12. DIHYDRIDOBIS(SOLVENT)BIS(TRIPHENYL-PHOSPHINE)IRIDIUM(III) TETRAFLUOROBORATES

Submitted by ROBERT H. CRABTREE,* MICHELLE F. MELLEA,* and JEAN M. MIHELCIC*
Checked by AYUSMAN SEN† and VENKATASURYANARYANA CHEBOLU†

Complexes between organometallic or metal hydride complexes and hard bases have attracted attention recently.[1] In many cases the reaction solvent plays the role of a Lewis base; such complexes can be catalytically active by solvent ligand dissociation.[2,3] We describe here the complexes $[IrH_2S_2(PPh_3)_2][BF_4]$, where $S = Me_2CO$,[4,5] H_2O,[5] or $S_2 = o$-diiodobenzene.[6] The oxygen-donor complexes are very labile[7] and are even capable of dehydrogenating alkanes.[8] The acetone complex seems to have been reported first by Araneo et al.[9] in 1965, but they formulated it as $[IrH_2L_2]^+$. The aqua[5] and halocarbon[6] complexes were reported by us much more recently, the latter being the first crystallographically characterized halocarbon complex.

Several of the syntheses described in this chapter start from the well-known $[Ir(cod)(PPh_3)_2][BF_4]$, (cod = 1,5-cyclooctadiene), which can easily be obtained[4,10] from the commercially available $[Ir(cod)Cl]_2$ (Strem Chemicals).

Our own synthesis is a modification of the ones that have been described but in which we use $Ag[BF_4]$ to abstract chloride ion, this prevents the formation of $Ir(cod)Cl(PPh_3)$ as a side product.

A. (η^4-1,5-CYCLOOCTADIENE)BIS(TRIPHENYLPHOSPHINE)-IRIDIUM(1+) TETRAFLUOROBORATE(1−)

$$[Ir(C_8H_{12})Cl]_2 + 2P(C_6H_5)_3 + Ag[BF_4] \rightarrow [Ir(C_8H_{12})(P(C_6H_5)_3)_2][BF_4] + AgCl$$

Precautions against the admission of air must be taken in this preparation; an N_2 or Ar atmosphere is satisfactory. A 100-mL Schlenk tube is equipped with a rubber septum and a magnetic stirrer. To an evacuated flask filled with N_2 were added 700 mg of dichlorobis(η^4-1,5-cyclooctadiene)diiridium(I) (1.04 mmol), 546 mg of triphenylphosphine (2.09 mmol), and 224 mg (1.15 mmol) of silver tetrafluoroborate. Degassed (but not necessarily dry) methanol (50 mL) is added and the mixture stirred at room temperature for

---

*Department of Chemistry, Yale University, 225 Prospect Street, New Haven, CT 06520.
†Department of Chemistry, Pennsylvania State University, University Park, PA 16820.

24 h. This step is best done in the dark by covering the flask with aluminum foil. The solvent is then completely removed *in vacuo* and 20 mL of degassed dichloromethane added to the solid residue. The resulting mixture is filtered in an inert atmosphere, preferably through Celite, and the filtrate collected. The dichloromethane is reduced to ~ 2 mL *in vacuo* and 20 mL of degassed diethyl ether slowly added through a syringe to precipitate the red product, which is filtered and washed with three portions of 5 mL of diethyl ether. The crude material is then recrystallized from the minimum volume of dichloromethane by the addition of diethyl ether to give a fine red powder, and filtered and washed as above. Yield: 1.8 g (90%).

Anal. Calcd. for $C_{44}H_{42}P_2BF_4Ir \cdot CH_2Cl_2$: C, 56.22; H, 4.61. Found: C, 56.10; H, 4.71.

Properties

The complex is air stable in the solid state but slightly air sensitive in solution. It is soluble in CH_2Cl_2, $CHCl_3$, and Me_2CO, but insoluble in H_2O, C_6H_6, and alkanes. Other characteristics are given in the literature.[4, 10]

B. BIS(ACETONE)DIHYDRIDOBIS(TRIPHENYL-PHOSPHINE)IRIDIUM(1+) TETRAFLUOROBORATE(1−)

$$[Ir(C_8H_{12})\{P(C_6H_5)_3\}_2][BF_4] + 2(CH_3)_2CO + 3H_2 \rightarrow$$
$$[IrH_2\{(CH_3)_2CO\}_2\{P(C_6H_5)_3\}_2][BF_4] + C_8H_{16}$$

Precautions against the admission of air were taken. A 100-mL Schlenk tube is equipped with a rubber septum and a magnetic stirrer, distilled (but not especially dried) acetone (10 mL) is then added, followed by 500 mg (0.543 mmol) of (η^4-1,5-cyclooctadiene)bis(triphenylphosphine)iridium(1+)-tetrafluoroborate(1−).[10] The flask is evacuated briefly (10 s) and N_2 introduced. The mixture is cooled to 0 °C by immersion in an ice bath, and magnetically stirred. A long steel needle is passed through the septum, and by this means a gentle stream of hydrogen (~ 10 mL min^{-1}) is introduced into the mixture, a second needle, (not dipping in the solution) serving as an exhaust for the spent gases. The red solution becomes colorless or yellow. After 30 min, this treatment is stopped, and the gas entry needle is removed. Without removing the flask from the ice bath, the hydrogen from the flask, or the exhaust needle from the septum, 25 mL of diethyl ether is slowly added to the stirred solution with a syringe. Over a few minutes the crude product precipitates. The mixture is left for 3 h to complete the precipitation. This light tan material is recrystallized by dissolving in degassed dichloromethane (10 mL) and adding diethyl ether (~25 mL) as an upper layer. If the flask is placed in a cold room for several days, crystals appear. Alternatively, one can stir the mixture, in which case a microcrystalline colorless powder precipitates. The product is removed by filtration, washed with diethyl ether (3 × 5 mL), and dried *in vacuo*. Yield: 510 mg (93%).*

Anal. Calcd. for $C_{42}H_{44}O_2P_2F_4BIr \cdot CH_2Cl_2$: C, 51.30; H, 4.60. Found: C, 51.25; H, 4.60%.

Properties

The complex is air stable and soluble in CH_2Cl_2, Me_2CO, and $CHCl_3$. It is insoluble in H_2O, C_6H_6, and alkanes. It is best identified by its strong (C=O) stretching absorptions at 1713 and 1666 cm^{-1} and the (Ir—H) stretching mode at 2257 cm^{-1} (Nujol). The ^{1}H NMR (CD_2Cl_2, 25 °C) shows an IrH resonance at -27.8δ (triplet, $^2J_{P,H} = 17$ Hz) and a Me_2CO resonance at 1.8 δ. The complex has also been crystallographically characterized.[11] In contrast to many other acetone complexes,[2] it is not moisture sensitive; water seems to bind more weakly than the ketone. It is catalytically active for alkene hydrogenation[12] and isomerization[13] and for cyclohexene aromatization.[14]

C. DIAQUADIHYDRIDOBIS(TRIPHENYLPHOSPHINE)-IRIDIUM(1+) TETRAFLUOROBORATE(1−)

$$[Ir(C_8H_{12})\{P(C_6H_5)_3\}_2][BF_4] + 2H_2O + 3H_2 \rightarrow$$
$$[IrH_2(H_2O)_2\{P(C_6H_5)_3\}_2][BF_4] + C_8H_{16}$$

To a 100-mL Schlenk tube equipped as in the preparation in Section B, are added distilled water (60 mL) and 500 mg of (η^4-1,5-cyclooctadiene)bis-(triphenylphosphine)iridium(1+) tetrafluoroborate(1−) (0.543 mmol). The flask is evacuated briefly and filled with N_2. Hydrogen is gently bubbled through the solution as in the previous preparation for up to 3 h or until the original red suspension has turned to a yellow suspension. The time taken for this heterogeneous reaction is very sensitive to the exact conditions. The yellow solid is removed by filtration and recrystallized from dichloromethane (10 mL). Droplets of water are often present at this stage. These are removed by careful decantation of the organic phase with a hypodermic syringe. To the resulting organic solution is added 20 mL of olefin-free pentanes, bp 30–40 °C

---

*The checkers found 47% yield.

(Aldrich). The white complex precipitates, is filtered, and washed with pentanes (3 × 5 mL). Yield: 325 mg (65%).*

Anal. Calcd. for $C_{36}H_{36}P_2F_4BO_2Ir \cdot CH_2Cl_2$: C, 47.96, H, 4.13. Found: C, 47.70; H, 4.13.

Properties

The aqua complex closely resembles the acetone complex described above. The chief distinguishing features are a triplet resonance at $-29.8 \delta (^2J_{P,H} = 16$ Hz), an H_2O resonance at 2.4 δ, and a v_{Ir-H} vibration in the IR at 2280 cm^{-1} (w). The H_2O vibrations appear at 3650 (m) and 1610 cm^{-1} (m).

D. (1,2-DIIODOBENZENE)DIHYDRIDOBIS(TRIPHENYL-PHOSPHINE)IRIDIUM(1+) TETRAFLUOROBORATE(1−)

$$[IrH_2\{(CH_3)_2CO\}_2\{P(C_6H_5)_3\}_2][BF_4] + C_6H_4I_2 \rightarrow$$
$$[IrH_2(C_6H_4I_2)\{P(C_6H_5)_3\}_2][BF_4] + 2(CH_3)_2CO$$

Bis(acetone)dihydridobis(triphenylphosphine)iridium(1+) tetrafluoroborate(1−) (40.5 mg, 0.044 mmol), as prepared in Section C, is dissolved in 20 mL of degassed dichloromethane in a 250-mL Schlenk tube. To this is added 1,2-diiodobenzene (degassed, 500 mg, 1.5 mmol) and the mixture stirred for 10 min at room temperature. Diethyl ether (50 mL) is added to precipitate the crude product, which is then washed with diethyl (3 × 5 mL) and dried *in vacuo*. This material is recrystallized by dissolving in degassed CH_2Cl_2 (5 mL) and adding degassed Et_2O (15 mL) slowly with stirring. Yield: 41.7 mg (79%).†

Anal. Calcd. for $C_{42}H_{36}I_2F_4BIr \cdot \frac{1}{4}CH_2Cl_2$: C, 43.87; H, 3.18. Found: C, 43.77; H, 3.23%.

Properties

The complex can be identified from the ^1H NMR resonance at -16.5δ (triplet $^2J_{P,H} = 13$ Hz) and by the IR absorption at 2217 cm^{-1} (w). It has been characterized crystallographically[6] and has normal Ir—I covalent bonds [2.726(2) and 2.745(1) Å]. More recently the analogous bis(iodomethane) complex[6] has also been crystallographically characterized.[15]

---
*The checkers obtained 71%.
† The checkers found 54% yield.

Acknowledgments

We thank the Petroleum Research Fund (M. F. M.) and National Science Foundation (J. M. M.) for funding, and Professor J. W. Faller and Dr. Brigitte Segmuller for performing the crystal structures of the complexes in Sections B and D.

References

1. W. Beck, *Z. Naturforsch Teil B*, **33,** 1214 (1978).
2. J. A. Davies and F. R. Hartley, *Chem. Rev.* **81,** 79 (1981).
3. A. Sen and T.-W. Lai, *J. Am. Chem. Soc.*, **103,** 4627 (1981).
4. J. R. Shapley, R. R. Schrock, and J. A. Osborn, *J. Am. Chem. Soc.*, **91,** 2816 (1969).
5. R. H. Crabtree, P. C. Demou, D. Eden, J. M. Mihelcic, P. Parnell, J. M. Quirk, and G. E. Morris, *J. Am. Chem. Soc.*, **104,** 6994 (1982).
6. R. H. Crabtree, J. W. Faller, M. F. Mellea, and J. M. Quirk, *Organometallics*, **1,** 1361 (1982).
7. O. W. Howarth, C. H. McAteer, P. Moore, and G. E. Morris, *J. Chem. Soc. Dalton Trans.*, **1981,** 1481.
8. R. H. Crabtree, J. M. Mihelcic, and J. M. Quirk, *J. Am. Chem. Soc.*, **101,** 7738 (1979).
9. A. Araneo, S. Martinengo, and P. Pasquale, *Rend. Ist. Lomb. Sci. Lett. Parte Sen. Atti. Uffi.*, **99,** 797 (1965).
10. M. Green, T. A. Kuc, and S. H. Taylor, *J. Chem. Soc. A*, **1971,** 2334.
11. R. H. Crabtree, G. G. Hlatky, C. P. Parnell, B. E. Segmuller, and R. J. Uriarte, *Inorg. Chem.*, **23,** 354 (1984).
12. R. H. Crabtree, H. Felkin, and G. E. Morris, *J. Organomet. Chem.*, **141,** 205 (1977).
13. D. Baudry, M. Ephritikine, and H. Felkin, *Nouv. J. Chim.*, **2,** 355 (1978).
14. R. H. Crabtree and C. P. Parnell, *Organometallics*, **4,** 519 (1985).
15. M. J. Burk, R. H. Crabtree, and B. Segmuller, *Organometallics* **6,** 2241 (1987).

13. BIS(BENZONITRILE)DICHLORO COMPLEXES OF PALLADIUM AND PLATINUM

Submitted by GORDON K. ANDERSON* and MINREN LIN*
Checked by AYUSMAN SEN† and EFI GRETZ†

Dichloropalladium and dichloroplatinum complexes containing weakly bound ligands represent convenient starting materials for the synthesis of a wide range of compounds of these two metals. The bis(acetonitrile) and bis(benzonitrile) species are commonly employed since they are easily prepared, the benzonitrile complexes having greater solubility in common organic solvents.

---

*Department of Chemistry, University of Missouri-St. Louis, St. Louis, MO 63121.
†Department of Chemistry, Pennsylvania State University, University Park, PA 16802.

Bis(benzonitrile)dichloropalladium was first prepared[1] in 1938 by the reaction of palladium(II) chloride with benzonitrile at 100 °C, and similar procedures have been described elsewhere.[2, 3] The analogous platinum complex has been synthesized in a related manner;[3-5] it has also been prepared, although in much lower yield, from potassium tetrachloroplatinate(II).[6, 7] The infrared spectra of both bis(benzonitrile) compounds have been described,[8, 9] and their ^{14}N NMR data[10] have been reported; ^{13}C NMR parameters have been reported for the platinum species.[5]

The following procedures for the preparation of bis(benzonitrile)dichloropalladium and -platinum involve slight modifications of the methods reported previously. Palladium(II) chloride and platinum(II) chloride were provided by Johnson Matthey. Benzonitrile was distilled immediately prior to use.

■ **Caution.** *Benzonitrile is a toxic liquid; contact with the skin or inhalation should be avoided. The preparations should be performed in an efficient fume hood.*

A. BIS(BENZONITRILE)DICHLOROPALLADIUM

$$PdCl_2 + 2PhCN \rightarrow PdCl_2(NCPh)_2$$

Procedure

Anhydrous palladium(II) chloride (2.00 g, 11.3 mmol) is loaded into a 200-mL round-bottomed flask, equipped with a stirring bar. Benzonitrile (60 mL) is added, and the mixture is stirred magnetically while being heated to 100 °C. After 20 min most of the palladium chloride has dissolved to give a red-brown solution. The solution is filtered while still hot, and the filtrate is poured into a 500-mL Erlenmeyer flask containing 350 mL of petroleum ether. A precipitate is formed immediately. The yellow-orange product is filtered through a fine frit, washed with petroleum ether (3 × 15 mL), and dried *in vacuo*. The yield is 4.12 g (95%), mp 126 °C.

Anal. Calcd. for $C_{14}H_{10}Cl_2N_2Pd$: C, 43.84; H, 2.63; N, 7.31. Found: C, 44.16; H, 2.57; N, 7.38%.

Properties

Bis(benzonitrile)dichloropalladium is a yellow-orange solid that is soluble in benzene, chloroform, and dichloromethane, but insoluble in water, alcohols, and diethyl ether. The IR spectrum (KBr pellet) exhibits C≡N stretching vibrations at 2235 and 2290 cm^{-1}. The ^{1}H NMR spectrum (CDCl$_3$) contains several overlapping multiplets in the δ 7.40–7.75 range. The ^{13}C{^{1}H} NMR spectrum ($-45\,°$C) exhibits resonances due to the trans isomer at 108.5 (C_1), 122.2 (CN), 129.4 (C_2), 133.2 (C_3), and 135.3 (C_4). On warming to ambient temperature the solution darkens, and resonances due to free benzonitrile (112.4, 118.8, 129.1, 132.1, 132.7) are also observed.

B. BIS(BENZONITRILE)DICHLOROPLATINUM

$$PtCl_2 + 2PhCN \rightarrow PtCl_2(NCPh)_2$$

Procedure

Benzonitrile (75 mL) is placed in a 250-mL round-bottomed flask and heated to 100 °C. Anhydrous platinum(II) chloride (2.66 g, 10.0 mmol) is added in small portions, while the mixture is stirred magnetically. A light brown solution is formed. Ten minutes after all the platinum chloride has been added, the solution is filtered while still hot, and the filtrate is collected in a 1-L Erlenmeyer flask. A precipitate begins to form on cooling, and the slow addition of 500 mL of petroleum ether completes the precipitation. The pale yellow product is filtered through a fine frit, washed with petroleum ether (3 × 15 mL), and dried *in vacuo*. The yield is 4.25 g (90%), mp 210 °C (decomp.).

Anal. Calcd. for $C_{14}H_{10}Cl_2N_2Pt$: C, 35.60; H, 2.13; N, 5.93. Found: C, 35.32; H, 2.02; N, 5.92%.

Properties

Bis(benzonitrile)dichloroplatinum is a pale yellow solid that is soluble in benzene, chloroform, and dichloromethane but insoluble in water, alcohols, and diethyl ether. The IR spectrum (KBr pellet) shows a C≡N stretching vibration at 2290 cm^{-1}, and two Pt—Cl stretching vibrations at 350 and 360 cm^{-1}. The ^{1}H NMR spectrum (CDCl$_3$) exhibits two multiplets centered at 7.56 and 7.77 ppm. The ^{13}C{^{1}H} NMR spectrum indicates that both isomers are present in solution. The trans isomer, which is the major component, exhibits resonances at 109.0 (C_1), 116.8 (J_{PtC} 290 Hz, CN), 129.5 (C_2), 133.7 (C_3), and 135.3 (C_4); the cis isomer gives rise to peaks at 109.4, 115.2, 129.6, 133.6, and 135.2.

References

1. M. S. Kharasch, R. C. Seyler, and F. R. Mayo, *J. Am. Chem. Soc.*, **60**, 882 (1938).
2. J. R. Doyle, P. E. Slade, and H. B. Jonassen, *Inorg. Synth.*, **6**, 218 (1960).

3. F. R. Hartley, *The Chemistry of Palladium and Platinum*, Applied Science, New York, 1973, p. 462.
4. M. S. Kharasch and T. A. Ashford, unpublished results quoted in Ref. 1.
5. T. Uchiyama, Y. Toshiyasu, Y. Nakamura, T. Miwa, and S. Kawaguchi, *Bull. Chem. Soc. Jpn.*, **54**, 181 (1981).
6. K. A. Hofmann and G. Bugge, *Chem. Ber.*, **40**, 1772 (1907).
7. L. Ramberg, *Chem. Ber.*, **40**, 2578 (1907).
8. R. A. Walton, *Spectrochim. Acta*, **21**, 1795 (1965).
9. R. A. Walton, *Can. J. Chem.*, **46**, 2347 (1968).
10. W. Becker, W. Beck, and R. Rieck, *Z. Naturforsch. B*, **25**, 1332 (1970).

14. ACETONITRILE COMPLEXES OF SELECTED TRANSITION METAL CATIONS

Submitted by RICHARD R. THOMAS* and AYUSMAN SEN*
Checked by WOLFGANG BECK† and REICH LEIDL†

A necessary requirement of homogeneous catalysts is that they have the ability to create vacant coordination sites by the dissociation of weakly held ligands, thereby allowing the metal to interact with the substrates. Transition metal cations incorporating weakly coordinating solvent molecules and having noncoordinating counteranions should meet this criterion. Indeed, the three cationic acetonitrile solvated transition metal complexes, the syntheses of which are described in this chapter, have been shown to catalyze a variety of organic transformations.[1] In addition, because of the high lability of the coordinated acetonitriles, these complexes serve as convenient synthetic precursors to other transition metal compounds. The synthesis of $[Pd(CH_3CN)_4][BF_4]_2$ has been previously reported.[2]

Since these cationic, weakly solvated transition metal complexes are reactive towards atmospheric moisture, all manipulations should be performed using standard inert atmosphere techniques.

A. TETRAKIS(ACETONITRILE)PALLADIUM(2+)-BIS[TETRAFLUOROBORATE(1−)]

$$Pd + 2(NO)[BF_4] + 4CH_3CN \rightarrow [Pd(CH_3CN)_4][BF_4]_2 + 2NO$$

■ **Caution.** *Since the preparation described here results in the evolution of gas from the reaction solution, the apparatus must contain a volume sufficient to*

---
*Department of Chemistry, Pennsylvania State University, University Park, PA 16802.
†Institut für Anorganische Chemie, Universität München 8000 München 2, Federal Republic of Germany.

accommodate this gas so that the internal pressure does not exceed atmospheric. In the present case, the gas was released through a mercury-filled bubbler into a well-ventilated hood.

Procedure

A Schlenk-type apparatus, illustrated in Fig. 1, is used for the synthesis. This apparatus is similar to that employed in other syntheses[3] but allows for multiple filtrations of the reaction solution and manipulation of the complex in a completely sealed system. In a nitrogen glove box, a 100-mL round-bottomed flask (B) is charged with 1.0 g (9.4 mmol) of palladium sponge (99.95%, Johnson Matthey) and 2.2 g (18.8 mmol) of nitrosyl tetrafluoroborate (Aldrich, sublimed 220°C, 10^{-3} torr). This flask is connected to a medium

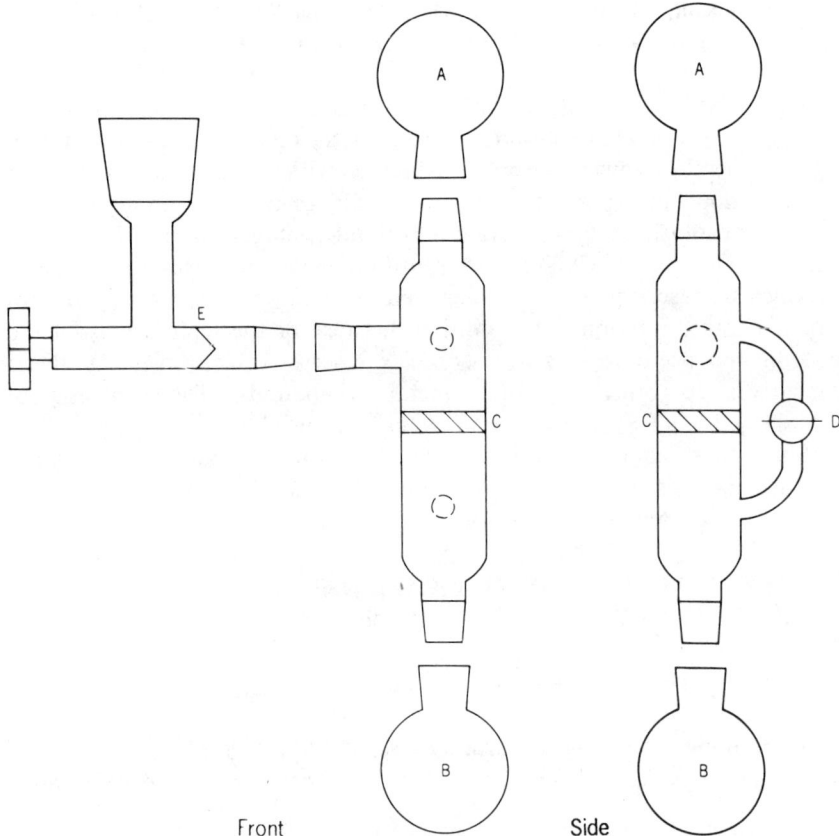

Fig. 1. Modified Schlenk apparatus, front and side views.

porosity filter frit (C) that is attached to another 100-mL round-bottomed flask, via a ground-glass joint. The entire apparatus is attached to a vacuum manifold through a high-vacuum Teflon valve (E) with a ground-glass joint. The apparatus is evacuated to 10^{-3} torr, and 50 mL of acetonitrile is added by vacuum distillation from a phosphorus(V) oxide slurry. On warming to 25°C, visible evolution of NO gas is noted from the metal surface.

The reaction is allowed to stir until no metal remains (12 h). The light yellow solution is filtered by closing the ground-glass valve (D), inverting the apparatus, and cooling the round-bottomed flask (A) to liquid nitrogen temperature. After filtration the ground-glass joint (D) is opened to allow for pressure equalization in the apparatus. The solution is then concentrated to 10 mL in vacuum to yield a light yellow precipitate. A 40-mL quantity of diethyl ether is added by vacuum distillation, from sodium benzophenone ketyl, to effect further precipitation. The suspension is filtered in a manner previously described and the solid is washed by further vacuum distillation of two 5-mL portions of diethyl ether. The precipitate is dried under vacuum (10^{-3} torr, 25°C) overnight to yield 3.0 to 4.0 g (75–96%) of pale yellow tetrakis(acetonitrile)palladium(2+) bis[tetrafluoroborate(1−)].

Anal. Calcd. for $C_8H_{12}N_4F_8B_2Pd$: C, 21.7; H, 2.7; N, 12.6. Found: C, 21.8; H, 2.9; N, 12.3 (authors); C, 21.6; H, 4.0; N, 12.5 (checkers).

Properties

The compound $[Pd(CH_3CN)_4][BF_4]_2$ is a moisture-sensitive pale yellow crystalline material. It is very soluble in acetonitrile and nitromethane but virtually insoluble in less polar organic solvents such as diethyl ether, chloroform, and benzene. The ^1H NMR spectrum in CD_3NO_2 exhibits a singlet at 2.65 ppm. The IR spectrum of its Nujol mull shows a coordinated C≡N band at 2335 cm^{-1}.

B. cis-TETRAKIS(ACETONITRILE)DINITROSYLMOLYBDE-NUM(2+) BIS[TETRAFLUOROBORATE(1−)]

$$Mo(CO)_6 + 2(NO)[BF_4] + 4CH_3CN \rightarrow$$
$$[Mo(NO)_2(CH_3CN)_4][BF_4]_2 + 6CO$$

■ **Caution.** *Since the preparation described here results in the evolution of gas from the reaction solution, the apparatus must contain a volume sufficient to accommodate this gas so that the internal pressure does not exceed atmospheric.*

In the present case, the gas was released through a mercury-filled bubbler into a well-ventilated hood.

Procedure

The apparatus and techniques used are similar to those for the synthesis of tetrakis(acetonitrile)palladium(2+) bis[tetrafluoroborate(1−)]. A 100-mL round-bottomed flask (B) is charged with 2.0 g (7.6 mmol) of hexacarbonylmolybdenum (Alfa) and 1.77 g (15.2 mmol) of nitrosyl tetrafluoroborate (Aldrich, sublimed 220°C, 10^{-3} torr). Dry acetonitrile (50 mL) is added and the solution is allowed to stir at 25°C until no hexacarbonylmolybdenum remains (6 h).

The resulting dark emerald green solution is filtered and the filtrate is concentrated to near dryness in vacuum. Diethyl ether (40 mL) is added and the solution is vigorously stirred to cause precipitation of the dark green molybdenum complex. The solution is filtered and the solid washed with two 5-mL portions of diethyl ether. The precipitate is dried in vacuum (10^{-3} torr, 25°C) to afford 3.36 g (90%) of dark emerald green *cis*-tetrakis(acetonitrile)dinitrosylmolybdenum(2+) bis[tetrafluoroborate(1−)].

Anal. Calcd. for $C_8H_{12}N_6O_2F_8B_2Mo$: C, 19.5; H, 2.5; N, 17.0. Found: C, 19.0; H, 2.4; N, 16.9.

Properties

The compound $[Mo(NO)_2(CH_3CN)_4][BF_4]_2$ is a moisture-sensitive dark crystalline emerald green substance. It is very soluble in acetonitrile and nitromethane but virtually insoluble in less polar organic solvents such as diethyl ether, chloroform, and benzene. The ^1H NMR spectrum in CD_3NO_2 exhibits two singlets of equal intensity at 2.65 and 2.55 ppm, respectively, which is consistent with a cis arrangement of the two NO groups on the metal atom. The IR spectrum of its Nujol mull shows absorptions at 2330 and 2300 cm^{-1} (coordinated C≡N), 1862, 1755, and 1724 cm^{-1} (coordinated NO) and may show weak absorptions at 2040 and 1957 cm^{-1} due to coordinated CO (impurity).

C. *cis*-TETRAKIS(ACETONITRILE)DINITROSYLTUNGSTEN(2+) BIS[TETRAFLUOROBORATE(1−)]

$$W(CO)_6 + 2(NO)[BF_4] + 4CH_3CN \rightarrow [W(NO)_2(CH_3CN)_4][BF_4]_2 + 6CO$$

■ **Caution.** *Since the preparation described here results in the evolution of gas from the reaction solution, the apparatus must contain a volume sufficient to* accommodate this gas so that the internal pressure does not exceed atmospheric. In the present case, the gas was released through a mercury-filled bubbler into a well-ventilated hood.

Procedure

In a procedure identical to that described in Section B, the reaction is performed using 2.5 g (7.1 mmol) of hexacarbonyltungsten (Alfa) and 1.66 g (14.2 mmol) of nitrosyl tetrafluoroborate (Aldrich, sublimed 220°C, 10^{-3} torr). An emerald green complex is isolated to yield 3.95 g (96%) of cis-tetrakis(acetonitrile)dinitrosyltungsten(2+) bis[tetrafluoroborate(1−)].

Anal. Calcd. for $C_8H_{12}N_6O_2F_8B_2W$: C, 16.4; H, 2.1. Found: C, 16.5; H, 2.3.

Properties

The compound $[W(NO)_2(CH_3CN)_4][BF_4]_2$ is a moisture-sensitive dark emerald green crystalline substance. It is very soluble in acetonitrile and nitromethane but virtually insoluble in less polar organic solvents such as diethyl ether, chloroform, and benzene. The ^1H NMR spectrum in CD_3NO_2 exhibits two singlets of equal intensity at 2.65 and 2.55 ppm, respectively, which is consistent with a cis arrangement of the two NO groups on the metal atom. The IR spectrum of its Nujol mull shows absorptions at 2332 and 2300 cm^{-1} (coordinated C≡N), 1862, 1820, 1772, and 1728 cm^{-1} (coordinated NO) and may show weak absorptions at 2030 and 1940 cm^{-1} due to coordinated CO (impurity).

References

1. (a) A. Sen and T.-W. Lai, *Inorg. Chem.*, **23**, 3257 (1984); (b) A. Sen and R. R. Thomas, *Organometallics*, **1**, 1251 (1982); (c) A. Sen and T.-W. Lai, *Organometallics*, **1**, 415 (1982); (d) A. Sen and T.-W. Lai, *J. Am. Chem. Soc.*, **103**, 4627 (1981); (e) W. A. Nugent and F. W. Hobbs, *J. Org. Chem.*, **48**, 5364 (1983).
2. R. F. Schramm and B. B. Wayland, *Chem. Commun.*, 898 (1968). $[Pd(CH_3CN)_4]^{2+}$ may also be generated by the reaction of $Pd(CH_3CN)_2Cl_2$ with two equivalents of $AgClO_4$, see: F. R. Hartley, S. G. Murray, W. Levason, H. E. Soutter, and C. A. McAuliffe, *Inorg. Chim. Acta*, **35**, 265 (1979).
3. For example, M. Ghedini and G. Dolcetti, *Inorg. Synth.*, **21**, 104 (1982). We thank Professor J. E. Bercaw of California Institute of Technology for providing us with the design of the present apparatus. For an early version of this apparatus, see: R. W. Parry, D. R. Schultz, and P. R. Girardot, *J. Am. Chem. Soc.*, **80**, 1 (1958).

15. TETRAKIS(ACETONITRILE)COPPER(1+) HEXAFLUOROPHOSPHATE(1−)

$$Cu_2O + 2HPF_6 \xrightarrow{CH_3CN} 2[Cu(CH_3CN)_4][PF_6] + H_2O$$

Submitted by G. J. KUBAS*
Checked by B. MONZYK† and A. L. CRUMBLISS†

Subsequent to the initial report of its preparation in *Inorganic Syntheses*,[1] [Cu(CH$_3$CN)$_4$][PF$_6$] has been widely used as a precursor for syntheses of macrocylic and other Cu(I) complexes. The [Cu(CH$_3$CN)$_4$]$^+$ cation was originally isolated as a nitrate salt in 1923 by the reduction of silver nitrate with copper powder in acetonitrile.[2] The cation is stabilized by large anions, and preparations of the perchlorate and tetrafluoroborate salts have since appeared in the literature.[3–5] The preparation of the [PF$_6$]$^-$ salt is described here, but conceivably any large-anion salt can be similarly prepared. Other synthetic options exist also (e.g., Cu + Ag[PF$_6$]), but the following method, based on Refs. 4 and 5, is simple, economical, and avoids Ag(I) contamination. The preparation can be carried out in glassware open to the atmosphere until final drying of the product.

Procedure

■ **Caution.** *The following procedures should be carried out in a well-ventilated hood because of the toxicity of acetonitrile and the HF fumes evolved from HPF$_6$.*

To a magnetically stirred suspension of 4.0 g (28 mmol) of copper(I) oxide in 80 mL of acetonitrile in a 125-mL Erlenmeyer flask is added 10 mL of 60–65% HPF$_6$ (about 113 mmol of HPF$_6$) in 2-mL portions. The reaction is very exothermic and may cause the solution to boil. However, the reaction temperature is not critical and the warming is beneficial in that the product remains dissolved. After addition of the final portion of HPF$_6$, the solution is stirred for about 3 min and is then filtered hot through a medium-porosity frit to remove small amounts of undissolved black solid (some white [Cu(CH$_3$CN)$_4$][PF$_6$] may begin to crystallize before filtration; if so, it is washed through the frit with a minimum amount of CH$_3$CN). The pale-blue solution is cooled in a freezer to about −20° for several hours (addition of an equal volume of diethyl ether and cooling to 0° yields equivalent results),

---

*Los Alamos National Laboratory, University of California, Los Alamos, NM 87545.
†Department of Chemistry, Duke University, Durham, NC 27706.

whereupon a blue-tinged white microcrystalline precipitate of [Cu(CH$_3$CN)$_4$][PF$_6$] forms. The solid is collected by filtration, washed with diethyl ether, and immediately redissolved in 100 mL of CH$_3$CN. A small amount of blue material, presumably a Cu^{2+} species, remains undissolved and is removed by filtration. To the filtrate (which may still retain a slight blue coloration) is added 100 mL of diethyl ether, and the mixture is allowed to stand for several hours at $-20°$. The precipitated complex may still retain a bluish cast, in which case a second recrystallization may be necessary if high purity is desired. This second recrystallization is carried out using 80 mL each of CH$_3$CN and diethyl ether. The product is pure white and is dried *in vacuo* for about 30 min immediately after being washed with diethyl ether. The yield is 12.5 g (60%) and is dependent on recrystallization losses.

Anal. Calcd. for C$_8$H$_{12}$N$_4$PF$_6$Cu: C, 25.8; H, 3.3; N, 15.0; P, 8.3; Cu, 17.0. Found: C, 25.9: H, 3.3; N, 15.1; P, 8.1; Cu, 16.7.

Properties

Tetrakis(acetonitrile)copper(1+) hexafluorophosphate(1−) is a free-flowing, white, microcrystalline powder that does not darken on long-term storage in an inert atmosphere. Exposure to air for longer than about 1 h results in minor surface oxidation, due to the slightly hygroscopic nature of the complex. The acetonitrile ligands are firmly bound and cannot be removed at a significant rate by pumping *in vacuo* at ambient temperature. Approximate CH$_3$CN dissociation pressures are 5 torr at 80° and 25 torr at 110°. The infrared spectrum in Nujol shows absorptions at 2277 (m) and 2305 (m) cm^{-1} due to CH$_3$CN, and at 850 (vs) and 557 (s) cm^{-1} due to [PF$_6$]$^-$.

The complex is moderately soluble in polar solvents and is remarkably stable to air oxidation in CH$_3$CN solution. It does not react with halide ions to give CuX (X=Cl, Br, I) in CH$_3$CN solution, but the coordinated acetonitriles can be displaced in other solvents or even in CH$_3$CN in certain cases. For example, copper(I) phenoxide, Cu[OC$_6$H$_5$] can be precipitated upon addition of ethanolic Na[OC$_6$H$_5$] to an acetonitrile solution of [Cu(CH$_3$CN)$_4$][PF$_6$].[6] Thus the complex is especially suited for non-aqueous-media syntheses of Cu(I) compounds. The X-ray structure[7] of [Cu(CH$_3$CN)$_4$][ClO$_4$] revealed nearly ideal tetrahedral coordination of copper by the almost linear CH$_3$CN molecules.

Acknowledgment

This work was funded by the U.S. Department of Energy, Division of Chemical Sciences, Office of Basic Energy Sciences.

References

1. G. J. Kubas, *Inorg. Synth.*, **19**, 90 (1979).
2. H. H. Morgan, *J. Chem. Soc.*, **1923**, 2901.
3. B. J. Hathaway, D. G. Holah, and J. D. Postlethwaite, *J. Chem. Soc.*, **1961**, 3215.
4. G. D. Davis and E. C. Makin, Jr., *Sep. Purif. Meth.*, **1**, 199 (1972).
5. P. Hemmerich and C. Sigwart, *Experentia*, **19**, 488 (1963).
6. P. G. Eller and G. J. Kubas, *J. Am. Chem. Soc.*, **99**, 4346 (1977).
7. I. Csoregh, P. Kierkegaard, and R. Norrestam, *Acta Cryst.*, **B31**, 314 (1975).

16. TRIFLUOROMETHANESULFONATES AND TRIFLUOROMETHANESULFONATO-O COMPLEXES

NICHOLAS E. DIXON,* GEOFFREY A. LAWRANCE,† PETER A. LAY,‡ ALAN M. SARGESON,* and HENRY TAUBE§

Trifluoromethanesulfonic acid (triflic acid) and its organic esters have found considerable use in organic chemistry.[1,2] The desirable qualities of these compounds include a high thermal stability, excellent leaving-group properties of the $[CF_3SO_3]^-$ group, and the ease of purification of the acid and of synthesis of its esters. Its use in synthetic inorganic chemistry has been somewhat hampered by the lack of convenient routes for the preparation of the desired complexes. Despite this, the excellent leaving-group properties of the triflato ligand were established a decade ago by Scott and Taube,[3] who showed that the $[Cr(OH_2)_5(OSO_2CF_3)]^{2+}$ and $[Cr(OH_2)_4(OSO_2CF_3)_2]^+$ complexes aquated rapidly. The preparative procedure utilized for these chromium complexes was to oxidize a Cr(II) solution in the presence of 6 M CF_3SO_3H followed by low-temperature cation exchange chromatographic separation of the various species. Subsequent to this, the $[CO(NH_3)_5(OSO_2CF_3)](CF_3SO_3)_2$ complex was synthesized by way of the nitrosation of $[Co(NH_3)_5(N_3)](CF_3SO_3)_2$ by $[NO][CF_3SO_3]$ in nonaqueous solvents.[4,5] More recently, a method of general utility was found by Dixon et al.[6] that involved heating chloro complexes in neat triflic acid, followed by precipitation of the resultant triflato complexes by diethyl ether. This procedure, or modifications of the procedure, has been utilized for the synthesis of amine complexes of Co(III),[6-10] Rh(III),[8,9] Ir(III),[8,9] Cr(III),[8,9]

---

*Research School of Chemistry, The Australian National University, G.P.O. Box 4, Canberra, A.C.T. 2601 Australia.
†Chemistry Department, The University of Newcastle, N.S.W. 2308 Australia.
‡Department of Inorganic Chemistry, The University of Sydney, N.S.W. 2006 Australia.
§Department of Chemistry, Stanford University, Stanford, CA 94305.

Ru(III)[8,9,11] Os(III),[12] Pd(II),[13] Pt(II),[13] and Pt(IV).[8,9] In addition, polypyridyl triflato complexes of Ru(II), Ru(III), Os(II), Os(III) have been synthesized.[14–16] Other useful techniques have appeared recently in the literature for the preparation of triflato complexes. These include heating triflate salts of aqua and other solvent complexes in the solid state under vacuum[17] and using trimethylsilyl triflate as a reagent instead of triflic acid in the above routes.[18]

Once prepared, the triflato complexes have many desirable properties. The most useful property in syntheses is the relatively high lability of the triflato group, which is substituted at a rate comparable to the best leaving groups known,[6] including the perchlorato[19] and fluoromethanesulfonato[20] ligands. This feature, combined with a high solubility in most polar organic solvents, a higher thermal stability, relatively low reactivity with atmospheric moisture, and simple and high-yielding preparative routes from readily available starting materials, makes these complexes extremely versatile synthetic intermediates en route to a large variety of important classes of transition metal complexes. Moreover, the use of the triflate anion for applications traditionally centered around the perchlorate anion is highly recommended. For instance, the explosive hazard, especially in organic solvents, is eliminated and the triflate anion often imparts higher solubility for the complexes in organic solvents.[21,22] Other commonly used anions that fall into the category of being poor nucleophiles, such as $[PF_6]^-$, $[BF_4]^-$, or $[BPh_4]^-$ do not have the thermal or photochemical stability exhibited by the $[CF_3SO_3]^-$ anion. Some chemical and physical properties of triflic acid are given in Refs. 1, 2, and 7.

Procedures

■ **Caution.** *Triflic acid is one of the strongest known protic acids. It is necessary to take adequate precautions to prevent contact with the skin and eyes. Precautions should also be taken to minimize inhalation of the corrosive vapors given off from the acid. Reactions with the neat acid must be conducted in a well-ventilated fume hood.*

■ **Caution.** *Under no circumstances should perchlorate salts be used in any of the reactions involving neat triflic acid. The anhydrous hot perchloric acid thus produced represents an extremely explosive hazard, especially in contact with transition metal complexes. Addition of anhydrous diethyl ether to such solutions would represent an additional explosive hazard.*

Trifluoromethanesulfonic acid, its salts, and its complexes are extremely stable thermally, and no explosive hazards are known. However, consideration should be given to the thermal stability of other components of the complex before any new reactions are attempted at elevated temperatures.

Vacuum distillation of triflic acid (Aldrich Chemical) is performed as described previously.[6,7] Use of the acid as supplied does not appear to affect yields or purity of products markedly, although the complexes may be contaminated by highly colored impurities on occasion if the distillation procedure is not adopted.

The major synthetic methods utilized are described in the following sections. Methods for the syntheses of triflate salts are also described, because of their general use in nonaqueous chemistry in place of the perchlorate salts. Syntheses of specific trifluoromethanesulfonato complexes are described in cited references, and in Volumes 22 and 24 of *Inorganic Syntheses*. A review of coordinated triflates has also appeared recently.[23]

A. TRIFLATE SALTS FROM CHLORIDE SALTS

$$MCl_n + nCF_3SO_3H \rightarrow M(CF_3SO_3)_n + nHCl \uparrow$$

Procedure

This procedure may be applied to metal salts generally, and similar procedures have been described.[24-26]

To MCl_n (1 g) contained in a two-necked round-bottomed flask fitted with a nitrogen bubbler is carefully added anhydrous triflic acid (3–5 mL).

■ **Caution.** *Triflic acid is one of the strongest known protic acids, and gaseous hydrogen chloride is produced rapidly in the reaction. It is necessary to take adequate precautions to protect the skin and eyes and to prevent inhalation of the corrosive vapors. These manipulations must be performed in a well-ventilated fume hood. Because of the initial rapid evolution of HCl, care must be taken not to add the triflic acid too rapidly.*

A steady stream of nitrogen is passed through the solution while it is warmed to $\sim 60°$. After 0.5–1 h, the heating is discontinued and the solution is cooled to $\sim 0°$ in an ice bath while an N_2 flow is maintained. In order to precipitate the complex, anhydrous diethyl ether is added cautiously to the rapidly stirred solution in a dropwise fashion.

■ **Caution.** *This is a very exothermic addition and due care must be taken not to add the diethyl ether too quickly. Diethyl ether is toxic and very flammable. The addition must be performed in a well-ventilated fume hood.*

The salt is filtered on a medium-porosity sintered-glass funnel (15 mL) initially under gravity, and then the filtration is completed using a water aspirator. The ethereal solutions of triflic acid obtained at this stage may be kept for recovery of triflic acid as the sodium salt.[6] The powdery solid is washed with copious amounts of anhydrous diethyl ether (4 × 20 mL) and air-dried after each washing. At this stage further purification is generally not necessary, since the complexes are normally analytically pure. Yields are essentially quantitative except for mechanical losses.

In some of the precipitation processes, $Et_2O \cdot CF_3SO_3H$ may be coprecipitated but may be removed readily by boiling the solid in chloroform for ~ 0.5 h after any solid lumps have been broken up using a mortar and pestle.

■ **Caution.** *Chloroform is toxic and a carcinogen; this procedure must be performed in a well-ventilated fume hood.*

The powder is collected on a medium-porosity frit and air-dried.

If crystalline material is required, most triflate salts are sparingly soluble in acetonitrile and may be recrystallized from hot solutions of this solvent.

■ **Caution.** *Acetonitrile is toxic and flammable. These crystallizations should be performed in a well-ventilated fume hood.*

The same procedures may be applied to other salts such as other halides and pseudohalides, carbonates, and acetates.

■ **Caution.** *Under no circumstances should perchlorate salts be used in any of the reactions involving neat triflic acid. The anhydrous hot perchloric acid thus produced represents an extremely explosive hazard, especially in contact with transition metal complexes. Moreover, the addition of anhydrous diethyl ether to such solutions would represent an additional explosive hazard.*

B. TRIFLATE SALTS FROM SULFATE SALTS

$$M(SO_4)_n + nBa(CF_3SO_3)_2 \rightarrow M(CF_3SO_3)_{2n} + nBaSO_4 \downarrow$$

Procedure

This procedure is convenient for acid-sensitive complexes and is similar to that described elsewhere.[17] To a solution of $M(SO_4)_n$ (1 g) dissolved in water (~ 10 mL), $Ba(CF_3SO_3)_2$ (n equiv) is added.

■ **Caution.** *Barium salts are extremely toxic. Avoid contact with skin.*

The solution is stirred, and the precipitated $BaSO_4$ is filtered using a medium-porosity sintered-glass filter. The solvent is removed by rotary evaporation to yield a powder, which may be recrystallized as before. For heat-sensitive complexes, the solvent is removed by freeze-drying techniques.

C. TRIFLATE SALTS USING SILVER TRIFLATE

$$MCl_n + nAgCF_3SO_3 \rightarrow M(CF_3SO_3)_n + nAgCl \downarrow$$

Procedure

This procedure[22] utilizes commercially available $AgCF_3SO_3$ (Alfa Products) and can also be used for other salts for which the AgX salts are very sparingly soluble.

■ **Caution.** *Silver salts are toxic and strong skin irritants. Avoid contact with skin.*

To MCl_n (1 g) dissolved in the minimum volume of water is added $AgCF_3SO_3$ (n equiv). The solution is stirred rapidly for 5–10 min, and the precipitated AgCl is removed by vacuum filtration through a bed of Hyflo Supercel (Gallard Schlesinger) on a sintered-glass funnel. Isolation and purification procedures are the same as described in Sections A and B.

D. TRIFLATO COMPLEXES FROM CHLORO COMPLEXES

$$[ML_xCl_y]Cl_n + (y+n)CF_3SO_3H \rightarrow$$
$$[ML_x(OSO_2CF_3)_y](CF_3SO_3)_n + (y+n)HCl \uparrow$$

Procedure

This procedure is a general procedure applicable to inert transition metal complexes.[6–9] To $[ML_xCl_y]Cl_n$ (1 g) contained in a two-necked round-bottomed flask (25 mL) connected with a nitrogen bubbler is cautiously added anhydrous triflic acid (~5 mL).

■ **Caution.** *Triflic acid is one of the strongest known protic acids, and gaseous hydrogen chloride is produced rapidly in the reaction. It is necessary to take adequate precautions to protect the skin and eyes and to prevent inhalation of the corrosive vapors. These manipulations must be performed in a well-ventilated fume hood. Because of the initial rapid evolution of HCl, care must be taken not to add the triflic acid too quickly.*

■ **Caution.** *Under no circumstances should perchlorate salts be used in any of the reactions involving neat triflic acid. The anhydrous hot perchloric acid thus produced represents an extremely explosive hazard, especially in contact with transition metal complexes. Addition of anhydrous diethyl ether to such solutions would represent an additional explosive hazard.*

A steady stream of nitrogen is passed through the solution, which is then lowered into a silicone oil bath preheated to 100–120°. Lower temperatures may be required for certain complexes [e.g., Cr(III) amines] where heating may cause decomposition; lower temperatures require extended reaction times. Specific examples of reactions performed at lower temperature appear in the literature.[9,13]

Evolution of HCl gas is monitored by periodically passing the effluent gas through an $AgNO_3$ bubbler. After the HCl evolution has ceased (1–20 h, depending on the complex and temperature), the flask is removed from the oil bath and allowed to cool to ~30° before cooling further in an ice bath. Ice cooling can be omitted for small-scale reactions, although boiling will occur on initial addition of diethyl ether. Use of a larger reaction vessel, vigorous mechanical stirring, and a well-ventilated fume hood are mandatory under these conditions. While the solution is rapidly stirred, diethyl ether (20 mL) is added carefully, in dropwise fashion, to precipitate the complex.

■ **Caution.** *This is a very exothermic addition, and due care must be exercised to avoid adding the diethyl ether too quickly. Diethyl ether is toxic and very flammable. Its addition must be performed in a well-ventilated fume hood.*

The complex is collected on a medium-porosity sintered glass funnel by initially allowing the solution to filter under gravity, then finally under vacuum. The filtrate may be saved for recovery of the triflic acid as $NaCF_3SO_3$.[6,7] The complex may be purified by boiling in chloroform as described in Section A. However, the complexes may not be recrystallized from acetonitrile, since the triflato ligands are readily substituted by acetonitrile molecules.

Similar procedures may be used for other complexes containing halo, pseudohalo, acetato, carbonato,[7] aqua, and many other ligands that are either relatively labile or are decomposed by strong acid.

Preparations of several triflate complexes of Co, Cr, Ru, Os, Rh, Ir, and Pt are given in *Inorganic Syntheses*, Volume 22, pages 103–107, and Volume 24, pages 250–306.

E. TRIFLATO COMPLEXES FROM SOLID STATE REACTIONS OF TRIFLATO SALTS

$$[ML_x(solvent)_y](CF_3SO_3)_n \rightarrow [ML_x(OSO_2CF_3)_y](CF_3SO_3)_{(n-y)} + y(solvent)$$

Procedure

This procedure has been described for inert aqua complexes,[17] and is applicable to many other solvent species. For aqua complexes, $[ML_x(OH_2)_y](CF_3SO_3)_n$ (1 g) is placed in a vacuum oven that has been preheated to 100–180° (depending on the lability of the complex). The solid is kept under vacuum for 8–24 h depending on the complex, at which stage the substitution is complete. Once dehydrated, the triflato complexes are relatively stable to atmospheric moisture and generally may be manipulated without precautions to exclude air, unless otherwise stated.

For complexes containing solvent ligands that are both poorly coordinating and volatile, this procedure may be carried out at room temperature or even lower temperatures using vacuum-line techniques. Such procedures are particularly useful for complexes that are thermally unstable toward isomerization or other chemical processes. Coordinated and ionic trifluoromethanesulfonate can usually be distinguished by infrared spectroscopy.[23]

F. REGENERATION OF TRIFLATO COMPLEXES

Procedure

Aged samples of triflato complexes may have undergone some aquation due to atmospheric moisture, although storage of the complexes in a desiccator over a suitable drying agent is generally sufficient to retain the integrity of the complexes for many months. The complexes are readily regenerated either by heating in the solid phase under vacuum (Section E) or by heating in neat triflic acid (Section D). Similar procedures may be used to regenerate the triflato complexes from a variety of product complexes.

References

1. J. B. Hendrickson, D. D. Stembach, and K. W. Bair, *Acc. Chem. Res.*, **10**, 306 (1977).
2. R. D. Howells and J. D. McCown, *Chem. Rev.*, **71**, 69 (1977).
3. A. Scott and H. Taube, *Inorg. Chem.*, **10**, 62 (1971).
4. P. J. Cresswell, Ph.D. thesis, The Australian National University, 1974.
5. D. A. Buckingham, P. J. Cresswell, W. G. Jackson, and A. M. Sargeson, *Inorg. Chem.*, **20**, 1647 (1981).
6. N. E. Dixon, W. G. Jackson, M. J. Lancaster, G. A. Lawrance, and A. M. Sargeson, *Inorg. Chem.*, **20**, 470 (1981).
7. N. E. Dixon, W. G. Jackson, G. A. Lawrance, and A. M. Sargeson, *Inorg. Synth.*, **22**, 103 (1983).
8. N. E. Dixon, G. A. Lawrance, P. A. Lay, and A. M. Sargeson, *Inorg. Chem.*, **22**, 846 (1983).
9. N. E. Dixon, G. A. Lawrance, P. A. Lay, and A. M. Sargeson, *Inorg. Chem.*, **23**, 2940 (1984).
10. N. J. Curtis, K. S. Hagen, and A. M. Sargeson, *J. Chem. Soc., Chem. Commun.*, **1984**, 1571.
11. B. Anderes, S. T. Collins, and D. K. Lavallee, *Inorg. Chem.*, **23**, 2201 (1984).
12. R. H. Magnuson, P. A. Lay, and H. Taube, *J. Am. Chem. Soc.*, **105**, 2507 (1983); P. A. Lay, R. H. Magnuson, J. Sen, and H. Taube, *J. Am. Chem. Soc.*, **104**, 7658 (1982).
13. C. Diver and G. A. Lawrance, *J. Chem. Soc. Dalton Trans.*, 931 (1988).
14. P. A. Lay, A. M. Sargeson, and H. Taube, *Inorg. Synth.*, **24**, 291 (1986).
15. D. StC. Black, G. B. Deacon, and N. C. Thomas, *Trans. Metal Chem.*, **5**, 317 (1980).
16. D. StC. Black, G. B. Deacon, and N. C. Thomas, *Aust. J. Chem.*, **35**, 2445 (1982).
17. W. C. Kupferschmidt and R. B. Jordan, *Inorg. Chem.*, **21**, 2089 (1982).
18. M. R. Churchill, H. J. Wasserman, H. W. Terner, and R. R. Schrock, *J. Am. Chem. Soc.*, **104**, 1710 (1982).
19. J. MacB. Harrowfield, A. M. Sargeson, B. Singh, and J. C. Sullivan, *Inorg. Chem.*, **14**, 2864 (1975).
20. W. G. Jackson, and C. M. Begbie, *Inorg. Chem.*, **20**, 1654 (1981).
21. T. Fujinaga and I. Sakamoto, *Pure Appl. Chem.*, **52**, 1389 (1980).
22. W. G. Jackson, G. A. Lawrance, P. A. Lay, and A. M. Sargeson, *Aust. J. Chem.*, **35**, 1561 (1982).
23. G. A. Lawrance, *Chem. Rev.*, **86**, 17 (1986).
24. J. S. Haynes, J. R. Sams, and R. C. Thompson, *Can. J. Chem.*, **59**, 669 (1981).
25. R. J. Batchelor, J. N. B. Ruddick, J. R. Sams, and F. Aubke, *Inorg. Chem.*, **16**, 1414 (1977).
26. A. M. Bond, G. A. Lawrance, P. A. Lay, and A. M. Sargeson, *Inorg. Chem.*, **22**, 2010 (1983).

Chapter Two

LOW-VALENT COMPLEXES OF Rh, Ir, Ni, Pd, AND Pt

17. CHLOROTRIS(TRIPHENYLPHOSPHINE)RHODIUM(I) (WILKINSON'S CATALYST)

Submitted by J. A. OSBORN*† and G. WILKINSON*
Checked by J. J. MROWCA‡*†

The discovery of chlorotris(triphenylphosphine)rhodium[1] and its utility as the first practical homogeneous catalyst for the hydrogenation of C=C and C≡C bonds opened up an enormous and still developing field of chemistry, not only in catalysis but in stoichiometric reactions.[2]

Procedure

$$RhCl_3 + 4P(C_6H_5)_3 \to RhCl[P(C_6H_5)_3]_3 + Cl_2P(C_6H_5)_3$$

$$Cl_2P(C_6H_5)_3 + H_2O \to OP(C_6H_5)_3 + 2HCl$$

Rhodium (III) chloride trihydrate§ (2 g) is dissolved in 70 mL of ethanol (95%) in a 500 mL round-bottomed flask fitted with gas inlet tube, reflux condenser, and gas exit bubbler. A solution of 12 g of triphenylphosphine

---

*Chemistry Department, Imperial College, London SW7 2AY.
†Now at Chemistry Department, University Louis Pasteur, Strasbourg, France.
‡Central Research Department, Experimental Station, E. I. du Pont de Nemours & Company, Wilmington, DE 19898.
§The commercial product usually corresponds closely to $RhCl_3 \cdot 3H_2O$ but small divergences from this stoichiometry are not significant in this preparation. The yield is calculated from the Rh content.

(freshly crystallized from ethanol to remove triphenylphosphine oxide) in 350 mL of hot ethanol is added and the flask purged with nitrogen. The solution is refluxed for about 2 h, and the crystalline product that forms is collected from the hot solution on a Büchner funnel or sintered-glass filter. The product is washed with small portions of 50 mL of anhydrous ether; yield 6.25 g (88% based on Rh). This crystalline product is deep red in color.

An isomeric species that is orange is obtained if the total volume of ethanol used is 200 mL or less and the solution is refluxed for a period of about 5 min. This substance often contains small amounts of the red product and, on continued refluxing, the orange crystals are slowly converted to the red form.

The excess triphenylphosphine used in the preparation can be recovered by addition of water to the ethanol filtrates until precipitation begins. After allowing the solutions to stand 2 to 3 days in a stoppered flask, the triphenylphosphine crystallizes out. Recrystallization from ethanol and ethanol–benzene (1:1) removes triphenylphosphine oxide contaminant.

Properties

The burgundy red (mp 157°C) and orange polymorphic forms of $RhCl[(P(C_6H_5)_3]_3$ have identical chemical properties. The complex is soluble in chloroform and dichloromethane to about 20 g L^{-1} at 25°. The solubility in benzene or toluene is about 2 g L^{-1} at 25° but is very much lower in acetic acid, acetone, and other ketones, methanol, and lower aliphatic alcohols. In alkanes and cyclohexane, the complex is virtually insoluble. Donor solvents such as pyridine, dimethyl sulfoxide, or acetonitrile dissolve the complex with reaction, initially to give complexes of the type $RhCl[P(C_6H_5)_3]_2L$, but further reaction with displacement of phosphine may occur.

The solutions are very air sensitive, giving soluble dioxygen compounds. Both solid and solutions should be handled under oxygen-free dinitrogen or argon. On heating benzene, toluene, or best, methyl ethyl ketone solutions (or suspensions) of $RhCl[P(C_6H_5)_3]_3$, salmon-pink crystals of the chlorine-bridged dimer $[(C_6H_5)_3P]_2RhCl_2Rh[P(C_6H_5)_3]_2$ are obtained essentially quantitatively. This dimer absorbs oxygen slowly even in the solid state. It may be reconverted to $RhCl[P(C_6H_5)_3]_3$ by cleavage with triphenylphosphine in refluxing ethanol.

The red solutions of $RhCl[P(C_6H_5)_3]_3$ absorb molecular hydrogen reversibly at 1 atm and 25°, becoming pale yellow; these solutions are highly effective for the catalytic homogeneous hydrogenation of compounds with C=C and C≡C bonds often selectivity depending on the nature of the substrate. Cationic species of the type $[Rh(PPh_3)_2(sol)_2]^+$, where sol is a solvent molecule and many other similar species with chelating phosphines are also effective.

References

1. J. A. Osborn, F. H. Jardine, F. H. Young, and G. Wilkinson, *J. Chem. Soc. A*, **1966**, 171.
2. F. H. Jardine, *Prog. Inorg. Chem.*, **28**, 63 (1981) (a review with 650 references on stoichiometric and catalytic reactions).

18. trans-CARBONYLCHLOROBIS(TRIPHENYL-PHOSPHINE)RHODIUM AND RELATED COMPLEXES

$$RhCl_3 \cdot 3H_2O + 2PPh_3 \xrightarrow{HCHO} RhCl(CO)(PPh_3)_2 *$$

Submitted by D. EVANS,[†] J. A. OSBORN,[†] and G. WILKINSON[†]
Checked by ROBERT PAINE, Jr.,[‡] and R. W. PARRY[‡]

Carbonylchlorobis(triphenylphosphine)rhodium and related compounds with substituted phosphine or arsine ligands have been made by interaction of the ligand with rhodium carbonyl chloride.[1] Alternative procedures starting with solutions of rhodium(III) chloride and the ligand and using carbon monoxide in hot alcohols, methanolic pentene, or ethanol and potassium hydroxide as reducing agents and sources of carbon monoxide are useful but slow, the reactions often taking several hours.[2,3] During studies of the rapid decarbonylation of aldehydes by chlorotris(triphenylphosphine)rhodium(I),[4] it was found that aqueous formaldehyde could be used for the preparation of the carbonyl complex directly from rhodium(III) chloride and that the method was also rapid and efficient when other phosphines and arsines were employed. With ortho-substituted aryl phosphines, bivalent rhodium species may be stabilized.[5]

Procedure

Rhodium(III) chloride 3-hydrate (2 g, 0.0076 mol) in 70 mL of absolute ethanol is slowly added to 300 mL of boiling absolute ethanol containing about a twofold excess§ of triphenylphosphine (7.2 g, 0.0275 mol). The solution, which may be turbid, becomes clear in about 2 min. Then sufficient (10–20 mL) 37% formaldehyde solution is added to cause the red solution to become pale yellow in about 1 min, and yellow microcrystals to precipitate.

---

*The stoichiometry of the reaction is uncertain.
†Imperial College of Science and Technology, London S.W. 7.
‡University of Michigan, Ann Arbor, MI.
§If a stoichiometric quantity is used, the product is contaminated and in low yield.

After cooling, the collected crystals are washed with ethanol and diethyl ether and dried in air or vacuum. They can be recrystallized from the minimum of hot benzene. Yield: 4.5 g (85% based on $RhCl_3 \cdot 3H_2O$). The triphenylarsine complex can be prepared in an analogous manner.

Properties

The properties of trans-$RhCl(CO)(PPh_3)_2$ (mp 195–197°) and trans-$RhCl(CO)(AsPh_3)_2$ (mp 242–244°) have been given.[1] The chlorides can be rapidly converted to the corresponding bromides, iodides, or thiocyanates by the interaction[3] in acetone solutions at room temperature with lithium bromide, sodium iodide, or potassium thiocyanate, respectively. Alternatively, rhodium(III) chloride can first be converted to the bromide or iodide by boiling the ethanolic solution with about a fivefold excess of lithium bromide or·iodide.

The complex trans-$RhCl(CO)(PPh_3)_2$ has been shown to act as a catalyst for the hydroformylation of olefins and acetylenes under mild conditions,[6] and for decarbonylation of aldehydes[7] and of acyl and aroyl halides.[8]

The properties of other complexes made in the above way and in yields exceeding 80%, together with original literature references, are as follows (R for PR_3 given): $p\text{-}CH_3C_6H_4$,[9] mp 195°, v_{CO} 1965 cm,$^{-1}$; $o\text{-}CH_3C_6H_4$,[10] mp 230°, v_{CO} 1962 cm^{-1}; $p\text{-}FC_6H_4$,[10] mp 175–180°, v_{CO} 1984 cm^{-1}. (Melting points uncorrected on Köfler hot stage). Infrared spectra were taken in Nujol mulls on a grating spectrophotometer.

References

1. J. A. McCleverty and G. Wilkinson, *Inorganic Syntheses*, **8**, 214 (1966).
2. R. F. Heck, *J. Am. Chem. Soc.*, **86**, 2796 (1964).
3. J. Chatt and B. L. Shaw, *J. Chem. Soc.*, **1966**, 1437.
4. J. A. Osborn, F. H. Jardine, J. F. Young, and G. Wilkinson, *J. Chem. Soc.*, 1711 (1966).
5. M. A. Bennett, R. Bramley, and P. A. Longstaff, *Chem. Commun.*, 806 (1966).
6. J. A. Osborn, G. Wilkinson, and J. F. Young, *Chem. Commun.*, 17 (1965); F. H. Jardine, J. A. Osborn, G. Wilkinson, and J. F. Young, *Chem. Ind.* (London), 560 (1965).
7. J. Blum, E. Oppenheimer, and E. D. Bergmann, *J. Am. Chem. Soc.*, **89**, 2338 (1967).
8. K. Ohno and J. Tsuji, *J. Am. Chem. Soc.*, **90**, 99 (1968).
9. L. Vallerino, *J. Chem. Soc.*, **1957**, 2287.
10. D. Evans, J. A. Osborn, and G. Wilkinson, *J. Chem. Soc.*, A, **1968**, 3133.

19. HYDRIDO PHOSPHINE COMPLEXES OF RHODIUM(I)

Submitted by N. AHMAD,* J. J. LEVISON,* S. D. ROBINSON,*
and M. F. UTTLEY*
Checked by E. R. WONCHOBA† and G. W. PARSHALL†

The syntheses of $RhH[P(C_6H_5)_3]_4$ and $RhH(CO)[P(C_6H_5)_3]_3$ described herein are members of a group of preparations each involving addition of a platinum metal halide or halo complex and sources of hydride, carbonyl, and/or nitrosyl ligands to a briskly boiling solution of triphenylphosphine in an alcoholic solvent.[1]

The reactions are performed in a 250-mL conical (Erlenmeyer) reaction flask surmounted by a quick-fit MA4/13 adapter bearing a 30-cm condenser, a nitrogen inlet, and a stoppered port for the introduction of reagents. The flask is situated on a magnetic stirrer–hot plate to ensure effective stirring and heating of the reaction solution. Vigorous boiling, efficient stirring, and rapid successive addition of reagent solutions are essential if clean products free from contamination by insoluble intermediates are to be obtained.

A. HYDRIDOTETRAKIS(TRIPHENYLPHOSPHINE)RHODIUM(I)

$$RhCl_3 \cdot 3H_2O + 4(C_6H_5)_3P \xrightarrow{KOH} RhH[P(C_6H_5)_3]_4$$

Hydridotetrakis(triphenylphosphine)rhodium(I) has previously been prepared by addition of triphenylphosphine to preformed hydridotris(triphenylphosphine)rhodium(I) in toluene solution[2,3] and by reaction of preformed chlorotris(triphenylphosphine)rhodium(I) with hydrazine and hydrogen in an ethanol–benzene medium containing excess triphenylphosphine.[2] Other syntheses employ aluminum alkyls,[4] Grignard reagents,[5] sodium propoxide,[6] and hydrogen under pressure[7] as reductants.

The procedure[8,9] given below affords an efficient, one-step synthesis of hydridotetrakis(triphenylphosphine)rhodium (I) from hydrated rhodium trichloride.

---

*Department of Chemistry, King's College, Strand, London WC2R 2LS.
Present address: Department of Chemistry, West Chester University, West Chester, PA (N. A.); International Nickel, Bashley Road, London N.W. 10 (J. J. L.).

†Central Research Department, E. I. du Pont de Nemours & Company, Wilmington, DE 19898.

Procedure

Hot solutions of rhodium trichloride 3-hydrate (0.26 g, 1.0 mmol) in ethanol (20 mL) and potassium hydroxide (0.4 g) in ethanol (20 mL) are added rapidly and successively to a vigorously stirred, boiling solution of triphenylphosphine (2.62 g, 10 mmol) in ethanol (80 mL). The mixture is heated under reflux for 10 min, and allowed to cool to 30°. The precipitated product is filtered; washed with ethanol, water, ethanol, and *n*-hexane; and dried *in vacuo*. Yield: 1.10 g, (97% based on $RhCl_3 \cdot 3H_2O$).

Anal. Calcd. for $C_{72}H_{61}P_4Rh$: C, 75.0; H, 5.33; P, 10.73. Found: C, 74.69; H, 5.31; P, 10.52.

Properties

Hydridotetrakis(triphenylphosphine)rhodium(I) forms yellow microcrystals that melt at 145–147° in air and at 154–156° in a capillary sealed under nitrogen. The IR spectrum shows a band at 2156 (m) cm^{-1} attributable to $\nu(RhH)$. The complex is soluble in benzene, chloroform, and dichloromethane forming highly air sensitive solutions.

B. CARBONYLHYDRIDOTRIS(TRIPHENYLPHOSPHINE)-RHODIUM(I)

$$RhCl_3 \cdot 3H_2O + 3(C_6H_5)_3P \xrightarrow{KOH, HCHO} RhH(CO)[P(C_6H_5)_3]_3$$

Carbonylhydridotris(triphenylphosphine)rhodium(I) was first prepared from $RhCl(CO)[P(C_6H_5)_3]_2$ by reduction with hydrazine in ethanolic suspension.[10] More recent syntheses involve reaction of $RhCl(CO)[P(C_6H_5)_3]_2$ with sodium tetrahydroborate[11] or triethylamine and hydrogen in ethanol[11] containing excess triphenylphosphine. Addition of ethanolic rhodium trichloride solution, aqueous formaldehyde, and ethanolic sodium tetrahydroborate to a boiling solution of triphenylphosphine in ethanol has also been employed to synthesize $RhH(CO)[P(C_6H_5)_3]_3$.[12] The following single-stage procedure[9] utilizes ethanolic potassium hydroxide in place of sodium tetrahydroborate.

Procedure

A solution of rhodium trichloride 3-hydrate (0.26 g, 1.0 mmol) in ethanol (20 mL) is added to a vigorously stirred, boiling solution of triphenylphosphine (2.64 g, 10 mmol) in ethanol (100 mL). After a delay of 15 s, aqueous formaldehyde (10 mL, 40% w/v solution) and a solution of potassium hydroxide (0.8 g) in hot ethanol (20 mL) are added rapidly and successively to the vigorously stirred, boiling reaction mixture. The mixture is heated under reflux for 10 min and then allowed to cool to room temperature. The bright yellow, crystalline product is filtered; washed with ethanol, water, ethanol, and n-hexane; and dried *in vacuo*. Yield is 0.85 g (94% based on $RhCl_3 \cdot 3H_2O$).

Anal. Calcd. for $C_{55}H_{46}OP_3Rh$: C, 71.90; H, 5.05; P, 10.11. Found: C, 72.11; H, 5.17; P, 9.86.

Properties

Carbonylhydridotris(triphenylphosphine)rhodium(I) forms yellow microcrystals that melt at 120–122° in air and at 172–174° in a capillary tube under nitrogen. The IR spectrum shows bands at 2041 (s) cm^{-1}, attributed to $\nu(RhH)$, and at 1918 (vs), attributed to $\nu(CO)$. The high-field NMR spectrum contains a single signal, δ −9.7 in $CDCl_3$ solution, broadened by ligand dissociation and exchange processes. The complex is soluble in benzene, chloroform, and dichloromethane.

References

1. N. Ahmad, J. J. Levison, S. D. Robinson and M. F. Uttley, *Inorg. Synth.*, **15**, 45 (1974).
2. K. C. Dewhirst, W. Keim, and C. A. Reilly, *Inorg. Chem.*, **7**, 546 (1968).
3. B. Ilmaier and R. S. Nyholm, *Naturwissenschaften*, **56**, 415 (1969).
4. A. Yamamoto, S. Kitazume, and S. Ikeda, *J. Am. Chem. Soc.*, **90**, 1089 (1968).
5. M. Takesada, H. Yamazaki and N. Hagihara, *J. Chem. Soc. Jpn.*, **89**, 1121 (1968).
6. G. Gregorio, G. Pregaglia, and R. Ugo, *Inorg. Chim. Acta*, **3**, 89 (1969).
7. M. Takesada, H. Yamazaki, and N. Hagihara, *J. Chem. Soc. Jpn.*, **89**, 1126 (1968).
8. J. J. Levison and S. D. Robinson, *J. Chem. Soc. A*, **1970**, 2947.
9. N. Ahmad, S. D. Robinson, and M. F. Uttley, *J. Chem. Soc. Dalton Trans.*, **1972**, 843.
10. S. S. Bath and L. Vaska, *J. Am. Chem. Soc.*, **85**, 3500 (1963).
11. P. S. Hallman, D. Evans, J. A. Osborn, and G. Wilkinson, *Chem. Commun.*, **1967**, 305.
12. D. Evans, G. Yagupsky, and G. Wilkinson, *J. Chem. Soc. A*, **1968**, 2660.

20. TETRACARBONYLDICHLORODIRHODIUM

$$2(RhCl_3 \cdot 3H_2O) + 6CO \rightarrow [Rh_2(CO)_4Cl_2] + 6H_2O + 2COCl_2$$

Submitted by J. A. McCLEVERTY* and G. WILKINSON*
Checked by LOREN G. LIPSON,† MICHAEL L. MADDOX,† and HERBERT D. KAESZ†

Tetracarbonyldichlorodirhodium has been obtained by the action of carbon monoxide at high temperature and pressure on a mixture of anhydrous rhodium(III) chloride and finely divided copper powder[1] and by reaction of rhodium(III) chloride 3-hydrate with carbon monoxide saturated with methanol at moderate temperatures and atmospheric pressure.[2] The preparation described here is a modification of the latter method, without use of methanol. This procedure is considerably simpler than the recently described preparation which involves adsorption of rhodium chloride on silica gel, chlorination, and subsequent carbonylation.[3]

Procedure

■ **Caution.** *The reaction must be carried out in a well-ventilated hood.*

A tube (20 cm long and 2 cm in diameter) with a porous disk (porosity 3 or medium) sealed in at one end and with an ungreased ground joint, is arranged as shown in Fig. 1. Rhodium(III) chloride 3-hydrate‡ (11.0 g; 0.042 mol) is pulverized and placed on top of the disk. The apparatus is then flushed with carbon monoxide and lowered into a paraffin-oil bath maintained at 100°. Bath temperatures above 100° should be avoided to prevent the formation of anhydrous rhodium(III) chloride, which is inert to carbon monoxide. Carbon monoxide is passed slowly through the system, a bubbler being attached to the end of the apparatus to indicate the rate of flow.

■ **Caution.** *The rate of flow of carbon monoxide through the apparatus must be determined by the capacity of the hood being used. If the flow is too rapid, the escaping noxious gases (carbon monoxide and phosgene) may not be removed entirely from the atmosphere in the hood and may escape into the laboratory.*

At hourly intervals, or more frequently if necessary, the water that condenses near the top of the tube is removed with adsorbent cotton. During

---

*Imperial College of Science and Technology, London.
†University of California, Los Angeles, CA.
‡The commercial salt is often not quite stoichiometric for $RhCl_3 \cdot 3H_2O$.

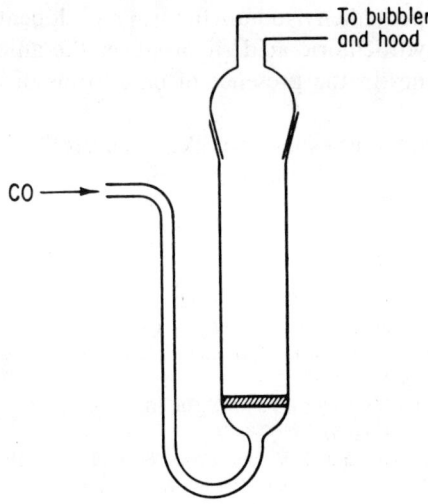

Fig. 1. Apparatus for the preparation of tetracarbonyldichlorodirhodium.

the reaction, the orange tetracarbonyldichlorodirhodium sublimes about halfway up the reaction tube.

When the reaction is complete (3–5 h), the apparatus is withdrawn from the oil bath. When the apparatus has been cooled, the orange-red crystals can be scraped from the reaction vessel and the last traces of compound washed out with dry benzene or hexane. The yield is 8.3 g (96%). The rhodium carbonyl chloride so obtained is pure enough for most purposes, but it may be recrystallized[4] from hexane or sublimed at 80° at a pressure of 0.1 mmHg.

The compound is stable in air but should be stored in a desiccator, for it is somewhat sensitive to moisture.

Properties

Tetracarbonyldichlorodirhodium is an orange-red crystalline solid very soluble in most organic solvents (except the aliphatic hydrocarbons) producing orange solutions. The compound has a melting point of 124–125°, carbonyl stretching frequencies[5] in petroleum ether solution (bp 40–60°) or hexane at 2105 (m), 2089 (s), about 2080 (vw), 2035 (s), and 2003 (w) cm^{-1}. It is quite volatile, forming a red crystalline sublimate. Although the pure compound is stable in dry air, its solutions in organic solvents decompose to insoluble brown materials when left exposed to air.

Tetracarbonyldichlorodirhodium reacts readily with ligands such as phosphines, arsines, stibines, and phosphites[4,5] to form mononuclear complexes.

It reacts with cyclopentadienylsodium to form π-cyclopentadienyldicarbonylrhodium.[7] With hydrochloric acid[8] it produces the anion $[Rh(CO)_2Cl_2]^-$, and with β-diketones in the presence of base forms dicarbonylrhodium β-diketonates.[9]

The complex has an unusual booklike structure[10] with chloride bridges, and there is evidence for a metal–metal bond.

References

1. W. Manchot and J. König, *Ber.*, **58**, 2173 (1925).
2. W. Hieber and H. Lagally, *Z. Anorg. u. Allgem. Chem.*, **251**, 96 (1943).
3. K. S. Brenner, E. O. Fischer, H. P. Fritz, and C. G. Kreiter, *Chem. Ber.*, **96**, 2632 (1963).
4. L. Vallarino, *J. Chem. Soc.*, **1957**, 2287.
5. C. W. Garland and J. R. Wilt, *J. Chem. Phys.*, **36**, 1094 (1962); A. C. Yang and C. W. Garland, *J. Phys. Chem.*, **61**, 1504 (1957).
6. W. Hieber, H. Heusinger, and O. Vohler, *Chem. Ber.*, **90**, 2425 (1957).
7. E. O. Fischer and H. P. Fritz, *Angew. Chem.*, **73**, 353 (1961).
8. A. Davison and G. Wilkinson, unpublished work.
9. F. Bonati and G. Wilkinson, *J. Chem. Soc.*, **1964**, 3156.
10. L. F. Dahl, C. Martell, and D. L. Wampler, *J. Am. Chem. Soc.*, **83**, 1761 (1961).

21. DI-μ-CHLORO-TETRAKIS(ETHENE)DIRHODIUM(I): $[Rh(Cl)(C_2H_4)_2]_2$

Submitted by RICHARD CRAMER*
Checked by J. A. McCLEVERTY† and J. BRAY†

A variety of organic syntheses that involve reactions of olefins are catalyzed by rhodium compounds. As a consequence, considerable attention has been given to the study of the properties of olefin complexes of rhodium. Moreover, since ethylene is very labile and volatile, a variety of compounds (including complexes of other olefins) are easily accessible from them by nucleophilic displacement of ethylene. Displacement of ethene from $Rh_2Cl_2(C_2H_4)_4$[1,2] makes it a useful synthetic intermediate.

$2RhCl_3 \cdot 3H_2O + 6C_2H_4 \longrightarrow$

$(C_2H_4)_2Rh \underset{Cl}{\overset{Cl}{<>}} Rh(C_2H_4)_2 + 4HCl + 2CH_3CHO + 4H_2O$

---

*Central Research Department, Experimental Station, E. I. du Pont de Nemours & Company, Wilmington, DE 19898.
†Department of Chemistry, The University, Sheffield S3 7HF, United Kingdom.

Procedure

"Rhodium trichloride trihydrate"* (10 g, 0.037 g-atom of Rh) is dissolved in 15 mL of water by warming on a steam bath. The solution is transferred to a 500-mL round-bottomed flask containing a Teflon-covered magnetic stirring bar and 250 mL of methanol. The flask is freed of oxygen by alternately evacuating (water pump) and repressuring with ethylene to 1 atm. The methanolic solution is stirred at room temperature under ethene at a pressure of about 1 atmosphere.† The product begins to precipitate as a finely divided solid after about an hour. It usually has the color of dichromate but occasionally is dark-rust-colored. After about 7 h, it is collected by filtration under vacuum on a sintered-glass funnel. It is best to decant most of the liquid before transferring the solid to the filter, because the product is sometimes so finely divided that filtration is slow. One should avoid drawing air through the solid. The product is washed with about 50 mL of methanol and dried *in vacuo* at room temperature. The yield is 4.8–5.0 g (60–65%).

Anal. Calcd. for C_4H_8ClRh: C, 24.70; H, 4.15; Cl, 18.23; Rh, 52.90. Found: C, 24.81; H, 4.17; Cl, 18.26; Rh, 51.23.

The product dissolves by reaction with HCl, and a second crop can be recovered by neutralizing the acid generated during synthesis. A solution of 1.5 g of NaOH in 3 mL of H_2O is added to the filtrate and washings from the first crop. The solution is treated with ethene as before to recover 1.0–1.5 g of $(C_2H_4)_2RhCl_2Rh(C_2H_4)_2$, giving a combined yield of about 6 g or 75% of theory. Attempts to get a third crop by a second treatment with NaOH give a small amount of inferior product.

The reaction has been run successfully, using 1–30 g of $RhCl_3 \cdot 3H_2O$ with a proportionate adjustment in the amount of solvents. Both the submitter and the checkers have found occasional samples of $RhCl_3 \cdot 3H_2O$ that gave only about half the normal yield. In these cases the yield has been improved by adding 20% aqueous NaOH to raise the initial pH of the reaction solution to 4 (pH indicator paper).

Properties

Di-μ-chloro-tetrakis(ethene)dirhodium does not melt. The product is best characterized by elemental analysis and by its infrared spectrum (KBr wafer), which has medium to strong absorptions at 3060, 2980, 1520, 1430, 1230, 1215, 999, 952, 930, and 715 cm^{-1}. It is sparingly soluble in all liquids and

---

*Engelhard Industries, 429 Delancy St., Newark, NJ 07105.

†Alternatively the reaction has been run by bubbling ethene through the stirred solution at the rate of about a bubble per second.

cannot be purified by recrystallization. The compound is relatively stable to air at room temperature, but stored samples develop the odor of acetaldehyde and darken superficially. It is preferred to store $[Rh_2Cl_2(C_2H_4)_4]$ at around 0°.

References

1. R. Cramer. *Inorg. Chem.*, **1**, 722 (1962).
2. R. Cramer, *J. Am. Chem. Soc.*, **86**, 217 (1964).

22. DI-μ-CHLORO-BIS(η^4-1,5-CYCLOOCTADIENE)-DIRHODIUM(I)

Submitted by G. GIORDANO* and R. H. CRABTREE†
Checked by R. M. HEINTZ,‡ D. FORSTER,‡ and D. E. MORRIS‡

Di-μ-chloro-bis(η^4-1,5-cyclooctadiene)dirhodium(I), $[RhCl(1,5-C_8H_{12})]_2$, has been prepared in 60% yield by reducing rhodium trichloride hydrate in the presence of excess olefin in aqueous ethanol.[1] In the present preparation the yield has been greatly increased (to 94%). Two related complexes, $[RhCl(1,5-C_6H_{10})]_2$[2] and $[RhCl(C_6H_{12})_2]_2$, are similarly prepared in high yield from 1,5-hexadiene and 2,3-dimethyl-2-butene, respectively.

Such diene complexes can be used to prepare homogeneous hydrogenation catalysts *in situ*, especially where a variable tertiary phosphine : rhodium ratio is required[3] or where an asymmetric tertiary phosphine is employed for asymmetric synthesis.[4] The cyclooctadiene complex is also the starting point for the preparation of a number of complexes of the type $[Rh(1,5-C_8H_{12})L_2]^+$ (L represents a variety of P— and N— donor ligands) of interest in homogeneous catalysis.[5]

Procedure

$$2RhCl_3 + 2C_8H_{12} + 2CH_3CH_2OH + 2Na_2CO_3 \rightarrow$$
$$[RhCl(C_8H_{12})]_2 + 2CH_3CHO + 4NaCl + 2CO_2 + 2H_2O$$

---

*Present address: Istituto di Chimica Generale, Università, Via Venezian 21, 20133 Milano, Italy.
†Sterling Chemistry Laboratory, Yale University, New Haven, CT 06520.
‡Monsanto Chemical Co, St Louis, MO 63166.

A 100-mL, two-necked, round-bottomed flask is fitted with a reflux condenser connected to a nitrogen bubbler. The flask is charged with 2.0 g (7.6 mmol) of rhodium trichloride trihydrate* (a generous loan of the Compagnie des Métaux Précieux) and 2.2 g (7.7 mmol) of sodium carbonate decahydrate.† Under nitrogen, 20 mL of deoxygenated ethanol-water (5:1) and 3 mL of 1,5-cyclooctadiene‡ are added and the mixture is then heated at reflux with stirring for 18 h, during which time the product precipitates as a yellow-orange solid. The mixture is cooled and immediately filtered and the product is washed with pentane and then with methanol–water (1:5) until the washings no longer contain chloride ion. The product is dried *in vacuo*. Yield 1.67 g (94%).

Anal. Calcd. for $C_{16}H_{24}Cl_2Rh_2$: C, 38.97; H, 4.91; Cl, 14.38; Rh, 41.74. Found: C, 39.01; H, 4.80; Cl, 14.08.

Properties

Di-μ-chloro-bis(η^4-1,5-cyclooctadiene)dirhodium(I) is a yellow-orange, air-stable solid. It can be used directly as obtained for preparative purposes[5] or as a precursor for homogeneous catalysts.[3,4] It can be recrystallized from dichloromethane–diethyl ether to give orange prisms. The compound is soluble in dichloromethane somewhat less soluble in acetone and insoluble in pentane and diethyl ether. Characteristic strong bands occur in the infrared spectrum at 819, 964, and 998 cm^{-1} (Nujol mull). The cyclooctadiene vinylic protons resonate in the ^{1}H NMR spectrum at τ 5.7 and the allylic protons at τ 7.4–8.3 (chloroform-*d* solution). Other physical properties are given by Chatt.[1]

Analogous Complexes

The 1,5-hexadiene complex, [RhCl(C_6H_{10})]$_2$, may be prepared by this method with a reaction time of 24 h. The temperature should not exceed 40° to avoid the deposition of metallic rhodium. Under these conditions the yield is 85% of analytically pure product.

---

*Available from Alfa Products, Ventron Corp., P.O. Box 299, Danvers, MA 01923.

†The checkers obtained an off-color (olive-green) product when using sodium carbonate. However, in the absence of sodium carbonate they repeatedly obtained the expected yellow-orange product in good yields (90–94%) and time periods (18 h). The primary authors believe that the Na_2CO_3 may be useful if the starting material is relatively acidic.

‡Available from Aldrich Chemical Co., 940 W. St. Paul Ave., Milwaukee, WI 53233.

References

1. J. Chatt and L. M. Venanzi, *J. Chem. Soc.*, **1957**, 4735.
2. G. Winkhaus and H. Singer, *Chem. Ber.*, **99**, 3602 (1966).
3. J. A. Osborn and G. Wilkinson, *J. Chem. Soc., A*, **1968**, 1054.
4. L. Horner, H. Buethe, and H. Siegel, *Tetrahedron Lett.*, 4023 (1968); H. B. Kagan, and T.-P. Dang, *J. Am. Chem. Soc.*, **94**, 6429 (1972); W. S. Knowles, J. J. Sabacky, and B. D. Vineyard, *Chem. Commun.*, **1972**, 10.
5. R. R. Schrock and J. A. Osborn, *J. Am. Chem. Soc.*, **93**, 2397 (1971).

23. CHLOROBIS(CYCLOOCTENE)RHODIUM(I) AND –IRIDIUM(I) COMPLEXES

Submitted by A. van der ENT* and A. L. ONDERDELINDEN†
Checked by ROBERT A. SCHUNN‡

The cyclooctene compounds $[MCl(C_8H_{14})_2]_n$, with M = Rh or Ir, are important starting materials for the preparation of rhodium(I) and iridium(I) complexes.[1,2] The compound $[RhCl(C_8H_{14})_2]_n$ can be separated in varying yields (35–60%) from solutions of rhodium(III) chloride 3-hydrate and cyclooctene in ethanol[3] after standing for 3–5 days. Di-μ-chloro-bis[bis(cyclooctene)iridium] can be prepared in 40% yield by refluxing chloroiridic(IV) acid and cyclooctene in 2-propanol.[4] The resulting product is always contaminated with an iridium hydride complex. The following modifications give better yields (70–80%) and an iridium(I) complex of higher purity.

A. CHLOROBIS(CYCLOOCTENE)RHODIUM(I)

$$RhCl_3 + 2C_8H_{14} + CH_3CH(OH)CH_3 \longrightarrow RhCl(C_8H_{14})_2 + CH_3COCH_3 + 2HCl$$

Procedure

In a 100-mL, three-necked, round-bottomed flask, 2 g (7.7 mmol) of rhodium(III) chloride 3-hydrate is dissolved in an oxygen-free mixture of

---

*Formerly Unilever Research Laboratorium Vlaardingen.
†Unilever Research Laboratorium Vlaardingen, P.O. Box 114, 3130 AC Vlaardingen, The Netherlands.
‡Central Research Department, E. I. du Pont de Nemours & Company, Wilmington, DE 19898.

40 mL of 2-propanol and 10 mL of water. Cyclooctene (6 mL) is added. The solution is stirred for about 15 min under nitrogen. The flask is then closed and allowed to stand at room temperature for 5 days. The resulting reddish-brown crystals are collected on a filter, washed with ethanol, dried under vacuum, and stored under nitrogen at $-5°C$. The yield is 2.0 g (74%).

Anal. Calcd. for $RhC_{16}H_{28}Cl$: Rh, 28.72; C, 53.56; H, 7.81; Cl, 9.91. Found: Rh, 28.55; C, 53.76; H, 7.89; Cl, 9.76.

Properties

The solubility of $[RhCl(C_8H_{14})_2]_n$ in benzene and chloroform is too low for molecular-weight measurements. Its reddish-brown color darkens slowly in air.

B. DI-μ-CHLORO-BIS[BIS(CYCLOOCTENE)IRIDIUM]

$$2(NH_4)_3IrCl_6 + 4C_8H_{14} + 2CH_3CH(OH)CH_3 \longrightarrow$$
$$[IrCl(C_8H_{14})_2]_2 + 6NH_4Cl + 2CH_3COCH_3 + 4HCl$$

Procedure

In a 250-mL, three-necked, round-bottomed flask, fitted with a nitrogen inlet and a reflux condenser, 6 g (0.01 mol) of ammonium hexachloroiridate(III)* (43.1% Ir) is suspended in an oxygen-free mixture of 30 mL of 2-propanol and 90 mL of water.† Cyclooctene (12 mL) is added. The mixture is refluxed on a water bath under a slow stream of nitrogen and with vigorous stirring for 3–4 h. After cooling, the alcohol–water mixture is decanted, the last few milliliters being pipetted off. The orange oil remaining in the flask is allowed to crystallize under ethanol at 0°C. The yellow crystals are collected on a filter, washed with cold ethanol, dried under vacuum, and stored under nitrogen at room temperature. The yield is 4.7 g (80%).‡

Anal. Calcd. for $Ir_2C_{32}H_{56}Cl_2$: C, 42.89; H, 6.25; Cl, 7.93. Found: C, 43.12; H, 5.97; Cl, 7.84.

---

*Available from Johnson, Matthey Company, Ltd., London.
†Similar results are obtained by using sodium or potassium chloroiridate(III).
‡Checkers found a yield of 74%.

Properties

The results of molecular-weight measurements on a freshly prepared solution in benzene suggest a dimeric structure (found: M = 886; calcd., M = 895). In the solid state, $[IrCl(C_8H_{14})_2]_2$ decomposes slowly under the influence of atmospheric moisture. The compound is moderately soluble in benzene, chloroform, and carbon tetrachloride, but in general, these solutions are unstable for long periods of time. In comparison with the corresponding rhodium complex, this compound is more reactive in oxidative addition reactions. This is demonstrated by the formation of iridium hydrides during reaction with hydrogen and hydrogen chloride, respectively.

References

1. S. Montelatici, A. van der Ent, J. A. Osborn, and G. Wilkinson, *J. Chem. Soc. A*, **1968**, 1054.
2. M. A. Bennett and D. L. Milner, *J. Am. Chem. Soc.*, **91**, 6983 (1969).
3. L. Porri, A. Lionetti, G. Allegra, and A. Immirzi, *Chem. Commun.*, **1965**, 336.
4. B. L. Shaw and E. Singleton, *J. Chem. Soc. A*, **1967**, 1683.

24. *trans*-CARBONYLCHLOROBIS-(TRIPHENYLPHOSPHINE)IRIDIUM

Submitted by J. P. COLLMAN,* C. T. SEARS, Jr.,* and M. KUBOTA*
Checked by ALAN DAVISON,† E. T. SHAWL,† JOHN R. SOWA, Jr.,‡ and ROBERT J. ANGELICI‡

Vaska and DiLuzio[1] prepared $(Ph_3P)_2(CO)IrCl$ from $IrCl_3(H_2O)_x$ or $(NH_4)_2IrCl_6$ and triphenylphosphine in various solvents such as aqueous 2-(2-methoxyethoxy)ethanol, ethylene glycol, diethylene glycol, triethylene glycol, with yields varying between 75 and 86%. Collman and Kang[2] have described a preparation which involves refluxing $IrCl_3(H_2O)_x$ and triphenylphosphine in N,N-dimethylformamide under nitrogen for 12 h. Upon cooling and adding excess methanol, yields of 85–90% are obtained.

$$IrCl_3 \cdot 3H_2O + Ph_3P \xrightarrow{N,N\text{-dimethylformamide}} (Ph_3P)_2(CO)IrCl$$

---

*University of North Carolina, Chapel Hill, NC 27514.
†Massachusetts Institute of Technology, Cambridge, MA 02139.
‡Iowa State University, Ames, IA 50011.

Procedure

A mixture of 3.52 g (0.010 mol) of iridium(III) chloride 3-hydrate, 13.1 g (0.05 mol) of triphenylphosphine, 4 mL of aniline, and 150 mL of N,N-dimethylformamide is heated at vigorous reflux for 12 h in a 250-mL round-bottomed flask.* The resulting red-brown solution is filtered while hot; 300 mL of warm methanol is rapidly added with stirring. After cooling the mixture in an ice bath, the yellow crystals are collected on a filter and washed with 25 mL of methanol and then 25 mL of diethyl ether. Yield: 6.8–7.0 g of product (87–90% yield). The purity of the compound can be ascertained by examining its infrared spectrum. An absorption band at 2000 cm^{-1} indicates the presence of some oxygen adduct. The pure complex may be regenerated from the oxygen adduct by heating the solid overnight at 100° *in vacuo*. The compound prepared by this method is pure, but further purification may be achieved by dissolving the compound under a nitrogen atmosphere in 100 mL of warm chloroform, filtering any undissolved impurity, and adding 300 mL of methanol to precipitate the product. The isostructural rhodium complex[3] may be prepared by this method,[4] except that the required heating time is only 2–3 h. The bromide and iodide complexes of iridium and rhodium may also be prepared by a similar procedure, except that oxygen must be excluded in their syntheses.

Properties

trans-Carbonylchlorobis(triphenylphosphine)iridium is a bright yellow crystalline solid (v_{CO} = 1961 cm^{-1} in Nujol mull), which is stable in air, but takes up oxygen in solution. The compound is soluble in benzene and chloroform but is insoluble in diethyl ether and alcohols. The compound was first prepared by Angoletta,[5] but it was correctly formulated by Vaska,[1] who first described its addition reactions. It is often referred to as "Vaska's compound."

This complex forms adducts with H_2, O_2, SO_2, CO, HCl, CS_2, CH_3I, allyl iodide, benzyl chloride, methyl iodoacetate, acetyl chloride, sulfonyl chlorides, trichlorosilane, olefins, electronegatively substituted acetylenes, metal halides, and nitrosylium and diazonium cations.[6,7] It catalyzes the hydrogenation of olefins. Treatment with organic azides affords the molecular nitrogen complex Ir(Ph$_3$P)$_2$N$_2$Cl, whereas reaction with sodium amalgam in tetrahydrofuran under carbon monoxide pressure gives an intermediate, Na[Ir(CO)$_3$Ph$_3$P], from which compounds with metal–metal bonds can be prepared. The chloride ion in Ir(Ph$_3$P)$_2$(CO)Cl can be replaced by other

---

*Although exclusion of air is not necessary, the checkers ran the reaction under an N$_2$ atmosphere.

halides, and the triphenylphosphine can be displaced by alkyl(phenyl)phosphines, but not by triphenylarsine. Treatment with 1,2-ethanediylbis(diphenylphosphine) (dppe) yields (dppe)$_2$IrCl.

References

1. L. Vaska and J. W. DiLuzio, *J. Am. Chem. Soc.*, **83**, 2784 (1961).
2. J. P. Collman and J. W. Kang, *J. Am. Chem. Soc.*, **89**, 844 (1967).
3. L. Vallarino, *J. Chem. Soc.*, **1957**, 2287.
4. A. Rusina and A. A. Vlček, *Nature*, **206**, 295 (1965).
5. M. Angoletta, *Gazz. Chim. Ital.*, **89**, 2359 (1959).
6. J. P. Collman and W. R. Roper, *Adv. Organomet. Chem.*, **10**, 101 (1967).
7. R. S. Dickson, *Organometallic Chemistry of Rhodium and Iridium*, Academic Press, Orlando, FL, 1983, Chapter 3.

25. BIS(1,5-CYCLOOCTADIENE)NICKEL(0)

$$Ni(C_5H_7O_2)_2 + 2C_8H_{12} + 2Al(C_2H_5)_3 \longrightarrow$$

$$Ni(C_8H_{12})_2 + 2Al(C_2H_5)_2(C_5H_7O_2) + C_2H_4 + C_2H_6$$

Submitted by R. A. SCHUNN, S. D. ITTEL, and M. A. CUSHING*
Checked by R. BAKER, R. J. GILBERT, and D. P. MADDEN†

Bis(1,5-cyclooctadiene)nickel(0) is useful for the synthesis of a variety of novel nickel complexes[1-6] since the cyclooctadiene ligands are easily displaced. The procedure given here is based on that described by Wilke;[7] butadiene is used to prevent the formation of nickel metal.[7]

Materials

Waters of hydration are removed from the light-blue complex, Ni(C$_5$H$_7$O$_2$)$_2$·2H$_2$O (Alfa Inorganics) by azeotropic distillation with toluene using a Dean–Stark apparatus to determine when water is no longer being evolved. The resulting dark-green slurry is filtered while still hot through Celite diatomaceous earth under nitrogen, and the filtrate is evaporated to dryness on a rotary evaporator. The green, anhydrous bis(2,4-pentanedionato)nickel(II) is crushed in a mortar and dried at 80°, 0.1 μm for

---

*Central Research and Development Department, Experimental Station, E. I. du Pont de Nemours & Company, Wilmington, DE 19880-0328. (R. A. S., retired.)
†Department of Chemistry, The University, Southampton, S09 5NH, United Kingdom.

16 h to give the desired material. It should then be handled and stored under an anhydrous atmosphere in a dry box or glove bag.

Peroxide impurities are removed from 1,5-cyclooctadiene (Aldrich Chemicals) by filtration through grade 1 neutral alumina; the filtration is repeated until the alumina is no longer colored yellow. Further purification is not necessary, but the liquid should be used immediately. Commercial butadiene is generally dry enough for this reaction, but may be dried by passing through a U trap containing type 3 Å molecular sieves, condensing the gas at $-78°$, and storing it in a small stainless steel cylinder.

■ **Caution.** *Triethylaluminum is a pyrophoric liquid that reacts violently with water or alcohols. It should be handled only in a rigorously oxygen and moisture free atmosphere, using face shield and gloves.*

The solution of triethylaluminum in toluene is most conveniently prepared in an inert-atmosphere box[8] and transferred to the addition funnel equipped with serum caps using a 100-mL hypodermic syringe.

Procedure

The entire procedure is performed in an anhydrous, oxygen-free atmosphere using anhydrous, deoxygenated solvents. Standard Schlenk techniques[8] for benchtop inert-atmosphere reactions are used in this procedure. A 1-L, four-necked, round-bottomed flask is equipped with a stopcock adapter having a hose end, a thermometer, and a Dry Ice condenser topped with a T tube that is connected to a mineral-oil bubbler. Ground-glass joints should be lubricated with a hydrocarbon grease because the product is decomposed catalytically by halocarbons. The center neck of the flask is left open. The flask is flushed thoroughly with nitrogen and charged through the center neck with 102.8 g (0.4 mol) of anhydrous bis(2,4-pentanedionato)nickel(II), 250 mL of toluene, and 216 g (2.0 mol) of 1,5-cyclooctadiene. The center neck is then equipped with a mechanical stirrer, and the mixture is stirred and cooled to $-10°$;* the condenser is filled with a mixture of Dry Ice and toluene.

Approximately 18 g (0.33 mol) of anhydrous 1,3-butadiene is admitted slowly to the flask through the stopcock adapter,† taking care that pressure does not build up in the system before the gas dissolves in the cold mixture. The stopcock adapter is then replaced with a nitrogen-flushed 250-mL pressure-equalizing dropping funnel having a Teflon stopcock, and the system

---

*Dry Ice is added to a toluene bath as necessary to obtain the desired temperature. A wet ice–methanol bath may also be used but is somewhat more hazardous because of the violent reaction of triethylaluminum with water and methanol.

† The weight transferred to the reaction flask is determined by supporting the cylinder on a balance while conducting the gas into the reaction flask through rubber tubing attached to the hose end of the stopcock adapter.

is slowly flushed with nitrogen by removing the stopper from the top of the addition funnel. A solution of 103 g (0.9 mol) of triethylaluminum in 100 mL of toluene is transferred into the addition funnel by cannula and added dropwise to the cold, stirred mixture. The temperature is maintained at -10 to $0°$; the addition is normally completed in 45 to 90 min. During the addition, the green slurry becomes yellow-brown and a yellow crystalline solid is formed. After being stirred at $0°$ for an additional 0.5 h, the mixture is allowed to warm to room temperature and is stirred for several hours or overnight.*
During this time, the Dry Ice condenser is allowed to come to room temperature and gas (predominantly ethylene and ethane) is evolved.

The yellow slurry is recooled to $-15°$ and slowly stirred for 2–3 h. With rapid nitrogen flushing, the addition funnel is replaced with an adapter attached to a nitrogen line, the thermometer and mechanical stirrer are replaced with stoppers, and all joints are secured with standard-taper-joint clips (A. H. Thomas Co.) or with electrical tape. The apparatus shown in

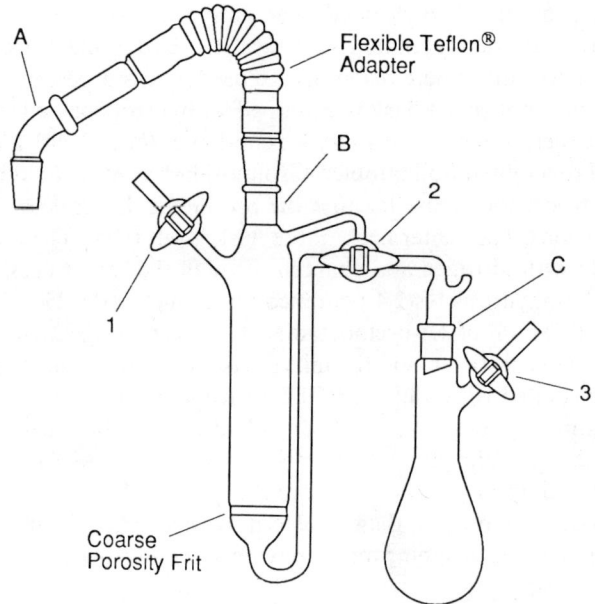

Fig. 1. Apparatus for the preparation of Ni(1,5-cyclooctadiene)$_2$.

*Since the product is often light sensitive, the reaction flask should be covered during this period.

Fig. 1* is assembled and filled with nitrogen. With nitrogen flushing through stopcock 1 and the reaction flask, the Dry Ice condenser is removed from the reaction flask and replaced by joint A; the joint is secured with a joint clip or tape. Care must be taken to support the reaction flask suitably, and it must be replaced in the cold bath. The filter is immersed in a cold bath ($-15°$), and the reaction solution is transferred to the filter in portions. The mixture is filtered by evacuating the reaction flask with stopcock 2 closed. Stopcock 3 is then closed and stopcock 2 is opened. The solution will then begin to filter into the receiving flask. The receiving flask should not be open to vacuum continuously, and the cake of product on the filter should not be allowed to go dry until all of the contents of the flask have been transferred to the filter. Alternatively, the filtration may be carried out by opening stopcock 1 and pressuring with nitrogen; the mineral-oil bubbler may be replaced by a mercury bubbler for this purpose.[8] Care must be taken to use sufficient pressure for filtering but to avoid popping joints or stoppers. If standard 500-mL Schlenk flasks are used as the receiving vessel, they will have to be changed several times during the filtration and washings; stopcock B should be closed during the change.† When the entire reaction mixture has been filtered, the product is washed by adding 100 mL of toluene to the reaction flask (via hypodermic needle through the septum joint), swirling to wash the walls of the flask, cooling it to $-15°$, and pouring it onto the filter. After three washes, the filtrate is pale yellow. A final wash with 100 mL of cold anhydrous diethyl ether removes the last traces of aluminum compounds and facilitates drying. With nitrogen flushing rapidly through stopcock 1, the adapter at joint B is replaced with a stopper which is secured and stopcock 2 is closed. The filter containing the product is evacuated through stopcock 1. After 0.5 h the filter is warmed to $25°$, and after 0.5 h the product is transferred to a vacuum-line flask and dried further (protect from light) at $25°$, 0.1 μm for 16 h to give 97.0 g (89%) of $Ni(1,5-C_8H_{12})_2$; decomposes $\sim 135-140°$.

Anal. Calcd. for $C_{16}H_{32}Ni$: C, 69.9; H, 8.8; Ni, 21.3. Found: C, 69.1; H, 8.7; Ni, 21.2.

---

*This apparatus is suitable for the filtration of large-scale preparations of air-sensitive compounds at temperatures down to $-78°$. The flexible, Teflon joint is available from Aldrich Chemical Co., Inc. Stainless steel versions of this flexible adapter are also available but are more expensive and do not allow visual inspection of the contents. For smaller-scale reactions, Schlenk techniques[7] may be used. The reaction mixture may also be filtered at room temperature, with a corresponding decrease in yield due to the increased solubility of the product.

†The filtrate contains highly reactive ethylaluminum compounds, which react violently with water or alcohols, and is most conveniently disposed of by incineration. Alternatively, the mixture may be decomposed by the careful, dropwise addition of 200 mL of ethanol to the cooled, stirred solution (much gas is evolved), followed by the cautious addition of water.

The solid complex decomposes after several minutes in air; solutions decompose in air more rapidly. It is moderately soluble in benzene and tetrahydrofuran, but heating these solutions above ~60° causes decomposition. It is nearly insoluble in diethyl ether and saturated hydrocarbons. The complex may be purified by extraction with toluene at 45–50°, addition of n-heptane to the yellow filtrate, and concentration of the mixture on a rotary evaporator. The complex is decomposed catalytically by halocarbons,[9] including halocarbon greases, even when present at very low concentrations so the use of "pesticide-grade" solvents and hydrocarbon greases often will result in higher-yield reactions.

^{1}H NMR (C_6D_6, TMS): δ 1.38 (s, 8, CH_2); δ 3.64 (s, bd, 4, CH).

References

1. J. Ashley-Smith, M. Green, and F. G. A. Stone, *J. Chem. Soc. A*, **1970**, 3161.
2. C. S. Cundy, M. Green, and F. G. A. Stone, *J. Chem. Soc. A*, **1970**, 1647.
3. J. Browning, C. S. Cundy, M. Green, and F. G. A. Stone, *J. Chem. Soc. A*, **1969**, 20.
4. D. H. Gerlach, A. R. Kane, G. W. Parshall, J. P. Jesson, and E. L. Muetterties, *J. Am. Chem. Soc.*, **93**, 3543 (1971).
5. U. Birkenstock, H. Bonnemann, B. Bogdanovic, D. Walter, and G. Wilke, *Adv. Chem. Ser.*, **70**, 250 (1968).
6. S. D. Ittel, *Inorg. Synth.*, **17**, 117 (1977).
7. B. Bogdanovic, M. Kroner, and G. Wilke, *Annalen*, **699**, 1 (1966).
8. D. F. Shriver and M. A. Drezdon, *The Manipulation of Air-Sensitive Compounds*, Wiley, Chichester, U.K., 1986.
9. C. A. Tolman, D. W. Reutter, and W. C. Seidel, *J. Organometal. Chem.*, **117**, C30 (1976).

26. COMPLEXES OF NICKEL(0)

Submitted by STEVEN D. ITTEL*
Checked by H. BERKE,† H. DIETRICH,† J. LAMBRECHT,† P. HÄRTER,† J. OPITZ,† and W. SPRINGER†

Complexes of Ni(0) have been investigated by a large number of workers.[1] They are interesting because they undergo oxidative addition,[2] are catalytically active,[3] and exist with a wide variety of ligands, including unsaturated molecules[4] and Group V donors. These complexes are easily prepared using bis(1,5-cyclooctadiene)nickel(0)[5] [Ni(1,5-C_8H_{12})$_2$] as a starting material. Examples of the preparation of the major classes of compounds,

---

*Central Research and Development Department, E. I. du Pont de Nemours & Co., P.O. Box 80328, Wilmington, DE 19880-0328.

†Fachbereich Chemie, Universität Konstanz, Postfach 233, Konstanz, West Germany.

NiL$_4$ (where L = phosphine, arsine, stibine, or alkylisocyanide) and "NiL$_2$" (where L = bidentate ligands) are presented.

General Procedure

■ **Caution.** *Isocyanide and phosphorus ligands are toxic and malodorous. The syntheses should therefore be carried out in a well-ventilated hood.*

The entire procedure must be performed in an anhydrous, oxygen-free atmosphere* using anhydrous, deoxygenated solvents.[6] The complexes rapidly decompose on exposure to air, either as solids or in solution. The reactions may be scaled down, but yields will be somewhat lower as a result of mechanical losses. The Schlenk apparatus shown in Fig. 1 is useful for these preparations and also for the preparation of the starting material Ni(1,5-C$_8$H$_{12}$)$_2$ on a scale reduced from that in the preceding synthesis,[5] because it minimizes contact with stopcock grease and allows one to perform all operations without a glove box. The frit with a Teflon plug allows one to rinse the product repeatedly or to prepare a solution by pulling a suitable solvent through the frit without danger of air leakage or contaminating it with grease.

A. TETRAKIS(*tert*-BUTYL ISOCYANIDE)NICKEL(0)

$$Ni(1,5\text{-}C_8H_{12})_2 + 4[(CH_3)_3CNC] \rightarrow Ni[(CH_3)_3CNC]_4 + 2C_8H_{12}$$

Procedure (see also General Procedure)

The apparatus is set up as shown in Fig. 1, with the dropping funnel in place. Bis(1,5-cyclooctadiene)nickel(0) (10.0 g, 36.4 mmol) and a magnetic stirring bar are transferred into flask A and the system is reevacuated and flushed with argon. Diethyl ether (125 mL) is injected into the dropping funnel and is allowed to run into flask A. The suspension is stirred and cooled with an ice bath. *tert*-Butyl isocyanide[7] (12.2 g; 146 mmol) in ether (20 mL) is injected into the dropping funnel and allowed to flow into flask A over a period of 5 min. The reaction mixture is stirred for 30 min. The pale-yellow product is filtered under vacuum and washed twice with 50-mL portions of hexane. The product (13.5 g, 95% yield), may be recrystallized from ethanol–diethyl ether (1:2 mixture), with loss in yield, to give pale-yellow crystals, mp 172° (decomposes). $v(N\equiv C) = 2000$ cm^{-1}.

---

*Either prepurified nitrogen or argon can be used. The author prefers the convenience of argon.

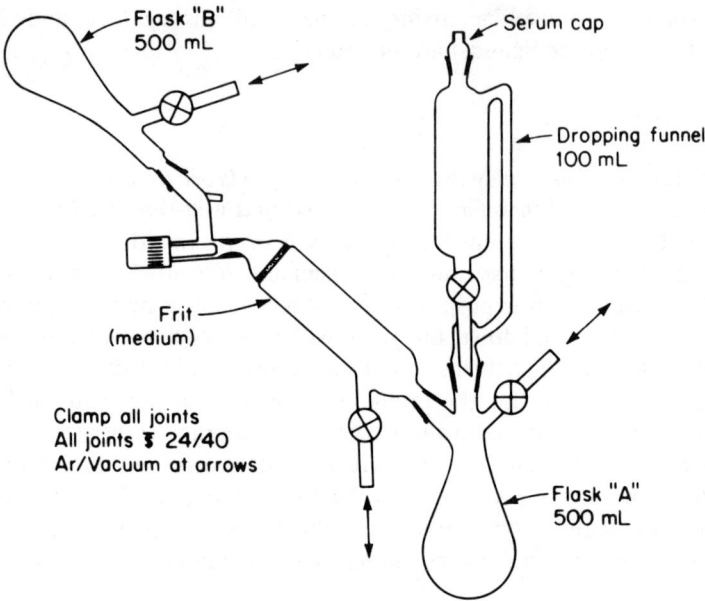

Fig. 1. Apparatus for the preparation of nickel(0) complexes.

Anal. Calcd. for $C_{20}H_{36}N_4Ni$: C, 61.2; H, 9.2; N, 14.0. Found: C, 61.7; H, 9.4; N, 13.3.

The compounds listed in Table I can be conveniently prepared by this procedure. The quantities of the appropriate ligands are listed, as are the preferred solvents for each reaction; melting points and colors are listed for characterization. Several of the products require cooling or removal of solvent to cause crystallization.

Properties

The air-stabilities of the complexes in Table I vary widely. Isocyanide complexes and complexes of the alkyl phosphines are very air sensitive or pyrophoric. The aryl phosphine complexes are moderately air sensitive; the phosphite complexes can be handled in air but should be stored in an inert atmosphere. The phosphite complexes are insoluble in polar solvents such as alcohols and water, and soluble in hydrocarbons. Aryl phosphine complexes are soluble in aromatic solvents or tetrahydrofuran and insoluble in alkanes and alcohols. The alkyl phosphine complexes are soluble in ethers and hydrocarbons; the triethylphosphine complex dissociates to the deep-purple tris complex, and is protonated by alcohols.

TABLE I. Synthesis and Properties of Selected Nickel(0) Complexes Prepared by Procedure A

						Analysis			
						Carbon		Hydrogen	
Compound	Ligand(g)	Solvent	Color	mp (°C)		Calcd.	Found	Calcd.	Found
Ni(C$_6$H$_{11}$NC)$_4$	16.0	Hexane	Light yellow	75		67.89	67.95	8.95	8.99
Ni[P(CH$_3$)$_3$]$_4$	11.2	Pentane	Yellow	198		39.70	39.29	10.00	9.81
Ni[P(C$_2$H$_5$)$_3$]$_4$	17.3	Pentane	Pale yellow	42		54.25	54.46	11.38	11.70
Ni[P(n-C$_4$H$_9$)$_3$]$_4$	29.7	Pentane	Off-white	50		55.65	55.64	10.51	10.31
Ni[P(CH$_3$)(C$_6$H$_5$)$_2$]$_4$	29.4	Ether	Red-orange	170		72.26	72.71	5.32	5.38
Ni[P(C$_2$H$_5$)$_2$(C$_6$H$_5$)]$_4$	24.4	Pentane	Yellow	87		66.40	66.49	8.36	8.47
Ni[P(OCH$_3$)$_3$]$_4$	18.2	Pentane	White	130		25.97	26.14	6.54	6.39
Ni[P(OC$_2$H$_5$)$_3$]$_4$	24.4	Pentane	White	108		39.86	39.71	8.36	8.26
Ni[P(O-i-C$_3$H$_7$)$_3$]$_4$	30.6	Pentane	White	187		48.50	48.59	9.50	9.41
Ni[P(OCH$_3$)(C$_6$H$_5$)$_2$]$_4$	31.7	Pentane	Yellow	185		67.63	67.58	5.68	5.78
Ni[P(OC$_6$H$_5$)$_3$]$_4$	45.5	Benzene	White	147		66.53	66.29	4.65	4.76
Ni[(CH$_3$)$_2$PC$_2$H$_4$P(CH$_3$)$_2$]$_2$	22.0	Ether	Pale yellow	121		40.15	40.00	8.99	9.17

B. TETRAKIS(TRIPHENYLPHOSPHINE)NICKEL(0)

$$Ni(1,5-C_8H_{12})_2 + 4P(C_6H_5)_3 \rightarrow Ni[P(C_6H_5)_3]_4 + 2C_8H_{12}$$

Procedure (see also General Procedure)

This synthesis is an alternative to the procedure given earlier.[8] If $Ni[P(C_6H_5)_3]_4$ is the only zero-valent nickel complex desired, the earlier synthesis will be preferred but if $Ni(1,5\text{-cod})_2$ is available or if several different nickel complexes are desired, this method will be superior.

A magnetic stirring bar is placed in flask A of the apparatus shown in Fig. 1. The dropping funnel is removed, and the serum-cap adapter is placed in the top of flask A before the entire system is evacuated and flushed with argon (or nitrogen). Bis(1,5-cyclooctadiene)nickel(0) (10.0 g, 36.4 mmol) is transferred into flask A, followed by finely powdered triphenylphosphine (38.2 g, 145.6 mmol). The system is reevacuated and flushed with argon; then flask A is cooled in an ice bath and 200 mL of hexane is injected through the serum cap. The stirred suspension begins to turn red at once. Stirring is continued, and the precipitate is allowed to come to room temperature. After 30 min, the ice bath is replaced; stirring is continued for 5 min, and then the entire apparatus is rotated to pour the product onto the frit. The red-brown product is filtered under vacuum and washed with one 50-mL portion of hexane and two portions of diethyl ether. The crude material may be dried under vacuum for 24 h to give about 39 g (96% yield) of product, mp 122–124°.

Anal. Calcd. for $C_{72}H_{60}NiP_4$: C, 78.3; H, 5.4. Found: C, 78.9; H, 5.7.

The material may be purified by extraction at 60° with 150 mL of toluene containing 10 g of triphenylphosphine. The toluene is removed on a rotary evaporator while *n*-heptane is added to keep the mixture at constant volume. The precipitated product is collected and washed with two 20-mL portions of diethyl ether.

This procedure is useful for the preparation of the compounds listed in Table II as well as many others where the ligands are solids and possibly insoluble in the reaction solvent. All the preparations are on a scale that starts with 10.0 g of $Ni(C_8H_{12})_2$. The quantities of various ligands for each complex are listed, as is the preferred solvent for the reaction. Colors and melting points are included for characterization.

Properties

All the compounds in Table II are soluble in aromatic solvents. They are all moderately air sensitive, and the arsine and stibine complexes are thermally

TABLE II. Synthesis and Properties of Selected Nickel(0) Complexes Prepared by Procedure B

Compound	Ligand (g)	Solvent	Color	mp (°C)	Analysis			
					Carbon		Hydrogen	
					Calcd.	Found	Calcd.	Found
Ni[$(C_6H_5)_2PC_2H_4P(C_6H_5)_2$]$_2$	29.0	hexane	orange	256	73.00	72.79	5.66	5.36
Ni[2,2'-bipyridine]$_2$	11.4	ether	dark-violet	155	64.74	64.71	4.35	4.39
Ni[1,10-phenanthroline]$_2$	13.2	ether	blue-black	280	68.78	68.67	3.85	3.72
Ni[As$(C_6H_5)_3$]$_4$	44.6	ether	orange-yellow	107	67.37	67.43	4.71	4.91
Ni[Sb$(C_6H_5)_3$]$_4$	51.4	ether	light-yellow	244	58.79	58.90	4.11	4.13
Ni[$(CH_3)_2AsC_6H_4As(CH_3)_2$]$_2$	31.8	ether	orange	195	38.08	38.18	5.11	5.27

unstable, decomposing on standing in solution—or more slowly in the solid state—to liberate nickel metal.

References

1. R. Ugo, *Coord. Chem. Rev.*, **3**, 319 (1968), or J. H. Nelson and H. B. Jonassen, *Coord. Chem. Rev.*, **6**, 27 (1971).
2. D. H. Fahey, *J. Am. Chem. Soc.*, **92**, 402 (1970).
3. S. Otsuka, A. Nakamura, and Y. Tatsuno, *Chem. Commun.*, **1967**, 836.
4. S. Otsuka, T. Yoshida, and Y. Tatsuno, *J. Am. Chem. Soc.*, **93**, 6462 (1971).
5. R. A. Schunn, S. D. Ittel, and M. A. Cushing, preceding synthesis.
6. D. F. Shriver, *The Manipulation of Air-Sensitive Compounds*, McGraw-Hill, New York, 1969.
7. J. Casanova, N. D. Werner, and R. E. Schuster, *J. Org. Chem.*, **31**, 3473 (1966).
8. R. A. Schunn, *Inorg. Synth.*, **13**, 124 (1972).

27. TETRAKIS(TRIETHYL PHOSPHITE)NICKEL(0), PALLADIUM(0), AND PLATINUM(0) COMPLEXES

Submitted by MAX MEIER* and FRED BASOLO*
Checked by W. R. KROLL,† D. MOY,† and M. G. ROMANELLI†

A. TETRAKIS(TRIETHYL PHOSPHITE)NICKEL(0)

$$NiCl_2 \cdot 6H_2O + 5P(OC_2H_5)_3 + 2(C_2H_5)_2NH \rightarrow$$
$$[Ni\{P(OC_2H_5)_3\}_4] + (C_2H_5O)_3PO + 2(C_2H_5)_2NH_2Cl + 5H_2O$$

Procedure

The triethyl phosphite (Eastman) used in these preparations was distilled in a vacuum prior to use, bp 51° at 13 mmHg. The preparations were carried out in air; an inert atmosphere was not necessary.

The nickel(0) compound can be prepared by the method of Vinal and Reynolds.[1] Nickel(II) chloride hexahydrate (5.0 g, 0.021 mol) is dissolved in 100 mL of methanol contained in a 250-mL round-bottomed flask. A stirring bar is placed in the flask, and it is put in an ice–water bath standing on a magnetic stirrer. The solution is allowed to cool with stirring for 10 min, and 18 mL of triethyl phosphite are then added over a 1-min period. On addition

---

*Department of Chemistry, Northwestern University, Evanston, IL 60201.
†Esso Research and Engineering Company, Corporate Research Laboratories, Linden, NJ 07036.

of $P(OC_2H_5)_3$ the solution turns dark red. With further cooling and vigorous stirring diethylamine is added dropwise from a syringe containing 5 mL of $(C_2H_5)_2NH$ over a period of 10 min. After about half the diethylamine has been added, white crystals of $[Ni\{P(OC_2H_5)_3\}_4]$ appear. Addition of diethylamine is stopped when the color of the liquid has faded to pink. [Further addition of diethylamine, until the liquid phase is yellow or green will cause contamination of the product with nickel(II) compounds.] The crystals are collected on a glass frit by means of a suction filter and washed with methanol which has previously been cooled in an ice–water bath. The washing is continued until the product is colorless. The product is transferred quickly to a drying vessel (e.g., a 50-mL round-bottomed flask) that is connected to a liquid-nitrogen-cooled trap and a vacuum pump (our pump was capable of producing a vacuum of 10^{-2} mmHg) and dried for 5 h at room temperature. The yield is typically about 40% using methanol as solvent. The checkers report that using acetonitrile yields of 55–60% can be obtained.

Anal. Calcd. for $C_{24}H_{60}O_{12}P_4Ni$: C, 39.83; H, 8.30. Found: C, 40.22; H, 8.58.

Properties

Tetrakis(triethyl phosphite)nickel(0) can be handled in air. It is best stored in an evacuated and sealed tube. On exposure to air for several hours the substance turns green. The compound is insoluble in water, somewhat soluble in methanol, and very soluble in hydrocarbons. It does not dissociate in hydrocarbon solutions.

B. TETRAKIS(TRIETHYL PHOSPHITE)PALLADIUM(0)

$$K_2[PdCl_4] + 5P(OC_2H_5)_3 + 2(C_2H_5)_2NH + H_2O \rightarrow$$
$$[Pd\{P(OC_2H_5)_3\}_4] + (C_2H_5O)_3PO + 2(C_2H_5)_2NH_2Cl + 2KCl$$

Procedure

A concentrated solution of potassium tetrachloropalladate(II), $K_2[PdCl_4]$ (0.330 g, 0.001 mol), is prepared by dissolving it in a minimum amount (~ 2.5 mL) of water at room temperature. A methanol (3 mL) solution of triethyl phosphite (0.831 g, 0.005 mol) is placed in a test tube containing a small stirring bar. The triethyl phosphite (Matheson, Coleman and Bell) was distilled before use and kept under nitrogen (bp 77°, 25 mm).

■ **Note.** *It is very important that the amount of triethyl phosphite not exceed the stoichiometrically required 0.005 mol.*

The test tube is placed in an ice–water bath standing on a magnetic stirrer, and the aqueous solution of chloropalladate is added with stirring.* A yellow solution results. If two liquid phases are formed, methanol is added dropwise until the solution is homogeneous. The yellow solution is cooled with stirring for 2 min, and then 0.21 mL of diethylamine is added to the solution from a small syringe, with further cooling and vigorous stirring. (The tip of the needle is immersed in the solution.) A white precipitate forms immediately, which is collected on a glass frit and washed quickly with a few milliliters of water. The product is rapidly transferred to a drying flask which is connected to a liquid-nitrogen-cooled trap and a vacuum pump and dried for 1 h at room temperature, then for an additional 3 h at 0°. The compound melts with decomposition at 112° under nitrogen, but at much lower temperatures in air.

Anal. Calcd. for $C_{24}H_{60}O_{12}P_4Pd$: C, 37.35; H, 7.87. Found: C, 36.48; H, 7.84.

Properties

The dry product, $[Pd\{P(OC_2H_5)_3\}_4]$, containing no adsorbed triethyl phosphite is air sensitive, turning black on exposures to air exceeding a few minutes. It should be handled in an inert atmosphere (glove box or a large beaker filled with argon). It is stored in an evacuated and sealed tube. The complex is insoluble in water, soluble in methanol, and very soluble in hydrocarbons, in which it does not dissociate.

C. TETRAKIS(TRIETHYL PHOSPHITE)PLATINUM(0)

$$K_2[PtCl_4] + 5P(OC_2H_5)_3 + 2KOH \rightarrow$$
$$[Pt\{P(OC_2H_5)_3\}_4] + (C_2H_5P)_3PO + 4KCl + H_2O$$

Procedure

The platinum(0) compound can be prepared by a method analogous to that of Malatesta and Cariello.[2] Powdered potassium hydroxide† (0.350–0.400 g, 0.006 mol) is dissolved in 10 mL of methanol contained in a large test tube (about 1-in. diameter). To this is added triethyl phosphite (2.5 g, 0.015 mol) and a small stirring bar, and then the test tube is placed in an oil bath kept at

---

* ■ *Note.* *The checkers recommend that the entire preparative procedure be carried out in an apparatus flushed with nitrogen.*

† The weight of potassium hydroxide depends on the assay of the material available: 0.006 mole of 100% KOH weighs 0.337 g. The weight range given is representative.

75° by a heater–stirrer plate. When the solution in the test tube has reached the temperature of 60°, a solution of potassium tetrachloroplatinate(II), $K_2[PtCl_4]$ (1.24 g, 0.003 mol), in about 20 mL of water is slowly added with stirring. Immediately or within a few minutes, colorless crystals separate. The crystals are collected on a glass frit, washed with a few milliliters of an ethanol–water mixture (50% by volume), and dried under vacuum for 4 h at room temperature. Yields vary from 0.45 to 0.58 g (52–67%), mp 114°.

Anal. Calcd. for $C_{24}H_{60}O_{12}P_4Pt$: C, 33.53; H, 7.03. Found: C, 31.0, H, 7.00.

Properties

The complex $[Pt\{P(OC_2H_5)_3\}_4]$ can be handled in air. It is stored in an evacuated and sealed tube. On exposure to air exceeding several hours the substance turns black. The compound is insoluble in water, somewhat soluble in methanol, and very soluble in hydrocarbons in which it does not dissociate.

References

1. R. S. Vinal and I. T. Reynolds, *Inorg. Chem.*, **3**, 1062 (1964).
2. L. Malatesta and C. Cariello, *J. Chem. Soc.*, **1958**, 2323.

28. TETRAKIS(TRIPHENYLPHOSPHINE)PALLADIUM(0)

$$2PdCl_2 + 8P(C_6H_5)_3 + 5NH_2NH_2 \cdot H_2O \rightarrow$$
$$2Pd\{P(C_6H_5)_3\}_4 + 4NH_2NH_2 \cdot HCl + N_2 + 5H_2O$$

Submitted by D. R. COULSON*
Checked by L. C. SATEK† and S. O. GRIM†

Preparations of phosphine and phosphite complexes of palladium(0) have been reported to result from reduction of palladium(II) complexes in the presence of the desired ligand.[1-5] The products are generally formulated as PdL_{4-n} (where $n = 0, 1$), depending on the nature and amount of the ligand used. A related complex, $[Pd\{P(C_6H_5)_3\}_2]_n$, has also been reported.[6]

*Central Research Department, Experimental Station, E. I. du Pont de Nemours & Company, Wilmington, DE 19880.
†University of Maryland, College Park, MD 20704.

Although this preparation is similar in concept to these previous ones, advantage is gained in being able to obtain a high yield of $Pd\{P(C_6H_5)_3\}_4$ in one step from palladium dichloride.

Proceudre

A mixture of palladium dichloride (17.72 g, 0.10 mol), triphenylphosphine (131 g, 0.50 mol), and 1200 mL of dimethyl sulfoxide is placed in a single-necked, 2-L, round-bottomed flask equipped with a magnetic stirring bar and a dual-outlet adapter. A rubber septum and a vacuum nitrogen system are connected to the outlets. The system is then placed under nitrogen with provision made for pressure relief through a mercury bubbler. The yellow mixture is heated by means of an oil bath with stirring until complete solution occurs ($\sim 140°$). The bath is then taken away, and the solution is rapidly stirred for approximately 15 min. Hydrazine hydrate (20 g, 0.40 mol) is then rapidly added over approximately 1 min from a hypodermic syringe. A vigorous reaction takes place with evolution of nitrogen. The dark solution is then immediately cooled with a water bath; crystallization begins to occur at $\sim 125°$. At this point the mixture is allowed to cool without external cooling. After the mixture has reached room temperature it is filtered under nitrogen on a coarse, sintered-glass funnel. The solid is washed successively with two 50-mL portions of ethanol and two 50-mL portions of diethyl ether. The product is dried by passing a slow stream of nitrogen through the funnel overnight. The resulting yellow crystalline product weighs 103.5–108.5 g (90–94% yield) (Note 1).

A melting point determination (Note 2) on a sample in a sealed capillary tube under nitrogen gave a decomposition point of 116° (uncorrected). This compares with a similar determination (115°) performed on the product prepared by the method of Malatesta and Angoletta.[1]

Anal. Calcd. for $C_{72}H_{60}PdP_4$: C, 75.88; H, 5.25; P, 10.75. Found: C, 75.3; H, 5.36; P, 10.7.

Properties

The $Pd\{P(C_6H_5)_3\}_4$ complex obtained by this procedure is a yellow, crystalline material possessing moderate solubilities in benzene (50 g L^{-1}), dichloromethane, and chloroform. The compound is less soluble in acetone, tetrahydrofuran and acetonitrile. Saturated hydrocarbon solvents give no evidence of solution. Although the complex may be handled in air, it is best stored under nitrogen to ensure its purity.

In benzene, molecular-weight measurements suggest substantial dissociation:[1,4]

$$Pd\{P(C_6H_5)_3\}_4 \rightleftarrows Pd\{P(C_6H_5)_3\}_{4-n} + nP(C_6H_5)_3$$

Solutions of the complex in benzene rapidly absorb molecular oxygen giving an insoluble, green oxygen complex, $Pd\{P(C_6H_5)_3\}_2O_2$.[7] This oxygen complex has been implicated as an intermediate in a catalytic oxidation of phosphines.[2,8]

Related displacements with acetylenes[9] and electrophilic olefins[6] have been reported to give complexes formulated as $[Pd\{P(C_6H_5)_3\}_2$ (olefin or acetylene)]. Also, oxidative additions of alkyl and aryl halides have been shown to occur giving palladium(II) complexes, $Pd\{P(C_6H_5)_3\}_2(R)Cl$.[10]

As a catalyst, the complex has been shown capable of dimerizing butadiene to give 1,3,7-octatriene.[11]

Notes

1. The checkers worked on one-third of the stated scale, obtaining a yield of 37.4 g (97%).
2. The checkers report that decomposition temperature does not appear to be a good criterion of identity or purity since it is not very reproducible.

References

1. L. Malatesta and M. Angoletta, *J. Chem. Soc.*, **1957**, 1186.
2. S. Takahashi, K. Sonogashira, and N. Hagihara, *Nippon Kagaku Zasshi*, **87**, 610 (1966); *Chem. Abstr.*, **65**, 14485 (1966).
3. T. Kruck and K. Baur, *Angew. Chem.*, **77**, 505 (1965).
4. E. O. Fischer and H. Werner, *Chem. Ber.*, **95**, 703 (1962).
5. J. Chatt, F. A. Hart, and H. R. Watson, *J. Chem. Soc.*, **1962**, 2537.
6. P. Fitton and J. E. McKeon, *Chem. Commun.*, **1968**, 4.
7. C. J. Nyman, C. T. Wymore, and G. Wilkinson, *J. Chem. Soc. A*, **1968**, 561.
8. G. Wilke, H. Schott, and P. Heimbach, *Angew. Chem. Int. Ed., Engl.*, **6**, 92 (1967).
9. S. Takahashi and N. Hagihara, *J. Chem. Soc. Jpn.* (Pure Chem. Sec.), **88**, 1306 (1967).
10. P. Fitton, M. P. Johnson, and J. E. McKeon, *Chem. Commun.*, **1968**, 6.
11. S. Takahashi, T. Shibano, and N. Hagihara, *Bull. Chem. Soc. Jpn.*, **41**, 454 (1968).

29. TETRAKIS(tert-BUTYL ISOCYANIDE)DI-μ-CHLORO-DIPALLADIUM(I)

$$Na_2[Pd_2Cl_6] + 4[CH_3CO_2]Na + 4C_{17}H_{14}O(dba)^* + 2CH_3OH \longrightarrow$$

$$2[Pd(dba)_2] + 6NaCl + 4CH_3COOH + 2CH_2O$$

$$[Pd(dba)_2] + [PdCl_2(C_6H_5CN)_2] + 4t\text{-}C_4H_9NC \xrightarrow{C_6H_5Cl}$$

$$[Pd_2Cl_2(t\text{-}C_4H_9NC)_4] + 2dba + 2C_6H_5CN$$

Submitted by M. F. RETTIG† and P. M. MAITLIS‡
Checked by F. A. COTTON§ and T. R. WEBB§

The unusual complex tetrakis(tert-butyl isocyanide)di-μ-chloro-dipalladium(I)·chlorobenzene has been prepared by Otsuka et al.[1] The preparation utilized a novel coupling reaction between bis(tert-butyl isocyanide)palladium(0) and cis-bis(tert-butyl isocyanide)dichloropalladium(II) in cold chlorobenzene. Although the reported[1] yield for the final step of this synthesis was good (70%), the preparation of the precursor bis(tert-butyl isocyanide)palladium(0) from (η^3-allyl)(η^5-cyclopentadienyl)palladium[2] is time-consuming and is accomplished in only ~50% yield. We have developed a greatly improved preparation of the title complex (chlorobenzene-free), which utilizes readily available Pd(0) and Pd(II) compounds as starting materials, namely bis(dibenzylideneacetone)palladium(0),[3] [Pd(dba)$_2$], and trans-bis(benzonitrile)dichloropalladium(II).[4] In addition, since full details of the preparation of Pd(dba)$_2$ have not been reported,[3] we give a complete account of the preparation here.¶

Tetrakis(tert-butyl isocyanide)dichlorodipalladium(I) is one of a very limited number of examples of the Pd(I) oxidation state. The convenient synthesis of this stable and easily soluble complex reported here should make it a useful starting material for continued study of the Pd(I) oxidation state. The synthesis may be successfully scaled down at least fivefold.

■ **Caution.** *Although tert-butyl isocyanide is apparently only moderately toxic,[7] its odor is unpleasant, and all operations involving isocyanides should be conducted in a fume hood.*

---

*dba = dibenzylideneacetone. (1,5-diphenyl-1,4-pentadien-3-one).
†Department of Chemistry, The University of California, Riverside, CA 92502.
‡Department of Chemistry, The University of Sheffield, Sheffield S3 7HF, United Kingdom.
§Department of Chemistry, Texas A. & M. University, College Station, TX 77843.
¶The exact nature of this complex, particularly in solution, is unclear (see Ref. 5), but a compound of this stoichiometry is obtained reproducibly by this method.

Procedure

Bis(dibenzylideneacetone)palladium(0), Pd(dba)$_2$, is first prepared by stirring 8.87 g of PdCl$_2$ (0.05 mol) and 2.92 g of NaCl (0.05 mol) in 250 mL of methanol at room temperature for 16 h. The resulting solution is filtered through a plug of cotton, the filtrate is diluted to ~1.5 L with methanol, and the solution is heated to 60°. Dibenzylideneacetone[6] (36.5 g, 0.15 mol) is added to the warm, stirred Na$_2$[Pd$_2$Cl$_6$] solution, and stirring is continued for 15 min, followed by addition of 75 g of anhydrous sodium acetate. The reaction commences at once, and the mixture is removed from the heat and stirred for ~1 h until it cools to room temperature. The dark-brown precipitate is filtered and washed successively with methanol (5 × 25 mL), water (5 × 50 mL), and acetone (5 × 15 mL). The product is air-dried. The yield is 23 g (80%).

Anal. Calcd. for C$_{34}$H$_{28}$O$_2$Pd: C, 66.82; H, 4.92. Found: C, 66.75; H. 5.06.

Tetrakis(*tert*-butyl isocyanide)di-μ-chloro-dipalladium(I), Pd$_2$Cl$_2$(*t*-C$_4$H$_9$-NC)$_4$, is then synthesized by placing 90 mL of redistilled chlorobenzene in a sidearm vessel (filter flask or Schlenk tube) and bubbling argon through the chlorobenzene for 15 min. The argon is then introduced through the side arm, and 5.40 g bis(dibenzylideneacetone)palladium(0) (9.4 mmol) is added, followed by 4.85 mL of *tert*-butyl isocyanide* (3.7 g, 44.5 mmol). The mixture is stirred until practically all of the dark solid dissolves to give a straw-colored solution† (~15 min), and 3.60 g of solid *trans*-bis(benzonitrile)dichloropalladium(II)[4] (9.4 mmol) is added in approximately five portions during a few minutes. The yellow-brown solution is stirred under argon for 45 min, during which time crystallization commences and the color of the solution darkens. The reaction mixture is stored under argon overnight in a freezer (~ −35°). The yellow to yellow-green precipitate is filtered and washed with 50-mL portions of anhydrous diethyl ether until the diethyl ether wash is colorless (some dibenzylideneacetone may precipitate; approximately three 50-mL washings with diethyl ether removes this impurity). The yellow to yellow-green‡ solid is air-dried. The crude product weighs 5.65 g (98% yield).

For purification, the crude product is dissolved in dichloromethane (8 mL g^{-1}) and is gravity-filtered through a fine-mesh paper (Whatman no.

---

* *tert*-Butyl isocyanide is readily prepared by the method of J. Casanova, Jr., N. D. Werner, and R. E. Schuster, *J. Org. Chem.*, **31**, 3473 (1966). The checkers report that the *t*-butylamine used in the synthesis of *tert*-butyl isocyanide should be dried, and that neutral alumina is an adequate drying-agent.

† Traces of Pd metal present in the Pd(dba)$_2$ can cause the solution to appear greenish. It is also responsible for the possible presence of a small amount of dark, insoluble deposit.

‡ Traces of palladium metal can cause a precipitate to appear greenish. Palladium is removed on recrystallization.

542 or equivalent).† The product is precipitated by dropwise addition of anhydrous diethyl ether to the dichloromethane solution, until 5 mL of the ether has been added for every milliliter of dichloromethane solution originally present. The mixture is chilled for several hours at $-35°$, filtered, washed with diethyl ether, and air-dried.

Traces of dichloromethane are removed by vacuum drying for 2 h at $56°$ 0.01 torr^{-1}. The yield is 5.03 g [87% based on Pd(dba)$_2$].

Anal. Calcd. for $C_{20}H_{36}Cl_2N_4Pd_2$: Pd, 34.55; Cl, 11.55; C, 38.98; H, 5.89; N, 9.09. Found: Pd, 34.8; Cl, 11.67, 11.50; C, 39.31, 38.77; H, 6.10, 6.13; N, 9.05, 9.33.

Properties

Tetrakis(*tert*-butyl isocyanide)di-μ-chloro-dipalladium(I) is a diamagnetic, bright-yellow powder when pure; it becomes orange above $155°$ and blackens above $210°$. It is stable in air for months at room temperature. Its solutions in common organic solvents are also stable to air if the temperature is not elevated. It is soluble in dichloromethane, chloroform, benzene, toluene, ethyl acetate, etc., and is insoluble in diethyl ether and petroleum ether. The near-infrared spectrum has major bands as follows (KBr disk, cm^{-1}): 2980 (m), 2935 (w), 2170 [s, $v(N\equiv C)$], 2160 [sh, $v(N\equiv C)$], 1475 (m), 1455 (br, m), 1400 (w), 1372 (m), 1235 (m), 1190 (s), 850 (w). The far-infrared spectrum has major absorption bands at (Nujol mull, cm^{-1}): 522 (m), 508 (m), 440 (m), 398 (m), 370 (w), 342 (w), 294 (m), and 258 (vs, possibly Pd–Cl). The PMR spectrum in CDCl$_3$ is a singlet at τ 8.47. The osmometrically determined molecular weight in chloroform (0.02 molal) is 614. (Calcd. 616.) The compound is conveniently converted to the bromine and iodine analogs[1] by treatment of acetone solutions with lithium bromide or lithium iodide, respectively.

References

1. S. Otsuka, Y. Tatsuno, and K. Ataka, *J. Am. Chem. Soc.*, **93**, 6705 (1971).
2. E. O. Fischer and H. Werner, *Chem. Ber.*, **95**, 703 (1962); M. Kh. Minasyants, S. P. Gubin, and Yu. T. Struchkov, *Zh. Strukt. Khim.*, **9**, 481 (1968).
3. Y. Takahashi, T. Ito, S. Sakai, and Y. Ishii, *Chem. Commun.*, **1970**, 1065.
4. M. S. Kharasch, R. C. Seyler, and F. R. Mayo, *J. Am. Chem. Soc.*, **60**, 882 (1938); J. R. Doyle, P. E. Slade, and H. B. Jonassen, *Inorg. Synth.* **6**, 218 (1960).
5. M. C. Mazza and C. G. Pierpont, *Chem. Commun.*, **1973**, 207; T. Ukai, H. Kawazura, Y. Ishii, J. J. Bonnet, and J. A. Ibers, *J. Organometal. Chem.*, **65**, 253 (1974); K. Moseley and P. M. Maitlis, *J. Chem. Soc. Dalton Trans.*, **1974**, 169.
6. C. R. Conard and M. A. Dolliver, in *Organic Syntheses*, A. H. Blatt (ed.), Coll. Vol. 2, p. 167, Wiley, New York, 1943.

7. (a) L. Malatesta and F. Bonati, *Isocyanide Complexes of Metals*, Wiley-Interscience, New York, 1969, pp. 5–6;
   (b) N. Irving Sax, *Dangerous Properties of Industrial Materials*, 6th ed., Van Nostrand Reinhold, New York, 1984, p. 589.

30. TWO-COORDINATE PHOSPHINE COMPLEXES OF PALLADIUM(0) AND PLATINUM(0)

Submitted by T. YOSHIDA* and S. OTSUKA*
Checked by D. G. JONES,† J. L. SPENCER,† P. BINGER,‡ A. BRINKMANN,‡ and P. WEDEMANN‡

Two-coordinate complexes still remain a rarity in transition metal chemistry. Recently a few bicoordinate complexes of palladium(0) and platinum(0), ML_2^{1-7} [M = Pd, Pt; L = PPh(t-Bu)$_2$ or P($cyclo$-C$_6$H$_{11}$)$_3$] have been reported, and X-ray studies reveal that they have almost linear structures.[1,2,6,8] The PdL$_2$ complexes were prepared by a reaction of the phosphine with (η^5-C$_5$H$_5$)(η^3-C$_3$H$_5$)Pd[1,2] or [η^3-(2-methylallyl)PdCl]$_2$.[3] The former reaction may involve an incipient formation of Pd(η^1-C$_3$H$_5$)(η^1-C$_5$H$_5$)L$_2$ (L = phosphines) followed by reductive elimination of organic moieties as C_8H_{10}.[9] The compound Pd[P($cyclo$-C$_6$H$_{11}$)$_3$]$_2$ was also prepared by removing the ethylene molecule from Pd(C$_2$H$_4$)[P($cyclo$-C$_6$H$_{11}$)$_3$]$_2$[7] or by reduction of Pd(acac)$_2$ (acac = acetylacetonato, 2,4-pentanedionato) with AlEt$_3$ in the presence of the phosphine.[4] The preparative procedure described here is a slight modification of the first method.[1,2]

The platinum analogues PtL$_2$ [L = PPh(t-Bu)$_2$ or P($cyclo$-C$_6$H$_{11}$)$_3$] may be prepared by reduction of the corresponding dichloro compounds with sodium amalgam or sodium naphthalene.[2] The compound Pt[P($cyclo$-C$_6$H$_{11}$)$_3$]$_2$ is also accessible from Pt(cod)$_2$ (cod = 1,5-cyclooctadiene) and P($cyclo$-C$_6$H$_{11}$)$_3$[5] or from Pt(n^3-allyl)[P($cyclo$-C$_6$H$_{11}$)$_3$]$_2^+$ and t-BuONa.[6]

■ **Caution.** *All the phosphines, (η^5-C$_5$H$_5$)(η^3-C$_3$H$_5$)Pd, and bicoordinate complexes are air sensitive. The phosphines and (η^5-C$_5$H$_5$)(η^3-C$_3$H$_5$)Pd are malodorous materials with unknown physiological effects. Therefore all the manipulations described here should be carried out under a dry nitrogen atmosphere using Schlenk-tube techniques*[10,11] *and in a well ventilated hood. A*

---

*Department of Chemistry, Faculty of Engineering Science, Osaka University, Toyonaka, Osaka, Japan 560.

†Department of Inorganic Chemistry, University of Bristol, Bristol BS8 1TS, United Kingdom; checked Sections A–C.

‡Max-Planck-Institute für Kohlenforschung, D-4330 Mülheim/Ruhr, Germany; checked Sections D and E.

hypodermic syringe is employed for weighing and transferring $PPh(t-Bu)_2$ and $P(t-Bu)_3$. All the solvents are dried with sodium metal (except methanol) and distilled under nitrogen.

A. BIS(DI-*tert*-BUTYLPHENYLPHOSPHINE)PALLADIUM(0)

$$(\eta^5\text{-}C_5H_5)(\eta^3\text{-}C_3H_5)Pd + 2PPh(t\text{-}Bu)_2 \rightarrow Pd[PPh(t\text{-}Bu)_2]_2 + C_8H_{10}$$

Procedure

A 50-mL Schlenk flask containing a magnetic stirring bar is charged with (η^3-allyl)(η^5-cyclopentadienyl)palladium[12] (0.40 g, 1.89 mmol), toluene (10 mL), and di-*tert*-butylphenylphosphine[13] (0.89 g, 4.00 mmol). The deep-red solution is heated at 70–75° for 1 h. The resulting pale-brown solution is concentrated *in vacuo* to dryness and the pale-brown crystals that separate are washed with methanol (five 3-mL portions). The crude crystals are dissolved in hot hexane (25 mL). The solution is filtered and the filtrate is concentrated *in vacuo* to a quarter of the original volume to give pale-yellow crystals. For complete crystallization the solution is cooled overnight at $-35°$. The mother liquor is removed with a syringe, and the crystals are washed at $-35°$ with cold hexane (two 5-mL portions) and dried *in vacuo*. Yield 0.91 g (86%), mp 160–162° (under N_2 in a sealed capillary tube).

Anal. Calcd. for $C_{28}H_{46}P_2Pd$: C, 61.06; H, 8.41. Found: C, 60.81; H, 8.46.

Properties

The properties of bis(di-*tert*-butylphenylphosphine)palladium(0) are described with those of the other bicoordinates complexes (see below).

B. BIS(TRICYCLOHEXYLPHOSPHINE)PALLADIUM(0)

$$(\eta^5\text{-}C_5H_5)(\eta^3\text{-}C_3H_5)Pd + 2P(cyclo\text{-}C_6H_{11})_3 \rightarrow Pd[P(cyclo\text{-}C_6H_{11})_3]_2 + C_8H_{10}$$

Procedure

In a 50-mL Schlenk flask containing a magnetic stirring bar are placed (η^3-allyl)(η^5-cyclopentadienyl)palladium[12] (0.34 g, 1.60 mmol) and a toluene solution (15 mL) of tricyclohexylphosphine*[14] (0.99 g, 3.54 mmol). The dark-red mixture is stirred with heating at 75–80° for 3 h. The brown solution is concentrated *in vacuo* to dryness. The brown crystalline solid is washed with MeOH (two 10-mL portions) to remove a slight excess of the phosphine. The solid is dissolved in hot toluene (5 mL), and methanol (5 mL) is added to give crystals. After standing in a freezer ($-35°$) overnight, the crystals are isolated by removing the mother liquor with a syringe, washed with MeOH (two 5-mL portions), and dried *in vacuo*. The off-white crystals thus obtained are pure enough to prepare organopalladium complexes. Yield: 0.84 g (79%). Analytically pure, colorless crystals can be obtained by recrystallization from a toluene (5 mL)–methanol (5 mL) mixture, mp 185–189° (under N_2 in a sealed capillary tube).

Anal. Calcd. for $C_{36}H_{66}P_2Pd$: C, 64.79; H, 9.99. Found: C, 64.76; H, 9.97.

Properties

The properties of bis(tricyclohexylphosphine)palladium(0) are described with those of the other bicoordinate complexes (see below).

C. BIS(TRI-*tert*-BUTYLPHOSPHINE)PALLADIUM(0)

$$(\eta^5\text{-}C_5H_5)(\eta^3\text{-}C_3H_5)Pd + 2P(t\text{-}Bu)_3 \rightarrow Pd[P(t\text{-}Bu)_3]_2 + C_8H_{10}$$

Procedure

This compound is prepared by a procedure similar to the one described above for the di-*tert*-butylphenylphosphine compound, employing (η^3-allyl)(η^5-cyclopentadienyl)palladium[12] (0.21 g, 1.0 mmol) and tri-*tert*-butylphosphine[15] (0.46 g, 2.3 mmol).† The product is obtained in 60% yields as colorless crystals, mp 150–153° (dec. in air).

Anal. Calcd. for $C_{24}H_{54}P_2Pd$: C, 56.40; H, 10.67. Found: C, 56.62; H, 10.73.

Properties

The properties of bis(tri-*tert*-butylphosphine)palladium(0) are described with those of the other bicoordinate complexes (see below).

---

*Tricyclohexylphosphine is available from Strem Chemicals Inc., Box 212, Danvers, MA 01923.

†Since tri-*tert*-butylphosphine is low melting (mp 30°), it is recommended to weigh the phosphine liquidfied at 50° employing a syringe preheated in a oven (60°). If the phosphine solidifies in the syringe, it can be melted by heating with a heat gun.

D. BIS(DI-*tert*-BUTYLPHENYLPHOSPHINE)PLATINUM(0)

$K_2[PtCl_4] + 2PPh(t\text{-}Bu)_2 \rightarrow trans\text{-}PtCl_2[PPh(t\text{-}Bu)_2]_2 + 2KCl$

$trans\text{-}PtCl_2[PPh(t\text{-}Bu)_2]_2 + 2Na(Hg) \rightarrow Pt[PPh(t\text{-}Bu)_2]_2 + 2NaCl$

Procedure

A 100-mL Schlenk flask containing a magnetic stirring bar is charged successively with $K_2[PtCl_4]$ (1.0 g, 2.3 mmol), deoxygenated water (5 mL), EtOH (10 mL), and di-*tert*-butylphenylphosphine (1.06 g, 4.8 mmol). The mixture is stirred at room temperature for 40 h, and the colorless solid is filtered in air, washed successively with H_2O (10 mL) and EtOH (20 mL), and dried *in vacuo*. A yield of 1.6 g (98%) is obtained. The crude *trans*-[dichlorobis(di-*tert*-butylphenylphosphine)platinum(II)][16] (0.95 g, 1.34 mmol) thus obtained, 1% sodium amalgam (50 g), and tetrahydrofuran (15 mL) are placed successively in a 100-mL Schlenk flask containing a magnetic stirring bar. The mixture is stirred vigorously at room temperature for 22 h. The gray suspension is transferred with a syringe to a filtration funnel[17] fitted to a 100-mL Schlenk flask and filtered through a filter paper. The sodium amalgam is washed with hexane (15 mL). The combined filtrate and washings are concentrated *in vacuo* to dryness. The solid residue is dissolved in hexane (30 mL) and the solution is filtered. The pale-yellow filtrate is concentrated to one-third of the original volume. After standing in a freezer ($-35°$), the colorless crystals are isolated by removing the solution with a syringe, washed at $-35°$ with cold hexane (5 mL), and dried *in vacuo*. A yield of 0.79 g (92%) is obtained, mp 171–174° (under N_2 in a sealed capillary tube).

Anal. Calcd. for $C_{28}H_{46}P_2Pt$: C, 52.57; H, 7.25. Found C, 52.68; H, 7.05.

Properties

The properties of bis(di-*tert*-butylphenylphosphine)platinum(0) are described with those of the other bicoordinate complexes (see below).

E. BIS(TRICYCLOHEXYLPHOSPHINE)PLATINUM(0)

$K_2[PtCl_4] + 2P(cyclo\text{-}C_6H_{11})_3 \rightarrow trans\text{-}PtCl_2[P(cyclo\text{-}C_6H_{11})_3]_2 + 2KCl$

$trans\text{-}PtCl_2[P(cyclo\text{-}C_6H_{11})_3]_2 + 2[C_{10}H_8^-]Na^+ \rightarrow$
$\qquad Pt[P(cyclo\text{-}C_6H_{11})_3]_2 + 2NaCl + 2C_{10}H_8$

Procedure*

A 100-mL Schlenk flask containing a magnetic stirring bar is charged with $K_2[PtCl_4]$ (1.0 g, 2.3 mmol), deoxygenated water (5 mL), and an ethanol solution (40 mL) of tricyclohexylphosphine[14] (1.5 g, 5.4 mmol). The mixture is stirred at room temperature for 15 h, and the colorless solid is filtered, washed successively with H_2O (10 mL) and EtOH (20 mL) in air, and dried *in vacuo*. A yield of 1.8 g (95%) is obtained. The crude *trans*-$PtCl_2[P(cyclo$-$C_6H_{11})_3]_2$[2] (1.0 g, 1.2 mmol) thus obtained is placed in a 50-mL Schlenk flask containing a magnetic stirring bar, and a 0.33 M tetrahydrofuran solution (15 mL) of sodium naphthalene prepared from sodium (0.5 g) and naphthalene (2.2 g) in THF (50 mL) is added. The mixture is stirred at room temperature for 5 h and the resulting brownish-green solution is concentrated *in vacuo*. The dark-brown solid residue is extracted with hot hexane (two 20-mL portions) (50–55°) and the extract is transferred into a filtration funnel[17] fitted with a sublimation apparatus and filtered through a filter paper.

■ **Caution.** *The residue, which is insoluble in hexane, is pyrophoric and should be treated with EtOH under a nitrogen atmosphere before it is discarded.*

The filtrate is concentrated *in vacuo* and the resulting solid is heated at 50–70° *in vacuo* (10^{-3} torr) for 10 hours to remove the naphthalene by sublimation. It is then dissolved in hot hexane (30 mL) and the solution is filtered as above. Concentration of the filtrate *in vacuo* to 5 mL gives pale-yellow crystals. After standing in a freezer ($-35°$) overnight, the crystals (0.40 g) are isolated by removing the mother liquor with a syringe, washed with hexane (three 2-mL portions), and dried *in vacuo*. Additional crystals (0.10 g) are obtained on concentration of the combined mother liquor and washings to 1 mL. Total yield 55%.† Colorless crystals are obtained by recrystallization from hexane; mp 204–208° (under N_2 in a sealed capillary tube).

Anal. Calcd. for $C_{36}H_{66}P_2Pt$: C, 57.19; H, 8.80. Found: C, 57.11; H, 8.98.

Properties

The bicoordinate complexes described here are soluble in benzene and hexane. The $Pd[P(t$-$Bu)_3]_2$ complex is stable in air in the solid state, whereas

---

*Sodium amalgam can also be employed as a reducing agent. However, in this case a prolonged heating (50 h at 55–60°) is required because of insolubility of *trans*-$PtCl_2[P(cyclo$-$C_6H_{11})_3]_2$ in tetrahydrofuran.

†The checkers found that for unknown reasons only two of five experiments gave this compound. In the other experiments they obtained $Pt[P(cyclo$-$C_6H_{11})_3]_3$ with metallic platinum.

Table I. ^{1}H NMR Spectra of Two-Coordinate Complexes

	Chemical Shift (ppm, Me$_4$Si)^a	$^3J_{HP}$ $+^5J_{HP}$	Area	Assignment
Pd[P(t-Bu)$_3$]$_2$	1.51 (t)	12.0		t-Bu
Pd[PPh(t-Bu)$_2$]$_2$^{b}	1.48 (t)	12.7	9	t-Bu
	8.40 (m)^c		1	o-H
	6.94–7.30 (m)		d	m- and p-H
Pd[P(cyclo-C$_6$H$_{11}$)$_3$]$_2$	0.70–2.60 (m)			cyclo-C$_6$H$_{11}$
Pt[PPh(t-Bu)$_2$]$_2$^{b}	1.56 (t)	13.5	9	t-Bu
	8.46 (m)^c		1	o-H
	6.90–7.30 (m)		d	m- and p-H
Pt[P(cyclo-C$_6$H$_{11}$)$_3$]$_2$	0.70–2.60 (m)			cyclo-C$_6$H$_{11}$

^{a}Measured in benzene-d_6 at 22.5°.
^{b}Measured in toluene-d_7 at 22.5°.
^{c}At $-71°$ the ortho proton signals of Pd[PPh(t-Bu)$_2$]$_2$ and Pt[PPh(t-Bu)$_2$]$_2$ are observed at δ 7.55 (m), 9.33 (m) and 7.40 (m), 9.38 ppm (m), respectively.
^{d}Because of overlap with the solvent signal, the area cannot be evaluated exactly.

the other complexes are unstable and give the dioxygen complexes MO$_2$L$_2$[18] [M = Pd, Pt; L = PPh(t-Bu)$_2$, P(cyclo-C$_6$H$_{11}$)$_3$]. In the case of palladium complexes, the formation of dioxygen complexes is readily detectable by the development of a pale-green color. All of the two-coordinate complexes can be stored under dry nitrogen for more than a year. The linear structure of the P(t-Bu)$_3$ and PPh(t-Bu)$_2$ complexes is readily deducible from a 1 : 2 : 1 triplet of the *tert*-butyl proton signal (Table I). Mass spectra of PtL$_2$ [L-PPh(t-Bu)$_2$ or P(cyclo-C$_6$H$_{11}$)$_3$] show the corresponding parent and fragment ions. [M-(R-1)]$^+$, [M-2(R-1)]$^+$, [M-3(R-1)]$^+$, and MP$_2^+$, where R is the alkyl substituent of the phosphine.[2] As expected from the high degree of coordinative unsaturation, the bicoordinate complexes, particularly the platinum complexes, show an enhanced reactivity toward small molecules and weak protonic acids, for example, alcohol, and π-acids like maleic anhydride.[18]

References

1. M. Matsumoto, H. Yoshioka, K. Nakatsu, T. Yoshida, and S. Otsuka, *J. Am. Chem. Soc.*, **96**, 3322 (1974).
2. S. Otsuka, T. Yoshida, M. Matsumoto, and K. Nakatsu, *J. Chem. Soc.*, **98**, 5850 (1976).
3. A. Musco, W. Kuran, A. Silvani, and M. W. Anker, *Chem. Commun.*, **1973**, 938.
4. K. Kudo, M. Hidai, and Y. Uchida, *J. Organomet. Chem.*, **56**, 413 (1973).
5. M. Green, J. A. Howard, J. L. Spencer, and F. G. A. Stone, *Chem. Commun.*, **1975**, 3.
6. A. Immirzi, A. Musco, and P. Zambelli, *Inorg. Chim. Acta*, **13**, L13 (1975).
7. R. van der Linde and R. O. der Jongh, *J. Chem. Soc. D*, **1971**, 563.
8. A. Immirzi and A. Musco, *Chem. Commun.*, **1974**, 400.

9. H. Werner and A. Kühn, *Angew. Chem. Int. Ed.*, **16**, 412 (1977).
10. D. F. Shriver, *The Manipulation of Air-Sensitive Compounds*, McGraw-Hill, New York, 1969.
11. R. B. King, in *Organometallic Syntheses*, Vol. 1, J. J. Eisch and R. B. King (eds.), Academic Press, New York, 1965.
12. Y. Tatsuno, T. Yoshida, and S. Otsuka, *Inorg. Synth.*, **19**, 220 (1979); also see section 88 in this volume.
13. B. E. Mann, B. L. Shaw, and R. M. Slade, *J. Chem. Soc. A*, **1971**, 2976.
14. K. Issleib and A. Brack, *Z. Anorg. Allg. Chem.*, **277**, 258 (1954).
15. H. Hofmann and P. Schellenbeck, *Chem. Ber.*, **100**, 692 (1967).
16. A. J. Cheney, B. E. Mann, B. L. Shaw, and R. M. Slade, *J. Chem. Soc. A*, **1971**, 3833.
17. A filtration funnel employed for preparation of $(\eta^5-C_5H_5)(\eta^3C_3H_5)Pd^{12}$ is satisfactory, see Fig. 1 in Ref. 12.
18. S. Otsuka and T. Yoshida, *J. Am. Chem. Soc.*, **99**, 2134 (1977).

31. THREE-COORDINATE PHOSPHINE COMPLEXES OF PLATINUM(0)

Submitted by T. YOSHIDA,* T. MATSUDA,* and S. OTSUKA*
Checked by G. W. PARSHALL† and W. G. PEET†,
P. BINGER,‡ A. BRINKMANN,‡ and P. WEDEMANN‡

Tris(triethylphosphine)platinum was originally prepared by vacuum thermolysis of $Pt(PEt_3)_4$, but the latter was synthesized from $Pt(B_3H_7)(PEt_3)_2$, which is an inconvenient starting material.[1,2] A direct synthesis from readily available starting materials is the reaction of $K_2[PtCl_4]$ with KOH and PEt_3 in alcohol,[3] but the product is difficult to isolate from the reaction mixture in pure form. The procedure described here is based on the vacuum thermolysis of $Pt(PEt_3)_4$, as obtained by a procedure outlined in this volume.[4]

The triisopropylphosphine analog, $Pt[P(i-Pr)_3]_3$, is obtained by reducing *trans*-$PtCl_2[P(i-Pr)_3]_2$ with sodium amalgam in the presence of $P(i-Pr)_3$[5] as described here.

■ **Caution.** *All zero-valent platinum compounds and trialkylphosphines employed here are extremely air sensitive and should be handled in a dry nitrogen or argon atmosphere. The trialkylphosphines are malodorous and toxic, and should be handled with care, in a well-ventilated hood. All solvents should be dried (except ethanol) and distilled under nitrogen.*

---

*Department of Chemistry, Faculty of Engineering Science, Osaka University, Toyonaka, Osaka, Japan 560.
†Central Research Department, E. I. du Pont de Nemours & Co., Wilmington, DE 19898; checked Section A.
‡Max-Planck-Institute für Kohlenforschung, D-4330 Mülheim/Ruhr, Germany; checked Section B.

A. TRIS(TRIETHYLPHOSPHINE)PLATINUM(0)

$$Pt(PEt_3)_4 \xrightarrow{in\ vacuo} Pt(PEt_3)_3 + PEt_3$$

Procedure

A 15-mL Schlenk flask is evacuated and refilled with nitrogen three times. Tetrakis(triethylphosphine)platinum[4] (0.66 g, 1 mmol) is charged by the Schlenk-tube techniques.[6] Under a nitrogen flow the flask is connected to a vacuum line through a liquid nitrogen U trap. The flask is heated at 50–60° at reduced pressure (5 torr) for 6 h to give an orange-red viscous oil. Yield 0.49 g (90%).

Anal. Calcd. for $C_{18}H_{45}P_3Pt$: C, 39.3; H, 8.3. Found: C, 38.9; H, 8.2.

Properties

The properties of tris(triethylphosphine)platinum(0) are described with those of tris(triisopropylphosphine)platinum(0).

B. TRIS(TRIISOPROPYLPHOSPHINE)PLATINUM(0)

$$K_2[PtCl_4] + 2P(i\text{-}Pr)_3 \rightarrow trans\text{-}PtCl_2[P(i\text{-}Pr)_3]_2 + 2KCl$$

$$trans\text{-}PtCl_2[P(i\text{-}Pr)_3]_2 + P(i\text{-}Pr)_3 + 2Na(Hg) \rightarrow Pt[P(i\text{-}Pr)_3]_3 + 2NaCl$$

Procedure

To a 50-mL nitrogen-flushed Schlenk flask containing a magnetic stirring bar is added K_2PtCl_4 (0.50 g, 1.2 mmol), deoxygenated water (3 mL), $P(i\text{-}Pr)_3$[7] (0.42 g, 2.6 mmol), and ethanol (3 mL). The mixture is stirred at room temperature for 2 h and the resulting colorless solid of $trans\text{-}PtCl_2[P(i\text{-}Pr)_3]_2$ is filtered, washed with ethanol, and dried *in vacuo*.

■ **Caution.** *The amalgamation of sodium is highly exothermic. Small pieces of sodium must be added to mercury behind a shield.*[8]

The crude $trans\text{-}PtCl_2[P(i\text{-}Pr)_3]_2$[4] (0.6 g, 1.0 mmol) is placed in a 50-mL Schlenk flask containing a stirring bar. A 20-g sample of 1% sodium amalgam and 10 mL of dried tetrahydrofuran[9] containing 0.24 g (1.5 mmol) of $P(i\text{-}Pr)_3$ are added successively. The mixture is stirred at room temperature for 10 h. The red solution is transferred with a syringe into a filtration funnel (see Fig. 1 in Ref. 10) and filtered through a filter paper. The sodium amalgam is washed with dried and degassed pentane (two 10-mL portions). The combined filtrate and washings are concentrated under reduced pressure (7 torr) to dryness. The solid residue is extracted with pentane (two 10-mL portions) and the extract is filtered as above. The filtrate is concentrated under reduced pressure to a quarter of the original volume. After standing at $-35°$ overnight, the pale-yellow crystals are isolated by removing the solution with a syringe, washing with pentane (three 2-mL portions) at $-78°$, and drying at $-35°$ under reduced pressure (7 torr). Yield 0.33–0.41 g (48–60%), mp 60–62° (under nitrogen in a sealed tube).

Anal. Calcd. for $C_{27}H_{63}P_3Pt$: C, 47.97; H, 9.39. Found: C, 48.09; H, 9.51.

Properties

The three-coordinate complexes PtL_3 [L = PEt_3, $P(i\text{-Pr})_3$] are extremely unstable toward air and should be kept under dry nitrogen in a freezer. They are readily soluble even in saturated aliphatic hydrocarbons. The ^1H NMR spectrum of $Pt(PEt_3)_3$ measured in benzene-d_6 shows two broad signals at $\delta 1.76$ (CH_2) and 1.16 ppm (CH_3), while that of $Pt[P(i\text{-Pr})_3]_3$ shows signals at $\delta 1.86$ (CH) and 1.24 ppm (CH_3). In contrast to $Pt(PEt_3)_3$, which does not dissociate the coordinate phosphine, $Pt[P(i\text{-Pr})_3]_3$ readily liberates $P(i\text{-Pr})_3$ even in the solid state ($K_d = 4.0 \times 10^{-2}$ M in heptane). They are strong nucleophiles and readily react with hydrogen and weak protonic acids, for example, C_2H_5OH and H_2O. With hydrogen $Pt(PEt_3)_3$ gives $PtH_2(PEt_3)_3$,[1] while $Pt[P(i\text{-Pr})_3]_3$ affords *trans*-$PtH_2[P(i\text{-Pr})_3]_2$.[11] Oxidative addition of alcohol to $Pt(PEt_3)_3$ is reversible to give $[PtH(PEt_3)_3]OC_2H_5$,[1] but with $Pt[P(i\text{-Pr})_3]_3$ it is irreversible and gives *trans*-$PtH_2[P(i\text{-Pr})_3]_2$.[11] They add H_2O reversibly to give the strong hydroxo bases $[PtH(PEt_3)_3]OH$[1] and *trans*-$PtH(OH)[P(i\text{-Pr})_3]_2$.[12]

References

1. D. H. Gerlach, A. R. Kane, G. W. Parshall, J. P. Jesson, and E. L. Muetterties, *J. Am. Chem. Soc.*, **93**, 3543 (1971).
2. L. J. Guggenberger, A. R. Kane, and E. L. Muetterties, *J. Am. Chem. Soc.*, **94**, 5665 (1972).
3. R. G. Pearson, W. Louw, and J. Rajaram. *Inorg. Chim. Acta.*, **9**, 251 (1974).
4. T. Yoshida, T. Matsuda, and S. Otsuka, *Inorg. Synth.*, **19**, 110 (1979).
5. S. Otsuka, T. Yoshida, M. Matsumoto, and K. Nakatsu, *J. Am. Chem. Soc.*, **98**, 5850 (1976).
6. D. F. Shriver, *Manipulation of Air-Sensitive Compounds*, McGraw-Hill, New York, 1969.
7. A. H. Cowley and M. W. Taylor, *J. Am. Chem. Soc.*, **91**, 2915 (1969).
8. W. B. Renfrow, Jr., and C. R. Hauser, *Org. Synth.*, Coll. Vol. 2, 609 (1943).
9. *Inorg. Synth.*, **12**, 317 (1970).
10. S. Otsuka, T. Yoshida, and Y. Tatsumo, *Inorg. Synth.*, **19**, 222 (1979).
11. S. Otsuka and T. Yoshida, *J. Am. Chem. Soc.*, **99**, 2134 (1977).
12. T. Yoshida, T. Matsuda, T. Okano, T. Kitani, and S. Otsuka, *J. Am. Chem. Soc.*, **101**, 2027 (1979).

32. TETRAKIS(TRIETHYLPHOSPHINE)PLATINUM(0)

$$K_2[PtCl_4] + 4PEt_3 + 2KOH + C_2H_5OH \rightarrow$$
$$Pt(PEt_3)_4 + 4KCl + CH_3CHO + 2H_2O$$

Submitted by T. YOSHIDA,* T. MATSUDA,* and S. OTSUKA*
Checked by G. W. PARSHALL† and W. G. PEET†

Tetrakis(triethylphosphine)platinum(0) has been prepared by two routes: (1) treatment of $Pt(B_3H_7)(PEt_3)_2$ with PEt_3[1,2] and (2) reduction of cis-$PtCl_2(PEt_3)_2$ with potassium[3] or sodium amalgam in the presence of PEt_3. The procedure described here is a direct synthesis from $K_2[PtCl_4]$, PEt_3, and potassium hydroxide in alcohol that was originally developed by Pearson et al.[4] for the preparation of $Pt(PEt_3)_3$.

Procedure

■ **Caution.** *Triethylphosphine and tetrakis(triethylphosphine)platinum(0) are extremely air-sensitive. The phosphine is malodorous and toxic. Therefore all manipulations should be carried out in a nitrogen atmosphere and in a well-ventilated hood. All solvents should be degassed with an inert gas.*

A 50-mL Schlenk flask containing a magnetic stirring bar is charged with a solution of KOH (0.7 g, 12.5 mmol) dissolved in a 30 : 1 EtOH–H_2O mixture (31 mL) and PEt_3 (3.0 mL, 20 mmol). To the mixture a solution of $K_2[PtCl_4]$ (1.5 g, 3.6 mmol) in H_2O (10 mL) is added dropwise by syringe over a period of 5 min. The mixture is stirred at room temperature for 1 h and then at 60° for 3 h. The colorless solution is concentrated to dryness *in vacuo* (5 torr) at room temperature under stirring. The reddish, oily–solid residue is extracted with hexane (two 15-mL portions) and the extract is filtered by the Schlenk-flask filtration method.[5] The orange filtrate is concentrated to a quarter of the original volume under reduced pressure (5 torr). The concentrate is treated with PEt_3 (0.5 mL, 3.4 mmol) and cooled at $-78°$ (Dry Ice–acetone) for 4 h. The colorless crystals that separate are isolated by removing the solution with a syringe, washing with hexane (two 3-mL portions) at $-78°$, and drying *in vacuo* (5 torr) at $-40°$. Yield: 2.0 g (85%); mp 47–48° (under nitrogen in a sealed capillary tube).

Anal. Calcd. for $C_{24}H_{60}P_4Pt$: C, 43.2; H, 9.1. Found: C, 42.6; H, 8.9.

---

*Department of Chemistry, Faculty of Engineering Science, Osaka University, Toyonaka, Osaka, Japan 560.
†Central Research Department, E. I. du Pont de Nemours & Co., Wilmington, DE 19898.

Properties

Tetrakis(triethylphosphine)platinum(0) is extremely air sensitive and readily soluble in saturated aliphatic hydrocarbons. The complex can be stored under dry nitrogen in a freezer ($-35°$) for several months. The complex readily loses one of the coordinated phosphine molecules to give $Pt(PEt_3)_3$[6] (dissociation constant (K_d) in heptane is 3.0×10^{-1}). The ^1H NMR spectrum measured in benzene-d_6 shows two multiplets at δ 1.56 (CH_2) and 1.07 ppm (CH_3). Tetrakis(triethylphosphine)platinum(0) is a strong nucleophile and reacts readily with chlorobenzene and benzonitrile to give σ-phenyl complexes $PtX(Ph)(PEt_3)_2$ (X = Cl, CN).[7] Oxidative addition of EtOH affords $[PtH(PEt_3)_3]^+$.

References

1. D. H. Gerlach, A. R. Kane, G. W. Parshall, J. P. Jesson, and E. L. Muetterties, *J. Am. Chem. Soc.*, **93**, 3543 (1971).
2. L. J. Guggenberger, A. R. Kane, and E. L. Muetterties, *J. Am. Chem. Soc.*, **94**, 5665 (1972).
3. R. A. Schunn, *Inorg. Chem.*, **15**, 208 (1976).
4. R. G. Pearson, W. Louw, and J. Rajaram, *Inorg. Chim. Acta.*, **9**, 251, (1974).
5. D. F. Shriver, *Manipulation of Air-Sensitive Compounds*, McGraw-Hill, New York, 1969.
6. T. Yoshida, T. Matsuda, and S. Otsuka, *Inorg. Synth.*, **19**, 108 (1979).
7. G. W. Parshall, *J. Am. Chem. Soc.*, **96**, 2360 (1974).

33. TRIS- AND TETRAKIS(TRIPHENYLPHOSPHINE)-PLATINUM(0)

Submitted by R. UGO,* F. CARIATI,* and G. LA MONICA*
Checked by JOSEPH J. MROWCA†

Tris- and tetrakis(triphenylphosphine) derivatives of zero-valent platinum can be obtained by reduction of platinum(II) triphenylphosphine compounds by alcoholic potassium hydroxide or hydrazine.[1] The tris(triphenylphosphine) derivative is easily obtained from the tetrakis derivative by treatment with hot ethanol.‡

---

*Dipartimento di Chimica Inorganica e Metallorganica CNR Center, Milan University, Italy.
†Central Research Department, E. I. du Pont de Nemours & Company, Wilmington, DE 19898.
‡The contributors are indebted to Professor G. Wilkinson (Imperial College, London SW7) for suggesting this simple method.

Tris- and tetrakis(triphenylphosphine)platinum(0) are the source of many platinum compounds. They react with carbon monoxide,[1,2] acids,[3] methyl iodide,[4] fluoroalkyl derivatives,[5] carbon disulfide,[6] oxygen,[7] chloro,[8] fluoro,[5] and activated olefins,[9,10] hydrogen sulfide and selenide,[11] and sulfur dioxide,[12] yielding many platinum(II) or platinum(0) compounds which cannot be obtained easily by other routes.

A. TETRAKIS(TRIPHENYLPHOSPHINE)PLATINUM(0)

$$K_2[PtCl_4] + 2KOH + 4PPh_3 + C_2H_5OH \rightarrow Pt(PPh_3)_4$$
$$+ 4KCl + CH_3CHO + 2H_2O$$

Procedure

Triphenylphosphine (15.4 g, 0.0588 mol) is dissolved in 200 mL of absolute ethanol at 65°. When the solution is clear, a solution of 1.4 g of potassium hydroxide in a mixture of 32 mL of ethanol and 8 mL of water is added. Then 5.24 g (0.0126 mol) of potassium tetrachloroplatinate(II) dissolved in 50 mL of water (0.0126 mol) is slowly added to the alkaline triphenylphosphine solution while stirring at 65°. The addition should be completed in about 20 min. A pale yellow compound begins to separate within a few minutes of the first addition. After cooling, the compound is recovered by filtration, washed with 150 mL of warm ethanol, then with 60 mL of cold water, and again with 50 mL of cold ethanol. The resulting pale yellow powder is dried *in vacuo* for 2 h. Yield 12.4 g, 79%. The compound must be stored under pure nitrogen.

Anal. Calcd. for $C_{72}H_{60}P_4Pt$: C, 69.50; H, 4.86; P, 9.96. Found: C, 69.32; H, 4.74; P, 10.0. Found by checker: C, 69.33; H, 4.99; P, 9.69.

Properties

Tetrakis(triphenylphosphine)platinum(0) is a pale yellow powder that decomposes in the air to a red liquid at 118–120° and melts *in vacuo* (1 mm) to a yellow liquid at 159–160°. The infrared spectrum in Nujol shows absorption maxima at 700 (vs), 737 (vs), 837 (w), 992 (s), 1022 (s), 1077 (vs), 1147 (m), 1162 (m), 1302 (w), and 1432 (vs) cm^{-1}. The compound is soluble in benzene with dissociation;[1] by leaving the benzene solution in the air, a white powder [the carbonatobis(triphenylphosphine)platinum(II) formed by action of oxygen and carbon dioxide] separates slowly.[13] The compound reacts with carbon tetrachloride, giving *cis*-dichlorobis(triphenylphosphine)platinum(II).[1]

B. TRIS(TRIPHENYLPHOSPHINE)PLATINUM(0)

$$Pt(PPh_3)_4 \rightarrow Pt(PPh_3)_3 + PPh_3$$

Procedure

Tetrakis(triphenylphosphine)platinum(0) (5.8 g, 0.00467 mol) is suspended in 250 mL of absolute ethanol under a nitrogen atmosphere and boiled, with stirring, for 2 h. The hot suspension is filtered, and the precipitate is washed with 30 mL of cold ethanol. The resulting yellow crystals are dried *in vacuo* for 2 h. Yield: 3.0 g, 66%. The compound must be stored under pure nitrogen.

Anal. Calcd. for $C_{54}H_{45}P_3Pt$: C, 66.0; H, 4.59; P, 9.48. Found: C, 65.90; H, 4.69; P, 9.55.

Properties

Tris(triphenylphosphine)platinum(0) is a yellow compound that can be crystallized from acetone in a nitrogen atmosphere. It decomposes in the air to a red liquid at 125–135° and melts *in vacuo* (1 mm) to a red liquid at 205–206°. The infrared spectrum in Nujol shows absorption maxima at 700 (vs), 742 (vs), 840 (w), 997 (s), 1023 (s), 1075 (vs), 1150 (m), 1177 (m), 1300 (w), and 1430 (vs) cm.$^{-1}$.

The compound dissolves in benzene, with dissociation. By leaving the benzene solution in the air, (carbonato)bis(triphenylphosphine)platinum(II) separates slowly.[13] It reacts easily with carbon tetrachloride giving *cis*-dichlorobis(triphenylphosphine)platinum(II).[1] The complex is tricoordinated in the solid state, and shows a nearly pure sp^2 hybridization of the platinum atom.[14]

References

1. L. Malatesta and C. Cariello, *J. Chem. Soc.*, 2323 (1958).
2. F. Cariati and R. Ugo, *Chim. Ind. (Milan)*, **48**, 1288 (1966).
3. F. Cariati, R. Ugo, and F. Bonati, *Inorg. Chem.*, **5**, 1128 (1966).
4. J. Chatt and B. L. Shaw, *J. Chem. Soc.*, 4020 (1959).
5. M. Green, R. B. L. Osborn, A. J. Rest, and F. G. A. Stone, *Chem. Commun.*, 502 (1966).
6. M. C. Baird and G. Wilkinson, *Chem. Commun.*, 514 (1966).
7. C. D. Cook and G. S. Janhal, *Inorg. Nucl. Chem. Lett.*, **3**, 31 (1967), and references therein.
8. W. J. Bland and R. D. W. Kemmitt, *Nature*, **211**, 963 (1966).
9. S. Cenini, R. Ugo, F. Bonati, and G. La Monica, *Inorg. Nucl. Chem. Lett.*, **3**, 191 (1967).
10. W. H. Baddley and L. M. Venanzi, *Inorg. Chem.*, **5**, 33 (1966).
11. D. Morelli, A. Segre, R. Ugo, G. La Monica, S. Cenini, F. Conti, and F. Bonati, *Chem. Commun.*, 524 (1967).

12. J. J. Levison and S. D. Robinson, *Chem. Commun.*, 198 (1967).
13. C. J. Nyman, C. E. Wymore, and G. Wilkinson, *Chem. Commun.*, 407 (1967); F. Cariati, R. Mason, G. B. Robertson, and R. Ugo, *ibid.*, 408 (1967).
14. V. Albano, P. L. Bellon, and W. Scatturin, *Chem. Commun.*, 507 (1966).

34. OLEFIN COMPLEXES OF PLATINUM

Significant advances in organonickel chemistry followed the discovery of *trans,trans,trans*-(1,5,9-cyclododecatriene)nickel, [Ni(cdt)], and bis(1,5-cyclooctadiene)nickel, [Ni(C_8H_{12})$_2$] by Wilke et al.[1] In these and related compounds, in which only olefinic ligands are bonded to the nickel, the metal is especially reactive both in the synthesis of other nickel compounds and in catalytic behavior. Extension of this chemistry to palladium and platinum requires convenient synthetic routes to zero-valent complexes of these metals in which mono- or diolefins are the only ligands. Here we describe the synthesis of tris(2-norbornene)platinum, bis(1,5-cyclooctadiene)platinum, tris(ethylene)platinum, and bis(ethylene)(tricyclohexylphosphine)platinum. The compound [Pt(C_8H_{12})$_2$] (C_8H_{12} = 1,5-cyclooctadiene) was first reported by Müller and Göser,[2] who prepared it by the following reaction sequence:

$$[PtCl_2(C_8H_{12})] \xrightarrow[\text{(b) MeOH, }-50°C]{\text{(a) }i\text{-PrMgBr, }-40°C} [Pt(i\text{-Pr})_2(C_8H_{12})] \xrightarrow[C_8H_{12}]{UV} [Pt(C_8H_{12})_2]$$

An alternative synthesis from [PtCl$_2$(C$_8$H$_{12}$)] using cobaltacene as reducing agent has been reported.[3] The method described below is easier than that which originally appeared in *Inorganic Syntheses*[4] and generally affords good yields.

A. BIS(1,5-CYCLOOCTADIENE)PLATINUM(0)

Submitted by LOUISE E. CRASCALL and JOHN L. SPENCER*
Checked by RUTH ANN DOYLE and ROBERT J. ANGELICI†

$$2Li + C_8H_8 \rightarrow Li_2C_8H_8$$

$$Li_2C_8H_8 + [PtCl_2(C_8H_{12})] + 3C_7H_{10} \rightarrow [Pt(C_7H_{10})_3] + 2LiCl$$
$$+ C_8H_8 + C_8H_{12}$$

$$[Pt(C_7H_{10})_3] + 2C_8H_{12} \rightarrow [Pt(C_8H_{12})_2] + 3C_7H_{10}$$

*Department of Chemistry and Applied Chemistry, University of Salford, Salford M5 4WT, United Kingdom.
†Department of Chemistry, Iowa State University, Ames, IA 50011.

Procedure

Dichloro(1,5-cyclooctadiene)platinum(II) may be prepared from hexachloroplatinic acid,[5] or by heating bis(benzonitrile)dichloroplatinum(II)[6] in 1,5-cyclooctadiene at 145°C for 5 min, or from potassium tetrachloroplatinate(2-).[7] The complex [$PtCl_2(C_8H_{12})$] has a very low solubility in the reaction mixture and must be finely ground to ensure complete reaction. The olefins 1,5-cyclooctadiene, bicyclo[2.2.1]hept-2-ene (2-norbornene), and 1,3,5,7-cyclooctatetraene and all solvents should be dried and freshly distilled under nitrogen. In particular, peroxide-free diethyl ether is first dried over sodium wire and then distilled under nitrogen from sodium–benzophenone.

1. (1,3,5,7-Cyclooctatetraene)dilithium ($Li_2C_8H_8$)

Lithium foil (1.0 g, 144 mmol) is suspended under nitrogen in dry diethyl ether (200 mL) in a magnetically stirred 250-mL, two-necked, round-bottomed flask at 0°C. A 5.0-g sample (48 mmol) of 1,3,5,7-cyclooctatetraene is added and the mixture is stirred for 16 h. The small quantity of white precipitate is allowed to settle, and an aliquot of the orange solution is removed with a syringe, and the molarity is checked by hydrolysis and titration against standard acid. A saturated solution of (1,3,5,7-cyclooctatetraene)dilithium is approximately 0.24 mol L^{-1}.

■ **Caution.** *The solution is no more flammable than diethyl ether but solid $Li_2C_8H_8$ is pyrophoric in air.*

2. Tris(bicyclo[2.2.1]heptene)platinum(0)

A 1000-mL three-necked round-bottomed flask, fitted with a pressure-equalized dropping funnel, a nitrogen inlet, and a magnetic stirring bar is placed in a cold bath at $-30°C$. Finely powdered [$PtCl_2(C_8H_{12})$] (13.1 g, 35 mmol) and bicyclo[2.2.1]hept-2-ene (25 g, 266 mmol) are added through the third neck. These are slurried with diethyl ether (40 mL) at $-30°C$. As freshly prepared solution of (1,3,5,7-cyclooctatetraene)dilithium (140 mL of a 0.24-mol dm^{-3} solution) is transferred with a syringe to the dropping funnel and then added over a 1-h period to the stirred slurry while the temperature is maintained at approximately $-30°C$. The tan reaction mixture is then allowed to warm to room temperature and the volatile material removed *in vacuo*. The tan residue is dried at 0.05 torr for 1 h to remove any traces of cyclooctatetraene. The flask is filled with nitrogen and the solid scraped from the walls before extraction with hexane (200 mL and then 2 × 50 mL). A few crystals of bicyclo[2.2.1]hept-2-ene are added to each extraction. The extract is filtered through an alumina pad (5-mL Brockman activity II) under a nitrogen atmosphere. A positive pressure may be applied to assist the filtration. The colorless or pale yellow filtrate obtained is evaporated *in vacuo*. A mass of fine needles precipitate as the volatiles are carefully removed. The product, which should be almost colorless, is dried at room temperature under vacuum (0.05 torr) for 10 min. Yield: 11.0–12.5 g (65–74%). The compound may be recrystallized as long fine colorless needles by slowly cooling a filtered saturated solution in hexane to $-20°C$.

Anal. Calcd. for $C_{21}H_{30}Pt$: C, 52.8; H, 6.3. Found: C, 53.1; H, 6.5%. Although tris(bicyclo[2.2.1]hept-2-ene)platinum(0) may be handled in air, it should be stored in a closed vessel with a small crystal of bicyclo[2.2.1]hept-2-ene. The properties of this compound have been reported.[8]

3. Bis(1,5-cyclooctadiene)platinum(0)

Tris(2-norbornene)platinum (1.2 g, 2.5 mmol) is dissolved in petroleum ether (bp 100–120°C) (25 mL) in a Schlenk tube[9] connected to a N_2 line and 1,5-cyclooctadiene (3 mL) is added. If necessary this solution should be filtered through a fine glass frit using standard Schlenk techniques into another tube. A small magnetic stirring bar is added to the tube, which is then connected to a vacuum line and placed in a water bath at 30°C on a magnetic stirring hotplate.

The volatile components are evaporated slowly (1 h) at reduced pressure until approximately 2 mL of solution remain. During the evaporation, off-white crystals of $[Pt(C_8H_{12})_2]$ are deposited. The mixture is cooled to 20°C, nitrogen is readmitted to the tube and the supernatant liquid is decanted with a syringe. A syringe is used to wash the crystals with hexane (2×1 mL) and they are then dried under vacuum (0.05 torr, 15 min). Yield: 0.8 g (78%).

Anal. Calcd. for $C_{16}H_{24}Pt$: C, 47.4; H, 5.9. Found: C, 47.7; H, 5.7%. An essentially quantitative yield may be obtained by evaporating the solution entirely but the product may then be contaminated with colored impurities. The success of the operation can be judged by the absence of infrared bands typical of $[Pt(C_7H_{10})_3]$ (e.g., at 1267, 1121, and 1064 cm^{-1}) in the Nujol mull spectrum. If necessary, the evaporation process may be repeated. As prepared above $[Pt(C_8H_{12})_2]$ is an off-white crystalline solid, and is pure enough for most purposes, however, pure white crystals may be obtained by dissolution in a large volume of petroleum ether (approximately 80 mL mmol^{-1}), filtration through alumina, and recrystallization at $-78°C$.

Properties

Bis(1,5-cyclooctadiene)platinum is appreciably more oxidatively and thermally stable than the nickel analog, and the dry solid may be handled safely in air.[8] The 1,5-cyclooctadiene groups are readily displaced by a range of other ligands, including phosphines, ethylene, strained olefins,[8] and isocyanides.[10] In several instances it has been found that the order of addition is important in these displacement reactions, the best results being obtained when the $[Pt(C_8H_{12})_2]$ is added slowly to a solution of the ligand.

The ^1H NMR spectrum (C_6D_6) shows resonances at δ 4.20 [m, 8H, CH, $J_{PtH} = 55$ Hz] and 2.19 ppm [m, 16H, CH$_2$]. The ^{13}C NMR spectrum $(C_6D_6$, proton decoupled) shows resonances at δ 73.3 [C=C, $J_{PtC} = 143$ Hz] and 33.2 ppm [CH$_2$, $J_{PtC} = 15$ Hz].

The crystal structure of $[Pt(C_8H_{12})_2]$ has been reported.[11]

B. TRIS(ETHENE)PLATINUM(0)

Submitted by J. L. SPENCER*
Checked by S. D. ITTEL and M. A. CUSHING†*

$$[Pt(C_8H_{12})_2] + 3C_2H_4 \rightarrow [Pt(C_2H_4)_3] + 2C_8H_{12}$$

Procedure

Ethene (ethylene) (CP grade) may be used directly from the cylinder without further purification.

■ **Caution.** *The entire preparation is performed under an atmosphere of ethene. Also, the physiological properties of the product, which is quite volatile, have not been investigated. Therefore all operations should be carried out in a well-ventilated hood.*

An 80-mL Schlenk tube,[9] equipped with a magnetic stirring bar and containing 20 mL of petroleum ether (bp 40–60°C), is flushed continuously with ethene and cooled in ice for 10 min. Bis(1,5-cyclooctadiene)platinum is added in 0.1-g portions, with rapid stirring, until no more dissolves readily (about 1.1 g). [If too much bis(1,5-cyclooctadiene)platinum is added, or if it is added too quickly, an insoluble white precipitate forms.] The tube is then temporarily sealed (to prevent large volumes of ethene from dissolving in the solution) and cooled with a toluene slush bath ($-96°C$) for 40 min, during which time a mass of white crystals of product forms. When crystallization is complete, ethene is readmitted, another Schlenk tube is attached to the first by means of a curved glass transfer tube, and the supernatant liquid is

---

*Department of Chemistry and Applied Chemistry, University of Salford, Salford M5 4WT, United Kingdom.

†Central Research and Development Department, E. I. du Pont de Nemours & Company, Inc, Experimental Station, Wilmington, DE 19898.

decanted into the second tube. The tubes are separated, the crude solid is redissolved in 9 mL of petroleum ether at room temperature, and the solution is filtered through a sintered-glass frit (porosity 3, fine) into a clean Schlenk tube. The product is again crystallized at $-96°C$, and the supernatant liquid is decanted into the tube containing the first mother liquor. One further recrystallization from pentane or light petroleum ether (5 mL of bp 30–40°C) gives, after drying at $-40°C$ (0.01 torr for 1 h), a 0.42-g (56%) yield of white crystals of tris(ethene)platinum(0).

Anal. Calcd. for $C_6H_{12}Pt$: C, 25.8; H, 4.3. Found: C, 25.2, H, 4.5%.

A considerable quantity of the $[Pt(C_8H_{12})_2]$ (20%) may be recovered by passing the combined mother liquors through a small pad of alumina and evaporating the filtrate to dryness at reduced pressure.

Properties

Tris(ethene)platinum(0) is stable for several hours at 20°C under 1 atm of ethene and keeps for many weeks at $-20°C$.[8] In the absence of ethene, decomposition to metallic platinum occurs in minutes at room temperature. The complex is quite volatile, with a vile smell, and sublimes slowly at 20°C in an atmosphere of ethene onto a cold finger at 0°C.

Despite the obvious lability of the complex, the ^1H NMR (C_6D_6, C_2H_4 1 atm) shows a well-defined singlet, δ 3.06 ppm, with ^{195}Pt satellites (J_{PtH} = 57 Hz), as well as the signal for free ethene at δ 5.28 ppm, although at 40°C both signals begin to broaden.

Other tris(olefin)platinum(0) complexes (where olefin represents a strained olefin such as bicyclo[2.2.1]heptene, dicyclopentadiene, or *trans*-cyclooctene) may be similarly obtained by direct displacement of 1,5-cyclooctadiene, often in quantitative yield.[8]

A neutron diffraction study[12] has confirmed a trigonal planar PtC_6 geometry for $[Pt(C_2H_4)_3]$.

C. BIS(ETHENE)(TRICYCLOHEXYLPHOSPHINE)PLATINUM(0)

$$[Pt(C_8H_{12})_2] + 3C_2H_4 \rightarrow [Pt(C_2H_4)_3] + 2C_8H_{12}$$
$$[Pt(C_2H_4)_3] + (C_6H_{11})_3P \rightarrow [Pt(C_2H_4)_2P(C_6H_{11})_3] + C_2H_4$$

Procedure

This preparation should be carried out in a well-ventilated hood.

To a 250-mL two-necked, round-bottomed flask, fitted with a magnetic stirring bar, is added 90 mL of petroleum ether (bp 30–40°C), cooled to 0°C.

The apparatus is flushed with ethene until the solvent is saturated. Bis(1,5-cyclooctadiene)platinum (3.8 g, 9.2 mmol) is added in small (0.2-g) portions, and each portion is allowed to dissolve before the next is added. A petroleum ether solution of tricyclohexylphosphine (2.6 g, 9.2 mmol in 15 mL) is added slowly. After a brief evolution of gas, crystallization of the product begins. The ethene source is removed, and the flask is flushed with a slow stream of nitrogen for 1 h. The supernatant liquid is decanted into a clean flask, and the crystals are washed well with cold petroleum ether (four 10-mL portions); the washings are added to the mother liquor. The white crystals may be dried at 20°C (0.05 torr) for 1 h.

Anal. Calcd. for $C_{22}H_{41}PPt$: C, 49.7; H, 7.8. Found: C, 49.5; H, 8.2%.

A second crop of crystals may be obtained as follows. The combined mother liquor and washings are evaporated to complete dryness at reduced pressure, and the residue is dissolved in petroleum ether (50–100 mL) under an atmosphere of ethene. This solution is filtered through a short column of alumina* (2 × 5 cm), under ethene, into a round-bottomed flask. Evaporating the solvent to a small volume gives the second crop of $[Pt(C_2H_4)_2P(C_6H_{11})_3]$; combined yield 4.2 g (85%).

Properties

Bis(ethene)(tricyclohexylphosphine)platinum(0) is a white crystalline solid that may be handled safely in air, but should be stored in an inert atmosphere. It is slightly soluble in petroleum ether and readily dissolves in aromatic solvents to give reasonably stable solutions, even in the absence of dissolved ethene. The ^1H NMR spectrum (C_6D_6) shows a sharp singlet with ^{195}Pt satellites at δ 2.75 ppm (C_2H_4, $J_{PtH} = 57$ Hz) and a broad featureless resonance centered at δ 1.5 ppm (C_6H_{11}). The complex undergoes a variety of reactions to afford organoplatinum compounds. With triorganosilanes and germanes, it gives binuclear complexes $[\{Pt(H)(MR_3)[P(C_6H_{11})_3]\}_2]$ (M = Si, Ge; R = Me, Et, Cl, OEt).[13] Other tertiary phosphine derivatives may be made by similar procedures,[14] although the yields and stability of the products are dependent on the nature of PR_3. In particular $[Pt(C_2H_4)_2(PBu^t_3)]$ is difficult to isolate and the bis(bicyclo[2.2.1]hept-2-ene)(tri-*tert*-butylphosphine)platinum is a more suitable source of the valuable $Pt(PBu^t_3)$ fragment.[15]

*BDH, Brockman activity II.

References

1. P. W. Jolly and G. Wilke, *The Organic Chemistry of Nickel*, Vols. 1,2, Academic Press, New York, 1974, 1975.
2. J. Müller and P. Göser, *Angew. Chem. Int. Ed.*, **6**, 364 (1967).
3. G. E. Herberich and B. Hessner, *Z. Naturforsch. B*, **34**, 638 (1979).
4. J. L. Spencer, *Inorg. Synth.*, **19**, 213, (1979).
5. D. Drew and J. R. Doyle, *Inorg. Synth.*, **13**, 48 (1972); also see section 89 in this volume.
6. F. R. Hartley, *Organomet. Chem. Rev. A*, **6**, 119 (1970).
7. J. X. McDermott, J. F. White, and G. M. Whitesides, *J. Am. Chem. Soc.*, **98**, 6521 (1976).
8. M. Green, J. A. K. Howard, J. L. Spencer, and F. G. A. Stone, *J. Chem. Soc. Dalton Trans.*, **1977**, 271; *J. Chem. Soc., Chem. Commun.*, **3**, 449, (1975).
9. D. F. Shriver and M. A. Drezdzon, *The Manipulation of Air Sensitive Compounds*, 2nd ed., Wiley-Interscience, New York, 1986.
10. M. Green, J. A. K. Howard, M. Murray, J. L. Spencer, and F. G. A. Stone, *J. Chem. Soc. Dalton Trans.*, **1977**, 1509.
11. J. A. K. Howard, *Acta Crystallogr. B*, **38**, 2896 (1982).
12. J. A. K. Howard, J. L. Spencer, and S. A. Mason, *Proc. Roy. Soc. Lond. A*, **386**, 145 (1983).
13. M. Ciriano, M. Green, J. A. K. Howard, J. Proud, J. L. Spencer, F. G. A. Stone, and C. A. Tsipis, *J. Chem. Soc. Dalton Trans.*, **1978**, 801.
14. N. C. Harrison, M. Murray, J. L. Spencer, and F. G. A. Stone, *J. Chem. Soc. Dalton Trans.*, **1978**, 1337.
15. P. W. Frost, J. A. K. Howard, J. L. Spencer, and D. E. Turner, *J. Chem. Soc. Chem. Commun.*, **1981**, 1104.

35. ETHENE COMPLEXES OF BIS(TRIALKYLPHOSPHINE)PLATINUM(0)

Submitted by R. A. HEAD*
Checked by D. M. ROUNDHILL† and D. HEDDEN†

A limited number of platinum(0) complexes that contain ethene and organophosphine ligands are known. The triphenylphosphine complex, $[Pt(C_2H_4)[P(C_6H_5)_3]_2]$, is readily prepared by the reaction of $[Pt[P(C_6H_5)_3]_2O_2]$ with ethene.[1] It is an excellent precursor to a range of platinum(0) complexes. Preparations of analogous compounds containing trialkylphosphine ligands are less convenient and include reaction of the very air sensitive $[Pt(C_2H_4)_3]$ with trialkylphosphines[2] and the thermal decomposition of $[Pt[P(C_2H_5)_3]_2(C_2H_5)_2]$.[3] The synthesis reported here is both simple and quick and gives stable solutions of these complexes, which should make them useful starting materials for new platinum(0) chemistry. Displacement of the

---

*New Science Group, Imperial Chemical Industries PLC, The Heath, Runcorn, Cheshire, WA7 4QE, United Kingdom.
†Department of Chemistry, Tulane University, New Orleans, LA 70118.

coordinated ethene by alkenes and ketones that contain electron-withdrawing groups is facile, and monomeric formaldehyde effects a slow displacement reaction to give solutions of $[Pt(CH_2O)[P(C_2H_5)_3]_2]$.[4]

A. (ETHENE)BIS(TRIETHYLPHOSPHINE)PLATINUM(0)

$$PtCl_2[P(C_2H_5)_3]_2 + 2NaC_{10}H_8 + C_2H_4 \rightarrow$$
$$Pt(C_2H_4)[P(C_2H_5)_3]_2 + 2NaCl + 2C_{10}H_8$$

Procedure

■ **Caution.** *All solvents should be dried thoroughly and air-free before use. The reactions must be performed under a rigorously oxygen-free atmosphere. Metallic sodium is exceptionally reactive, especially toward moisture, and should be used with utmost care. Excess sodium can be destroyed by careful reaction with 2-propanol.*

Freshly distilled oxygen-free tetrahydrofuran (120 mL) and naphthalene (2.2 g) are added to a 250-mL, three-necked, round-bottomed flask fitted with a nitrogen bubbler and containing a magnetic stirring bar. Sodium wire (0.5 g) is then added, and the solution is stirred vigorously for 5 h, during which time the sodium dissolves to give an intense green solution of $NaC_{10}H_8$. The molarity of this solution is then determined by removing an accurately known volume (~5 mL) and quenching into water. Titration of this aqueous solution to neutrality with 0.1 N HCl, using phenolphthalein as the indicator, allows the concentration of $NaC_{10}H_8$ in tetrahydrofuran to be calculated. If the above procedure is observed, the concentration of $NaC_{10}H_8$ is approximately 0.11 N.

To a 250-mL round-bottomed flask fitted with an ethene inlet, a bubbler, and a rubber septum and containing a magnetic stirring bar is added dichlorobis(triethylphosphine)platinum(II)* (0.62 g, 1.2 mmol) and dry, oxygen-free tetrahydrofuran (50 mL). The reaction works equally well with both cis and trans isomers of the platinum complex. A tetrahydrofuran solution of $NaC_{10}H_8$ (24 mL, 0.1 N) prepared as above is then added over a 15–20-min period to the stirred suspension at room temperature by means of a gastight syringe (Fig. 1). On mixing the two solutions, the deep green color of $NaC_{10}H_8$ is rapidly destroyed, and the platinum complex gradually dissolves, eventually giving a slightly cloudy solution of (ethene)bis(triethylphosphine)platinum(0). Addition of excess reducing agent gives an intense red

---

*Syntheses of $PtCl_2(PR_3)_2$ complexes are best carried out using the procedure outlined for $PtCl_2(PPr_3^i)_2$ by Yoshida et al. in Ref. 5.

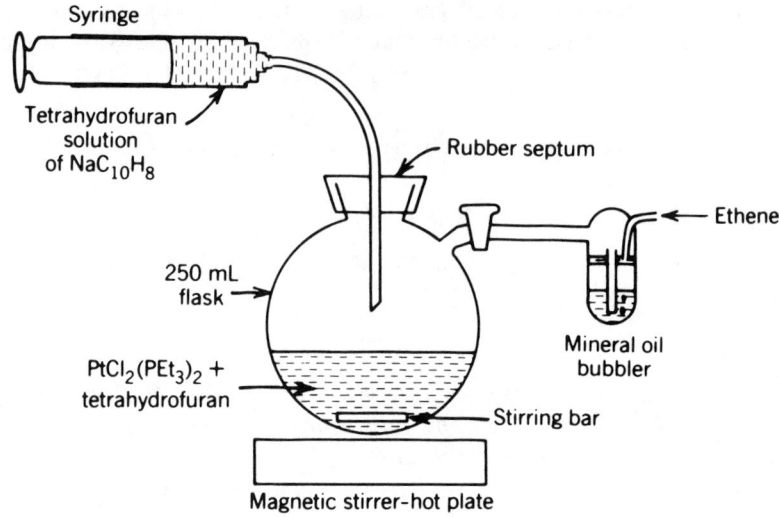

Fig. 1. Apparatus for platinum(0) complexes.

solution, but this is easily reversed by adding a crystal of $PtCl_2[P(C_2H_5)_3]_2$. The yield of product is essentially quantitative, as evidenced by ^{31}P NMR spectroscopy. A single line is observed at 20.4 ppm, relative to 85% H_3PO_4, with $^1J_{^{195}Pt-P} = 3520$ Hz.

B. (ETHENE)BIS(PHOSPHINE)PLATINUM(0)

$$PtCl_2(PR_3)_2 + 2NaC_{10}H_8 + C_2H_4 \rightarrow Pt(C_2H_4)(PR_3)_2 + 2NaCl + 2C_{10}H_8$$

Phosphine, PR_3* = tris(1-methylethyl)phosphine; diethylphenylphosphine; triphenylphosphine; $\frac{1}{2}$[1,2-ethanediylbis(diphenylphosphine)]

Procedure

The method for preparation of the zero-valent ethene complexes is identical to that described in Section A. For less soluble starting materials the reducing agent is best added over a longer period of time (up to 30 min). All complexes

---

*All phosphines are available through Strem Chemicals, Inc., P.O. Box 108, Newburyport, MA 01950, who also supply certain Pt(II) complexes including $PtCl_2(PPh_3)_2$.

Table I. ^{31}P NMR Data for [Pt(C$_2$H$_4$)(PR$_3$)$_2$] Complexes

PR$_3$	Chemical shift (ppm)^a	$^1J_{195Pt-P}$ (Hz)
P(C$_2$H$_5$)$_3$	20.4	3520
P(i-C$_3$H$_7$)$_3$	53.4	3657
P(C$_2$H$_5$)$_2$C$_6$H$_5$	23.2	3574
P(C$_6$H$_5$)$_3$	32.0	3660
$\frac{1}{2}$(C$_6$H$_5$)$_2$PCH$_2$CH$_2$P(C$_6$H$_5$)$_2$	54.5	3300

^{a}Relative to 85% H$_3$PO$_4$.

are stable in solution. They are prepared in nearly quantitative yield as shown by ^{31}P NMR spectroscopy (see Table I).

Properties

Solutions of the complexes are stable for several hours at room temperature under an ethene atmosphere. Decomposition takes place if the compound is heated (> ~65°). If the solvent is removed under reduced pressure, a red-brown oil of uncertain composition is formed.

References

1. F. R. Hartley, *The Chemistry of Platinum and Palladium with Particular Reference to Complexes of the Elements*, Halstead, New York, 1973.
2. M. Berry, J. A. K. Howard, and F. G. A. Stone, *J. Chem. Soc. Dalton Trans.*, **1980**, 1601, 1609; J. Spencer, *Inorg. Synth.*, **19**, 216 (1979).
3. R. G. Nuzzo, T. J. McCarthy, and G. M. Whitesides, *Inorg. Chem.*, **20**, 1312 (1981).
4. R. A. Head, *J. Chem. Soc. Dalton Trans.*, **1982**, 1637.
5. T. Yoshida, T. Matsuda, and S. Otsuka, *Inorg. Synth.*, **19**, 108 (1979); also see section 30 in this volume.

Chapter Three

SUBSTITUTED METAL CARBONYL COMPLEXES

36. (η^6-ARENE)TRICARBONYLCHROMIUM COMPLEXES

Submitted by C. A. L. MAHAFFY* and P. L. PAUSON*
Checked by M. D. RAUSCH† and W. LEE†

Most simple arenes react smoothly with $Cr(CO)_6$ to give hexahapto complexes (η^6-ArH)Cr(CO)$_3$. The reactions have been conducted under a wide variety of conditions,[1] and the chief problems encountered arise from (a) the volatility of $Cr(CO)_6$, (b) slowness of the reaction, and (c) difficulty in removing high-boiling solvents or excess arene from the product. Sublimation of $Cr(CO)_6$ from the reaction vessel has been overcome by the use of rather complex apparatus,[2] but the following procedure shows this to be unnecessary. In inert solvents [decahydronaphthalene (decalin) has been widely used], reaction is excessively slow. Donor solvents (sol) lead to appreciably more rapid reaction, probably by way of intermediates $Cr(CO)_{6-n}(sol)_n$ (where $n = 1$–3); but if the donor is too good (e.g., C_6H_5CN), it may compete, especially with the less reactive arenes (e.g., C_6H_5Cl), leading to incomplete complex formation. Alkylpyridines have been recommended,[3,4] but ethers have been much more widely used. Tetrahydrofuran (THF), a good donor, allows reaction to proceed cleanly[5] but too slowly because of its low boiling point. Dibutyl ether also leads to rather slow reaction, probably because of its weak donor properties, whereas bis(2-methoxyethyl) ether (diglyme), which is better in this respect, is relatively difficult to remove. The following procedure therefore uses a mixture of

---

*Department of Pure and Applied Chemistry, University of Strathclyde, Thomas Graham Building, Cathedral Street, Glasgow, G1 1XL.
†Department of Chemistry, University of Massachusetts, Amherst, MA 01002.

dibutyl ether with sufficient THF to "catalyze" the reaction and to wash back most of the $Cr(CO)_6$ that sublimes into the condenser, but not enough to lower the boiling point too much.

A. (η^6-ANISOLE)TRICARBONYLCHROMIUM

Procedure

■ **Caution.** *The reaction should be carried out in a well-ventilated hood, as hexacarbonylchromium is toxic and carbon monoxide is evolved during the reaction. Both ether solvents peroxidize; they should be carefully freed from peroxide and dried (conveniently by distilling from lithium tetrahydridoaluminate or from sodium) before use. Benzene is toxic; contact with the liquid or vapor should be avoided.*

In a 250-mL round-bottomed flask fitted with a gas inlet and a simple reflux condenser [not spiral or similar type from which subliming $Cr(CO)_6$ is washed back less efficiently] are placed hexacarbonylchromium (4 g, 18 mmol), anisole (25 mL), dibutyl ether (120 mL), and tetrahydrofuran (10 mL). A bubbler is placed at the top of the condenser to prevent access of air. The apparatus is thoroughly purged with nitrogen. The nitrogen stream is stopped and the mixture is then heated at reflux for 24 h (the checkers found stirring beneficial). The yellow solution is cooled in ice (most of the unreacted $Cr(CO)_6$ separates out in 10–15 min) and filtered through kieselguhr (diatomaceous earth) or a similar material (the checkers used Celite or, preferably, a small pad of anhydrous silica gel) on a sintered-glass filter, which is then washed with a little additional solvent. The solvents are distilled off on a rotary evaporator from a water bath held at 60° (an oil pump may be required to remove the solvents completely); a deep-yellow oil remains to which dry light petroleum ether (bp 40–60°) or hexane (20 mL) is added. Crystalline (η^6-anisole)tricarbonylchromium [4.1 g (92%), mp 83–84°] separates. A small amount of unreacted $Cr(CO)_6$ may be recoverable from the condenser; the remainder distills off with the solvent. If a very pure product is required the compound may be recrystallized by dissolving it in benzene or in diethyl ether and adding light petroleum ether to give 3.53 g (80%), mp 84–85°; lit.[6] mp 84–85°.

Anal. Calcd. for $C_{10}H_8CrO_4$: C, 49.2; H, 3.3. Found: C, 49.5; H, 3.4.

Solutions of this compound and other arenetricarbonylchromiums can be handled for only very brief periods in air. The work-up procedures are therefore best carried out in an inert atmosphere throughout, but they can be conducted in air if done rapidly and efficiently.

Properties

(η^6-Anisole)tricarbonylchromium is a yellow crystalline solid. Its NMR spectrum[8] (CDCl$_3$ solution) shows a singlet at δ 3.6 for the CH$_3$O group and three well-resolved signals for the metal-bound ring at δ 4.77 (1H, t) 5.03 (2H, d) and 5.4 (2H, t). Its solutions are air sensitive, forming greenish precipitates, but the pure solid is stable. However, refrigeration is recommended for prolonged storage.

B. OTHER ARENETRICARBONYLCHROMIUM COMPLEXES

The general applicability of the above method is illustrated by the examples in Table I. With the lowest boiling arenes, subliming Cr(CO)$_6$ is completely washed back, and none is recovered at the end of the reaction. However, the reduced boiling point of the mixture necessitates a longer reaction time. For arenes of higher boiling points, it is preferable not to use a large excess, since it would be difficult to remove at the end of the reaction. This leads to a somewhat less complete reaction, but since much of the unchanged Cr(CO)$_6$ is readily recovered from the condenser and the distillate, yields in the table are based on unrecovered carbonyl. The result with anisole when only a small excess is used is included for comparison. The arenes for which results are tabulated are liquids, but several solid arenes have been used similarly. Thus the checkers obtained an 81% yield of tricarbonyl(triphenylene)chromium using a 1:1 molar ratio of the hydrocarbon to Cr(CO)$_6$.

Reactivity

The chemistry of arenetricarbonylchromium complexes has been extensively studied and reviewed.[10,11] Compared to the uncomplexed arenes they show greatly enhanced reactivity towards nucleophiles.[12]

Addendum

Since its original publication this procedure has been successfully used for a very wide range of both mono- and polycyclic arenes. Provided that air is rigorously excluded anhydrous conditions are not always essential as shown by the use[13] of water–THF (4:1) for the formation of tricarbonylchromium complexes of several phenylalanine derivatives. Professor G. Jaouen reports that the method fails for oestradiol and for the relatively weakly bound condensed arenes (naphthalene, etc.), probably because THF can displace such arenes from their complexes; he recommends the use of dibutyl ether–heptane in such cases. For the formation of highly sensitive complexes,

TABLE I. Reaction of Arenes with 4 g of $Cr(CO)_6$ to Give the Corresponding Complexes $(\eta^6\text{-Arene})Cr(CO)_3$[a]

Arene	Volume (mL)	Reflux Time (h)	$Cr(CO)_6$ Recovered (g)	Yield g	Yield %	Product mp[b](°C)	ν_{CO} peaks in C_6H_{12}[c] (cm^{-1})
$C_6H_5(OCH_3)$	5	42	—	3.77	85	83–84 (84–85, Ref. 6)	1980, 1908
C_6H_6	20	40	—	3.44	89[d]	159–160 (161.5–163, Ref. 6)	1982, 1915
C_6H_5F	16	42	—	3.79	90	108[e] (116–117, Ref. 7)	1990, 1929, 1926
C_6H_5Cl	20	20	0.35	2.6	64[f]	101–102 (102–103, Ref. 6)	1991, 1929, 1925
$C_6H_5(NMe_2)$	5[g]	19	0.79	3.2	85	144 (145–146, Ref. 6)	1969, 1894, 1888
C_6H_5COOMe	10	20	0.70	3.63	89	92–93 (95.5–96, Ref. 6)	1990, 1927

[a]The procedure given for $[\eta^6\text{-}C_6H_5(OCH_3)]Cr(CO)_3$ was used, except for the conditions noted.
[b]Melting point of the product as initially isolated, before recrystallization; literature melting point of the pure compound in parentheses.
[c]Values from Ref. 9.
[d]A 19-h run [0.21 g recovered $Cr(CO)_6$] gave 2.3 g (62%).
[e]This melting point is exceptionally sensitive to the rate of heating: an analytically pure sample of mp 116° was obtained after two crystallizations.
[f]Slight decomposition during the reaction indicated by formation of gray-green precipitates was found difficult to avoid with chlorobenzene and some other arenes. This appears to become progressive and reaction should be stopped and solution filtered when such precipitation is observed.
[g]Solvent: dibutyl ether (60 mL) + tetrahydrofuran (5 mL); most of the product from this reaction crystallized from the solution when cooled in ice.

indirect methods are preferable. Of these, the most successful employs $(NH_3)_3(CO)_3Cr$;[14] the iodine-catalyzed arene-exchange reaction[15] also deserves attention for this purpose.

References

1. Comprehensive references may be found in *Gmelin's Handbuch der anorganischen Chemie*, Supplement to the 8th ed., Vol. 3, **1971**, pp. 181–289.
2. W. Strohmeier, *Chem. Ber.*, **94**, 2490 (1961); A. T. T. Hsieh, W. C. Matchan, H. van den Bergen, and B. B. West, *Chem. and Ind.* (London), **1974**, 114.
3. R. L. Pruett, J. E. Wyman, D. R. Rink, and L. Parts, U.S. Patent 3378569 (1968); *Chem. Abstr.*, **69**, 77512 (1968); U.S. Patent 3382263 (1968); *Chem. Abstr.*, **69**, 59376 (1968).
4. M. D. Rausch, *J. Org. Chem.*, **39**, 1787 (1974).
5. W. P. Anderson, N. Hsu, C. W. Stanger, and B. Munson, *J. Organomet. Chem.*, **69**, 249 (1974).
6. W. R. Jackson, B. Nicholls, and M. C. Whiting, *J. Chem. Soc.*, **1960**, 469.
7. J. F. Bunnett and H. Hermann, *J. Org. Chem.*, **36**, 4081 (1971); V. S. Khandkarova, S. P. Gubin, and B. A. Kvasov, *J. Organomet. Chem.*, **23**, 509 (1970).
8. W. McFarlane and S. O. Grim, *J. Organomet. Chem.*, **5**, 147 (1966); A. Mangini and F. Taddei, *Inorg. Chim. Acta*, **2**, 8 (1968).
9. R. D. Fischer, *Chem. Ber.*, **93**, 165 (1960); D. A. Brown and H. Sloan, *J. Chem. Soc.*, **1963**, 4389; D. A. Brown and J. R. Raju, *J. Chem. Soc. A*, **1966**, 1617; D. M. Adams and A. Squire. *J. Chem. Soc. Dalton Trans.*, **1974**, 558.
10. R. P. A. Sneeden, *Organochromium Compounds*, Academic Press, New York, 1975.
11. W. E. Silverthorn, *Adv. Organomet. Chem.*, **13**, 47 (1975).
12. M. F. Semmelhack, in *New Applications of Organometallic Reagents in Organic Synthesis*, D. Seyferth, ed., Elsevier, Amsterdam, 1976, p. 361.
13. C. Sergheraert, J.-C. Brunet, and A. Tartar, *J. Chem. Soc. Chem. Commun.*, **1982**, 1417.
14. J. Vebrel, R. Mercier, and J. Belleney, *J. Organomet. Chem.*, **235**, 197 (1982).
15. J. J. Harrison, *J. Am. Chem. Soc.*, **106**, 1487 (1984).

37. ZERO-VALENT ISOCYANIDE COMPLEXES OF CHROMIUM, MOLYBDENUM, AND TUNGSTEN

Submitted by MICHEL O. ALBERS* and NEIL J. COVILLE†
Checked by HAEWON L. UHM‡ and IAN S. BUTLER‡

The hexacarbonyl complexes of the Group 6 metals are typically fairly inert to substitution reactions. In particular it is well documented that the direct replacement of carbonyl ligands on $[M(CO)_6]$ (M = Cr, Mo, W) by iso-

---

*Department of Chemistry, University of Pretoria, Pretoria 0002, Republic of South Africa.
†Department of Chemistry, University of the Witwatersrand, 1 Jan Smuts Avenue, Johannesburg 2001, Republic of South Africa.
‡Department of Chemistry, McGill University, 801 Sherbrooke, St. W, Montreal, Quebec, Canada H3A 2K6.

cyanides can only be achieved with some difficulty.[1,2] Thus, under thermal reaction conditions, useful yields of monosubstituted products [M(CO)$_5$(CNR)] can be obtained, whereas only low to moderate yields of disubstituted products have been reported. Further, in only one instance has the direct thermal replacement of three carbonyl ligands by isocyanide groups been achieved. In this study,[3] [W(CO)$_3$(CNBu^t)$_3$] was obtained by reacting [W(CO)$_6$] in Bu^{t}NC as solvent for 4 days (yield 80%). Consequently it is somewhat surprising that photochemical methods of substituting [M(CO)$_6$] complexes have not been more extensively investigated,[4] and also that the use of phase-transfer catalysis[5] has not, apparently, gained popularity. Rather, the most general synthetic procedures for obtaining the complexes [M(CO)$_{6-n}$(CNR)$_n$] have depended on indirect methods of substitution. These have normally entailed the preparation of synthetic precursors containing labile ligand groups (e.g., NH$_3$, cyclohepta-1,3,5-triene), which may then subsequently be readily substituted in a controlled manner by isocyanide nucleophiles:[1,2,6]

$$[(\eta^6\text{-}C_7H_8)Cr(CO)_3] + 3\text{MeNC} \rightarrow fac\text{-}[Cr(CO)_3(CNMe)_3] + C_7H_8$$

(C$_7$H$_8$ = cyclohepta-1,3,5-triene)

Typically the desired isocyanide substitution products are obtained in good yield from such reactions. However, a major drawback of this strategy entails the prior preparation and purification of the labile precursor complexes. The transition metal *catalyzed* substitution of the metal hexacarbonyls [M(CO)$_6$] (M = Cr, Mo, W) by isocyanides has, by way of contrast, proved to be a direct and rapid synthetic route that reliably gives the products [M(CO)$_{6-n}$(CNR)$_n$] (n = 1–3) in high yield.[7–9] We describe here general procedures that illustrate the methods involved and that highlight the synthetic utility of these reactions.

General Procedure

Chromium, molybdenum and tungsten hexacarbonyl, and palladium oxide were purchased from Strem Chemicals (7 Mulliken Way, Dexter Industrial Park, P.O. Box 108, Newburyport, MA 01950). *tert*-Butylisocyanide is available commercially from Aldrich Chemical Co. (P.O. Box 355, Milwaukee, WI 53201) or it may be synthesized in 50–60-g quantities by the method of Gokel et al.[10] The catalyst CoCl$_2\cdot$2H$_2$O was obtained by heating CoCl$_2\cdot$6H$_2$O under vacuum (0.1 torr) at 50° for approximately 5 h.[8] This catalyst can be readily handled and stored in air. It does, however, slowly reabsorb atmospheric moisture over a period of several months, necessitating occasional regeneration. Reagent-grade toluene was deoxygenated prior to use by distillation under nitrogen or by bubbling a rapid stream of nitrogen through the solvent for several minutes. Silica gel (23–600 μm) was used for both product purification and column chromatography. Thin-layer chromatography plates were of silica gel and contained a fluorescent indicator (Merck "Silica Gel 60F$_{254}$"). All reactions were routinely carried out under a nitrogen atmosphere in a two-necked, round-bottomed flask. One neck was fitted with a water-cooled condenser and the other with a rubber septum or a glass stopper. The reaction solution was heated by a paraffin-oil bath preset at 110° and magnetically stirred.

■ **Caution.** *Chromium, molybdenum and tungsten hexacarbonyls are volatile solids (Cr $\gg$ Mo > W) and like all metal carbonyl compounds should be considered to be toxic. tert-Butylisocyanide has a pungent odor, and although many isocyanides are reported to exhibit no appreciable toxicity to mammals,[11] it should still be handled with care. Carbon monoxide is evolved in these reactions and being an odorless, toxic gas, care should be exercised to carry out the reactions in an efficient ventilation hood with the apparatus venting into a well-ventilated region of the hood.*

A. [M(CO)$_{6-n}$(CNBu^t)$_n$] (M = Cr, Mo, W; n = 1–3) using Cobalt(II) Chloride as Catalyst[8]

$$[M(CO)_6] + n\text{Bu}^t\text{NC} \xrightarrow{\text{CoCl}_2} [M(CO)_{6-n}(\text{CNBu}^t)_n] + n\text{CO}$$

(M = Cr, Mo, W; n = 1–3)

Procedure

The catalyst CoCl$_2 \cdot$2H$_2$O (0.033 g; 0.2 mmol), metal hexacarbonyl (3.0 mmol; M = Cr, 0.660 g; M = Mo, 0.792 g; M = W, 1.056 g) and toluene (10 mL) are added to the reaction flask. The stirred mixture is heated to reflux and the appropriate amount of *tert*-butylisocyanide required to achieve monosubstitution (3.6 mmol, 0.403 mL), disubstitution (6.6 mmol, 0.704 mL), or trisubstitution (9.6 mmol, 1.076 mL) is added by microsyringe to the hot reactants. This gives an immediate blue coloration due to the formation of a cobalt chloride–isocyanide complex.

■ **Caution.** *It is essential that this sequence of events be strictly adhered to. In particular, the isocyanide must be added to the hot solution of metal hexacarbonyl and catalyst, otherwise a competing reaction, the cobalt chloride-catalyzed polymerization of isocyanide, occurs, leading to catalyst deactivation and incomplete substitution reactions.*

TABLE I. Reaction Times and Yields for the Catalyzed Synthesis of the Complexes [M(CO)$_{6-n}$(CNBu^t)$_n$] (M = Cr, Mo, W; n = 1–3)

Complex	Cobalt Chloride		Palladium Oxide	
	Reaction Time^a (min)	Yield^b (%)	Reaction Time^a (min)	Yield^b (%)
[Cr(CO)$_5$(CNBu^t)]	30	75	3	85
[Mo(CO)$_5$(CNBu^t)]	3	84	3	92
[W(CO)$_5$(CNBu^t)]	30	82	3	90
cis-[Cr(CO)$_4$(CNBu^t)$_2$]	40	80	3	93
cis-[Mo(CO)$_4$(CNBu^t)$_2$]	3	90	3	92
cis-[W(CO)$_4$(CNBu^t)$_2$]	40	90	3	93
fac-[Cr(CO)$_3$(CNBu^t)$_3$]	180	80	10	90
fac-[Mo(CO)$_3$(CNBu^t)$_3$]	3	88	3	90
fac-[W(CO)$_3$(CNBu^t)$_3$]	50	84	3	92

^{a}As estimated by IR spectroscopy.
^{b}Isolated yields.

Continued reflux results in the solution turning bright green. Completion of the reaction (Table I) is indicated by monitoring the reaction by IR spectroscopy or by thin-layer chromatography [mobile phase: hexane (n = 1); hexane–diethyl ether 1:1 (n = 2); hexane–diethyl ether 1 : 2 (n = 3)]. The mixture is cooled to room temperature, silica gel (10–20 g) is added in order to adsorb the catalyst, and the product is extracted with 5–10, 10-mL portions of toluene. The extracts are filtered, giving colorless or pale yellow solutions of the products. Solvent removal is effected under vacuum giving the crude products as white or pale yellow solids. Purification is achieved by recrystallization (*n*-hexane for the mono- and disubstituted products and CH$_2$Cl$_2$–hexane 1 : 5 for the trisubstituted products) or by chromatography on silica gel (30 × 2-cm column; *n*-hexane for monosubstituted products and diethyl ether–hexane mixtures for the di- and trisubstituted products, 1 : 4 and 1 : 1 diethyl ether–hexane, respectively). The products are obtained from these procedures in 75–90% yields* as white or pale yellow crystalline solids.

---

*The checkers report lower yields (40–65%) and suggest extended reaction times. The major impurity observed was the next lower derivative; for example, in the synthesis of [W(CO)$_4$(CNBu^t)$_2$] a 45-min reaction gave a yield of 47% [W(CO)$_4$(CNBu^t)$_2$] and 18% [W(CO)$_5$(CNBu^t)].

B. $[M(CO)_{6-n}(CNBu^t)_n]$ (M = Cr, Mo, W; $n = 1-3$) using Palladium(II) Oxide as Catalyst

$$[M(CO)_6] + nBu^tNC \xrightarrow{PdO} [M(CO)_{6-n}(CNBu^t)_n] + nCO$$

(M = Cr, Mo, W; $n = 1-3$)

The catalyst PdO (0.020 g; 0.16 mmol), $[M(CO)_6]$ (2.0 mmol; M = Cr, 0.440 g; M = Mo, 0.528 g; M = W, 0.704 g), and toluene (40 mL) are combined in the reaction flask. The mixture is heated to boiling and the appropriate amount of *tert*-butylisocyanide required to achieve mono- (2.05 mmol, 0.230 mL), di- (4.1 mmol, 0.460 mL), or trisubstitution (6.15 mmol, 0.690 mL) is added by microsyringe. The reaction mixture is heated under reflux until completion of the reaction as monitored by IR spectroscopy or thin layer chromatography (Table I). The mixture is allowed to cool to room temperature and filtered through a fluted filter paper to separate the catalyst. The solvent is removed under vacuum to give the crude products. Product purification employed the same procedures detailed above for the cobalt(II) chloride-catalyzed reaction.*

C. Other Zero-Valent Isocyanide Complexes of Chromium, Molybdenum, and Tungsten

Complexes of the type $[M(CO)_{6-n}(CNR)_n]$ (M = Cr, Mo, W; $n = 1-3$) employing a range of isocyanides (e.g., R = 2,6-$Me_2C_6H_3$, benzyl) are readily obtainable by minor variation in the procedures outlined above.[8,9,12] Catalytic methods of synthesis have also been applied to the preparation of higher substituted derivatives $[M(CO)_{6-n}(CNR)_n]$ ($n = 4, 5$) and homoleptic isocyanide systems ($n = 6$).[9]

Properties

The complexes $[M(CO)_{6-n}(CNBu^t)_n]$ ($n = 1-3$) and those of the type $[M(CO)_{6-n}(CNR)_n]$ ($n = 1-3$) in general are air-stable crystalline solids soluble in most common organic solvents. Other physical and spectroscopic properties of these compounds have been extensively documented.[1,2,6,8]

---

*The checkers report that within the reaction time suggested lower yields of the trisubstituted products were obtained; for instance, $[Mo(CO)_3(CNBu^t)_3]$ was isolated in 58% yield after a 3-min reaction.

References

1. L. Malatesta and F. Bonati, *Isocyanide Complexes of Metals*, Wiley, London, 1969.
2. R. B. King and M. S. Saran, *Inorg. Chem.*, **13**, 74 (1974).
3. E. B. Dreyer, C. T. Lam, and S. J. Lippard, *Inorg. Chem.*, **18**, 1904 (1979).
4. W. Strohmeier, J. F. Guttenberger, and F. J. Müller, *Z. Naturforsch.*, **22B**, 1091 (1967); W. Strohmeier and H. Hellman, *Ber. Bunsengeselch. Physik. Chem.*, **68**, 481 (1964).
5. S. A. Al-Jibori and B. L. Shaw, *J. Organomet. Chem.*, **192**, 83 (1980).
6. J. A. Connor, E. M. Jones, G. K. McEwen, M. K. Lloyd, and J. A. McCleverty, *J. Chem. Soc. Dalton Trans.*, **1972**, 1246.
7. M. O. Albers and N. J. Coville, *Coord. Chem. Rev.*, **53**, 227 (1984).
8. M. O. Albers, N. J. Coville, T. V. Ashworth, E. Singleton, and H. E. Swanepoel, *J. Organomet. Chem.*, **199**, 55 (1980).
9. N. J. Coville and M. O. Albers, *Inorg. Chim. Acta*, **65**, L7 (1982).
10. G. W. Gokel, R. P. Widera, and W. P. Weber, *Org. Synth.*, **55**, 96 (1976).
11. J. A. Green and P. T. Hoffman, in *Isonitrile Chemistry*, I. Ugi, (ed.), Academic Press, New York, 1971, p. 2.
12. M. O. Albers, Ph.D. thesis, University of the Witwatersrand, Johannesburg, 1981.

38. cis-BIS(DIETHYLDITHIOCARBAMATO)DINITROSYL-MOLYBDENUM, DIBROMOTETRACARBONYL-MOLYBDENUM(II), AND DI- AND TRICARBONYL-BIS(DIETHYLDITHIOCARBAMATO)MOLYBDENUM(II)

$$Mo(CO)_6 + Br_2 \rightarrow MoBr_2(CO)_4 + 2CO$$

$$MoBr_2(CO)_4 + 2NaS_2CN(C_2H_5)_2 \rightarrow$$
$$Mo(CO)_2[S_2CN(C_2H_5)_2]_2 + 2CO + 2NaBr$$

$$Mo(CO)_2[S_2CN(C_2H_5)_2]_2 + 2NO \rightarrow Mo(NO)_2[S_2CN(C_2H_5)_2]_2 + 2CO$$

Submitted by J. A. BROOMHEAD,* J. BUDGE,* and W. GRUMLEY*
Checked by J. W. McDONALD†

Molybdenum complexes with sulfur-donor atoms play an important role in the nitrogenase enzyme system, and the study of their chemistry is of continuing interest. The present synthesis concerns cis-dinitrosylbis(N,N-diethyldithiocarbamato)molybdenum, originally described by Johnson et al.[1] Their method of synthesis involves the conversion of molybdenum hexacarbonyl to the unstable $MoBr_2(NO)_2$ using NOBr, followed by reaction

---
*Chemistry Department, Faculty of Science, Australian National University, Canberra, A.C.T. 2600, Australia. The authors wish to thank Dr. B. Tomkins for helpful discussions and the Australian Research Grants Committee for supporting the work.
†Kettering Scientific Research Laboratory, Yellow Springs, OH 45387.

with the N,N-diethyldithiocarbamate [$S_2CN(C_2H_5)_2$] ligand. Yields are 50–60% based on molybdenum hexacarbonyl. Both NOBr and $MoBr_2(NO)_2$ are unstable and must be freshly prepared. The method given here avoids these intermediates and makes use of carbonyl complexes instead. Molybdenum hexacarbonyl is treated with bromine followed by sodium N,N-diethyldithiocarbamate to yield $Mo(CO)_2[S_2CN(C_2H_5)_2]_2$. Subsequent reaction with NO affords cis-$Mo(NO)_2[S_2CN(C_2H_5)_2]_2$ in 75% yield.

Procedure

- **Caution.** *A fume hood must be used.*

Molybdenum hexacarbonyl (5 g, 0.019 mol) is ground to a fine powder and placed in a 250-mL, two-necked, round-bottomed flask fitted with a gas inlet tube, dropping funnel, and magnetic stirrer. The reaction vessel is cooled to $-10°$ in an acetone–Dry Ice bath. Dichloromethane (100 mL) is added, and a stream of dry nitrogen is maintained throughout the subsequent reaction procedure. Bromine (3 g, 0.019 mol) is added dropwise and gas evolution is evident. When the addition of bromine is complete, the cooling bath is removed, and the reaction mixture is allowed to evaporate to dryness. The effluent gas is tested with aqueous silver nitrate–nitric acid solution until there is complete removal of bromine. This takes about $\frac{3}{4}$ h. Methanol (50 mL, deoxygenated) is then added followed 10 min later by a solution of sodium N,N-diethyldithiocarbamate (8.55 g, 0.038 mol) in methanol (50 mL). A mixture of the orange tricarbonyl complex and purple dicarbonyl complex precipitates.[3,*] After a further 5 min, dichloromethane (100 mL) is added to the stirred suspension to dissolve the molybdenum-containing products (NaBr formed does not dissolve), and nitrogen oxide is passed through the solution for 30 min. The color changes to a deep brown during this time.† At this stage and during subsequent purification the reaction vessel is covered with aluminum foil to minimize exposure to light. The product complex is photosensitive, especially in solution, and gives the seven-coordinate tris(N,N-diethyldithiocarbamato)nitrosylmolybdenum complex along with other unidentified products.[2] The dark-brown solution is next evaporated to

---

*The checker reports a 3.2-g yield of carbonyl complexes here based on 2.0 g of $Mo(CO)_6$. The mixture of carbonyl complexes can be filtered off and stored under nitrogen and is useful for the synthesis of other bis(dithiocarbamato)molybdenum compounds.

†Thin-layer chromatography using a drop of the reaction mixture spotted onto Bakerflex silica gel IB-F strips (20 × 10 cm) and eluted with dichloromethane (in the air) readily monitors the disappearance of the purple intermediate complexes. Brown cis-$Mo(NO)_2[S_2CN(C_2H_5)_2]_2$ moves with the solvent front while the purple complexes have R_F values of ~0.5.

dryness with a rotary evaporator, and the residue is extracted with chloroform (50 mL). Further purification is best effected by column chromatography. It is not necessary to exclude air, but photochemical reactions should be avoided by wrapping the column with dark paper or aluminum foil. A 50 × 3-cm column is prepared by suspending silica gel powder in 1:1 n-hexane–chloroform. One-half of the chloroform solution of the crude product is placed on the column and eluted with 1:1 n-hexane–chloroform ($\sim$ 500 mL). The complex readily elutes as a brown band and is recovered by evaporation of the solvent. The remainder of the crude material is chromatographed similarly, and the combined products are recrystallized from hot methanol (200 mL). The yield is 6.5 g (0.014 mol).

Anal. Calcd. for $C_{10}H_{20}N_4S_4O_2Mo$: C, 26.55; H, 4.66; N, 12.38; S, 28.36. Found: C, 26.52; H, 4.54; N, 12.29; S, 28.70.

Properties

cis-Bis(diethyldithiocarbamato)dinitrosylmolybdenum is a brown crystalline solid stable in air but undergoing slow photochemical decomposition in solution.[2] It is soluble in organic solvents such as acetone, chloroform, and methanol. Solutions are unstable with respect to atmospheric oxidation, although the reaction is slow. The nmr and infrared spectra of the compound have been reported, and the cis geometry was assigned on the basis of these measurements.[1] For evaporated chloroform films and KBr pellets, the characteristic strong infrared bands are at 1755, 1642 [$\gamma(N\equiv O)$], and 1505 cm^{-1} [$\gamma(C=N)$]. The mass spectrum gives a strong parent molecular ion and fragmentation consistent with loss of NO and breakdown of the diethyldithiocarbamate ligand.[1] The stoichiometry of the present preparation remains obscure.

References

1. B. F. G. Johnson, K. H. Al-Obaidi, and J. A. McCleverty, *J. Chem. Soc. A*, **1969**, 1668.
2. J. A. Broomhead and W. Grumley, unpublished work.
3. R. Colton, G. R. Scollary, and I. B. Tomkins, *Aust. J. Chem.*, **21**, 15 (1968).

39. HEXACARBONYLBIS(η^5-CYCLOPENTADIENYL)DI-CHROMIUM, MOLYBDENUM, AND TUNGSTEN AND THEIR ANALOGS, $M_2(\eta^5\text{-}C_5H_4R)_2(CO)_6$ (M = Cr, Mo, and W; R = H, Me or PhCH$_2$).

$$2C_5H_6 + 2Na \rightarrow 2Na[C_5H_5] + H_2$$

$$Na[C_5H_5] + M(CO)_6 \rightarrow Na[M(\eta\text{-}C_5H_5)(CO)_3] + 3CO$$

$$2Na[M(\eta\text{-}C_5H_5)(CO)_3] + 2Fe(III) \rightarrow M_2(\eta\text{-}C_5H_5)_2(CO)_6$$
$$+ 2Fe(II) + 2Na^+$$

Submitted by A. R. MANNING,* PAUL HACKETT,* RALPH BIRDWHISTELL,* and PAUL SOYE*
Checked by DAVID C. MILLER† and ROBERT J. ANGELICI†

The direct reaction of dicyclopentadiene with $M(CO)_6$ at elevated temperatures provided an early route to the $M_2(\eta\text{-}C_5H_5)_2(CO)_6$ complexes where M = Mo or W.[1,2] Subsequently a number of other routes to this series of compounds were devised using the readily available hexacarbonylmetals as precursors: (a) the oxidation of $[Cr(\eta\text{-}C_5H_5)(CO)_3]^-$ with C_7H_7Br (M = Cr),[3] (b) the oxidation of $Mo(\eta\text{-}C_5H_5)(CO)_3H$ with air (M = Mo),[3] (c) the thermolysis of $Cr(\eta\text{-}C_5H_5)(CO)_3H$ (M = Cr),[4] and (d) the photolysis of $M(\eta\text{-}C_5H_5)(CO)_3CH_2Ph$ (M = Cr, Mo, and W). All suffer from the disadvantage that they proceed by way of intermediates that have to be isolated and purified, and, with the exception of (d), give the desired products in low yields.

The procedure described here is simple, quick, and straightforward. It is carried out in a single reaction vessel. No intermediates need be isolated. The products are usually obtained in high yields (90–98% when M = Mo or W and 60–85% when M = Cr) and are sufficiently pure (>95%) for most purposes.

Procedure

■ **Caution.** *The toxic and inflammable nature of some of the products requires that these reactions be performed in a well-ventilated fumehood.*

All chemicals were purchased from the usual commercial sources. Unless it is stated otherwise, all procedures were carried out under an atmosphere of

---

*Department of Chemistry, University College, Belfield, Dublin 4.
†Department of Chemistry, Iowa State University, Ames, IA 50011.

nitrogen. Diglyme* was dried by refluxing over calcium hydride or sodium benzophenone ketyl for 4 h and distilled from it prior to use. Dicyclopentadiene was cracked as described.[6]

First, 0.575 g (25 mmol) of sodium metal was cut into small pieces and added to a solution of 6 mL (70 mmol) of freshly cracked cyclopentadiene in 150 mL of dry diglyme in a 500-mL, three-necked, round-bottomed flask fitted with a Liebig condenser. The metal dissolved within 30 min to give a pale pink solution of sodium cyclopentadienide. To it was added 25 mmol of $M(CO)_6$ (5.5 g when M = Cr, 6.6 g when M = Mo and 8.8 g when M = W) and the mixture heated at reflux for 40 min.† During this time its color changed to bright yellow. $M(CO)_6$ sublimed into the condenser and was returned to flask at intervals using a glass rod. When no more sublimed, the reaction was complete, and the condenser was replaced by a clean one. The reaction mixture was cooled to room temperature and stirred using a magnetic stirrer. To it were added slowly 5 mL of methanol to destroy any unreacted sodium and 5 mL of water to destroy any unreacted sodium cyclopentadienide. Then a filtered solution in 250 mL of water and 15 mL of acetic acid of 25 g of iron(III) ammonium sulfate, $Fe_2(SO_4)_3 \cdot (NH_4)_2SO_4 \cdot 24H_2O$ or 20 g of hydrated iron(III) sulfate, $Fe_2(SO_4)_3 \cdot 9H_2O$ was added. The color of the reaction mixture changed to green when M = Cr and purple when M = Mo or W, and fine crystals precipitated. When M = Mo or W these were filtered off in air using a Büchner funnel and flask, washed with 250 mL of distilled water, 25 mL of cold methanol, and 25 mL of pentane, and dried. When M = Cr the finely divided product caught fire in air, so the filtration, washing, and drying had to be carried out under nitrogen using a G3 glass filter.

The yields of $M_2(\eta-C_5H_5)_2(CO)_6$ at this stage are 90–98% when M = Mo or W and 60–85% when M = Cr. Product purity is generally >95% and there are no carbonyl-containing impurities.

Anal. Calcd. for $C_{16}H_{10}Mo_2O_6$: C, 39.2; H, 2.0. Found: C, 38.5; H, 1.9. Calcd. for $C_{16}H_{10}W_2O_6$: C, 28.8; H, 1.5. Found: C, 27.5; H. 1.5.

Further purification if required may be effected by recrystallization under nitrogen from acetone (M = Mo and W) or tetrahydrofuran (M = Cr) with very little diminution in yield. The well-formed crystals of the chromium complex that can thus be obtained are no longer pyrophoric, but they are still very air-sensitive.

The procedure may be scaled up with no loss in product yield and without the necessity of increasing the volume of diglyme or iron(III) solution in

*1,1′-oxybis(2-methoxyethane)
†The checkers added 5 mL of hexanes to the solution, which washed any sublimed $M(CO)_6$ back into the reaction flask.

proportion. Cyclopentadiene may be replaced by substituted cyclopentadienes and $M_2(\eta\text{-}C_5H_4R)_2(CO)_6$ have been prepared for R = Me and $PhCH_2$ and M = Cr, Mo, and W. The method cannot be used to obtain indenyl complexes; $[M(\eta\text{-}C_9H_7)(CO)_3]^-$ are formed but are not oxidized to $M_2(\eta^5\text{-}C_9H_7)_2(CO)_6$.[7]

Properties

$Mo_2(\eta\text{-}C_5H_5)_2(CO)_6$ and its W analog are purple crystalline solids which are indefinitely stable in air at room temperature, provided they are stored in the dark. The chromium complex, in contrast, is a green crystalline solid that is very air sensitive. It must be stored under an atmosphere of nitrogen and, preferably, at $-20\,°C$. All three are soluble in the usual organic solvents and insoluble in water. $Cr_2(\eta\text{-}C_5H_5)_2(CO)_6$ reacts with strong donor solvents.[7]

Infrared spectra in chloroform: M = Cr, 1890 (sh), 1913 (sh), 1925 (8.3), 1948 (10), and 2012 (3.9). For M = Mo, 1905 (sh), 1916 (6.3), and 1959 (10). For M = W, 1895 (sh), 1908 (6.3), 1956 (10); cm^{-1} with relative peak heights in parentheses.

^1H NMR spectra: M = Cr; δ 5.84 (toluene-d_8);[8] M = Mo, δ 5.30 ($CDCl_3$); M = W, δ 5.39 ($CDCl_3$).

References

1. G. Wilkinson, *J. Am. Chem. Soc.*, **76**, 209 (1954).
2. R. B. King, *Organomet. Synth.*, **1**, 109 (1965).
3. R. B. King and F. G. A. Stone, *Inorg. Synth.*, **7**, 115 (1963).
4. S. A. Keppie and M. F. Lappert, *J. Chem. Soc. A*, **1971**, 3216.
5. D. S. Ginley, C. R. Bock, and M. S. Wrighton, *Inorg. Chim. Acta*, **23**, 85 (1977).
6. M. Korach, D. R. Nielsen, and W. H. Rideout, *Org. Synth.*, Coll. Vol. 5, **414**, (1973).
7. R. Birdwhistell, P. Hackett, and A. R. Manning, *J. Organomet. Chem.*, **157**, 239, (1978).
8. R. D. Adams, D. E. Collins, and F. A. Cotton, *J. Am. Chem. Soc.*, **96**, 749 (1974).

40. CYCLOPENTADIENYL METAL CARBONYL DIMERS OF MOLYBDENUM AND TUNGSTEN

Submitted by M. DAVID CURTIS* and MICHAEL S. HAY*
Checked by MOON GUN CHOI† and ROBERT J. ANGELICI†

The thermolysis of $Mo(CO)_6$ with neat dicyclopentadiene, or the oxidation of $(C_5H_5)Mo(CO)_3H$ with air[2a] or Fe^{3+} [2b] have been reported as routes to

---

*Department of Chemistry, University of Michigan, Ann Arbor, MI 48109–1055.
†Department of Chemistry, Iowa State University, Ames, IA 50011.

$(C_5H_5)_2Mo_2(CO)_6$(Mo—Mo). The first two methods[1, 2a] produce low to moderate yields of the hexacarbonyl dimer when run on a large (>20g) scale. The method presented herein for the preparation of $(C_5H_5)_2Mo_2(CO)_6$ gives consistently high yields of the dimer in a two-step, one-pot synthesis. This procedure is an improved variation of the synthesis reported by Keppie and Lappert.[3]

The methods for the preparation of the triply bonded dimers, $(C_5H_5)_2M_2(CO)_4$, of molybdenum and tungsten involve either the photolysis or thermolysis of $(C_5H_5)_2M_2(CO)_6$.[4, 5] The photolysis route is appropriate only for the preparation of very small amounts of $(C_5H_5)_2Mo_2(CO)_4$. An improved version of the thermolysis route is presented that shortens the reaction time and produces $(C_5H_5)_2M_2(CO)_4$ in high yield and purity.[6]

These compounds are of interest in catalysis,[7] as precursors to sulfided clusters,[8] and in preparative chemistry due to the broad reactivity of the metal–metal triple bond.[9]

All operations in the procedures described must be carried out under dry and oxygen-free conditions using standard Schlenk techniques. Solvents should be dried and purified by conventional methods and stored under nitrogen prior to use.*

A. Hexacarbonylbis(η^5-cyclopentadienyl)dimolybdenum(Mo—Mo)

$$Mo(CO)_6 + 3CH_3CN \xrightarrow{\Delta} Mo(CO)_3(CH_3CN)_3 + 3CO$$

$$2Mo(CO)_3(CH_3CN)_3 + 2C_5H_6 \xrightarrow{\Delta} (C_5H_5)_2Mo_2(CO)_6 + H_2 + 6CH_3CN$$

Procedure

Molybdenum hexacarbonyl (20.0 g, 0.076 mol) is added to a 500-mL Schlenk flask fitted with a reflux condenser (straight-bore Liebig type), a large magnetic stirring bar, and a glass rod [to prevent $Mo(CO)_6$ from plugging the condenser]. The glass rod should be about one-half the inner diameter of the condenser and be long enough to extend from the bottom of the flask 8–10 cm into the condenser. The system is filled with nitrogen and 200 mL of dry, degassed CH_3CN is added through the top of the condenser. A three-way stopcock is then fitted to the top of the condenser and the stopcock is attached to an oil bubbler and a source of nitrogen in order to maintain a

---
*The checkers carried out all of the reactions on one-tenth the scale described.

blanket of nitrogen over the reaction vessel. The lower nitrogen inlet is closed and the solution refluxed for 4 h. Occasionally, it may be necessary to wash the $Mo(CO)_6$ that sublimes into the condenser back into the flask with small portions of acetonitrile.

After the solution has cooled, the nitrogen inlet on the Schlenk flask is reopened and a good flow of nitrogen is established through the flask. The condenser is then replaced by a connection to a cold trap while maintaining the flow of nitrogen. The cold trap is immersed in liquid nitrogen and connected to a vacuum pump. The solvent is removed from the solution under vacuum at room temperature, leaving the $Mo(CO)_3(CH_3CN)_3$ as a yellow powder. The $Mo(CO)_3(CH_3CN)_3$ is quite air sensitive, so appropriate precautions must be observed.

Freshly distilled cyclopentadiene (50 mL) is then added directly to the $Mo(CO)_3(CH_3CN)_3$, the reflux condenser with oil bubbler and nitrogen inlet is replaced, and the mixture is refluxed for 2 h. Care should be taken during the initial period of refluxing as the evolution of hydrogen may become vigorous and carry the solution up the condenser. On cooling, $(C_5H_5)_2Mo_2(CO)_6$ crystallizes and is collected by filtration through a fritted funnel. The product is washed with hexane to remove cyclopentadiene dimer, and air-dried. The yield is 16.5 g (89%). Some additional $Cp_2Mo_2(CO)_6$ may be recovered from the initial filtrate by concentrating, adding hexane, and cooling to 0°C. Scale-up of this procedure to 50–100 g quantities of $Mo(CO)_6$ is readily achieved.

Properties

The compound $(C_5H_5)_2Mo_2(CO)_6$ is a dark red crystalline solid that is air stable in the solid state, but in solution slowly decomposes when exposed to air. The IR spectrum $v(CO)$ in CH_2Cl_2 shows strong absorptions at 1910 and 1957 cm^{-1}. The ^1H NMR spectrum in benzene-d_6 shows a single resonance at 4.67 (s) ppm.

B. Tetracarbonylbis(η^5-cyclopentadienyl)dimolybdenum(Mo≡Mo)

$$(C_5H_5)_2Mo_2(CO)_6 \xrightarrow[\Delta]{\text{diglyme}} (C_5H_5)_2Mo_2(CO)_4 + 2CO$$

Procedure

A 1000-mL, three-necked, round-bottomed flask is fitted with a Friedrichs condenser, a stopper, and a nitrogen inlet that impinges above the level of the liquid. (A long, 18-gauge needle inserted through a rubber septum will suffice in place of the nitrogen inlet, although this arrangement is not as efficient.) A large, magnetic stirring bar and 40 g (0.082 mol) of $(C_5H_5)_2Mo_2(CO)_6$ are placed in the flask, which is then evacuated and backfilled with nitrogen. Diglyme (400 mL, previously distilled from sodium–benzophenone ketyl) is then transferred into the flask. The mixture is heated to reflux while maintaining a slow nitrogen sweep across the surface of the solution and out the condenser until carbon monoxide evolution is complete. The rate of carbon monoxide evolution is checked periodically by turning off the nitrogen flow and observing the oil bubbler. The reaction is complete in 1.5–3 h depending on the efficiency of nitrogen sweep. The color of the solution changes from cherry red to brick red with an orange cast.

■ **Caution.** *Diglyme vapors are extremely flammable.*

The reaction is worked up in the following manner. While the solution is still warm, the condenser is removed while maintaining a flow of nitrogen through the flask. The reaction vessel is then connected to a cold trap, and the diglyme is distilled under vacuum into the cold trap. The residue is dissolved in CH_2Cl_2 (~ 150 mL) and filtered through a fritted funnel (1-in. diameter) containing 1 in. of dry, degassed Florisil. The Florisil is washed with CH_2Cl_2 (~ 100 mL) until the washings are lightly colored. Some black material is left at the top of the Florisil. After adding 75 mL of hexane, the volume of the solution is reduced under vacuum to approximately 100 mL, after which the mixture is cooled in an ice bath to 0°C. The crystalline $(C_5H_5)_2Mo_2(CO)_4$ is collected by filtration under nitrogen and the filtrate concentrated further to yield a second crop of crystals. The yield is 29.9 g (84%).

Anal. Calcd. for $C_{14}H_{10}Mo_2O_4$: C, 38.74; H, 2.33; Mo, 44.20. Found: C, 38.70; H, 2.43; Mo, 44.51.

The use of $(C_5H_5)_2W_2(CO)_6$ in place of $(C_5H_5)_2Mo_2(CO)_6$ allows comparable yields of $(C_5H_5)_2W_2(CO)_4$ to be prepared by essentially the same procedure as stated above. Longer reaction times are usually needed for complete conversion, however; so the progress of the decarbonylation should be monitored by taking IR or NMR spectra of small aliquots.

Anal. Calcd. for $C_{14}H_{10}O_4W_2$: C, 27.57; H, 1.65; W, 60.29. Found: C, 27.40; H, 1.56; W. 60.44.

Properties

The complex $(C_5H_5)_2Mo_2(CO)_4$ is a dark crystalline compound. In the solid state the dimer is air stable for short periods of time but should be stored under nitrogen. Solutions of $(C_5H_5)_2Mo_2(CO)_4$ slowly decompose when exposed to the atmosphere. The complex also reacts with chloroform and carbon tetrachloride. The IR spectrum $v(CO)$ in CH_2Cl_2 shows absorptions at 1958 (m), 1890 (s), and 1857 (s) cm^{-1}. The ^{1}H-NMR spectrum in benzene-d_6 shows a resonance at 4.61 (s) ppm.

The compound $(C_5H_5)_2W_2(CO)_4$ should be handled in the same manner as the molybdenum analog. The IR spectrum $v(CO)$ in CH_2Cl_2 shows absorptions at 1954 (m) cm^{-1}, 1892 (s) cm^{-1}, and 1833 (s) cm^{-1}. The ^{1}H-NMR spectrum in benzene-d_6 shows a resonance at 4.87 (s) ppm.

References

1. R. B. King, *Organometallic Syntheses*, Academic Press, New York, 1965, p. 109.
2. (a) R. B. King and F. G. A. Stone, *Inorg. Synth.*, **7**, 107 (1963); (b) R. Birdwhistell, P. Hackett, and A. R. Manning, *J. Organomet. Chem.*, **157**, 239 (1978).
3. S. A. Keppie and F. Lappert, *J. Chem. Soc. A*, **1971**, 3216.
4. D. A. Ginley, C. R. Bock, and M. S. Wrighton, *Inorg. Chim. Acta*, **23**, 85 (1977).
5. M. D. Curtis and R. J. Klinger, *J. Organomet. Chem.*, **161**, 23 (1978).
6. M. D. Curtis, N. A. Fotinos, L. Messerle, and A. P. Sattelberger, *Inorg. Chem.*, **22**, 1559 (1983).
7. R. F. Gerlach, D. N. Duffy, and M. D. Curtis, *Organometallics*, **2**, 1172 (1983).
8. P. D. Williams and M. D. Curtis, *Inorg. Chem.*, **22**, 2661 (1983).
9. M. D. Curtis, *Polyhedron*, **6**, 759 (1987).

41. PENTACARBONYLMANGANESE HALIDES

The pentacarbonylmanganese halides have been known for some time[1,2] but are still the subject of intense interest. Numerous vibrational analyses have been carried out,[3] as well as synthetic and kinetic studies of carbonyl substitution reactions.[4] Halide substitution, which requires elevated temperatures,[5] leads to charged species. Diazocyclopentadienes insert into the manganese–halogen bond.[6]

The best method for preparation of pentacarbonylchloromanganese is the reaction of chlorine with dimanganese decacarbonyl in carbon tetrachloride solution. Earlier versions of this procedure[2] often give low yields; however, the present version is reliable and produces a good yield.

The action of CCl_4 alone on $Mn_2(CO)_{10}$ also gives $MnCl(CO)_5$,[7] but the reaction is too slow to be of synthetic utility. Recent reports[8] suggest that photolysis enhances both the rate and yield of this reaction.

A commonly used preparation of $MnBr(CO)_5$ is the reaction of $Mn_2(CO)_{10}$ with Br_2 in CCl_4.[2,9] However, the product is frequently contaminated with $MnCl(CO)_5$ owing to reaction of the decacarbonyl with the solvent.[7] A substantial increase in the yield and purity of this compound can be achieved by the use of CS_2 as the reaction solvent, as in the procedure given here.

The iodo analogue has been prepared by thermal[1,7] and photochemical[8] reactions of $Mn_2(CO)_{10}$ with I_2. The former method, which is simple to perform, is described in Section D. The reaction of $MnH(CO)_5$ and I_2 also has been reported to give $MnI(CO)_5$.[10] Another procedure utilizes the nucleophilic character of the anion $Mn(CO)_5^-$. This reaction, which forms the basis of the procedure in Section C, was mentioned briefly some years ago,[11] but no experimental detail was provided. Advantages of the method are speed, high yield, and absence of product contamination by unreacted $Mn_2(CO)_{10}$.

Commercial $Mn_2(CO)_{10}$ should be purified by sublimation (70°, 0.05 torr) before use in these preparations. Solutions of $MnX(CO)_5$ must not be heated during evaporations, owing to the facile decarbonylation of these compounds to $[MnX(CO)_4]_2$[2,12] (especially for X = Cl). Evaporations can be assisted by immersion of the flask in a *room-temperature* water bath.

A. PENTACARBONYLCHLOROMANGANESE, $MnCl(CO)_5$

$$Mn_2(CO)_{10} + Cl_2 \rightarrow 2MnCl(CO)_5$$

Submitted by KENNETH J. REIMER* and ALAN SHAVER†
Checked by MICHAEL H. QUICK‡ and ROBERT J. ANGELICI‡

The reaction is conducted in a 100-mL, three-necked, round-bottomed flask fitted with a nitrogen inlet (attached to a nitrogen line equipped with a mercury bubbler),[13] a magnetic stirrer, and an equipressure dropping funnel. Finely ground $Mn_2(CO)_{10}$ (4.0 g, 0.01 mol) is dissolved in a minimum of degassed carbon tetrachloride at 0° in the nitrogen-filled flask.

■ **Caution.** *Carbon tetrachloride and chlorine gas are highly toxic and should be handled in an efficient fume hood.*

Another sample of carbon tetrachloride (25 mL) in a dropping funnel is saturated by means of a stream of chlorine and the yellow solution is added dropwise over 30 min to the cooled, stirred solution of $Mn_2(CO)_{10}$. As the addition proceeds, some $MnCl(CO)_5$ begins to precipitate. After complete addition, the reaction mixture is allowed to warm to room temperature and is then stirred for 4 h. The yellow precipitate is filtered in air and washed several times with carbon tetrachloride. The yield of crude product is 4.04 g (86%).

---

*Department of Chemistry, Royal Roads Military College, FMO, Victoria, B.C., Canada, V0S 1B0.
†Department of Chemistry, McGill University, Montreal, Quebec, Canada, H3A 2K6.
‡Department of Chemistry, Iowa State University, Ames, IA 50011.

The compound is contaminated with only small amounts of $[MnCl(CO)_4]_2$ and white insoluble material. These impurities are negligible and usually no further purification is undertaken. If pure $MnCl(CO)_5$ is required, it is easily obtained by sublimation (40°, 0.1 torr, yield 65%) [based on $Mn_2(CO)_{10}$].

Anal. Calcd. for $MnCl(CO)_5$: C, 26.06; Cl, 15.39. Found: C, 26.08; Cl, 15.44.

B. BROMOPENTACARBONYLMANGANESE, $MnBr(CO)_5$

$$Mn_2(CO)_{10} + Br_2 \rightarrow 2MnBr(CO)_5$$

Submitted by MICHAEL H. QUICK* and ROBERT J. ANGELICI*
Checked by KENNETH J. REIMER† and ALAN SHAVER‡

A Schlenk tube of about 100-mL capacity, equipped with a magnetic stirring bar, is flushed with N_2. Dimanganese decacarbonyl (2.00 g, 5.13 mmol) and CS_2 (50 mL) are then added, under a nitrogen flush, and the mixture is stirred for about 10 minutes under nitrogen.

■ **Caution.** *Carbon disulfide is highly flammable and toxic and should be vented through an efficient fume hood.*

Some of the $Mn_2(CO)_{10}$ remains undissolved, but this does not affect the reaction. A solution of Br_2 (0.35 mL, 1.0 g, 6.3 mmol) in 20 mL of CS_2 is then added under nitrogen from an equipressure dropping funnel over a period of 20–30 min; some of the product begins to precipitate during this time. Stirring is continued for an additional hour. The dark-red mixture is then evaporated under reduced pressure to give an orange powder.

The crude product is stirred with 150 mL of CH_2Cl_2, and the orange solution is filtered. Hexane (60 mL) is added, and the solution is slowly evaporated under reduced pressure (50–60 torr) to about 30 mL. The yellow-orange, microcrystalline $MnBr(CO)_5$ is filtered off, washed with cold (0°) pentane, and dried *in vacuo*. Yield 2.53 g (90%). The product thus obtained is sufficiently pure for most purposes, but it can be sublimed (50°, 0.05 torr) with only a slight decrease in yield if further purification is desired.

The reaction may also be carried out in dichloromethane or cyclohexane. However, the yield and purity of the product are generally superior in CS_2.

Anal. Calcd. for $MnBr(CO)_5$: C, 21.8; Br, 29.1. Found: C, 21.7; Br, 28.7.

*Department of Chemistry, Iowa State University, Ames, IA 50011.
†Department of Chemistry, Royal Roads Military College, FMO, Victoria, B.C., Canada, V0S 1B0.
‡Department of Chemistry, McGill University, Montreal, Quebec, Canada, H3A 2K6.

C. PENTACARBONYLIODOMANGANESE, MnI(CO)$_5$

$$Mn_2(CO)_{10} + 2Na \rightarrow 2Na[Mn(CO)_5]$$
$$Na[Mn(CO)_5] + I_2 \rightarrow MnI(CO)_5 + NaI$$

Submitted by MICHAEL H. QUICK* and ROBERT J. ANGELICI*
Checked by KENNETH J. REIMER† and ALAN SHAVER‡

The reaction is conducted in a 100-mL three-necked amalgam reduction flask, which is described elsewhere.[14] The mixture is agitated with a paddle stirrer driven by a small electric motor. (The checkers found a Teflon-coated stirring bar and magnetic stirrer adequate.) The apparatus is flushed with nitrogen, after which 6 mL of 1% Na(Hg) and a solution of $Mn_2(CO)_{10}$ (2.00 g, 5.13 mmol) in 40 mL of tetrahydrofuran (distilled from $LiAlH_4$; see Ref. 15 for the distillation procedure) are added. The mixture is stirred vigorously for 30–40 min, after which the excess amalgam is drained off. The flask and solution are then "washed" for a few minutes by vigorous stirring with 3–4 mL of fresh mercury, which is then removed as before. A somewhat cloudy greenish-gray solution of $Na[Mn(CO)_5]$ is obtained.[14]

A solution of I_2 (2.65 g, 10.4 mmol) in 20 mL of THF (tetrahydrofuran) is then added from an equipressure dropping funnel to the stirred solution over a period of 15–20 min. The solution, now clear and deep red-orange, is stirred for an additional 5 min and then evaporated to dryness under reduced pressure. The residue is extracted with 120 mL of 1 : 1 hexane–dichloromethane, the mixture is filtered, and the filtrate is evaporated to dryness. Sublimation (50°, 0.05 torr) or recrystallization from hexane (about 40 mL per gram of crude product) at $-20°$ gives red-orange crystals of pure $MnI(CO)_5$. Yield 2.50–2.72 g (75–82%).

Anal. Calcd. for $MnI(CO)_5$: C, 18.6; I, 39.4. Found: C, 18.4; I, 40.3.

---

*Department of Chemistry, Iowa State University, Ames, IA 50011.
†Department of Chemistry, Royal Roads Military College, FMO, Victoria, B.C., Canada, V0S 1B0.
‡Department of Chemistry, McGill University, Montreal, Quebec, Canada, H3A 2K6.

D. PENTACARBONYLIODOMANGANESE, ALTERNATE PROCEDURE

$$Mn_2(CO)_{10} + I_2 \rightarrow 2MnI(CO)_5$$

Submitted by KENNETH J. REIMER* and ALAN SHAVER†
Checked by MICHAEL H. QUICK‡ and ROBERT J. ANGELICI‡

Stoichiometric quantities of $Mn_2(CO)_{10}$ and I_2 are introduced into a thick-walled Carius tube (the checkers used a capable glass pressure bottle[16]), mixed thoroughly, and cooled to 0° and the tube is sealed *in vacuo*. The mixture is uniformly heated in an oven at 90° until the color is uniform (12 h is convenient for a reaction employing about 1 g of $Mn_2(CO)_{10}$).

■ **Caution.** *Care should be taken when handling sealed reaction vessels.*[1] *Face and hands should be protected from the possibility of flying glass and chemicals.*

The residue is placed in a sublimation apparatus fitted with a water-cooled probe. Sublimation (40°, 0.1 torr) separates the product from a nonvolatile powder, but the sublimate is usually contaminated with $Mn_2(CO)_{10}$. Purification may be accomplished by fractional sublimation or, more conveniently, by recrystallization from pentane to give an overall 50% yield of ruby-red crystals.

Properties of $MnX(CO)_5$, X = Cl, Br, I

The yellow $MnCl(CO)_5$ is slightly to moderately soluble in nonpolar organic solvents. Although stable in air, prolonged exposure to light results in the formation of $[MnCl(CO)_4]_2$. In a closed vessel an equilibrium exists between the monomer and dimer in benzene[17] and pentane.[6b] Unstoppered reaction vessels allow loss of CO and subsequent dimer formation. Heating solutions or solid samples of $MnCl(CO)_5$ accelerates dimer formation.

Bromopentacarbonylmanganese is an air-stable yellow-orange crystalline solid. Unlike the chloro analog, $MnBr(CO)_5$ is not particularly light sensitive. Its properties, such as solubility in organic solvents and ease of decarbonylation,[12] are intermediate between those of the chloro and iodo compounds.

The red crystalline (orange powder) $MnI(CO)_5$ is more stable in common organic solvents than the chloro analogue. It dimerizes slowly in solution.

---

*Department of Chemistry, Royal Roads Military College, FMO, Victoria, B.C., Canada, V0S 1B0.
†Department of Chemistry, McGill University, Montreal, Quebec, Canada, H3A 2K6.
‡Department of Chemistry, Iowa State University, Ames, IA 50011.

TABLE I. IR Spectra in Carbonyl Stretching Region

Compound	$\nu(CO)$ (cm^{-1}) in CCl$_4$[a]			
MnCl(CO)$_5$	2140 (w)	2054 (s)	2021 (vw)	1998 (m)
MnBr(CO)$_5$	2134 (w)	2051 (s)	2020 (vw)	2000 (m)
MnI(CO)$_5$	2127 (w)	2045 (s)	2016 (vw)(sh)	2005 (m)

[a] Referenced to 2147 cm^{-1} band of CO gas.

The compound is best stored at 5°, since iodine is liberated on prolonged standing at room temperature.

The infrared spectrum in the carbonyl stretching region is very useful in characterizing these complexes (Table I). Three infrared active bands are predicted;[18] however, limited solubility may preclude observation of the weaker bands. Dimer formation is easily detected by the presence of characteristic bands.[12]

References

1. E. O. Brimm, M. A. Lynch, Jr., and W. J. Sesny, *J. Am. Chem. Soc.*, **76**, 3831 (1954).
2. E. W. Abel and G. Wilkinson, *J. Chem. Soc.*, **1959**, 1501.
3. L. M. Haines and M. H. B. Stiddard, *Adv. Inorg. Radiochem.*, **12**, 53 (1969) and references therein.
4. (a) G. R. Dobson, *Acc. Chem. Res.*, **9**, 300 (1976); (b) D. A. Brown, *Inorg. Chim. Acta*, **1**, 35 (1967); (c) R. J. Angelici, *Organomet. Chem. Rev.*, **3**, 173 (1968).
5. R. H. Reimann and E. Singleton, *J. Organomet. Chem.*, **59**, C24 (1973).
6. (a) K. J. Reimer and A. Shaver, *Inorg. Synth.*, **20**, 188 (1980); (b) K. J. Reimer and A. Shaver, *Inorg. Chem.*, **14**, 2707 (1975), *J. Organomet. Chem.*, **93**, 239 (1975).
7. J. C. Hilemann, D. K. Huggins, and H. D. Kaesz, *Inorg. Chem.*, **1**, 933 (1962).
8. (a) S. A. Hallock and A. Wojcicki, *J. Organomet. Chem.*, **54**, C 27 (1973); (b) M. S. Wrighton and D. S. Ginley, *J. Am. Chem. Soc.*, **97**, 2065 (1975).
9. R. B. King, *Organomet. Synth.*, **1**, 174 (1965).
10. I. G. DeJong, S. C. Srinivasan, and D. R. Wiles, *J. Organomet. Chem.*, **26**, 119 (1971).
11. M. F. Farona and L. M. Frazee, *J. Inorg. Nucl. Chem.*, **29**, 1814 (1967).
12. F. Zingales and U. Satorelli, *Inorg. Chem.*, **6**, 1243 (1967).
13. D. F. Shriver, *The Manipulation of Air-Sensitive Compounds*, McGraw-Hill, New York, 1969.
14. (a) R. B. King, *Organomet. Synth.*, **1**, 149 (1965); (b) R. B. King and F. G. A. Stone, *Inorg. Synth.*, **7**, 198 (1963).
15. *Inorg. Synth.*, **12**, 111, 317 (1970).
16. See Ref. 13, p. 157.
17. C. H. Bamford, J. W. Burley, and M. Coldbeck, *J. Chem. Soc. Dalton Trans.*, **1972**, 1846.
18. F. A. Cotton, *Inorg. Chem.*, **3**, 702 (1964) and references therein.

42. PENTACARBONYLRHENIUM HALIDES

Submitted by STEVEN P. SCHMIDT,* WILLIAM C. TROGLER,† and FRED BASOLO*
Checked by MICHAEL A. URBANCIC‡ and JOHN R. SHAPLEY‡

The pentacarbonylrhenium halides, first prepared by Hieber,[1,2] are starting materials for the syntheses of many novel rhenium carbonyl compounds.[3-7] Photochemical,[8,9] vibrational,[10,11] and kinetic[12-15] properties of these molecules have been studied. A rhenium carbonyl halide–alkyl aluminum halide system polymerizes acetylene and is a useful olefin-metathesis catalyst.[16-18]

One member of the halo family, bromopentacarbonylrhenium, is conveniently prepared by direct reaction of an excess of bromine with dirhenium decacarbonyl. Existing literature methods report high yields and facile syntheses.[4,10,19] The chloro analog has been prepared in a similar manner by the action of chlorine on dirhenium decacarbonyl in carbon tetrachloride. Although no explicit mention has been made in the literature of this procedure for the synthesis of $[ReCl(CO)_5]$, it is a straightforward modification of the preparation reported[20] for pentacarbonylchlorotechnetium. This synthesis readily yields the product in acceptable purity and yield.

An alternative procedure involves the photochemical cleavage of the rhenium–rhenium bond in dirhenium decacarbonyl in the presence of a chlorocarbon solvent.[21,22] This method suffers from contamination by the tetracarbonylchlororhenium dimer, which forms in large quantities during photolysis.[21,22] To prevent formation of dimer, the photolysis is performed under an atmosphere of carbon monoxide. The yield of $ReCl(CO)_5$ is essentially quantitative using this procedure.

Existing preparations of the iodo analog involve high temperature[11,23,24] and high pressure.[1,2,25] Photolysis in the presence of iodine under a carbon monoxide atmosphere provides an alternative method of synthesis. Advantages of the photochemical reaction are speed, high yield, and the absence of product contamination by dimer. All reactions may be monitored conveniently by infrared spectroscopy.

---

*Department of Chemistry, Northwestern University, Evanston, IL 60201.
†Present Address: Department of Chemistry, University of California, San Diego, La Jolla, CA 92093.
‡Chemistry Department, University of Illinois, Urbana, IL 61801.

Procedure

A. PENTACARBONYLCHLORORHENIUM, [ReCl(CO)$_5$]

$$[Re_2(CO)_{10}] \xrightarrow[CCl_4]{h\nu} 2[ReCl(CO)_5]$$

■ **Caution.** *Carbon monoxide is a toxic gas and should be used in an efficient fume hood. Avoid looking directly at the ultraviolet light source. Protective goggles should be worn. Carbon tetrachloride is a carcinogen and must be handled with care in a hood.*

Reagent-grade CCl$_4$ is stirred for 2 h over P$_4$O$_{10}$ (50 mL CCl$_4$/0.5 g P$_4$O$_{10}$) and then distilled (25 mL) into a 50-mL Schlenk flask (Pyrex), equipped with a Teflon-coated stir bar under an N$_2$ atmosphere. Dirhenium decacarbonyl (Strem Chemicals—used as received—0.97 g, 1.49 mmol) is dissolved in the CCl$_4$. The solution is then saturated with CO (Matheson, technical grade) by bubbling with a syringe needle (through a rubber septum) for 10 min.

The photolysis set-up employs a standard 450-W medium-pressure Hanovia mercury arc lamp surrounded by a water-cooled immersion well. The light is directed on the Pyrex Schlenk flask by partially wrapping the well with aluminum foil. The distance from lamp to flask is approximately 3.5 cm. A small magnetic stirrer placed below the flask provides stirring, which is very important to ensure complete reaction and to prevent a buildup of precipitate on the walls of the flask. During photolysis, white [ReCl(CO)$_5$] crystallizes from solution. Photolysis is terminated when the IR bands due to [Re$_2$(CO)$_{10}$] (2073, 2016, and 1976 cm^{-1}) disappear (about 140 min). Aliquots for spectroscopic analysis are withdrawn from the reaction mixture by syringe. A slight positive pressure may accumulate during the reaction.

After solvent is removed under continuous vacuum (0.1 mm) at room temperature, the crude, slightly brownish residue is transferred to a sublimator and sublimed at 80–85° and 0.1 mm. Yield 1.10 g [94.1%, based on Re$_2$(CO)$_{10}$] [Checkers' yield: 85%].

Anal. Calcd. for [ReCl(CO)$_5$]: C, 16.60; Cl, 9.80. Found: C, 16.63; Cl, 9.84.

B. PENTACARBONYLCHLORORHENIUM (ALTERNATIVE PROCEDURE)

$$[Re_2(CO)_{10}] + Cl_2 \xrightarrow{CCl_4} 2[ReCl(CO)_5]$$

Reagent-grade CCl$_4$ is stirred for 2 h over P$_4$O$_{10}$ (50 mL CCl$_4$/0.5 g P$_4$O$_{10}$), and then distilled (25 mL) into a 50-mL Schlenk flask equipped with a Teflon-coated stir bar under a N_2 atmosphere. The $[Re_2(CO)_{10}]$ (0.32 g, 0.49 mmol) is dissolved in the CCl_4, and the solution is saturated with a stream of chlorine from a syringe needle for one minute. The solution color changes from colorless to yellow-green, and, after 2 min of stirring, a white precipitate forms. The reaction is terminated when the IR bands due to $[Re_2(CO)_{10}]$ disappear ($\sim$140 min).

The solvent and excess chlorine are removed under continuous vacuum (0.1 mm) at room temperature. The white solid is transferred to a sublimator and sublimed at 80–85° and 0.1 mm.* Yield 0.30 g (84.4%, based on $[Re_2(CO)_{10}]$ [Checkers' yield: 75%].

Anal. Calcd. for $[ReCl(CO)_5]$: C, 16.60; Cl, 9.80. Found: C, 16.56; Cl, 9.57.

C. BROMOPENTACARBONYLRHENIUM, [ReBr(CO)$_5$]

$$[Re_2(CO)_{10}] + Br_2 \xrightarrow{\text{hexane}} 2[ReBr(CO)_5]$$

Hexane (30 mL freshly distilled from sodium benzophenone ketyl) is transferred to a 50-mL Schlenk flask equipped with a Teflon-coated stir bar. The $[Re_2(CO)_{10}]$ (0.89 g, 1.36 mmol) is added under a stream of N_2, and bromine (0.24 g, 1.50 mmol) is added to the solution by means of a syringe.† Immediately on stirring at room temperature, a precipitate forms in the flask. Stirring is continued for 30 min, and almost all of the orange bromine color disappears, along with the IR bands of $[Re_2(CO)_{10}]$.

Volatiles are removed under continuous vacuum (0.1 mm) at room temperature. The white powder is transferred to a sublimator and sublimed at 85–90° and 0.2 mm. Yield: 1.04 g (91.3%, based on $[Re_2(CO)_{10}]$).

Anal. Calcd. for $[ReBr(CO)_5]$: C, 14.79; Br, 19.67. Found: C, 14.98; Br, 17.72.

---

*Alternatively, the checkers have found that the compound can be purified by dissolving the crude solid in a minimum of warm acetone, adding two volumes of methanol, and recrystallizing at $-10°$.

† The checkers have found that in dichloromethane, $[Re_2(CO)_{10}]$ reacts instantaneously with bromine. Hence, the carbonyl compound can be "titrated" with bromine until the slightest yellow color persists. This minimizes the amount of residual bromine that must be removed. After solvent removal, the crude solid can be recrystallized by dissolving it in a minimum of warm acetone, adding two volumes of methanol, and cooling to $-10°$.

D. PENTACARBONYLIODORHENIUM, [ReI(CO)$_5$]

$$[Re_2(CO)_{10}] + I_2 \xrightarrow[\text{hexane}]{hv} 2[ReI(CO)_5]$$

■ **Caution.** *See Section A.*

Hexane (30 mL freshly distilled from sodium benzophenone ketyl) is transferred to a 50-mL Pyrex Schlenk flask containing a Teflon-coated stir bar. The [Re$_2$(CO)$_{10}$] (0.74 g, 1.13 mmol) is added under a N$_2$ stream and dissolved. An excess of iodine (0.435 g, 1.71 mmol) is then added under a N$_2$ stream, and the solution is purged with CO for 10 min.

The photolysis procedure follows that given in the chloro derivative synthesis (Section 11.A). During irradiation, a white precipitate of [ReI(CO)$_5$] appears in the violet solution. The reaction is terminated when the IR bands due to [Re$_2$(CO)$_{10}$] (2073, 2016, and 1976 cm^{-1}) disappear (about 140 min).*

The solvent is removed under continuous vacuum (0.1 mm) at *room temperature*, and most of the excess I$_2$ can be pumped off at this point. To remove residual amounts of I$_2$,† the crude solid is transferred to a sublimator and gently heated from 25 to 50° under continuous vacuum with no working cold finger. When the solid appears nearly white, the temperature is raised to 85°, and the pure [Re(CO)$_5$I] is sublimed at 0.1 mm. The compound sublimes to a milky-white crystalline solid. Yield: 0.82 g (80.2%) based on [Re$_2$(CO)$_{10}$].

Anal. Calcd. for [ReI(CO)$_5$]: C, 13.25; I, 28.00. Found: C, 13.37; I, 28.18.

Properties

Pentacarbonylrhenium halides are white, crystalline solids that exhibit moderate solubility in nonpolar organic solvents. The iodo derivative is most soluble, and the chloro analog least soluble. All the complexes are stable in air at room temperature; however, heating the solutions or solids results in formation of the tetracarbonylrhenium halide dimers.

---

*The checkers have found that this reaction was considerably slower than that for the synthesis of the chloro derivative, and thus required a longer period of irradiation (320 min).

†Alternatively, the residual I$_2$ can be removed by dissolving it in 10 mL of ethanol, and then adding 0.30 g of Na$_2$S$_2$O$_3$ in 10 mL of water. The crude product is then filtered, washed with water, and dried under vacuum before being sublimed. If desired, the sublimate can be recrystallized by dissolving it in a minimum of THF, adding two volumes of methanol, and cooling to $-10°$.

TABLE I. Carbonyl Stretching Modes (cm^{-1})[a]

Compound		In CCl$_4$		
[Re(CO)$_5$Cl]	NO[b]	2049 s	2020 w	1985 m
[Re(CO)$_5$Br]	2154 w	2048 s	2018 s	1987 m
[Re(CO)$_5$I]	2149 w	2047 s	2018 w	1991 s
[Re(CO)$_4$Cl]$_2$[26]	2125 w	2040 s	2005 m	1990 s
[Re(CO)$_4$Br]$_2$[26]	2115 w	2037 s	2000 m	1960 m
[Re(CO)$_4$I]$_2$[26]	2111 w	2032 s	2004 m	1968 m
[Re$_2$(CO)$_{10}$]	2073	2016	1976	

[a]Calibrated against polystyrene.
[b]NO = not observed.

As mentioned previously, the infrared spectrum in the carbonyl stretching region of the reaction mixtures is very useful in determining the extent of reaction. The presence of dimeric halide impurities is also easily detected (Table I).

Note. Weak peaks at higher frequency may not appear in solution IR spectra as a result of limited solubility.

References

1. W. Hieber and H. Schulten, *Z. anorg. allg. Chem.*, **243**, 164 (1939).
2. W. Hieber, R. Schuh, and H. Fuchs, *Z. anorg. allg. Chem.*, **248**, 243 (1941).
3. W. Hieber and W. Opavsky, *Chem. Ber.*, **101**, 2966 (1968).
4. R. J. Angelici and A. E. Kruse, *J. Organomet. Chem.*, **22**, 461 (1970).
5. R. B. King and R. H. Reimann, *Inorg. Chem.*, **15**, 179 (1976).
6. K. P. Darst and C. M. Lukehart, *J. Organomet. Chem.*, **171**, 65 (1979).
7. L. H. Staal, A. Oskam, and K. Vrieze, *J. Organomet. Chem.*, **170**, 235 (1979).
8. D. R. Tyler and D. P. Petrylak, *Inorg. Chim. Acta*, **53**, L185 (1981).
9. D. M. Allen, A. Cox, T. J. Kemp, Q. Sultana, and R. B. Pitts, *J. Chem. Soc. Dalton Trans.*, **1976**, 1189.
10. H. D. Kaesz, R. Bau, D. Hendrickson, and J. M. Smith, *J. Am. Chem. Soc.*, **89**, 2844 (1967).
11. G. Keeling, S. F. A. Kettle, and I. Paul, *J. Chem. Soc. Dalton Trans.*, **1971**, 3143.
12. F. Zingales, M. Graziani, F. Faraone, and U. Belluco, *Inorg. Chim. Acta*. **1**, 172 (1967).
13. D. A. Brown and R. T. Sane, *J. Chem. Soc. A*, **1971**, 2088.
14. M. J. Blandamer, J. Burgess, S. J. Cartwright, and M. Dupree, *J. Chem. Soc. Dalton Trans.*, **1976**, 1158.
15. J. Burgess and A. J. Duffield, *J. Organomet. Chem.*, **177**, 435 (1979).
16. W. S. Greenlee and M. F. Farona, *Inorg. Chem.*, **15**, 2129 (1976).
17. C. D. Tsonis and M. F. Farona, *J. Polym. Sci., Polym. Chem. Ed.*, **17**, 1779 (1979).
18. C. Tsonis and M. F. Farona, *J. Polym. Sci., Polym. Chem. Ed.*, **16**, 185 (1979).
19. E. Horn and M. R. Snow, *Aust. J. Chem.*, **33**, 2369 (1980).

20. J. C. Hileman, D. K. Huggins, and H. D. Kaesz, *Inorg. Chem.*, **1**, 933 (1962).
21. M. Wrighton and D. Bredesen, *J. Organomet. Chem.*, **50**, C35 (1973).
22. M. S. Wrighton and D. S. Ginley, *J. Am. Chem. Soc.*, **97**, 2065 (1975).
23. M. A. Lynch, Jr., W. J. Sesny, and E. O. Brimm, *J. Am. Chem. Soc.*, **76**, 3831 (1954).
24. E. W. Abel, M. M. Bhatti, K. G. Orell, and V. Sik, *J. Organomet. Chem.*, **208**, 195 (1981).
25. L. Vancea and W. A. G. Graham, *J. Organomet. Chem.*, **134**, 219 (1977).
26. R. Colton and J. E. Knapp, *Aust. J. Chem.*, **25**, (1972).

43. PENTACARBONYLHYDRIDORHENIUM

Submitted by MICHAEL A. URBANCIC and JOHN R. SHAPLEY*
Checked by NANCY N. SAUER and ROBERT J. ANGELICI†

$$Re(CO)_5Br \xrightarrow{Zn-H^+} ReH(CO)_5$$

Hieber and coworkers first prepared the pentacarbonyl hydride complex of rhenium[1] by protonation of the corresponding pentacarbonyl anion, $[Re(CO)_5]^-$. Modified versions of the original syntheses have been reported,[2] but these procedures still require extensive vacuum-line manipulations. Furthermore, the yield of $ReH(CO)_5$ is only ~30% based on pure $Na[Re(CO)_5]$,[1,3] which means that the overall yield from $Re_2(CO)_{10}$ is much lower.

Pentacarbonylhydridorhenium can be prepared readily in solution in ~85% yield by the reaction of $Re(CO)_5Br$ with zinc and acetic acid in methanol.[4] This procedure has been modified to allow the convenient isolation of pure $ReH(CO)_5$ in high yield.

■ **Caution.** *Volatile metal carbonyls are highly toxic and should be handled in a well-ventilated hood.*

Procedure

Into a 100-mL single-necked (14/20) round-bottomed Schlenk flask are placed a Teflon-coated magnetic stirring bar, 2.50 g (6.16 mmol) of powdered $Re(CO)_5Br$,[5] and 2.5 g (38 mmol) of zinc dust. The flask is then capped with a rubber serum stopper and evacuated. Into another Schlenk flask is placed 50 mL of 2, 5, 8, 22, 14-pentaoxapentadecane (tetraglyme, Aldrich) and 4 mL of 85% phosphoric acid. This flask is also capped with a serum stopper and the solution is degassed *in vacuo* for 2 h at room temperature. At this point

*Department of Chemistry and the Materials Research Laboratory, University of Illinois, Urbana, IL 61801.
†Department of Chemistry, Iowa State University, Ames, IA 50011.

the acid solution is transferred under nitrogen to the flask containing the solids via a double-tipped needle (cannula). The flask is wrapped with aluminum foil for protection from light, and the mixture is stirred under nitrogen until no solid $Re(CO)_5Br$ remains. This should take ~24 h.

When the reaction is complete, 5 g of P_4O_{10} is added to the flask and the mixture is stirred vigorously to remove the water in the solution. After 30 min, the product is ready for collection by vacuum distillation at room temperature. This is accomplished by using the apparatus shown in the Fig. 1, which is assembled inside a nitrogen-filled glove bag. The reaction flask is attached to a Schlenk filter tube by means of a U-shaped adapter. The adapter should be as small as possible, preferably not larger than 7 × 10 cm.

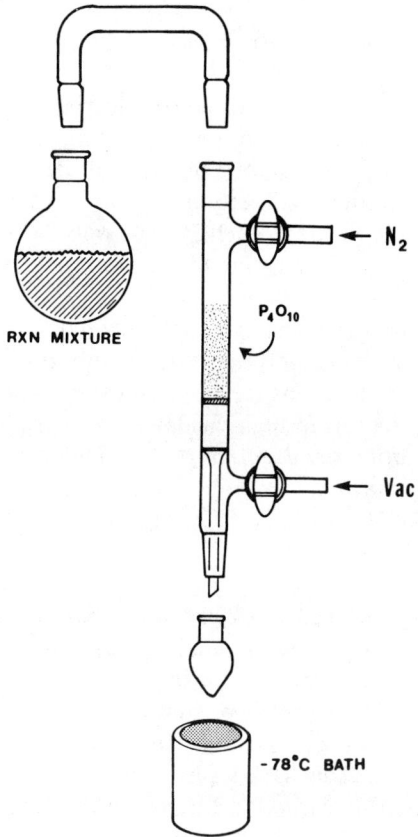

Fig. 1. Apparatus for vacuum distillation of the reaction mixture containing $HRe(CO)_5$.

A column of P_4O_{10} (10 × 25 mm) on the frit serves to remove any last traces of water during the distillation. A layer of glass wool (~2 mm thick) on the frit beneath the P_4O_{10} helps to keep the frit unclogged. A 5-mL pear-shaped flask is used as the receiver and is shielded from light by aluminum foil. The receiver is cooled at $-78°C$, and the apparatus is evacuated with a vacuum pump. The reaction mixture bubbles slowly as the hydride distills out.

If the top of the column of P_4O_{10} becomes syrupy, it may prevent flow through the column during the distillation. In this case, the top portion of the column should be removed and placed into the reaction flask along with some additional P_4O_{10} to react with the remaining water.

Normally, three fractions are collected over a 12-h period to complete the distillation. Yield: 1.67 g, 5.10 mmol 84%.

Anal. Calcd. for $ReH(CO)_5$: C, 18.35; 0.31. Found: C, 18.25; H, 0.21.

Properties

Pure $ReH(CO)_5$ is a colorless liquid with a density of 2.30 g mL^1 (determined at 24°C) and a melting point of 12.5°C.[1b] It is weakly acidic, more soluble in nonpolar than polar solvents, and practically insoluble in water.[1] The equilibrium vapor pressure (6–100°C) is given by the equation.[1b]

$$\log_{10} P(\text{mm}) = 8.598 - 2353.6/T$$

Three major infrared absorptions occur in the carbonyl stretching region near 2014, 2005, and 1982 cm^{-1} (cyclohexane)[3,6] and the molar absorptivities at these positions are 86, 22, and 2.7 L g^{-1} cm^{-1}, respectively.* The metal–hydrogen stretching absorptions are at 1882 cm^{-1} for the hydride and 1313 cm^{-1} for the deuteride (cyclohexane).[3] A singlet is observed in the ^{1}H NMR spectrum at δ -5.88 (hexane).[6] The gas-phase IR[1b] and Raman[7] spectra as well as the UV–Vis spectrum[8] have also been reported.

Contrary to early reports,[1b] pentacarbonylhydridorhenium is air stable.[6] The neat liquid, however, is moderately sensitive to light, turning yellow with formation of $Re_3H(CO)_{14}$.[9] Pentacarbonylhydridorhenium is stable up to 100°C, whereupon it decomposes to form $Re_2(CO)_{10}$ and H_2.[1] Substitution of carbonyl groups in $ReH(CO)_5$ by tertiary phosphines and other Lewis bases is well known and apparently occurs by a radical chain pathway.[6] More detailed surveys of the reactivity of $ReH(CO)_5$ have been summarized elsewhere.[10,11]

---

*Determined in cyclohexane solutions from 0.4 to 1 mg mL^{-1} at 2014 cm^{-1} and from 0.4 to 4 mg mL^{-1} at 2005 and 1982 cm^{-1}.

References and Notes

1. (a) W. Hieber and G. Braun, *Z. Naturforsch.*, **14b**, 132 (1959); (b) W. Beck, W. Hieber, and G. Braun, *Z. Anorg. Allg. Chem.*, **308**, 23 (1961).
2. (a) R. B. King and F. G. A. Stone, *Inorg. Synth.*, **7**, 198 (1963); (b) J. J. Eisch and R. B. King (eds.), *Organometallic Synthesis*, Vol. I, Academic Press, New York, 1965, p. 158; (c) W. F. Edgell and W. M. Risen, Jr., *J. Am. Chem. Soc.*, **88**, 5451 (1966).
3. P. S. Braterman, R. W. Harrill, and H. D. Kaesz, *J. Am. Chem. Soc.*, **89**, 2851 (1967).
4. J. R. Shapley, G. A. Pearson, M. Tachikawa, G. E. Schmidt, M. R. Churchill, and F. J. Hollander, *J. Am. Chem. Soc.*, **99**, 8064 (1977).
5. $Re(CO)_5Br$ was prepared by "titrating" a dichloromethane solution of $Re_2(CO)_{10}$ with liquid bromine at room temperature and recrystallizing from 1:2 acetone–methanol. See: S. P. Schmidt, W. C. Trogler, and F. Basolo, *Inorg. Chem.*, **23**, 41 (1985).
6. B. H. Byers and T. L. Brown, *J. Am. Chem. Soc.*, **99**, 2527 (1977).
7. A. Davison and J. W. Faller, *Inorg. Chem.*, **6**, 845 (1967).
8. G. B. Blakney and W. F. Allen, *Inorg. Chem.*, **10**, 2763 (1971).
9. Ultraviolet photolysis of neat $HRe(CO)_5$ yields $HRe_3(CO)_{14}$ in high yield; Mass spectrum, m/z 954 (M^+, ^{187}Re) and $(M - xCO)^+$, $x = 1$–14; IR (cyclohexane) ν_{CO} 2100 (w), 2047 (vs), 2014 (m), 1996 (s), 1979 (m), 1969 (w), 1933 (m) cm^{-1}. Compare: W. Fellmann and H. D. Kaesz, *Inorg. Nucl. Chem. Lett.*, **2**, 63 (1966).
10. (a) A. P. Ginsberg, in *Transition Metal Chemistry*, Vol. I, R. L. Carlin (ed.), Marcel Dekker, New York, 1965, p. 111; (b) H. D. Kaesz and R. B. Saillant, *Chem. Rev.*, **72**, 231 (1972).
11. (a) D. Giusto, *Inorg. Chim. Acta Rev.*, **6**, 91 (1972); (b) N. M. Boag and H. D. Kaesz, in *Comprehensive Organometallic Chemistry*, Vol. IV, G. Wilkinson, F. G. A. Stone, and E. W. Abel (eds.); Pergamon, New York, 1982, p. 172.

44. TETRACARBONYLIRON(0) COMPLEXES CONTAINING GROUP V DONOR LIGANDS

Submitted by MICHEL O. ALBERS,* ERIC SINGLETON,* and NEIL J. COVILLE†
Checked by CARLTON E. ASH,‡ CHRISTINE C. KIM,‡ and MARCETTA Y. DARENSBOURG‡

Since the first reported synthesis of $Fe(CO)_4(PPh_3)$ by Reppe and Schweckendiek in 1948[1] there have been numerous attempts to prepare this complex in high yield and in particular, free from contamination with $Fe(CO)_3(PPh_3)_2$. Direct procedures have included the thermal[2] as well as the photochemical[3] reaction between triphenylphosphine and iron pentacar-

---

*National Chemical Research Laboratory, Council for Scientific and Industrial Research, P.O. Box 395, Pretoria 0001, Republic of South Africa.
†Department of Chemistry, University of the Witwatersrand, 1 Jan Smuts Avenue, Johannesburg 2001, Republic of South Africa.
‡Department of Chemistry, Texas A & M University, College Station, TX 77843-3255.

bonyl. Indirect synthetic methods have included the reaction between triphenylphosphine and triiron dodecacarbonyl,[4] and the reduction of iron(II) carbonyl halide complexes with phenyllithium in the presence of triphenylphosphine.[5] More recent synthetic procedures have included the use of a combination of high-temperature and photochemical irradiation,[6] main group metal hydrides as promoters of carbonyl substitution on iron pentacarbonyl,[7] the use of Wilkinson's catalyst, $RhCl(PPh_3)_3$, as a stoichiometric decarbonylation reagent,[8] and iron carbonyl anions as CO substitution catalysts.[9] A similar checkered variety of synthetic methods also exists for the synthesis of other $Fe(CO)_4L$ complexes, where L is a large range of phosphine, phosphite, arsine, and stibine ligands.[10] In general, however, these methods all suffer from a number of disadvantages that include long reaction times, forcing conditions, the use of expensive reagents, and most significantly, the formation of mixtures of $Fe(CO)_4L$ and $Fe(CO)_3L_2$ with concomitant low yields of $Fe(CO)_4L$.

In this chapter, we describe the high-yield, selective synthesis of the complex $Fe(CO)_4(PPh_3)$ directly from iron pentacarbonyl and triphenylphosphine in boiling toluene using cobalt(II) chloride as a catalyst.[11] The method may also be generalized to a large variety of other Group V donor ligands and we provide a guide to a suitable choice of catalysts for such reactions.[11, 12]

General Procedure

■ **Caution.** *The use of the volatile, toxic iron pentacarbonyl necessitates that all manipulations be carried out in a well-ventilated hood. In addition, carbon monoxide, evolved in the reaction, is an odorless, extremely toxic gas, and care must be exercised that the apparatus vents into the best ventilated region of the hood.*

All reactions are routinely carried out under an inert atmosphere and in a well-ventilated fume hood. Standard ground-glass apparatus is used throughout.

All solvents are of analar grade and are freshly dried and distilled under nitrogen prior to use.[13] The catalysts $CoX_2 \cdot nH_2O$ (X = Cl, I) are dried under vacuum (0.1 torr, 50°C, 5 h) to give materials that are pink, analyzing approximately as $CoCl_2 \cdot 2H_2O$ or $CoI_2 \cdot 4H_2O$.[11] These turn color when added to the reaction mixtures, turning blue (X = Cl) or brown (X = I). The catalyst $[Fe(\eta^5\text{-}C_5Me_5)(CO)_2]_2$ was purchased from Strem and is used without further purification.

The catalyst $[Fe(\eta^5\text{-}C_5Me_5)(CO)_2]_2$ is synthesized by the method of King and Bisnette.[14] Pentamethylcyclopentadiene was purchased from Strem. Column chromatography utilized 63 to 200 μm of silica gel and 63 to 200 μm of neutral alumina activity grade 1. Melting points are all recorded in air in a well-ventilated hood, and are uncorrected. The reactions are all routinely monitored by infrared (IR) spectroscopy using a 0.05-mm pathlength cell.

A. TETRACARBONYL(TRIPHENYLPHOSPHINE)IRON(0)

■ *See cautionary note under General Procedure.*

$$Fe(CO)_5 + PPh_3 \xrightarrow{CoCl_2} Fe(CO)_4(PPh_3) + CO$$

Procedure

A 50-mL two-necked flask containing a magnetic stirrer bar is equipped with a reflux condenser and connected to a check-valve oil bubbler to observe evolution of gases. The flask is charged with PPh_3 (2.26 g, 10.0 mmol), $CoCl_2 \cdot 2H_2O$ (0.050 g, 0.3 mmol) and toluene (30 mL). The stirred solution is brought to reflux using an oil bath temperature controlled at 120°C, with a magnetic stirrer–heater device. When reflux is achieved, pentacarbonyliron (3.92 g, 20 mmol) is added. The course of the reaction is followed by monitoring the changes in the region from 1900 to 2100 cm^{-1} of the IR spectrum. Reflux is continued for 2 h or until the spectrum remains invariant with time. The reaction solution is cooled, and catalyst and traces of excess ligand are removed by eluting the reaction solution through a column (21 × 1, 5 cm) containing three layers: $CoCl_2 \cdot 6H_2O$ (5 g), neutral alumina (20 g), and silica gel (20 g). Benzene is used as eluant.

■ **Caution.** *Vapors of benzene are harmful, and toluene should be substituted if possible. The procedure must be carried out in a well-ventilated hood.*

The solvent and excess pentacarbonyliron are removed under reduced pressure and the product is recrystallized from dichloromethane–hexane. Yield: 3.56 g (83% based on consumed pentacarbonyliron).

Anal. Calcd. for $C_{22}H_{15}O_4PFe$: C, 61.43, H, 3.52. Found: C, 61.31, H, 3.53.

B. OTHER TETRACARBONYL(GROUP V DONOR LIGAND)IRON(0) COMPLEXES

The general applicability of the reaction catalyzed by cobalt(II) halide is illustrated by the examples in Table I. An important feature of the catalyzed reaction is the ability to prepare $Fe(CO)_4L$ in high yield with little contamination by $Fe(CO)_3L_2$. Typically the final reaction product contains <5% $Fe(CO)_3L_2$ except for L = $P(OEt)_3$, where ~10% $Fe(CO)_3[P(OEt)_3]_2$ is produced. Excess $Fe(CO)_5$ is used in the reactions to ensure that $Fe(CO)_4L$ is

TABLE I. Reaction Conditions and Spectroscopic Data for the Complexes Fe(CO)$_4$L [L = PPh$_3$, AsPh$_3$, SbPh$_3$, PMePh$_2$, PMe$_2$Ph, PCy$_3$, P(n-Bu)$_3$, P(OPh)$_3$, P(OEt)$_3$, P(OMe)$_3$] Using CoX$_2$·nH$_2$O (X = Cl, I) as Catalyst.[a]

	CoCl$_2$		CoI$_2$		IR Frequencies ν_{CO}[b] (cm^{-1})				Melting Point[c] (°C)	Ref.
	Yield (%)	Time[d] (h)	Yield (%)	Time[d] (h)						
Fe(CO)$_4$(PPh$_3$)	76–83[e]	2	95	0.5	2052	1978	1940		204–205	1–9
Fe(CO)$_4$(AsPh$_3$)	86	2	99[e]	0.5	2048	1972	1942		178–179	4, 6, 7
Fe(CO)$_4$(SbPh$_3$)	90	5	97[e]	1	2045	1970	1938		133–135	4, 6, 7
Fe(CO)$_4$(PMePh$_2$)	98[e]	1	78	1.5	2058	1977	1947		Oil	16
Fe(CO)$_4$(PMe$_2$Ph)	91[e]	1.3	93	4	2055	1973	1937		49–50	16
Fe(CO)$_4$(PCy$_3$)	45–60[e]	3	58	5	2043	1964	1930		174–175	3
Fe(CO)$_4$[P(n-Bu)$_3$]	38	6	50	6	2045	1968	1932		Oil	3, 6
Fe(CO)$_4$[P(OPh)$_3$]	79	4	75–95[e]	0.5	2067	1997	1960		68–69	7, 17
Fe(CO)$_4$[P(OEt)$_3$]	15[f]	6	59	6	2062	1983	1950		Oil	7, 17
Fe(CO)$_4$[P(OMe)$_3$]	nr[g]		nr	6	2053	1986	1949		44–45	6

[a] See Sections A and B and Ref. 11.
[b] Recorded in CHCl$_3$.
[c] Recorded in air, uncorrected.
[d] Monitored by IR spectroscopy until the spectrum remained invariant with time or for a period of 6 h.
[e] Verified by checkers.
[f] Maximum yield estimated by IR.
[g] NR = no reaction.

the major product. The excess iron pentacarbonyl is readily removed together with the solvent, and if stored under nitrogen, may be reused. Yields are thus based on consumed $Fe(CO)_5$ or alternatively, on added ligand. Purification of the product requires the removal of both unreacted ligand and catalyst. This is achieved by the use of a chromatographic column made up of three layers.[11] The top layer consists of $CoCl_2 \cdot 6H_2O$ and is used to remove the small amounts of unreacted ligand via the reaction $CoX_2 + 2L \rightarrow CoX_2L_2$.[15] This is followed by a layer of silica gel and a layer of alumina that remove the catalyst from the product.

Other catalysts may also be used for the syntheses of $Fe(CO)_4L$ complexes. Of particular use are $[Fe(\eta^5\text{-}C_5H_5)(CO)_2]_2$ and $[Fe(\eta^5\text{-}C_5Me_5)(CO)_2]_2$, which have proved in many ways to be complementary to the cobalt(II) halide catalysts.[12] Details are outlined in Table II. An identical synthetic procedure to that outlined above for $CoCl_2 \cdot 2H_2O$ as catalyst is used except that 0.1 mmol of catalyst is used and the reaction is carried out in 20 mL of toluene as solvent. Once the reaction is complete, the solvent is removed and the crude product is chromatographed on silica gel.[12] There is usually little, if any, unreacted ligand present, obviating the need for a $CoCl_2 \cdot 6H_2O$ layer. These reactions may all be scaled up or down in size without affecting the yields of products noticeably.

TABLE II. Reaction Conditions for the Synthesis of the Complexes $Fe(CO)_4L$ [L = PPh_3, $AsPh_3$, $SbPh_3$, $PMePh_2$, PMe_2Ph, PCy_3, $P(n\text{-}Bu)_3$, $P(OEt)_3$, $P(OMe)_3$] Using $[Fe(\eta^5\text{-}C_5H_5)(CO)_2]_2$ and $[Fe(\eta^5\text{-}C_5Me_5)(CO)]_2$ as Catalysts[a].

	$[Fe(\eta^5\text{-}C_5H_5)(CO)_2]_2$		$[Fe(\eta^5\text{-}C_5Me_5)(CO)_2]_2$	
	Yield (%)	Time[b] (h)	Yield (%)	Time[b] (h)
$Fe(CO)_4(PPh_3)$	90	5	97	1.25
$Fe(CO)_4(AsPh_3)$	88	5	94	1.25
$Fe(CO)_4(SbPh_3)$	92	2.75	97	1.25
$Fe(CO)_4(PMePh_2)$	98	1.75	96	1.25
$Fe(CO)_4(PMe_2Ph)$	93	0.5	91	1.25
$Fe(CO)_4(PCy_3)$	69	5	90	1.5
$Fe(CO)_4[P(n\text{-}Bu)_3]$	98–100[c]	1	93	1.25
$Fe(CO)_4[P(OPh)_3]$	83–93	4	94	1.5
$Fe(CO)_4[P(OEt)_3]$	93[c]	1.5	90	1.25
$Fe(CO)_4[P(OMe)_3]$	56–83[d]	0.75	86	1.5

[a]See Section B and Ref. 12.
[b]Monitored by IR spectroscopy until the reaction had gone to completion.
[c]Verified by checkers.
[d]The checkers were unable to repeat these yields. Crude yields were 56% and the yield of isolated, purified product was ~15%. The synthesis of the checkers' choice for $Fe(CO)_4[P(OMe)_3]$ is that of Ref. 9.

Properties

The properties of the complexes $Fe(CO)_4L$ [L = PPh_3, $AsPh_3$, $SbPh_3$, $PMePh_2$, PMe_2Ph, PCy_3, $P(n$-$Bu)_3$, $P(OPh)_3$, $P(OEt)_3$, and $P(OMe)_3$] have been extensively studied and reviewed.[10, 16, 17] The complexes are all air stable.

References

1. W. Reppe and W. J. Schweckendiek, *Liebigs Ann. Chem.*, **560**, 104 (1948).
2. F. A. Cotton and R. V. Parish, *J. Chem. Soc.*, **1960**, 1440.
3. W. Strohmeier and F.-J. Müller, *Chem. Ber.*, **102**, 3613 (1969); M. O. Albers and N. J. Coville, unpublished results.
4. A. F. Clifford and A. K. Mukherjee, *Inorg. Chem.*, **2**, 151 (1963); A. F. Clifford and A. K. Mukherjee, *Inorg. Synth.*, **8**, 185 (1966).
5. T. A. Manuel, *Inorg. Chem.*, **2**, 854 (1963).
6. H. L. Condor and M. Y. Darensbourg, *J. Organometal. Chem.*, **67**, 93 (1974).
7. W. O. Siegel, *J. Organometal. Chem.*, **92**, 321 (1975).
8. Yu. S. Varshavsky, E. P. Shestakova, N. V. Kiseleva, T. G. Cherkasova, N. A. Buzina, L. S. Bresler, and V. A. Kormer, *J. Organometal. Chem.*, **170**, 81 (1979).
9. S. B. Butts and D. F. Shriver, *J. Organometal. Chem.*, **169**, 191 (1979).
10. Gmelin, *Handbuch der Anorganische Chemie*, Eisen-Organische Verbindungen, Teil B, U. Kruerke (ed.), Springer Verlag, Berlin, 1978.
11. M. O. Albers, N. J. Coville, T. V. Ashworth, and E. Singleton, *J. Organometal. Chem.*, **217**, 385 (1981).
12. M. O. Albers, N. J. Coville, and E. Singleton, *J. Organometal. Chem.*, **232**, 261 (1982).
13. D. D. Perrin, W. L. F. Armarego, and D. R. Perrin, *Purification of Laboratory Chemicals*, 2nd ed., Pergamon, Oxford, 1980.
14. R. B. King and M. B. Bisnette, *J. Organometal. Chem.*, **8**, 287 (1967).
15. F. A. Cotton, O. D. Faut, D. M. L. Goodgame, and R. H. Holm, *J. Am. Chem. Soc.*, **83**, 1780 (1961).
16. P. M. Treichel, W. M. Douglas, and W. K. Dean, *Inorg. Chem.*, **11**, 1615 (1972).
17. J. D. Cotton and R. L. Heazelwood, *Aust. J. Chem.*, **22**, 2673 (1969).

45. BIS(PHOSPHINE) DERIVATIVES OF IRON PENTACARBONYL AND TETRACARBONYL (TRI-*tert*-BUTYLPHOSPHINE)IRON(0)

Submitted by MICHAEL J. THERIEN† and WILLIAM C. TROGLER*
Checked by ROSALICE SILVA† and MARCETTA Y. DARENSBOURG†

Bis(phosphine) derivatives of pentacarbonyliron are starting materials for the synthesis of several organometallic iron complexes.[1-7] Iron carbonyl phosphine complexes have attracted attention [8-11] because of their relevance to

---

*Department of Chemistry, D-006, University of California at San Diego, La Jolla, CA 92093.
†Department of Chemistry, Texas A&M University, College Station, TX 77843.

photochemical catalysis of olefin hydrosilation. Though $Fe(CO)_3(PR_3)_2$ complexes are used widely in organotransition metal chemistry, an efficient preparation of these compounds has not been reported. Clifford and Mukherjee[12] describe two methods for the synthesis of tricarbonyl-bis(triphenylphosphine)iron(0). They report that direct reaction between $Fe_3(CO)_{12}$ and triphenylphosphine in THF (tetrahydrofuran) solvent gives a mixture of $Fe(CO)_3[P(C_6H_5)_3]_2$ (27%) and $Fe(CO)_4[P(C_6H_5)_3]$ (34%). The second method [the reaction between $Fe(CO)_5$ and $P(C_6H_5)_3$ in cyclohexanol] gives both $Fe(CO)_3[P(C_6H_5)_3]_2$ and $Fe(CO)_4[P(C_6H_5)_3]$ in 15% yield. Again, the mono- and bissubstituted compounds have to be separated by vacuum sublimation.

A better synthesis (89% yield) of $Fe(CO)_3(PPh_3)_2$ is reported[13] from $[PPN]_2[Fe_4(CO)_{13}]$, where $PPN^+ = \mu$-nitridobis(triphenylphosphorus)(1+). The $CoX_2(X=Cl, Br, I)$-catalyzed substitution of CO in $Fe(CO)_5$ is reported[14] to yield $Fe(CO)_4L$ species in 15 to 99% yield and $Fe(CO)_3(PPh_3)_2$ was prepared (net 62% yield) from $Fe(CO)_5$ in a two-step procedure that requires a chromatographic separation. Strohmeier and Muller[15] report that irradiation of $Fe(CO)_5$ in the presence of several phosphines produces $Fe(CO)_3L_2$ and $Fe(CO)_4L$ complexes in yields that range from 13% for the synthesis of $Fe(CO)_3[P(n\text{-}Bu)_3]_2$ to 35% for $Fe(CO)_3[P(cyclo\text{-}C_6H_{11})_3]_2$. For some of the compounds synthesized, vacuum sublimation is necessary to separate the $Fe(CO)_3L_2$ species from $Fe(CO)_4L$. The one-step photochemical procedure we report here employs cyclohexane as a solvent. That enables unreacted phosphine. $Fe(CO)_5$, and $Fe(CO)_4L$ to remain in solution while pure $Fe(CO)_3L_2$ precipitates. It is essential that the phosphines used in these reactions be free of phosphine oxides, which labilize[16] CO and yield products other than $Fe(CO)_3(PR_3)_2$ complexes.

General Procedure

Method 1. For large-scale reactions, 300–450 mL of cyclohexane is heated at reflux over sodium, distilled, and transferred under N_2 into a photochemical reaction vessel (500 mL) that is fitted around a 450-W Hanovia mercury arc lamp contained in a water-cooled quartz immersion well (see Fig. 1). This apparatus is available from Ace Glass.* A septum-capped sidearm permits the system to be flushed with N_2 through a syringe needle. The carbon monoxide evolved during the reaction is collected by venting the airtight irradiation vessel through a mineral oil bubbler. The outlet of the bubbler is connected via Tygon tubing, to release gas into the bottom of an inverted 2000- or 3000-mL graduated cylinder that is filled with water and contained

*Ace Glass Co., P.O. Box 688, 1430 Northwest Blvd., Vineland, NJ 08360.

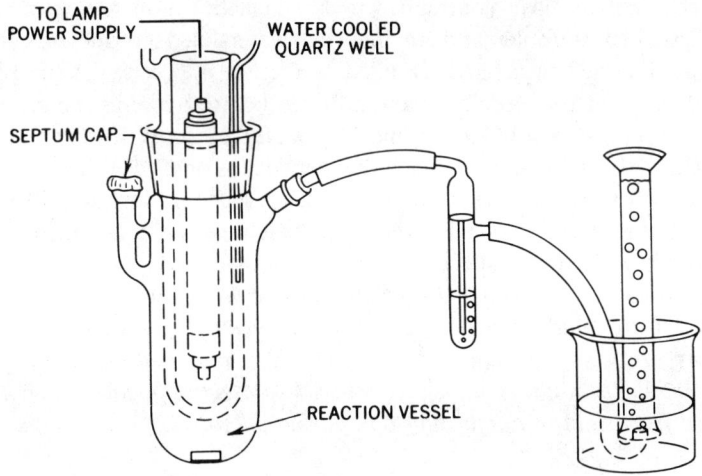

Fig. 1. Experimental setup for large-scale photochemical syntheses.

in a partially filled 5000-mL beaker. The volume of CO gas produced is measured as it displaces water in the graduated cylinder. The oil bubbler prevents exposure of the reaction solution to water vapor and serves as a safety valve if a slight back pressure develops. Then 2–10 mL of $Fe(CO)_5$ (152–760 mmol) is introduced by syringe into the cyclohexane solvent along with 3–5 equiv of phosphine through the septum capped side arm. Photolysis is begun with stirring, and the volume of gas evolved is monitored as light yellow $Fe(CO)_3L_2$ precipitates during the reaction. Irradiation is stopped when the volume of CO gas evolved equals the theoretical amount or when the reaction mixture ceases to evolve CO. At this point one must carefully remove the immersion well from the center of the reaction vessel (under an N_2 flush) without dislodging solid decomposition products that adhere to the surface of the inner quartz well. The solution is filtered under N_2, and the recovered $Fe(CO)_3(PR_3)_2$ is dried under vacuum. The volume of the filtrate is reduced by 50% and then filtered to yield more $Fe(CO)_3(PR_3)_2$. The recovered $Fe(CO)_3(PR_3)_2$ complexes give satisfactory elemental analyses and IR spectra and require no further purification.

Method 2. A convenient method for small scale preparations of these compounds uses Schlenk techniques. To a 100-mL quartz Schlenk tube is added 0.5–1.0 mL (38–76 mmol) of $Fe(CO)_5$, 3–5 equiv of phosphine, and 80 mL of dry deoxygenated cyclohexane. The Schlenk tube is placed adjacent to the mercury arc lamp contained in the quartz immersion well. The side arm of the Schlenk flask is attached to N_2 flushed Tygon tubing, which is connected to the bubbler and an inverted water-filled graduated cylinder (500 mL) contained in a partially filled beaker. It is best to let the photochemical reaction proceed for a few minutes before opening the stopcock. This allows a pressure of CO to build up; this forces air in the Tygon tubing to the other side of the bubbler when the stopcock is opened. When complete, the reaction is worked up as in the large-scale procedure just described. The checkers found reduced yields when the reactions were conducted on a smaller scale without the specified excess of phosphine ligand.

■ **Caution.** *The compound $Fe(CO)_5$ is toxic and should be used only in a fume hood. Care should be used when handling trimethyl- and triethylphosphine since they are toxic and ignite readily. The UV lamp will cause severe eye damage or blindness if it is viewed without UV-protective goggles. It should be concealed from sight in a lighttight box during photolysis.*

A. TRICARBONYLBIS(TRIPHENYLPHOSPHINE)IRON(0)

$$Fe(CO)_5 + 2P(C_6H_5)_3 \xrightarrow{h\nu} Fe(CO)_3[P(C_6H_5)_3]_2 + 2CO$$

Method 1 is used for the reaction between 10 mL of $Fe(CO)_5$ and 95 g of $P(C_6H_5)_3$. After 15 h of irradiation, 2.95 L of gas is evolved, and the reaction is stopped. The first filtration yields 28.1 g of product. Reducing the volume by 50% yields an additional 5.2 g of $Fe(CO)_3[P(C_6H_5)_3]_2$. Yield: 33.3 g, 66%

Anal. Calcd. for $Fe(CO)_3[P(C_6H_5)_3]_2$: C, 70.50; H, 4.52; P, 9.34. Found: C, 70.54; H, 4.98; P, 9.23.

B. TRICARBONYLBIS(TRICYCLOHEXYLPHOSPHINE)IRON(0)

$$Fe(CO)_5 + 2P(c\text{-}C_6H_{11})_3 \xrightarrow{h\nu} Fe(CO)_3[P(cyclo\text{-}C_6H_{11})_3]_2 + 2CO$$

The photochemical reaction between 0.90 mL of $Fe(CO)_5$ (6.84×10^{-3} mol) and 9.3 g (3.32×10^{-2} mol) of $P(cyclo\text{-}C_6H_{11})_3$ in 300 mL of cyclohexane (method 1) went to completion in 6 h with the evolution of 320 mL of gas. After the initial filtration, 2.65 g of product is recovered. Reducing the volume of cyclohexane and refrigerating the solution yields an additional 0.75 g of product. Yield: 3.4 g; 71%.

Anal. Calcd. for $Fe(CO)_3[P(cyclo\text{-}C_6H_{11})_3]_2$: C, 66.86; H, 9.43; P, 8.86. Found: C, 67.45; H, 9.52; P, 8.50.

C. TRICARBONYLBIS(TRI-n-BUTYLPHOSPHINE)IRON(0)

$$Fe(CO)_5 + 2P(n\text{-}Bu)_3 \xrightarrow{h\nu} Fe(CO)_3[P(n\text{-}Bu)_3]_2 + 2CO$$

Using method 1, a mixture of 5 mL of $Fe(CO)_5$ (3.83×10^{-2} mol) and 21.2 g of $P(n\text{-}Bu)_3$ (1.04×10^{-1} mol) is irradiated. The reaction is complete after 12 h. The initial filtration yields 3.7 g of product. After reducing the volume of the solution, an additional 2.2 g is obtained. Yield: 5.9 g; 28%.

Anal. Calcd. for $Fe(CO)_3[P(n\text{-}Bu)_3]_2$: C, 59.58; H, 9.93; P, 11.40. Found: C, 59.32; H, 9.86; P, 11.15.

D. TRICARBONYLBIS(TRIMETHYLPHOSPHINE)IRON(0)

$$Fe(CO)_5 + 2PMe_3 \xrightarrow{h\nu} Fe(CO)_3(PMe_3)_2 + 2CO$$

To 1 mL of $Fe(CO)_5$ (7.61×10^{-3} mol) was added 2.76 g of PMe_3 (3.63×10^{-2} mol) in 80 mL of dry cyclohexane (method 2). The reaction vessel is cooled in a quartz beaker containing an ice–salt mixture to avoid loss of volatile PMe_3 during the irradiation even though the cooling bath scatters much of the light. After 5 h the reaction is complete. The initial filtration yields 1.23 g of product. After reducing the volume of solvent by 50%, an additional 0.54 g recovered. Yield: 1.77 g, 80%.

Anal. Calcd. for $Fe(CO)_3[P(CH_3)_3]_2$: C, 37.02; H, 6.21; P, 21.21. Found: C, 36.65; H, 6.27; P, 21.13.

E. TETRACARBONYL(TRI-tert-BUTYLPHOSPHINE)IRON(0)

$$Fe(CO)_5 + P(t\text{-}Bu)_3 \xrightarrow{h\nu} Fe(CO)_4[P(t\text{-}Bu)_3]$$

The reaction between 1.3 mL of $Fe(CO)_5$ (9.69×10^{-3} mol) and 6.31 g of $P(t\text{-}Bu)_3$ in 350 mL of dry cyclohexane (method 1) produces 475 mL of CO after 8 h of irradiation. The reaction mixture is filtered to yield 1.51 g of yellow product. An additional 1.05 g of solid is obtained by reducing the volume of solution. The combined yield (2.56 g) proves to be exclusively the monosubstituted product. Apparently, the conditions of the reaction do not allow two bulky tri-*tert*-butylphosphine ligands to replace two carbonyl groups on $Fe(CO)_5$. Yield: 2.56 g; 72%.

Anal. Calcd. for Fe(CO)$_4$[P(*t*-Bu)$_3$]: C, 51.91; H, 7.35; P, 8.37. Found: C, 51.43; H, 7.23; P, 8.32.

Properties

Both the Fe(CO)$_3$(PR$_3$)$_2$ complexes and Fe(CO)$_4$[P(*t*-Bu)$_3$] are soluble in organic solvents such as CH$_2$Cl$_2$, benzene, toluene, THF, acetone, and hot heptane or cyclohexane, with the degree of solubility varying with the type of phosphine (the PPh$_3$ derivative is least soluble). All the complexes are air stable in the solid state. Prolonged exposure to light darkens the surfaces of these compounds. When left in solution for long periods of time and exposed to air, the compounds decompose slowly; those complexes containing small phosphine ligands seem to be the most sensitive. These compounds should be kept cold for long-term storage. Physical properties are listed in the following table:

Compound	mp (°)	v_{CO} (cm^{-1})[a]	^{31}P (ppm)[b]
Fe(CO)$_4$[P(*t*-Bu)$_3$]	170 dec	2040, 1960, 1920	126.9
Fe(CO)$_3$(PMe$_3$)$_2$	195	1863	42.4
Fe(CO)$_3$[P(*n*-Bu)$_3$]$_2$	55	1855	66.0
Fe(CO)$_3$[P(*cyclo*-C$_6$H$_{11}$)$_3$]$_2$	228	1846	89.1
Fe(CO)$_3$[PPh$_3$]$_2$	272	1878	78.2

[a] Solution IR spectra used CH$_2$Cl$_2$ as the solvent.
[b] NMR spectra were recorded on a Nicolet 200-MHz instrument in CDCl$_3$ solvent. Chemical shifts are parts per million (ppm) downfield from 85% H$_3$PO$_4$.

References

1. P. K. Baker, N. S. Connelly, B. M. R. Jones, J. P. Maher, and K. R. Somers, *J. Chem. Soc. Dalton Trans.*, **1980**, 579.
2. W. E. Carroll and F. J. Lalor, *J. Chem. Soc. Dalton Trans.*, **1973**, 1754.
3. G. R. Crooks and B. F. G. Johnson, *J. Chem. Soc. A*, **1968**, 1238.
4. R. K. Kummer and W. A. G. Graham, *Inorg. Chem.*, **7**, 1208, (1968).
5. A. Davison, W. McFarlane, L. Pratt, and G. Wilkinson, *J. Chem. Soc.*, **1962**, 3653.
6. K. Farmey and M. Kilner, *J. Chem. Soc. A*, **1970**, 634.
7. M. J. Therien, C.-L. Ni, F. C. Anson, J. G. Osteryoung, and W. C. Trogler, *J. Am. Chem. Soc.*, **108**, 4037, (1986).
8. R. D. Sanner, R. G. Austin, M. S. Wrighton, W. D. Honnick, and C. U. Pittman, *Inorg. Chem.*, **18**, 928, (1979).
9. D. K. Liu, C. G. Brinkley, and M. S. Wrighton, *Organometallics*, **3**, 1449, (1984).
10. J. L. Graff, R. D. Sanner, and M. S. Wrighton, *Organometallics*, **1**, 837, (1982).
11. D. K. Liu, M. S. Wrighton, D. R. McKay, and G. R. Maciel, *Inorg. Chem.*, **23**, 212, (1984).
12. A. F. Clifford and A. K. Mukherjee, *Inorganic Synth.*, **8**, 184, (1966).

13. S. B. Butts and D. F. Shriver, *J. Organomet. Chem.*, **169**, 191, (1979).
14. M. O. Albers and N. J. Coville, *J. Organomet. Chem.*, **17**, 385, (1981).
15. W. Strohmeier and F. J. Muller, *Chem. Ber.*, **102**, 3613, (1969).
16. D. J. Darensbourg, M. Y. Darensbourg, and N. Walker, *Inorg. Chem.*, **20**, 1918, (1981).

46. ZERO-VALENT IRON ISOCYANIDE COMPLEXES

Submitted by MICHEL O. ALBERS,* ERIC SINGLETON,* and
NEIL J. COVILLE†
Checked by ARTHUR Y-J. CHEN,‡ RUSTY BLANSKI,‡ and
HERBERT D. KAESZ‡

A recent review has highlighted the extensive and interesting chemistry of metal isocyanide complexes.[1] Although synthetic procedures are varied, a vast number are based on substitution in metal carbonyl complexes by isonitriles. Such procedures are, however, not always successful. This is especially so in cases where multiple substitution of CO is required, as in the syntheses of homoleptic isocyanide complexes. Many of the inherent difficulties are illustrated by the reaction of iron pentacarbonyl with isocyanides.

Pentacarbonyl iron is fairly inert to substitution reactions, and attempts to prepare $Fe(CO)_{5-n}(CNR)_n$ ($n = 1-5$) by the direct reaction of $Fe(CO)_5$ with isocyanides in Carius tubes has produced only the complexes $Fe(CO)_{5-n}(CNR)_n (n = 1$ and $2)$.[2] The products were obtained as mixtures that required separation. Other syntheses, including photochemical[3] and trimethylamine N-oxide-promoted[4a] displacement of carbonyl groups, or other means,[4b] give the same products in variable yield. Procedures based on diiron nonacarbonyl[5] and triiron dodecacarbonyl[6] have produced similar results. The only zero-valent iron complex $Fe(CO)_{5-n}(CNR)_n$ where $n > 2$ is the complex $Fe(CNR)_5$ prepared either by metal vapor synthesis techniques[7] or by sodium amalgam reduction of iron(II) bromide in the presence of isocyanide.[8]

In this contribution we describe facile, high-yield syntheses of the series of zerovalent iron isocyanide complexes $Fe(CO)_{5-n}(CNC_6H_3Me_2-1,3)_n$ ($n = 1-5$). The starting material is iron pentacarbonyl, and cobalt(II) chloride

---

*National Chemical Research Laboratory, Council for Scientific and Industrial Research, P.O. Box 395, Pretoria 0001, Republic of South Africa.
†Department of Chemistry, University of the Witwatersrand, 1 Jan Smuts Avenue, Johannesburg 2001, Republic of South Africa.
‡Department of Chemistry and Biochemistry, University of California, Los Angeles, CA 90024-1569.

is used as a catalyst to achieve the stepwise replacement of carbonyl groups by 2-isocyano-1,3-dimethylbenzene.[4,9]

Although the synthetic procedure is exemplified by the use of 2-isocyano-1,3-dimethylbenzene, the series of complexes $Fe(CO)_{5-n}(CNR)_n (n = 1-5)$ can be obtained for other aromatic isonitriles, and for some alkyl isocyanides (e.g., $n = 1-4$, $R = t$-Bu; $n = 1-3$, $R = Me, C_6H_{11}, C_6H_5CH_2$) using a similar technique.[4]

General Procedure

■ **Caution.** *The use of the volatile, toxic iron pentacarbonyl necessitates that all manipulations be carried out in a well-ventilated hood. In addition, the carbon monoxide evolved in the reaction is an odorless, extremely toxic gas, and care should be exercised that the apparatus vents into the best ventilated region of the hood. The compound 2-isocyano-1,3-dimethylbenzene is a vile smelling, volatile solid that is best handled in a well-ventilated hood, using protective gloves.*

Unless otherwise stated, all manipulations are performed under nitrogen using standard Schlenk techniques.[10] Solvents are all of analar grade and were dried and distilled under nitrogen prior to use.[11] The 2-isocyano-1,3-dimethylbenzene was purchased from Fluka AG and is used without further purification. The cobalt(II) chloride catalyst is obtained by heating $CoCl_2·6H_2O$ under vacuum at ~50°C for 5 h. The material so obtained is blue-pink in color and analyzed approximately as $CoCl_2·2H_2O$.[4] Column chromatography on 60–200 μm of silica gel is used throughout. Melting points are all recorded in air in a well-ventilated hood, and are uncorrected.

A. TETRACARBONYL(2-ISOCYANO-1,3-DIMETHYLBENZENE)IRON(0)

■ *See cautionary note under General Procedure.*

1. $Fe(CO)_5 + 1,3\text{-}Me_2C_6H_3NC \longrightarrow Fe(CO)_4(CNC_6H_3Me_2\text{-}1,3) + CO$

Procedure

A 250-mL two-necked reaction flask containing a magnetic stirrer bar is equipped with a reflux condenser attached to a check-valve oil bubbler to observe venting of gases. The flask is purged with dry, oxygen-free nitrogen and charged with pentacarbonyl iron (11.76 g, 60.0 mmol), 2-isocyano-1,3-dimethylbenzene (3.93 g, 30.0 mmol), and toluene (80 mL). The yellow reaction solution is brought to 85°C using an oil bath and a magnetic stirrer–heater. Heating is continued at 85°C until CO evolution ceases, generally requiring between 15 and 30 min. It is important to stop heating the reaction mixture as soon as the reaction is complete. Otherwise, a darkening occurs that imparts an undesirable color to the product. Purification to remove this discoloration has proved to be difficult.[4]

The reaction mixture is cooled to room temperature and the solvent and excess $Fe(CO)_5$ are removed under vacuum. The product is obtained as a yellow, crystalline solid. Yield: 8.6 g (96%). This material is sufficiently pure for most purposes, but it may be recrystallized from pentane.

Anal. Calcd. for $C_{13}H_9NO_4Fe$: C, 52.2; H, 3.05; N, 4.7. Found: C, 52.45; H, 3.15; N, 4.7, mp 82–83°C.

2. $Fe(CO)_5 + 1,3\text{-}Me_2C_6H_3NC \xrightarrow{CoCl_2} Fe(CO)_4(CNC_6H_3Me_2\text{-}1,3) + CO$

Procedure

■ *See cautionary note under General Procedure.*

The same equipment is used as described in the preceding procedure; this method has the advantage that stoichiometric quantities of reagents are used. The reaction flask (250 mL) containing a magnetic stirrer bar is charged with pentacarbonyl iron (3.92 g, 20.0 mmol), $CoCl_2 \cdot 2H_2O$ (0.050 g, 0.3 mmol), and toluene (100 mL). The reaction mixture is heated to 85°C, 1,3-$Me_2C_6H_3NC$ (2.62 g, 20.0 mmol) is then added and the reaction mixture is maintained there until CO evolution ceases (~5 min). The green reaction solution is cooled to 0°C, filtered, and passed down a short silica gel column (1 × 10 cm) in order to separate remaining traces of catalyst, indicated by a green coloration. Further small quantities of toluene are used as eluant. The solvent is then removed from the eluant under vacuum to give the product as a yellow, crystalline solid. Yield: 5.4 g (90%).

Anal. Calcd. for $C_{13}H_9NO_4Fe$: C, 52.2; H, 3.05; N, 4.7. Found: C, 52.32; H, 3.01; N, 4.69, mp 82–83°C.

B. TRICARBONYLBIS(2-ISOCYANO-1,3-DIMETHYLBENZENE)IRON(0)

■ *See cautionary note under General Procedure.*

$Fe(CO)_4(CNC_6H_3Me_2\text{-}1,3) + 1,3\text{-}Me_2C_6H_3NC$

$\xrightarrow{CoCl_2} Fe(CO)_3(CNC_6H_3Me_2\text{-}1,3)_2 + CO$

Procedure

A 100-mL two-necked reaction flask containing a magnetic stirrer bar is equipped with a reflux condenser connected to a check-valve oil bubbler to observe evolution of gases. The flask is charged with $Fe(CO)_4(CNC_6H_3Me_2\text{-}1,3)$ (1.79 g, 6.0 mmol), $CoCl_2 \cdot 2H_2O$ (0.034 g, 0.2 mmol), and toluene (30 mL). The reaction mixture is brought to 85°C using an oil bath and a magnetic stirrer–heater device. When the reaction mixture has reached 85°C, 2-isocyano-1,3-dimethylbenzene (0.786 g, 6.0 mmol) is added to give an immediate green coloration. Vigorous CO evolution occurs for a period of ~10 min. The completion of the reaction is indicated by the cessation of CO evolution, but is best confirmed by IR spectroscopy (Table I). The reaction mixture is cooled to 0°C, filtered to remove the catalyst, and then passed down a short silica gel column (~20 g, wrapped in foil to minimize photochemical decomposition) in order to remove any remaining traces of catalyst. Portions of toluene may be used as eluant. The solvent is removed under reduced pressure to give the product as a yellow, crystalline solid. Yield: 2.12 g (88%). This material is sufficiently pure for most purposes, but it may be recrystallized from dichloromethane–pentane to give the analytically pure product.

Anal. Calcd. for $C_{21}H_{18}N_2O_3Fe$: C, 62.7; H, 4.50; N, 6.95. Found: C, 62.3; H, 4.60; N, 6.95, mp 132–134°C (dec).

C. DICARBONYLTRIS(2-ISOCYANO-1,3-DIMETHYL-BENZENE)IRON(0)

■ *See cautionary note under General Procedure.*

$$Fe(CO)_4(CNC_6H_3Me_2\text{-}1,3) + 2(1,3\text{-}Me_2C_6H_3NC)$$

$$\xrightarrow{CoCl_2} Fe(CO)_2(CNC_6H_3Me_2\text{-}1,3)_3 + 2CO$$

Procedure

The same equipment is used as described in the procedure in Section B. The flask is charged with $Fe(CO)_4(CNC_6H_3Me_2\text{-}1,3)$ (1.79 g, 6.0 mmol), $CoCl_2 \cdot 2H_2O$ (0.034 g, 0.2 mmol), and toluene (30 mL), and the mixture is heated to 85°C with stirring. To the hot solution is added 2-isocyano-1,3-dimethylbenzene (1.57 g, 12.0 mmol). It is crucial to add the isocyanide to the hot solution. Otherwise, catalyst deactivation (believed to be due to isocyanide polymerization) occurs. This results in a sluggish and incomplete reaction.[4,12]

The reaction begins immediately as evidenced by a green coloration and vigorous evolution of CO. Continued heating at 85°C (~5 min) gives an orange reaction solution. The end of the reaction is indicated when the evolution of CO ceases, but is best confirmed by IR spectroscopy (see Table I). Cooling to 0°C, followed by filtration to remove the catalyst, gives a clear orange solution. The volume of the solution is reduced under vacuum to 10–15 mL. Addition of pentane (15–30 mL) (with cooling to −20 or −78°C if necessary) gives the product as a yellow, crystalline solid. Further cycles of solvent removal and addition of pentane followed by crystallization, may be necessary. Yield: 2.1 g (70%).

Anal. Calcd. for $C_{29}H_{27}N_3O_2Fe$: C, 68.91; H, 5.35; N, 8.32. Found: C, 68.82; H, 5.51; N, 8.40.

D. CARBONYLTETRAKIS(2-ISOCYANO-1,3-DIMETHYL-BENZENE)IRON(0)

■ *See cautionary note under General Procedure.*

$$Fe(CO)_4(CNC_6H_3Me_2\text{-}1,3) + 3(1,3\text{-}Me_2C_6H_3NC)$$

$$\xrightarrow{CoCl_2} Fe(CO)(CNC_6H_3Me_2\text{-}1,3)_4 + 3CO$$

Procedure

The same equipment is used as described in the procedure in Section B. The reaction flask is charged with $Fe(CO)_4(CNC_6H_3Me_2\text{-}1,3)$ (1.79 g, 6.0 mmol), $CoCl_2 \cdot 2H_2O$ (0.034 g, 0.2 mmol), and toluene (30 mL), and these components are heated to 85°C with magnetic stirring, using an oil bath. To the hot mixture is added 2-isocyano-1,3-dimethylbenzene (2.36 g, 18.0 mmol). There is an immediate green coloration and vigorous evolution of CO. Completion of the reaction is indicated by the cessation of CO evolution, but is best confirmed by IR spectroscopy (see Table I). Filtration of the cold (0°C) reaction solution gives a clear, orange solution. The volume is reduced under vacuum to ~10–15 mL. Addition of pentane (15–30 mL) gives the product as an orange, crystalline solid. If the product does not crystallize immediately, solvent should be removed under vacuum to reduce the volume again to ~10–15 mL, followed by addition of pentane (15–30 mL) and cooling to −20 or −78°C. Repeat this cycle if necessary. Yield: 2.8 g (76%).

Anal. Calcd. for $C_{37}H_{36}N_4OFe$: C, 73.03; H, 5.92; N, 9.21. Found: C, 72.26; H, 5.99; N, 8.99.

E. PENTAKIS(2-ISOCYANO-1,3-DIMETHYLBENZENE)IRON(0)

■ *See cautionary note under General Procedure.*

$$Fe(CO)_4(CNC_6H_3Me_2\text{-}1,3) + 4(1,3\text{-}Me_2C_6H_3NC)$$

$$\xrightarrow{CoCl_2} Fe(CNC_6H_3Me_2\text{-}1,3)_5 + 4CO$$

Procedure

The same equipment is used as described in the procedure in Section B. The flask is charged with $Fe(CO)_4(CNC_6H_3Me_2\text{-}1,3)$ (1.79 g, 6.0 mmol), $CoCl_2 \cdot 2H_2O$ (0.034 g, 0.2 mmol), and toluene (30 mL). These components are heated to 85°C, and 2-isocyano-1,3-dimethylbenzene (3.15 g, 24.0 mmol) is added to the hot solution. There is an immediate green coloration and vigorous evolution of CO. Completion of the reaction is indicated by the cessation of CO evolution, but is best confirmed by IR spectroscopy (see Table I). Cooling to 0°C followed by filtration to remove the catalyst gives a clear, red solution of the product. The volume of the solution is reduced under vacuum to ~10–15 mL. Addition of pentane (15–30 mL) followed by cooling to -20 or $-78°C$, gives the product as a red crystalline solid. Repeated toluene–pentane crystallization cycles at -20 or $-78°C$ may be necessary if the first obtained product is an oil. Yield: 2.9 g (68%).

Anal. Calcd. for $C_{45}H_{45}N_5Fe$: C, 75.95; H, 6.33; N, 9.85. Found: C, 75.33; H, 6.15; N, 9.05.

Properties

The complexes $Fe(CO)_4(CNC_6H_3Me_2\text{-}1,3)$ and $Fe(CO)_3(CNC_6H_3Me_2\text{-}1,3)_2$ are yellow, crystalline solids.[4] They are air stable and mildly light sensitive. The compounds are soluble in most common organic solvents. The complexes $Fe(CO)_{5-n}(CNC_6H_3Me_2\text{-}1,3)_n (n = 3\text{-}5)$[4] are yellow ($n = 3$), orange ($n = 4$), and red ($n = 5$), air- and light-sensitive materials, susceptible to oxidation by O_2 and by solvents such as $CHCl_3$ and CH_2Cl_2. The compounds are soluble in, and best handled in hydrocarbon or ether solvents.

The compounds $Fe(CO)_{5-n}(CNC_6H_3Me_2\text{-}1,3)_n$ ($n = 1\text{-}5$) are characterized by IR vibrational spectroscopy (v_{CO} and v_{NC}, 1800–2200-cm^{-1} region) and by NMR spectroscopy (aromatic and methyl protons). Selected spectroscopic data for these complexes are given in the Table I. Infrared spectroscopy may conveniently be used for the monitoring of the progress of the substitution reaction, and NMR spectroscopy for an estimate of product purity. The complexes may also be characterized by mass spectrometry

TABLE I. Selected Spectroscopic Data for the Complexes
$Fe(CO)_{5-n}(2\text{-}CNC_6H_3Me_2\text{-}1,3)_n$ ($n = 1\text{-}5$) see Ref. 4)

Complex	IR[a] Frequencies (cm^{-1})		NMR[b] δ Values
	ν_{NC}	ν_{CO}	CH_3
$Fe(CO)_4(2\text{-}CNC_6H_3Me_2\text{-}1,3)$	2151	2051, 1999, 1975	1.93
$Fe(CO)_3(2\text{-}CNC_6H_3Me_2\text{-}1,3)_2$	2108	2000, 1938	2.08
$Fe(CO)_2(2\text{-}CNC_6H_3Me_2\text{-}1,3)_3$	2065, 2045 (sh)	1940, 1906	2.17
$Fe(CO)(2\text{-}CNC_6H_3Me_2\text{-}1,3)_4$	2045, 1990	1903	2.31
$Fe(2\text{-}CNC_6H_3Me_2\text{-}1,3)_5$	2028, 1960, 1920 (sh)		2.39

[a] $n = 1$ recorded in hexane; $n = 2$ recorded in $CHCl_3$; $n = 3\text{-}5$ recorded in C_6H_6.
[b] $n = 1\text{-}5$ recorded in C_6D_6 relative to TMS.

(operating temperatures 25–200°C); a molecular ion is observed in each case.[4] This series of complexes have been shown to be precursors to a varied and rich chemistry, particularly involving the isonitrile ligand and the ease of oxidation of the electron-rich iron(0) complexes.[4, 13, 14] In addition, the spectroscopic and structural details of these complexes have attracted attention within the context of five coordination at iron(0) centers.[9, 13, 15]

References

1. E. Singleton and H. E. Oosthuizen, Adv. Organometal. Chem., **22**, 209 (1983).
2. W. Hieber and D. von Pigenot, Chem. Ber., **89**, 193 (1956).
3. W. Strohmeier and F. J. Müller, Chem. Ber., **102**, 3613 (1969).
4a M. O. Albers, N. J. Coville, and E. Singleton, J. Chem. Soc. Dalton Trans., **1982**, 1069.
4b. S. B. Butts and D. F. Shriver, J. Organometal. Chem., **169**, 191 (1979).
5. A. Reckziegel and M. Bigorgne, J. Organometal. Chem., **3**, 341 (1965).
6. S. Grant, J. Newman, and A. R. Manning, J. Organometal. Chem., **96**, C11 (1975).
7. D. Gladkowski and F. R. Scholer, Abstracts of Papers from the Centennial American Chemical Society Meeting, New York, 1976, INOR 133.
8. J. M. Basset, D. E. Berry, G. K. Barker, M. Green, J. A. K. Howard, and F. G. A. Stone, J. Chem. Soc. Dalton Trans., **1979**, 1003.
9. M. O. Albers, N. J. Coville, T. V. Ashworth, E. Singleton, and H. E. Swanepoel, J. Chem. Soc. Chem. Commun., **1980**, 489.
10. D. F. Shriver and M. A. Drezdzon, The Manipulation of Air-Sensitive Compounds, 2nd ed., McGraw-Hill, New York, 1986.
11. D. D. Perrin, W. L. F. Armarego, and D. R. Perrin, Purification of Laboratory Chemicals, 2nd ed., Pergamon Press, Oxford, 1980.
12. M.O. Albers, N. J. Coville, T. V. Ashworth, E. Singleton, and H. E. Swanepoel, J. Organometal. Chem., **199**, 55 (1980).
13. M. O. Albers, E. Singleton, and N. J. Coville, unpublished results.

14. J. M. Bassett, M. Green, J. A. K. Howard, and F. G. A. Stone, *J. Chem. Soc. Dalton Trans.*, **1980**, 1779; J. M. Bassett, L. J. Farrugia, and F. G. A. Stone, *J. Chem. Soc. Dalton Trans.*, **1980**, 1789.
15. G. W. Harris, J. C. A. Boeyens, and N. J. Coville, *Acta Crystallogr. Sect. C*, **39**, 1180 (1983).

47. DICARBONYL(η^5-CYCLOPENTADIENYL)-(THIOCARBONYL)IRON(1 +) TRIFLUOROMETHANE-SULFONATE(1 −) AND DICARBONYL(η^5-CYCLO-PENTADIENYL)[(METHYLTHIO)THIOCARBONYL]IRON

$$[(\eta^5\text{-}C_5H_5)_2Fe_2(CO)_4] + Na/Hg \rightarrow 2[(\eta^5\text{-}C_5H_5)Fe(CO)_2]^-$$

$$[(\eta^5\text{-}C_5H_5)Fe(CO)_2]^- + CS_2 \rightarrow [(\eta^5\text{-}C_5H_5)Fe(CO)_2(CS_2)]^-$$

$$[(\eta^5\text{-}C_5H_5)Fe(CO)_2(CS_2)]^- + CH_3I \rightarrow$$

$$(\eta^5\text{-}C_5H_5)Fe(CO)_2[(CS)SCH_3] + I^-$$

$$(\eta^5\text{-}C_5H_5)Fe(CO)_2[(CS)SCH_3] + CF_3SO_3H \rightarrow$$

$$[(\eta^5\text{-}C_5H_5)Fe(CO)_2(CS)]^+[SO_3CF_3]^- + CH_3SH$$

Submitted by B. DUANE DOMBEK,* MOON-GUN CHOI,* and ROBERT J. ANGELICI*
Checked by IAN S. BUTLER† and DANIEL COZAK†

The dicarbonyl(η^5-cyclopentadienyl)(thiocarbonyl)iron(1 +) cation, $[(\eta^5\text{-}C_5H_5)Fe(CO)_2(CS)]^+$, has been prepared by the reaction of the $[(\eta^5\text{-}C_5H_5)Fe(CO)_2]^-$ ion with $ClC(S)OCH_3$ or $ClC(S)OC_2H_5$ followed by treatment of the resulting intermediate with hydrogen chloride.[1,2] A second procedure[2] involved reaction of the $[(\eta^5\text{-}C_5H_5)Fe(CO)_2]^-$ ion with carbon disulfide, followed by methyl iodide to give the dithioester, dicarbonyl(η^5-cyclopentadienyl)[(methylthio)(thiocarbonyl)]iron, $(\eta^5\text{-}C_5H_5)Fe(CO)_2[C(S)SCH_3]$. On reaction with hydrogen chloride, the $[(\eta^5\text{-}C_5H_5)Fe(CO)_2(CS)]^+$ ion was produced. The low yields obtained by this latter route made it of limited synthetic utility. Now we report an improved procedure for this reaction which provides a convenient, high-yield synthesis of the $[(\eta^5\text{-}C_5H_5)Fe(CO)_2(CS)]^+$ ion by using trifluoromethanesulfonic acid.

---

*Department of Chemistry, Iowa State University, Ames, IA 50011.
†Department of Chemistry, McGill University, Montreal, Quebec, Canada H3A 2K6.

Procedure

■ **Caution.** *The drying of tetrahydrofuran (THF) may be accompanied by serious explosions under certain conditions.* [*See Inorg. Synth.*, **12**, *111, 317 (1970)*].

The starting material, $[(\eta^5\text{-}C_5H_5)Fe(CO)_2]_2$, may be prepared[3] from iron pentacarbonyl and cyclopentadiene dimer, or it may be purchased commercially. Before use, dicarbonyl(η^5-cyclopentadienyl)iron dimer may be recrystallized by dissolving it in a minimum amount of dichloromethane, filtering the saturated solution through Celite filter aid, adding an equal volume of *n*-hexane to the solution, and cooling to $-20°$.

A 500-mL three-necked flask[4] with a stopcock fused to the bottom and equipped with a mechanical stirrer (Fig. 1), is flushed well with nitrogen, and 35 mL of mercury is added. While rapidly stirring the mercury, 2.0 g (87 mmol) of sodium is added to the flask against a countercurrent of nitrogen, in pieces of about 0.4 g cut under a hydrocarbon solvent. After the amalgam has cooled, dry tetrahydrofuran (250 mL) is added to the flask, followed by 10.0 g (28.3 mmol) of dicarbonyl(η^5-cyclopentadienyl)iron dimer. The mixture is stirred *vigorously* for ~ 1 hr.

The sodium amalgam is then drained out of the flask through the bottom stopcock. The mercury may be washed with ethanol and water and reused.[5] Carbon disulfide (5 mL, 73 mmol) is added rapidly to the solution under a nitrogen atmosphere, and the mixture is stirred for 20–30 s. *Longer reaction times reduce the yield.* Methyl iodide (5 mL, 79 mmol) is then added immediately. The mixture is stirred for ~ 5 min under a nitrogen atmosphere. The

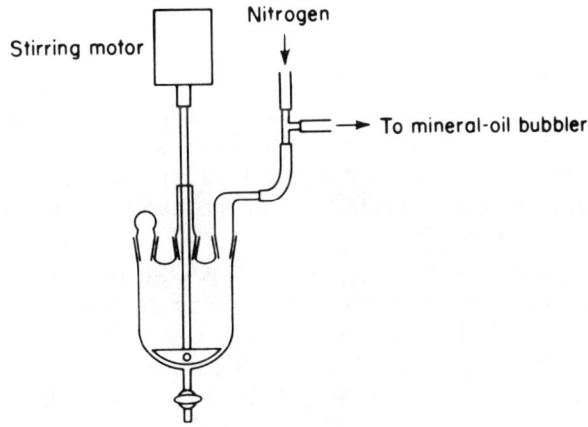

Fig. 1. Apparatus for the preparation of dicarbonyl(η^5-cyclopentadienyl)-[(methylthio)thiocarbonyl]iron.

operations beyond this point may be done in air. The reaction mixture is then decanted into a 1000-mL round-bottomed flask and evaporated to dryness on a rotary evaporator.

The $(\eta^5\text{-}C_5H_5)Fe(CO)_2[C(S)SCH_3]$ may be isolated at this stage by extracting the residue with diethyl ether until the wash is colorless. The solution is filtered through Celite filter aid and concentrated by evaporation under vacuum. Cooling to $-20°$ gives 7.5–9.5 g of brown crystals of $(\eta^5\text{-}C_5H_5)Fe(CO)_2[C(S)SCH_3]$ (49–62% yield).

Anal. Calcd. for $C_9H_8FeO_2S_2$: C, 40.3; H, 2.98; S, 23.8. Found: C, 40.4; H, 2.95; S, 22.7.

In the direct preparation of the $[(\eta^5\text{-}C_5H_5)Fe(CO)_2(CS)]^+$ ion, the dithioester is not normally isolated. After the volume of the extracted Et_2O solution is reduced to about 100 mL, trifluromethanesulfonic acid (5.6 mL, 63 mmol in 60 mL of Et_2O) is added slowly and dropwise (~ 1 drop s^{-1}) by using a dropping funnel. As the reaction mixture is stirred for an additional 2 h, a yellow-brown precipitate of the product is formed. The precipitate is filtered off and washed with Et_2O.

The crude product is then dissolved in 150 mL of acetone, and 350 mL of hexane or diethyl ether is added to reprecipitate the product. The yellow-brown powder is filtered and washed successively with Et_2O (~ 50 mL), THF (~ 30 mL), Et_2O (~ 50 mL) to remove the brown impurity, yielding 10.4 g (50% yield) of the bright yellow product.

Anal. Calcd. for $C_9H_5F_3FeO_5S_2$: C, 29.21; H, 1.36. Found: C, 29.21; H, 1.38.

Properties

The dithioester complex $(\eta^5\text{-}C_5H_5)Fe(CO)_2[C(S)SCH_3]$, is an air-stable, brown solid with a melting point of 72°, soluble in polar and nonpolar organic solvents. The infrared spectrum of its hexane solutions shows two strong carbonyl absorptions at 2035 and 1988 cm^{-1}. Its ^1H NMR spectrum taken in CDCl$_3$ shows a singlet C_5H_5 resonance at δ 4.91 and singlet —SCH$_3$ resonance at δ 2.63 ppm.

The thiocarbonyl complex $[(\eta^5\text{-}C_5H_5)Fe(CO)_2(CS)]^+[SO_3CF_3]^-$ crystallizes as air-stable yellow crystals from acetone–diethyl ether. The infrared spectrum in acetonitrile shows two strong carbonyl absorptions (2105 and 2071 cm^{-1}) and a C—S stretching band at 1353 cm^{-1}. An X-ray crystallographic study of $[(\eta^5\text{-}C_5H_5)Fe(CO)_2(CS)]PF_6$ confirms its formulation as a thiocarbonyl complex.[6] Its ^1H NMR spectrum taken in acetone-d_6 solvent exhibits a singlet C_5H_5 resonance at δ 6.05 ppm. The thiocarbonyl ligand in this complex is very susceptible to attack by nucleophiles.[7] The photolysis of

$[(\eta^5\text{-}C_5H_5)Fe(CO)_2(CS)]^+[SO_3CF_3]^-$ yields an unstable but useful intermediate, $(\eta^5\text{-}C_5H_5)Fe(CO)(CS)(SO_3CF_3)$, for the synthesis of other substituted products.[8] Conversion to the $[(\eta^5\text{-}C_5H_5)Fe(CO)_3]^+$ ion has been observed in moist acetone or acetonitrile.[9] The analogous $[(\eta^5\text{-}C_5Me_5)Fe(CO)_2(CS)]BF_4$ has also been prepared.[10]

References

1. L. Busetto and R. J. Angelici, *J. Am. Chem. Soc.*, **90**, 3283 (1968).
2. L. Busetto, U. Belluco, and R. J. Angelici, *J. Organometal. Chem.*, **18**, 213 (1969).
3. R. B. King and F. G. A. Stone, *Inorg. Synth*, **7**, 110 (1963).
4. R. B. King, *Organometallic Syntheses*, Vol. 1, Academic Press, Inc., New York, 1965, p. 149.
5. Ibid., p. 150.
6. J. W. Richardson, Jr., R. J. Angelici, and R. A. Jacobson, *Inorg. Chem.*, **26**, 452 (1987).
7. L. Busetto, M. Graziani, and U. Belluco, *Inorg. Chem.*, **10**, 78 (1971).
8. R. J. Angelici and J. W. Dunker, *Inorg. Chem.*, **24**, 2209 (1985).
9. B. D. Dombek and R. J. Angelici, unpublished results.
10. M. G. Choi, L. M. Daniels, and R. J. Angelici, *J. Organometal, Chem.*, **383**, 321 (1990).

48. TETRACARBONYLBIS(η^5-CYCLOPENTADIENYL)-DIRUTHENIUM

Submitted by N. M. DOHERTY,* S. A. R. KNOX,* and M. J. MORRIS*
Checked by C. P. CASEY† and G. T. WHITEKER†

The generation and interconversion of hydrocarbon fragments on metal surfaces is an important aspect of transition metal catalysis.[1] In an effort to model and understand these transformations, much attention has been focused on the synthesis and reactivity of organic species coordinated at polynuclear transition metal centers.[2,3] Organodiruthenium complexes have provided a particularly rich area of study. The availability of a variety of organometallic derivatives of the bis(η^5-cyclopentadienyl)diruthenium carbonyl system has allowed extensive examination of the reactivity of a range of hydrocarbon ligands.[4,5]

The starting material for preparation of these derivatives, $[Ru_2(CO)_4(\eta^5\text{-}C_5H_5)_2]$, has previously been obtained by the reaction of sodium cyclopentadienide with a dihaloruthenium(II) carbonyl $\{[Ru(CO)_2I_2]$ (Ref. 6) or $[Ru(CO)_3Cl_2]_2$ (Ref. 7)$\}$ prepared by carbonylation of the corresponding ruthenium(III) trihalide. A more facile synthesis was later reported involving

---

*Department of Inorganic Chemistry, The University, Bristol BS8 1TS, United Kingdom.
†Department of Chemistry, University of Wisconsin, Madison, WI 53706.

the reaction of triruthenium dodecacarbonyl with cyclopentadiene.[8] The procedure described herein represents a modification of this second method, resulting in an improved yield.

$$\tfrac{2}{3}[Ru_3(CO)_{12}] + 2C_5H_6 \longrightarrow$$

$$2[RuH(CO)_2(\eta^5\text{-}C_5H_5)] \xrightarrow{O_2} [Ru_2(CO)_4(\eta^5\text{-}C_5H_5)_2]$$

Procedure

A three-necked, 500-mL, round-bottomed flask equipped with a nitrogen bypass, a reflux condenser, and a magnetic stirring bar is charged with triruthenium dodecacarbonyl (8.5 g, 0.013 mol) (best prepared by the carbonylation of $RuCl_3$ in methanol[9]), 350 mL of dry, deoxygenated heptane, and freshly distilled cyclopentadiene (17.5 g, 0.265 mol) [prepared by cracking dicyclopentadiene over iron filings under a nitrogen atmosphere, and collecting the cyclopentadiene distillate (40–45°) from a 12-in fractionating column]. The mixture is heated at reflux for 1 h, producing $[RuH(CO)_2(\eta^5\text{-}C_5H_5)]$. A stopper is then removed from the flask and the volume of solvent is reduced to 50 mL by continued heating at reflux under a brisk flow of nitrogen, allowing the heptane to boil away. (This procedure is performed in a fume hood.) At this point orange product begins to crystallize from the reaction mixture. An additional 300 mL of untreated heptane, obtained directly from the reagent bottle, is added to the flask, the stopper is replaced, and the solution is heated at reflux for a further 2 h. On cooling to room temperature the reaction mixture affords orange crystals of the product. The solid is collected by decantation and washed three times with 30-mL portions of hexane. After drying under vacuum the yield is 7.15 g. Additional product is obtained from the decanted solution and hexane washings by evaporation of the solvent followed by chromatography on a 3 × 20 cm alumina (Brockman grade 2) column. Elution with dichloromethane–hexane (1:3) removes unreacted starting materials and impurities as yellow bands. Elution with dichloromethane–hexane (1:1) develops a yellow band from which 0.77 g of product is obtained. Overall yield of product is 90–95% by this procedure.

Anal. Calcd. for $C_{14}H_{10}O_4Ru_2$: C, 37.8; H, 2.3. Found: C, 37.6; H, 2.1.

Properties

The compound is an air-stable, orange crystalline solid, soluble in common organic solvents. Solutions decompose slowly in air on exposure to light.

The IR spectrum (in CH_2Cl_2) shows bands at 2003 (s), 1966 (s), 1934 (m), and 1771 (s) cm^{-1} due to the carbonyl ligands.

References

1. E. L. Muetterties and J. Stein, *Chem. Rev.*, **79**, 479 (1979).
2. See, for example, W. A. Herrmann, *Angew. Chem., Int. Ed. Engl.*, **21**, 117 (1982).
3. "Reactivity of Hydrocarbyl Ligands," in *Polyhedron* (Symposium in print No. 5), M. H. Chisholm (ed.), **7**, 757 (1988).
4. S. A. R. Knox, *Pure Appl. Chem.*, **56**, 81 (1984) and references cited therein.
5. G. S. Lewandos, N. M. Doherty, S. A. R. Knox, K. A. Macpherson, and A. G. Orpen, *Polyhedron*, **7**, 837 (1988) and references cited therein.
6. E. O. Fischer and A. Vogler, *Z. Naturforsch.*, **B17**, 421 (1962).
7. T. Blackmore, M. I. Bruce, and F. G. A. Stone, *J. Chem. Soc. A*, **1968**, 2158.
8. A. P. Humphries and S. A. R. Knox, *J. Chem. Soc. Dalton Trans.*, **1975**, 1710.
9. M. I. Bruce and C. H. Hameister, see *Comprehensive Organometallic Chemistry*, Vol. 4, G. Wilkinson, F. G. A. Stone, and E. W. Abel (eds.), Pergamon Press, New York 1982, p. 664.

Chapter Four

METAL CARBONYL ANION COMPLEXES

49. SODIUM HEXACARBONYLNIOBATE(1−)

Submitted by FAUSTO CALDERAZZO* and GUIDO PAMPALONI*
Checked by J. E. ELLIS†

$$2NbCl_5 + 12CO + 6 Mg \xrightarrow{pyridine} Mg[Nb(CO)_6]_2 + 5MgCl_2$$

$$Mg[Nb(CO)_6]_2 + 2NaOH \longrightarrow 2Na[Nb(CO)_6] + Mg(OH)_2$$

$$Na[Nb(CO)_6] \xrightarrow{THF} Na[Nb(CO)_6] \cdot THF$$

The literature covering the preparative methods of the hexacarbonylniobate-(1−) anion has been reviewed.[1] Most of the known methods require reductive carbonylation of $NbCl_5$ at elevated CO pressures,[1,2] with yields as high as 14% obtained with Na–K alloy.[2] Recently, a 30% yield of $[Na(CH_3OCH_2CH_2)_2O)_2][Nb(CO)_6]$ has been achieved[3] by carbonylating $NbCl_5$ at atmospheric pressure of CO by using sodium naphthalene as reducing agent in 1,2-dimethoxyethane at −80°C.

By using pyridine as the solvent and magnesium–zinc as a reducing agent, $NbCl_5$ can be carbonylated to the $[Nb(CO)_6]^−$ anion at room temperature and at *atmospheric pressure* of carbon monoxide. The method does not require the use of the hazardous alkali metals and gives better yields than the ones reported earlier.

---

*Dipartimento di Chimica e Chimica Industriale, Sezione di Chimica Inorganica, Via Risorgimento 35, University of Pisa, 56126 Pisa, Italy.
†Department of Chemistry, University of Minnesota, Minneapolis, MN 55455.

Procedure

■ **Caution.** *All operations must be carried out in an efficient fume hood, due to the poisonous nature of carbon monoxide. Use care when handling CaH_2 and avoid flame.*

Sodium Hexacarbonylniobate(1 −) Tetrahydrofuran

Purification of the pyridine solvent is critical, and must be carried out with great care, to eliminate water. It is first dried over KOH pellets, refluxed over CaH_2 under prepurified nitrogen for about 36 h and then distilled from it. The distillation heads must be tested for the presence of water by addition of sodium and naphthalene, and are discarded until a persistent violet color is observed. The solvent to be used in the reaction (1100 mL) is then directly distilled into the reaction flask containing the magnetically stirred mixture of magnesium (12 g, 0.49 mol) and zinc (20 g, 0.31 mol) powders.* At the end of the distillation, the flask is evacuated and is then filled with carbon monoxide.† In a stream of carbon monoxide, freshly sublimed‡ niobium pentachloride (18.26 g, 67.6 mmol) is added. The reaction flask is then connected to a gas buret containing carbon monoxide over mercury, and the reaction mixture is vigorously stirred magnetically. The temperature is maintained at 16° by an external water bath operated by a thermostat.

Reduction of $NbCl_5$ occurs rapidly, as denoted by the change in color of the reaction mixture from yellow-orange to violet and then blue. When the reaction mixture is greenish, a fast carbon monoxide absorption starts, and continues at an approximate rate of $2 \, L \, h^{-1}$ up to a molar CO/Nb ratio of about 3.§ After that, a considerable decrease in the gas absorption rate is noted. The reaction mixture is vigorously stirred for about 30 h when the molar CO/Nb ratio reaches approximately 6. The red-brown suspension is

---

*Magnesium and zinc powders are reagent grade and are used without further purification. Activation of magnesium by iodine can be used, but is unnecessary.

†The submitters used either pure, dry carbon monoxide, free from molecular hydrogen, or commercially available carbon monoxide containing 1–3% of molecular hydrogen, with equally satisfactory results.

‡Commercially available $NbCl_5$ must be purified by double sublimation at 100–110° $\sim 10^{-2}$ torr. In the case of particularly inpure samples, the chloride must be treated with refluxing sulfinyl chloride[4] followed by evaporation of the solvent and double sublimation of the resulting solid residue.

§The checker did not measure CO absorption and maintained the CO pressure at approximately 800 torr throughout the reaction period ($\sim$ 30hr). CP-grade CO was purified of CO_2 by passing through a column of Ascarite, molecular sieves, and activated copper.

decanted after about 12 h, and then filtered under an atmosphere of carbon monoxide.*

The filtered solution is evaporated under vacuum to dryness† at room temperature (17°, 10^{-1} torr), and the resulting deep brown residue is treated rapidly with 500 mL of a 1.2 M aqueous solution of NaOH while the temperature is controlled with an external ice bath. The resulting orange mixture is transferred into a separatory funnel equipped with an upper stopcock for operating under an inert atmosphere, and extracted several times with diethyl ether free of peroxides (distilled over Li[AlH$_4$]). The nature of the suspension is such that the precipitate of Mg(OH)$_2$ can easily be eliminated by way of the lower exit of the separatory funnel, together with the aqueous layer.

The combined diethyl ether extracts are evaporated to dryness initially by a water pump and finally by a mechanical pump. The oily yellow residue becomes a microcrystalline powder by addition of 50 mL of prepurified tetrahydrofuran (THF). After cooling the resulting suspension to Dry Ice temperature, the crude hexacarbonylniobate(1 −) is collected by filtration and then recrystallized by dissolution in 300 mL of tetrahydrofuran at 40°, filtration, and cooling to room temperature. Completion of the crystallization is achieved by cooling the mother liquor to Dry Ice temperature overnight. The sodium derivative is collected by filtration and dried under vacuum for about 5 h. Yield: 9.4 g, corresponding to 39.1%‡.

Anal. Calcd. for [Na(C$_4$H$_8$O)][Nb(CO)$_6$]; C$_{10}$H$_8$NaNbO$_7$: CO, 47.2; Nb, 26.1. Found: CO, 48.0; Nb, 25.0.

Properties

The sodium derivative, Na[Nb(CO)$_6$], stabilized by tetrahydrofuran, is a yellow-orange solid that is extremely sensitive to oxygen, and it can contain variable amounts of tetrahydrofuran. Some attempts to reduce the amount of tetrahydrofuran present in the solid (by evacuation at 20°) have resulted in

---

* Gas-volumetric measurements of the filtered solution by decomposition with iodine showed that the carbonylation yields vary by 44–60%, depending on the operating conditions. Atomic absorption analyses of the cationic metals contained in the solution give magnesium–zinc molar ratios ranging from 9 to 24, showing that magnesium is mainly responsible for the reduction. However, the simultaneous presence of both metals gives the best results. The checker used a medium porosity filtration unit and filtered under an atmosphere of nitrogen.

†From now on, the operations are carried out under reduced pressure, as indicated, or under an atmosphere of prepurified argon.

‡Yields as high as 47% of recrystallized product based on initial niobium have been obtained.

complete decomposition. When a product with a low content of tetrahydrofuran is desired, the treatment of the solid under vacuum should be done carefully. If the solid shows signs of blackening, the operation should be discontinued immediately. The sodium derivative is quite soluble in diethyl ether, sparingly soluble in tetrahydrofuran at room temperature, very slightly soluble in dichloromethane, and substantially insoluble in hydrocarbons. It is quite soluble in water, in which it forms stable, yellow-orange solutions under a carbon monoxide atmosphere, at a pH of 7–8 or higher. At pH values even slightly lower than 7, rapid decomposition occurs with evolution of carbon monoxide and molecular hydrogen. The aqueous solution is characterized by a single CO stretching vibration at 1875 cm^{-1} (CaF$_2$, 0.01-mm cell), which should be compared with the 1862 cm^{-1} value for [V(CO)$_6$]$^-$ under similar conditions. The tetrahydrofuran solution has a main band at 1860 cm^{-1} and a shoulder at 1887 cm^{-1}, presumably caused by distortion of the octahedral structure by the countercation.

From the aqueous solution of [Nb(CO)$_6$]$^-$, the nickel–1,10-phenanthroline derivative [Ni(phen)$_3$][Nb(CO)$_6$]$_2$ can be precipitated as a brick-red, microcrystalline solid. From a dichloromethane suspension of Na[Nb(CO)$_6$], the μ-nitrido-bis(triphenylphosphorus)(1 +) derivative [(PPh$_3$)$_2$N][Nb(CO)$_6$], is obtained as a yellow solid, soluble in dichloromethane, after addition of (PPh$_3$)$_2$NCl and recrystallization from dichloromethane–diethyl ether. The crystal and molecular structures of ionic [(PPh$_3$)$_2$N][Nb(CO)$_6$], determined by X-ray diffraction methods, show that [Nb(CO)$_6$]$^-$ has an almost perfect octahedral geometry.[5]

References

1. J. E. Ellis and A. Davison, *Inorg. Synth.*, **16**, 68 (1976).
2. (a) R. P. M. Werner and H. E. Podall, *Chem. Ind. (London)*, 144 (1961); (b) R. P. M. Werner, A. H. Filbey, and S. A. Manastyrskyi, *Inorg. Chem.*, **3**, 298 (1964).
3. C. G. Dewey, J. E. Ellis, K. L. Fjare, K. M. Pfahl, and G. F. P. Warnock, *Organometallics*, **2**, 388 (1983).
4. C. Brauer, *Handbook of Preparative Inorganic Chemistry*, Vol. 2, 2nd ed., Academic Press, New York, 1963, p. 1303.
5. F. Calderazzo, U. Englert, G. Pampaloni, G. Pelizzi and R. Zamboni, *Inorg. Chem.*, **22**, 1865 (1983).

50. DICARBONYL(η^5-CYCLOPENTADIENYL)NITROSYL COMPLEXES OF CHROMIUM, MOLYBDENUM, AND TUNGSTEN

Submitted by TEEN T. CHIN,* JAMES K. HOYANO,* PETER LEGZDINS,* and JOHN T. MALITO*
Checked by THOMAS ARNOLD† and BASIL I. SWANSON†

$$NaC_5H_5 + M(CO)_6 \longrightarrow Na[(\eta^5\text{-}C_5H_5)M(CO)_3] + 3CO \quad (1)$$

$$Na[(\eta^5\text{-}C_5H_5)M(CO)_3] + p\text{-}CH_3C_6H_4SO_2N(NO)CH_3 \xrightarrow{THF}$$

$$(\eta^5\text{-}C_5H_5)M(CO)_2(NO) + CO + p\text{-}CH_3C_6H_4SO_2N(CH_3)Na \quad (2)$$

The dicarbonyl(η^5-cyclopentadienyl)nitrosyl complexes of chromium, molybdenum, and tungsten, $(\eta^5\text{-}C_5H_5)M(CO)_2(NO)$ (M = Cr, Mo, W), are convenient synthetic precursors to a variety of organometallic nitrosyl complexes.[1] The compound $(\eta^5\text{-}C_5H_5)Cr(CO)_2NO$ may be prepared in good yield by the action of nitric oxide on $[(\eta^5\text{-}C_5H_5)Cr(CO)_3]_2$,[2] but the latter reagent can be obtained only in low yields and with the expenditure of much effort.[3] Preparative routes leading to the analogous carbonyl nitrosyl complexes of Mo and W have involved treatment of (a) aqueous solutions of $Na[(\eta^5\text{-}C_5H_5)M(CO)_3]$ (M = Mo or W) with nitric oxide,[4] (b) $[(\eta^5\text{-}C_5H_5)M(CO)_3]_2$ (M = Mo or W) with NO(g) under UV irradiation,[5] and (c) $(\eta^5\text{-}C_5H_5)Mo(CO)_3H$ with $p\text{-}CH_3C_6H_4SO_2N(NO)CH_3$ (Diazald) in THF.[6–8] These methods suffer either from low yields of the products or the non-ready availability of the requisite reagents. The procedures described below are of general applicability to all three compounds and provide high yields of the desired materials by utilizing commercially available or easily preparable starting materials.

The $(\eta^5\text{-}C_5H_5)M(CO)_2(NO)$ (M = Cr, Mo, W) compounds are obtained by treatment of their corresponding anions, $[(\eta^5\text{-}C_5H_5)M(CO)_3]^-$, with $p\text{-}CH_3C_6H_4SO_2N(NO)CH_3$ (Diazald), a cleaner nitrosylating agent than NO or ClNO in this case.[9] To effect the preceding reactions in high yield, it is of paramount importance that the $Na[(\eta^5\text{-}C_5H_5)M(CO)_3]$ salts prepared in reaction (1) do not contain a large excess of unreacted NaC_5H_5. The molybdenum and tungsten salts are obtained according to the published procedure,[6] although refluxing for 3 days is advised for the preparation of the

---

*Department of Chemistry, The University of British Columbia, Vancouver, B.C., Canada, V6T 1Y6.
†Department of Chemistry, The University of Texas at Austin, Austin, TX 78712.

tungsten compound to ensure complete conversion. The chromium analogue is best prepared in the following manner.

Procedure

■ **Caution.** *Diazald* [p-$CH_3C_6H_4SO_2N(NO)CH_3$] *is a severe irritant, and any skin contact with it should be avoided. Also, $M(CO)_6$ ($M = Cr, Mo, W$) compounds are highly volatile and toxic.*

All reactions and manipulations are carried out under nitrogen in a well-ventilated fume hood. A 200-mL three-necked flask is fitted with a nitrogen inlet and stirrer and is thoroughly flushed with prepurified nitrogen. Into the flask is syringed a tetrahydrofuran (THF) solution containing 4.18 g (47.5 mmol) of NaC_5H_5.[3] The THF is removed *in vacuo*, and 11.00 g (50.0 mmol) of $Cr(CO)_6$ (Pressure Chemical Co.) and 100 mL of anhydrous dibutyl ether ([$CH_3(CH_2)_3]_2O$, 99 + %, Aldrich Chemical Co.; deaerated with N_2) are added. The flask is then equipped with a Liebig condenser, and the reaction mixture is refluxed with vigorous stirring for 12 h. During this time the reaction vessel is shaken occasionally to reintroduce any sublimed $Cr(CO)_6$ into the refluxing reaction mixture. The final reaction mixture is allowed to cool to room temperature and filtered, and the pale-yellow solid thus collected is washed with dibutyl ether (3 × 10 mL) followed by anhydrous hexanes (2 × 10 mL) and dried *in vacuo* (0.005 torr). The excess $Cr(CO)_6$ and any dibutyl ether remaining in this solid are removed by sublimation at 90°, 0.005 torr onto a water-cooled probe. The $Na[(\eta^5\text{-}C_5H_5)M(CO)_3]$ complexes of molybdenum and tungsten[6] are freed of any unreacted hexacarbonyl in a similar manner, and all three sodium salts are used without further purification.

The preparations of all three $(\eta^5\text{-}C_5H_5)M(CO)_2(NO)$ complexes from their corresponding $[(\eta^5\text{-}C_5H_5)M(CO)_3]^-$ anions are similar. The experimental procedure, using the tungsten complex [dicarbonyl(η^5-cyclopentadienyl)-nitrosyltungsten] as a typical example, is given below.

A 300-mL three-necked flask is equipped with a nitrogen inlet, an addition funnel, and a stirrer. It is charged with 17.3 g (48.5 mmol) of $Na[(\eta^5\text{-}C_5H_5)W(CO)_3]$ and 120 mL of THF (Fisher Scientific Co., reagent grade, dried by distillation from sodium benzophenone and deaerated with nitrogen). A THF solution (50 mL) containing 10.4 g (48.6 mmol) of Diazald (N-methyl-N-nitroso-p-toluenesulfonamide, Eastman Kodak reagent grade) is syringed into the addition funnel. The solution of Diazald is added dropwise over a period of 15 min to the stirred reaction mixture. Gas evolution occurs and an orange-brown solid precipitates. The mixture is stirred for an additional 15 min, and the solvent is then removed *in vacuo*.

Sublimation of the resulting brown residue at 50–60°, 0.005 torr onto a water-cooled probe for 3 days affords 13.6 g (84% yield) of (η^5-C_5H_5)W(CO)$_2$(NO). The corresponding chromium and molybdenum complexes are obtained similarly in yields of 60 and 93%, respectively.

Anal. Calcd. for $C_5H_5Cr(CO)_2(NO)$: C, 41.39; H, 2.48; N, 6.90. Found: C, 41.40; H, 2.60; N, 6.70.
Calcd. for $C_5H_5Mo(CO)_2(NO)$: C, 34.03; H, 2.04; N, 5.67. Found: C, 34.28, H, 2.24; N, 5.54.
Calcd. for $C_5H_5W(CO)_2(NO)$: C, 25.10; H, 1.50; N, 4.18. Found: C, 25.29; H, 1.70; N, 4.13.

Properties

The (η^5-C_5H_5)M(CO)$_2$(NO) (M = Cr, Mo, W) complexes are orange to orange-red solids that are readily soluble in common organic solvents to afford bright orange, relatively air stable solutions. In the solid state, they are also stable in air for short periods of time, but are best stored under nitrogen at 0°C. Other physical properties of these compounds are as follows:

Complex	IR (Nujol) (cm^{-1})		^{1}H NMR (CDCl$_3$) δ(C$_5$H$_5$), (ppm)	mp (°C) (under N$_2$, Uncorrected)
	ν_{CO}	ν_{NO}		
(η^5-C_5H_5)Cr(CO)$_2$(NO)	2024 (s), 1954 (s)	1713 (s)	5.03	67–68
(η^5-C_5H_5)Mo(CO)$_2$(NO)	2018 (s), 1944 (s)	1686 (s)	5.53	85–86
(η^5-C_5H_5)W(CO)$_2$(NO)	2012 (s), 1933 (m)	1678 (m)	5.60	103–104

References

1. See, for example: (a) J. K. Hoyano, P. Legzdins, and J. T. Malito, *Inorg. Synth.*, **18**, 127 (1978); (b) A. D. Hunter, P. Legzdins, J. T. Martin, and L. Sánchez, *Organomet. Synth.*, **3**, 58 (1987).
2. E. O. Fischer and K. Plesske, *Chem. Ber.*, **94**, 93 (1961).
3. R. B. King and F. G. A. Stone, *Inorg. Synth.*, **7**, 99 (1963).
4. E. O. Fischer, O. Beckert, W. Hafner, and H. O. Stahl, *Z. Naturforsch.*, **10b**, 598 (1955).
5. A. M. Rosan and J. W. Faller, *Synth. React. Inorg. Metal–Org. Chem.*, **6**, 357 (1976).
6. T. S. Piper and G. Wilkinson, *J. Inorg. Nucl. Chem.*, **3**, 104 (1956).
7. R. B. King and M. B. Bisnette, *Inorg. Chem.*, **6**, 469 (1967).
8. D. Seddon, W. G. Kita, J. Bray, and J. A. McCleverty, *Inorg. Synth.*, **16**, 24 (1976).
9. A. E. Crease and P. Legzdins, *J. Chem. Soc. Dalton Trans.*, 1501 (1973).

51. ACETYLPENTACARBONYLMANGANESE AND ACETYLPENTACARBONYLRHENIUM

Metal carbonyl acyl complexes have been an important class of organometallic compound because of the fundamental need to understand the structure and bonding of metal–carbon linkages and the chemistry of organic functional groups that are attached to metal moieties via carbon–donor atoms. These compounds have played a particularly important role in our understanding of alkyl-migration reactions and other types of ligand transformations.

The title complexes are prepared from acetylation of the corresponding pentacarbonylmetalate anions as described in the original literature.[1,2] Acetyl chloride is used as the acetylation reagent. The required pentacarbonylmetalate anions are formed by sodium amalgam reduction of the appropriate metal carbonyl dimer. Good yields of the pentacarbonylmanganate anion have also been obtained by reductive cleavage of the metal-metal bond of $Mn_2(CO)_{10}$ by trialkylborohydride reagents.[3]

A. ACETYLPENTACARBONYLMANGANESE

$$Mn_2(CO)_{10} + 2Na/Hg \rightarrow 2Na[Mn(CO)_5]$$

$$Na[Mn(CO)_5] + CH_3C(O)Cl \rightarrow [[CH_3C(O)]Mn(CO)_5] + NaCl$$

Submitted by C.M. LUKEHART,* G. PAULL TORRENCE,*
and JANE V. ZEILE*
Checked by B. DUANE DOMBEK† and ROBERT J. ANGELICI†

Procedure

A 100-mL two-necked flask is removed from a 130° drying oven and flushed well with prepurified nitrogen, after a gas inlet is attached to the side neck. This flask is charged with 4 mL of mercury and a stirring bar, and then 0.50 g (21.7 mmol) of sodium metal is added to the stirred mercury puddle in small pieces, one at a time, under a continuous nitrogen flush.

■ **Caution.** *This operation should be done in the hood since the dissolution of the sodium metal in mercury is a highly exothermic reaction.*

---

*Department of Chemistry, Vanderbilt University, Nashville, TN 37235.
†Department of Chemistry, Iowa State University, Ames, IA 50010.

After the sodium amalgam has cooled to room temperature, 50 mL of freshly distilled tetrahydrofuran[4] is introduced, followed by the addition of 3.0 g (7.7 mmol) of decacarbonyldimanganese.*

■ **Caution.** *Metal carbonyl compounds are extremely toxic and should be handled in an efficient hood.*

The yellow solution is stirred at 25° under nitrogen for 75 min.

After this time the solution of $Na[Mn(CO)_5]$ is transferred by means of a syringe into another 100-mL two-necked flask that contains a stirring bar and has been flushed well with nitrogen. The solution is cooled to $-78°$ (Dry Ice–acetone bath), and 1.2 mL (17.0 mmol) of acetyl chloride is added with a syringe from a freshly opened bottle. The reaction solution is stirred at $-78°$ for 1 h, the bath is removed, and the stirring is continued for 1 h more. The solvent is removed at reduced pressure (5 torr, 25°), and the solid residue is stirred with 100 mL of hexane for 30 min at 25°. The hexane solution is filtered through a Schlenk frit,[5] and the filtrate is cooled at $-20°$ for 16 h. The crystallized solid is collected on a glass frit in air and dried briefly, using a water aspirator, giving 1.75 g (48%) of an off-white solid. A second crop of slightly impure product (about 0.5 g) is obtained similarly by concentrating the hexane filtrate to one-half the original volume, followed by cooling at $-78°$ for 4 h.

Anal. Calcd. for $C_7H_3MnO_6$: C, 35.29; H, 1.26. Found: C, 35.10; H, 1.35.

Properties

Acetylpentacarbonylmanganese is a moderately volatile, white solid, mp 54.5–56°. It is air stable over at least a 2-day period. It has excellent solubility in most organic solvents. The infrared spectrum in cyclohexane* shows $v(C\equiv O)$ bands at 2114(w), 2049(w), 2011(vs), 2002(s) and a v(acyl) band at 1663(s) cm^{-1}. The ^1H NMR spectrum ($CDCl_3$ vs. TMS) shows a singlet for the methyl resonance at $\delta\,2.57$.

---

*Decacarbonyl dimanganese was purchased from Pressure Chemical Co., Pittsburgh, PA 15201.

B. ACETYLPENTACARBONYLRHENIUM

$$Re_2(CO)_{10} + 2Na/Hg \rightarrow 2Na[Re(CO)_5]$$
$$Na[Re(CO)_5] + CH_3C(O)Cl \rightarrow [[(CH_3C(O)]Re(CO)_5] + NaCl$$

Submitted by K. P. DARST,* C. M. LUKEHART,* L. T. WARFIELD,* and JANE V. ZEILE*
Checked by J. A. GLADYSZ† and J. C. SELOVER†

Procedure

A 100-mL two-necked flask is removed from a 130° drying oven and is flushed well with prepurified nitrogen, while a gas inlet is attached to the side neck. This flask is charged with 4 mL of mercury and a stirring bar, and then 0.35 g (15.2 mmol) of sodium metal is added to the stirred mercury puddle in small pieces, one at a time, under a continuous nitrogen flush.

■ **Caution.** *This operation should be performed in a hood since the dissolution of the sodium metal is a highly exothermic reaction.*

After the sodium amalgam has cooled to room temperature, 50 mL of freshly distilled tetrahydrofuran[4] is introduced into the flask followed by the addition of 4.0 g (6.1 mmol) of dirhenium decacarbonyl.‡

■ **Caution.** *Metal carbonyl compounds are toxic chemicals and they should be handled in the hood.*

The red-orange solution is stirred at 25° under nitrogen for 1 h.

After this time, the solution of $Na[Re(CO)_5]$ is transferred under a nitrogen flush, by means of a syringe into another 100-mL, two-necked flask containing a stirring bar and fitted with a gas inlet. The solution of the anion is cooled to $-78°$ (Dry Ice–acetone bath), and 0.86 mL (12.1 mmol) of acetyl chloride is added dropwise by means of a syringe. The reaction solution is stirred at $-78°$ for 5 min, at which time the bath is removed and stirring is continued for 15 min more at $-20°$. The solvent is removed at reduced pressure (5 torr) using a 0° ice–water bath, after which the solid residue is stirred with 75 mL of hexane for 20 min at 30°. The hexane solution is filtered through a Schlenk frit,[5] and the filtrate is cooled at $-20°$ for 16 h. The residue is extracted with 40 mL more of hexane at 30° and filtered through a Schlenk frit. The two crops of crystallized solid are collected by syringing the supernatant solutions into a third flask, followed by drying the solids at

---

*Department of Chemistry, Vanderbilt University, Nashville, TN 37235.
†Department of Chemistry, University of California, Los Angeles, CA 90024.
‡Dirhenium decacarbonyl was purchased from Strem Chemicals Inc., Danvers, MA 01923.

reduced pressure (5 torr, 25°) for 5 min. The volume of the combined supernatants is reduced to 20 mL at reduced pressure, and this solution is cooled to $-20°$ for 16 h. The third crop of product is collected by the above procedure. The total yield of product is 1.98 g (44%).

Anal. Calcd. for $C_7H_3O_6Re$: C, 22.76; H, 0.82. Found: C, 22.60; H, 0.70.

Properties

Acetylpentacarbonylrhenium is a slightly volatile, pale-yellow solid, mp 79.5–80.5°. It is air stable over at least a 2-day period, and it has excellent solubility in most organic solvents. The infrared spectrum in cyclohexane shows $v(C\equiv O)$ bands at 2130 (w), 2020 (vs), 2005 (s) cm^{-1} and a v(acyl) band at 1622 (m) cm^{-1}. The ^{1}H NMR spectrum (CS$_2$, vs. TMS) shows a singlet for the methyl resonance at δ 2.44.

References

1. T. H. Coffield, J. Kozikowski, and R. D. Closson, *J. Org. Chem.*, **22**, 598 (1957).
2. W. Hieber, G. Braun and W. Beck, *Chem. Ber.*, **93**, 901 (1960).
3. J. A. Gladysz, G. M. Williams, W. Tam, D. L. Johnson, D. W. Parker, and J. C. Selover, *Inorg. Chem.*, **18**, 553 (1979).
4. Appendix, *Inorg. Synth.*, **12**, 317 (1970).
5. R. B. King, in *Organometallic Syntheses*, Vol. 1, J. J. Eisch and R. B. King (eds.), Academic Press, New York, 1965, p. 56.

52. SODIUM CARBONYL FERRATES, $Na_2[Fe(CO)_4]$, $Na_2[Fe_2(CO)_8]$, AND $Na_2[Fe_3(CO)_{11}]$. BIS[μ-NITRIDO-BIS(TRIPHENYLPHOSPHORUS)(1 +)] UNDECA-CARBONYLTRIFERRATE(2 −), $[(Ph_3P)_2N]_2[Fe_3(CO)_{11}]$

$$Fe(CO)_5 + 2NaC_{10}H_8 \xrightarrow{THF} Na_2[Fe(CO)_4] + 2C_{10}H_{18} + CO$$

$$2Fe(CO)_5 + 2NaC_{10}H_8 \xrightarrow{THF} Na_2[Fe_2(CO)_8] + 2C_{10}H_8 + 2CO$$

$$Fe_3(CO)_{12} + 2NaC_{10}H_8 \xrightarrow{THF} Na_2[Fe_3(CO)_{11}] + C_{10}H_8 + CO$$

$$Na_2[Fe_3(CO)_{11}] + 2[(Ph_3P)_2N]Cl \xrightarrow{MeOH} [(Ph_3P)_2N]_2[Fe_3(CO)_{11}] + 2NaCl$$

Submitted by HENRY STRONG,* PAUL J. KRUSIC,† and JOSEPH SAN FILIPPO, Jr.*
Checked by SCOTT KEENAN‡ and RICHARD G. FINKE‡

Carbonylferrates have been the subject of many studies. The well-defined mono-, di-, and trinuclear species $[Fe(CO)_4]^{2-}$, $[Fe_2(CO)_8]^{2-}$, and $[Fe_3(CO)_{11}]^{2-}$ have been obtained by a variety of methods[1-3] in varying yields and degrees of convenience. The procedures described here provide uniform, convenient, high-purity, high-yield syntheses of the sodium salts of these three important reagents. In addition, the preparation of the bis [μ-nitrido-bis(triphenylphosphorus)(1 +)] salt of $[Fe_3(CO)_{11}]^{2-}$ by metathesis of $[(Ph_3P)_2N]Cl$ with $Na_2[Fe_3(CO)_{11}]$ is presented.

Procedure

■ **Caution.** *The toxic nature of the reagents and products requires that these reactions be performed in a well-ventilated fume hood.*

■ **Caution.** *The use of $Li[AlH_4]$ in purifying THF is dangerous. It should not be attempted until it is ascertained that the THF is peroxide-free and also not grossly wet.*[4]

---

*Department of Chemistry, Rutgers University, New Brunswick, NJ 08903.
†Central Research and Development Department, E. I. du Pont de Nemours & Co., Wilmington, DE 19898.
‡Department of Chemistry, University of Oregon, Eugene, OR 97403.

Peroxide-free tetrahydrofuran (THF) is distilled under nitrogen from lithium tetrahydroaluminate. Pentane is distilled under nitrogen from P_4O_{10}. Iron pentacarbonyl (Pressure Chemical) is freshly distilled prior to each use.[5] Commercial triiron dodecacarbonyl (Alfa Products) is obtained wet with methanol, which is removed by subjecting the sample to vacuum (0.1 torr) overnight. Commercial naphthalene (recrystallized quality) is used without further purification. All manipulations are carried out in a nitrogen-flushed dry box or in standard Schlenk apparatus under a nitrogen atmosphere.

Disodium tetracarbonylferrate(2 −), $Na_2[Fe(CO)_4]$, is prepared in a 1-L three-necked, round-bottomed flask, one arm of which is modified to permit the contents of the flask to be filtered under an inert atmosphere.[5] The flask is equipped with a Teflon-coated magnetic stirrer bar and a 200-mL addition funnel. All remaining inlets are sealed with rubber serum stoppers (Ace Scientific), and the vessel is flame-dried.* Under a flush of nitrogen the flask is charged with a weighed quantity (3.45 g, 75.0 mmol) of sodium dispersion (20 μm, 50% by weight) in paraffin (Alfa Products). It is then placed in an ice bath, and a solution of naphthalene (9.90 g, 77.0 mmol) in tetrahydrofuran (500 mL) is added via a stainless steel cannula. The contents of the flask are stirred for 2 h at 0° and the resulting deep-green solution of (naphthalene)-sodium is chilled at $\lesssim -70°$ in a Dry Ice–acetone bath. A solution of freshly distilled iron pentacarbonyl (7.02 g, 36.0 mmol) in tetrahydrofuran (100 mL) is added slowly over a 30-min period, attended by vigorous stirring. Failure to use freshly distilled $Fe(CO)_5$ leads to diminished yields and purity. The deep-green color is gradually replaced by a persistent beige color. At this point, addition is discontinued, and the resulting mixture is stirred an additional hour before being permitted to warm to ambient temperature.

Pentane† (200 mL) is added by cannula to the reaction mixture, which is then stirred for an additional 30 min before the flask is tilted and its contents filtered under a positive pressure of nitrogen through a coarse-frit glass disk filter. The collected snow-white precipitate of $Na_2[Fe(CO)_4]$ is rinsed with two 100-mL portions of pentane,† and the flask is transferred to the dry box, where the contents are dried under vacuum (0.1 torr) for 4 h to give 7.39 g [96%‡ based on $Fe(CO)_5$] of disodium tetracarbonylferrate 2(−).§ Approximate elapsed time for total synthesis is 12 h.

■ **Caution.** *Solid $Na_2[Fe(CO)_4]$ is an exceedingly pyrophoric material.*

Disodium octacarbonyldiferrate(2 −), $Na_2[Fe_2(CO)_8]$, is prepared by a procedure similar to that described above. Thus, a solution of iron pentacar-

---

*Checkers dried the flask in an oven at 150° and flushed it immediately with dry nitrogen.
†Checkers used hexane.
‡Checkers obtained a yield of 83%. They report difficulty in dissolving all of the sodium in THF.
§$Na_2[Fe(CO)_4] \cdot 1.5$ dioxane is available commercially [Aldrich].[2a]

bonyl (14.04 g, 72.0 mmol) in tetrahydrofuran (200 mL) is added over a 30-min period with stirring to the previously described solution of (naphthalene) sodium. Following a work-up procedure equivalent to that described above, the orange precipitate that is obtained is rinsed with three 200-mL portions of pentane. On drying under vacuum the orange solid yields 13.6 g (99%* based on iron pentacarbonyl) of bright yellow $Na_2[Fe_2(CO)_8]$. The addition and removal of THF causes a reversible color change.[2b] Elapsed time for the total synthesis is ~ 3 h.

- **Caution.** *Dry $Na_2[Fe_2(CO)_8]$ is a pyrophoric substance.*

Disodium undecacarbonyltriferrate(2 −), $Na_2[Fe_3(CO)_{11}]$, is prepared by a modification of the above procedure. Thus, in a nitrogen flushed dry box, a 250-mL, single-necked, round-bottomed flask equipped with a Teflon-coated stirring bar is charged with 1.84 g (44 mmol) of sodium dispersion and capped with a rubber septum stopper. The flask is removed from the dry box and cooled in an ice bath; a solution of naphthalene (6.00 g, 47.0 mmol) in tetrahydrofuran (150 mL) is introduced by cannula, and the resulting mixture is stirred for 2 h.

A modified (see above) 1-L three-necked flask equipped with addition funnel and Teflon-coated stirrer bar is charged with 10.07 g (20.0 mmol) of $Fe_3(CO)_{12}$, capped with a rubber septum, and flushed with nitrogen before adding THF (125 mL). The flask is placed in a Dry Ice–acetone bath, and the solution of sodium naphthalene is transferred through a cannula into the 200-mL addition funnel. This solution is added slowly over a period of 1 h to the chilled, well-stirred solution of $Fe_3(CO)_{12}$ in THF. This order of addition is essential; reversal of the indicated order leads to substantial contamination of the product by unidentified side products. The resulting mixture is stirred for an additional 2 h before it is permitted to warm to ambient temperature. The flask is then transferred to the dry box, and the contents are concentrated under vacuum to dryness. The remaining dark red-brown solid is rinsed with three 200-mL portions of pentane and dried under vacuum once again. The isolated yield of $Na_2[Fe_3(CO)_{11}]$ is 10.2 g [98%†based on $Fe_3(CO)_{12}$]. Approximate elapsed time for total synthesis is 4 h.

- **Caution.** *Dry $Na_2[Fe_3(CO)_{11}]$ is a pyrophoric substance.*

Bis[μ-nitrido-bis(triphenylphosphorus)(1 +)] undecacarbonyltriferrate(2 −) is obtained by treating a solution of $Na_2[Fe_3(CO)_{11}]$ (1.2 g, 2.3 mmol) in 25 mL of anhydrous methanol, which is distilled from $Mg(OCH_3)_2$ and is contained in a 250-mL, single-necked, round-bottomed flask with a solution of $[(Ph_3P)_2N]Cl$[6] (Aldrich) (3.0 g, 5.2 mmol) in methanol (25 mL.) The dark red-brown solid that precipitates is collected by suction filtration on a

---

*Checkers obtained a yield of 88% after having initial difficulties in dissolving sodium.
†Checkers obtained 57% yield.

medium-porosity frit under an inert atmosphere. Recrystallization from dichloromethane as previously described[3] yields 2.9 g (81%)* of crystalline, dark red-brown crystals of $[(Ph_3P)_2N]_2[Fe_3(CO)_{11}]$.

Properties

Disodium tetracarbonylferrate(2 −) is a snow-white solid that is extremely sensitive to oxygen. It has a reported[2] solubility of 7×10^{-3} M in THF and can be stored for moderate periods of time in an inert atmosphere, at room temperature, if kept in the dark. The IR spectrum, recorded in N,N-dimethylformamide (DMF), exhibits a stretching frequency at 1730 cm^{-1}, consistent with previous literature reports.[7] The structure of the $[Fe(CO)_4]^{2-}$ anion has been established by X-ray,[8] and the utility of this reagent has been discussed.[9]

Disodium octacarbonyldiferrate(2 −) has been reported previously.[2b] This extremely air-sensitive solid is largely insoluble in most organic solvents, with only marginal solubility in THF. Its IR spectrum, recorded in DMF, shows the following CO stretching vibrations: 1835 (w), 1860 (s), and 1910 (m) cm^{-1}, consistent with previously reported values.[2b] A single-crystal X-ray structure determination of the $[Fe_2(CO)_8]^{2-}$ anion has been carried out.[10]

Disodium undecacarbonyltriferrate(2 −) is a well-known substance that has also been characterized structurally.[11] The IR spectrum of this material, recorded in DMF, shows CO stretching bands at 1940 (s), 1915 (m), and 1880 (w) cm^{-1}, consistent with values observed previously.[7,11]

References

1. R. B. King and F. G. A. Stone, *Inorg. Synth.* **7**, 197 (1963).
2. (a) J. P. Collman, R. G. Finke, J. N. Cawse, and J. I. Brauman, *J. Am. Chem. Soc.*, **99**, 2515 (1977); (b) J. P. Collman, R. G. Finke, P. L. Matlock, R. Wahren, R. G. Komoto, and J. I. Brauman, *ibid.*, **100**, 1119 (1978).
3. H. A. Hodali and D. F. Shriver, *Inorg. Synth.*, **20**, 222 (1980).
4. Anon., *Inorg. Synth.*, **12**, 317 (1970).
5. D. F. Shriver, *Manipulation of Air-Sensitive Compounds*, McGraw-Hill, New York, 1969.
6. J. K. Ruff and W. J. Schlientz, *Inorg. Synth.*, **15**, 84 (1974).
7. W. F. Edgell, M. T. Yang, B. J. Bulkin, R. Bayer, and N. Koizumi, *J. Am. Chem. Soc.*, **87**, 3080 (1965).
8. R. G. Teller, R. G. Finke, J. P. Collman, H. B. Chin, and R. Bau, *J. Am. Chem. Soc.*, **99**, 1104 (1977); J. P. Collman, R. G. Finke, P. L. Matlock, and J. I. Brauman, *ibid.*, **98**, 4685 (1976).
9. J. P. Collman, *Accounts Chem. Res.*, **8**, 342 (1975). A preparative scale synthesis of the solvated complex $Na_2Fe(CO)_4 \cdot 1.5$ dioxane was previously reported; see *Org. Synth.*, **59**, 102 (1980).

*Checkers obtained 71% yield.

10. H. B. Chin, M. B. Smith, R. D. Wilson, and R. Bau, *J. Am. Chem. Soc.*, **96**, 5285 (1974).
11. F. Y.-K. Lo, G. Longoni, P. Chini, L. D. Lower, and L. F. Dahl, *J. Am. Chem. Soc.*, **102**, 7691 (1980).

53. DICARBONYL(η^5-CYCLOPENTADIENYL) (2-METHYL-1-PROPENYL-κC^1)IRON AND DICARBONYL(η^5-CYCLOPENTADIENYL)(η^2-2-METHYL-1-PROPENE)IRON(1 +) TETRAFLUOROBORATE(1 −)

Submitted by MYRON ROSENBLUM,* WARREN P. GIERING,† and SARI-BETH SAMUELS‡
Checked by PAUL J. FAGAN§

Cationic olefin complexes of dicarbonyl(η^5-cyclopentadienyl) iron have been of wide interest in syntheses for a number of years. These complexes, generally isolated as their tetrafluoroborate or hexafluorophosphate salts, have been prepared by the reaction of $Fe(\eta^5-C_5H_5)(CO)_2Br$ with simple olefins in the presence of Lewis acid catalysts,[1] by protonation of allyl ligands in $Fe(\eta^5-C_5H_5)(CO)_2[(allyl)\kappa C^1]$ complexes,[2] or by treatment of these with cationic electrophiles,[3] by hydride abstraction from $Fe(\eta^5-C_5H_5)(CO)_2(alkyl)$ complexes,[4] through reaction of epoxides with $Fe(\eta^5-C_5H_5)(CO)_2$ anion followed by protonation,[5] or by thermally induced ligand exchange between $[Fe(\eta^5-C_5H_5)(CO)_2(\eta^2$-2-methyl-1-propene$)][BF_4]$[5-7] or $[Fe(\eta^5-C_5H_5)(CO)_2(tetrahydrofuran)][BF_4]$[8] and excess olefin.

The latter two methods are often the most convenient. Dicarbonyl(η^5-cyclopentadienyl)(η^2-2-methyl-1-propene)iron(1 +) tetrafluoroborate(1 −) is a readily synthesized crystalline solid that can be stored indefinitely at −20°. When solutions of the salt are heated in 1,2-dichloroethane (65–70°, 5–10 min) or in dichloromethane (40°, 3–4 h) in the presence of 2–3 M equivalents of an olefin, ligand exchange occurs, yielding the derived $[Fe(\eta^5-C_5H_5)(CO)_2(olefin)][BF_4]$ complex.[3,5] The exchange reaction is limited to the preparation of those complexes that are thermodynamically more stable than the 2-methyl-1-propene complex itself under the conditions of the exchange reaction. These generally include terminal, alkyl-substituted olefins, 1,2-dialkyl-substituted olefins, and cycloalkenes. Heteroatoms such as O, N,

---
*Department of Chemistry, Brandeis University, Waltham, MA 02254.
†Department of Chemistry, Boston University, Boston, MA 02215.
‡Union Carbide Corp., Bound Brook, NJ 08805.
§Central Research and Development Department, E. I. du Pont de Nemours & Co., Wilmington, DE 19898.

and S present in the olefin may interfere with formation of the olefin complex through competitive complexation.

The procedure given here can be completed easily within a day. Although specific for the preparation of the 2-methyl-1-propene complex, it can be adapted readily as an alternative method for the preparation of $[\text{Fe}(\eta^5\text{-}C_5H_5)(CO)_2(\eta^2\text{-olefin})][BF_4]$ complexes through metallation of an allyl halide or tosylate, followed by protonation of the monohapto(allyl)iron complex.

A. DICARBONYL(η^5-CYCLOPENTADIENYL)-(2-METHYL-1-PROPENYL-κC^1)IRON

$$[\text{Fe}(\eta^5\text{-}C_5H_5)(CO)_2]_2 + 2\text{Na(Hg)} \xrightarrow{\text{THF}} 2\text{Na}[\text{Fe}(CO)_2(\eta^5\text{-}C_5H_5)]^* + 2\text{Hg}$$

$$\text{Na}[\text{Fe}(CO)_2(\eta^5\text{-}C_5H_5)] + C_4H_7Cl \xrightarrow{\text{THF}}$$
$$\text{Fe}(\eta^5\text{-}C_5H_5)(CO)_2(2\text{-MeC}_3H_4\text{-}\kappa C^1) + \text{NaCl}$$

Procedure

■ **Caution.** *Care should be exercised in the preparation of the mercury amalgam because the initial reaction is highly exothermic. This and all subsequent operations should be carried out in a well-ventilated fume hood.*

A 500-mL three-necked flask with a stopcock at the bottom is fitted with a nitrogen inlet and a motor-driven mechanical stirrer with a Teflon paddle. The flask is flushed thoroughly with nitrogen while being flame-dried, and then 30 mL of mercury is introduced. A pan may be placed under the flask in case of breakage. The mercury is stirred vigorously as 4.5 g (0.196 mol) of sodium metal, cut into small pieces, is slowly added to it. The flask is capped with a rubber septum.

■ **Caution.** *The amalgamation of sodium is highly exothermic. Small pieces of sodium must be added to mercury behind a shield.*

After the resulting hot amalgam has cooled to room temperature, 200 mL of tetrahydrofuran (THF), which is predried over KOH and then freshly distilled under a nitrogen atmosphere from sodium benzophenone ketyl, is

---

*For smaller-scale preparations, $K[\text{Fe}(CO)_2(\eta^5\text{-}C_5H_5)]$, available from Alfa Division of Ventron Corporation or preparable by reduction of $[\text{Fe}(\eta^5\text{-}C_5H_5)(CO)_2]_2$ with potassium metal,[9] may be used.

The product may be stored indefinitely under nitrogen at $-20°$ without decomposition. It is soluble in dichloromethane, acetone, and nitromethane but insoluble in hydrocarbons and in diethyl ether.

References

1. E. O. Fischer and K. Fichtel, *Chem. Ber.*, **94**, 1200 (1961); **95**, 2063 (1962); E. O. Fischer and E. Moser, *Inorg. Synth.*, **12**, 38 (1970).
2. M. L. H. Green and P. L. I. Nagy, *J. Chem. Soc.*, **1963**, 189.
3. A. Cutler, D. Ehntholt, P. Lennon, K. Nicholas, D. F. Marten, M. Madhavarao, S. Raghu, A. Rosan, and M. Rosenblum, *J. Am. Chem. Soc.*, **97**, 3149 (1975).
4. M. L. H. Green and P. L. I. Nagy, *J. Organometal. Chem.*, **1**, 58 (1963).
5. W. P. Giering, M. Rosenblum, and J. Tancrede, *J. Am. Chem. Soc.*, **94**, 7170 (1972); A. Cutler, D. Ehntholt, W. P. Giering, P. Lennon, S. Raghu, A. Rosan, M. Rosenblum, J. Tancrede, and D. Wells, *ibid.*, **98**, 3495 (1976).
6. W. P. Giering and M. Rosenblum, *J. Chem. Soc., Chem. Commun.*, **1971**, 441.
7. B. Foxman, D. Marten, A. Rosan, S. Raghu, and M. Rosenblum, *J. Am. Chem. Soc.*, **99**, 2160 (1977).
8. D. L. Reger and C. Coleman, *J. Organometal. Chem.*, **131**, 153 (1977).
9. J. S. Plotkin and S. G. Shore, *Inorg. Chem.*, **20**, 284 (1981).
10. J. J. Eisch and R. B. King (eds.), *Organometallic Syntheses*, Vol. 1, Academic Press, New York, p. 114.
11. M. Nitay and M. Rosenblum, *J. Organometal. Chem.*, **186**, C23 (1977).
12. D. F. Shriver, *The Manipulation of Air-Sensitive Compounds*, McGraw-Hill, New York, 1969, p. 147.

54. μ-NITRIDO-BIS(TRIPHENYLPHOSPHORUS) (1 +) TETRACARBONYLRHODATE(1 −) AND μ-NITRIDO-BIS(TRIPHENYLPHOSPHORUS)(1 +) TETRACARBONYLIRIDATE(1 −), [(PPh$_3$)$_2$N][M(CO)$_4$] (M = Rh, Ir)

$$MCl_3 + 8KOH + 6CO \rightarrow K[M(CO)_4] + 2K_2CO_3 + 4H_2O + 3KCl*$$

$$K[M(CO)_4] + [(PPh_3)_2N]Cl \rightarrow [(PPh_3)_2N][M(CO)_4] + KCl$$

Submitted by L. GARLASCHELLI,[†] **R. DELLA PERGOLA,**[†] **and S. MARTINENGO**[†]
Checked by J. E. ELLIS[‡]

The $[M(CO)_4]^-$ (M = Rh, Ir) anions were originally prepared by reduction of the $[M_4(CO)_{12}]$ carbonyls, or, in the case of Rh, also of $[Rh_2(CO)_4Cl_2]$,

---

*This balanced equation does not represent the real course of the reaction, which involves formation of many intermediate products. For more details, see Ref. 3.
[†]Dipartimento di Chimica Inorganica e Metallorganica dell'Universita' e Centro C.N.R. sui Bassi Stati di Ossidazione, Via G. Venezian 21, 20133 Milan, Italy.
[‡]Department of Chemistry, University of Minnesota, Minneapolis, MN 55455.

with sodium in tetrahydrofuran (THF).[1,2] The yields were rather low, with formation of large amounts of by-products. More recently we reported a method that started from the above carbonyls or directly from the hydrated metal trichlorides, which gave better yields.[3] The method involved carbonylation of the substrates in the presence of powdered KOH in solvents such as dimethyl sulfoxide (DMSO) or N,N-dimethylformamide. The synthesis that we describe here uses this method starting from the soluble hydrated trichlorides and KOH in DMSO, and is complete in about 2 days.

General Procedure

■ **Caution.** *Because of the use of the highly poisonous carbon monoxide, all operations should be carried out in an efficient fume hood.*

All the operations must be carried out in the absence of oxygen using the Schlenk-tube technique.[4] Carbon monoxide and nitrogen should contain less than 2 ppm of oxygen by volume.* The solvents are laboratory grade and after purification are stored under nitrogen. DMSO is dried by distillation under vacuum over KOH, discarding the first fraction of the distillate, which contains the water. THF is distilled under nitrogen over sodium benzophenone ketyl. Isopropyl alcohol is distilled under nitrogen. Water is degassed by pumping in vacuum and stored under nitrogen.

The $RhCl_3$ trihydrate may be a commercial product or prepared according to the literature,[5] while the hydrated $IrCl_3$ is a commercial product. μ-Nitrido-bis(triphenylphosphorus)(1+) chloride is prepared according to the literature.[6] Potassium hydroxide is powdered immediately prior to use by milling KOH pellets in a mortar.† The CO and N_2 lines are attached to a low-pressure (2 atm) manifold through fine-needle valves, and include, before the reaction vessel, a mineral-oil bubbler and a pressure-release safety bubbler filled with an 80-mm mercury column.

The CO is not flowed through the reaction vessel, but it is added at such a rate that some continuously escapes from the pressure-release bubbler; in this way a CO pressure slightly higher than atmospheric is maintained in the apparatus, and only the CO necessary to the reaction enters in the vessel, thus minimizing the effect of its residual oxygen content. Great attention is required to ensure that, when the reaction vessel is opened for the addition of the reagents or solvents, the CO stream is sufficient to prevent back-diffusion of air into the vessel.

---

*The checker used 99.9% pure CO (ultra-high-purity grade) without further purification.

†The checker recommends that the powdering be done in a glove bag or dry box if the humidity is high. Otherwise the KOH powder will be clumpy and less effective. Useful observations concerning the preparation of powdered KOH have been reported by W. L. Jolly, *Inorg. Syn.*, **11**, 114 (1968).

A. μ-NITRIDO-BIS(TRIPHENYLPHOSPHORUS)-(1 +) TETRACARBONYLRHODATE(1 −)

Procedure

A sample of 0.28 g of $RhCl_3$ trihydrate (Rh 39.11%, 1.06 mmol) is placed in a 500-mL Schlenk tube equipped with a magnetic stirring bar. The tube is evacuated and filled with CO twice, then 20 mL of DMSO is added, and the mixture is stirred until the rhodium chloride is dissolved. To the red-brown solution powdered KOH (0.75 g) is then added, and the mixture is vigorously stirred for 24 h. During the reaction the originally red solution turns first violet, then green, then orange-red, and finally nearly colorless or slightly brown. At this point the IR spectrum should show only the characteristic bands of the $[Rh(CO)_4]^-$ anion at 2000 (vw) and 1895 (vs) cm^{-1}. However, because of the very low solubility of the KOH in DMSO, the reaction time is very dependent on the fineness of the KOH powder, its degree of dispersion in the medium, and stirring efficiency; thus it may happen that after 24 h the solution is still green or orange-red, indicating an incomplete reaction. In this case other additions of KOH (0.3 g each) followed by a few hours stirring may be necessary to render the solution colorless. At the end of the reaction, the product is precipitated by addition, while stirring, of a solution of $[(PPh_3)_2N]Cl$ (2 g) in a mixture of 2-propanol (15 mL) and water (100 mL). The white precipitate is collected on a fritted-glass filter, then washed three times with 10 mL of water, and finally is vacuum-dried. The subsequent operations may be carried out under nitrogen. The product is purified by extraction from the filtration apparatus with the minimum amount of THF added in 10-mL portions (total ~30 mL), which leaves some insoluble material on the frit. The extract is treated with 2-propanol (75 mL) and concentrated in vacuum with stirring, until most of the THF is volatilized. The crystalline precipitate is filtered, washed with 2-propanol (3 × 20 mL), and then vacuum dried. Yield 0.57 g, 71%.

Anal. Calcd. for $C_{40}H_{30}NO_4P_2Rh$: C, 63.76; H, 4.01; N, 1.85; Rh, 13.65. Found: C, 63.20; H, 3.85; N, 1.86; Rh, 13.02.

Properties

Pure $[(PPh_3)_2N][Rh(CO)_4]$ is a white crystalline product. It may be also ivory or very pale green-colored, without detectable changes in the IR spectrum or in analysis, and may be considered sufficiently pure also in this case. The IR spectrum in THF exhibits a very weak band at 2000 cm^{-1} and a characteristic very strong band at 1895 cm^{-1}. All manipulations involving the product should be carried out under an inert atmosphere; in fact, while crystalline samples are stable in air for a few minutes, the solutions are extremely air sensitive. For a short time, crystalline [(PPh$_3$)$_2$N][Rh(CO)$_4$] may be left on the filtering septum, while for longer periods it is better to seal it under nitrogen in glass ampoules. [(PPh$_3$)$_2$N][Rh(CO)$_4$] is quite soluble in THF, acetone, and acetonitrile, slightly soluble in methanol, sparingly soluble in 2-propanol, and insoluble in water and hexane.

The use of bulky cations other than [(PPh$_3$)$_2$N]Cl for the precipitation of the [Rh(CO)$_4$]$^-$ anion is to be avoided, because in most cases only tacky materials are obtained, or a reaction occurs between the cation and the anion.[7]

The [Rh(CO)$_4$]$^-$ anion has been used as building block in the synthesis of various rhodium clusters, such as [Rh$_7$(CO)$_{16}$]$^{3-}$,[8] [Rh$_5$(CO)$_{15}$]$^-$,[9] [Rh$_{11}$(CO)$_{23}$]$^{3-}$,[10] and [Rh$_7$N(CO)$_{15}$]$^{2-}$.[11] It has also been used in the synthesis of rhodium cluster carbides isotopically enriched on the interstitial carbon atom.[12]

B. μ-NITRIDO-BIS(TRIPHENYLPHOSPHORUS)-(1 +) TETRACARBONYLIRIDATE(1 −)

Procedure

The synthesis of [(PPh$_3$)$_2$N][Ir(CO)$_4$] is carried out exactly as described for the rhodium derivative, but starting from a sample of 0.7 g of commercial IrCl$_3$ trihydrate (Ir 54.44%, 1.98 mmol), and using 1.2 g of powdered KOH. During the reduction the starting red solution changes first to yellow, then to red, and then to brown, and finally is colorless or slightly brown. When the reaction is complete, the IR spectrum of the solution should show only the two characteristic bands of the [Ir(CO)$_4$]$^-$ anion at 2000 (vw) and 1895 (vs) cm^{-1}. As in the case of the rhodium derivative, further additions of KOH may be necessary to complete the reaction. The product is precipitated and purified exactly as in the Rh case. Yield 1.25 g, 75%.

Anal. Calcd. for C$_{40}$H$_{30}$IrNO$_4$P$_2$: C, 57.00; H, 3.68; N, 1.66; Ir, 22.80. Found: C, 56.75; H, 3.41; N, 1.66; Ir, 22.0.

Properties

Pure [(PPh$_3$)$_2$N][Ir(CO)$_4$] is a white crystalline product. The IR spectrum in THF solution shows bands at 2000 (vw) and 1895 (vs) cm^{-1}. The crystals are stable in air for a few minutes, while the solutions are rapidly oxidized. The same storage precautions described for the rhodium derivative are recommended. Also the solubility is the same.

The $[Ir(CO)_4]^-$ anion is used as building block for the synthesis of cluster anions such as $[Ir_6(CO)_{15}]^{2-}$,[2] and of the mixed-metal species $[Rh_6IrN(CO)_{15}]^{2-}$,[11] $[Rh_4Ir(CO)_{15}]^-$,[13] $[Ni_6Ir_3(CO)_{17}]^{3-}$,[14] and $Ir_2Rh_2(CO)_{12}$.[15]

References

1. P. Chini and S. Martinengo, *Inorg. Chim. Acta*, **3**, 21 (1969).
2. M. Angoletta, L. Malatesta, and G. Caglio, *J. Organometal. Chem.*, **94**, 99 (1975).
3. L. Garlaschelli, P. Chini, and S. Martinengo, *Gazz. Chim. Ital.*, **112**, 285 (1982).
4. D. F. Shriver, *The Manipulation of Air-Sensitive Compounds*, McGraw-Hill, New York, 1969.
5. S. N. Sanderson and F. Basolo, *Inorg. Synth.*, **7**, 214 (1963).
6. J. K. Ruff and W. J. Schlientz, *Inorg. Synth.*, **15**, 84 (1974).
7. J. E. Ellis, *J. Organometal. Chem.*, **111**, 331 (1976).
8. S. Martinengo and P. Chini, *Gazz. Chim. Ital.*, **102**, 344 (1972).
9. A. Fumagalli, T. F. Koetzle, F. Takusagawa, P. Chini, S. Martinengo, and B. T. Heaton, *J. Am. Chem. Soc.*, **102**, 1740 (1980).
10. A. Fumagalli, S. Martinengo, G. Ciani, and A. Sironi, *J. Chem. Soc., Chem. Commun.*, **1983**, 453.
11. S. Martinengo, G. Ciani, and A. Sironi, *J. Chem. Soc., Chem. Commun.*, **1984**, 1577.
12. V. G. Albano, P. Chini, S. Martinengo, D. J. A. McCaffrey, D. Strumolo, and B. T. Heaton, *J. Am. Chem. Soc.*, **96**, 8106 (1974).
13. A. Fumagalli, R. Della Pergola, L. Garlaschelli, and A. Sironi, *3rd International Conference on the Chemistry of the Platinum Group Metals*, Sheffield, July 12–17, 1987.
14. R. Della Pergola, L. Garlaschelli, F. Demartin, M. Manassero, N. Masciocchi and G. Longoni, *J. Chem. Soc., Dalton Trans.*, **1988**, 201.
15. S. Martinengo, P. Chini, V. G. Albano, F. Cariati, and T. Salvatori, *J. Organometal. Chem.*, **59**, 379 (1973).

Chapter Five

METAL CLUSTER COMPLEXES

55. POLYNUCLEAR RUTHENIUM COMPLEXES

A. DODECACARBONYLTRIRUTHENIUM, $Ru_3(CO)_{12}$

Submitted by M. I. BRUCE,* C. M. JENSEN,† and N. L. JONES‡
Checked by GEORG SÜSS-FINK,§ GERHARD HERRMANN,§ and VERA DASE§

$$RuCl_3 \cdot nH_2O + CO \rightarrow Ru_3(CO)_{12}$$

Dodecacarbonyltriruthenium can be prepared by several methods. Johnson and Lewis[1] have reported a procedure in which ruthenium trichloride hydrate is converted to tris(2,4-pentanedionato)ruthenium(III), which in turn is reacted with hydrogen and carbon monoxide. Reaction pressure and temperature are high (160 atm and 165°C) and the yield is in the range from 50 to 55%.

James and coworkers[2] reported a method for the synthesis of dodecacarbonyltriruthenium from hexakis(μ-acetato)trisaquaoxotriruthenium(III) acetate, which requires only ambient pressures of carbon monoxide. The reaction time can be long and the yield is 59% based on the starting triruthenium complex.

Mantovani and Cenini[3] have also reported a two-step ambient pressure synthesis of dodecacarbonyltriruthenium starting with ruthenium trichloride

---

*Department of Physical and Inorganic Chemistry, University of Adelaide, Adelaide, South Australia 5000.
†Department of Chemistry, University of Hawaii, 2545 The Mall, Honolulu, HI 96822.
‡Department of Chemistry, La Salle University, Philadelphia, PA 19141.
§Laboratorium für Anorganische Chemie, Universität Bayreuth, Universitätstrasse 30, D-8580 Bayreuth, Federal Republic of Germany.

hydrate resulting in a 50–60% yield, but the product requires recrystallization.

We give here details of a one-step, high-yield (70% or greater), medium-pressure (65 atm) synthesis of $Ru_3(CO)_{12}$ from $RuCl_3 \cdot nH_2O$.[4] No solvent purification is necessary and this synthesis can be completed in 1 day.

Procedure

■ **Caution.** *All manipulations with carbon monoxide should be carried out in a well-ventilated area.*

A mixture of $RuCl_3 \cdot nH_2O$ (Strem or Aldrich) (25.4 g) and anhydrous methanol (Mallinckrodt, fresh bottle with no further drying or deaerating) (300 mL) is pressurized to ∼1000 psi (65 atm) with carbon monoxide in a 1-L autoclave. The autoclave is heated at 250°F (125°C) with stirring. After 8 h of heating the autoclave is cooled and then vented in a well-ventilated hood.

■ **Caution.** *Highly toxic carbonyl chloride (phosgene) may be formed as a by-product and therefore use of an efficient fume hood is mandatory.*

The crude orange crystalline dodecacarbonyltriruthenium is separated by filtration in air on a Büchner funnel. The crude product is extracted into dichloromethane (3.5–4 L), leaving a blue-black solid behind. The amount of blue-black solid formed varies and based on Ru and Cl elemental analyses it is identified as RuO_2.

Anal. Calcd. for RuO_2: Ru, 75.95%, Cl, 0.0%; Found: Ru, 76.30%, Cl < 0.1%.

The solution is concentrated by rotary evaporation at room temperature. Spectroscopically pure orange crystalline $Ru_3(CO)_{12}$ is isolated by filtration. Yields vary slightly from preparation to preparation and are typically ∼70% (15.4 g) but sometimes can be as high as 92%.

Anal. Calcd. for $Ru_3(CO)_{12}$: C, 22.54: Found: C, 22.50.

Properties

Dodecacarbonyltriruthenium is an orange, air- and light-stable crystalline solid. It is soluble in most organic solvents. Its infrared (IR) spectrum in hexane displays three bands attributable to terminal CO ligands: 2061 (vs), 2031 (s), and 2011 (m) cm^{-1}. No band assignable to a bridging carbonyl ligand is observed.

Note. The mother liquor can be recycled, an amount of $RuCl_3 \cdot nH_2O$ equal to the amount of $Ru_3(CO)_{12}$ formed in the previous preparation being added to the solution. We have successfully operated this process for up to four successive preparations, with essentially quantitative conversion of the added $RuCl_3 \cdot nH_2O$ to $Ru_3(CO)_{12}$.

In some instances, particularly when the ruthenium trichloride sample contains more than the usual amount of water (this may occur, e.g., with old samples or on long exposure to moist air), the isolated product may be a mixture of $Ru_3(CO)_{12}$ and $Ru_4(\mu\text{-}H)_4(CO)_{12}$ (as indicated by the IR ν_{CO} spectrum). In such cases, depending on the final product required (a) the product may be used directly as in the synthesis of $Ru_4(\mu\text{-}H)_4(CO)_{12}$ described below, when conversion to the cluster carbonyl hydride is completed by reaction with H_2; or (b) treatment of the product with CO for 1 h while suspended in refluxing octane, using the apparatus depicted in Fig. 1, results in conversion of any $Ru_4(\mu\text{-}H)_4(CO)_{12}$ to $Ru_3(CO)_{12}$.

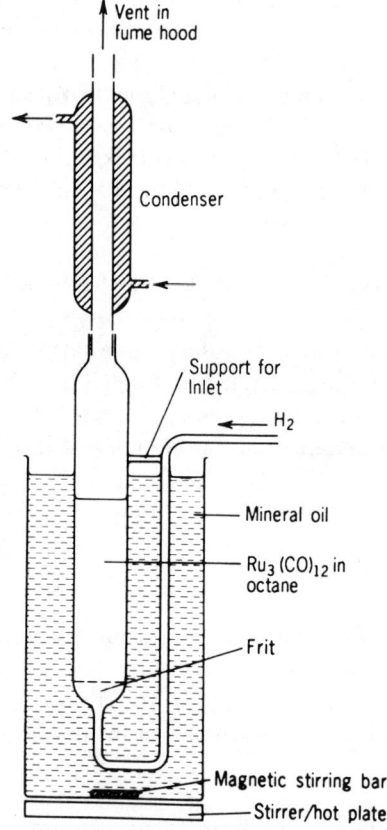

Fig. 1. Apparatus for preparation of $Ru_4H_4(CO)_{12}$.

B. DODECACARBONYLTETRA(μ-HYDRIDO)-TETRARUTHENIUM, $Ru_4(\mu\text{-}H)_4(CO)_{12}$

$$4Ru_3(CO)_{12} + 6H_2 \rightarrow 3Ru_4(\mu\text{-}H)_4(CO)_{12} + 12CO$$

Submitted by MICHAEL I. BRUCE* and MICHAEL L. WILLIAMS*
Checked by GUY LAVIGNE† and THÉRÈSE ARLIGUIE†

This tetranuclear ruthenium carbonyl hydride was described on several occasions,[5] but early preparations were usually contaminated with $Ru_3(CO)_{12}$, giving rise to suggestions of the existence of two isomeric forms. The situation was clarified by the work of Kaesz and coworkers,[6] who discovered the direct route from $Ru_3(CO)_{12}$ and hydrogen, which is described below. The compound is often obtained from reactions between $Ru_3(CO)_{12}$ and substrates containing hydrogen (hydrocarbons, ethers, alcohols, water, etc.) and by acidification of anionic ruthenium cluster carbonyls.[7]

■ **Caution.** *Because of the highly toxic nature of carbon monoxide and flammable nature of hydrogen gas, this procedure must be carried out in a well-ventilated hood; the autoclave room must also be well ventilated.*

Procedure

1. $Ru_4(\mu\text{-}H)_4(CO)_{12}$ can be obtained from $Ru_3(CO)_{12}$ by the original method,[6] in which hydrogen is passed through a solution of $Ru_3(CO)_{12}$ in refluxing octane. Finely powdered $Ru_3(CO)_{12}$ (250 mg, 0.35 mmol) is suspended in octane (80 mL), which has been washed successively with concentrated H_2SO_4 and water and distilled, contained in the apparatus shown in the figure. A gentle stream of H_2 is passed through the solution while it is being heated at the reflux point in the oil bath. After 1 h, or when the ν_{CO} band of $Ru_3(CO)_{12}$ at 2061 cm^{-1} is no longer present, the solution is filtered hot through a short (10 × 5 cm) column of silica gel. Reduction of volume to ~10 mL (rotary evaporator) results in deposition of $Ru_4(\mu\text{-}H)_4(CO)_{12}$ as a yellow powder, which is filtered, washed with cold petroleum ether (2 × 2 mL), and dried. Yield: 180 mg (90%). The cluster hydride can be recrystallized from a dichloromethane–hexane mixture; however, it is generally pure enough for further reactions.

Note. This reaction does not give high yields if more concentrated solutions of $Ru_3(CO)_{12}$ are used.

*Department of Physical and Inorganic Chemistry, University of Adelaide, South Australia 5000.
†CNRS Laboratoire de Chimie de Coordination, 31077 Toulouse, France.

2. Alternatively, this hydrogenation may be carried out in a small laboratory autoclave [$Ru_3(CO)_{12}$ in cyclohexane solution (1 g 50 mL^{-1}); H_2 at 25 atm, 120°C, for 2 h]. Finely powdered $Ru_3(CO)_{12}$ (1.0 g, 1.4 mmol) is added to octane (50 mL), purified as in (a), contained in the close-fitting glass liner of a 100-mL capacity stainless steel laboratory autoclave (Röth). After initial pressurization with hydrogen and venting, the autoclave is charged with H_2 (25 atm) and heated at 120°C for 2 h. After cooling and venting, the solution is removed from the autoclave and filtered; removal of solvent (rotary evaporator) gives $Ru_4(\mu\text{-H})_4(CO)_{12}$. Yield: 0.70 g (90%).

Properties

The compound $Ru_4(\mu\text{-H})_4(CO)_{12}$ is obtained as a yellow air-stable powder, which is soluble in most organic solvents, but insoluble in water. The IR spectrum contains ν_{CO} bands at 2081 (s), 2067 (vs), 2030 (m), 2024 (s), and 2009 (w) cm^{-1} (cyclohexane solution); the ^{1}H NMR spectrum has a resonance at δ -17.98 (CDCl$_3$ solution). The molecular structure of $Ru_4(\mu\text{-H})_4(CO)_{12}$ has been determined by X-ray diffraction: the four hydrogen atoms bridge the edges of the tetrahedral Ru_4 core in a D_{2d} arrangement, while three CO ligands are terminally bonded to each ruthenium.[8] The deuterated complex $Ru_4(\mu\text{-D})_4(CO)_{12}$ can be prepared in the same way if D_2 is used in place of H_2.[6]

References

1. B. F. G. Johnson and J. Lewis, *Inorg. Synth.*, **13**, 92 (1972) and references cited therein.
2. B. R. James, G. L. Rempel, and W. K. Teo, *Inorg. Synth.*, **16**, 45 (1975).
3. A. Mantovani and S. Cenini, *Inorg. Synth.*, **16**, 47 (1975).
4. M. I. Bruce, J. G. Matisons, R. C. Wallis, J. M. Patrick, B. W. Skelton, and A. H. White, *J. Chem. Soc. Dalton Trans.*, **1983**, 2365.
5. J. W. S. Jamieson, J. V. Kingston, and G. Wilkinson, *Chem. Commun.*, **1966**, 569; B. F. G. Johnson, R. D. Johnston, J. Lewis, and B. H. Robinson, *Chem. Commun.*, **1966**, 851; H. Pichler, H. Meier zu Kocker, W. Gabler, R. Gartner, and D. Kioussis, *Brennstoff Chem.*, **48**, 266 (1967); F. Piacenti, M. Bianchi, P. Frediani, and E. Benedetti, *Inorg. Chem.*, **10**, 2759 (1971).
6. S. A. R. Knox, J. W. Koepke, M. A. Andrews, and H. D. Kaesz, *J. Am. Chem. Soc.*, **97**, 3942 (1975).
7. B. J. G. Johnson, R. D. Johnston, J. Lewis, B. H. Robinson, and G. Wilkinson, *J. Chem. Soc. A.*, **1958**, 2856.
8. R. D. Wilson, S. M. Wu, R. A. Love, and R. Bau, *Inorg. Chem.*, **17**, 1271 (1987).

56. TRI- AND TETRANUCLEAR CARBONYL-RUTHENIUM CLUSTER COMPLEXES CONTAINING ISOCYANIDE, TERTIARY PHOSPHINE, AND PHOSPHITE LIGANDS. RADICAL ION-INITIATED SUBSTITUTION OF METAL CLUSTER CARBONYL COMPLEXES UNDER MILD CONDITIONS

Submitted by MICHAEL I. BRUCE,* BRIAN K. NICHOLSON,† and MICHAEL L. WILLIAMS*
Checked by THÉRÈSE ARLIGUIE‡ and GUY LAVIGNE‡

Replacement of CO by tertiary phosphines, arsines, and similar ligands in metal cluster carbonyls is a reaction that is highly dependent on the metal carbonyl. In some cases, such as $Co_4(CO)_{12}$ or $Ir_4(CO)_{12}$, the reaction proceeds readily at room temperature or on gentle warming, and the stoichiometric products can be obtained fairly easily.[1] In other cases, such as $Ru_3(CO)_{12}$, $Ru_4H_4(CO)_{12}$, or $Os_3(CO)_{12}$, the reactions proceed only under vigorous conditions, and usually result in (a) polysubstitution of the cluster carbonyl, (b) the formation of a mixture of products, or (c) further reaction of the initial substitution product. Examples of these situations are the formation of $Ru_3(CO)_9(PPh_3)_3$ as the sole product when mixtures of $Ru_3(CO)_{12}$ and PPh_3 are heated in refluxing benzene, even when a deficiency of the phosphine is present;[2] the formation of $Ru_4H_4(CO)_{12-n}[P(OMe)_3]_n$ (n = 0–4) on heating mixtures of $P(OMe)_3$ and $Ru_4H_4(CO)_{12}$; individual complexes are separated with difficulty and in low overall yield,[3] and the formation of P—C and C—H bond cleavage products among the nine complexes isolated after heating mixtures of $Os_3(CO)_{12}$ and PPh_3 in xylene.[4]

The current interest in cluster complexes as possible catalysts has made desirable a method of synthesis of selectively substituted derivatives, which does not suffer the disadvantages just outlined. Cluster-bound molecules have high and often unique reactivity, and the introduction of tertiary phosphines containing functional groups is often difficult.

It was earlier observed that radical anions of binuclear metal carbonyls, generated electrochemically, were unusually susceptible to attack by nucleophilic ligands.[5] Specific substitution of CO ligands was thus achieved. This was followed by the demonstration of substitution reactions initiated by

---

*Department of Physical and Inorganic Chemistry, University of Adelaide, South Australia, 5001.
†Department of Chemistry, University of Waikato, Hamilton, New Zealand.
‡Laboratoire de Chimie de Coordination, CNRS, 31077 Toulouse, France.

radical anions, such as sodium benzophenone ketyl.[6] A range of metal cluster carbonyls can be used in the reactions, which proceed under mild conditions and in high yields. Development of the method led to syntheses of specifically substituted compounds containing two or more different ligands.[7] It has resulted in the possibility of the syntheses of complexes that are chiral by virtue of having four different ligands on a tetrahedral M_4 core, such as $Ru_4H_4(CO)_9(PMe_2Ph)[P(OMe)_2Ph][P(OCH_2)_3CEt]$.[8]

Under these conditions, it is also possible to prepare (a) complexes in which bidentate ligands contain one cluster moiety attached to each donor atom, such as $[Ru_3(CO)_{11}]PPh_2(CH_2)_2PPh_2[Ru_3(CO)_{11}]$ (Ref. 9) or $[Ru_3(CO)_{11}]PPh_2C\equiv CPPh_2[Ru_3(CO)_{11}]$,[10] or (b) derivatives that readily rearrange thermally, such as $Ru_3(CO)_{10}(Ph_2PC_6H_4CH=CH_2)$, which is converted to $Ru_3H_2(CO)_8(Ph_2PC_6H_4C\equiv CH)$ at 40°C,[11] or $Ru_3(CO)_{10}$-$[(Ph_2P)_2CH_2]$;[10] the products obtained from the thermal reaction between $Ru_3(CO)_{12}$ and $CH_2(PPh_2)_2$ are $Ru_3(CO)_8[(Ph_2P)_2CH_2]_2$ (in xylene, at 80–85°C, 73%) and $Ru_3(CO)_7(\mu_3\text{-}PPh)(\mu\text{-}CHPPh_2)[(Ph_2P)_2CH_2]$ (in xylene, > 130°C, 42%).[12]

The following syntheses describe the method as applied to $Ru_3(CO)_{12}$ and $Ru_4H_4(CO)_{12}$, the experimental application of which is quite general. Brief accounts of similar reactions with $Co_3(\mu_3\text{-}CR)(CO)_9$, $Rh_6(CO)_{16}$, or $Os_3(CO)_{12}$ have been given.[13] Some of these substitution reactions may also be catalyzed by various $[(Ph_3P)_2N]^+$ salts.[14]

Reagents

Dodecacarbonyltriruthenium, $Ru_3(CO)_{12}$, is available commercially from Strem, Pressure Chemicals, or Aldrich; it may also be made by any of the methods described in previous volumes of this series,[15] or by the modification described in Ref. 16. The synthesis of $Ru_4(\mu\text{-}H)_4(CO)_{12}$ is also described elsewhere in this volume.[17]

The tertiary phosphines PMe_2Ph, $CH_2(PPh_2)_2$, and $C_2(PPh_2)_2$, the phosphite $P(OC_6H_4Me\text{-}p)_3$, and the isocyanide $CNBu^t$ are available from Strem or Pressure Chemicals.

A. SODIUM BENZOPHENONE KETYL SOLUTION

$$Ph_2CO + Na \rightarrow Na[Ph_2CO]$$

■ **Caution.** *Sodium metal is highly reactive. It should be handled under a covering layer of mineral oil and an atmosphere of dry nitrogen.*

Procedure

A 50 mL, two-necked, round-bottomed flask is removed from a hot oven and flushed with nitrogen. A sample of benzophenone (0.091 g, 0.5 mmol) together with 20 mL of freshly distilled and deoxygenated tetrahydrofuran (THF) and a stirring bar are placed in the flask. A small lump of sodium metal (~ 0.5 g) is cut into smaller pieces directly into the flask against the emergent stream of nitrogen. The second neck of the flask is then closed with a septum cap. The mixture is stirred for 0.5–2 h in the stoppered flask under nitrogen until the initial ultramarine blue is replaced by an intense blue-purple color. The resulting solution is ~ 0.025 mol L^{-1} in $Na[Ph_2CO]$. This solution may decolorize in 1 or 2 days' time but it can be regenerated by addition of more benzophenone.

Properties

A solution of sodium benzophenone ketyl is exceedingly reactive, and is immediately decolorized by air or traces of water, alcohols, and so on. The solution should therefore be kept under nitrogen at all times. Aliquots may be removed using a dry syringe via the septum cap. Under these conditions, the solution appears to be stable for ~ 2 days.

B. UNDECACARBONYL(DIMETHYLPHENYLPHOSPHINE)-TRIRUTHENIUM, $Ru_3(CO)_{11}(PMe_2Ph)$

$$Ru_3(CO)_{12} + PMe_2Ph \xrightarrow{Na[Ph_2CO]} Ru_3(CO)_{11}(PMe_2Ph) + CO$$

■ **Caution.** *Because of evolution of highly toxic carbon monoxide (a colorless and odorless gas), and also because of the toxic nature and bad odor of tertiary phosphine, phosphite, or isocyanide ligands, these reactions should be performed in a well-ventilated hood.*

■ **Caution.** *The procedure for obtaining accurately small amounts of PMe_2Ph (or similar toxic liquids with bad odor) is to weigh ~ 0.4 g of the ligand into a stoppered, nitrogen-flushed flask. Sufficient dry THF is then added to make a solution of 1.0 mmol mL^{-1}. A calibrated syringe can be used to transfer the required volume of this standard solution.*

Procedure

A 50-mL Schlenk tube, previously kept overnight in a hot oven, is flushed with nitrogen while cooling. After attaching a gas inlet to the side arm, the flask is charged with $Ru_3(CO)_{12}$ (0.20 g, 0.31 mmol), 25 mL of dry, deoxygenated THF, 0.33 mL of a solution of PMe_2Ph in THF (~ 1.0 mmol mL^{-1}), and a magnetic stirring bar, and is then closed with a rubber septum cap. The Schlenk tube is placed in an oil bath on a stirrer hot plate, and warmed to $\sim 40°C$ to dissolve the $Ru_3(CO)_{12}$. A dry 1-mL syringe is loaded with the $Na[Ph_2CO]$ solution, which is added dropwise to the reaction mixture through the septum cap until the solution rapidly darkens in color (typically 5–10 drops; see cautionary note immediately above). To check that reaction is complete, an aliquot of the reaction mixture is removed by a second syringe to an IR solution cell; the 2061 cm^{-1} ν_{CO} band of $Ru_3(CO)_{12}$ is monitored. If still present, several more drops of the $Na[Ph_2CO]$ are added until disappearance of the band. The reaction mixture is then transferred into a 100-mL round-bottomed flask, and solvent is removed (rotary evaporator) to leave an oily residue, which is extracted with two 15-mL portions of warm ($\sim 40°C$) petroleum ether (boiling range 40–60°C). The combined extracts are filtered, concentrated to ~ 10 mL, and cooled overnight ($-10°C$) to give red-orange crystals, which are collected on a sintered-glass filter and air-dried. Yield: 0.18 g (78%).

Anal. Calcd. for $C_{19}H_{11}O_{11}PRu_3$: C, 30.46; H, 1.48. Found: C, 30.26; H, 1.23.

Properties

The complex $Ru_3(CO)_{11}(PMe_2Ph)$ is a reasonably air stable red-orange crystalline solid (mp 104–106°C), which dissolves readily in organic solvents. The IR spectrum (hexane) shows ν_{CO} bands at 2096 (m), 2044 (s), 2028 (s), 2016 (s), 2000 (w, sh), and 1987 (w) cm^{-1}. The ^1H NMR spectrum ($CDCl_3$) shows a broad multiplet at δ 7.52 (Ph) and a doublet at δ 1.97 (J_{HP} 10 Hz, Me).

C. DECACARBONYL(DIMETHYLPHENYLPHOSPHINE)-(2-ISOCYANO-2-METHYLPROPANE)TRIRUTHENIUM, $Ru_3(CO)_{10}(CNBu^t)(PMe_2Ph)$

$$Ru_3(CO)_{11}(PMe_2Ph) + CNBu^t \xrightarrow{Na[Ph_2CO]} Ru_3(CO)_{10}(CNBu^t)(PMe_2Ph) + CO$$

■ *See cautionary notes, Section B.*

Procedure

A dry-nitrogen-flushed 50-mL Schlenk tube is set up as described in Section B, and charged with 0.10 g (0.133 mmol) $Ru_3(CO)_{11}(PMe_2Ph)$, 10 mL dry, deoxygenated THF, a stirring bar and 0.14 mL of a solution of 2-isocyano-2-methylpropane ($CNBu^t$, *t*-butyl isocyanide) in THF (1.0 mmol mL^{-1}). The tube is then sealed with a septum cap. A solution of $Na[Ph_2CO]$ is added dropwise by syringe as before, $\sim$ 5–10 drops usually being required.

The amount of $Na[Ph_2CO]$ solution required depends on how dry the reaction mixture is. When the reaction appears visually to be complete, a small aliquot should be examined (solution IR spectrum) to ensure that the precursor cluster complex has reacted completely [using the 2096 cm^{-1} ν_{CO} band of $Ru_3(CO)_{11}(PMe_2Ph)$]. Alternatively, the progress of the reaction may be monitored by TLC (thin-layer chromatography) (silica gel, petroleum ether eluant).

After the reaction is determined to be complete, solvent is removed (rotary evaporator), and the resulting deep red oil is extracted with two 10-mL aliquots of warm ($\sim 40°C$) petroleum ether (boiling range, 40–60°C). The extracts are filtered, concentrated to ~ 5 mL under vacuum, and cooled at $-10°C$ overnight to give $Ru_3(CO)_{10}(CNBu^t)(PMe_2Ph)$ (0.052 g, 42%) as deep red crystals.

Anal. Calcd. for $C_{23}H_{20}NO_{10}PRu_3$: C, 34.33; H, 2.51; N, 1.74. Found: C, 33.82; H, 2.37; N, 1.59.

Properties

The deep red crystals (mp 128–132 °C) of $Ru_3(CO)_{10}(PMe_2Ph)(CNBu^t)$ are reasonably air stable, although they darken in color after several days in air. An IR spectrum (hexane) shows a ν_{CN} band at 2168 cm^{-1}, and ν_{CO} bands at 2098 (m), 2068 (w), 2046 (m), 2029 (s), 2016 (s), 2004 (w), and 1988 (w) cm^{-1}. The ^1H NMR spectrum ($CDCl_3$) contains resonances at δ 7.42 (multiplet, Ph), 1.88 (J_{HP}, 9 Hz, P*Me*) and 1.49 (singlet, C*Me*$_3$).

D. DECACARBONYL[METHYLENEBIS(DIPHENYL-PHOSPHINE)]TRIRUTHENIUM, $Ru_3(CO)_{10}[(Ph_2P)_2CH_2]$

$$Ru_3(CO)_{12} + CH_2(PPh_2)_2 \xrightarrow{Na[Ph_2CO]} Ru_3(CO)_{10}[(Ph_2P)_2CH_2] + 2CO$$

■ See cautionary notes, Section B.

Procedure

A dry, nitrogen-flushed 100-mL Schlenk tube is charged with $Ru_3(CO)_{12}$ (1.00 g, 1.56 mmol), methylenebis(diphenylphosphine) (0.62 g, 1.61 mmol) and 70 mL of freshly distilled, deoxygenated THF. A magnetic stirring bar is added and the mixture is warmed to about 40 °C to dissolve the $Ru_3(CO)_{12}$. A solution of $Na[Ph_2CO]$ is added dropwise as before, carbon monoxide being evolved and the solution rapidly darkening. An aliquot is withdrawn by syringe to ensure that the $Ru_3(CO)_{12}$ is consumed by monitoring the 2061 cm^{-1} v_{CO} band in the IR spectrum. The reaction mixture is then transferred into a 100-mL round-bottomed flask and concentrated (rotary evaporator) to ~3mL. Addition of 40 mL of methanol and cooling in a freezer overnight gives orange-red crystals, which are collected on a sintered-glass filter and air-dried. Yield: 1.37 g (91%).

Anal. Calcd. for $C_{35}H_{22}O_{10}P_2Ru_3$: C, 43.44; H, 2.29. Found: C, 43.76; H, 2.05.

Properties

The complex forms orange-red, air-stable crystals (mp 180–181 °C), which readily dissolve in acetone, dichloromethane, chloroform, THF, and benzene. It is sparingly soluble in cyclohexane, petroleum ether, and methanol. The IR spectrum (cyclohexane) shows v_{CO} bands at 2086 (m), 2024 (sh), 2018 (s), 2005 (s), 1991 (w), 1968 (m), 1965 (m), and 1947 (w) cm^{-1}. The ^{1}H NMR spectrum (CDCl$_3$) shows a broad multiplet at δ 7.37 (Ph) and a triplet at δ 4.29 (J_{HP}, 10.5 Hz; CH$_2$).

E. [μ-ETHYNEDIYLBIS(DIPHENYLPHOSPHINE)]-BIS[UNDECACARBONYLTRIRUTHENIUM], [Ru$_3$(CO)$_{11}$]$_2$[μ-C$_2$(PPh$_2$)$_2$]

$$2Ru_3(CO)_{12} + C_2(PPh_2)_2 \xrightarrow{Na[Ph_2CO]} [Ru_3(CO)_{11}]_2[\mu\text{-}C_2(PPh_2)_2] + 2CO$$

■ See cautionary notes, Section B.

Procedure

A dry, nitrogen flushed 50-mL Schlenk tube is set up as described in Section B, and charged with $Ru_3(CO)_{12}$ (0.15 g, 0.235 mmol), and $C_2(PPh_2)_2$ (0.047 g, 0.12 mmol) and 20 mL of freshly distilled THF. A magnetic stirring bar is added and the mixture is warmed to $\sim 30\,°C$ to dissolve the $Ru_3(CO)_{12}$. A solution of $Na[Ph_2CO]$ solution is added dropwise from a syringe as in Section B or C, ~ 5–10 drops being required. The progress of the reaction is assessed by monitoring the v_{CO} band of $Ru_3(CO)_{12}$ at 2061 cm^{-1}. When the reaction is complete the solution is transferred to a 50-mL round-bottomed flask and solvent is removed (rotary evaporator). The residue is dissolved in 1 mL of dichloromethane, and 5 mL of hexane is added. Cooling in a freezer for 24 h gives the complex as an orange powder, which is collected on a sintered-glass filter and air-dried. Yield: 0.135 g (71%).

Anal. Calcd. for $C_{48}H_{20}O_{22}P_2Ru_6$: C, 35.65; H, 1.25. Found: C, 35.56; H, 1.21.

Properties

The complex is obtained as an orange microcrystalline solid [mp $> 150\,°C$ (dec)], which is readily soluble in acetone, dichloromethane, chloroform, and THF, but only sparingly soluble in light petroleum ether, cyclohexane, and methanol. The IR spectrum (cyclohexane) shows v_{CO} bands at 2102 (m), 2068 (sh), 2052 (vs), 2034 (s), 2021 (vs), 2005 (sh), 1997 (w), 1987 (sh), and 1970 (w) cm^{-1}. A broad multiplet at δ 7.50 (Ph) is the only resonance in the ^{1}H NMR spectrum (CDCl$_3$).

F. UNDECACARBONYLTETRAHYDRIDO[TRIS(4-METHYL-PHENYL)PHOSPHITE]TETRARUTHENIUM, $Ru_4H_4(CO)_{11}[P(OC_6H_4Me\text{-}p)_3]$

$$Ru_4H_4(CO)_{12} + P(OC_6H_4Me\text{-}p)_3$$

$$\xrightarrow{Na[Ph_2CO]} Ru_4H_4(CO)_{11}[P(OC_6H_4Me\text{-}p)_3] + CO$$

■ *See cautionary notes, Section B.*

Procedure

A dry, nitrogen-flushed 100-mL Schlenk tube is charged with $Ru_4H_4(CO)_{12}$ (0.21 g, 0.28 mmol), 20 mL of freshly distilled, deoxygenated THF, and 0.3 mL of a solution of tris(4-methylphenyl)phosphite in THF (1.0 mmol mL^{-1}). A stirring bar is added and the mixture is warmed to about 40 °C to dissolve the $Ru_4H_4(CO)_{12}$. A solution of $Na[Ph_2CO]$ is next added as described in Section B or C. The reaction is monitored by following the disappearance of the 2081 cm^{-1} v_{CO} band of the starting cluster complex. An additional 0.4 to 0.5 mL of the Na[Ph$_2$CO] solution may be required for complete reaction.

The reaction mixture is transferred to a 50-mL round-bottomed flask and solvent is removed (rotary evaporator). The oily residue is dissolved in 2 mL of benzene and 8 mL of absolute ethanol is added. The mixture is kept in a freezer for 24 h. The bright yellow crystals that form are collected on a glass sinter and vacuum-dried (1 torr, 25 °C). Yield: 0.20 g (66%).

Anal. Calcd. for C$_{32}$H$_{25}$O$_{14}$PRu$_4$: C, 35.97; H, 2.35; P, 2.90. Found: C, 36.41; H, 2.31; P, 2.94.

Properties

The golden yellow crystals [mp 130–133 °C (dec)] of Ru$_4$H$_4$(CO)$_{11}$-[P(OC$_6$H$_4$Me-*p*)$_3$] are apparently stable in air indefinitely. A hexane solution shows v_{CO} bands at 2098 (m), 2071 (s), 2061 (s), 2037 (s), 2028 (w), 2016 (s), 2001 (w), and 1983 (w) cm^{-1}. The ^{1}H NMR spectrum [(CD$_3$)$_2$CO] has peaks at δ 7.87 (multiplet, C$_6$H$_4$) and 2.33 (multiplet, CH$_3$), with the metal hydride resonance at δ − 17.7 [d, J_{HP} 7 Hz].

G. DECACARBONYL(DIMETHYLPHENYLPHOSPHINE)-TETRAHYDRIDO[TRIS(4-METHYLPHENYL)-PHOSPHITE]TETRARUTHENIUM, Ru$_4$H$_4$(CO)$_{10}$ (PMe$_2$Ph)[P(OC$_6$H$_4$Me-*p*)$_3$]

$$Ru_4H_4(CO)_{11}[P(OC_6H_4Me\text{-}p)_3] + PMe_2Ph$$

$$\xrightarrow{Na[Ph_2CO]} Ru_4H_4(CO)_{10}(PMe_2Ph)[P(OC_6H_4Me\text{-}p)_3] + CO$$

■ See cautionary notes, Section B.

Procedure

A dry, nitrogen-flushed 100-mL Schlenk tube is loaded with Ru$_4$H$_4$(CO)$_{11}$[P(OC$_6$H$_4$Me-*p*)$_3$] (0.18 g, 0.17 mmol), 0.17 mL of a solution of PMe$_2$Ph in THF (1.0 mmol mL^{-1}), and 8 mL of dry, deoxygenated THF as in Sections B or C. A solution of Na[Ph$_2$CO] is next added dropwise from a syringe through the septum cap, again as described in Sections B or C. Typically, 5 to 10 drops are sufficient for complete reaction as evidenced by the disappearance of the 2097 cm^{-1} ν_{CO} band of the precursor cluster. The reaction mixture is evaporated under vacuum and the residue is dissolved in 1 mL of benzene. Addition of 3 mL of petroleum ether (boiling range 40–60 °C), and cooling in a freezer overnight gives the product as a yellow-brown powder (0.14 g, 64%), which is collected on a glass sinter and air-dried.

Anal. Calcd. for $C_{39}H_{36}O_{13}P_2Ru_4$: C, 39.74; H, 3.08. Found: C, 40.53; H, 2.96.

Properties

The compound $Ru_4H_4(CO)_{10}(PMe_2Ph)[P(OC_6H_4Me-p)_3]$ is a yellow-brown powder (mp 136–137 °C), which is stable in air and very soluble in organic solvents. The IR spectrum (hexane) has ν_{CO} bands at 2079 (m), 2059 (vs), 2043 (w), 2029 (s), 2017 (s), 2002 (m), 1998 (sh), and 1956 (w) cm^{-1}. The ^{1}H NMR spectrum contains singlet resonances at δ 7.43 (Ph), 7.11 (C_6H_4), and 2.30 (C_6H_4Me), a doublet at δ 1.87 (J_{HP} 9 Hz, P*Me*), and a broad, poorly resolved triplet at δ −19 (J_{HP} 10 Hz, RuH).

References

1. D. J. Darensbourg and M. J. Incorvia, *Inorg. Chem.*, **19**, 2585 (1980); D. Sonnenberger and J. D. Atwood, *Inorg. Chem.*, **20**, 3243 (1981).
2. M. I. Bruce, G. Shaw, and F. G. A. Stone, *J. Chem. Soc. Dalton Trans.*, **1972**, 2094, and references cited therein.
3. S. A. R. Knox and H. D. Kaesz, *J. Am. Chem. Soc.*, **93**, 4594 (1971).
4. C. W. Bradford, R. S. Nyholm, G. J. Gainsford, J. M. Guss, P. R. Ireland, and R. Mason, *J. Chem. Soc. Chem. Commun.*, **1972**, 87; G. J. Gainsford, J. M. Guss, P. R. Ireland, R. Mason, C. W. Bradford, and R. S. Nyholm, *J. Organomet. Chem.*, **40**, C70 (1972).
5. M. Arewgoda, B. H. Robinson, and J. Simpson, *J. Am. Chem. Soc.*, **105**, 1893 (1983).
6. M. I. Bruce, J. G. Matisons, and B. K. Nicholson, *J. Organomet. Chem.*, **247**, 321 (1983).
7. M. I. Bruce, J. G. Matisons, B. K. Nicholson, and M. L. Williams, *J. Organomet. Chem.*, **236**, C57 (1982).
8. M. I. Bruce, B. K. Nicholson, J. M. Patrick, and A. H. White, *J. Organomet. Chem.*, **254**, 361 (1983).
9. M. I. Bruce, T. W. Hambley, B. K. Nicholson, and M. R. Snow, *J. Organomet. Chem.*, **235**, 83 (1982).
10. M. I. Bruce, M. L. Williams, J. M. Patrick, and A. H. White, *J. Chem. Soc. Dalton Trans.*, **1985**, 1229; M. I. Bruce, M. L. Williams, J. M. Patrick, B. W. Skelton and A. H. White, *J. Chem. Soc. Dalton Trans.*, **1986**, 2557.
11. M. I. Bruce, B. K. Nicholson, and M. L. Williams, *J. Organomet. Chem.*, **243**, 69 (1983).
12. G. Lavigne and J.-J. Bonnet, *Inorg. Chem.*, **20**, 2713 (1981).
13. M. I. Bruce, D. C. Kehoe, J. G. Matisons, B. K. Nicholson, P. H. Rieger, and M. L. Williams, *J. Chem. Soc. Chem. Commun.*, **1982**, 442.
14. G. Lavigne and H. D. Kaesz, *J. Am. Chem. Soc.*, **106**, 4647 (1984).

15. B. F. G. Johnson and J. Lewis, *Inorg. Synth.*, **13**, 92 (1972); B. R. James, G. L. Rempel, and W. K. Teo, *Inorg. Synth.*, **16**, 45 (1976); A. Mantovani and S. Cenini, *Inorg. Synth.*, **16**, 47 (1976).
16. (a) M. I. Bruce, J. G. Matisons, R. C. Wallis, J. M. Patrick, B. W. Skelton, and A. H. White, *J. Chem. Soc. Dalton Trans.*, **1983**, 2365; (b) M. I. Bruce, N. L. Jones, and C. M. Jensen, *Inorg. Synth.*, **26**, 259 (1989). See also section 55 in this volume.
17. M. I. Bruce and M. L. Williams, *Inorg. Synth.*, **26**, 262 (1989). See also section 55 in this volume.

57. DODECACARBONYLTRIOSMIUM

$$3OsO_4 + 24CO \rightarrow Os_3(CO)_{12} + 12CO_2$$

Submitted by SIMON R. DRAKE* and PAUL A. LOVEDAY*
Checked by RUTH ANN DOYLE† and ROBERT J. ANGELICI†

The cluster dodecacarbonylosmium[1] has been known for over forty years. Originally reported by Manchot et al.[2] as $Os_2(CO)_9$, its molecular structure was determined by X-ray single-crystal analysis and the trimeric formulation $Os_3(CO)_{12}$ established.[3] Although formally analogous to $Fe_3(CO)_{12}$, this cluster has a different structure consisting of a triangular arrangement of metal atoms with all the carbonyl groups terminal.

Previous syntheses of this compound often gave low yields from the reaction of OsO_4 with carbon monoxide in the gas phase[1] or by the reaction of OsO_4 with carbon monoxide in xylene under high pressure (128 atm) and temperatures (175 °C). Improved synthesis of this cluster was obtained by the use of OsO_4 in MeOH under carbon monoxide pressure, with the white oxycarbonyl $Os_4O_4(CO)_{12}$ as a side product.[4] We now describe an improved low-pressure synthesis of $Os_3(CO)_{12}$ from OsO_4 in EtOH in very high yield.[5]

■ **Caution.** *Metal carbonyls are toxic and should be handled in a well-ventilated fume hood.*

Reactions are carried out in a 1-L magnetically stirring autoclave (Baskerville and Lindsay) capable of withstanding pressure up to 300 atm and operating at maximum temperatures of 300 °C.‡ All solvents must be freshly distilled before use and all manipulations carried out in a well-ventilated fume hood.

---

*University Chemical Laboratory, Lensfield Road, Cambridge, CB2 1EW, United Kingdom.
†Department of Chemistry, Iowa State University, Ames, IA 50011.
‡The checkers carried out the reaction in a 250-mL stainless steel high-pressure Parr autoclave without stirring.

Procedure

■ **Caution.** *Because of the toxicity of osmium tetraoxide, all operations must be carried out in a well-ventilated hood using positive breathing apparatus. The toxicity of this material cannot be understated.*

Osmium tetraoxide (10 g) and ethyl alcohol (200 mL) are placed in the autoclave in that order.* Carbon monoxide (CP grade) (80 atm) is then added and the reaction mixture is heated at 175 °C for 7 h. When cold (~ 12 h) the pressure is released (behind an explosive proof shield). The large hexagonal yellow crystals of dodecacarbonyltriosmium are separated by filtration and washed with ice-cold ethanol (3 × 10-mL portions). The product obtained is dried under vacuum for at least 3 h (20 °C, 10^{-1} mmHg). The sample thus obtained is sufficiently pure for most purposes. Further purification may be carried out by recrystallization from hot benzene. Yield: 9.7–10.0 g, 97–99.5%. Other minor products from this reaction are the trinuclear derivatives $HOs_3(CO)_{10}(OMe)$ and $Os_3(CO)_{10}(OMe)_2$.[6,7]

Properties

Dodecacarbonyltriosmium is a yellow crystalline solid that shows only very limited solubility in organic solvents and is volatile. The infrared spectrum of this cluster is given and is useful for identification and for checking purity. v_{CO} cm^{-1} (*n*-hexane) 2070 (s), 2036 (s), 2003 (s). In the solid state this cluster is reasonably stable over a period of several months.

References

1. W. Heiber and H. Stallman, *Z. Electrochem.*, **49**, 228, (1943).
2. W. Manchot and W. J. Manchot, *Z. Anorg. Allgem. Chem.*, **226**, 385, (1936).
3. E. R. Corey and L. F. Dahl, *J. Am. Chem. Soc.*, **88**, 1821, (1966).
4. C. W. Bradford and R. S. Nyholm, *J. Chem. Soc., Chem. Commun.*, **1967**, 384.
5. B. F. G. Johnson and J. Lewis, *Inorg. Synth.*, **13**, 93, (1972).
6. B. F. G. Johnson, P. A. Kilty, and J. Lewis, *J. Chem. Soc. A*, **1968**, 2859.
7. M. A. Pearsall, Ph.D. thesis, University of Cambridge, 1984.

*The checkers dissolved the OsO_4 in EtOH and syringed the solution into the autoclave.

58. SOME USEFUL DERIVATIVES OF DODECA-CARBONYLTRIOSMIUM

Submitted by J. N. NICHOLLS* and M. D. VARGAS*†
Checked by A. J. DEEMING† and S. E. KABIR‡

The preparation of substituted derivatives of $Os_3(CO)_{12}$ is hampered by the reluctance of this cluster complex to undergo replacement of carbonyl ligands by other donor ligands.[1] There is, however, an alternative route to such species via the derivatives $Os_3(CO)_{11}(CH_3CN)$, **1**, and $Os_3(CO)_{10}(CH_3CN)_2$, **2**, the acetonitrile ligands of which are readily displaced by other nucleophiles such as PR_3, C_5H_5N, HX (X = Cl, Br, I), C_2H_4, and so on.[2,3] Much milder conditions are needed to replace the acetonitrile ligands in the cluster complexes **1** or **2** than the CO ligands in $Os_3(CO)_{12}$. Simple substitution products such as $Os_3(CO)_{11}(C_2H_4)$, and $Os_3(CO)_{11}(C_5H_5N)$, for instance, can thus be prepared only from the monoacetonitrile derivative, **1**. By contrast, the direct reaction of $Os_3(CO)_{12}$ with C_2H_4 or C_5H_5N proceeds to give $HOs_3(CO)_{10}(C_2H_3)$ or $HOs_3(CO)_{10}(C_5H_4N)$, respectively.[2]

The preparation of complexes **1** or **2** involves the reaction of $Os_3(CO)_{12}$ in acetonitrile with trimethylamine oxide, which removes CO ligands as CO_2; the vacant site(s) on the cluster are then filled by acetonitrile rather than the resulting trimethylamine, which is a poorer coordinating ligand. Complexes **1** or **2** are easily isolated in high yields. Their syntheses are reported in detail here, together with the synthesis of a typical derivative $Os_3(CO)_{11}(C_5H_5N)$.

A. (ACETONITRILE)UNDECACARBONYLTRIOSMIUM

$$Os_3(CO)_{12} + Me_3NO \xrightarrow{CH_3CN} Os_3(CO)_{11}CH_3CN + Me_3N + CO_2$$

■ **Caution 1.** *The metal carbonyls described in this synthesis are toxic by absorption through the skin and should therefore be handled with care. All procedures should be carried out in a well-ventilated fume hood.*

■ **Caution 2.** *Dry trimethylamine oxide can be explosive; it should be handled and stored only in small quantities.*

---

*University Chemical Laboratory, Lensfield Road, Cambridge, CB2 1EW, United Kingdom.
†Address correspondence to Professor M. D. Vargas, Universidade Estadual de Campinas, Instituto de Quimica, Caixa Postal 6154, 13081, Campinas, S. Paulo, Brasil.
‡Department of Chemistry, University College, London, WC1H OAJ, United Kingdom.

Trimethylamine oxide (Aldrich) must be freshly sublimed for the procedure that follows. Typically, 2 g of trimethylamine oxide is sublimed in a cold-finger apparatus (60°C, 0.05 torr, 16 h). Dry trimethylamine oxide is very hygroscopic. It is important to handle it in a dry atmosphere using suitably dried solvents.

In a 1-L, three-necked, round-bottomed flask and under an atmosphere of nitrogen, a suspension of $Os_3(CO)_{12}$ (0.5 g, 0.55 mmol)[4] is prepared in acetonitrile (750 mL), freshly distilled from a slurry of phosphorus pentoxide under nitrogen. To ensure dissolution of the maximum amount of $Os_3(CO)_{12}$, the suspension is heated under reflux for 1 h. The solution is then allowed to cool to room temperature, and under a nitrogen atmosphere, a solution of freshly sublimed, dry trimethylamine oxide (42 mg, 0.55 mmol) in acetonitrile (200 mL) is added to the stirred suspension over a period of 4 h using a pressure-equalized dropping funnel.

After being stirred for a total of 12 h, the mixture is filtered through silica to remove any excess amine oxide. The solvent is then removed at room temperature and the yellow solid is taken up in dichloromethane (120 mL) from which it is crystallized at 0°C. Yield: 406 mg (80%). Checkers noted that the product may alternatively be recrystallized from acetonitrile (15 mL) at 0°C.

Properties

The air-stable yellow product $Os_3(CO)_{11}(CH_3CN)$ is soluble in a variety of organic solvents apart from hydrocarbons in which it is only sparingly soluble. Its purity can be checked by IR spectrum: (CH_2Cl_2 solution) ν_{CO}, 2107 (w), 2054 (vs), 2040 (vs), 2017 (s, sh), 2008 (vs), 1981 (m) cm^{-1}.

It seems to be virtually impossible to obtain 100% pure $Os_3(CO)_{11}(MeCN)$ via this route. We generally prefer to use exactly one equivalent of amine oxide in the preparation. Although traces of $Os_3(CO)_{12}$ may then be present in the product, this compound is much less reactive and therefore less likely to interfere with subsequent reactions than $Os_3(CO)_{10}(MeCN)_2$, the impurity found if excess amine oxide is used. Possible impurities are ν_{CO} in CH_2Cl_2, cm^{-1}: $Os_3(CO)_{12}$: 2066 (s), 2032 (s), 2011 (w), 1997 (w); $Os_3(CO)_{10}(CH_3CN)_2$: 2079 (w), 2025 (s, sh), 2021 (vs), 1983 (m), 1960 (w).

Anal. Calcd. for $C_{13}H_3NO_{11}Os_3$: C, 16.95; H, 0.35; N, 1.50. Found: C, 17.2; H, 0.5; N, 1.2.

A typical acetonitrile replacement reaction on $Os_3(CO)_{11}(CH_3CN)$ is that with pyridine, described in Section B.

B. UNDECACARBONYL(PYRIDINE)TRIOSMIUM

$$Os_3(CO)_{11}(CH_3CN) + C_5H_5N \xrightarrow{\text{cyclohexane}} Os_3(CO)_{11}(C_5H_5N) + CH_3CN$$

Procedure

■ See cautionary note 1 in Section A. Pyridine must be handled in a well-ventilated fume hood.

In a 1.5-L, three-necked, round-bottomed flask and under a nitrogen atmosphere, a quantity of $Os_3(CO)_{11}(CH_3CN)$ (700 mg, 0.76 mmol) is dissolved in degassed cyclohexane (1 L). Excess pyridine (0.5 mL, 6.2 mmol) is then added. The solution is stirred at 60°C for 15 min, after which time the reaction is complete. The progress of the reaction should be monitored by IR spectroscopy so that heating is not continued after completion. This could lead to orthometallation of the pyridine derivative to $HOs_3(CO)_{10}(C_5H_4N)$.[2]

The solvent and excess pyridine are removed by a fast stream of nitrogen (in a well-ventilated fume hood), and the yellow-green residue is crystallized from dichloromethane–diethyl ether at −10°C. The product is obtained as yellow-orange crystals. A green powder might precipitate. This is somewhat less pure $Os_3(CO)_{11}(C_5H_5N)$. Yield: 537 mg (74%).

Anal. Calcd. for $[C_{16}H_5NO_{11}Os_3]CH_2Cl_2$: C, 20.60; H, 0.60; N, 1.35, Found: C, 20.55; H, 0.60; N, 0.90.

Properties

The compound $Os_3(CO)_{11}(C_5H_5N)$ is very soluble in acetonitrile, dichloromethane, acetone, and methanol, and its solutions are stable to air. It is sparingly soluble in hydrocarbons. Its purity can be checked by IR spectroscopy (dichloromethane solution): v_{CO} cm^{-1}: 2106 (w), 2052 (s), 2035 (vs), 2008 (s), 1976 (m). Other physical properties have been reported.[2] The vacuum pyrolysis of this compound provides a high-yield route to the carbido-dianion $[Os_{10}C(CO)_{24}]^{2-}$.[5]

C. BIS(ACETONITRILE)DECACARBONYLTRIOSMIUM

$$Os_3(CO)_{12} + 2(CH_3)_3NO$$
$$\xrightarrow{CH_3CN} Os_3(CO)_{10}(CH_3CN)_2 + 2Me_3N + 2CO_2$$

■ See cautionary notes in Section A.

Using the apparatus and sublimation procedure for trimethylamine oxide described in Section A a suspension of $Os_3(CO_{12})$ (0.5 g, 0.55 mmol) is prepared in acetonitrile, (700 mL) freshly distilled from phosphorous pentoxide under nitrogen. In order to dissolve the maximum amount of $Os_3(CO)_{12}$, the suspension is heated under reflux for 1 h. The solution is then cooled to 40°C. Slightly more than two equivalents of trimethylamine oxide (100 mg, 1.31 mmol) in acetonitrile (200 mL) is added under nitrogen over a period of 2 h. The excess Me_3NO does not react further with $Os_3(CO)_{10}(CH_3CN)_2$ under these conditions. The mixture is left stirring at this temperature for a further 2 h. The dark yellow solution is filtered through silica to remove excess trimethylamine oxide; the solvent is then removed under vacuum at room temperature to yield a brown-yellow solid.

This solid is suspended in 2 mL of acetonitrile and transferred by pipette to a 25-mL round-bottomed flask. The supernatant liquid is returned to the original flask and the procedure repeated as many times as is necessary to transfer all the solid. At this stage, evaporation of the solvent gives slightly impure $Os_3(CO)_{10}(CH_3CN)_2$ in yields between 85 and 95%. The trace brown impurity does not impair the reactivity of the compound even in sensitive reactions. We therefore generally use it without going through the purification process described next.

Purification of the product can be achieved by washing it with acetonitrile (2 mL), leaving the suspension overnight at $-25°C$, and then twice decanting the brown solution. Pure $Os_3(CO)_{10}(CH_3CN)_2$ is thus obtained in $\sim 60\%$ yield. The supernatant of the brown solution may be purified by thin-layer chromatography, using Merck plates precoated with silica gel 60F-254, and 50% $CHCl_3$, 45% hexane and 5% MeCN as eluant. (R_f 0.4.) Additional product is obtained by this procedure.

Anal. Calcd. for $C_{14}H_6N_2O_{10}Os_3$: C, 18.0; H, 0.6; N, 3.0. Found: C, 18.2; H, 0.6; N, 3.2.

Properties

Infrared absorptions in the carbonyl region for $Os_3(CO)_{10}(CH_3CN)_2$ in CH_2Cl_2 are 2079 (w), 2025 (sh), 2021 (vs), 1983 (m), 1960 (w) cm^{-1}. The acetonitrile ligands are readily displaced by other donor ligands; see Ref. 2.

References

1. B. F. G. Johnson and J. Lewis, *Gazz. Chim. Ital.*, **109**, 271 (1979).
2. B. F. G. Johnson, J. Lewis, and D. A. Pippard, *J. Chem. Soc. Dalton Trans.*, **1981**, 407.
3. Johnson et al., *J. Chem. Soc. Dalton Trans.*, **1981**, 407.
4. B. F. G. Johnson and J. Lewis, *Inorg. Synth.*, **13**, 93 (1972). See also section 57 in this volume.
5. P. F. Jackson, B. F. G. Johnson, J. Lewis, W. J. H. Nelson, and M. McPartlin, *J. Chem. Soc. Dalton Trans.*, **1982**, 2099.

59. μ-NITRIDO-BIS(TRIPHENYLPHOSPHORUS) (1 +)
μ-CARBONYL-DECACARBONYL-
μ-HYDRIDOTRIOSMATE(1 −)

Submitted by K. BURGESS* and R. P. WHITE*
Checked by S. BASSNER,† G. L. GEOFFROY,† R. L. GRAY,‡ and D. J. DARENSBOURG‡

The starting material for most of the osmium cluster chemistry published to date, $Os_3(CO)_{12}$,[1] is quite stable and relatively unreactive. Hydroxide ions remove one carbonyl ligand from $Os_3(CO)_{12}$ and a triosmium anion results that is far more reactive than the cluster from which it is produced. Indeed, this anion, $[Os_3(\mu\text{-H})(\mu\text{-CO})(CO)_{10}]^-$, combines under mild conditions with a range of inorganic,[2,3] organic,[4,5] and organometallic,[6,7] electrophiles providing a route to functionalized triosmium complexes and mixed-metal clusters. A reliable and convenient modification of the original synthesis of $[Os_3(\mu\text{-H})(\mu\text{-CO})(CO)_{10}]^-$ (Ref. 8) is presented here.

$$Os_3(CO)_{12} + KOH + (PPh_3)_2NCl \longrightarrow$$
$$[(PPh_3)_2N][Os_3(\mu\text{-H})(\mu\text{-CO})(CO)_{10}] + KCl + CO_2$$

■ **Caution.** *Owing to the high toxicity of carbon monoxide and osmium carbonyls these reactions should be carried out in an efficient fume hood.*

Powdered samples of 0.454 g (0.5 mmol) of $Os_3(CO)_{12}$ (Ref. 1) and 1.403 g (25 mmol) of KOH are placed in a 50-mL Schlenk tube with a magnetic stirrer. The reaction vessel is sealed with a septum and alternately evacuated and filled with nitrogen three times to remove oxygen. The Schlenk tube and its contents are cooled in liquid nitrogen, 25 mL of methanol (distilled from magnesium and taken directly from a still collector vessel under nitrogen) is added by syringe, and the mixture is degassed using three freeze-thaw cycles. The reaction mixture is then warmed to 25° and vigorously stirred for 6 h, during which time the solution changes from yellow to red. A degassed solution of 0.32 g (0.56 mmol) of $(PPh_3)_2NCl$ (Aldrich Chemical Co.) in 3 mL of methanol is added all at once and the stirring is continued for 15 min. Degassed distilled water (~ 2 mL required) is carefully added dropwise, with stirring, over 5 min until a red precipitate appears and persists on stirring for

---

*University Chemical Laboratory, Cambridge, CB2 1EW, United Kingdom. Address correspondence to Kevin Burgess, Chemistry, Box 1892, Rice University, Houston TX 77251.
†Department of Chemistry, The Pennsylvania State University, University Park, PA 16802.
‡Department of Chemistry, Texas A&M University, College Station, TX 77843.

2 min. The water should be added with care since too much will precipitate some $(PPh_3)_2NCl$. The suspension is stirred further for 1 h. The precipitate is then filtered in air, washed with 5 mL of distilled water and 5 mL of cold methanol, and dried *in vacuo* overnight to give 58%, 0.414 g, (0.292 mmol) of μ-nitrido-bis(triphenylphosphorus)(1 +) μ-carbonyl-decarbonyl-μ-hydridotriosmate(1 −).

Anal. Calcd. for $C_{47}H_{31}NO_{11}P_2Os_3$: C, 39.80; H, 2.20; N, 0.99; P, 4.37. Found: C, 40.05, H, 2.28; N, 1.10; P, 4.45.

Properties

μ-Nitrido-bis(triphenylphosphorus)(1 +) μ-carbonyl-decarbonyl-μ-hydridotriosmate(1 −), $[(PPh_3)_2N][Os_3(\mu\text{-}H)(\mu\text{-}CO)(CO)_{10}]$ is obtained as a light red powder. The solid is air stable but solutions decompose in ∼ 5 min when exposed to the atmosphere. It is soluble in tetrahydrofuran, dichloromethane, acetonitrile, methanol, and diethyl ether, and insoluble in hydrocarbon solvents. The IR spectrum of the compound contains four CO stretching vibrations for terminal carbonyls and one bridging carbonyl stretch (CH_2Cl_2, cm^{-1}): 2038 (w), 2021 (s), 1996 (s), 1951 (ms), and 1667 (w), respectively. The ^1H NMR spectrum (80 MHz, chloroform-d_1, δ in ppm downfield from tetramethylsilane, ambient) shows a broad multiplet at 7.5, due to the protons attached to the aromatic rings in the cation, and a sharp metal hydride signal at − 13.8.

References

1. B. F. G. Johnson and J. Lewis, *Inorg. Synth.*, **13**, 93 (1972). See also section 57 in this volume.
2. B. F. G. Johnson, P. R. Raithby, and C. Zuccaro, *J. Chem. Soc. Dalton Trans.*, **1980**, 99.
3. B. F. G. Johnson, J. Lewis, P. R. Raithby, and S. W. Sankey, *J. Organomet. Chem.*, **228**, 135 (1982).
4. J. B. Keister, *J. Chem. Soc. Chem. Commun.*, **979**, 214
5. C. E. Kampe, N. M. Boag, and H. D. Kaesz, *J. Am. Chem. Soc.*, **105**, 2896 (1983).
6. B. F. G. Johnson, D. A. Kaner, J. Lewis, and P. R. Raithby, *J. Organomet. Chem.*, **215**, C33 (1981).
7. M. Fajardo, H. D. Holden, B. F. G. Johnson, J. Lewis, and P. R. Raithby, *J. Chem. Soc. Chem. Commun.*, **1984**, 24.
8. C. R. Eady, B. F. G. Johnson, J. Lewis, and M. C. Malatesta, *J. Chem. Soc. Dalton Trans.*, **1978**, 1358.

60. DECACARBONYLDI-μ-HYDRIDOTRIOSMIUM: $Os_3(\mu\text{-H})_2(CO)_{10}$

Submitted by H. D. KAESZ*
Checked by G. N. GLAVEE and R. J. ANGELICI†

$$Os_3(CO)_{12} + H_2(1 \text{ atm}) \xrightarrow[\text{octane}]{120°C, 1.5 h} Os_3(\mu\text{-H})_2(CO)_{10} + 2CO$$

The complex $Os_3(\mu\text{-H})_2(CO)_{10}$ contains 46 valence electrons and is electronically unsaturated.[1] It is a key reagent in the preparation of a variety of hydrocarbon complexes of the Os_3 cluster,[2] and is in general more reactive than the electronically saturated ($48e^-$) parent carbonyl, $Os_3(CO)_{12}$, from which it is prepared.[3]

Procedure

■ **Caution.** *Use of a stream of hydrogen gas, and evolution of carbon monoxide, require this reaction to be run in a well-ventilated hood.*

Dodecacarbonyltriosmium, $Os_3(CO)_{12}$, is commercially available (Strem Chemicals) or may be prepared from OsO_4 and CO.[4] Solvents should be dried over CaH_2 and distilled under an atmosphere of N_2 before use.

■ **Caution.** *Osmium tetroxide, OsO_4, is a severe poison and must be handled in a well-ventilated hood.*

A quantity of $Os_3(CO)_{12}$ (200 mg, 0.22 mmol) is placed into a three-necked flask (100-mL capacity) equipped with a reflux condenser and a gas inlet tube (constricted to a fine tip as for a medicine dropper). Octane (40 mL) is distilled from CaH_2 under reduced pressure (100 torr) and added to the reaction flask. Air is displaced from the flask by bubbling a slow stream of hydrogen while still at room temperature, after which the contents are warmed to 120°C for a period of 4 h. The solution becomes deep red and IR indicates little starting material remains. The checkers have found that yield and reaction rate depend on the particle size; finely ground starting material is recommended.

The reaction mixture is allowed to cool to room temperature, transferred to a single-necked flask, and the solution concentrated to a few milliliters under reduced pressure (15 torr, assisted by gentle heating, $\sim 50°C$). The remaining liquid is placed at the top of a silica gel column (20 × 2 cm, 60-

---

*Department of Chemistry and Biochemistry, University of California, Los Angeles, CA 90024-1569.
†Department of Chemistry, Iowa State University, Ames, IA 50011.

mesh, EM Reagents), and eluted with hexane. The product $Os_3(\mu\text{-H})_2(CO)_{10}$ elutes as a deep purple band (130 mg, 0.15 mmol) in 73% yield, followed by a light yellow band of unreacted starting material ($\sim$ 9 mg). IR (hexane), cm^{-1}: 2074 (s), 2063 (m), 2024 (s), 2009 (s), 1990 (w), and 1971 (vw).[3,4] ^{1}H NMR: δ(ppm) 11.6(C_6H_6).[4]

Anal. Calcd. for $C_{10}H_2O_{10}Os_3$: 14.0 %C, 0.23% H. Found, 14.3%C, 0.3%H.

Properties

The complex $Os_3(\mu\text{-H})_2(CO)_{10}$ is moderately air stable in the solid, but its solutions are best handled under an atmosphere of N_2. The complex is soluble in most common organic solvents such as hydrocarbons and ethers, however, extensive contact with chlorinated solvents should be avoided because of possible halogenation reaction of the metal complex. Infrared, Raman and mass spectra have been recorded,[3-5] and the structure determined both by X-ray[6] and neutron[7] diffraction. The reactions of this complex, including those with unsaturated hydrocarbons, have been reviewed.[2]

References

1. (a) D. M. P. Mingos, *Nature* (London) *Phys. Sci.*, **236**, 99 (1972); (b) J. W. Lauher, *J. Am. Chem. Soc.*, **100**, 5305 (1978); (c) D. M. P. Mingos, *Acc. Chem. Res.*, **17**, 311 (1984).
2. (a) A. P. Humphries and H. D. Kaesz, *Progr. Inorg. Chem.*, **25**, 145 (1979); (b) R. D. Adams and J. P. Selegue, in *Comprehensive Organometallic Chemistry*, Vol. 4, G. Wilkinson, F. G. A. Stone, and E. W. Abel (eds.), Pergamon, Oxford, U.K., 1982, pp. 1023ff.
3. S. A. R. Knox, J. W. Koepke, M. A. Andrews, and H. D. Kaesz, *J. Am. Chem. Soc.*, **97**, 3942 (1975).
4. (a) B. F. G. Johnson, J. Lewis, and P. Kilty, *J. Chem. Soc. A*, 2859 (1968); (b) S. R. Drake and P. A. Loveday, this volume.
5. M. A. Andrews, S. W. Kirtley, and H. D. Kaesz, *Adv. Chem. Ser.*, **167**, 215 (1978).
6. M. R. Churchill, F. J. Hollander, and J. P. Hutchinson, *Inorg. Chem.*, **16**, 2697 (1977).
7. R. W. Broach and J. M. Williams, *Inorg. Chem.*, **18**, 314 (1979).

61. DODECACARBONYLTETRA-μ-HYDRIDO--*tetrahedro*-TETRAOSMIUM

Submitted by CAMILO ZUCCARO*
Checked by G. PAMPLONI and F. CALDERAZZO†

$$4Os_3(CO)_{12} + 6H_2 \rightleftarrows 3Os_4H_4(CO)_{12} + 12CO$$

The complex $Os_4H_4(CO)_{12}$ was first synthesized in low yield by acidification of the anions formed in the reaction of methanolic potassium hydroxide and $Os_3(CO)_{12}$.[1] In 1975, Kaesz and coworkers synthesized $Os_4H_4(CO)_{12}$ in ~30% yield using atmospheric hydrogenation of $Os_3(CO)_{12}$ in refluxing octane over a period of 50 h.[2] In 1978, Lewis and coworkers synthesized $Os_4H_4(CO)_{12}$ in the range from 60 to 70% yield using 24 h of reaction at 120 atm of hydrogen and 100°C.[3] Bau and coworkers[4] found that in the direct hydrogenation procedure of Kaesz[2] the yield could be doubled by raising the temperature from that of refluxing octane (120°C) to that of a mixture of refluxing xylenes (137–144°C). The compound $Os_4H_4(CO)_{12}$ so obtained is found to be contaminated with many uncharacterized impurities, which make its purification tedious. A high-yield synthetic method is reported here to facilitate further study of the chemistry of $Os_4H_4(CO)_{12}$.

Procedure

A suspension of $Os_3(CO)_{12}$ (2 g) in hexane (60 mL) is placed into a stirred stainless steel autoclave (450-mL total capacity), under a nitrogen atmosphere.[5] After the autoclave is sealed, the gases are evacuated and the autoclave is pressurized to 25 atm of hydrogen. The autoclave is then heated with stirring to 140°C for 24 h. After cooling, the gas is slowly vented. After coming to atmospheric pressure the autoclave is flushed with an atmosphere of nitrogen, and opened.

■ **Caution.** *The vented gases may contain CO; ventilation of hydrogen gas into the atmosphere must be carried out with care in a well-ventilated hood.*

Pale yellow crystals of $Os_4H_4(CO)_{12}$ are separated by decantation of the supernatant solution. The sample thus obtained is sufficiently pure for most purposes. The yield of the $Os_4H_4(CO)_{12}$ is usually in the range from 80 to 85% (1.8 g). Checkers report a yield of 45% for a smaller-scale reaction,

---

*Departamento de Química, Facultad de Ciencias, Universidad de Los Andes, Mérida 5101, Venezuela.

†Departmento di Chimica e Chimica Industriale, Via Risorgimento, 35-56100 Pisa, Italy.

namely, 1 g $Os_3(CO)_{12}$ in 30 mL of solvent in an autoclave of 125-mL total capacity.

Further purification may be carried out by extracting the solids with chloroform (200 mL) in a Soxhlet extractor.

Properties

The complex $Os_4H_4(CO)_{12}$ is a pale yellow solid, insoluble in common organic solvents. The mass spectrum shows the parent ion peak and other fragments arising from successive loss of 12 CO groups. The IR spectrum (in cyclohexane solution) in the carbonyl region consists of four peaks 2085 (m), 2036 (s), 2020 (s), 1997 (w) cm^{-1}. The proton NMR spectrum in CD_2Cl_2 shows a sharp singlet at -20.5δ throughout the temperature range from 30 to $-50°C$, consistent with the fact that the hydrogen atoms are magnetically equivalent as is shown in the X-ray structure.[6]

Anal. Calcd.: C, 13.00%; H, 0.46%.

References

1. B. F. G. Johnson, J. Lewis, and P. A. Kilty, *J. Chem. Soc. A*, **1968**, 2889.
2. S. A. R. Knox, J. W. Keople, M. A. Andrews, and H. D. Kaesz, *J. Am. Soc.*, **97**, 3942 (1975).
3. B. F. G. Johnson, J. Lewis, P. R. Raithby, G. M. Sheldrick, K. Wong, and M. McPartlin, *J. Chem. Soc. Dalton Trans.*, **1978**, 673.
4. C.-Y. Wey, L. Garlaschelli, R. Bau, and T. F. Koretzli, *J. Organomet. Chem.*, **213**, 63 (1981).
5. J. R. Norton, J. P. Collman, G. Dolcetti, and W. T. Robinson, *Inorg. Chem.* **11**, 382 (1972); $Os_3(CO)_{12}$ may also be purchased from Strem Chemicals.
6. B. F. G. Johnson, J. Lewis, P. R. Raithby, and C. Zuccaro, *Acta Cryst.*, **B34**, 1728 (1981).

62. TRI-μ-CARBONYL-NONACARBONYLTETRA-RHODIUM, $Rh_4(CO)_9(\mu\text{-}CO)_3$

$$RhCl_3 + 2Cu + 4CO + NaCl \rightarrow Na[Rh(CO)_2Cl_2] + 2Cu(CO)Cl$$
$$4Na[Rh(CO)_2Cl_2] + 6CO + 2H_2O \rightarrow Rh_4(CO)_{12} + 2CO_2$$
$$+ 4NaCl + 4HCl$$

Submitted by S. MARTINENGO,* G. GIORDANO,* and P. CHINI*
Checked by G. W. PARSHALL† and E. R. WONCHOBA†

Two general methods have been given for the preparation of $Rh_4(CO)_{12}$: high-pressure methods that employ $RhCl_3$ or $Rh_2(CO)_4Cl_2$ with metal powders as halogen acceptors,[1] and atmospheric pressure syntheses starting with $K_3[RhCl_6]\cdot H_2O$ and copper powder in water,[2] or with $Rh_2(CO)_4Cl_2$ and $NaHCO_3$ in hexane.[3] The synthesis described here is conducted at atmospheric pressure with $RhCl_3\cdot 3H_2O$ as the starting material rather than $K_3[RhCl_6]\cdot H_2O$. The overall procedure requires about 30 h, with a yield of 80 to 90%.

The compound $Rh_4(CO)_{12}$ has a large number of applications. It is used in catalysis directly or as a catalyst precursor, and it is the starting material both for substitution reactions and for the synthesis of other rhodium carbonyl clusters. In particular it is easily converted into $Rh_6(CO)_{16}$, thus providing a method for the synthesis of this carbonyl in a highly pure state.[4]

Procedure

The starting material, $RhCl_3\cdot 3H_2O$, may be obtained commercially,‡ or prepared as described in the literature.[5] Copper powder is first activated by washing with a mixture of concentrated hydrochloric acid and acetone and then rinsing with acetone and drying *in vacuo*. The solvent CH_2Cl_2 may be laboratory grade and is distilled from anhydrous sodium carbonate before use.

■ **Caution.** *The operations must be carried out in a well-ventilated hood because carbon monoxide is used.*

---

*Centro del C. N. R. per lo Studio dei Composti di Coordinazione nei Bassi Stati di Ossidazione. Via G. Venezian, 21, 20133 Milan, Italy.
†Central Research and Development Department, E. I. du Pont de Nemours & Co., Wilmington, DE 19898.
‡Available from Alfa Products, Ventron Corp., P.O. Box 299, Danvers, MA 92923.

A two-necked 1-L flask, equipped with a stopcock on the side neck and a magnetic stirring bar, is charged with activated copper powder (1.5 g, 24 mmol), sodium chloride (0.6 g, 10 mmol), and water (200 mL).

A 100-mL dropping funnel with a pressure-equalizing tube is then placed on the central neck. The funnel is stoppered, and the entire apparatus is pumped *in vacuo* until the water is well degassed. It then is filled through the side stop-cock with carbon monoxide from a CO line, including a mineral oil bubbler and a pressure-release bubbler with an 80-mm mercury column. The dropping funnel is opened while CO is passed through it, and it is charged with a degassed solution of $RhCl_3 \cdot 3H_2O$ (2.6 g, Rh 39.6%, 9.0 mmol) in water (50 mL). The funnel is closed, and the solution is dropped into the flask at a rate of about 1 mL min^{-1} while the mixture is stirred vigorously. During the addition, care must be taken that the pressure in the flask never lowers, for the lack of CO can cause the partial formation of rhodium metal. The best conditions for the reaction are achieved by adding the carbon monoxide at such a rate that a small portion of the gas continuously escapes from the pressure-release bubbler. For the same reason, the stirring must be vigorous enough to ensure good contact between the gaseous and aqueous phases. The rate of absorption of the CO is initially slow, but gradually increases, reaching a maximum when about half the $RhCl_3$ solution has been added.

At the end of the addition, the last traces of the $RhCl_3$ solution in the funnel are washed down into the flask with 10 mL of degassed water, and the mixture is stirred for 2 h. During this time, some $Rh_4(CO)_{12}$ begins to separate as an orange powder. The dropping funnel is then opened in a CO stream and charged with a degassed solution of disodium citrate (0.4 *M*, 50 mL). The funnel is stoppered, and the solution of citrate is dropped into the yellow mixture during 50 min. The mixture is then stirred for 20 h, and the resulting orange suspension of the carbonyl is filtered under carbon monoxide atmosphere.* The solid is washed with water to remove all the mother liquor and then dried *in vacuo* ($\sim 10^{-1}$ torr) at room temperature.

The product is extracted on the filter under CO with the minimum amount of CH_2Cl_2 (about 25 mL, in 5-mL portions). The solution is quickly evaporated to dryness *in vacuo* on a water bath at room temperature, and the resulting crystalline mass is dried *in vacuo* for 2 h to give 1.55–1.75 g of $Rh_4(CO)_{12}$ with a yield of 80–90%.†

Anal. Calcd. For $Rh_4(CO)_{12}$: Rh, 55.08%. Found: Rh, 54.74%. This product is sufficiently pure for most purposes, the impurity being a trace of

---

*The CO atmosphere is required to prevent the precipitation of the insoluble CuCl by decomposition of Cu(CO)Cl.

†The checkers report that a ninefold scale-up of this synthesis proceeds satisfactorily.

$Rh_6(CO)_{16}$. A $Rh_6(CO)_{16}$-free product can be obtained by extraction, under carbon monoxide with pentane, followed by crystallization at $-70°C$, filtration at this temperature, and drying in a stream of CO.

Properties

The compound $Rh_4(CO)_{12}$ is a dark-red solid when crystalline and orange when powdered. It decomposes under nitrogen at 130–140°, giving $Rh_6(CO)_{16}$ (lit. 150°).[1a, 1c] The same decomposition also occurs at room temperature, but it is very slow. For this reason $Rh_4(CO)_{12}$ is best stored under a CO atmosphere. It is stable in air, although on prolonged standing decomposition occurs. It is readily soluble in CH_2Cl_2 and $CHCl_3$, soluble in pentane ($\sim 12\,g\,L^{-1}$ at 25°), toluene and tetrahydrofuran (THF) ($\sim 10\,g\,L^{-1}$), and acetone, and less soluble in cyclohexane and methanol.

Under nitrogen the solutions slowly decompose to $Rh_6(CO)_{16}$. This decomposition is faster at higher temperatures and provides a method for the synthesis of $Rh_6(CO)_{16}$. Very pure crystalline $Rh_6(CO)_{16}$ is formed in high yields on heating a saturated solution of $Rh_4(CO)_{12}$, or a suspension, in heptane at 80–90° under nitrogen until the orange-red color has disappeared.

The solutions in polar solvents (THF, acetone) are unstable toward water, which causes transformation into $Rh_6(CO)_{16}$ through the violet $[H_3O]_2$-$[Rh_{12}(CO)_{30}]$[6].

The compound $Rh_4(CO)_{12}$ does not react with CO at room temperature at 300 atm. At higher temperature it is transformed into $Rh_6(CO)_{16}$, even under high CO pressures (80–120°, 360–420 atm).[7] More recently IR evidence has been reported for the formation of $Rh_2(CO)_8$ by the reaction of $Rh_4(CO)_{12}$ with CO at 430 atm and $-19°$.[8]

The IR spectrum in heptane solution shows bands at 2101 (w), 2074.5 (s), 2069 (s), 2059 (w, sh), 2044.5 (m), 2041.5 (m, sh), 2001 (w), 1918.4 (w), 1882 (s), 1848 (w) cm^{-1}.[9]

Tri-μ-carbonyl-nonacarbonyltetrarhodium is the starting material for the synthesis of a large number of rhodium carbonyl cluster compounds.[10]

References

1. (a) W. Hieber and H. Lagally, *Z. anorg. allg. Chem.*, **251**, 96 (1943). (b) S. H. Chaston and F. G. A. Stone, *J. Chem. Soc. A*, **1969**, 500. (c) B. L. Booth, M. J. Else, R. Fields, H. Goldwhite, and R. N. Haszeldine, *J. Organomet. Chem.*, **14**, 417 (1968).
2. S. Martinengo, P. Chini, and G. Giordano, *J. Organomet. Chem.*, **27**, 389 (1971).
3. G. Giordano, S. Martinengo, D. Strumolo, and P. Chini, *Gazzetta*, **105**, 613 (1975).
4. P. Chini and S. Martinengo, *Inorg. Chim. Acta*, **3**, 315 (1969).
5. S. N. Anderson and F. Basolo, *Inorg. Synth.*, **7**, 214 (1963).
6. P. Chini, S. Martinengo, and G. Garlaschelli, *Chem. Commun.*, **1972**, 709.

7. P. Chini and S. Martinengo, *Inorg. Chim. Acta*, **3**, 21 (1969).
8. R. Whyman, *J. Chem. Soc. Dalton Trans.*, **1972**, 1375.
9. F. Cariati, P. Fantucci, V. Valenti, and P. Barone, *Rend. ist. Lomb., Sci.*, **105**, 122 (1971).
10. P. Chini, G. Longoni, and V. G. Albano, *Adv. Organomet. Chem.*, **XIV**, 285 (1976).

63. DODECACARBONYLTETRAIRIDIUM: $Ir_4(CO)_{12}$

$$IrCl_3 + 3CO + H_2O \rightarrow [Ir(CO)_2Cl_2]^- + Cl^- + 2H^+ + CO_2$$

$$4[Ir(CO)_2Cl_2]^- + 6CO + 2H_2O \rightarrow Ir_4(CO)_{12} + 2CO_2 + 4H^+ + 8Cl^-$$

Submitted by ROBERTO DELLA PERGOLA,* LUIGI GARLASCHELLI,* and SECONDO MARTINENGO*
Checked by STEPHEN SHERLOCK† and WAYNE GLADFELTER†

The compound $Ir_4(CO)_{12}$ is prepared by several methods, most of which involve the use of high-pressure equipment. The only high-yield synthesis that works at atmospheric pressure is that reported by Malatesta[1] starting from $Na_2[IrCl_6]$. The synthesis that we describe here is a modification of another mild conditions synthesis that we recently adopted.[2] The synthesis starts from hydrated iridium trichloride and involves two steps. In the first step the $IrCl_3$ is transformed into $[Ir(CO)_2Cl_2]^-$ with a high-temperature carbonylation, and in the second step $[Ir(CO)_2Cl_2]^-$ is converted to $Ir_4(CO)_{12}$ at room temperature by partial buffering of the acidity formed. Though the reaction time and yields are comparable to those of the synthesis reported in Ref. 1, the synthesis here described requires fewer operations. The overall procedure requires 2–3 days.

Procedure

■ **Caution.** *With the use of the highly poisonous carbon monoxide, all operations should be carried out in an efficient fume hood.*

2-Methoxyethanol is Merck analytical grade and is used as received. Tetrahydrofuran (THF) is distilled under nitrogen over sodium benzophenone ketyl. Distilled water is degassed by pumping in vacuum and stored under nitrogen. Aqueous 1 *M* disodium citrate is prepared from citric acid and NaOH and is degassed as the water. The hydrated $IrCl_3$ is commercial product. Carbon monoxide should contain less than 2 ppm of oxygen by

---

*Dipartimento di Chimica Inorganica e Metallorganica dell'Universita' e Centro C.N.R. sui Bassi Stati di Ossidazione, Via G. Venezian 21, 20133 Milano, Italy.
†Department of Chemistry, University of Minnesota, Minneapolis, MN 55455-0431.

volume and is supplied through a gas line including a mineral oil bubbler and a pressure-release safety bubbler filled with an 80-mmHg column.

A sample of 3.1 g of $IrCl_3$ trihydrate (Ir 54.44%, 8.79 mmol) is placed in a 250-mL, two-necked, round-bottomed flask containing a 4-cm magnetic stirring bar. Then 10 mL of water is added and the mixture heated until the $IrCl_3$ is dissolved. The clear solution is left to cool, then 100 mL of 2-methoxyethanol is added; the resulting solution should be clear. The central neck of the flask is stoppered and the side neck is fitted with a stopcock. The solution is briefly degassed by pumping in vacuum, then CO is introduced. The flask is opened while passing a CO stream, and a reflux condenser with a mineral oil bubbler on the top is placed on the central neck. The apparatus is flushed with CO for a few minutes, then the stopcock is quickly substituted with a gas dispersion tube dipping into the solution. The stream of CO is regulated at about two bubbles per second; this flow rate is maintained during all the synthesis, and is temporarily increased whenever the flask is opened for the addition of the reagents. The flask is placed in a controlled-temperature oil bath at 110°C, and the magnetic stirring is started. The carbonylation is complete in about 4.5 h. At the end the solution is yellow and the IR shows two strong bands at 2055 and at 1970 cm^{-1}, with the latter being slightly stronger; if the band at 2055 cm^{-1} is stronger, the carbonylation is to be continued for an additional hour. The flask is removed from the oil bath without interrupting the CO stream, and left to cool, then the condenser is substituted with a mineral-oil bubbler. During all the subsequent operations and till the end of the reaction, a vigorous stirring of the mixture is necessary, in order to have efficient CO saturation of the solution, because a lack of CO could cause formation of some metal. To the stirred mixture, 25 mL of the 1 M disodium citrate solution is added in portions of 5 mL each, one every 30 min. At the end of the additions, the carbonylation is left to continue overnight. At this point an IR check of the solution should show only traces of the two bands of the $[Ir(CO)_2Cl_2]^-$ anion. If this is not the case, the carbonylation is continued further. The suspension of the carbonyl is filtered in air on a weighed porous porcelain fine frit and the solid washed well with two portions of 10 mL each of water, acetone, and methanol, and dried first briefly in air, then in a desiccator under vacuum. The crude yellow-green carbonyl weighs 1.85 g, 76%, and is sufficiently pure for most purposes. A better-looking highly pure yellow crystalline product can be obtained by purification in the following way: the crude product together with 0.9 g of LiI is placed in a 100-mL, two-necked, round-bottomed flask equipped with a gas inlet tube, a reflux condenser, and a magnetic stirring bar. The flask is purged with nitrogen then 50 mL of THF is added; the mixture is refluxed under a nitrogen stream for 2–3 h. The carbonyl slowly dissolves to give a dark red solution containing the

[Ir$_4$(CO)$_{11}$I]$^-$ anion.[2] The solution is cooled at room temperature, then filtered under nitrogen through a fine fritted glass filter into another flask. The nitrogen is removed under vacuum and CO is introduced. The solution is stirred under flowing CO for 6 h to give an almost quantitative precipitation of the yellow crystalline iridium carbonyl that is separated from the reddish mother liquor by filtration in air through a fine glass frit. The carbonyl is washed two times with 10 mL each of water, methanol, and acetone, and dried in vacuum. Yield: 1.75 g, 72%.

Anal. Calcd. for C$_{12}$Ir$_4$O$_{12}$:C, 13.04; Ir, 69.58. Found: C, 13.10; Ir, 69.40.

Properties

The purified Ir$_4$(CO)$_{12}$ is canary-yellow, while the crude product varies in color from yellow-green to ivory-yellow, but no significant differences in analytical data are observed. It is stable for a very long period of time in the air and is insensitive to moisture. Practically it is sparingly soluble in all solvents. A saturated solution in dichloromethane (0.1-mm cells) shows IR bands at 2067 (m) and 2027 (w) cm^{-1} [3] while a saturated solution in THF has bands at 2060 (m) and 2020 (w) cm^{-1}. An IR spectrum in the solid state (KBr pellets) shows bands at 2084, 2053, and 2020 cm^{-1}.[1]

Dodecacarbonyltetrairidium is the starting material for the synthesis of a large number of substitution products and of most anionic iridium carbonyl cluster compounds.[4-6] Possible uses in catalysis of carbonyl and its substituted derivatives is also emerging.[7]

References

1. L. Malatesta, G. Caglio, and M. Angoletta, *Inorg. Synth*, **13**, 95 (1972).
2. R. Della Pergola, L. Garlaschelli, and S. Martinengo, *J. Organometal. Chem.*, **331**, 271 (1987).
3. F. Cariati, V. Valenti, and G. Zerbi, *Inorg. Chim. Acta*, **3**, 378 (1969).
4. R. Ros, A. Scrivanti, V. G. Albano, D. Braga, and L. Garlaschelli, *J. Chem. Soc. Dalton Trans.*, **1986**, 2411 and references cited therein.
5. H. Vahrenkamp, *Adv. Organomet. Chem.*, **32**, 169 (1983) and references cited therein.
6. R. S. Dickson, *Organometallic Chemistry of Rhodium and Iridium*, Academic Press, London, 1983.
7. B. C. Gates, L. Guczi, and H. Knozinger, *Metal Clusters in Catalysis*, Elsevier, Amsterdam, 1986.

Chapter Six
CYCLOPENTADIENYL COMPLEXES

64. DICARBONYLBIS(η^5-CYCLOPENTADIENYL) COMPLEXES OF TITANIUM, ZIRCONIUM, AND HAFNIUM

Submitted by DAVID J. SIKORA,* KEVIN J. MORIARTY,* and MARVIN D. RAUSCH*
Checked by A. RAY BULLS,† JOHN E. BERCAW,† VIKRAM D. PATEL,‡ and ARTHUR J. CARTY‡

Dicarbonylbis(η^5-cyclopentadienyl)titanium Ti(CO)$_2$(η^5-C$_5$H$_5$)$_2$, was first synthesized in 1959. It was the first carbonyl complex of a Group 4 metal.[1,2] Since the original synthesis, several preparations of Ti(CO)$_2$(η^5-C$_5$H$_5$)$_2$ have been reported.[3-9] The corresponding zirconium and hafnium analogs of Ti(CO)$_2$(η^5-C$_5$H$_5$)$_2$ were not described until 1976. The preparation of Zr(CO)$_2$(η^5-C$_5$H$_5$)$_2$ was reported independently by three different research groups, each employing different methods.[7,8,10] One of the groups also described the only known procedure for the synthesis of Hf(CO)$_2$(η^5-C$_5$H$_5$)$_2$.[7]

Because of our interest in the structure and reactivity of M(CO)$_2$(η^5-C$_5$H$_5$)$_2$ (M = Zr, Hf),[11,12] we sought a more convenient method of preparation for these compounds, since existing syntheses were of low yield and/or required severe reaction conditions.[7,8,10] We have subsequently developed facile routes to the syntheses of M(CO)$_2$(η^5-C$_5$H$_5$)$_2$ (M = Ti, Zr, Hf) utilizing the reductive carbonylation of MCl$_2$(η^5-C$_5$H$_5$)$_2$. The synthesis of Ti(CO)$_2$(η^5-C$_5$H$_5$)$_2$ described here is a modification of the convenient procedure reported originally by Demerseman et al.[6] Our procedure involves

---

*Department of Chemistry, University of Massachusetts, Amherst, MA 01003.
†Department of Chemistry, California Institute of Technology, Pasadena, CA 91125.
‡Department of Chemistry, University of Waterloo, Waterloo, Ontario, Canada N2L 3Gl.

magnesium metal activated *in situ* by mercury(II) chloride. The advantages of this method are as follows: (1) no high-pressure apparatus is required, since the reaction occurs readily at 1 atm CO pressure;[8,10] (2) the reaction does not involve large amounts of elemental mercury or sodium amalgam;[7] (3) each synthesis is a one-step procedure that avoids the preparation of intermediates;[8] (4) the yields are reproducible and give ample amounts of product for further studies;[7] and (5) the starting metallocene dichlorides are commercially available.

Because of the current interest in the structure and chemistry of the corresponding pentamethylcyclopentadienyl analogs $M(CO)_2(\eta^5-C_5Me_5)_2$ (M = Ti, Zr, Hf),[13-15] we have extended our procedure to the syntheses of $Ti(CO)_2(\eta^5-C_5Me_5)_2$,[16,17] $Zr(CO)_2(\eta^5-C_5Me_5)_2$,[18] and $Hf(CO)_2(\eta^5-C_5Me_5)_2$.[19,20] Dichlorobis($\eta^5$-pentamethylcyclopentadienyl)titanium and -zirconium are readily reduced with activated magnesium powder. The hafnium analog $HfCl_2(\eta^5-C_5Me_5)_2$, however, is resistant to reduction under these conditions. A more reactive form, Rieke magnesium[21] activated with mercury(II) chloride, is therefore utilized. A main advantage of these procedures for the preparation of $M(CO)_2(\eta^5-C_5Me_5)_2$ (M = Ti, Zr, Hf) is that $MCl_2(\eta^5-C_5Me_5)_2$ is used directly in the synthesis without the necessity of preparing reaction intermediates.[16,18,20]

All procedures are performed using standard Schlenk tube techniques[22] under an atmosphere of dry oxygen-free argon. All glassware is oven-dried and then flame-dried under vacuum and allowed to cool to room temperature while under argon. The tetrahydrofuran (THF) and 1,2-dimethoxyethane (DME) are predried over potassium hydroxide flakes, further dried over sodium wire, and finally distilled under argon from the sodium ketyl of benzophenone. Hexane is dried over calcium hydride and freshly distilled under argon prior to use. CAMAG neutral grade alumina (Alfa Products) is heated by means of a heat gun on a rotary evaporator operating at 10^{-2} torr (vacuum pump) for 2 h and then allowed to cool under an argon atmosphere to room temperature. Five percent by weight of degassed water is added to the alumina, the mixture is shaken until thoroughly mixed, and the alumina is stored under argon. Water is purged with argon for 15 min, heated at reflux under argon for 12 h, and then allowed to cool under argon.

■ **Caution.** *Because of the known toxicity of carbon monoxide and metal carbonyls, all preparations must be carried out in an efficiently operating fume hood. The residual material contained in the reaction vessel after the initial filtration includes activated magnesium and metallic mercury. The magnesium metal may be decomposed by the careful addition of water (100 mL) (except for Rieke magnesium, which should be decomposed with 2-propanol). Although this reaction vigorously evolves hydrogen, at no time has ignition been observed. The metallic mercury should be disposed of properly.*[22]

A. DICARBONYLBIS(η^5-CYCLOPENTADIENYL)TITANIUM

$$TiCl_2(\eta^5\text{-}C_5H_5)_2 + Mg + 2CO \rightarrow Ti(CO)_2(\eta^5\text{-}C_5H_5)_2 + MgCl_2$$

Procedure

Dichlorobis(η^5-cyclopentadienyl)titanium (Alfa Products) (5.00 g, 20.1 mmol), magnesium turnings* (1.62 g, 66.6 mmol), and 100 mL of THF are placed into a 250-mL Schlenk tube and stirred magnetically. Recrystallization of the titanium compound from xylene is recommended. The tube is flushed with carbon monoxide for 5 min. Mercury(II) chloride (3.60 g, 13.3 mmol) is then added while carbon monoxide is allowed to flow slowly over the solution through the side-arm stopcock and out to a mercury overpressure valve. A small amount of heat is generated during the amalgamation of the magnesium. Although this does not seem to have any detrimental effects on the reduction, it is recommended that the Schlenk tube be cooled (water bath) during the addition of mercury(II) chloride and subsequent amalgamation. After 5 min of stirring, the bath may be removed and the reaction run at room temperature. The reaction mixture is stirred in the carbon monoxide atmosphere for 12 h at room temperature, during which time the color changes from bright red to dark green and finally to dark red. The reaction vessel is then flushed with argon, and the solution is poured into a fritted funnel[22] containing a plug (12 × 4 cm) of 5% deactivated alumina covered with 1.5 cm of sea sand. The reaction mixture is allowed to pass through the plug in order to remove the magnesium chloride formed in the reaction. Hexane is subsequently used to elute the remaining material until the eluate is colorless. The THF–hexane solution is concentrated to dryness under reduced pressure, leaving the crude dark red $Ti(CO)_2(\eta^5\text{-}C_5H_5)_2$. The product is purified by dissolving it in ~100 mL of hexane and passing this solution through another plug (8 × 4 cm) of deactivated alumina. The plug is eluted with fresh hexane until the solution emerging from the fritted funnel is colorless. The solvent is then removed under reduced pressure, leaving 4.1 g (87%) of $Ti(CO)_2(\eta^5\text{-}C_5H_5)_2$. The purity of this material is satisfactory for further reactions. However, the product can be conveniently recrystallized from hexane at $-20°$.

Properties

Dicarbonylbis(η^5-cyclopentadienyl)titanium is a maroon red air-sensitive solid that is soluble in both aliphatic and aromatic solvents. The ^1H NMR

---

*Fisher Brand magnesium turnings—for Grignard reaction (Fisher Scientific).

spectrum (C_6D_6) exhibits a singlet at δ 4.62 ppm (external tetramethylsilane, TMS). The IR spectrum shows two metal carbonyl stretching vibrations at 1977 and 1899 cm^{-1} in hexane, or at 1965 and 1883 cm^{-1} in THF.

B. DICARBONYLBIS(η^5-CYCLOPENTADIENYL)ZIRCONIUM

$$ZrCl_2(\eta^5\text{-}C_5H_5)_2 + Mg + 2CO \rightarrow Zr(CO)_2(\eta^5\text{-}C_5H_5)_2 + MgCl_2$$

Procedure

Dichlorobis(η^5-cyclopentadienyl)zirconium (Alfa Products) (2.00 g, 6.84 mmol) together with magnesium turnings (Fisher) (0.83 g, 34.2 mmol) and 50 mL of THF are placed in a 100-mL Schlenk tube and magnetically stirred. On dissolution of the $ZrCl_2(\eta^5\text{-}C_5H_5)_2$, mercury(II) chloride (1.85 g, 6.81 mmol) is added to the mixture, at which time carbon monoxide is allowed to flow slowly over the solution through the side-arm stopcock and out to a mercury overpressure valve. The solution is stirred in the carbon monoxide atmosphere for 24 h at room temperature, during which time the solution changes from colorless to dark green and finally to dark red. The reaction vessel is flushed with argon, and the solution is poured into a fritted funnel[22] containing a plug (10 × 3 cm) of 5% deactivated alumina covered with 1.5 cm of sea sand. The reaction mixture is allowed to pass through the plug in order to remove the magnesium chloride formed in the reaction. The plug is then eluted with hexane until the eluate is colorless. The dark red reaction solution appears green when passing through the plug and dark reddish-green on exiting, depending on how the solution is viewed. The THF–hexane solution is then concentrated to dryness, leaving a dark solid. The $Zr(CO)_2(\eta^5\text{-}C_5H_5)_2$ is purified by dissolving it in ~75 mL of hexane and passing the solution through another plug (5 × 3 cm) of 5% deactivated alumina. The plug is eluted with fresh hexane until the solution emerging from the fritted funnel is colorless. The hexane is then concentrated under vacuum until black needle-like crystals begin to form. At this point the solution is cooled to $-20°$, resulting in further crystal formation. The hexane is decanted from the crystals into another Schlenk tube and further concentrated and cooled. The resulting black crystal crops are dried under vacuum and combined, yielding 1.00 g (53%) of $Zr(CO)_2(\eta^5\text{-}C_5H_5)_2$.*

---

*The checkers report that the same percentage yield is obtained when the amounts of the starting materials are doubled.

Properties

Dicarbonylbis(η^5-cyclopentadienyl)zirconium is a black air-sensitive solid that on dissolution in aliphatic or aromatic solvents yields dark reddish-green solutions. The ^{1}H NMR spectrum (C_6D_6) exhibits a singlet at δ 4.95 ppm (external TMS). The IR spectrum displays two metal carbonyl stretching vibrations at 1975 and 1885 cm^{-1} in hexane, or at 1967 and 1872 cm^{-1} in THF.

C. DICARBONYLBIS(η^5-CYCLOPENTADIENYL)HAFNIUM

$$HfCl_2(\eta^5\text{-}C_5H_5)_2 + Mg + 2CO \rightarrow Hf(CO)_2(\eta^5\text{-}C_5H_5)_2 + MgCl_2$$

Procedure

Dichlorobis(η^5-cyclopentadienyl)hafnium (Alfa Products) (2.00 g, 5.27 mmol), magnesium powder [RMC-50/100-UM($\sim$50–100 mesh), Reade Manufacturing] (0.50 g, 20.6 mmol), and 50 mL of THF are placed into a 100-mL Schlenk tube and stirred magnetically. The use of magnesium powder for this preparation as opposed to magnesium turnings is critical to the success of the reaction. The use of magnesium powder for the preparation of $Zr(CO)_2(\eta^5\text{-}C_5H_5)_2$, however, was found to result in lower yields of product. When the solution is stirred, a deep narrow vortex is desirable. This can be accomplished with a small stirring bar. On dissolution of the HfCl$_2(\eta^5$-$C_5H_5)_2$, mercury(II) chloride (1.00 g, 3.68 mmol) is added to the mixture, at which time carbon monoxide is allowed to flow slowly over the solution through the side-arm stopcock and out to a mercury overpressure valve. The solution is stirred in the carbon monoxide atmosphere of 24 h, during which time the color changes from colorless to dark green and finally to dark red. The reaction vessel is flushed with argon, and the solution is poured into a fritted funnel[22] containing a plug (10 × 3 cm) of 5% deactivated alumina covered with 1.5 cm of sea sand. The reaction mixture is allowed to pass through the plug in order to remove the magnesium chloride formed in the reaction. Hexane is then used to elute the remaining material until the eluate is colorless. The THF–hexane solution is concentrated to dryness, leaving a purple solid. The Hf(CO)$_2(\eta^5$-$C_5H_5)_2$ is purified by dissolving it in $\sim$75 mL of hexane and passing this solution through another plug (5 × 3 cm) of 5% deactivated alumina. This plug is eluted with fresh hexane until the solution emerging from the fritted funnel is colorless. The hexane is concentrated under vacuum until purple needle-like crystals begin to form. The solution is then cooled to $-20°$, resulting in further crystal formation. The hexane is decanted from the crystals into another Schlenk tube and further concentrated and cooled. The resulting crystal crops are dried under vacuum and combined, yielding 0.58 g (30%) of $Hf(CO)_2(\eta^5\text{-}C_5H_5)_2$.

Properties

Dicarbonylbis(η^5-cyclopentadienyl)hafnium is a purple air-sensitive solid that is soluble in both aliphatic and aromatic solvents. The ^1H NMR spectrum (C_6D_6) exhibits a singlet at δ 4.81 ppm (external TMS). The IR spectrum displays two metal carbonyl stretching vibrations at 1969 and 1878 cm^{-1} in hexane, or at 1960 and 1861 cm^{-1} in THF.

D. DICARBONYLBIS(η^5-PENTAMETHYLCYCLOPENTADIENYL)TITANIUM

$$TiCl_2(\eta^5\text{-}C_5Me_5)_2 + Mg + 2CO \rightarrow Ti(CO)_2(\eta^5\text{-}C_5Me_5)_2 + MgCl_2$$

Procedure

Dichlorobis(η^5-pentamethylcyclopentadienyl)titanium[16] (Strem Chemicals) (2.00 g, 5.14 mmol), magnesium powder [RMC-50/100-UM(~50–100 mesh), Reade] (0.62 g, 25.5 mmol), and 50 mL of THF are placed in a 100-mL Schlenk tube and stirred magnetically. On dissolution of the $TiCl_2(\eta^5\text{-}C_5Me_5)_2$, mercury(II) chloride (1.39 g, 5.12 mmol) is added to the mixture, at which time carbon monoxide is allowed to flow slowly over the solution through the side-arm stopcock and out to a mercury overpressure valve. The solution is stirred under a carbon monoxide atmosphere for 24 h at room temperature, during which time the color changes from red to green and finally to red. The THF is removed under reduced pressure, leaving a red residue. The residue is extracted with hexane until the extracts are colorless. The hexane solution is poured into a fritted funnel[22] containing a plug (5 × 3 cm) of 5% deactivated alumina covered with 1.5 cm of sea sand. The plug is eluted with hexane until the eluate is colorless. The resulting red solution is then concentrated under reduced pressure to approximately one-fourth of its original volume and cooled to $-20°$. On crystal formation, the hexane is decanted into another Schlenk tube and further concentrated and cooled. The resulting crystal crops are dried under vacuum and combined, yielding 1.25 g (65%) of $Ti(CO)_2(\eta^5\text{-}C_5Me_5)_2$ as lustrous brick-red needles.

Properties

Dicarbonylbis(η^5-pentamethylcyclopentadienyl)titanium is moderately stable in air, but it is best handled and stored under an inert atmosphere. It is extremely soluble in both aliphatic and aromatic solvents. The ^{1}H NMR spectrum (C_6D_6) exhibits a singlet at δ 1.67 ppm (external TMS). The IR spectrum displays two metal carbonyl stretching vibrations at 1940 and 1858 cm^{-1} (hexane). The mass spectrum shows a parent ion at m/e 374.

E. DICARBONYLBIS(η^5-PENTAMETHYLCYCLOPENTADIENYL)ZIRCONIUM

$$ZrCl_2(\eta^5\text{-}C_5Me_5)_2 + Mg + 2CO \rightarrow Zr(CO)_2(\eta^5\text{-}C_5Me_5)_2 + MgCl_2$$

Procedure

Dichlorobis(η^5-pentamethylcyclopentadienyl)zirconium[23] (Strem Chemicals) (0.80 g, 1.85 mmol), magnesium powder [RMC-50/100-UM($\sim$50–100 mesh), Reade] (0.22 g, 9.05 mmol), and 20 mL of THF are placed in a 50-mL Schlenk tube and stirred magnetically such that a deep narrow vortex is formed. On dissolution of the $ZrCl_2(\eta^5\text{-}C_5Me_5)_2$, mercury(II) chloride (0.50 g, 1.84 mmol) is added to the mixture, at which time carbon monoxide is allowed to flow slowly over the solution through the side-arm stopcock and out to a mercury overpressure valve. The solution is stirred under a carbon monoxide atmosphere for 24 h at room temperature, during which time the color changes from pale yellow to dark reddish-green. The THF is removed under reduced pressure, leaving a dark-colored solid. This solid is extracted with hexane until the extracts are colorless. The hexane solution is poured into a fritted funnel[22] containing a plug (5 × 3 cm) of 5% deactivated alumina covered with 1.5 cm of sea sand. The plug is eluted with hexane until the eluate is colorless. The resulting dark reddish-green solution is then concentrated under reduced pressure to approximately one-fourth of its original volume and cooled to $-20°$. On crystal formation, the hexane is decanted into another Schlenk tube and further concentrated and cooled. The resulting crystal crops are dried under vacuum and combined, yielding 0.62 g (80%) of $Zr(CO)_2(\eta^5\text{-}C_5Me_5)_2$ as lustrous black needles.

Properties

Dicarbonylbis(η^5-pentamethylcyclopentadienyl)zirconium, unlike its titanium analog, is highly air-sensitive. It is easily soluble in both aliphatic and aromatic solvents. The ^{1}H NMR spectrum (C_6D_6) exhibits a singlet at δ 1.73 ppm (external TMS), and the IR spectrum displays two metal carbonyl stretching vibrations at 1945 and 1852 cm^{-1} (hexane). The mass spectrum shows a parent ion at m/e 416.

F. DICARBONYLBIS-(η^5-PENTAMETHYLCYCLOPENTADIENYL)HAFNIUM

$$2K + MgCl_2 \rightarrow Mg + 2KCl$$
$$HfCl_2(\eta^5\text{-}C_5Me_5)_2 + Mg + 2CO \rightarrow Hf(CO)_2(\eta^5\text{-}C_5Me_5)_2 + MgCl_2$$

Procedure

A 100-mL, three-necked, round-bottomed flask is fitted with a reflux condenser and gas inlet and outlet tubes. Approximately 60 mL of THF is added to the flask together with freshly cut potassium metal (1.11 g, 28.4 mmol), anhydrous $MgCl_2$ (1.90 g, 20.0 mmol), and KI (0.50 g, 3.0 mmol).

■ **Caution.** *Potassium metal reacts explosively on contact with water and may also form potentially dangerous superoxides. Consult Ref. 24 for proper handling procedure.*

The reaction mixture is stirred and slowly heated to reflux, at which point the potassium metal melts and the reaction commences. After refluxing for 3 h, a very finely divided dark-gray suspension of magnesium metal can be observed.[21] The reaction mixture is then cooled to room temperature and purged with carbon monoxide for 5 min. Dichlorobis(η^5-pentamethylcyclopentadienyl)hafnium[20] (Strem Chemicals) (1.50 g, 2.89 mmol) is added to the magnesium slurry, and the mixture is allowed to stir under a carbon monoxide atmosphere for 12 h, with a slow, continuous flow of carbon monoxide through the flask. Mercury(II) chloride (0.50 g, 1.84 mmol) is added, and the mixture is allowed to stir further under a stream of carbon monoxide for an additional 12 h. The dark reaction mixture is then filtered through a fritted funnel[22] in order to separate it from the magnesium. This filtration may be slow, because of the magnesium particles clogging the pores of the frit. The use of deoxygenated Celite helps alleviate this problem. (Celite was deoxygenated in a manner analogous to the deoxygenation of alumina; however, it was not deactivated with water.) The frit is then washed with fresh THF, and the dark red filtrate is concentrated to dryness under reduced pressure, leaving a solid residue. The residue is extracted with hexane until the extracts are colorless. The hexane solution is poured into a fritted funnel[22] containing a plug (5 × 3 cm) of 5% deactivated alumina covered with 1.5 cm of sea sand. The plug is eluted with hexane until the eluate is colorless. The resulting purple solution is concentrated to approximately one-fourth of its original volume and cooled to $-20°$. On crystal formation, the hexane is decanted into another Schlenk tube and is further concentrated and cooled. The resulting crystal crops are dried under vacuum and combined, yielding 0.36 g (25%) of $Hf(CO)_2(\eta^5\text{-}C_5Me_5)_2$ as lustrous purple needles.

Anal. Calcd. for $C_{22}H_{30}HfO_2$: C, 52.32; H, 5.99. Found: C, 52.19; H, 5.95.

■ **Caution.** The residual material contained in the reaction flask and fritted funnel includes highly activated magnesium metal and metallic mercury. The residual material must not come into contact with water, since the hydrogen that is generated would ignite. The magnesium should be decomposed by reaction with 2-propanol. The metallic mercury should be disposed of properly.[22]

Properties

Dicarbonylbis(η^5-pentamethylcyclopentadienyl)hafnium is a purple air-sensitive solid that is very soluble in both aliphatic and aromatic solvents. The ^{1}H NMR spectrum (C_2D_6) exhibits a singlet at δ 1.74 ppm (external TMS), and the IR spectrum displays two metal carbonyl stretching vibrations at 1940 and 1844 cm^{-1} (hexane). The mass spectrum shows a parent ion at m/e 506.

References

1. J. G. Murray, *J. Am. Chem. Soc.*, **81**, 752 (1959).
2. J. G. Murray, *J. Am. Chem. Soc.*, **83**, 1287 (1961).
3. D. J. Sikora, D. W. Macomber, M. D. Rausch, *Adv. Organomet. Chem.*, **25** (1986), 317.
4. M. D. Rausch and H. Alt, *J. Am. Chem. Soc.*, **96**, 5936 (1974).
5. B. Demerseman, G. Bouquet, and M. Bigorgne, *J. Organomet. Chem.*, **93**, 199 (1975).
6. B. Demerseman, G. Bouquet, and M. Bigorgne, *J. Organomet. Chem.*, **101**, C24 (1975).
7. J. T. Thomas and K. T. Brown, *J. Organomet. Chem.*, **111**, 297 (1976).
8. G. Fachinetti, G. Fochi, and C. Floriani, *J. Chem. Soc., Chem. Commun.*, **1976**, 230.
9. B. Demerseman, G. Bouquet, and M. Bigorgne, *J. Organomet. Chem.*, **145**, 41 (1978).
10. B. Demerseman, G. Bouquet, and M. Bigorgne, *J. Organomet. Chem.*, **107**, C19 (1976).
11. D. J. Sikora, M. D. Rausch, R. D. Rogers, and J. L. Atwood, *J. Am. Chem. Soc.*, **101**, 5079 (1979).
12. D. J. Sikora and M. D. Rausch, *J. Organomet. Chem.*, **276**, 21 (1984).
13. P. T. Wolczanski and J. E. Bercaw, *Acc. Chem. Res.*, **13**, 121 (1980).
14. D. J. Sikora, M. D. Rausch, R. D. Rogers, and J. L. Atwood, *J. Am. Chem. Soc.*, **103**, 982 (1981).
15. D. J. Sikora, M. D. Rausch, R. D. Rogers, and J. L. Atwood, *J. Am. Chem. Soc.*, **103**, 1265 (1981).
16. J. E. Bercaw, R. H. Marvich, L. G. Bell, and H. H. Brintzinger, *J. Am. Chem. Soc.*, **94**, 1219 (1972).
17. B. Demerseman, G. Bouquet, and M. Bigorgne, *J. Organomet. Chem.*, **132**, 223 (1977).
18. J. M. Manriquez, D. R. McAlister, R. D. Sanner, and J. E. Bercaw, *J. Am. Chem. Soc.*, **98**, 6733 (1976).
19. J. A. Marsella, C. J. Curtis, J. E. Bercaw, and K. G. Caulton, *J. Am. Chem. Soc.*, **102**, 7244 (1980).
20. D. M. Roddick, M. D. Fryzuk, P. F. Seidler, G. L. Hillhouse, and J. E. Bercaw, *Organometallics*, **4**, 97, 1694 (1985).

21. R. E. Rieke and S. E. Bales, *J. Am. Chem. Soc.*, **96,** 1775 (1974).
22. D. F. Shriver, *The Manipulation of Air-Sensitive Compounds*, McGraw-Hill, New York, 1978.
23. J. M. Manriquez, D. R. McAlister, E. Rosenberg, A. M. Shiller, K. L. Williamson, S. I. Chan, and J. E. Bercaw. *J. Am. Chem. Soc.*, **100,** 3078 (1978).
24. L. F. Fieser and M. Fieser, *Reagents for Organic Synthesis*, Wiley, New York, 1967, p. 950.

65. (η^5-CYCLOPENTADIENYL)HYDRIDOZIRCONIUM COMPLEXES

Submitted by P. C. WAILES* and H. WEIGOLD*
Checked by JEFFREY SCHWARTZ† and CHU JUNG†

The dihydridozirconium complex, $ZrH_2(\eta^5\text{-}C_5H_5)_2$, was prepared originally[1] by the action of trialkylamines on the bis(tetrahydroborate), $Zr(BH_4)_2(\eta^5\text{-}C_5H_5)_2$. Below are described more convenient methods for the preparation of the dihydride and of the chlorohydrido derivative, $ZrHCl(\eta^5\text{-}C_5H_5)_2$, from readily available starting materials. These compounds are useful intermediates in the preparation of alkyl- and alkenylzirconium compounds,[2,3] and under hydrogen pressure they are active catalysts for hydrogenation of olefins and acetylenes.[2] The "hydrozirconation" reactions require $ZrHCl(\eta^5\text{-}C_5H_5)_2$ as starting material.[4]

A. BIS(η^5-CYCLOPENTADIENYL)DIHYDRIDOZIRCONIUM

$$2ZrCl_2(\eta^5\text{-}C_5H_5)_2 + H_2O \rightarrow [ZrCl(\eta^5\text{-}C_5H_5)_2]_2O + 2HCl$$

$$[ZrCl(\eta^5\text{-}C_5H_5)_2]_2O + Li[AlH_4] \rightarrow 2ZrH_2(\eta^5\text{-}C_5H_5)_2 + LiCl$$
$$+ \tfrac{1}{2}[AlOCl]_2$$

1. μ-Oxo-bis[chlorobis(η^5-cyclopentadienyl)zirconium]

The method of preparation described here is an adaption of the original,[5,6] which gives consistently high yields.

Procedure

To a solution of $ZrCl_2(\eta^5\text{-}C_5H_5)_2$ (29 g, 0.1 mol) in dichloromethane (250 mL) in a conical flask is added aniline (10 mL) and water (1.3 mL) with

---
*Division of Chemicals and Polymers CSIRO, Private Bag 10, Clayton, Victoria, Australia 3168.
†Department of Chemistry, Princeton University, Princeton, NJ, 08540.

shaking. A white precipitate of aniline hydrochloride forms immediately. After it is chilled for several hours in a refrigerator, the suspension is filtered. Occasionally large crystals of the product are present at this stage and these are dissolved by addition of more CH_2Cl_2. The filtrate is evaporated to a small volume, and light petroleum ether (100 mL, bp 30–40°) is added to precipitate the product. After filtration and washing with petroleum ether, white crystals of the oxo-bridged compound are obtained (25.5 g, 97% yield); these slowly turn pink on storage. The melting point varies with the rate of heating, but if the sample in an evacuated capillary is placed in a melting point apparatus preheated to 260°, a melting point around 305° should be obtained.

2. Bis(η^5-cyclopentadienyl)dihydridozirconium

In this reaction purified tetrahydrofuran (THF) is required, as well as a standardized solution of $Li[AlH_4]$ in tetrahydrofuran. The THF from freshly opened bottles is distilled from $Li[AlH_4]$.

(■ **Caution.** *$Li[AlH_4]$ is a hazardous material and must be handled in dry conditions and in small quantities. Serious explosions can occur when impure THF is purified if it contains peroxides (see Ref. 7).*

Standardized solutions of $Li[AlH_4]$ are prepared by stirring $Li[AlH_4]$ in purified THF for several hours under nitrogen or argon and filtering through Celite (previously baked out at 140° and degassed under vacuum) using the apparatus shown in Fig. 1, followed by hydrolysis of an aliquot of this solution with dilute acid and accurate measurement of the hydrogen evolved. This is best done on a vacuum line using a Töpler pump, but simpler methods should also suffice.

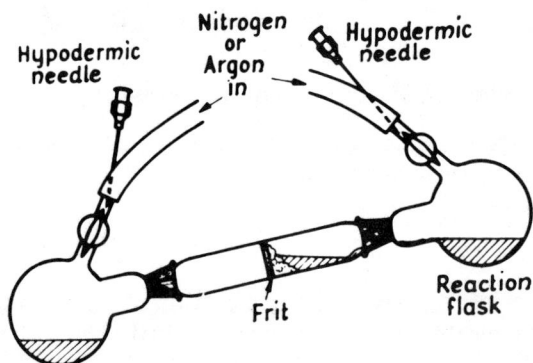

Fig. 1. Simple apparatus for filtering under inert atmosphere. The flow of N_2 or Ar is controlled by a finger over the base of the needle.

Procedure

To a solution of $[ZrCl(\eta^5\text{-}C_5H_5)_2]_2O$ (17.7 g, 33.4 mmol) in purified THF (200 mL) in a 500-mL flask of the type shown in Fig. 1, a clear solution of $Li[AlH_4]$ in THF (20 mL of 1.7 M, 34 mmol) is added dropwise from a hypodermic syringe with stirring. An atmosphere of nitrogen or argon is maintained at all times. A white precipitate slowly appears, but precipitation is not complete for several hours. The mixture is set aside overnight and then filtered anaerobically as in Fig. 1, giving the dihydrido complex as an almost colorless microcrystalline solid (yield 10.1 g, 66%).

Anal. Calcd. for $C_{10}H_{12}Zr$: ash (ZrO_2), 55.14%; hydrolyzable H, 2.00 g-atoms mol^{-1}. Found: ash, 55.5%; hydrolyzable H, 1.89 g-atom mol^{-1}. The yield of dihydride never exceeds 66%; the remainder of the zirconium is believed to be a complex with an aluminum compound (see Properties section).

B. CHLOROBIS(η^5-CYCLOPENTADIENYL)HYDRIDOZIRCONIUM

$$ZrCl_2(\eta^5\text{-}C_5H_5)_2 + Li[AlH(t\text{-}BuO)_3] \rightarrow ZrHCl(\eta^5\text{-}C_5H_5)_2 + LiCl$$
$$+ Al(t\text{-}BuO)_3$$

Procedure*

The apparatus and procedures are similar to those in the preparation above, and a 1-L flask is used. A solution of lithium tri-*tert*-butoxyhydridoaluminate†[9] (28.6 g, 113 mmol) in purified THF (100 mL) is added slowly to a solution of $ZrCl_2(\eta^5\text{-}C_5H_5)_2$ (32.9 g, 113 mmol) in THF (500 mL) with stirring. After complete addition, stirring is continued for 1 h, after which the mono-hydrido complex is collected by anaerobic filtration (Fig. 1) and washed with THF (yield 26.3 g, 90%).

Anal. Calcd. for $C_{10}H_{11}ClZr$: ash (ZrO_2), 47.77%; hydrolyzable H, 1.00 g-atom mol^{-1}; Cl, 13.75. Found: ash, 47.0%; hydrolyzable H, 1.02 g-atoms mol^{-1}; Cl, 13.4. One-quarter mol of $Li[AlH_4]$ may be used instead of the tri-*tert*-butoxy hydrido complex, but the essential control of stoichiometry is more difficult (see Properties section).

---

*An interesting alternative procedure is to stir $ZrCl_2(\eta^5\text{-}C_5H_5)_2$ in tetrahydrofuran with 0.5 g-atom of magnesium turnings. A red color develops and after 3–5 days a 30% yield of $ZrHCl(\eta^5\text{-}C_5H_5)_2$ can be recovered by filtration.

†Available from Aldrich Chemical Co.

Other hydridozirconium derivatives that have been prepared from aluminum hydrides are $ZrH(CH_3)(\eta^5\text{-}C_5H_5)_2$, $ZrH(AlH_4)(\eta^5\text{-}C_5H_5)_2$, the complex $ZrH_2(\eta^5\text{-}C_5H_5)_2 \cdot [ZrH(\eta\text{-}C_5H_5)_2]_2O$, and deuterido derivatives corresponding to all of these hydrido complexes.[8]

Properties

The bis(η^5-cyclopentadienyl)zirconium hydrides are colorless solids that hydrolyze in water. Accurate measurement of the hydrogen thus evolved is a sensitive method of analysis. Alternatively, reaction with CH_2Cl_2 in a stoppered NMR tube and quantitative estimation of the CH_3Cl so formed can be used. The compounds are associated through bridging hydrido ligands, which explains their insolubility and the low infrared frequencies of the metal-hydrogen bands [1520 and 1300 cm^{-1} for $ZrH_2(\eta^5\text{-}C_5H_5)_2$ and 1390 cm^{-1} for $ZrHCl(\eta^5\text{-}C_5H_5)_2$]. All dissolve readily in excess $Li[AlH_4]$ solution, probably forming zirconium-aluminum complexes with bridging hydrido ligands. All but the dihydrido complex slowly develop a pink color when exposed to light and therefore appear to be photosensitive.

References

1. R. K. Nanda and M. G. H. Wallbridge, *Inorg. Chem.*, **3**, 1798 (1964).
2. P. C. Wailes, H. Weigold, and A. P. Bell, *J. Organomet. Chem.*, **43**, C32 (1972).
3. P. C. Wailes, H. Weigold, and A. P. Bell, *J. Organomet. Chem.*, **27**, 373 (1971).
4. J. Schwartz and J. A. Labinger, *Angew. Chem. Int. Ed. Engl.*, **15**, 333 (1976).
5. E. Samuel and R. Setton, *C. R.*, **256**, 443 (1963).
6. A. F. Reid, J. S. Shannon, J. M. Swan, and P. C. Wailes, *Aust. J. Chem.*, **18**, 173 (1965).
7. *Inorg. Synth.*, **12**, 317 (1970).
8. P. C. Wailes and H. Weigold, *J. Organomet. Chem.*, **24**, 405 (1970).
9. H. C. Brown and R. F. McFarlin, *J. Am. Chem. Soc.*, **80**, 5372 (1958).

66. CYCLOPENTADIENYL COMPLEXES OF TITANIUM(III) AND VANADIUM(III)

Submitted by L. E. MANZER*
Checked by E. A. MINTZ† and T. J. MARKS†

The chlorodicyclopentadienyl complexes of Ti(III) and V(III) are useful synthetic reagents for the synthesis of a variety of paramagnetic organo-

*Central Research and Development Dept., E. I. du Pont de Nemours & Co., Experimental Station, Wilmington, DE 19898.
†Dept. of Chemistry, Northwestern University, Evanston, IL 60201.

metallic and coordination compounds.[1-4] $TiCl(\eta^5\text{-}C_5H_5)_2$ has been prepared a number of ways, including reduction of $TiCl_2(\eta^5\text{-}C_5H_5)_2$ with zinc dust[5] and reaction of $TiCl_3$ with $Mg(C_5H_5)_2$.[6] Previous routes to $VCl(\eta^5\text{-}C_5H_5)_2$ include the reaction of VCl_4 with $Na(C_5H_5)$[7,8] and the oxidation of $V(\eta^5\text{-}C_5H_5)_2$ with HCl,[8,9] $PhCH_2Cl$,[9] or CH_3Cl.[9] These methods suffer from either low yields or the use of nonreadily available reagents. The procedures described below provide high yields of $MCl(C_5H_5)_2$ (M = Ti, V) using readily available reagents. $Tl(C_5H_5)$ should be sublimed prior to use. All solvents were dried over 4 Å molecular sieves and purged with nitrogen prior to use.

■ **Caution.** *All reactions and manipulations should be performed under an atmosphere of dry nitrogen, either in a dry box or using Schlenk-tube techniques. Thallium compounds are extremely toxic and should be handled with care.*

A. CHLOROBIS(η^5-CYCLOPENTADIENYL)TITANIUM(III)

$$TiCl_3 + 2Tl(C_5H_5) \xrightarrow{C_4H_8O} TiCl(\eta^5\text{-}C_5H_5)_2 + 2TlCl$$

■ **Caution.** *See the note above concerning the use of inert atmosphere. $TiCl_3$ as a dry powder is pyrophoric in air.*

Procedure

A single-necked, round-bottomed, 100-mL flask equipped with a reflux condenser and magnetic stirring bar is charged with 2.0 g (13 mmol) of titanium chloride, 30 mL of dry tetrahydrofuran, and 6.99 g (26 mmol) of thallium(I) cyclopentadienide. The solution is heated to reflux for 1 h, cooled to room temperature, and filtered through a medium frit. The precipitated thallium(I) chloride is washed with tetrahydrofuran until the washings are colorless. The filtrate and washings are combined and stripped by rotary evaporation. Yield: 2.59 g of green-yellow crystals (93.8%).

Anal. Calcd. for $C_{10}H_{10}ClTi$: C, 56.25; H, 4.72; Cl, 16.60. Found: C, 56.00; H, 4.86; Cl, 16.05.

Properties

The product is a very air sensitive, green-yellow crystalline solid. It does not melt below 250°. Its infrared spectrum contains a strong, sharp peak at 1015 cm^{-1} and strong, broad peaks at 795 and 815 cm^{-1}.

B. CHLOROBIS(η^5-CYCLOPENTADIENYL)VANADIUM(III)

$$VCl_3 + 2Tl(C_5H_5) \xrightarrow{C_4H_8O} VCl(\eta^5\text{-}C_5H_5)_2 + 2TlCl$$

■ **Caution.** *See the note above concerning the use of inert atmosphere.*

Procedure

A single-necked, round-bottomed, 100-mL flask equipped with a reflux condenser and a magnetic stirring bar is charged with 2.06 g (13 mmol) of vanadium trichloride, 60 mL of anhydrous tetrahydrofuran, and 7.10 g (26 mmol) of thallium(I) cyclopentadienide. The solution is then heated at reflux for 2 h, cooled to room temperature, and filtered through a medium frit. The precipitated thallium(I) is washed with tetrahydrofuran until the washings are colorless. The filtrate and washings are combined and stripped by rotary evaporation. The resulting blue-black solid is dissolved in a minimum amount of dichloromethane, and the suspension is filtered through a fine frit. The dichloromethane is then removed by rotary evaporation to give blue-black crystals. Yield: 2.76 g (97.5%).

Anal. Calcd. for $C_{10}H_{10}ClV$: C, 55.46; H, 4.65; Cl, 16.37. Found: C, 56.38; H, 4.91, Cl, 16.24.

Properties

The product is an extremely air- and moisture-sensitive blue-black, crystalline solid. It has a melting point of 203–205°. Its infrared spectrum contains two strong, sharp peaks at 1020 and 1010 cm^{-1} and a strong, broad peak at 810 cm^{-1}. It is soluble in THF and CH_2Cl_2 and insoluble in hydrocarbons. It readily sublimes at 10^{-4} torr and 100°.

References

1. P. C. Wailes, R. S. P. Coutts, and H. Weigold, *Organometallic Chemistry of Titanium, Zirconium and Hafnium*, Academic Press, New York, 1974.
2. H. Bowman and J. Teuben, *J. Organomet. Chem.*, **110**, 327 (1976).
3. J. H. Teuben and H. J. De Liefde Meijer, *J. Organometal. Chem.*, **46**, 313 (1972).
4. L. B. Manzer, *J. Am. Chem. Soc.*, **99**, 276 (1977).
5. M. L. H. Green and C. R. Lucas, *J. Chem. Soc. Dalton Trans.*, **1972**, 1000.
6. A. F. Reid and P. C. Wailes, *Aust. J. Chem.*, **18**, 9 (1965).
7. S. Vigoureux and P. Keizel, *Chem. Ber.*, **93**, 701 (1960). *Chem. Abstr.*, **54**, 15347 (1960).
8. H. J. De Liefde Meijer, M. H. Jansen, and G. J. M. Van der Kerk, *Chem. Ind.* (London), 119 (1960).

9. H. J. De Liefde Meijer, M. H. Jansen, and G. J. M. Van den Kerk, *Recl. Trav. Chim. Pays-Bas*, **80**, 831 (1961).
10. T. L. Brown and J. A. Ladd, *Adv. Organometal. Chem.*, **2**, 373 (1964).

67. VANADOCENE, BIS(η^5-CYCLOPENTADIENYL)VANADIUM

$$[(THF)_3V(\mu\text{-}Cl)_3V(THF)_3]_2[Zn_2Cl_6] + 8NaCp \xrightarrow{THF} 4Cp_2V$$

$$(Cp = \eta^5\text{-}C_5H_5)$$

Submitted by CARLO FLORIANI* and VLADIMIR MANGE*
Checked by[1] E. KENT BAREFIELD† and MICHAEL W. CARRIS†

Vanadocene is usually prepared by direct reaction of VCl_3 + 3NaCp to generate $(\eta^5\text{-}Cp)_2(\eta^1\text{-}Cp)V$;[2,3] subsequent thermolysis of this product yields vanadocene. Moderate yields (< 50%) are obtained by this procedure and several undesirable by-products are formed, which are very difficult to remove. The synthesis[4] described below provides a higher yield route (80%) for the preparation of vanadocene in larger quantities ($\sim$ 50 g). The vanadium(II) precursor $[(V_2Cl_3)(THF)_6]_2[Zn_2Cl_6]$[5] is a very useful starting material for various vanadium organometallic syntheses.

■ **Caution.** *Vanadocene, as well as all the vanadium(II) complexes $[(V_2(\mu\text{-}Cl_3)(THF)_6]_2[Zn_2Cl_6]$ and sodium sand, react exothermically with oxygen and water. Therefore, all manipulations should be carried out in a nitrogen-purified atmosphere and in a well-ventilated hood.*

Procedure

A. **Preparation of $[(V_2Cl_3)(THF)_6]_2[Zn_2Cl_6]$:** 100 g of commercial VCl_3 (0.63 mol) (Fluka purum, $\sim$ 96%) is placed in 600 mL of THF[6] in a 1-L flask with a nitrogen inlet. The flask is fitted with a reflux condenser and the mixture refluxed for 12 h to form the red $VCl_3 \cdot 3THF$. The flask is then cooled to room temperature, the condenser replaced with a stopper, and degassed zinc powder (72.5 g, 1.11 mol) added and stirred for 24 h. After 3–4 h the red color of $VCl_3 \cdot 3THF$ starts changing to form the green vanadium(II) adduct (using Zn pellets instead of Zn powder lengthens the

---

*Section de Chimie, Université de Lausanne, Place du Château 3, CH-1005 Lausanne, Switzerland.
†School of Chemistry, Georgia Institute of Technology, Atlanta, GA 30332.

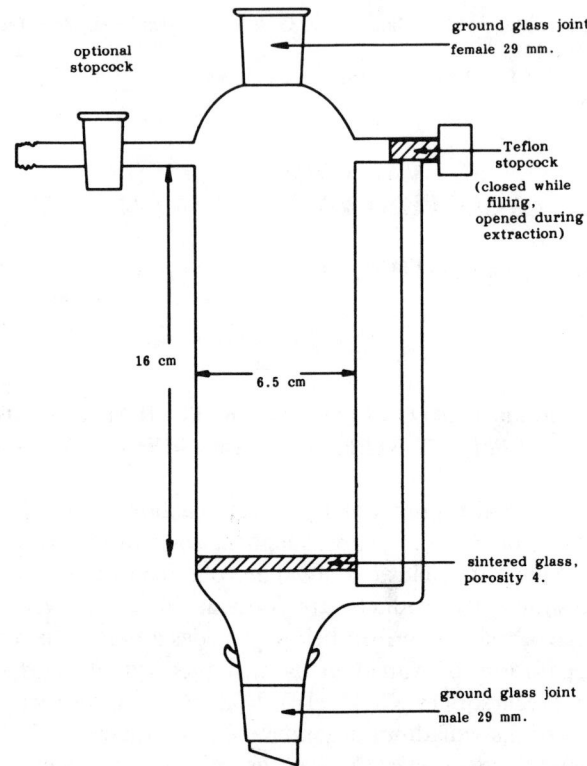

Fig. 1. Extractor.

stirring time). The product is transferred into an extractor (Fig. 1) and the solvent is filtered off. The color of the filtrate depends on the quality of the starting VCl_3. When the filtrate is dark, the solid is washed with clean THF until the filtrate is almost colorless. The product is then extracted with 500 mL of fresh THF. Green air-sensitive crystalline powder of $[(V_2Cl_3)(THF)_6]_2[Zn_2Cl_6]$ is obtained, which is then filtered and dried under vacuum (10^{-2} torr) (226 g, yield 88%).

B. A 1-L flask with nitrogen inlet (similar to that shown in Fig. 2) is charged with a suspension of sodium sand[7,8] (16.8 g, 0.72 mol) in 600 mL of THF. Cyclopentadiene[9] (66 mL, 0.80 mol) is added dropwise to the suspension, with care taken to prevent the reaction from becoming too vigorous. There is an evolution of hydrogen as the reaction proceeds. After all the cyclopentadiene has been added, the funnel is removed and replaced with a

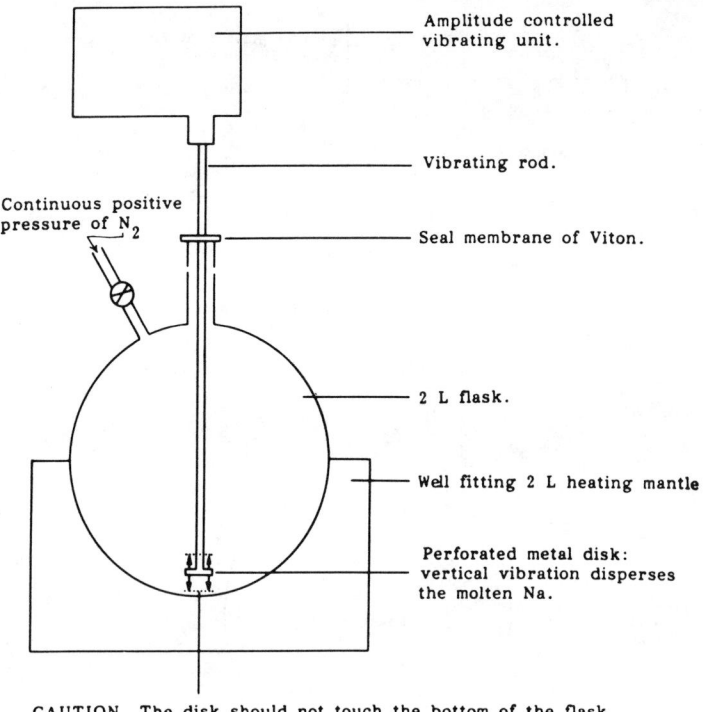

Fig. 2. Vibromixer apparatus.

reflux condenser and the reactants are refluxed until all the remaining sodium is consumed. The solution is colorless if prepared properly.

The reflux condenser is then removed and replaced with a stopper and the solution cooled in an acetone–Dry Ice bath where the cyclopentadienide sodium partially precipitates. The air-sensitive $[V_2(\mu\text{-Cl})_3(THF)_6]_2[Zn_2Cl_6]$ (135 g, equivalent to 0.33 mol of vanadium) is added under nitrogen. The cold bath is removed and the reactants are allowed to warm to room temperature. The green vanadium(II) adduct, $[V_2(\mu\text{-Cl})_3(THF)_6]_2[Zn_2Cl_6]$, starts reacting at about -30 to $-20°$, forming the deep violet vanadocene. A reflux condenser is attached and the solution refluxed for one hour to complete the reaction.

The condenser is removed and the solution concentrated to 300–400 mL, then transferred into a sublimator (Fig. 3) and evaporated to dryness. As it is impossible to avoid bumping onto the walls of the sublimator, we start

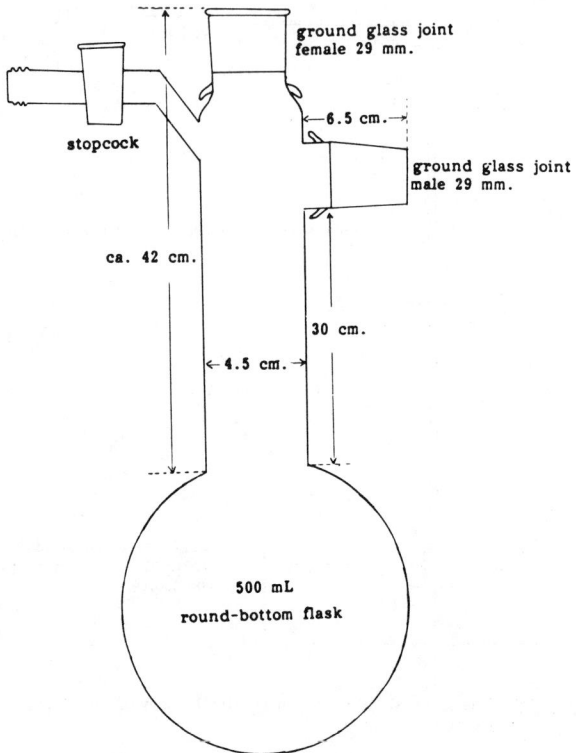

Fig. 3. Sublimator.

heating at 80–100° under vacuum then scrape[10] back the first part of sublimed material and finally clean carefully the walls with paper wetted with THF. Violet crystals of vanadocene sublime on the walls of the sublimator when heated under vacuum ($\sim 10^{-1}$ torr) at 120–140°. The sublimed product is scraped[12] off the walls of the sublimator under a flow of nitrogen into a Schlenk flask attached to the side arm (Fig. 3). This operation must be repeated several times every 2–4 h until no more vanadocene sublimes. It is advisable to scrape the sublimed vanadocene frequently, otherwise the crust will become too hard and difficult to break. The very air sensitive product is transferred to glass tubes, which are then sealed. A total of 47.4 g (0.26 mol) of vanadocene is collected (yield 80%, based on vanadium).

Anal. Calcd. for $C_{10}H_{10}V$: C, 66.31; H, 5.56; V, 28.12. Found: C, 66.08; H, 5.42; V, 28.35.

References and Notes

1. The checkers performed the synthesis on one-half scale.
2. E. O. Fischer and S. Vigoureux, *Chem. Ber.*, **91**, 2205 (1958).
3. G. Wilkinson, F. A. Cotton, and J. M. Birmingham, *J. Inorg. Nucl. Chem.*, **2**, 95 (1956).
4. F. H. Köhler, and W. Prössdorf, *Z. Naturforsch.*, **32** (B), 1026 (1977).
5. V. M. Hall, C. D. Schmulbach, and W. N. Soby, *J. Organomet. Chem.*, **209**, 69 (1981); F. A. Cotton, S. A. Duraj, M. W. Extine, G. E. Lewis, W. J. Roth, C. D. Schmulbach, and W. Schwotzer, *J. Chem. Soc., Chem. Commun.*, **1983**, 1377; R. J. Bouma, J. H. Teuben, W. R. Beukema, R. L. Bansemer, J. C. Huffman, and J. C. Caulton, *Inorg. Chem.*, **23**, 2715 (1984).
6. Tetrahydrofuran (THF) is distilled over K-benzophenone ketyl.
7. To a 2-L Schlenk flask are added 1.2 L of dry degassed xylene under N_2. Freshly cut pieces of sodium metal (30–40 g) are immersed in the xylene. The Schlenk flask is then equipped with the vibromixer (Vibromixer Model E1, by Chemap AG, Hölzliwisenstr. 5, CH-8604 Volketswil, Switzerland) as shown in Fig. 2. It is very important to align the vibrating rod properly and firmly adjust it to the correct height above the bottom of the Schlenk flask, as shown. The flask is then slowly heated in a 2 L heating mantle until the Na is completely molten, after which heating is stopped. The vibromixer is then switched on and the amplitude of vibration slowly increased to a maximum. The vibration is maintained at a maximum for about one minute until a fine Na sand is obtained. The vibromixer is then switched off and the heating mantle is removed being careful not to agitate the molten Na sand to prevent fusing. The mixture is then allowed to cool to room temperature and the vibromixer then removed and replaced with a stopper. The Na sand is filtered and dried under vacuum. The checkers used a commercial Na dispersion for the preparation of sodium cyclopentadienide. The oil was washed from this material with THF in a pressure filtration device, and the Na was washed from the filter into the reaction vessel with THF.
8. R. B. King and F. G. A. Stone, *Inorg. Synth.*, **7**, 99 (1963).
9. Freshly cracked cyclopentadiene is collected in a dropping funnel with Dry Ice jacket.
10. A long strong spatula flattened, rounded, and hooked at one end: long enough to fit into the sublimator ($\geqslant$ 50 cm), flat part 1–2 cm large.

68. DICHLOROBIS(η^5-CYCLOPENTADIENYL)NIOBIUM(IV)

Submitted by C. R. LUCAS*
Checked by J. A. LABINGER† and J. SCHWARTZ†

The observation of catalytic behavior exhibited by lower-valent compounds of Group 5 elements[1,2] makes it desirable to continue to expand our knowledge of the chemistry of these and neighboring elements. The organic chemistry of the early transition elements is relatively less explored in comparison with that of later members of the series. This situation has arisen

---

*Department of Chemistry, Memorial University of Newfoundland, St. John's, Newfoundland A1B 3X7, Canada.
†Department of Chemistry, Princeton University, Princeton, NJ 08540.

for a number of reasons, some at least of which are related to the lack of good preparative methods for starting materials. Preparations of dichlorobis(η-cyclopentadienyl)niobium[3,4] are described in the literature, but give low yields and involve tedious procedures or the use of toxic thallium reagents. In contrast, the preparation reported below provides a relatively rapid route to good yields of a suitable starting material for the investigation of organoniobium chemistry. For example, approximately 40 g of dichlorobis(η-cyclopentadienyl)niobium can be prepared in 1 day by this method.

$$NbCl_5 + 6NaC_5H_5 \longrightarrow 5NaCl + (C_5H_5)_4Nb + \text{organic products}$$

$$2(C_5H_5)_4Nb + 4HCl + [O] \xrightarrow[O_2]{Br_2} [\{(\eta\text{-}C_5H_5)_2NbCl\}_2O]Cl_2^* + 2C_5H_6$$

$$5Cl^- + [\{(\eta\text{-}C_5H_5)_2NbCl\}_2O]^{2+} + SnCl_3^- + 2H^+ \longrightarrow$$

$$2(\eta\text{-}C_5H_5)_2NbCl_2 + SnCl_6^{2-} + H_2O$$

Procedure

The reaction is carried out in a 3-L three-necked flask. Through the central neck, a mechanical stirrer is fitted, and through another is added the niobium pentachloride.† To the third neck is attached a T piece, one side of which is connected to an inert-gas supply and the other to a bubbler.

Under an atmosphere of dry nitrogen, dry cyclopentadienylsodium‡ (100 g, 1.14 mol) is prepared and suspended at room temperature in dry

---

* The ionic substances containing niobium(V) that exist in these red solutions have not all been identified. In addition to the ion indicated, species of the form $[(\eta\text{-}C_5H_5)_2Nb(X)Y]^+$ (X, Y = Cl, Br, OH, or O—) are probably present in an unknown equilibrium with each other.

† Niobium pentachloride is available commercially from Alfa Inorganics, Ltd. The checkers found that sublimation of the commercial material *in vacuo* was necessary before use in the preparation.

‡ White cyclopentadienylsodium can be prepared by cracking dicyclopentadiene (for details, see *Organometallic Syntheses*, J. J. Eisch and R. B. King (eds.), Vol. 1, pp. 64ff., Academic Press, New York, 1965) under an atmosphere of dry nitrogen directly into a suspension of sodium sand in dry tetrahydrofuran from which dissolved oxygen has been removed. The apparatus for this operation should be assembled so that the distillate of cyclopentadiene monomer is added directly under nitrogen to the sodium. The distillation should proceed at a rate of ~3 mL min^{-1} until almost all the sodium has dissolved. The bulk of volatile substances may then be removed *in vacuo* on a hot-water bath to give a sticky white residue. Final removal of the last traces of tetrahydrofuran from this product is essential. This can be accomplished by heating the residue to 100° *in vacuo* until no more liquid distils into a liquid-nitrogen trap. Breaking up large lumps assists solvent removal. Large batches (300 g) may require heating for up to 72 h.

benzene* (400 mL). With vigorous mechanical stirring, powdered niobium pentachloride (57 g, 0.21 mol) is added in small portions, with care being taken to prevent the reaction temperature from exceeding $\sim 70°$ after each addition. After 1 h the suspension has cooled to room temperature and is a uniform purple-brown color. It is stirred for another $\frac{1}{2}$ h and then poured in air onto concentrated hydrochloric acid (1 L). The resulting mixture is heated gently (in the hood) with occasional swirling until all the benzene has been evaporated. To the boiling brown suspension, small quantities of bromine are then carefully added (~ 60 mL in all) until the solid phase no longer acts like a sticky lump (some scraping of the vessel walls may be necessary) but as a suspended powder.

■ **Caution.** *Bromine attacks respiratory passages and should only be handled in an efficient hood.*

Excess bromine is then boiled away. The hot red† liquid is decanted in air through a filter; extraction with hot concentrated hydrochloric acid and bromine is repeated until the extracts are nearly colorless. After about 10 extractions, the collected filtrates are brought to the boiling point to redissolve any precipitated matter, and they are placed under an inert atmosphere.‡ To the hot solution is added under an inert atmosphere a hot solution of tin dichloride dihydrate (47.5 g, 0.21 mol) in concentrated (12 M) hydrochloric acid (200 mL), and the mixture is left standing overnight. The brown tabular crystals of dichlorobis(η-cyclopentadienyl)niobium are separated by filtration, and the product is washed several times with water and once with acetone (50 mL) before being dried *in vacuo*. Yield is 45 g (75% based on $NbCl_5$).

Anal. Calcd. for $C_{10}H_{10}Cl_2Nb$: C, 40.8; H, 3.4; Cl, 24.1; Nb, 31.6. Found: C, 40.8; H, 3.5; Cl, 24.2; Nb, 31.3.

Properties

Crystals of the product obtained by this method may be weighed and handled in air although they should be stored and used in further reactions under an inert atmosphere. The compound is only very sparingly soluble in common organic solvents. It is paramagnetic and its electron spin resonance spectrum at room temperature and 77 K has been reported.[5] Its infrared spectrum

---

*Freshly distilled from calcium hydride and free from dissolved oxygen.
†If the cyclopentadienylsodium is not dry and white, the hydrochloric acid extracts will be pink or yellow instead of red, and the major product of this preparation will be a white precipitate of Nb_2O_5.
‡The checkers find that concentration of the extracts improves the yield.

(Nujol) is 3090 (m), 1011 (m), 822 (s), 725 (m), 308 (m), 290 (s), and 268 (s) cm^{-1}.

References

1. F. N. Tebbe and G. W. Parshall, *J. Am. Chem. Soc.*, **93**, 3793 (1971).
2. M. D. Curtis, L. G. Bell, and W. M. Butler, *Organometallics*, **4**, 701 (1985).
3. F. W. Siegert and H. J. de Liefde Meijer, *J. Organomet. Chem.*, **23**, 177 (1970).
4. W. E. Douglas and M. L. H. Green, *J. Chem. Soc. Dalton Trans.*, **1972**, 1796.
5. A. T. Casey and J. B. Raynor, *J. Chem. Soc. Dalton Trans.*, **1983**, 2057.

69. CHLORO(η^5-CYCLOPENTADIENYL)BIS(TRIPHENYL-PHOSPHINE)RUTHENIUM(II): RuCl(PPh$_3$)$_2$(η^5-C$_5$H$_5$)

Submitted by M. I. BRUCE,* C. HAMEISTER,* A. G. SWINCER,* and R. C. WALLIS*
Checked by S. D. ITTEL†

The chemistry of RuCl(PPh$_3$)$_2$(η^5-C$_5$H$_5$) differs markedly from that of the corresponding dicarbonyl complex, RuCl(CO)$_2$(η^5-C$_5$H$_5$).[1] The triphenylphosphine complex was first described by Gilbert and Wilkinson,[2] who obtained it from a 2-day reaction between RuCl$_2$(PPh$_3$)$_3$ and cyclopentadiene in benzene. An improved synthesis was described later,[3] from RuCl$_2$(PPh$_3$)$_3$ and thallium(I) cyclopentadienide. Both reactions are difficult to employ if large amounts of this complex are required; the competing dimerization of RuCl$_2$(PPh$_3$)$_3$ to [RuCl$_2$(PPh$_3$)$_2$]$_2$ reduces yields in the first reaction, especially at temperatures above 20°, while the bulk of the thallium compounds makes the second reaction less convenient. The present method employs a slight modification of the one-pot reaction described earlier.[4]

The most significant differences in the chemistries of RuCl(PPh$_3$)$_2$(η^5-C$_5$H$_5$) and RuCl(CO)$_2$(η^5-C$_5$H$_5$) are the ready displacement of chloride from the former complex by neutral ligands, L, to form cationic complexes [RuL(PPh$_3$)$_2$(η^5-C$_5$H$_5$)]$^+$ [3,5] and exchange of one or both triphenylphosphine ligands for other neutral ligands. The chloro complex is the precursor of choice for syntheses of related complexes containing other monodentate or chelating bidentate tertiary phosphines, including those with optically active

---

*Department of Physical and Inorganic Chemistry, University of Adelaide, Adelaide, South Australia 5000.
†Central Research and Development Department, E. I. du Pont de Nemours & Co., Wilmington, DE 19898.

phosphines; replacement of only one PPh_3 ligand affords complexes which are chiral at the ruthenium center.

In methanol, the equilibrium lies predominantly to

$$RuCl(PPh_3)_2(\eta^5\text{-}C_5H_5) + MeOH \rightarrow [Ru(MeOH)(PPh_3)_2(\eta^5\text{-}C_5H_5)]^+ + Cl^-$$

the right, and salts with large anions can be readily isolated.[6] Several novel unsaturated ligands have been made by reactions of alkynes with the hydride or alkyls, including 1,3,4-η^3-butadienyl,[7] allenyl,[8] cumulenyl,[9] and η^5-pentadienyl[9] species. Terminal acetylenes (1-alkynes) react with the chloride affording complexes containing substituted vinylidene ligands, which are readily deprotonated to the corresponding η^1-acetylides on treatment with bases, such as sodium carbonate, butyllithium, or even alumina.[10]

The chloro complex is prepared by adding a mixture of ruthenium trichloride and cyclopentadiene to an excess of triphenylphosphine, which acts as the reducing agent as well as a ligand:

$$RuCl_3 + 2PPh_3 + C_5H_6 \rightarrow RuCl(PPh_3)_2(\eta^5\text{-}C_5H_5) + \cdots$$

Procedure

The reaction is carried out in a 2-L, two-necked, round-bottomed flask equipped with a 500-mL dropping funnel and a reflux condenser topped with a nitrogen bypass. The apparatus is purged with nitrogen. Triphenylphosphine (21.0 g, 0.08 mol) is dissolved in 1 L of ethanol by heating. (If the solution is not clear, it should be filtered before proceeding further.) Hydrated ruthenium trichloride (5.0 g, 0.02 mol) is dissolved in ethanol (100 mL) by bringing the mixture to the boil and then allowing the solution to cool. Freshly distilled cyclopentadiene (10 mL, 8.0 g, 0.12 mol) is added to the ruthenium trichloride solution, and the mixture is transferred to the dropping funnel. The dark-brown solution is then added to the triphenylphosphine solution over a period of 10 min while maintaining the temperature at the reflux point. After the ruthenium trichloride/cyclopentadiene solution has been added, the mixture has a dark-brown color, which after 1 h has lightened to a dark red-orange. The solution, which can now be exposed to air, is filtered quickly while hot and cooled overnight at $-10°$. (The product may also be isolated as fine crystals by evaporating this solution to about one-third the volume on a rotary evaporator. Subsequent work-up is as described below.) Orange crystals separate, leaving a pale yellow-orange supernatant.

The crystals are collected on a sintered-glass filter, washed with ethanol (4 × 25 mL) and with light petroleum (4 × 25 mL), and dried *in vacuo*.

Yield: ~14 g, 90–95%. The yield depends on the composition of the ruthenium trichloride used. Although nominally a trihydrate, commercially available hydrated ruthenium trichloride contains varying amounts of water. The yield quoted by the submitter was obtained using material of approximate composition $RuCl_3 \cdot 2H_2O$. This reaction has been run successfully using material containing up to 4 molecules of water per $RuCl_3$ unit, in which case ~12 g of complex was obtained. The checker obtained 12.2 g from $RuCl_3 \cdot 3.2H_2O$.

Anal. Calcd. for $RuCl(PPh_3)_2(\eta^5\text{-}C_5H_5)$: C, 67.7; H, 4.8. Found: C, 67.8; H, 5.3%.

Properties

The complex forms orange crystals mp 130–133°C (dec., sealed tube) (checker 134–136°C) which are stable in air for prolonged periods. It is insoluble in light petroleum and water, slightly soluble in cold methanol or ethanol, diethyl ether, or cyclohexane, more soluble in chloroform, carbon tetrachloride, dichloromethane, carbon disulfide, and acetone, and very soluble in benzene, acetonitrile, and nitromethane. The ^1H NMR spectrum contains a sharp singlet at δ 4.01 for the C_5H_5 protons and a broad signal at δ 7.16 for the aromatic protons (in $CDCl_3$ solution).

References

1. (a) M. A. Bennett, M. I. Bruce, and T. W. Matheson, in *Comprehensive Organometallic Chemistry*, Vol. 4, G. Wilkinson, F. G. A. Stone, and E. W. Abel (eds.), Pergamon, Oxford, U.K., 1982, Section 32.3, p. 775; (b) M. O. Albers, D. J. Robinson, and E. Singleton, *Coord. Chem. Rev.*, **79,** 1 (1987); (c) G. Consiglio and F. Morandini, *Chem. Rev.*, **87,** 761 (1987).
2. J. D. Gilbert and G. Wilkinson, *J. Chem. Soc. A*, **1969,** 1749.
3. T. Blackmore, M. I. Bruce, and F. G. A. Stone, *J. Chem. Soc. A*, **1971,** 2376.
4. M. I. Bruce and N. J. Windsor, *Aust. J. Chem.*, **30,** 1601 (1977).
5. G. S. Ashby, M. I. Bruce, I. B. Tomkins, and R. C. Wallis, *Aust. J. Chem.*, **32,** 1003 (1979).
6. R. J. Haines and A. L. du Preez, *J. Organometal. Chem.*, **84,** 357 (1975).
7. T. Blackmore, M. I. Bruce, and F. G. A. Stone, *J. Chem. Soc. Dalton Trans.*, **1974,** 106.
8. M. I. Bruce, R. C. F. Gardner, J. A. K. Howard, F. G. A. Stone, M. Welling, and P. Woodward, *J. Chem. Soc. Dalton Trans.*, **1977,** 621.
9. M. I. Bruce, R. C. F. Gardner, and F. G. A. Stone, *J. Chem. Soc. Dalton Trans.*, **1979,** 906.
10. M. I. Bruce and R. C. Wallis, *J. Organometal. Chem.*, **161,** C1 (1978); *Aust. J. Chem.*, **32,** 1471 (1979); (b) M. I. Bruce and A. G. Swincer, *Adv. Organometal. Chem.*, **22,** 59 (1983).

70. (η^5-PENTAMETHYLCYCLOPENTADIENYL)COBALT COMPLEXES

Submitted by STEVEN A. FRITH and JOHN L. SPENCER*‡
Checked by W. E. GEIGER, Jr. and J. EDWIN†

Pentamethylcyclopentadienyl complexes exhibit a chemistry complementary to that of the unsubstituted cyclopentadienyl analogs, with differences in behavior that may be ascribed to the greater bulk or the greater electron-releasing ability of the pentamethylcyclopentadienyl ligand.[1] The following syntheses lead to a group of complexes that have given access to rich areas of organocobalt chemistry. The syntheses have been designed such that later members of the group may be prepared without lengthy purification of the intermediates.

A. DICARBONYL-(η^5-PENTAMETHYLCYCLOPENTADIENYL)COBALT(I)

$$Co_2(CO)_8 + 2C_5Me_5H + C_6H_8 \rightarrow 2[Co(\eta^5\text{-}C_5Me_5)(CO)_2] + C_6H_{10} + 4CO$$

Originally [Co(η^5-C$_5$Me$_5$)(CO)$_2$] was prepared[2] by the photolysis of Co$_2$(CO)$_8$ and C$_5$Me$_5$H and subsequently[3] by the reaction of Co$_2$(CO)$_8$ and C$_5$Me$_5$C(O)Me in refluxing cyclohexane. The method described here is a substantial modification of that used by Byers and Dahl[4] and gives excellent yields from both precursors.

Procedure

■ **Caution.** *Both dicobalt octacarbonyl and carbon monoxide are poisonous and therefore the reaction should be carried out in a well-ventilated hood.*

A 100-mL, two-necked, round-bottomed flask is fitted with a reflux condenser and a magnetic stirring bar. A T junction, placed on top of the

---

*Department of Inorganic Chemistry, University of Bristol, Bristol BS8 ITS, United Kingdom.
†Department of Chemistry, University of Vermont, Burlington, VT 05405.
‡Present address: Department of Chemistry and Applied Chemistry, University of Salford, Salford M5 4WT, United Kingdom.

condenser, is connected to a supply of dry nitrogen and a mineral oil bubbler, and the flask is flushed for several minutes. Dry, deoxygenated dichloromethane (50 mL), octacarbonyldicobalt(0) (6.0 g, 17.5 mmol), pentamethylcyclopentadiene* (3 g, 22.1 mmol) and 1,3-cyclohexadiene (2.5 mL) are added through the second neck, which is then closed with a stopper. The mixture is stirred and heated to maintain a very gentle reflux. After 45 min, a further 2.3 g (16.9 mmol) of pentamethylcyclopentadiene is added and heating is continued for another 1.75 h. The reaction may be conveniently monitored, using IR spectroscopy, by the appearance of v_{CO} bands at 1999 and 1935 cm^{-1} (CH_2Cl_2). When reaction is complete, there may also be weak absorptions due to traces of $Co_4(CO)_{12}$ (2065, 2055, and 1858 cm^{-1}) or $[Co(\eta^3-C_6H_9)(CO)_3]$ [2047 and 1983 (sh) cm^{-1}].

The condenser is replaced with a stopcock adapter and the volatile components are evaporated at reduced pressure and ambient temperature. If the temperature is kept at or below 20°, the oily residue will crystallize spontaneously. This crude material may be used in the synthesis of $[CoI_2(\eta^5-C_5Me_5)(CO)]$, or it may be purified as follows. Nitrogen is readmitted to the flask, the crude crystalline mass is dissolved in dry deoxygenated hexane (30 mL), and the solution is applied to a column of alumina (Brockman, activity II) (20 × 2.5 cm), previously prepared under nitrogen and washed with deoxygenated hexane (200 mL). The product is eluted as an orange-brown band with hexane and collected under nitrogen. Evaporation of the solvent under reduced pressure yields deep-red crystals of $[Co(\eta^5-C_5Me_5)(CO)_2]$, 7.9 to 8.3 g [90–95% based on $Co_2(CO)_8$]. Further purification may be effected by crystallization from hexane solution at low temperatures or by sublimation (40°, 0.01 torr).

Anal. Calcd. for $C_{12}H_{15}O_2Co$: C, 57.61; H, 6.04. Found: C, 57.51; H, 6.26 (mp 58°, sealed tube).

Properties

Dicarbonyl(η^5-pentamethylcyclopentadienyl)cobalt(I) is a deep red-brown crystalline solid that may be handled briefly in air but should be stored in an inert atmosphere. It dissolves readily in dry, deoxygenated organic solvents to give stable solutions but, in the presence of air or water, it slowly decomposes. The ^1H NMR spectrum ($CDCl_3$) shows a sharp singlet at δ 1.98 ppm.

*Available from Strem. Chem., P.O. Box 212, Danvers, MA 01923 or Aldrich Chemical Co., 940 W. St. Paul Ave., Milwaukee, WI 53233.

B. CARBONYLDIIODO-(η^5-PENTAMETHYLCYCLOPENTADIENYL)COBALT(III)

$$[Co(\eta^5\text{-}C_5Me_5)(CO)_2] + I_2 \rightarrow [CoI_2(\eta^5\text{-}C_5Me_5)(CO)] + CO$$

The method described here is based on that of King and coworkers.[5] Roe and Maitlis[6] followed a similar procedure but used an alkane solvent.

Procedure

■ **Caution.** *A large volume of carbon monoxide is released in a short time. The reaction should be carried out in a well-ventilated hood.*

A 250-mL two-necked flask is fitted with a magnetic stirrer bar and a T piece connected to a nitrogen supply and a mineral oil bubbler. The flask is purged with nitrogen, and sodium–dry diethyl ether (150 mL) is added, followed by 8.9 g of iodine (35.0 mmol). The second neck is then closed with a rubber septum and a solution of $[Co(\eta^5\text{-}C_5Me_5)(CO)_2]$ (8.75 g, 35.0 mmol, or all of the crude material from the previous synthesis) in deoxygenated diethyl ether (20 mL) is added dropwise from a syringe to the rapidly stirred reaction mixture. Addition may be completed in 5–10 min. A black, crystalline precipitate forms rapidly with the evolution of carbon monoxide gas. Stirring is continued for another hour. The mixture is filtered in the air using a sintered-glass funnel and the crude product is washed with diethyl ether (three 20-mL portions) and dried in an oven at 80°. The material so-formed is sufficiently pure for the preparation of $[\{CoI_2(\eta^5\text{-}C_5Me_5)\}_2]$. However, it may be purified by dissolution in a large volume of dichloromethane to give a deep purple solution, which is filtered through a glass sinter (porosity 3) and evaporated to small volume using a rotary evaporator. Lustrous black crystals of $[CoI_2(\eta^5\text{-}C_5Me_5)(CO)]$ are filtered off and dried under vacuum. Yield: 15.0–15.8 g (90–95%).

Anal. Calcd. for $C_{11}H_{15}OI_2Co$: C, 27.76; H, 3.18. Found: C, 27.70; H, 3.45.

Properties

Carbonyldiiodo(η^5-pentamethylcyclopentadienyl)cobalt(III) is a black, crystalline, air-stable solid. It dissolves readily in polar solvents forming deep purple solutions but is virtually insoluble in diethyl ether, alkanes, and aromatic solvents. The ^1H spectrum ($CDCl_3$) shows a singlet at δ 2.20 ppm. The IR spectrum exhibits a single metal–carbonyl stretching band at

2053 cm^{-1} (CH$_2$Cl$_2$). Removal of the iodide ligands by Ag$^+$ provides a facile route to cationic organometallic derivatives of Co(III).[7]

C. DI-μ-IODO-BIS[IODO(η^5-PENTAMETHYLCYCLOPENTADIENYL)COBALT(III)]

$$2[CoI_2(\eta^5\text{-}C_5Me_5)(CO)] \xrightarrow{\Delta} [\{CoI_2(\eta^5\text{-}C_5Me_5)\}_2] + 2CO$$

The method described here is based on that of Roe and Maitlis.[6] An analogous dichloro(η-ethyltetramethylcyclopentadienyl) derivative, [{CoCl$_2$(η^5-C$_5$Me$_4$Et)}$_2$], has been prepared[8] by the extraction of CoCl$_2$ from [Co$_3$Cl$_6$(η^5-C$_5$Me$_4$Et)$_2$] in turn made from [Sn(n-Bu)$_3$(η^1-C$_5$Me$_4$Et)] and CoCl$_2$.

Procedure

■ **Caution.** *The reaction releases toxic carbon monoxide and should be carried out in a well-ventilated hood.*

A 250-mL, two-necked, round-bottomed flask is charged with 150 mL of dry n-octane and 15.5 g of powdered [CoI$_2$(η^5-C$_5$Me$_5$)(CO)] and equipped with a magnetic stirring bar, a nitrogen inlet, and a reflux condenser. The top of the condenser is connected to a mineral-oil bubbler venting into the hood. The rapidly stirred mixture is heated at reflux for 5 h under a slow N$_2$ purge. During the reaction, the microcrystalline suspension is converted to a suspension of lustrous, black crystals. After cooling to room temperature, the crude [{CoI$_2$(η^5-C$_5$Me$_5$)}$_2$] is collected in a sintered-glass filtration funnel, washed with pentane, and dried in an oven at 80°. The product may be purified by dissolving it in a large volume of dichloromethane, filtering, and removing most of the solvent at reduced pressure on a rotary evaporator, or by using the continuous extraction apparatus shown in Fig. 1. The crude material is transferred to the paper thimble, a loose plug of glass wool is packed above it, and dry dichloromethane (120 mL), contained in the flask, is boiled at a sufficient rate that condensate drips into the thimble. This process may be run conveniently overnight. The flask is then detached and cooled to 0° for 3 h. The product, as lustrous, black crystals, is recovered by filtration and dried under vacuum (0.05 torr, 1 h). Yield: 11.5 g. A further crop ($\sim$ 1.3 g) may be obtained by reducing the volume of the filtrate to 30 mL, but this material should be checked by IR to ensure that it contains no [CoI$_2$(η^5-C$_5$Me$_5$)(CO)]. Total yield of [{CoI$_2$(η^5-C$_5$Me$_5$)}$_2$] is 12.8 g (88%).

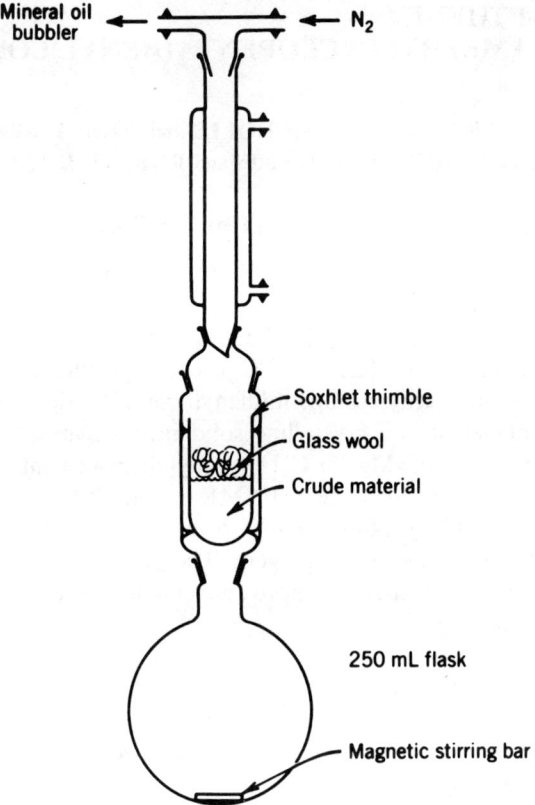

Fig. 1. Continuous extraction apparatus.

Anal. Calcd. for $C_{20}H_{30}I_4Co_2$: C, 26.81; H, 3.38. Found: C, 26.71; H, 3.50.

Properties

Di-μ-iodo-bis[iodo(η^5-pentamethylcyclopentadienyl)cobalt(III)] is a black, crystalline, air-stable solid. It dissolves in polar solvents, forming dark green solutions, but is virtually insoluble in diethyl ether, alkanes, and aromatic solvents. The ^{1}H NMR spectrum (CDCl$_3$) exhibits a singlet at δ 1.80 ppm.

D. BIS(η-ETHENE)(η^5-PENTAMETHYLCYCLOPENTADIENYL)COBALT(I)

Submitted by JULIAN C. NICHOLLS and JOHN L. SPENCER*
Checked by JOHN R. SOWA, Jr. and ROBERT J. ANGELICI[†]

$$[\{CoI_2(\eta^5\text{-}C_5Me_5)\}_2] + 4C_2H_4 + Zn(Hg) \rightarrow 2[Co(\eta^5\text{-}C_5Me_5)(\eta\text{-}C_2H_4)_2] + 2ZnI_2$$

Green and Pardy[8] prepared $[Co(\eta^5\text{-}C_5Me_4Et)(\eta^2\text{-}C_2H_4)_2]$ by the sodium amalgam reduction of $[\{CoCl_2(\eta^5\text{-}C_5Me_4Et)\}_2]$ under ethene (toluene, 110°C). The pentamethylcyclopentadienyl complex described here offers several advantages: it is a crystalline solid rather than an oil; compounds derived from $[Co(\eta^5\text{-}C_5Me_5)(\eta\text{-}C_2H_4)_2]$ by displacement of ethylene are frequently solids also; and the ^{1}H NMR is simpler. Frith and Spencer[9] prepared $[Co(\eta^5\text{-}C_5Me_5)(\eta\text{-}C_2H_4)_2]$ from $[\{CoI_2(\eta^5\text{-}C_5Me_5)\}_2]$ using Na/Hg amalgam as the reducing agent. The procedure described here uses activated zinc and is cleaner and more convenient although a longer reaction time is required.

Procedure

■ **Caution.** *The large volumes of ethylene used in this synthesis are a fire hazard. The preparation should be carried out in a well-ventilated hood in the absence of naked flames. Drying tetrahydrofuran is hazardous and should be carried out with due precaution.*[10]

A two-necked, round-bottomed flask is flushed with nitrogen and charged with zinc dust (20 g). Dilute hydrochloric acid (25 mL of 2 M) is added and allowed to react for 5 min, before being decanted. A solution of $HgCl_2$ (1.2 g) in 2 M HCl (25 mL) is added, allowed to react for 5 min, and then decanted. The activated zinc is washed with deionised water (2 × 25 mL), ethanol (2 × 25 mL), and diethyl ether (2 × 25 mL) and dried *in vacuo* 0.1 torr for 3 h at room temperature. The finely divided powder may form a solid mass on standing and should be used soon after preparation.

A 250-mL, two-necked, round-bottomed flask fitted with a magnetic stirrer is flushed with ethene and charged with 20 g of activated zinc and dry deoxygenated tetrahydrofuran (100 mL). The mixture is rapidly stirred under

---

*Department of Chemistry and Applied Chemistry, University of Salford, Salford M5 4WT, United Kingdom.
[†]Department of Chemistry, Iowa State University, Ames, IA 50011.

a gentle flow of ethene as pure* solid [{CoI$_2$(η^5-C$_5$Me$_5$)}$_2$] (10.0 g, 11.15 mmol) is added, and stirring is continued for 1.5 h, during which time the color of the solution changes from dark green to orange-brown.

A syringe is used to transfer the solution to a nitrogen-filled 250 mL, three-necked, round-bottomed flask connected by stopcock adapters to nitrogen and vacuum. Solvent is evaporated at reduced pressure and nitrogen is readmitted to the flask. The residue is dissolved in 40 mL of dry deoxygenated hexane and the solution transferred by a syringe to a N$_2$-filled filtration apparatus containing a 1-cm pad of kieselguhr. The red-brown filtrate is reduced in volume to 20 mL by evaporation at reduced pressure, and the product is crystallized at $-78°C$ under nitrogen. The supernatant liquid is decanted with a syringe and the orange-brown crystals dried at 20°C, 0.05 torr for 1 h. A second crop of crystals may be obtained from the mother liquor by filtering, evaporating, and cooling as before. The combined yield is 3.9–4.5 g (70–85%).

Properties

Bis(η-ethene)(η^5-pentamethylcyclopentadienyl)cobalt(I) is an orange-brown, volatile, crystalline solid. It may be handled briefly in air but if it is to be kept for long periods, it should be under an inert atmosphere at $-20°C$. It is soluble in organic solvents but the solutions decompose in the presence of air and water. In coordinating solvents decomposition occurs slowly. It is best therefore to allow it to react in alkane, diethyl ether, or aromatic solvents. The ^{1}H NMR spectrum (C$_6$D$_6$) consists of a singlet at δ 1.43 [15H,C$_5$Me$_5$] and an [AB]$_2$ pattern for the C$_2$H$_4$ protons, δ_A 1.60, δ_B 0.86 ppm ($|J(AB) + J(AB')|$ 13 Hz). The IR spectrum (Nujol) has strong bands at 1197 and 1180 cm^{-1}.

References

1. (a) R. B. King, *Coord. Chem. Rev.*, **20**, 155 (1976); (b) P. M. Maitlis, *Chem. Soc. Rev.*, **10**, 1 (1981); (c) T. P. Wolczanski and J. E. Bercaw, *Acc. Chem. Res.*, **13**, 121 (1980); (d) J. M. Manriquez et al., *Inorg. Synth.*, **21**, 181 (1982).
2. R. B. King and M. B. Bisnette, *J. Organomet. Chem.*, **8**, 287 (1967).
3. R. B. King and A. Efraty, *J. Am. Chem. Soc.*, **94**, 3773 (1972).
4. L. R. Byers and L. F. Dahl, *Inorg. Chem.*, **19**, 277 (1980).
5. R. B. King, A. Efraty, and W. M. Douglas, *J. Organomet. Chem.*, **56**, 345 (1973).
6. D. M. Roe and P. M. Maitlis, *J. Chem. Soc. A*, **1971**, 3173.
7. G. Fairhurst and C. White, *J. Chem. Soc. Dalton Trans.*, **1979**, 1524, 1531.

*If any trace of [CoI$_2$(η-C$_5$Me$_5$)(CO)] is present reduction will give the intense green compound [{Co(η-C$_5$Me$_5$)(μ-CO)}$_2$], which will mask the progress of the reaction and may be difficult to separate later.

8. M. L. H. Green and R. B. A. Pardy, *J. Chem. Soc. Dalton Trans.*, **1979**, 355.
9. S. A. Frith and J. L. Spencer, *Inorg. Synth.*, **23**, 15 (1985).
10. R. W. Parry, *Inorg. Synth.*, **12**, 317 (1970).

71. CYCLOPENTADIENYLBIS(TRIMETHYLPHOSPHINE) AND CYCLOPENTADIENYLBIS(TRIMETHYLPHOSPHITE) COMPLEXES OF Co AND Rh

Submitted by H. WERNER,* R. FESER,* V. HARDER,* W. HOFMANN,* and H. NEUKOMM*
Checked by W. D. JONES†

The title complexes, $CpM(PMe_3)_2$ and $CpM[P(OMe)_3]_2$, are electron-rich half-sandwich complexes that have been demonstrated to be valuable precursors to a large number of organometallic derivatives.[1] They behave as Lewis bases and react with a wide variety of electrophiles E or EX to form new metal–element bonds. They have also been shown to be valuable starting materials for the syntheses of heterometallic di- and trinuclear complexes via their reactions with unsaturated transition metal compounds.[1] The only viable syntheses of these compounds now known are those reported here.

A. (η^5-CYCLOPENTADIENYL)BIS-(TRIMETHYLPHOSPHINE)RHODIUM(I)[2]

$$[(C_8H_{14})_2RhCl]_2 + 4PMe_3 + 2LiC_5H_5 \rightarrow$$
$$2Rh(\eta\text{-}C_5H_5)(PMe_3)_2 + 2LiCl + 2C_8H_{14}$$

Procedure

In a 125-mL Schlenk tube, equipped with a nitrogen inlet and a magnetic stirring bar, 3.7 g (5.15 mmol) $[(C_8H_{14})_2RhCl]_2$ (Ref. 3) is dissolved in 40 mL of tetrahydrofuran (THF), freshly distilled over Na and benzophenone. The solution is treated dropwise with 2.1 mL (21.0 mmol) PMe_3 and then stirred for 1 h at room temperature. A 1.01-g (14.0-mmol) quantity of LiC_5H_5, freshly prepared from equimolar amounts of *n*-BuLi and C_5H_6 in hexane,[4] is added and the reaction mixture is stirred for 2 h at room temperature. The

---
*Institut für Anorganische Chemie der Universität, Am Hubland, D-8700 Würzburg, Federal Republic of Germany.
†Department of Chemistry, The University of Rochester, Rochester, NY 14627.

suspension is filtered, and the red-brown filtrate is evaporated to dryness under reduced pressure. The solid residue is extracted with pentane (2 × 20 mL). The pentane solution is filtered, and the filtrate is concentrated *in vacuo* to ~5 mL. After cooling to −78°, red-brown air-sensitive crystals are obtained, which must be stored under thoroughly purified nitrogen or argon (preferably in a refrigerator). Yield: 2.37 g (72%), mp 85°.

Anal. Calcd. for $C_{11}H_{23}P_2Rh$: C, 41.26; H, 7.24; P, 19.35; Rh, 32.13; MW, 320.2. Found: C, 41.04; H, 7.09; P, 19.62; Rh, 32.38; MW, 320 (mass spectroscopy).

Note: In the original procedure,[2] the 1,5-cyclooctadiene complex $[C_8H_{12}RhCl]_2$ was used as starting material. The present method avoids the isolation of the intermediate $[Rh(PMe_3)_4]Cl$.

Properties

(η-Cyclopentadienyl)bis(trimethylphosphine)rhodium(I) is a red-brown crystalline product that remains unchanged when stored under an inert atmosphere at 0–10°. It is soluble in hydrocarbon solvents but decomposes in chloroform and carbon tetrachloride. In dichloromethane, an oxidative addition reaction occurs to give the cation $[Rh(\eta-C_5H_5)(PMe_3)_2CH_2Cl]^+$, which can be isolated as the PF_6^- salt.[5] The ^1H NMR spectrum of the title compound [60 MHz, benzene-d_6, δ in ppm downfield from tetramethylsilane (TMS)] shows two signals at 5.27 (triplet, J_{PH} = 0.6 Hz, C_5H_5 protons) and 1.16 (doublet of virtual triplets, J_{RhH} = 1.2 Hz, N = 8.4 Hz, PMe_3 protons).

The complex is a useful starting material for the preparation of other cyclopentadienylrhodium complexes, for example, $[Rh(\eta-C_5H_5)(PMe_3)_2R]PF_6$ (R = H, Me, Et, COMe, COPh, $GeMe_3$, $SnMe_3$, Cl, Br, I),[2] $[Rh(\eta-C_5H_5)(PMe_3)_2(CH_2PMe_3)]I_2$,[5] $[Rh(\eta-C_5H_5)(PMe_3)(CH_2PMe_3)I]I$,[5] and $[Rh(\eta-C_5H_5)(PMe_3)_2(Al_2Me_4Cl_2)]$.[6]

B. (η^5-CYCLOPENTADIENYL)BIS-(TRIMETHYLPHOSPHINE)COBALT(I)[7]

$$2CoCl_2 + Mg + 6PMe_3 \rightarrow 2CoCl(PMe_3)_3 + MgCl_2$$

$$CoCl(PMe_3)_3 + LiC_5H_5 \rightarrow Co(\eta-C_5H_5)(PMe_3)_2 + PMe_3 + LiCl$$

The original procedures for $CoCl(PMe_3)_3$ (Ref. 8) and $C_5H_5Co(PMe_3)_2$ (Ref. 9) were modified as described next.

Procedure

(a) $CoCl(PMe_3)_3$. In a 125-mL Schlenk tube, equipped with a nitrogen inlet, a solution of 7 mL (73.7 mmol) of PMe_3 (Ref. 10) in 50 mL of THF, freshly distilled over Na and benzophenone, is treated with 2.0 g (15.4 mmol) of anhydrous $CoCl_2$, 600 mg of Mg turnings, freshly cut, and 50 mg (0.39 mmol) of anthracene. The reaction mixture is placed for 20 min in an ultrasonic bath, which leads to a color change from violet to dark brown. The solvent is removed *in vacuo*, and the brown solid residue is extracted with diethyl ether ($\sim 3 \times 20$ mL). The ether solution (which is *very* air sensitive), together with 2.6 mL (27.3 mmol) of PMe_3,[10] is added to a Schlenk tube, equipped with a nitrogen inlet, and containing 1.4 g (10.8 mmol) of anhydrous $CoCl_2$. The reaction mixture is stirred for 2 h at room temperature. The blue precipitate formed is filtered off, washed with diethyl ether, and dried *in vacuo*. The product must be stored under thoroughly purified nitrogen or argon at low temperature. Yield: 5.49 g (65%).

(b) $Co(\eta\text{-}C_5H_5)(PMe_3)_2$. In a 125-mL Schlenk tube, equipped with a nitrogen inlet and a magnetic stirring bar, 5.96 g (18.5 mmol) of $CoCl(PMe_3)_3$ is dissolved in 50 mL of THF, freshly distilled over Na and benzophenone, and the resulting solution cooled to $-50°$. To the solution, a quantity of 1.4 g (19.5 mmol) of LiC_5H_5 is added in small portions. The reaction mixture is stirred for 30 min at $-50°$ and then slowly warmed to room temperature. While keeping the temperature at 20–25°, the solvent is removed *in vacuo*, and the brown solid residue is extracted with hexane (3×20 mL). The hexane solution is filtered and concentrated *in vacuo* to ~ 5 mL. After cooling at $-78°$ dark brown air-sensitive crystals are isolated, which must be stored under thoroughly purified nitrogen or argon (preferably at low temperature). Yield: 4.76 g (93%), mp 55–57°.

Anal. Calcd. for $C_{11}H_{23}CoP_2$: C, 47.85; H, 8.34; Co, 21.36; P, 22.45; MW, 275.9. Found: C, 47.87; H, 8.12; Co, 21.10; P, 22.28; MW, 276 (mass spectrometry).

Properties

(η-Cyclopentadienyl)bis(trimethylphosphine)cobalt(I) is a dark brown crystalline product that remains unchanged when stored under an inert atmosphere at 0–10°. It is soluble in hydrocarbon solvents but decomposes rapidly in chloroform. In methanol, slow reaction takes place that leads to the formation of the cation $[Co(\eta\text{-}C_5H_5)(PMe_3)_2H]^+$. This cation (with PF_6^- as the anion) is formed more easily from $[Co(\eta\text{-}C_5H_5)(PMe_3)_2]$ and NH_4PF_6. The ^1H NMR spectrum of the title compound (60 MHz, benzene-d_6, δ in ppm downfield from TMS) shows two signals at 4.51 (triplet, $J_{PH} = 1.4$ Hz, C_5H_5 protons) and 1.07 (virtual triplet, $N = 7.6$ Hz, PMe_3 protons).

The complex is a useful starting material for the preparation of other cyclopentadienylcobalt complexes, for example, [Co(η-C_5H_5)(PMe_3)$_2$R]X (R = H, Me, Et, COMe, COPh, $SnCl_3$, $SnMe_3$, $SnPh_3$; X = I or PF_6),[9,11] [Co(η-C_5H_4R)(PMe_3)$_2$] (R = $CHMe_2$, CMe_3, CMe_2Et),[12] [Co(η-C_5H_5)(PMe_3)(η^2-CS_2)],[13] [Co(η-C_5H_5)(PMe_3)(η^2-CSSe)],[14] [Co(η-C_5H_5)(PMe_3)(η^2-CSe_2)],[14] [Co(η-C_5H_5)(PMe_3)(CE)] (E = S, Se),[14] [Co(η-C_5H_5)(PMe_3)(CNR)] [R = Me, CMe_3, Ph),[15] [Co(η-C_5H_5)(PMe_3)(η^2-S_5)],[16] [(η-C_5H_5)(PMe_3)Co(η-CO)$_2$Mn(CO)(η-C_5H_4Me)].[17] The (carbon disulfide)-cobalt complex has further been used for the synthesis of Co_3 clusters containing a bridging thiocarbonyl ligand.[18]

C. (η^5-CYCLOPENTADIENYL)BIS(TRIMETHYL PHOSPHITE)COBALT(I)[19]

$$Co(\eta\text{-}C_5H_5)_2 + 2P(OMe)_3 \rightarrow [Co(\eta\text{-}C_5H_5)(P(OMe)_3)_2] + \{C_5H_5\}$$

Procedure

In a 20-mL Schlenk tube, equipped with a nitrogen inlet, a reflux condenser, and a magnetic stirring bar, 2.09 g (11.06 mmol) of Co(η-C_5H_5)$_2$ (Ref. 20) and 7 mL (59.4 mmol) of $P(OMe)_3$ are heated under reflux for 3 days. After cooling to room temperature, excess of $P(OMe)_3$ is removed *in vacuo* (~ 15 torr), and the oily residue is distilled at $\sim 149°$ and $\sim 10^{-4}$ torr using a short path distillation apparatus. The distilled product is dissolved in pentane (~ 10 mL), and the pentane solution is concentrated *in vacuo*. Cooling to $\sim -30°$ gives red-brown air-sensitive crystals that are filtered off, washed with small amounts of cold pentane, and dried *in vacuo*. Yield: 2.2 g (56%), mp 38°.

Anal. Calcd. for $C_{11}H_{23}CoO_6P_2$: C, 35.50; H, 6.23; Co, 15.84; MW, 372.2. Found: C, 35.39; H, 6.01; Co, 15.87; MW, 372 (mass spectrometry).

Properties

(η-Cyclopentadienyl)bis(trimethyl phosphite)cobalt(I) is a red-brown crystalline low-melting product that is air sensitive and should be stored under an inert atmosphere at 10–20°. It is soluble in hydrocarbon solvents, but decomposes in chloroform and carbon tetrachloride. The ^{1}H NMR spectrum

(60 MHz, acetone-d_6, δ in ppm downfield from TMS) shows two signals at 4.63 (singlet, C_5H_5 protons) and 3.48 [virtual triplet, $N = 12.0$ Hz, $P(OMe)_3$ protons]. The UV spectrum (in hexane, λ_{max} in cm^{-1}) shows three maxima at 23,920 (log ε 2.7), 38,910 (log ε 4.7), and 43,860 (log ε 5.0).

The complex is a good nucleophile and reacts with acids (e.g., CF_3CO_2H) and methyl iodide to form the corresponding salts of the cations [Co(η-C_5H_5)(P(OMe)$_3$)$_2$R]$^+$(R = H, Me).[21] Thermolysis gives the "super-sandwich" complex [Co$_3$(η-C_5H_5)$_2$(μ-P(O)(OMe)$_2$)$_6$],[22] which is an important starting material for the syntheses of heterometallic di- and trinuclear phosphonatemetal complexes.[23, 24]

D. (η^5-CYCLOPENTADIENYL)BIS(TRIMETHYL PHOSPHITE)RHODIUM(I)[25]

$$[Rh(P(OMe)_3)_2Cl]_2 + 2NaC_5H_5 \rightarrow [Rh(\eta-C_5H_5)(P(OMe)_3)_2]$$
$$+ 2NaCl$$

Procedure

In a 125-mL Schlenk tube, equipped with a nitrogen inlet and a magnetic stirring bar, 1.0 g (2.03 mmol) of [Rh(C$_8$H$_{12}$)Cl]$_2$ (Ref. 26) is dissolved in 50 mL of CH$_2$Cl$_2$. The solution is treated dropwise with 1 mL (8.1 mmol) of P(OMe)$_3$ and stirred for 2 h at room temperature. The solvent and excess phosphite are removed *in vacuo*. The residue is dissolved in 60 mL of THF, and 450 mg (5.1 mmol) of NaC$_5$H$_5$ (Ref. 20) is added. The reaction mixture is stirred for 24 h at room temperature, filtered, and concentrated *in vacuo*. The oily residue is dissolved in 10 mL of hexane, and the solution is chromatographed on Al$_2$O$_3$ (neutral, activity III) using diethyl ether as eluant. The major yellow fraction is collected, the solvent is removed, and the resultant oily residue is dried for 24 h in high vacuum ($\sim 10^{-4}$ torr). After storing the product for 3–4 weeks under an inert atmosphere without additional solvent at $-10°$, yellow crystals are formed. Yield: 605 mg (36%), mp 31°.

Anal. Calcd. for C$_{11}$H$_{23}$O$_6$P$_2$Rh: C, 31.74; H, 5.57; P, 14.88; MW, 416.15. Found: C, 31.84; H, 5.73; P, 14.72; MW, 416 (mass spectrometry).

Properties

(η-Cyclopentadienyl)bis(trimethyl phosphite)rhodium(I) is a yellow crystalline low-melting product that is air sensitive and should be stored under an inert atmosphere at 10–20°. It is soluble in hydrocarbon solvents but decomposes in chloroform and carbon tetrachloride. The ^{1}H NMR spectrum (60 MHz, benzene-d_6, δ in ppm downfield from TMS) shows two signals at 5.32 (doublet of triplets, $J_{\text{RhH}} = 0.6$ Hz, $J_{\text{PH}} = 1.2$ Hz, C_5H_5 protons) and 3.40 [virtual triplet, $N = 12.2$ Hz, P(OMe)$_3$ protons].

The complex is a good nucleophile and can be used as starting material for the syntheses of cationic complexes [Rh(η-C$_5$H$_5$)(P(OMe)$_3$)$_2$R]$^+$ (R = H, Me).[21, 27] It also reacts smoothly with alkali metal iodides MI (M = Li, Na, K) in two stages, via the intermediate [Rh(η-C$_5$H$_5$)CH$_3$(P(OMe)$_3$)P(O)(OMe)$_2$], to the corresponding rhodiumbis(phosphonate) complexes [Rh(η-C$_5$H$_5$)CH$_3$(P(O)(OMe)$_2$)$_2$]M.[28]

References

1. H. Werner, *Angew. Chem. Int. Ed. Engl.*, **22**, 927 (1983).
2. H. Werner, R. Feser, and W. Buchner, *Chem. Ber.*, **112**, 834 (1979).
3. A. van der Ent and L. Onderdelinden, *Inorg. Synth.*, **14**, 92 (1973).
4. M. A. Lyle and S. R. Stobart, *Inorg. Synth.*, **17**, 178 (1977).
5. H. Werner, L. Hofmann, R. Feser, and W. Paul, *J. Organomet. Chem.*, **281**, 317 (1985).
6. J. M. Mayer and J. C. Calabrese, *Organometallics*, **3**, 1292 (1984).
7. H. Otto, Ph.D. thesis, Universität Würzburg, 1986, p. 239.
8. H. F. Klein and H. H. Karsch, *Chem. Ber.*, **108**, 944 (1975).
9. H. Werner and W. Hofmann, *Chem. Ber.*, **110**, 3481 (1977).
10. R. T. Markham, E. A. Dietz Jr., and D. R. Martin, *Inorg. Synth.*, **16**, 153 (1976).
11. K. Dey and H. Werner, *Chem. Ber.*, **112**, 823 (1979).
12. H. Werner and W. Hofmann, *Chem. Ber.*, **114**, 2681 (1981).
13. H. Werner, K. Leonhard, and C. Burschka, *J. Organomet. Chem.*, **160**, 291 (1978).
14. O. Kolb and H. Werner, *J. Organomet. Chem.*, **268**, 49 (1984).
15. H. Werner, S. Lotz., and B. Heiser, *J. Organomet. Chem.*, **209**, 197 (1981).
16. C. Burschka, K. Leonhard, and H. Werner, *Z. Anorg. Allg. Chem.*, **464**, 30 (1980).
17. K. Leonhard and H. Werner, *Angew. Chem. Int. Ed. Engl.*, **16**, 649 (1977).
18. H. Werner, K. Leonhard, O. Kolb, E. Röttinger, and H. Vahrenkamp, *Chem. Ber.*, **113**, 1654 (1980).
19. V. Harder, J. Müller, and H. Werner, *Helv. Chim. Acta*, **54**, 1 (1971).
20. R. B. King and F. G. A. Stone, *Inorg. Synth.*, **7**, 99 (1963).
21. H. Werner, H. Neukomm, and W. Kläui, *Helv. Chim. Acta*, **60**, 326 (1977).
22. V. Harder, E. Dubler, and H. Werner, *J. Organomet. Chem.*, **71**, 427 (1974).
23. W. Kläui and H. Werner, *Angew. Chem. Int. Ed. Engl.*, **15**, 172 (1976).
24. W. Kläui and K. Dehnicke, *Chem. Ber.*, **111**, 451 (1978).
25. H. Neukomm and H. Werner, *Helv. Chim. Acta*, **57**, 1067 (1974).
26. J. Chatt and L. M. Venanzi, *J. Chem. Soc.*, **1957**, 4735.
27. H. Neukomm and H. Werner, *J. Organomet. Chem.*, **108**, C 26 (1976).
28. H. Werner and R. Feser, *Z. Anorg. Allg. Chem.*, **458**, 301 (1979).

Chapter Seven

LANTHANIDE AND ACTINIDE COMPLEXES

72. LANTHANIDE TRICHLORIDES BY REACTION OF LANTHANIDE METALS WITH MERCURY(II) CHLORIDE IN TETRAHYDROFURAN

Submitted by GLEN B. DEACON,* TRAN D. TUONG,* and
DALLAS L. WILKINSON*†
Checked by TOBIN MARKS‡

Anhydrous lanthanide trihalides, particularly the trichlorides, are important reactants for the formation of a variety of lanthanide complexes, including organometallics. Routes for the syntheses of anhydrous lanthanide trihalides generally involve high-temperature procedures or dehydration of the hydrated halides.[1-5] The former are inconvenient and complex for small-scale laboratory syntheses, while dehydration methods may also be complex[4] and have limitations, for example, use of thionyl chloride.[1,5] Moreover, the products from these routes may require purification by vacuum sublimation at elevated temperatures.[3,4] Redox transmetalation between lanthanide metals and mercury(II) halides was initially carried out at high temperatures.[2,3] However, this reaction can be carried out in tetrahydrofuran (THF) to give complexes of lanthanide trihalides with the solvent.[6] These products are equally as suitable as reactants for synthetic purposes as the uncomplexed trihalides. Other workers have used this transmetalation reaction for activation of the lanthanide metals.[7]

---

*Chemistry Department, Monash University, Clayton, Victoria, Australia, 3168.
†The submitters are grateful to the Australian Research Grants Scheme for support and to Rare Earth Products for a gift of REACTON ytterbium.
‡Department of Chemistry, Northwestern University, Evanston, IL 60201.

Detailed syntheses of four representative lanthanide trihalide–tetrahydrofuran (THF) complexes, $YbCl_3(THF)_3$, $ErCl_3(THF)_{3.5}$, $SmCl_3(THF)_2$, and $NdCl_3(THF)_{1.5}$ by redox transmetalation are described. The first three compounds have previously been prepared by direct reaction of anhydrous lanthanide trihalides with THF,[8,9] but the composition of the last differs slightly from that reported,[9] namely, $NdCl_3(THF)_2$, for the product from reaction of $NdCl_3$ with THF. Although the method below describes isolation of the complex trichlorides, the THF solutions can be used *in situ* for further reactions.[7] The method can also be used for other trichlorides[7,10] and other trihalides.[6]

General Procedure

The lanthanide trichlorides described here are moisture sensitive, but they can be stored indefinitely under purified nitrogen or argon at room temperature. All operations are carried out under nitrogen or argon, which is purified by passage through BASF R3/11 oxygen removal catalyst and molecular sieves (see ref. 11 for a discussion of inert atmosphere techniques). Tetrahydrofuran is dried by distillation from sodium benzophenone ketyl under nitrogen.

■ **Caution.** *Only fresh, peroxide-free THF should be distilled.*

The solvent is stored under nitrogen in Schlenk vessels equipped with high-vacuum Teflon taps (e.g., Young or Rotaflo). Syringes (preflushed with nitrogen) are used for transfers of solvent. For reliable results, an oven-dried greaseless Schlenk apparatus incorporating polytetrafluoroethylene O-rings and taps (Young or Rotaflo) should be used, and the lanthanide metal powder should be stored, weighed, and handled under nitrogen. The metal powder can be obtained from Research Chemicals. Alternatively, RE-ACTON distilled lanthanide metals from Rare Earth Products, can be crushed under nitrogen to a powder. Dried reagent-grade mercury(II) chloride can be used without purification.

The apparatus used in all reactions is shown in Fig. 1. Mercury(II) chloride (1.084 g, 4.00 mmol), an excess of the lanthanide metal powder (6.00 mmol), and a magnetic stirring bar are placed in the lower 100-mL Schlenk flask. Onto this flask are attached (in order) a condenser,* a Schlenk filter [covered with a Whatman microfiber glass filter paper, a layer at least 6 mm thick of dried diatomaceous earth (Sigma grade 1, D-3877) and a second glass filter paper], and a preweighed 100-mL Schlenk flask. The Schlenk apparatus is purged of air, and an inert atmosphere is established by evacuation of the apparatus to 10^{-3} torr and backfilling with nitrogen or argon at least three

---

*If heating is carefully controlled, the condenser may be omitted.

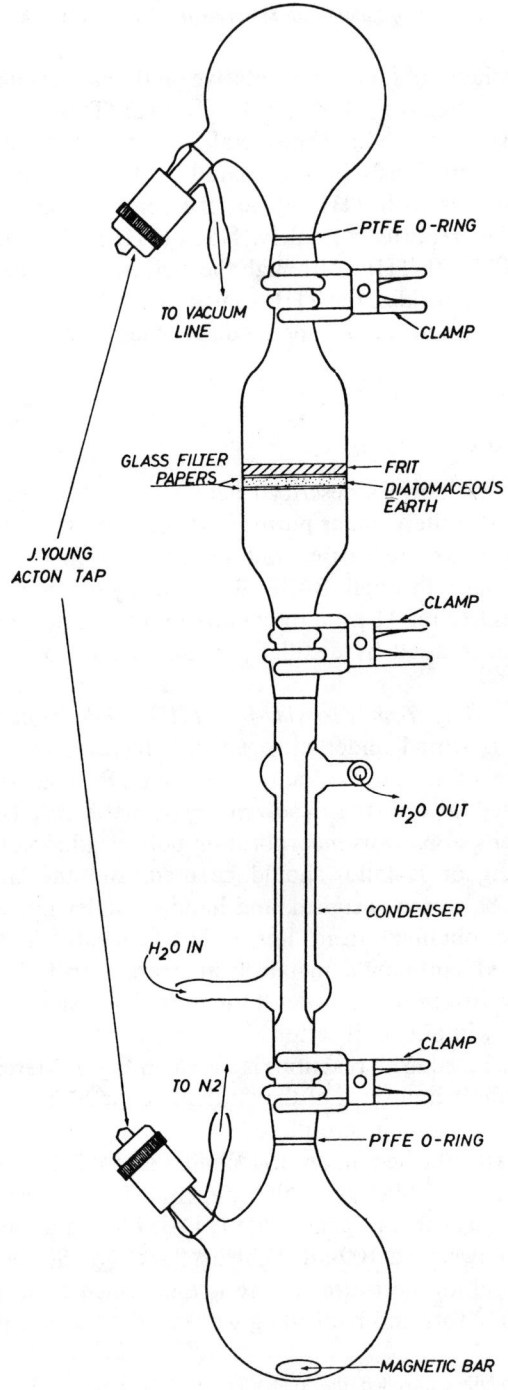

Fig. 1. Reaction apparatus.

times.* Tetrahydrofuran (25 mL) is then added, stirring is commenced, and the reaction mixture is heated to 65°C. After stirring and heating for 1.5 h the apparatus is inverted, and the reaction mixture is filtered into the second Schlenk flask under reduced pressure. The residue on the Schlenk filter is washed with THF (25 mL). Evaporation of the filtrate and washings (10^{-3} torr, room temperature) gives the lanthanide trichloride, which is dried under vacuum (10^{-3} torr) for 1–2 h at room temperature and weighed.

■ **Caution.** *Finely divided metal residues from the Schlenk filter can be pyrophoric.*

The composition of the THF complexes of the lanthanide trichlorides is established by dissolution of a weighed amount of each compound in water and determination of chloride potentiometrically with aqueous silver nitrate.[12] Analysis for the lanthanide is effected by titration with disodium ethylenediaminetetraacetate using xylenol orange indicator after appropriate buffering of the solution.[13,14] The IR spectra of all the trichlorides show intense bands at ~ 1020 and 880 cm^{-1} (see below), indicative of coordinated THF.[15]

A. YTTERBIUM TRICHLORIDE–TETRAHYDROFURAN(1/3)

$$2\text{Yb} + 3\text{HgCl}_2 + 6\text{THF} \xrightarrow{\text{THF}} 2\text{YbCl}_3(\text{THF})_3 + 3\text{Hg}\downarrow$$

■ **Caution.** *Mercury(II) chloride and mercury metal are toxic. Protective rubber gloves should be used. Operations should be carried out in a well-ventilated hood.*

Procedure

Mercury(II) chloride (1.084 g, 4.00 mmol) and ytterbium metal (1.038 g, 6.00 mmol) are allowed to react as described previously, giving the white title compound. Yield: 1.16 g (88%).

Anal. Calcd. for $C_{12}H_{24}Cl_3O_3Yb$: Cl, 21.5; Yb, 34.9. Found: Cl, 21.4; Yb, 34.9%.

IR (Nujol mull, prepared under N_2), 1055 (sh), 1018 (s), 925 (m), 880 (s), 860 (sh), and 735 (w) cm^{-1}.

---

*Vacuum–nitrogen (argon) lines are connected to both Schlenk flasks. Initially the reaction flask is attached to the nitrogen line and the product flask is attached to the vacuum line.

B. ERBIUM TRICHLORIDE–TETRAHYDROFURAN (2/7)

$$2Er + 3HgCl_2 + 7THF \xrightarrow{THF} 2ErCl_3(THF)_{3.5} + 3Hg\downarrow$$

■ **Caution.** See Section A.

Procedure

Mercury(II) chloride (1.08 g, 4.00 mmol) and erbium metal (1.004 g, 6.00 mmol) are allowed to react as indicated above to give the pink title compound. Yield: 1.22 g (87%).

Anal. Calcd. for $C_{28}H_{56}Cl_6Er_2O_7$: Cl, 20.2; Er, 31.8. Found: Cl, 20.1; Er, 31.9%.

IR (Nujol mull, prepared under N_2): 1191 (m), 1021 (s), 931 (m), 871 (s), and 737 (w) cm^{-1}.

C. SAMARIUM TRICHLORIDE–TETRAHYDROFURAN (1/2)

$$2Sm + 3HgCl_2 + 4THF \xrightarrow{THF} 2SmCl_3(THF)_2 + 3Hg\downarrow$$

■ **Caution.** See Section A.

Procedure

Mercury(II) chloride (1.084 g, 4.00 mmol) and samarium metal (0.902 g, 6.00 mmol) are allowed to react as described above, giving the white title compound. Yield: 0.520 g (49%).

Anal. Calcd. for $C_8H_{16}Cl_3O_2Sm$: Cl, 26.5; Sm, 37.5. Found: Cl, 25.6; Sm, 37.5%.

IR (Nujol mull, prepared under N_2): 1020 (s), 926 (m), 870 (s), and 746 (m) cm^{-1}.

D. NEODYMIUM TRICHLORIDE–TETRAHYDROFURAN (2/3)

$$2Nd + 3HgCl_2 + 3THF \xrightarrow{THF} 2NdCl_3(THF)_{1.5} + 3Hg\downarrow$$

■ **Caution.** See Section A.

Procedure

Mercury(II) chloride (1.084 g, 4.00 mmol) and neodymium metal (0.865 g, 6.00 mmol) are allowed to react as described above, giving the pale blue title compound. Yield: 0.560 g (58%).

Anal. Calcd. for $C_{12}H_{24}Cl_6Nd_2O_3$: Cl, 29.6; Nd, 40.2. Found: Cl, 29.0; Nd, 39.8%.

IR (Nujol mull, prepared under N_2): 1022 (s), 930 (m), 871 (s), and 736 (m) cm^{-1}.

References

1. M. D. Taylor, *Chem. Rev.*, **62**, 503 (1962).
2. L. B. Asprey, T. K. Keenan, and F. H. Kruse, *Inorg. Chem.*, **3**, 1137 (1964).
3. F. L. Carter and J. F. Murray, *Mater. Res. Bull.* **7**, 519 (1972).
4. K. E. Johnson and K. R. MacKenzie, *J. Inorg. Nucl. Chem.*, **32**, 43 (1970).
5. J. H. Freeman and M. L. Smith, *J. Inorg. Nucl. Chem.*, **7**, 224 (1958).
6. G. B. Deacon and A. J. Koplick, *Inorg. Nucl. Chem. Lett.*, **15**, 263 (1979).
7. G. Z. Suleimanov, T. Kh. Kurbanov, Yu. A. Nuriev, L. F. Rybakova, and I. P. Beletskaya, *Dokl. Chem.*, **265**, 254 (1982).
8. K. Rossmanith and C. Auer-Welsbach, *Monatsh. Chem.*, **96**, 602 (1965).
9. K. Rossmanith, *Monatsh. Chem.*, **100**, 1484 (1969).
10. A. A. Passynskii, I. L. Eremenko, G. Z. Suleimanov, Yu. A. Nuriev, I. P. Beletskaya, V. E. Shklover, and Yu. T. Struchkov, *J. Organomet. Chem.*, **266**, 45 (1984).
11. J. J. Eisch, *Organometallic Syntheses*, Vol. 2, *Nontransition-Metals*, J. J. Eisch and R. B. King (eds.), Academic Press, New York, 1981.
12. A. I. Vogel, *A Text Book of Quantitative Inorganic Analysis*, 3rd ed., Longmans, Green, London, **1961**, p. 950.
13. J. Korbl and R. Pribil, *Chem. Anal*, **45**, 102 (1956).
14. J. L. Atwood, W. E. Hunter, A. L. Wayda, and W. J. Evans, *Inorg. Chem.*, **20**, 4115 (1981).
15. J. Lewis, J. R. Miller, R. L. Richards, and A. Thompson, *J. Chem. Soc.*, **1965**, 5850. R. J. H. Clark, J. Lewis, D. J. Machin, and R. S. Nyholm, *J. Chem. Soc.*, **1963**, 379.

73. (η^5-CYCLOPENTADIENYL)LANTHANIDE COMPLEXES FROM THE METALLIC ELEMENTS

Submitted by GLEN B. DEACON,* GEOFF N. PAIN,* and TRAN D. TUONG*
Checked by WILLIAM J. EVANS,† KEITH R. LEVAN,† and RAUL DOMINGUEZ†

Tris(η^5-cyclopentadienyl)lanthanides were the first authentic organolanthanides to be prepared[1] and bis(η^5-cyclopentadienyl)lanthanide(II) compounds

---
*Department of Chemistry, Monash University, Clayton, Victoria, Australia, 3168.
†Department of Chemistry, University of California, Irvine, CA 92717.

have played a germinal part in the development of lower oxidation state organolanthanide chemistry.[2] These cyclopentadienyls are sources of coordination compounds of structural interest and are reagents for the synthesis of other organolanthanides, for example, bis- and mono(η^5-cyclopentadienyl)lanthanide(III) derivatives.[2]

Early syntheses of tris(η^5-cyclopentadienyl)lanthanides employed reactions of anhydrous lanthanide trihalides with alkali metal, magnesium, or beryllium cyclopentadienides, generally followed by sublimation of the crude product.[1,2] Bis(η^5-cyclopentadienyl)europium(II) and bis(η^5-cyclopentadienyl)ytterbium(II) have been obtained by reactions of the free metals with cyclopentadiene in liquid ammonia, again with sublimation work-up.[3-5] Recently, reductive transmetalation reactions of thallium(I)[6,7] and mercury(II)[8,9] cyclopentadienides with lanthanide elements have been developed as a route to trivalent and divalent (η^5-cyclopentadienyl)lanthanide complexes. These methods avoid the need for anhydrous lanthanide halides, highly air-sensitive metal cyclopentadienide reagents, and use of liquid ammonia. For transmetallation, thallium(I) cyclopentadienide is the more attractive reagent, since it is more readily prepared, is less heat and light sensitive, and has better storage characteristics than bis(cyclopentadienyl)mercury.[10] Tetrahydrofuran and 1,2-dimethoxyethane are the most convenient solvents for transmetallation (see below), but the scope of reactions with $Tl(C_5H_5)$ can be increased by use of pyridine.[11] Thus, Eu gives $Eu(C_5H_5)_n(py)$ (n = 2 or 3) in this solvent,[11] but no reaction occurs in tetrahydrofuran.

In the following sections we detail transmetalation syntheses from $Tl(C_5H_5)$ (see Ref. 12) of three (η^5-cyclopentadienyl)lanthanide complexes, which can be isolated analytically pure in nearly quantitative yields simply by evaporation of the filtered reaction solvent. These simplify and optimize our reported syntheses.[6,7]

General Comments

The organolanthanides described are very sensitive to oxygen and water both in solution and as solids, but they can be handled and stored under purified nitrogen or argon (BASF R3/11 oxygen removal catalyst and molecular sieves), and in dry solvents. For reliable results, "grease-free' Schlenk apparatus incorporating polytetrafluoroethylene O-rings and "Rotaflo," "Young," or similar taps should be used. (See Refs. 13 and 14 for general techniques in organometallic chemistry.)

A small amount of mercury metal aids initiation of transmetalation, but is not essential. The apparatus used in all three reactions is shown in Fig. 1 of synthesis number (Section) 72. This consists of a 100-mL Schlenk flask containing a magnetic stirring bar, on top of which are attached (in order) a condenser, a Schlenk filter [covered with a Whatman Microfibre filter paper, a thin (~3 mm) layer of dried diatomaceous earth (Sigma, grade 1) and a second glass fiber filter paper], and a further Schlenk flask (100 mL, preweighed). Because of the use of poisonous materials (see Caution note in procedure in Section A), the apparatus should be placed in a well-ventilated fume hood. Because of the air-sensitive nature of the products, the apparatus should be thoroughly flushed with purified nitrogen or argon.

The powdered lanthanide element, thallium(I) cyclopentadienide and a trace of mercury metal are placed into the first flask. An inert atmosphere is established by evacuation of the apparatus and backfilling with N_2 or Ar at least three times. Dried and degassed solvent is then added to the reagents by syringe and the reaction is carried out as described in each of the following procedures. After reaction and cooling, the apparatus is inverted, and the reaction mixture is filtered into the second flask by gravity or by reduced pressure. The filter is washed with solvent (2 × 5 mL) added by syringe, and the filtrate is evaporated slowly to dryness under vacuum at room temperature yielding the pure cyclopentadienyllanthanide. All operations including handling of solids must be carried out under purified argon or nitrogen.

Lanthanide metals were obtained from Research Chemicals as powders. The metals can be weighed and stored in air, but for long-term storage, especially of neodymium, use of a dry box is recommended.

■ **Caution.** *Finely divided residues after reaction can be pyrophoric.*

Peroxide-free tetrahydrofuran (THF) (predried first over KOH pellets followed by sodium wire) and 1,2-dimethoxyethane (DME) (predried over sodium wire) are distilled from sodium benzophenone ketyl and stored under purified nitrogen in Schlenk vessels equipped with Teflon Rotaflo or Young taps.

A. TRIS(η^5-CYCLOPENTADIENYL)(TETRAHYDROFURAN)-NEODYMIUM(III)

$$Nd + 3Tl(C_5H_5) + C_4H_8O \xrightarrow{\text{THF}} Nd(C_5H_5)_3(C_4H_8O) + 3Tl \downarrow$$

■ **Caution.** *Compounds of thallium and thallium metal are extremely poisonous. Protective rubber gloves should be used in handling thallium compounds and their solutions, and all operations should be carried out in a well-ventilated fume hood. Disposal of thallium and mercury-containing residues must be handled according to procedures established for poisonous metals.*

Procedure

The Schlenk apparatus is charged with Nd (1.45 g, 10.05 mmol), Hg (0.170 g, 0.85 mmol), Tl(C_5H_5) (1.090 g, 4.04 mmol), and THF (25 mL). The mixture is heated at reflux point for 20 h with magnetic stirring. After filtration, washing of the filter, and evaporation of solvent, the pure complex is obtained as pink-blue crystals (pink in daylight, blue in artificial light). [The color appears to be dependent on crystal size, since the checkers report that, isolated as a microcrystalline solid, the compound is light blue in natural light, but lavender under a light bulb, whereas larger crystals (from THF) appear pink in sunlight.] Yield: 0.500 g (90%). (The checkers report 73% when the product was scraped from the flask in a dry box.)

Anal. Calcd. for $C_{19}H_{23}NdO$: C, 55.4; H, 5.6. Found: C, 55.6; H, 5.6%.

Properties

The complex $Nd(\eta^5\text{-}C_5H_5)_3(C_4H_8O)$ is unstable to air both in solution and as the solid.[7] It must be handled using techniques for air-sensitive compounds.[13,14] The IR spectrum of a Nujol mull prepared under inert atmosphere shows absorptions at 1335 (m), 1254 (m), 1009 (vs), 925 (m), 858 (m) (characteristic of coordinated THF), 795 (vs), 768 (s), 740 (s, sh), and 670 (m) cm^{-1}. The Vis–near IR absorptions (in THF solution) are λ_{max} (ε): 323 (75), 528 (8), 531 (7), 576 (21), 582 (22), 588 (123), 591 (94), 594 (133), 597 (92), 600 (23), 605 (15), 618 (5), 622 (4), 733 (6), 737 (8), 740 (12), 743 (5), 755 (8), 792 (9), 800 (6), 802 (4) nm.

The mass spectrum is identical with that previously reported for the unsolvated compound.[15]

B. TRIS(η^5-CYCLOPENTADIENYL)(TETRAHYDROFURAN)-SAMARIUM(III)

$$Sm + 3Tl(C_5H_5) + C_4H_8O \xrightarrow{THF} Sm(C_5H_5)_3(C_4H_8O) + 3Tl\downarrow$$

■ *See cautionary note, Section A.*

Procedure

The Schlenk apparatus is charged with Sm (2.52 g, 16.75 mmol), Hg (0.10 g, 0.5 mmol), Tl(C_5H_5) (1.980 g, 7.35 mmol), and THF (50 mL). The mixture is heated at reflux point for 4 h with magnetic stirring. After work-up, as described previously, the pure complex is obtained as bright yellow crystals. Crystal size may be increased by slow evaporation of the solvent, and this can give crystals suitable for X-ray characterization.[7] Yield: 0.979 g (96%). [The checkers' isolation method (see Section A) gave 72%.]

Anal. Calcd. for $C_{19}H_{23}OSm$: C, 54.6; H, 5.5. Found: C, 54.9; H, 5.5%.

Properties

The complex $Sm(\eta^5\text{-}C_5H_5)_3(C_4H_8O)$ is unstable to air both in solution and as the solid.[7] Crystallographic data[7] show the complex to be isostructural with $Gd(\eta^5\text{-}C_5H_5)_3(THF)$, hence samarium has pseudotetrahedral stereochemistry and formal 10 coordination. The IR spectrum of a Nujol mull prepared under inert atmosphere shows absorptions at 1335 (m), 1257 (m), 1009 (vs), 858 (m) (characteristic of coordinated THF), 790 (vs), 773 (s, br), 750 (s, sh), and 667 (m) cm^{-1}. The Vis–near IR absorptions (in THF solution) are $\lambda_{max}(\varepsilon)$: 296 (1584), 424 (134), 429 (134), 1050 (4), 1075 (5), 1104 (4), 1206 (15), 1227 (19), 1268 (8), 1320 (14) nm. The mass spectrum agrees with that previously reported for the unsolvated compound.[15] Heating at 140–150°C under vacuum (10^{-3} torr) gives orange tris(η^5-cyclopentadienyl)-samarium(III) with IR absorption as previously reported.[16]

C. BIS(η^5-CYCLOPENTADIENYL)(1,2-DIMETHOXYETHANE)-YTTERBIUM(II)

$$Yb + 2Tl(C_5H_5) + C_4H_{10}O_2 \xrightarrow{DME} Yb(C_5H_5)_2(C_4H_{10}O_2) + 2Tl\downarrow$$

■ *See cautionary note, Section A.*

Procedure

The Schlenk apparatus is charged with Yb (3.003 g, 17.35 mmol), Hg (0.1 g, 0.5 mmol), $Tl(C_5H_5)$ (2.000 g, 7.42 mmol), and 1,2-dimethoxyethane (50 mL). The mixture is heated for 2 h, during which time the color of the reaction mixture changes from an olive green to the final blue-green. Monitoring the reaction by Vis–near IR spectroscopy reveals tris(η^5-cyclopentadienyl)ytterbium(III) (olive green) to be the first detectable organometallic product. This is then reduced to bis(η^5-cyclopentadienyl)ytterbium(II) (blue-green) by the excess metallic ytterbium.[7]

Work-up, as above, gives the pure complex as green crystals. Yield: 1.40 g (96%). (The checkers' isolation method gave 68%).

Anal. Calcd. for $C_{14}H_{20}O_2Yb$: C, 42.8; H, 5.1. Found: C, 42.9; H, 4.9%.

Properties

The complex $Yb(\eta^5\text{-}C_5H_5)_2(C_4H_{10}O_2)$ is very air sensitive both in solution and in the solid state.[7] The IR spectrum as a Nujol mull prepared under inert atmosphere shows absorptions at 3070 (m), 1276 (m), 1240 (m), 1190 (m), 1108 (m), 1090 (m), 1058 (vs), 1002 (vs), 854 (s), 825 (m), 772 (s, sh), 750 (vs), and 738 (vs) cm^{-1}. Visible–near IR spectrum (in 1,2-dimethoxyethane), $\lambda_{max}(\varepsilon)$: 382 (582), 624 (226) nm. The complex dissolves in THF to form a red-purple solution, and attempted dissolution in benzene results in precipitation of unsolvated $Yb(C_5H_5)_2$, and formation of a green solution with a dme:$Yb(C_5H_5)_2$ ratio (by ^1H NMR) of >1:1. (The checkers were able to dissolve a sample completely in benzene.) In a sealed capillary under nitrogen, the compound turns yellow at 140°C and brick red at 150°C, the green color being restored on cooling. On being heated under vacuum at 140 to 150°C, unsolvated bis(η^5-cyclopentadienyl)ytterbium is obtained. The ^1H NMR spectrum of $Yb(C_5H_5)_2$ in C_4D_8O solution (acetone-d_6 lock tube) shows a cyclopentadienyl resonance at δ 5.64 (s) downfield from external Me_4Si. The crystal and molecular structure of $Yb(\eta^5\text{-}C_5H_5)_2(C_4H_{10}O_2)$ has been determined.[17]

References

1. G. Wilkinson and J. M. Birmingham, *J. Am. Chem. Soc.*, **76**, 6210 (1954); **78**, 42 (1956).
2. T. J. Marks and R. D. Ernst, "Scandium, Yttrium, and the Lanthanides and Actinides," in *Comprehensive Organometallic Chemistry*, Vol. 3, G. Wilkinson, F. G. A. Stone, and E. W. Abel (eds.), Pergamon, Oxford, 1982, Chapter 21; H. Schumann, *Angew. Chem. Int. Ed. Engl.*, **23**, 474 (1984).
3. E. O. Fischer and H. Fischer, *J. Organomet. Chem.*, **3**, 181 (1965).
4. R. G. Hayes and J. L. Thomas, *Inorg. Chem.*, **8**, 2521 (1969).
5. F. Calderazzo, R. Pappalardo, and S. Losi, *J. Inorg. Nucl. Chem.*, **28**, 987 (1966).
6. G. B. Deacon, A. J. Koplick, and T. D. Tuong, *Polyhedron*, **1**, 423 (1982).
7. G. B. Deacon, A. J. Koplick, and T. D. Tuong, *Aust. J. Chem.*, **37**, 517 (1984).
8. G. Z. Suleimanov, L. F. Rybakova, Ya. A. Nuriev, T. Kh. Kurbanov, and I. P. Beletskaya, *J. Organomet. Chem.*, **235**, C19 (1982).
9. G. Z. Suleimanov, T. Kh. Kurbanov, Ya. A. Nuriev, L. F. Rybakova, and I. P. Beletskaya, *Dokl. Chem.*, **265**, 254 (1982).
10. F. A. Cotton and G. Wilkinson, *Advanced Inorganic Chemistry*, 4th ed., Wiley, New York, 1980, p. 1163; G. Wilkinson and T. S. Piper, *J. Inorg. Nucl. Chem.*, **2**, 32 (1955).
11. G. B. Deacon, C. M. Forsyth, R. H. Newnham, and T. D. Tuong, *Aust. J. Chem.*, **40**, 895 (1987).

12. A. J. Nielson, C. E. F. Rickard, and J. M. Smith, *Inorg. Synth.*, **24**, 97 (1986).
13. D. F. Shriver, *The Manipulation of Air-Sensitive Compounds*, McGraw-Hill, New York, 1969.
14. R. B. King, "Transition-Metal Compounds," in *Organometallic Syntheses*, Vol. I, J. J. Eisch and R. B. King (eds.), 1965; Vol. 2, J. J. Eisch, "Nontransition-Metal Compounds," 1981; Academic Press, New York.
15. J. L. Thomas and R. G. Hayes, *J. Organomet. Chem.*, **23**, 487 (1970).
16. G. W. Watt and E. W. Gillow, *J. Am. Chem. Soc.*, **91**, 775 (1969).
17. G. B. Deacon, P. I. MacKinnon, T. W. Hambley, and J. C. Taylor, *J. Organomet. Chem.*, **259**, 91 (1983).

74. BIS(η^5-PENTAMETHYLCYCLOPENTADIENYL)-BIS(TETRAHYDROFURAN)SAMARIUM(II)

$$SmI_2(THF)_2 + 2KC_5Me_5 \longrightarrow Sm(C_5Me_5)_2(THF)_2 + 2KI$$

THF (ligand) and THF (solvent) = tetrahydrofuran

Submitted by WILLIAM J. EVANS* and TAMARA A. ULIBARRI
Checked by HERBERT SCHUMANN† and SIEGBERT NICKEL

In recent years, the organometallic chemistry of the lanthanide metals in low oxidation states has been actively investigated. These low-valent studies have involved the zero-valent metals in the elemental state, using metal–vapor techniques, as well as the complexes of the three lanthanide metals that have divalent states readily accessible under "normal" solution reaction conditions, that is, Eu, Yb, and Sm. Although Sm(II) is the most reactive of these divalent lanthanides [Sm(III) + e → Sm(II): − 1.5 V vs. NHE],[1] its chemistry in organometallic systems had not been investigated previously because the only known divalent organosamarium complexes, $[Sm(C_5H_5)_2(THF)_x]_y$,[2,3] and $[Sm(CH_3C_5H_4)_2(THF)_x]_y$,[4] are insoluble. Recently, however, the first soluble organosamarium(II) complex, $Sm(C_5Me_5)_2(THF)_2$, was synthesized by metal–vapor techniques.[5,6] Although the original synthesis was achieved on a preparative scale, a rotary metal vaporization reactor was required. The following solution synthesis[7] of the title compound is a more generally available route to this soluble divalent organosamarium complex.

---

*Department of Chemistry, University of California, Irvine, Irvine, CA 92717.
†Institut für Anorganische and Analytische Chemie der Technischen Universitat, Berlin, D-1000 Berlin 12, Federal Republic of Germany.

General Procedure

The complexes described below are extremely air and moisture sensitive. Therefore, the syntheses are conducted under nitrogen with rigorous exclusion of air and water by using Schlenk, vacuum-line, and glove-box techniques.[8]

- **Caution.** *Tetrahydrofuran and toluene are harmful if inhaled or absorbed through the skin. They should be handled in a well-ventilated fume hood, and gloves should be worn. THF forms explosive peroxide; only fresh, peroxide-free material should be distilled. Potassium hydride removed from oil suspension and KC_5Me_5 are pyrophoric; they should be manipulated in an inert atmosphere only.*[8] *The compound 1,2-$C_2H_4I_2$ is heat and light sensitive, as well as sublimable.*

Toluene, hexane, and THF are distilled under nitrogen from sodium–benzophenone. The compound C_5Me_5H (Strem, 95%) is dried with 4 Å molecular sieves and degassed by repeated freeze–pump–thaw cycles. The compound KC_5Me_5 is prepared by slowly adding ~5 to 7 g of C_5Me_5H (3% molar excess) to a vigorously stirring suspension of the appropriate amount of KH in ~40 mL of THF. Immediately on addition of the C_5Me_5H, H_2 evolution is evident. The reaction mixture is stirred for 10 h or until H_2 evolution stops. The solution is filtered through a medium-porosity fritted funnel (10–20 μm) to isolate the insoluble white KC_5Me_5. The KC_5Me_5 is washed with three 5-mL aliquots of hexane and dried by rotary evaporation.*
The complex $SmI_2(THF)_2$ is prepared from excess Sm metal (Research Chemicals) and 1,2-$C_2H_4I_2$ in a THF solution.[3,7] In the air in a fume hood, 1,2-$C_2H_4I_2$ (Aldrich, 97%) is dissolved in diethyl ether, washed with a sodium thiosulfate solution, and then washed with water. The ether solution is dried with $MgSO_4$, the ether is removed by rotary evaporation, and 1,2-$C_2H_4I_2$ is recovered as a white powder. To remove all traces of H_2O, the 1,2-$C_2H_4I_2$ is placed in a round-bottomed flask and dried on a vacuum line. Excess samarium pieces are placed in a 500-mL round-bottomed flask equipped with a Teflon stirring bar. Approximately 2–3 g of 1,2-$C_2H_4I_2$ and 50 to 75 mL of THF are added to the samarium. The solution is stirred until it is a homogeneous dark blue color without any trace of the insoluble, yellow, samarium triiodide. The solution is then filtered through a medium-porosity fritted funnel to remove the excess metal pieces, which can be saved and recycled. The dark blue solution is dried by rotary evaporation to yield the dark blue solid, $SmI_2(THF)_2$. This reaction is quantitative in 1,2-$C_2H_4I_2$. The compound $SmI_2(THF)_2$ is very sensitive to oxidation. Solid $SmI_2(THF)_2$

---

*The checkers prepared KC_5Me_5 from C_5Me_5H and KNH_2 in THF or with potassium in 1,2-dimethoxyethane (DME).

is stable if stored under N_2. Solutions should be stored with a small amount of metal present.*

Procedure

In the glove box,† KC_5Me_5 (2.01 g, 11.5 mmol) and $SmI_2(THF)_2$ (3.06 g, 5.58 mmol) are placed in a 125-mL Erlenmeyer flask equipped with a Teflon stirring bar. The powder is mixed thoroughly, and while stirring, 25 mL of THF is added. The mixture is allowed to stir for ~12 h. The purple solution is filtered through a medium-porosity fritted funnel (10–20 μm). The potassium salts are washed with 5 mL of THF and discarded. The 5 mL of wash solution is combined with the filtrate and the THF is removed by rotary evaporation to yield a purple solid. The purple solid is dissolved in 30 mL of toluene, and the resulting solution is allowed to stir for 6 h. The purple solution is filtered through a fine-porosity fritted funnel (4–8 μm). The purple insoluble materials are washed with two 5-mL portions of toluene. The toluene washes are combined with the filtrate, and the toluene is removed by rotary evaporation to yield a greenish-purple solid. The toluene extraction procedure is repeated (this is important to insure that all iodide containing species are excluded). The resulting solid is dissolved in ~5 mL of THF. The THF is removed from the resulting purple solution by rotary evaporation to yield the purple solid, $Sm(C_5Me_5)_2(THF)_2$. Yield: 2.29 g (73%).

Anal. Calcd. for $SmO_2C_{28}C_{46}$: Sm, 26.61. Found: 26.1.

Properties

The compound $Sm(C_5Me_5)_2(THF)_2$ is a purple air-sensitive solid that is soluble in both aromatic hydrocarbons and ethers. The monosolvate, $Sm(C_5Me_5)_2(THF)$, is a green solid; it is obtained by repeated evaporation of toluene solutions of the disolvate. The degree of solvation is easily monitored by integration of the absorptions in the ^1H NMR spectrum in benzene-d_6. The ^1H NMR shifts of these samarium(II) complexes are concentration and solvent dependent. The C_5Me_5 signal is typically located between 2.0 and 3.0 ppm in solutions of $Sm(C_5Me_5)_2(THF)_2$ in benzene-d_6. The two THF signals are found between 12.0–16.0 and 1.0–2.0 ppm. The most common

---

*The checkers recommended that the $SmI_2(THF)_2$ be freshly prepared.

†All of the following procedures were done in the nitrogen atmosphere of a Vacuum/Atmospheres HE-553 Dri-Lab glove box equipped with an oxygen- and water-getter and an atmosphere recirculation system. The checkers used Schlenk technique with 100-mL flasks for the entire procedure.

impurity formed as a result of partial decomposition is $[Sm(C_5Me_5)_2]_2$ $(\mu\text{-}O)$,[9] which has a distinctive ^1H NMR signal in benzene-d_6 at 0.05–0.06 ppm.

References

1. L. R. Morss, *Chem. Rev.*, **76**, 827 (1976).
2. G. W. Watt and E. W. Gillow, *J. Am. Chem. Soc.*, **91**, 775 (1969).
3. J. L. Namy, P. Girard, H. B. Kagan, and P. E. Caro, *Nouv. J. Chim.*, **5**, 479 (1981).
4. W. J. Evans and H. A. Zinnen, unpublished results.
5. W. J. Evans, I. Bloom, W. E. Hunter, and J. L. Atwood, *J. Am. Chem. Soc.*, **103**, 6507 (1981).
6. W. J. Evans, I. Bloom, W. E. Hunter, and J. L. Atwood, *Organometallics*, **4**, 112 (1985).
7. W. J. Evans, J. W. Grate, H. W. Choi, I. Bloom, W. E. Hunter, and J. L. Atwood, *J. Am. Chem. Soc.*, **107**, 941 (1985).
8. D. F. Shriver and M. A. Drezdzon, *The Manipulation of Air-Sensitive Compounds*, 2nd. ed. Wiley, New York, 1986.
9. W. J. Evans, J. W. Grate, I. Bloom, W. E. Hunter, and J. L. Atwood, *J. Am. Chem. Soc.*, **107**, 405 (1985).

75. CHLOROTRIS(η^5-CYCLOPENTADIENYL)COMPLEXES OF URANIUM(IV) AND THORIUM(IV)

$$UCl_4 + 3C_5H_5Tl \xrightarrow{DME} (\eta\text{-}C_5H_5)_3UCl + 3TlCl$$

$$ThCl_4 + 3C_5H_5Tl \xrightarrow{DME} (\eta\text{-}C_5H_5)_3ThCl + 3TlCl$$

Submitted by T. J. MARKS,* A. M. SEYAM,*† and W. A. WACHTER*‡
Checked by G. W. HALSTEAD§ and K. N. RAYMOND§

The compounds $(\eta\text{-}C_5H_5)_3UCl$ and $(\eta\text{-}C_5H_5)_3ThCl$ are useful precursors for the synthesis of a large number of organoactinides,[1] such as tris(cyclopentadienyl)metal alkoxides,[2] tetrahydroborates,[3,4] alkyls and aryls,[5-9] halides,[10] amides,[11] and other derivatives.[12] The chlorouranium compound was first synthesized by Reynolds and Wilkinson[13] in 1956 from uranium tetrachloride and sodium cyclopentadienide in tetrahydrofuran. Vacuum sublimation, in our hands, invariably gives the pure product in low yield. The thorium analog was originally prepared in a similar manner,[14]

---

*Department of Chemistry, Northwestern University, Evanston, IL 60201.
†UNESCO Fellow, on leave from the University of Jordan.
‡NSF Predoctoral Fellow.
§Department of Chemistry, University of California, Berkeley, CA 94720.

except that the rapid rate of attack by thorium tetrachloride on tetrahydrofuran requires that dimethoxyethane be used as solvent. A pure product is again obtained only on sublimation, and in low yield.

We describe here improved syntheses of $(\eta\text{-}C_5H_5)_3UCl$ and $(\eta\text{-}C_5H_5)_3ThCl$. The newer procedures have a number of attractive features. Cyclopentadienylthallium is used as the cyclopentadienylating reagent.[3,15] Unlike cyclopentadienylsodium, it is air stable and can be stored and handled without special precautions. In addition, it generally gives cleaner reactions (less reduction) and is readily separated from the reaction solution by careful filtration.[16] Dimethoxyethane is used as a solvent for both preparations. The above innovations yield products on solvent removal that are sufficiently pure for most synthetic work; sublimation, which is accompanied by extensive thermal decomposition, is unnecessary.

In both syntheses, air and moisture are excluded from reagents at all times. Hence, all operations are performed under dry, prepurified nitrogen, and all glassware is oven-dried, or, when possible, flame-dried. All solvents are distilled from Na/K alloy–benzophenone under nitrogen. For general discussions of techniques and apparatus employed in the synthesis of air-sensitive organometallics, the reader is referred to the excellent books by Shriver[17] and by Eisch and King.[18]

A. CHLOROTRIS(η^5-CYCLOPENTADIENYL)URANIUM(IV)

Procedure

A 1-L three-necked flask is charged with 21.6 g (0.057 mol) of uranium tetrachloride* and a large magnetic stirring bar in a glove bag or glove box.† The flask is fitted with a gas inlet valve and stoppers and is removed to the benchtop. With a strong flow of nitrogen gas through the flask, it is fitted with a condenser also equipped with a gas inlet. Next, the vessel is filled with 700 mL of 1,2-dimethoxyethane, freshly distilled; with strong stirring, 46.0 g (0.171 mol) of cyclopentadienylthallium‡ is added.

---

*The uranium tetrachloride(anhydrous) was purchased from Cerac Chemical Co., Milwaukee, Wisconsin 53201. It can be synthesized by the procedure of J. A. Hermann and J. F. Suttle, *Inorg. Synth.*, **5**, 143 (1957).

†This compound is hygroscopic and should be protected from the atmosphere. Storage in a Schlenk tube is recommended.

‡Cyclopentadienylthallium was purchased from Aldrich Chemical Co., Milwaukee, Wisconsin 53233. It can be easily synthesized by the method of C. C. Hunt and J. R. Doyle, *Inorg. Nucl. Chem. Lett.*, **2**, 283 (1966); H. Meister, *Angew. Chem.*, **69**, 533 (1957) (recommended by the checkers); or A. J. Nielson, C. E. F. Rickard, and J. M. Smith, *Inorg. Synth.*, **24**, 97 (1986) (also in section 78 of this volume).

■ **Caution.** *This compound is toxic and should be handled in a hood.*

The milky-tan suspension is stirred at room temperature for 24 h. After this time, the mixture will be suction-filtered by pouring it through a glass connecting tube into a fritted Schlenk funnel.* The frit assembly rests on the top of a 1-L three-necked flask fitted with a nitrogen inlet valve and a valve connected to a vacuum line. The sintered disk is covered with a layer of diatomaceous earth. The apparatus is evacuated and filled with nitrogen several times before filtration is performed. The reaction mixture is slowly poured into the Schlenk frit, while gentle suction is applied. The filtration residue is then washed with 50 mL of 1,2-dimethoxyethane, and suction is applied again. Next, the dark red-brown filtrate is stripped of solvent via trap-to-trap distillation to yield a golden-brown solid as the product. Under a nitrogen flow, this is scraped to the bottom of the flask and washed with 50 mL of freshly distilled *n*-hexane. The washings are removed with a syringe, and the product is dried for 8 h under high vacuum. Yield 20.0 g (75%).

Anal. Calcd. for $C_{15}H_{15}UCl$: C, 38.45; H, 3.23. Found: C, 38.30; H, 3.31.

Properties

Chlorotris(η-cyclopentadienyl)uranium(IV) is an oxygen-sensitive brown solid. It can be handled in air for brief periods of time with minimal oxidation, which is evidenced by darkening of the color. The compound is soluble in ethereal and aromatic solvents but only sparingly soluble in aliphatic hydrocarbons. Solutions are exceedingly air sensitive. The ^{1}H NMR spectrum in benzene exhibits a sharp singlet 9.60 ppm to high field of the solvent (δ − 1.90). The infrared spectrum (Nujol mull) exhibits typical π-cyclopentadienyl bands at 1013 (m) and 784 (s) cm^{-1}. Oxidation is evidenced by the appearance of the antisymmetric v(OUO) stretch of the uranyl group at 930 cm^{-1}.

B. CHLOROTRIS(η^5-CYCLOPENTADIENYL)THORIUM(IV)

Procedure

In the same manner as described in Section A, a 1-L three-necked flask is charged with 16.0 g (0.043 mol) of anhydrous thorium tetrachloride.† Next,

---

*See Ref. 17, p. 150, for a diagram of a typical setup. Similar glassware can be purchased from Ace Glass, Inc., Vineland, NJ 08360, and from Kontes Glass Co., Vineland, NJ 08360.

† This reagent was purchased from Cerac Chemical Co. or prepared by passing CCl_4 over ThO_2 at 450–500 °C [P. J. Fagan, J. M. Manriquez, E. A. Maatta, A. M. Seyam, and T. J. Marks, *J. Am. Chem. Soc.*, **103**, 6650 (1981)].

the flask is evacuated and gently flamed with a torch.* After cooling, a magnetic stirring bar and 250 mL of freshly distilled 1,2-dimethoxyethane are added to the flask under a strong flush of nitrogen. The flask is then fitted with a condenser equipped with a gas inlet, and, with rapid stirring, 36.0 g (0.133 mol) of cyclopentadienylthallium is added. The mixture is refluxed for 2 days by heating in an oil bath on a hot plate with magnetic stirring. The solvent is then stripped from the tan suspension by trap-to-trap distillation, and the residue is dried under high vacuum for 1 h.

The slate-gray residue is next transferred, in a glove bag or glove box, to an evacuable Soxhlet extractor.† After the apparatus has been evacuated and filled with nitrogen several times, 60 mL of freshly distilled benzene is introduced via syringe into the receiving flask. Next, all inlet valves are closed, and the benzene is frozen in liquid nitrogen. The apparatus is subsequently evacuated to 10^{-3} mm, all valves are closed, and the benzene is thawed. This cycle is repeated, and then extraction is allowed to take place under high vacuum by closing the valves, allowing the apparatus to warm to room temperature, and introducing cool water into the condenser.‡ In some cases it may be necessary to warm the receiving flask to $\sim 40°$. After 2 days, the benzene is removed from the apparatus by trap-to-trap distillation. The resulting flesh-colored extracted solid is scraped to the bottom of the receiving flask and washed with three 10-mL portions of freshly distilled hexane. The washings are removed with a syringe. The product is then dried for 8 h under high vacuum. Yield: 13.0 g (65%) of light-flesh-colored product.

Anal. Calcd. for $C_{15}H_{15}ThCl$: C, 38.90; H, 3.27. Found: C, 38.0; H, 3.00. A purer, white crystalline product in considerably lower yield can be obtained by vacuum sublimation at 175°.

Properties

Chloro(η-cyclopentadienyl)thorium(IV) is a white solid, less oxygen sensitive than the uranium analog but considerably more moisture sensitive. Exposure to air causes the appearance of an ocher color. The compound is less soluble

---

*Samples of thorium tetrachloride are sometimes contaminated with thionyl chloride, which can be removed by this procedure.

†The apparatus employed was similar to Kontes unit K-212800 (Kontes Glass Co., Vineland, NJ 08360), except that both receiving flask and condenser were attached by greaseless O-ring joints. The receiving flask was equipped with a gas inlet and an O-ring stopper. All gas inlets were fitted with Teflon needle valves.

‡Soxhlet extraction can also be carried out at atmospheric pressure in a less sophisticated apparatus. The product may be slightly less pure but is suitable for most preparative purposes.

in all organic solvents than the uranium analog; solutions are extremely air sensitive. The ^{1}H NMR spectrum in benzene-d_6 exhibits a sharp singlet at δ 6.19. The infrared spectrum (Nujol mull) shows π-cyclopentadienyl bands at 1016 (m), 812 (sh), and 788 (s) cm^{-1}.

References

1. (a) T. J. Marks and A. Streitwieser, Jr., in *The Chemistry of the Actinide Elements*, 2nd ed., J. J. Katz, G. T. Seaborg, and L. R. Morss (eds.), Chapman and Hall, London, 1986, Chapter 22; (b) T. J. Marks, *ibid.*, Chapter 23; (c) T. J. Marks, and R. D. Ernst, in *Comprehensive Organometallic Chemistry*, G. W. Wilkinson, F. G. A. Stone, and E. W. Abel (eds.), Pergamon Press, Oxford, U.K., 1982; Chapter 21; (d) T. J. Marks, *Progr. Inorg. Chem.*, **25**, 223 (1979).
2. R. von Ammon, R. D. Fischer, and B. Kanellakopulos, *Chem. Ber.*, **105**, 45 (1972).
3. T. J. Marks and J. R. Kolb, *J. Am. Chem. Soc.*, **97**, 27 (1975).
4. T. J. Marks, W. J. Kennelly, J. R. Kolb, and L. A. Shimp, *Inorg. Chem.*, **11**, 2540 (1972).
5. T. J. Marks and A. M. Seyam, *J. Am. Chem. Soc.*, **94**, 6545 (1972).
6. A. E. Gabala and M. Tsutsui, *J. Am. Chem. Soc.*, **95**, 91 (1973).
7. G. Brandi, M. Brunelli, G. Lugli, and A. Mazzei, *Inorg. Chim. Acta*, **7**, 319 (1973).
8. T. J. Marks, A. M. Seyam, and J. R. Kolb, *J. Am. Chem. Soc.*, **95**, 5529 (1973).
9. T. J. Marks and W. A. Wachter, *J. Am. Chem. Soc.*, **98**, 703 (1976).
10. R. D. Fischer, R. von Ammon, and B. Kanellakopulos, *J. Organomet. Chem.*, **25**, 123 (1970).
11. R. E. Cramer, V. Engelhardt, K. T. Higa, and J. W. Gilje, *Organometallics*, **6**, 41 (1987).
12. J. Takats, in *Fundamental and Technological Aspects of Organo-f-Element Chemistry*, T. J. Marks and I. Fragala (eds.), Reidel Publishing Co.: Dordrecht, The Netherlands, 1985; Chapter 5.
13. L. T. Reynolds and G. Wilkinson, *J. Inorg. Nucl. Chem.*, **2**, 246 (1956).
14. N. TerHarr and M. Dubeck, *Inorg. Chem.*, **3**, 1649 (1964).
15. J. Leong, K. O. Hodgson, and K. N. Raymond, *Inorg. Chem.*, **12**, 1329 (1973).
16. R. B. King, *Inorg. Chem.*, **9**, 1936 (1970).
17. D. F. Shriver, *The Manipulation of Air-Sensitive Compounds*, McGraw-Hill, New York, 1969.
18. J. J. Eisch and R. B. King (eds.), *Organometallic Syntheses*, Vol. 1, Part I, Academic Press, New York, 1965.

Chapter Eight

LIGANDS AND OTHER TRANSITION METAL COMPLEXES

76. TRIMETHYLPHOSPHINE

$$P(OC_6H_5)_3 + 3CH_3MgBr \xrightarrow{Bu_2O} P(CH_3)_3 + 3Mg(OC_6H_5)Br$$

**Submitted by M. L. LUETKENS, Jr.,* A. P. SATTELBERGER,*†
H. H. MURRAY,‡ J. D. BASIL,‡ and J. P. FACKLER, Jr.‡
Checked by R. A. JONES§ and D. E. HEATON§**

Trimethylphosphine is a ligand of proven utility in organometallic chemistry. However, because of its expense when purchased from commercial sources (Strem Chemicals) or its poor[1] to moderate[2] yields when isolated as the silver iodide adduct [AgI(PMe$_3$)]$_4$, its potential is not fully realized. Frequently, large quantities of the phosphine are required, for example, as a reactive solvent[3] or in the field of lanthanide and actinide chemistry, wherein the lability of the phosphine ligand may require crystallization from neat trimethylphosphine.[4] Similar considerations apply in exploratory synthetic early transition metal chemistry.[5] In transition metal ylide chemistry,[6] access to quantities of PMe$_3$ is also very desirable, particularly when excess ylide is required. The volatility of PMe$_3$ facilitates work-up of reaction mixtures and

---

*Department of Chemistry, The University of Michigan, Ann Arbor, MI 48109.
†Present address: Los Alamos National Laboratory, Los Alamos, NM 87545.
‡Department of Chemistry, Texas A & M University, College Station, TX 77843.
§Department of Chemistry, University of Texas, Austin, TX 78712.

the ligand provides convenient proton NMR signals for product characterization. Trimethylphosphine derivatives usually have good solubility properties and show enhanced crystallizability relative to analogs containing other tertiary phosphines.

The following procedure is a hybrid of several trimethylphosphine syntheses.[1a, 7, 8] The key features of the new procedure are the use of triphenyl phosphite[8] in place of phosphorus(III) halides[1a, 7] and dibutyl ether in place of diethyl ether.[1a, 9] The present procedure has the following advantages over the preparations reported previously in *Inorganic Syntheses*:[2, 7] (a) large quantities (~ 1 mol) of PMe_3 can be made in 1 day in a "one-pot synthesis" from an easily handled and fairly innocuous and inexpensive phosphorus compound, $P(OC_6H_5)_3$; and (b) the lower boiling PMe_3 (bp 39–40°C) is easily separated from the reaction mixture (dibutyl ether, bp 142°C) by distillation at atmospheric pressure giving the phosphine in high yield and purity.

Starting Materials

Dibutyl ether (Aldrich, 99%) is dried and freed from dissolved molecular oxygen by distillation under nitrogen from a solution of the solvent and sodium benzophenone ketyl.

■ **Caution.** *Dibutyl ether should be stored under nitrogen or argon, over sodium wire to prevent the formation of peroxides.*

Triphenyl phosphite (Aldrich, 97%) is purified by fractional vacuum distillation (bp 181–183°C, 1 torr) and stored in the dark under nitrogen. A 2 M solution of CH_3MgBr in dibutyl ether may be purchased from Alfa Inorganics, or prepared according to the following procedure. If the CH_3MgBr solution is prepared in the laboratory, bromomethane (Air Products) is used without further purification.

Procedure

■ **Caution.** *Bromomethane and trimethylphosphine are volatile, toxic materials. This reaction must be carried out in an efficient fume hood. Trimethylphosphine may ignite spontaneously in air^2 and must be handled under an inert atmosphere at all times.*

All operations are performed under an atmosphere of prepurified nitrogen. An oven-dried, 3-L, three-necked, round-bottomed flask equipped with an N_2 gas inlet, a precision mechanical stirrer, and a jacketed 250-mL pressure-equalizing dropping funnel (Fig. 1) vented through an oil bubbler, is charged with Grignard grade magnesium turnings (101 g, 4.15 mol) and ~ 1.5 L of purified dibutyl ether under a countercurrent of N_2. The mixture is then

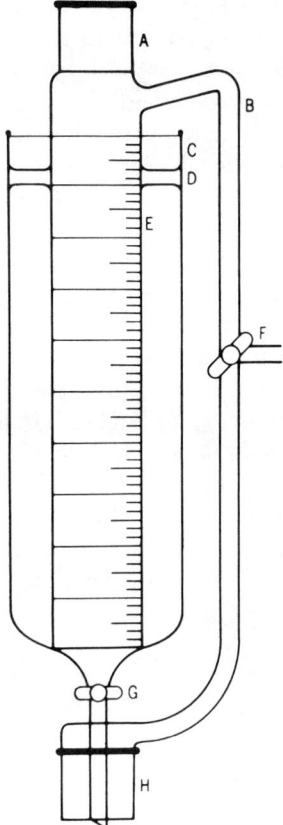

Fig. 1. Jacketed 250-mL pressure-equalizing dropping funnel: (A) 24/40 female ground joint, (B) 10-mm glass tubing, (C) 80-mm tubing, (D) solid glass braces, (E) 250-mL graduated cylinder, (F) three-way, 2-mm glass stopcock, (G) 2-mm glass stopcock, (H) 24/40 male ground joint.

cooled to 0°C in a large tub of ice water. The jacket of the dropping funnel is filled with a mixture of powdered Dry Ice–acetone, and bromomethane (200 mL, 365.6 g, 3.85 mol) is carefully poured into the dropping funnel. If a metal gas cylinder equipped with a manual control (needle) valve is the source of bromomethane, liquid can be withdrawn by *inverting* the cylinder. Tygon tubing is a suitable material for liquid transfer to the dropping funnel. The cold bromomethane is added dropwise to the stirred suspension. The Grignard reaction usually starts after the first 5–10 mL of bromomethane is added. If it does not, a few small crystals of iodine should be added before resuming the addition of bromomethane. The remaining bromomethane is added after the Grignard reaction has been initiated, as evidenced by a shiny surface appearing on the magnesium turnings. The addition of bromomethane should be carried out over a 2-h period.

After the addition of bromomethane is complete, the dropping funnel is replaced with a stopper, the ice tub is removed, and the Grignard solution is stirred at room temperature for 4 h (or overnight if this is more convenient). The stopper is then replaced with a 500-mL pressure-equalizing dropping funnel and the white Grignard suspension is cooled to $\sim -5°C$ with a large ice–salt bath. A solution of triphenylphosphite (300 mL, 355.2 g, 1.145 mol) and dibutyl ether (300 mL) is added dropwise to the Grignard solution, with vigorous stirring, over the course of 2.5 h. The ice–salt bath should be replenished periodically to maintain the temperature of this exothermic reaction below $\sim 10°C$. After the phosphite addition is complete, the cooling bath is removed and the dropping funnel is replaced with an N_2 purged distillation assembly (Fig. 2) under N_2 flow. The latter is connected via a glass "tee" to the nitrogen source and a mineral oil bubbler. A slow flow of N_2 is maintained during the following operation. The 200-mL receiving flask is cooled to $-78°C$ (Dry Ice–acetone) and the contents of the reaction flask are brought to a *gentle* reflux using a 3-L heating mantle. Ideally, at the early

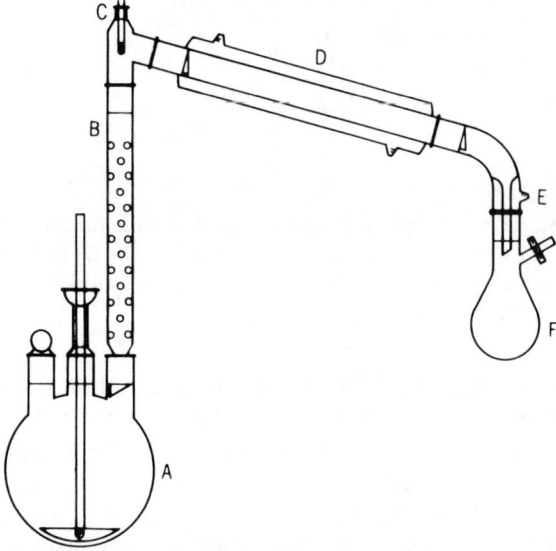

Fig. 2. First distillation assembly: (A) 3-L reaction flask, (B) 15-in Vigreux column, (C) 24/40 distillation head with 150°C thermometer, (D) water-cooled condenser, (E) to glass "tee," which is also connected to an N_2 source and a mineral oil bubbler, (F) 200-mL Schlenk receiver flask.

stages of the distillation, the butyl ether condenses about one-quarter of the way up the Vigreux column. A "cloud" of trimethylphosphine vapor appears on the upper part of the column, condenses, and collects in the receiver.

Product is collected until the still head temperature reaches 110°C. At this point the heating mantle is removed, the nitrogen flow is increased slightly, and the distillation flask is allowed to cool. When the latter is cool enough to touch, the receiving flask is carefully disconnected under a positive N_2 flow, and capped. The *warm* residue in the distillation flask (which solidifies on cooling to room temperature) may be poured into waste solvent bottles for disposal. The Schlenk flask, which contains ~150 mL of liquid (PMe_3 and Bu_2O), is fitted with a magnetic stirring bar and a clean N_2-purged distillation assembly, which consists of a 6-in. Vigreux column, a distillation head with a thermometer [a Kontes Vigreux Distillation Head (K-287450) works well here], a water-cooled condenser, and a 100-mL Schlenk receiving flask. As before, the distillation assembly is connected via a "tee" to the nitrogen source and a mineral-oil bubbler. The flask is placed in an oil bath (atop a stirrer hot plate) and the receiver is cooled to $-78°C$. The oil bath is heated to 60°C and pure trimethylphosphine distills at 39–40°C from the stirred solution. The temperature of the oil bath is slowly raised to ~110°C during the latter stages of the distillation. When the distillation is complete, the oil bath is removed, the nitrogen flow is increased slightly; and the distillation flask is allowed to cool to room temperature. At this point, the receiver is carefully disconnected under a positive N_2 flow and tightly capped. Any residual trimethylphosphine in the distillation flask is destroyed by adding Clorox bleach (dilute aqueous sodium hypochlorite). The yield of PMe_3 is typically 70–74 g or 80–85% based on $P(OC_6H_5)_3$. The ^1H NMR (C_6D_6, 360 MHz): δ 0.79 (d, $^2J_{HP} = 2.75$ Hz).

Properties

Trimethylphosphine is a colorless, pyrophoric[2] liquid with a very unpleasant odor. It forms air-stable phosphonium salts with a variety of alkyl halides, and the corresponding oxide and sulfide on reaction with O_2 and sulfur.

Trimethylphosphine may be stored as the air-stable silver iodide complex $[AgI(PMe_3)]_4$.[7] Other physicochemical properties are described in previous volumes of this series.[2,7]

References

1. (a) A. B. Burg and R. I. Wagner, *J. Am. Chem. Soc.*, **75**, 3872 (1953); (b) J. G. Evans, P. L. Goggin, R. J. Goodfellow, and J. G. Smith, *J. Chem. Soc. A*, **1968**, 464.
2. R. T. Markham, E. A. Dietz, Jr., and D. R. Martin, *Inorg. Synth.*, **16**, 153 (1974).

3. V. C. Gibson, C. E. Graimann, P. M. Hare, M. L. H. Green, J. A. Bandy, P. D. Grebenik, and K. Prout, *J. Chem. Soc. Dalton Trans.*, **1985**, 2025.
4. P. G. Edwards, R. A. Andersen, and A. Zalkin, *J. Am. Chem. Soc.*, **103**, 7792 (1981).
5. (a) G. A. Rupprecht, L. W. Messerle, J. D. Fellmann, and R. R. Schrock, *J. Am. Chem. Soc.*, **102**, 6236 (1980); (b) J. D. Fellmann, R. R. Schrock, and G. A. Rupprecht, *J. Am. Chem. Soc.*, **103**, 5752 (1981); (c) A. P. Sattelberger, R. B. Wilson, Jr., and J. C. Huffman, *J. Am. Chem. Soc.*, **102**, 7111 (1980); (d) M. L. Luetkens, Jr., J. C. Huffman, and A. P. Sattelberger, *J. Am. Chem. Soc.*, **105**, 4474 (1983); (e) K. W. Chiu, G. G. Howard, H. S. Rzepa, R. N. Sheppard, G. Wilkinson, A. M. R. Galas, and M. B. Hursthouse, *Polyhedron*, **1**, 441 (1982); (f) G. S. Girolami, V. V. Mainz, R. A. Andersen, S. H. Vollmer, and V. W. Day, *J. Am. Chem. Soc.*, **103**, 3953 (1981); (g) P. R. Sharp and R. R. Schrock, *J. Am. Chem. Soc.*, **102**, 1430 (1980).
6. H. Schmidbaur and R. Franke, *Inorg. Chim. Acta*, **13**, 79 (1975).
7. R. Thomas and K. Eriks, *Inorg. Synth.*, **9**, 59 (1967).
8. W. Wolfsberger and H. Schmidbaur, *Synth. Inorg. Met. Org. Chem.*, **4**, 149 (1974).
9. P. R. Sharp, Ph.D. thesis, Massachusetts Institute of Technology, 1980.

77. PHOSPHORUS TRIFLUORIDE

$$SbF_3 + PCl_3 \xrightarrow{SbCl_5} PF_3 + SbCl_3$$

Submitted by RONALD J. CLARK* and HELEN BELEFANT*
Checked by STANLEY M. WILLIAMSON†

Phosphorus trifluoride is a ligand that is used extensively in coordination chemistry. It substitutes readily into various metal carbonyl complexes using either thermal or photochemical techniques. As a ligand, it is unique in its similarity to carbon monoxide in lower-valent organometallic compounds. In its role as a model for CO, a number of studies are possible that cannot be done on the carbonyls themselves.[1] The name normally used for PF_3 in complexes is trifluorophosphine.

Although PF_3 is inherently an inexpensive material, the fact that it currently has no industrial uses results in a high cost from speciality suppliers. It can be prepared in a number of ways, almost all of which are based on the interaction of phosphorus trichloride with some mild fluorinating agent.[2] A convenient procedure is the interaction of PCl_3 with solid ZnF_2.[3] However, the yield of PF_3 is strongly dependent on the reactivity of the ZnF_2, which can vary greatly depending on the process of manufacture. The use of AsF_3 as fluorinating agent has obvious disadvantages particularly when sizable quantities of PF_3 are needed.[4] The use of NaF as the fluorinating agent has

---
*Department of Chemistry, Florida State University, Tallahassee, FL 32306.
†Department of Chemistry, University of California, Santa Cruz CA 95064.

many attractive features but the reaction does not progress with consistent rates, yields varying from 20 to 80% in 48 h. Another method found in the literature is to bubble HF gas through a tower of liquid PCl_3.[5] However, one not only has to scrub 3 mol of HCl from every mole of the PF_3, but the use of HF is not desirable except in the most skilled of hands. (Hydrogen chloride and PF_3 have almost identical boiling points so they cannot be separated by vacuum techniques.)

Our resolution to this dilemma has been to work out a compact procedure based upon the fluorinating agent SbF_3 catalyzed by $SbCl_5$ using acetonitrile as a solvent (i.e., the Swarts reaction).[6]

Procedure

■ **Caution.** *Because of the scale of the reaction, care must be exercised regarding free flow of gas through the apparatus, especially the collection of product in the cold trap. Mineral oil bubblers are used to monitor gas flow and ensure that the product collecting in the cold trap has not plugged the flow.*

Because of the toxicity of PF_3, this reaction should be carried out in a well-ventilated hood.

The heart of the apparatus is a 1-L, three-necked flask equipped with a dropping funnel, gas purge tube, a low-temperature reflux cold finger, and a strong magnetic stirrer (see Fig. 1). Alternatively, a motor-driven stirrer and a gastight gland can be used. The dropping funnel is arranged so that the PCl_3 and gas purge enter below the surface of the solvent. Following the cold finger are two traps cooled to $-78°C$, a 4 Å molecular sieve trap ~ 40 cm in length, and a final $-196°C$ product trap.

Oil-filled bubblers are placed at the start of the gas train and after the PF_3 condenser to monitor the gas flow. About 300 mL of acetonitrile, from a fresh bottle with a water content of no more than 0.05%, is admitted to flask. The acetonitrile can be dried over 3 Å molecular sieves. As purge gas, helium is preferred but nitrogen can also be used. Under a gas purge, 150 g (0.84 mol) of SbF_3 is dropped in over a period of a few minutes with vigorous stirring. If the order of addition is reversed, the SbF_3 tends to *form clumps and resists* stirring. Three milliliters of (0.02 mol) $SbCl_5$ is added. The cold finger is cooled to $\sim -40°C$ by the judicious addition of Dry Ice to acetone. A suitable thermometer or thermocouple device should be used to monitor this bath temperature. Finally, 60 mL (95 g, 0.69 mol) of PCl_3 is placed in the dropping funnel and is added to the flask over the course of ~ 1 h. If excess SbF_3 is used and the PCl_3 is introduced below the surface of the solvent, the tendency to form $PClF_2$ and PCl_2F is kept quite low. These by-products are monitored by chloride test of the product,[7] which should be negative (i.e., no visible chloride in the test).

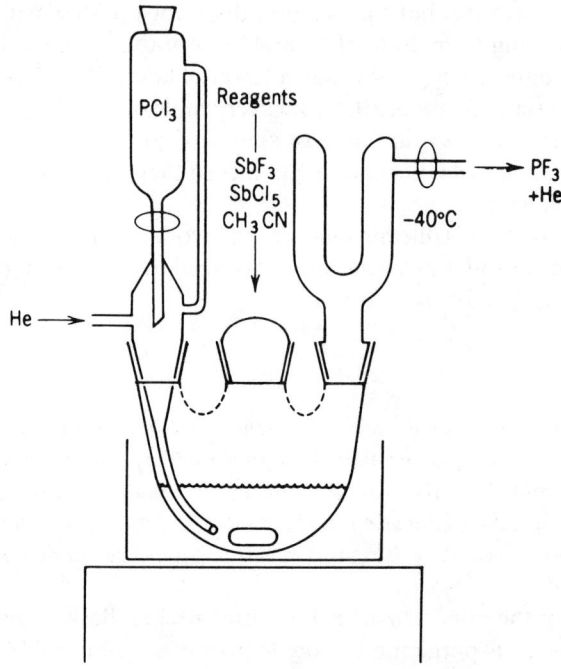

Fig. 1. Apparatus for the fluorination of PCl_3.

The reaction is mildly exothermic; a room temperature water bath should be placed around the reaction flask to control the reaction temperature and particularly the vapor pressure of CH_3CN to reduce the problems of product clean-up. The PF_3 collects as a fluffy solid in the $-196°C$ trap and occupies a volume of about five times its liquid volume. It is essential that this trap be fairly large. Typical dimensions for the large trap are 27×5 cm; the inner tube reaches no closer than 10 cm of the bottom. The acetonitrile traps are 13×2.5 cm. In addition, it is extremely important that gases come in the sidearms of all the traps rather than the central tube in order to avoid plugging. This can occur either with CH_3CN transported into the $-78°C$ trap, or with PF_3 collecting in the $-196°C$ trap. Liquid nitrogen should be added to the $-196°C$ trap frequently and in small increments to avoid large pressure surges.

If nitrogen is being used as a purge gas, a moderate amount becomes trapped in the lattice of the frozen PF_3 at $-196°C$ and its removal necessitates a series of subsequent freeze–pump–thaw cycles. After completion of the PCl_3 addition, a gas purge of 20 min is used to sweep the PF_3

out of the reaction flask into the $-196°C$ trap. The PF_3 is then isolated from the rest of the system without the removal of the liquid nitrogen. Then it is connected to a vacuum system and the helium or nitrogen pumped off. Once it is under vacuum, the liquid nitrogen can be removed and the snowy PF_3 allowed to become warm enough to melt. It is advisable to monitor the pressure during the thaw cycle, however, PF_3 has wide liquid range (-152 to $-102°C$), and will not build too much pressure as long as some solid PF_3 is in evidence, or shortly after the disappearance of the last solid PF_3. The liquid should be refrozen by liquid nitrogen. If the system manometer returns to zero pressure, the product is free of purge gas. If not, additional freeze–pump–thaw cycles are needed. Yield: ~ 50 g or 83%.

If the PF_3 is not to be used immediately, it can be condensed into a metal–pressure vessel at liquid nitrogen temperature using vacuum techniques.

■ **Caution.** *Only stainless steel (i.e., 316 SS) or aluminum pressure vessels approved for cryogenic service should be used. The use of lecture bottles of mild steel must be avoided, since such lecture bottles lose their strength at cryogenic temperatures.*

The $-78°C$ trap and molecular sieve trap are designed to remove CH_3CN. Since traces of moisture are inevitably present, this means that HCl is also likely to be produced; the molecular sieve trap will also remove it. An alternate method of purification is to pass the product through a water tower packed with glass beads as shown in Fig. 2.

The HCl dissolves quickly in the water but the PF_3 hydrolysis is slow. The product is then given a final purification from water and other materials by a vacuum distillation from a methylcyclohexane slush bath ($-128°C$). However, the molecular sieve trap works better and avoids the 10–20% loss that occurs from PF_3 hydrolysis. The product should give a negative chloride test[7] indicating absence of HCl and/or by-products $PClF_2$ or PCl_2F.

Properties

The gas-phase infrared (IR) spectrum contains two strong vibrational stretching bands in the region from 800 to 1000 cm^{-1}, which are indicative of PF_3. However, the gas-phase IR absorptivity of the P—F bands is so much stronger than the bands of impurities such as HCl and CH_3CN that IR spectroscopy is not a good method to check purity unless one is careful. To detect the presence of HCl or CH_3CN at approximately the 1% level by IR requires gas pressures of ~ 700 torr in the 10-cm cell. At that pressure, the P—F stretching bands are totally off scale and various broad combination and overtone bands are seen as strong absorptions in the regions of 1700 and 1200 cm^{-1}.

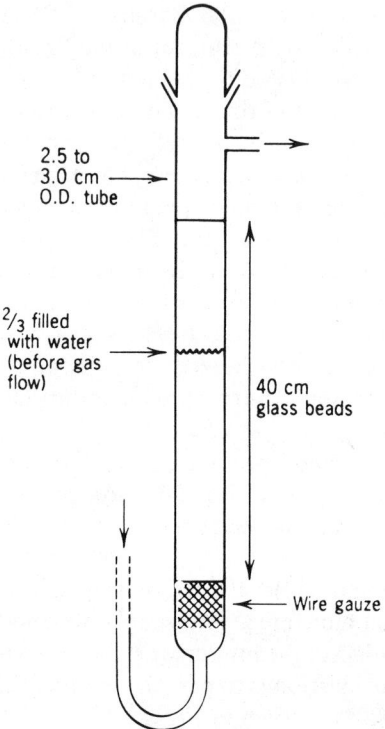

Fig. 2. Apparatus for alternate method of purification of PF_3.

The gas has a boiling point of $-102°C$ and a freezing point of $-152°C$. In high-pressure cylinders, the deviation from PVT ideality is quite significant. At higher pressures, room temperature gauge readings are about one-half to two-thirds of what one would calculate based on the mass of PF_3 present. Thus pressure is not a good measure of yield unless the data are carefully corrected.

Phosphorus trifluoride does not have a particularly sharp odor that one would expect of an acid gas until the concentration becomes higher than is safe. It combines with hemoglobin[8] in a manner comparable to carbon monoxide. It should also be assumed that the subsequent hydrolysis products are highly undesirable in the body. Thus, one is strongly advised to work with PF_3 only with a good hood system or a well-designed vacuum system. Pressure storage vessels should be examined for slow leaks before use in storage of PF_3.

References

1. R. J. Clark and M. A. Busch, *Acc. Chem. Res.*, **6**, 246 (1973).
2. R. Schmutzler, *Advances in Fluorine Chemistry*, Vol. 5, M. Stacey, J. C. Tatlow, and A. G. Sharpe (eds.), Butterworths, Washington, DC, 1965, p. 37.
3. A. A. Williams, *Inorg. Synth.*, **5**, 95 (1957).
4. C. J. Hoffman, *Inorg. Synth.*, **4**, 149 (1953).
5. G. Brauer, *Handbook of Preparative Inorganic Chemistry*, Vol. 1, 2nd ed., Academic Press, New York, 1963, p. 189.
6. F. Swarts, *Bull. Acad. R. Belg.*, **24**, 309 (1892).
7. The chloride test is performed by bubbling a sample of the product into a $AgNO_3/HNO_3$ solution. Alternately, vacuum techniques can be used to transfer a sample of the product into a flask followed by the admission of the silver nitrate solution. T. L. Brown and H. E. LeMay. *Qualitative Inorganic Chemistry*, Prentice Hall, Englewood Cliffs, NJ, 1984.
8. J. W. Irvine and G. Wilkinson, *Science*, **113**, 742 (1951).

78. CYCLOPENTADIENYLTHALLIUM

$$Tl_2SO_4 + 2NaOH \rightarrow 2TlOH + Na_2SO_4$$

$$2TlOH + 2C_5H_6 \rightarrow 2TlC_5H_5 + 2H_2O$$

Submitted by A. J. NIELSON,[*] C. E. F. RICKARD,[*] and J. M. SMITH[*]
Checked by BRUCE N. DIEL[†]

Reagents used for introducing the cyclopentadienyl ligand (Cp) into metal complexes[1] often give low yields of product or suffer from difficulties in handling. The frequently used and easily prepared NaCp in tetrahydrofuran solution[2] is extremely sensitive to traces of oxygen and does not store well. Product yields are usually lower with $CpMgBr$[3] or Cp_2Mg,[4] both of which are air sensitive, with the latter involving a tedious preparation. Biscyclopentadienylmercury[5] has low thermal stability and does not store well.

Cyclopentadienylthallium[6] is a superior reagent for many syntheses. It is easily prepared and stored, it may be handled in air, and the insolubility of thallium halide leads to high yields of product that may be filtered easily and worked up. The reagent has been used for preparing a variety of actinide,[7] lanthanide,[8] and transition metal[9] cyclopentadiene complexes.

■ **Caution.** *Thallium compounds are extremely toxic. Avoid inhalation of dust and contact with skin. All operations should be carried out in a fume hood with the use of neoprene gloves. Wastes should be stored in bottles or disposed of*

---

[*] Department of Chemistry, University of Auckland, Private Bag, Auckland, New Zealand.
[†] Department of Chemistry, University of Idaho, Moscow, ID 83843.

by thoroughly mixing with a large amount of sand and burying the mixture in a safe, open area.

Procedure

Thallium(I) sulfate (25 g, 0.0495 mol) and sodium hydroxide (8 g, 0.2 mol) are dissolved in 250 mL of water in a 500-mL round-bottomed flask. Freshly cracked cyclopentadiene[2,10] (8.2 mL) is added, the flask is stoppered, and the solution is stirred magnetically for 12 h. The brown precipitate is removed by filtering, washed with 50 mL of water followed by 50 mL of methanol, and dried under vacuum for 2 h. The solid is transferred to a water-cooled sublimation apparatus. A plug of glass wool is placed on top, and the TlCp is sublimed at 90–100° at 10^{-3} torr. The sublimation should be continued until no further product sublimes onto the cold finger after the bulk of the TlCp has been scraped away. A brown powdery residue remains after the sublimation is completed. Yield: 24.4 g, (91%*).

Anal. Calcd. for C_5H_5Tl: C, 22.3; H, 1.9. Found: C, 22.5; H, 2.2.

Properties

Cyclopentadienylthallium is a light yellow solid that decomposes at 230°.[6] It is stable for several months if stored under N_2 in a Schlenk flask kept in the dark. Slow decomposition takes place in the air and light, whereupon the solid turns brown, but pure product may be sublimed away from this material. The IR spectrum shows absorption bands at 3022, 1584, 995, and 725 cm^{-1}. Cyclopentadienylthallium is insoluble in most organic solvents. It may be handled in air, and it is normally heated at reflux in THF solution with the substrate for about 2 h under an inert gas atmosphere. When allowed to react with metal halides, the insoluble thallium halide produced may be filtered from the reaction solution. Most cyclopentadienyl complexes are air- and moisture-sensitive and should be handled with appropriate techniques.[11]

References

1. J. M. Birmingham, *Adv. Organometal. Chem.*, **2**, 365 (1964).
2. G. Wilkinson, *Org. Synth.*, **36**, 31 (1956).
3. G. Wilkinson, F. A. Cotton, and J. M. Birmingham, *J. Inorg. Nucl. Chem.*, **2**, 95 (1956).
4. W. A. Barber, *Inorg. Synth.*, **6**, 11 (1960).

*Checker obtained 94.4% yield.

5. G. Wilkinson and T. S. Piper, *J. Inorg. Nucl. Chem.*, **2**, 32 (1956).
6. H. Meister, *Angew. Chem.*, **69**, 533 (1957).
7. P. Zanella, S. Faleschini, L. Doretti, and G. Faraglia, *J. Organometal. Chem.*, **26**, 353 (1971); **43**, 339 (1972).
8. E. O. Fischer and H. Fischer, *J. Organometal. Chem.*, **3**, 181 (1965).
9. L. E. Manzer, *Inorg. Synth.*, **21**, 84 (1982).
10. R. B. King and F. G. A. Stone, *Inorg. Synth.*, **7**, 99 (1963).
11. D. F. Shriver, *The Manipulation of Air-Sensitive Compounds*, McGraw-Hill, New York, 1969.

79. 1,2,3,4,5-PENTAMETHYLCYCLOPENTADIENE

Submitted by JUAN M. MANRIQUEZ,* PAUL J. FAGAN,* LARRY D. SCHERTZ,* and TOBIN J. MARKS*
Checked by JOHN BERCAW† and NANCY MCGRADY†

The 1,2,3,4,5-pentamethylcyclopentadienyl ligand is a complexing agent of great utility in main-group,[1,2] transition metal,[3-5] and f-element[6] organometallic chemistry. In comparison to the unsubstituted cyclopentadienyl ligand, it is a greater donor of electron density[3,4] while imparting considerably enhanced solubility and crystallizability to the metal complexes it forms. In several cases, peralkylcyclopentadienyl incorporation also increases the thermal stability of metal-to-carbon σ-bonded derivatives.[4,7] The latter characteristic appears to reflect the well-known instability of the unsubstituted η^5-C_5H_5 moiety with respect to ring hydrogen atom transfer.[4,6d] The steric influence of the pentamethylcyclopentadienyl ligand is also important, and for actinides it leads to coordinative unsaturation by restricting the number of cyclopentadienyl groups that can be bound to the metal ion.[6]

The first detailed study of pentamethylcyclopentadienyl organometallic chemistry was carried out by King and Bisnette.[3] Pentamethylcyclopentadiene was prepared by the laborious (6 steps) and expensive procedure of de Vries.[8] This method was latter improved somewhat by substituting CrO_3–pyridine for MnO_2 in the step involving oxidation of 3,5-dimethyl-2,5-heptadien-4-ol(di-sec-2-butenylcarbinol).[4a] More recently, Burger, et al. reported two alternative syntheses of pentamethylcyclopentadiene. The shortest approach (3 steps) begins with expensive hexamethyl Dewar benzene, while the second procedure involves 5 steps and proceeds in rather low overall yield. A three-step preparation by Feitler and Whitesides[10] is more efficient than any of the above methods. Even simpler and more straightforward is the adaptation of the procedure of Sorensen et al.[4c] by Threlkel and

*Department of Chemistry, Northwestern University, Evanston, IL 60201.
†Department of Chemistry, California Institute of Technology, Pasadena, CA 91125.

Bercaw,[4d] which begins with a commercially available mixture of *cis*- and *trans*-2-bromo-2-butene. The present approach (shown below) is based on the latter procedure but introduces several important modifications which result both in increased economy and the possibility of larger-scale syntheses.

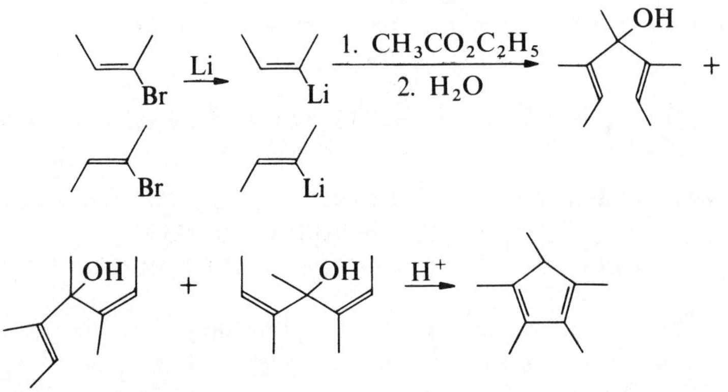

$$\text{(2-bromo-2-butene)} \xrightarrow{\text{Li}} \text{(2-lithio-2-butene)} \xrightarrow[\text{2. H}_2\text{O}]{\text{1. CH}_3\text{CO}_2\text{C}_2\text{H}_5} \text{(dienol)} + \text{(dienol)} \xrightarrow{\text{H}^+} \text{(pentamethylcyclopentadiene)}$$

Thus, reduced quantities of solvents, different addition procedures for reagents, and streamlined work-up methods are employed.

Procedure

■ **Caution.** *2-Bromo-2-butene and p-toluenesulfonic acid are considered hazardous. Avoid skin contact and inhalation.*

A mixture of *cis*- and *trans*-2-bromo-2-butene was prepared from *cis*- and *trans*-2-butene (Matheson Co., East Rutherford, NJ) by the literature procedure.[11] (Alternatively, this mixture can be purchased from Pfaltz and Bauer, Inc., Stanford, CT; for large-scale syntheses, it is more economical to prepare it.) A 5-L three-necked reaction flask equipped with a 500-mL pressure-equalizing addition funnel, reflux condenser, efficient magnetic stirring bar, and gas inlet is heated with a hot air gun while flushing with argon. (For further scale-up, a mechanical stirrer is recommended.) Next, 58.0 g (8.36 mol) of 3.2-mm-diameter Li (0.02% Na) wire (Alfa Division, Ventron Corp., Danvers, MA) is cut into the flask in ~5-mm pieces under a flush of argon. Then, 1600 mL of diethyl ether (freshly distilled from Na/K–benzophenone) is added to the flask under argon. The composition of the sodium-potassium alloy is ~3:1 K/Na by weight. The addition funnel is charged with 120 g (0.88 mol, 90.4 mL) 2-bromo-2-butene, which has been dried over Davison 4 Å molecular sieves. A 10-mL aliquot of 2-bromo-2-butene is added to the lithium–ether mixture with stirring. As the reaction begins,* cloudiness due to suspended LiBr should be observable. The 2-bromo-2-butene is added dropwise at a rate sufficient to maintain reflux of the diethyl ether.

After the addition of the 2-bromo-2-butene is complete (~ 1.5 h), a mixture of 166 g (1.88 mol, 184 mL) of ethyl acetate (dried over Davison 4 Å molecular sieves) and 430 g (3.18 mol, 324 mL) 2-bromo-2-butene is added dropwise over a period of 4–5h, maintaining reflux. Since the reaction rate often slows markedly midway through the addition, the addition rate should be increased if necessary and then slowed once the reflux becomes vigorous again. After the addition of this mixture is complete, stirring is continued for 15 min. Next, the addition funnel is charged with a further 50 g (55.4 mL) of dry ethyl acetate, and only a portion of this reagent (usually 10–20 g) is added dropwise to the reaction mixture until refluxing of the ether ceases. No further ethyl acetate should be added. After the reaction mixture has cooled for 4 h, 1200 mL of a saturated NH_4Cl solution is added dropwise to hydrolyze the remaining lithium. The diethyl ether layer is isolated using a separatory funnel, and the aqueous layer is extracted with three 200-mL portions of diethyl ether. The diethyl ether solutions are combined and concentrated to ~ 350 mL on a rotary evaporator.

The diethyl ether concentrate is added to a slurry of 36 g p-toluenesulfonic acid monohydrate (Aldrich Chemical Co., Milwaukee, WI) in 500 mL of diethyl ether contained in a 2-L, three-necked flask equipped with a reflux condenser and magnetic stirring bar. The rate of addition should be sufficiently slow that the diethyl ether gently refluxes. The reaction mixture is stirred for 5 min after refluxing ceases and is then poured into 1200 mL of a saturated $NaHCO_3$ solution containing 19 g of Na_2CO_3. The yellow aqueous phase is removed and extracted with three 200-mL portions of diethyl ether. The combined diethyl ether solutions are dried over Na_2SO_4 and then concentrated to ~ 250–300 mL on a rotary evaporator. The crude product is trap-to-trap distilled *in vacuo* [with the aid of a warm water bath (35–40°)], and the resulting yellow liquid (85% pure by gas chromatography) is then vacuum-distilled under N_2 using a 50-cm Vigreux column. The fraction boiling from 65–70°, 20 torr is collected to yield 142 g (53% overall yield based on ethyl acetate) 1,2,3,4,5-pentamethylcyclopentadiene as a pale-yellow liquid [92% pure by gas chromatography (GC)]. An additional fraction boiling at 70–75°, 20 torr can also be collected (85% pure by GC) and represents an additional 15 g (5%) of product.†

---

*Experience has shown that rigorously anhydrous and anaerobic conditions are necessary for the success of this reaction. All reagents must be dried appropriately and degassed.

† The checkers report three fractions selected in this sequence: 46 g (67% purity), 129 g (88% purity), and 12 g (74% purity).

Properties

1,2,3,4,5-Pentamethylcyclopentadiene is a colorless to pale-yellow liquid with a sweet olefinic odor. It should be stored under nitrogen in a freezer. The ^{1}H NMR spectrum (CCl$_4$) exhibits a multiplet at $\delta = 2.4$ (1H), a broadened singlet at $\delta = 1.75$ (12H), and a sharp doublet at $\delta = 0.95$ (3H, $^3J_{H-H} = 8$ Hz). The infrared spectrum (neat liquid) exhibits significant transitions at 2960 (vs), 2915 (vs), 2855 (vs), 2735 (w), 1660 (m), 1640 (w), 1390 (s), 1355 (m), 1150 (w), 1105 (mw), 1048 (w), 840 (mw), 668 (w) cm^{-1}. For the preparation of metal complexes, pentamethylcyclopentadiene can be converted to the lithium reagent by treatment with lithium alkyls[4,10] or to the Grignard reagent by refluxing with isopropyl magnesium chloride in toluene.[6c]

Acknowledgements

This research was supported by NSF Grants CHE76-84494A01 and CHE8009060.

References

1. A. Davison and P. E. Rakita, *Inorg. Chem.*, **9**, 289 (1970).
2. P. Jutzi, F. Kohl, and C. Krüger, *Angew. Chem. Int. Ed.*, **18**, 59 (1979).
3. R. B. King and M. B. Bisnette, *J. Organometal. Chem.*, **8**, 287 (1967).
4. (a) J. E. Bercaw, R. H. Marvich, L. G. Bell, and H. H. Brintzinger, *J. Am. Chem. Soc.*, **94**, 1219 (1972); (b) J. E. Bercaw, *J. Amer. Chem. Soc.*, **96**, 5087 (1974); (c) P. H. Campbell, N. W. K. Chiu, K. Deugan, I. J. Miller, and T. S. Sorensen, *J. Am. Chem. Soc.*, **91**, 6404 (1969); (d) R. S. Threlkel and J. E. Bercaw, *J. Organometal. Chem.*, **136**, 1 (1977) and references therein.
5. (a) P. M. Maitlis, *Acc. Chem. Res.*, **11**, 301 (1978) and references therein; (b) D. P. Freyberg, J. L. Robbins, K. N. Raymond, and J. C. Smart, *J. Am. Chem. Soc.*, **101**, 892 (1979); (c) S. J. McLain, R. R. Schrock, P. R. Sharp, M. R. Churchill, and W. J. Youngs, *J. Am. Chem. Soc.*, **101**, 263 (1979).
6. (a) P. J. Fagan, E. A. Maatta, and T. J. Marks, *ACS Symp. Series*, **152**, 52 (1981); (b) P. J. Fagan, J. M. Manriquez, S. H. Vollmer, C. S. Day, V. W. Day, and T. J. Marks, *J. Am. Chem. Soc.* (in press); (c) P. J. Fagan, J. M. Manriquez, E. A. Maatta, A. M. Seyam, and T. J. Marks, *J. Am. Chem. Soc.*, **103**, 6650 (1981) and references therein; (d) J. M. Manriquez, P. J. Fagan, T. J. Marks, C. S. Day, and V. W. Day, *J. Am. Chem. Soc.*, **100**, 7112 (1978); (e) P. J. Fagan, J. M. Manriquez, and T. J. Marks, in *Organometallics of the f-Elements*, T. J. Marks and R. D. Fischer (eds.), Reidel Publishing Co., Dordrecht, The Netherlands, 1979, Chapter 4.
7. F. H. Köhler, W. Prössdorf, U. Schubert, and D. Neugebauer, *Angew. Chim. Int. Ed.*, **17**, 850 (1978).
8. L. deVries, *J. Org. Chem.*, **25**, 1838 (1960).
9. U. Burger, A. Delay, and F. Mazenod, *Helv. Chim. Acta*, **57**, 2106 (1974).
10. D. Feitler and G. M. Whitesides, *Inorg. Chem.*, **15**, 466 (1976).
11. F. G. Bordwell and P. S. Landis, *J. Am. Chem. Soc.*, **79**, 1593 (1957).

80. ANHYDROUS METAL CHLORIDES

$$MCl_n \cdot xH_2O + xSOCl_2 \rightarrow MCl_n + xSO_2 + 2xHCl$$

Submitted by ALFRED R. PRAY*
Checked by RICHARD F. HEITMILLER, † STANLEY STRYCKER,† VICTOR D. AFTANDILIAN,† T. MUNIYAPPAN,† D. CHOUDHURY,† and MILTON TAMRES†

General methods for the preparation of anhydrous chlorides have been described by Tyree.[1] Of the methods listed for dehydrating metal chlorides, that involving treatment with thionyl chloride[2] has the advantages of convenience and simplicity and requires no special apparatus. This method is generally useful regardless of the periodic group in which the metal appears.

Procedure

First, 20 g of the finely ground metal chloride hydrate is put into a round-bottomed flask, and 50 mL of freshly distilled‡ thionyl chloride§ (~ 0.64 mol) is added at room temperature. Evolution of sulfur dioxide and hydrogen chloride begins at once. After bubbling stops, the flask is equipped with a reflux condenser, and the slurry is refluxed for 1–2 h. The condenser is then arranged for distillation, and the excess thionyl chloride is removed *in vacuo* using a dry nitrogen bleed. The flask containing the product is transferred immediately to a vacuum desiccator containing potassium hydroxide and stored for at least 12 h to remove remaining thionyl chloride. The product is then transferred in a dry box to a suitable container that can be sealed. Except for mechanical losses, the yield is quantitative. Results for a series of typical chlorides are as shown in the table.

Properties

Lithium chloride is a colorless, crystalline, hygroscopic, and deliquescent salt; its melting point is 613° and its boiling point is 1360°. Its specific gravity is 2.068. The compound is somewhat hydrolyzed in aqueous solution and is soluble in a number of organic liquids.

Copper(II) chloride is a brown solid, melting at 498° with some decomposition to copper(I) chloride. The compound is hygroscopic and deliquescent.

*University of Minnesota, Minneapolis, MN.
†University of Illinois, Urbana, IL.
‡The reagent must be freshly distilled.
§Sulfinyl chloride.

		Analyses			
Chloride[a]	Starting Material	Metal		Chlorine	
		Calcd.	Found	Calcd.	Found
LiCl	LiCl + 20% H_2O	16.37	16.3	83.63	83.5
$CuCl_2$	$CuCl_2 \cdot 2H_2O$	47.27	47.3	52.73	52.7
$ZnCl_2$	$ZnCl_2$ + 10% H_2O	47.97	47.5	52.03	51.5
$CdCl_2$	$CdCl_2 \cdot 2\frac{1}{2}H_2O$	61.32	61.3	38.68	38.7
$ThCl_4$[b]	$ThCl_4 \cdot 8H_2O$	62.05	62.13	37.92	
$CrCl_3$	$CrCl_3 \cdot 6H_2O$	32.84	31.5	67.16	64.5
$FeCl_3$	$FeCl_3 \cdot 6H_2O$	34.43	34.3	65.57	65.3
$CoCl_2$	$CoCl_2 \cdot 6H_2O$	45.39	45.3	54.61	54.4
$NiCl_2$	$NiCl_2 \cdot 6H_2O$	45.29	44.3	54.71	53.5

[a] The method has also been applied to magnesium and neodymium chlorides by the checkers, but the products were not analyzed.
[b] Data obtained by checker T. Muniyappan (master's dissertation, University of Illinois, 1955).

It dissolves in water and the alcohols but is insoluble in many organic solvents.

Zinc chloride is a white, crystalline, deliquescent solid (mp $\sim 290°$, bp 730°). It dissolves readily in water to give solutions that are extensively hydrolyzed. It is also soluble in alcohols, in acetone and in diethyl ether.

Cadmium chloride is a hygroscopic, translucent, white, crystalline solid, melting at 568° and boiling at 960°. It is somewhat less soluble than zinc chloride in water and the alcohols and is insoluble in diethyl ether.

Thorium(IV) chloride is a white, hygroscopic solid, which dissolves readily in water, in the lower alcohols, and in ethylenediamine.[3]

The properties of chromium(III) chloride have been summarized by Heisig, et al.[4] The compound apparently exists in two forms. The hygroscopic, soluble form results in this synthesis. This form has a peach-blossom color.

The properties of iron(III) chloride have been described by Tarr.[5]

Cobalt(II) chloride is a blue, crystalline solid, which gives the red hexahydrate on exposure to moist air. The compound melts at 724° and boils at 1049°. It dissolves readily in water, methanol, and ethanol but is sparingly soluble in most organic solvents.

Nickel(II) chloride is a yellow, crystalline, scale-like solid, which absorbs water from moist air to give the green hexahydrate. The boiling point is 973°. The salt dissolves readily in water and the alcohols but is insoluble in acetone and the esters.

References

1. S. Y. Tyree, Jr., *Inorg. Synth.*, **4**, 104 (1953).
2. H. Hecht, *Z. anorg. Chem.*, **254**, 37 (1947).
3. T. Muniyappan, master's dissertation, University of Illinois, 1955.
4. G. B. Heisig, B. Fawkes, and R. Hedin, *Inorg. Synth.*, **2**, 193 (1946).
5. B. R. Tarr: *Inorg. Synth.*, **3**, 191 (1950).

81. TUNGSTEN AND MOLYBDENUM TETRACHLORIDE OXIDES

Submitted by A. J. NIELSON*
Checked by R. A. ANDERSEN†

$$H_2WO_4 + SOCl_2 \rightarrow WOCl_4$$

$$MoO_3 + SOCl_2 \rightarrow MoOCl_4$$

Tungsten tetrachloride oxide is a convenient starting material for the synthesis of a variety of tungsten complexes containing oxo[1] and imido[2] ligands. It has been prepared by refluxing tungsten trioxide with octachlorocyclopentane,[3] by sealed-tube reactions of tungsten hexachloride with WO_3,[4] or dry sulfur dioxide[5] and by chlorination of WO_3.[6] A simple preparation involves refluxing WO_3 with sulfinyl chloride,[7] but as commercial samples of WO_3 may be relatively unreactive, the use of H_2WO_4[7,8] is preferred.

Molybdenum tetrachloride oxide has been used in the synthesis of oxo complexes of molybdenum(VI) and (V).[9] It has been prepared by reaction of $MoCl_5$ with MoO_2Cl_2[10] or MoO_3,[11] and by subliming it away from the solid material obtained by treating carbon tetrachloride solutions of $MoCl_5$ with oxygen.[7] The compound is also prepared by refluxing MoO_3,[12] sodium molybdate or MoO_2Cl_2 with sulfinyl chloride.[7]

For large-scale preparations of the tetrachloride oxides, reactions using sulfinyl chloride are more convenient. The procedures outlined below give details for obtaining the compounds, particularly tungsten tetrachloride oxide, in excess of 50-g quantities. The reactions may be scaled up or down without deleterious effect.

---

*Department of Chemistry, University of Auckland, Auckland, New Zealand.
†Department of Chemistry, University of California, Berkeley, CA 94720.

Materials and General Procedure

Tungstic acid and molybdenum trioxide are available from most chemical suppliers. Commercial sulfinyl chloride should be distilled prior to use; otherwise, brown-colored solutions result. All reactions should be carried out in an efficient fume hood, and all manipulations involving the solid oxytetrachlorides should be carried out under dry, oxygen-free nitrogen, using normal techniques for air-sensitive compounds.

Procedure

■ **Caution.** *Some of the substances in this synthesis (e.g., sulfinyl chloride) are volatile and toxic, and must be handled with care in an efficient fume hood.*

A. TUNGSTEN TETRACHLORIDE OXIDE

Finely ground tungstic acid (H_2WO_4), (40 g, 0.16 mol) is placed in a three-necked 2-L flask and 1 L of sulfinyl chloride is added. The flask is fitted with a stopcock, a stopper, and two spiral reflux condensers in tandem with all glass attachments (wired down to prevent forceful ejection). The mixture is refluxed at a rate such that violent bumping is avoided, until most of the H_2WO_4 is consumed, (3–5 days). If bumping proves troublesome, a mechanical stirrer should be used. More sulfinyl chloride is added if the volume of solution decreases during the reflux; the condenser mouth is wiped periodically to prevent accumulating droplets from falling into the reaction mixture.

The heat is turned down and any unreacted H_2WO_4 is allowed to sink to the bottom of the orange-red solution. While the solution is warm, the reflux condensers are removed, a rubber septum is inserted, and the flask is flushed with nitrogen. The liquid is transferred to another 2-L two-necked flask (previously fitted with a stopcock, serum cap, and vent needle, and filled with nitrogen), by way of a stainless steel transfer tube. The transfer is carried out carefully, using a positive nitrogen pressure, until the solution begins to carry H_2WO_4 with it. The receiving flask should be kept warm with a heating mantle, since cooling causes precipitation of $WOCl_4$. If H_2WO_4 contaminates the warm solution, it should be allowed to settle, and the solution should be transferred to another flask. (Alternatively, the solution is filtered under N_2 while hot. Cold glass filtering devices will cause precipitation of $WOCl_4$ and clogging of the sinter.)

The solvent is removed under vacuum, leaving the compound as orange-red crystals. When dry, the solid is finely ground with a mortar and pestle in a glove bag, and placed in a Schlenk flask, which is then evacuated for several hours. The yield of tungsten oxytetrachloride is about 48–51 g, 88–93%. The solid may smell slightly of sulfinyl chloride, but it is of sufficient purity for most uses. A higher-purity material may be obtained by sublimation at 140° and 5–10 mmHg.

Anal. Calcd. for $WOCl_4$: Cl, 41.5; W, 53.7. Found: Cl, 41.2; W, 53.1.

B. MOLYBDENUM TETRACHLORIDE OXIDE

Molybdenum trioxide (30 g, 0.21 mol) is refluxed under nitrogen with 500 mL of sulfinyl chloride for 8–12 h, or until no white solid remains. The cooled solution is filtered under nitrogen through a sintered-glass filter, and the solvent is removed under vacuum, to give the compound as a dark-green solid. The material is held under vacuum for several hours to remove vestiges of sulfinyl chloride, and is stored in a Schlenk flask under nitrogen at $-20°$. The yield is approximately 51 g, 97%. The compound is sufficiently pure for most uses. A higher-purity material may be obtained by sublimation at 50–60° and 10^{-3} torr.

Anal. Calcd. for $MoOCl_4$: Cl, 55.9; Mo, 37.8. Found: Cl, 55.4; Mo, 37.3.

Properties

Tungsten tetrachloride oxide is an orange-red, air- and moisture-sensitive solid melting at 211°.[3] It is insoluble in aliphatic hydrocarbons, but dissolves appreciably in aromatic and chlorinated solvents, to give dark-red solutions. Its 1:1 adducts form in coordinating solvents, such as tetrahydrofuran and MeCN. The compound does not mull in Nujol, and reacts with NaBr and CsI. The crystal structure[13] shows that $WOCl_4$ is tetragonal, with $WOCl_4$ square pyramids weakly associated into chains through W· · · ·O interactions through the basal plane. The compound may be stored at room temperature under nitrogen for many months, but vacuum grease in glass joints should be renewed periodically.

Molybdenum tetrachloride oxide is a very moisture sensitive, green solid, melting at 101 to 103° to give a brown liquid. A red-brown vapor is given off at about 120°.[12] The compound is soluble in aromatic and chlorinated solvents, forming adducts with coordinating solvents. It does not mull in Nujol, and reacts with NaBr or CsI. In carbon tetrachloride, $v(MoO)$ occurs at 1009 cm^{-1}.[14] The crystal structure[15] determination shows an $MoOCl_4$ square pyramid, weakly associated with one other $MoOCl_4$, through Mo· · · ·Cl interactions through the basal plane, suggesting incipient dimer formation. The $MoOCl_4$ is reduced photolytically and thermally to $MoOCl_3$ and chlorine,[14] and should be stored at $-20°$ in the dark. Long storage should be avoided, and fresh samples should be prepared prior to use.

References

1. (a) S. P. Anand, R. K. Multani, and B. D. Jain, *J. Organomet. Chem.*, **17**, 423 (1969); (b) H. Funk, W. Weiss, and G. Mohaupt, *Z. anorg. allg. Chem.*, **304**, 238 (1960); (c) A. V. Butcher, J. Chatt, G. J. Leigh, and P. L. Richards, *J. Chem. Soc. Dalton*, **1972**, 1064; (d) S. P. Anand, R. K. Multani, and B. D. Jain, *Bull. Chem. Soc. Jpn.*, **42**, 3459 (1969).
2. Imido ligands are not included in this synthesis herewith.
3. S. E. Feil, S. Y. Tyree, and F. N. Collier, *Inorg. Synth.*, **9**, 123 (1967).
4. J. Tillack, *Inorg. Synth.*, **14**, 109 (1973).
5. G. W. A. Fowles and J. L. Frost, *J. Chem. Soc. A*, **1967**, 671.
6. A. Michael and A. Murphy, *Am. Chem. J.*, **44**, 382 (1910).
7. R. Colton and I. B. Tomkins, *Aust. J. Chem.*, **18**, 447 (1965).
8. R. H. Crabtree and G. G. Hlatky, *Polyhedron*, **4**, 521 (1985).
9. (a) N. K. Kaushik, R. P. Singh, H. S. Sangari, and G. S. Sodhi, *Syn. React. Inorg. Metal Org. Chem.*, **10**, 617 (1980); (b) S. K. Anand, R. K. Multani, and B. D. Jain, *J. Ind. Chem. Soc.*, **45**, 1130 (1968).
10. W. Puttbach, *Liebig's Ann.*, **201**, 125 (1880).
11. I. A. Glukov and S. S. Eliseev, *Zhur. Neorg. Khim.*, **7**, 81 (1962).
12. R. Colton, I. B. Tomkins, and P. W. Wilson, *Aust. J. Chem.*, **17**, 496 (1964).
13. H. Hess and H. Hartung, *Z. anorg. allg. Chem.*, **344**, 157 (1966).
14. M. L. Larson and F. W. Moore, *Inorg. Chem.*, **5**, 801 (1966).
15. J. C. Taylor and A. B. Waugh, *J. Chem. Soc. Dalton*, **1980**, 2006.

82. TUNGSTEN CHLORO PHOSPHINE COMPLEXES

Submitted by PAUL R. SHARP,* JEFFREY C. BRYAN,† and JAMES M. MAYER†
Checked by JOSEPH L. TEMPLETON‡ and SHAOGUANG FENG‡

The lability of phosphine ligands and their ability to stabilize transition metals in a variety of oxidation states makes transition metal phosphine complexes versatile synthetic reagents. Tungsten(II) chlorophosphine complexes, $WCl_2(PR_3)_4$ (R = alkyl and aryl) fall into this category. They have been prepared by reduction of tungsten(IV) chlorophosphine complexes, $WCl_4(PR_3)_n$ (n = 2 or 3) in the presence of excess phosphine[1-3] or by the reduction of WCl_6 in neat PMe_3.[4] The tungsten(IV) chlorophosphine complexes, $WCl_4(PR_3)_n$, have been prepared by reduction of WCl_6 with Zn amalgam followed by the addition of phosphine,[5] by phosphine reduction of WCl_6,[6] by direct reaction of $[WCl_4]_x$ with phosphine,[1,2,7] by displacement

---
*Department of Chemistry, University of Missouri-Columbia, Columbia, MO 65211.
†Department of Chemistry, University of Washington, Seattle, WA 98195.
‡Department of Chemistry, University of North Carolina at Chapel Hill, Chapel Hill, NC 27599.

of other ligands (L), from WCl_4L_2 complexes,[7] or by chlorine oxidation of $W(CO)_4(PR_3)_2$.[8] The Zn amalgam reduction gives poor yields. The displacement reaction and the chlorine reaction require additional steps for the preparation of WCl_4L_2 and $W(CO)_4(PR_3)_2$. The phosphine reduction has the advantage of being a simple one-step procedure but the disadvantage of consuming additional phosphine for the reduction. The commercial availability of $[WCl_4]_x$ (Aldrich) also makes the direct reaction a one-step procedure and requires no additional phosphine.

General

The procedures described are conducted under an inert atmosphere using either Schlenk–vacuum-line or dry-box techniques.[9] An argon atmosphere is recommended for the reduction reactions because of the facile formation of low valent tungsten dinitrogen complexes. Solvents must be free of oxygen and water. Toluene and tetrahydrofuran (THF) are distilled from CaH_2 or sodium benzophenone. Pentane is distilled from *n*-butyl lithium or CaH_2. Dichloromethane is distilled from CaH_2 or P_4O_{10}.

A. TETRACHLOROTRIS(TRIMETHYLPHOSPHINE)TUNGSTEN(IV)

$$[WCl_4]_x + 3PMe_3 \rightarrow WCl_4(PMe_3)_3$$

Procedure

■ **Caution.** *Phosphines are noxious and toxic. Trimethylphosphine is spontaneously flammable in air.*

In a dry box trimethylphosphine (18.8 mL, 198 mmol) is added to a stirred suspension of $[WCl_4]_x$ (19.5, 60 mmol) in CH_2Cl_2 (250 mL) at 0° in a 500-mL single-necked flask. The stirred mixture is allowed to warm to room temperature and stirring is continued for 10 h. A ^1H NMR aliquot is removed to check for the presence of $WCl_4(PMe_3)_2$ (singlet at -24.5 ppm, $\Delta v_{1/2} \sim 4$ Hz) and more PMe_3 is added as needed to convert any $WCl_4(PMe_3)_2$ to $WCl_4(PMe_3)_3$. The solution is then filtered to remove green solids and the filtrate is reduced *in vacuo* to 20 mL. The resulting red crystals (~ 23 g) are removed by filtration, washed with cold ($-40°$) dichloromethane (1×5 mL), and dried *in vacuo*. A second crop (~ 1 g) is obtained by combining the wash and filtrate and reducing the solution to 5 mL *in vacuo*. Total yields are 60–75%.

Anal. Calcd. (found): C, 19.52 (19.71); H, 4.91 (4.96).

Properties

The red crystalline product is sensitive to air in solution but much less so in the solid state. Solid samples show little decomposition after several days in the air but decompose to a white solid after several weeks' exposure. The X-ray crystal structure shows an approximate capped octahedral geometry.[10] The product is paramagnetic but gives a relatively sharp ^{1}H NMR (CD$_2$Cl$_2$) signal at -7.9 ppm ($\Delta v_{1/2} \sim 4$ Hz). It is soluble in dichloromethane, slightly soluble in THF and aromatic hydrocarbons, but insoluble in saturated hydrocarbons and ether. It reacts with LiBH$_4$[11] and LiAlH$_4$[12] to give borohydride and aluminohydride complexes and with Na/Hg or Na in the presence of H$_2$,[13] N$_2$,[14,15] phenylacetylene,[16] or methylisocyanide[16] to give hydride, dinitrogen, phenylacetylene, and methylisocyanide complexes, respectively.

B. TETRACHLOROBIS(DIPHENYLMETHYLPHOSPHINE)-TUNGSTEN(IV)

$$WCl_6 + 3PMePh_2 \rightarrow WCl_4(PMePh_2)_2 + Cl_2PMePh_2$$

Procedure

■ **Caution.** *See Synthesis A.*

Methyldiphenylphosphine (14 mL, 77 mmol) is added by syringe to a suspension of tungsten(VI) chloride (10 g, 26 mmol) in toluene (200 mL) in a 500-mL Schlenk flask carrying a reflux condenser. The reaction mixture is magnetically stirred and heated to reflux under an atmosphere of N$_2$. Stirring can become difficult early in the reaction because of the temporary formation of a viscous brown oil. After 4 h of reflux, the reaction mixture is cooled to room temperature and concentrated to ~ 30 mL. The resulting slurry is filtered in air and washed with acetone (reagent grade, used without further purification) until the washings are no longer green ($\sim 7 \times 50$ mL), yielding 11 g (57%) of yellow-orange solids.

Properties

The yellow-orange product is slightly hygroscopic, but can be handled in air for brief periods or stored in a desiccator for extended periods with minimal decomposition. It is best stored in a dry box. Infrared spectra (Nujol) show v_{W-Cl} at 320 cm^{-1}. The ^{1}H NMR (C$_6$D$_6$) spectrum exhibits paramagnetically shifted peaks at 11.28 (d, $J = 7$ Hz), 7.95 (t, $J = 7$ Hz), 7.56 (t, $J = 7$ Hz), and

−27 (br, s) ppm. A similar spectrum is obtained in $CDCl_3$.[5] The X-ray crystal structure indicates a trans geometry.[17]

C. DICHLOROTETRAKIS(TRIMETHYLPHOSPHINE)-TUNGSTEN(II)

$$WCl_4(PMe_3)_3 + 2Na/Hg + PMe_3 \rightarrow WCl_2(PMe_3)_4 + 2NaCl + Hg$$

Procedure

■ **Caution.** *See Synthesis A. Proper procedures for the disposal of Hg waste should be followed.*

This procedure is for an inert atmosphere dry box. For the Schlenk–vacuum-line procedure, see Synthesis E. The reaction vessel is a 100-mL single-necked flask. Sodium amalgam (0.4%, 20 mmol Na, 112 g) is added to a stirred suspension of $WCl_4(PMe_3)_3$ (5.90 g, 10 mmol) in a mixture of PMe_3 (1.1 mL, 11 mmol) and 60 mL of THF. The reaction mixture is stirred vigorously for 2 h. The solution is removed from above the Hg pool and filtered through a 1 cm plug of Celite over a medium sintered-glass filter. The Hg and Celite are washed with fresh THF and the wash is added to the filtrate. The resulting pale-brown solution is reduced *in vacuo* to 10 mL and a first crop of orange crystals is obtained by filtering, washing with cold (−30°) THF (1 × 5 mL), and drying *in vacuo* for 2 h. The wash and filtrate are combined and all volatiles are removed *in vacuo*. The solid residue is dissolved in a minimum volume of toluene and one volume of pentane is added. Cooling to −30° for 12 h gives a second crop of orange crystals after filtering, washing with pentane, and drying. A third crop may be obtained in a similar manner. Total yields range from 60 to 90% and appear to depend on the purity of the THF. High-purity THF such as HPLC (high-performance liquid chromatography) grade is recommended. The reaction may also be conducted in toluene (see Synthesis E).

Anal. Calcd. (found): C, 25.78 (26.20); H, 6.49 (6.48).

Properties

The orange product is nicely crystalline. The solid may be handled briefly in air, but solutions rapidly become green on exposure to air. It is soluble in THF and aromatic solvents and slightly soluble in diether. The ^1H NMR spectrum (C_6D_6) exhibits a broad peak ($\Delta v_{1/2} \sim 40$ Hz) at 3.8 ppm that shifts to lower field as the temperature is lowered. The solution (C_6H_6) magnetic susceptibility at 35° is $2.3\mu_B$. The X-ray crystal structure of the Mo analog indicates a trans geometry.[14] The mass spectrum shows peaks due to the loss of one and two PMe_3 ligands. Incomplete reduction can result in contamination of the product with brown $WCl_3(PMe_3)_3$.[2] This impurity is readily detected by its ^1H NMR (C_6D_6) spectrum [-11 (v, br, s, $v_{1/2} \sim 80$ Hz), -15 (v, br, s, $v_{1/2} \sim 80$ Hz) ppm] and is converted to the product by the addition of Na/Hg and PMe_3. The product reacts with H_2,[18] methanol,[16] and aluminum hydrides[19] to give hydride complexes, with trimethylaluminum[20] to give a methylidyne complex, with ethene[2] to give ethene complexes, and with CO[21] to give carbon monoxide complexes and can be thermally converted to a dimer with a W–W quadruple bond.[1]

D. DICHLOROTETRAKIS(DIMETHYLPHENYLPHOSPHINE)-TUNGSTEN(II)

$$WCl_4(PMe_2Ph)_2 + 2Na/Hg + 2PMe_2Ph \rightarrow WCl_2(PMe_2Ph)_4$$
$$+ 2NaCl + Hg$$

Procedure

■ **Caution.** *See Syntheses A and C.*

The general procedure is given in Synthesis C or E. Residual phosphine sometimes interferes with the isolation of the solid product (oily materials are obtained). In this case it is best to remove all volatiles and the excess phosphine by warming (40°) under vacuum (0.1 torr). Crystallizing the residue from toluene–pentane as described for the second crop in Synthesis C gives the product as orange crystals in 40–70% yield. Alternatively, the product may be purified by washing with methanol.[3]

Properties

The orange crystalline product is air sensitive, although it is reportedly stable to water and methanol at room temperature.[3] The solubility properties are similar to those of $WCl_2(PMe_3)_4$. Characterization data have been previously reported[3] [IR, $v_{W-Cl} = 280$ cm^{-1}; mp = 186–188 (decomp.); μ (solid, 20°) = $2.30\mu_B$]. A cryoscopic molecular-weight measurement indicates extensive dissociation in solution.[3] The product reacts with ethene[2] to give an ethene complex and with CO[3] to give a CO complex.

E. DICHLOROTETRAKIS(METHYLDIPHENYLPHOSPHINE)-TUNGSTEN(II)

$$WCl_4(PMePh_2)_2 + 2Na/Hg + 2PMePh_2 \rightarrow WCl_2(PMePh_2)_4$$
$$+ 2NaCl + Hg$$

Procedure

■ **Caution.** *See Syntheses A and C.*

Methyldiphenylphosphine (1.7 mL, 9.0 mmol) is added with a syringe through a septum to a suspension of tetrachlorobis(methyldiphenylphosphine) tungsten(IV) (4.0 g, 5.5 mmol) in toluene (60 mL) in a 100-mL three-necked flask. Sodium amalgam (0.4%, 70.3 g, 12 mmol Na) is added to the reaction flask from an addition funnel. The reaction mixture is stirred vigorously for 2 h under an argon atmosphere. The solution above the amalgam is transferred with a syringe to a Schlenk-type filtering apparatus and filtered through roughly 1 cm of Celite. The filtrate is concentrated *in vacuo* to 10 mL and pentane is added to aid the precipitation of the product. The resulting orange solids are isolated by filtration, washed with pentane (1 × 10 mL), and dried *in vacuo* (10^{-3} torr, 20°) for 3 h (3.0 g, 52%).

Properties

The orange product is sensitive to both oxygen and water, especially in solution. Solid samples may be handled briefly in air. The IR spectrum (Nujol) shows v_{W-Cl} at 280 cm^{-1}. The ^{1}H NMR (C_6D_6) spectrum of the paramagnetic product exhibits broad peaks from 7.5 to 10.0 ppm. Toluene is often observed in the ^{1}H NMR spectra of the isolated solids and can serve as a convenient internal standard for experiments monitored by NMR. The product is soluble in THF and moderately soluble in benzene, toluene, or diethyl ether but reacts with dichloromethane. It reacts with ethene[2] to give an ethene complex; with CO_2 and related heterocumulenes[22] to give W(IV) oxo, imido, and sulfido complexes; and with cyclopentanone[23] to give a W(VI) oxo alkylidene complex. An X-ray crystal structure determination indicates a trans geometry.[24]

References

1. P. R. Sharp and R. R. Schrock, *J. Am. Chem. Soc.*, **102**, 1430 (1980).
2. P. R. Sharp, *Organometallics*, **3**, 1217 (1984).
3. B. Bell, J. Chatt, and G. J. Leigh, *J. Chem. Soc. Dalton Trans.*, **1972**, 2492.

4. V. C. Gibson, C. E. Graimann, P. M. Hare, M. L. H. Green, J. A. Bandy, P. D. Grebenik, and K. Prout, *J. Chem. Soc. Dalton Trans.*, **1985**, 2025.
5. A. V. Butcher, J. Chatt, G. J. Leigh, and P. L. Richards, *J. Chem. Soc. Dalton Trans.*, **1972**, 1064.
6. E. Carmona, L. Sanchez, M. L. Poveda, R. A. Jones, and J. G. Hefner, *Polyhedron*, **2**, 797 (1983).
7. M. A. Schaefer King and R. E. McCarley, *Inorg. Chem.*, **12**, 1972 (1973).
8. J. R. Moss and B. L. Shaw, *J. Chem. Soc. A*, **1970**, 595.
9. D. F. Shriver, *The Manipulation of Air-Sensitive Compounds*, McGraw-Hill, New York, 1969.
10. R. D. Rogers, E. Carmona, A. Galindo, J. L. Atwood, and L. G. Canada, *J. Organomet. Chem.*, **277**, 403 (1984).
11. A. R. Barron, J. E. Salt, G. Wilkinson, M. Motevalli, and M. B. Hursthouse, *Polyhedron*, **5**, 1833 (1986).
12. A. R. Barron, D. Lyons, G. Wilkinson, M. Motevalli, A. J. Howes, and M. B. Hursthouse, *J. Chem. Soc. Dalton Trans.*, **1986**, 279; A. R. Barron, M. B. Hursthouse, M. Motevalli, and G. Wilkinson, *J. Chem. Soc., Chem. Commun.*, **1985**, 664.
13. D. Lyons and G. Wilkinson, *J. Chem. Soc. Dalton Trans.*, **1985**, 587.
14. E. Carmona, J. M. Marin, M. L. Poveda, J. L. Atwood, and R. D. Rogers, *Polyhedron*, **2**, 185 (1983).
15. E. Carmona, J. M. Marin, M. L. Poveda, R. D. Rogers, and J. L. Atwood, *J. Organomet. Chem.*, **238**, C63 (1982).
16. K. W. Chiu, D. Lyons, G. Wilkinson, M. Thornton-Pett, and M. B. Hursthouse, *Polyhedron*, **2**, 803 (1983).
17. L. Aslanov, R. Mason, A. G. Wheeler, and P. O. Whimp, *J. Chem. Soc., Chem. Commun.* **1970**, 30.
18. P. R. Sharp and K. G. Frank, *Inorg. Chem.*, **24**, 1808 (1985).
19. A. R. Barron, J. E. Salt, and G. Wilkinson, *J. Chem. Soc. Dalton Trans.*, **1986**, 1329.
20. P. R. Sharp, S. J. Holmes, R. R. Schrock, M. R. Churchill, and H. J. Wasserman, *J. Am. Chem. Soc.*, **103**, 965 (1981).
21. E. Carmona, K. Doppert, J. M. Marin, M. L. Poveda, L. Sanchez, and R. Sanchez-Delgado, *Inorg. Chem.*, **23**, 530 (1984).
22. J. C. Bryan, S. J. Geib, A. L. Rheingold, and J. M. Mayer, *J. Am. Chem. Soc.*, **109**, 2826 (1987).
23. J. C. Bryan and J. M. Mayer, *J. Am. Chem. Soc.*, **109**, 7213 (1987).
24. J. M. Mayer, unpublished results.

83. TETRABUTYLAMMONIUM OCTACHLORODIRHENATE(III)

$$2[(n\text{-}C_4H_9)_4N]ReO_4 + 8C_6H_5COCl \rightarrow [(n\text{-}C_4H_9)_4N]_2[Re_2Cl_8]$$

Submitted by T. J. BARDER* and R. A. WALTON*
Checked by F. A. COTTON† and G. L. POWELL†

The octachlorodirhenate(III) anion is the key starting material for entry into the chemistry of multiply bonded complexes of dirhenium.[1-3] To date, the

*Department of Chemistry, Purdue University, West Lafayette, IN 47907.
†Department of Chemistry, Texas A&M University, College Station, TX 77843.

most convenient synthetic route for the synthesis of this complex has been through the hypophosphorous acid reduction of $KReO_4$.[4,5] However, the yield of the complex is low and variable, rarely exceeding 40%. Other synthetic routes[5-8] require either high-pressure conditions or the use of starting materials that are best prepared from the $[Re_2Cl_8]^{2-}$ anion itself, and they offer no significant advantages over the hypophosphorous acid reduction of $KReO_4$. We describe here a simple, quick, and high-yield synthesis of $[(n-C_4H_9)_4N]_2[Re_2Cl_8]$, which has many advantages over existing procedures.[9]

Procedure

The starting material $[(n-C_4H_9)_4N]ReO_4$* (3.0 g, 6.1 mmol) is placed in a 250-mL round-bottomed flask fitted with a reflux condenser. The condenser is connected to a mercury bubbler system, which consists of a cylindrical glass reservoir (22-mm internal diameter) containing 40 mL of mercury and fitted with a 4-mm-diameter gas inlet tube extending 95 mm into the mercury pool. The reaction vessel is purged with N_2 and benzoyl chloride† (30 mL, 26 mmol) is then syringed into the reaction flask. The resulting mixture is refluxed for 1.5 h under a positive pressure of N_2. Under these conditions, the boiling point of the benzoyl chloride should be very close to 209°.‡ The resulting dark green solution is allowed to cool and is then treated with a solution of $[(n-C_4H_9)_4N]Br$ (5.0 g, 16 mmol) in ethanol (75 mL) that has been saturated with hydrogen chloride gas. This mixture is then refluxed for an additional hour, and the resulting solution is evaporated to approximately one half its original volume under a stream of N_2. The resulting blue crystals are filtered off, washed with three 10-mL portions of ethanol, then with diethyl ether, and dried under vacuum. Yield: 94% (3.28 g).

Anal. Calcd. for $C_{32}H_{72}Cl_8N_2Re_2$: C, 33.69; H, 6.36. Found: C, 33.97; H, 6.12.

Properties

The complex $[(n-C_4H_9)_4N]_2[Re_2Cl_8]$ is soluble in acetone, acetonitrile, methanol, and many other nonaqueous solvents. It can be recrystallized by dissolution of the complex in boiling methanol, followed by filtration into an

---

*This salt can be prepared in essentially quantitative yield by the slow addition of a hot, aqueous solution of $[(n-C_4H_9)_4N]Br$ to one of $KReO_4$. The resulting white precipitate is washed with water and dried in vacuum.

†Benzoyl chloride can be used as received from a commercial source.

‡This is a critical factor in ensuring a high yield of the desired product. For example, with a mineral-oil bubbler, the boiling point of benzoyl chloride was ~198°, and the yield of $[(n-C_4H_9)_4N]_2Re_2Cl_8$ fell from ~90 to 60%.

equal volume of concentrated HCl. Subsequent evaporation of the solution to remove the methanol gives dark blue crystals. The complex is quite stable in air and can be stored indefinitely without special precautions.

This complex possesses a metal–metal quadruple bond, and can be converted readily into other dirhenium species containing multiple bonds. Solutions of $[Re_2Cl_8]^{2-}$ react with concentrated HBr to form the analogous bromo-anion $[Re_2Br_8]^{2-}$,[4] and with acetic acid–acetic anhydride mixtures to produce the acetate-bridged dirhenium(III) complex $Re_2(O_2CCH_3)_4Cl_2$.[10] The $[(n-C_4H_9)_4N]_2[Re_2Cl_8]$ complex reacts with phosphines (PR_3) to yield complexes of stoichiometry $Re_2Cl_6(PR_3)_2$, $Re_2Cl_5(PR_3)_3$, and $Re_2Cl_4(PR_3)_4$.[11]

References

1. F. A. Cotton and R. A. Walton, *Multiple Bonds Between Metal Atoms*, Wiley-Interscience, New York, 1982.
2. F. A. Cotton and R. A. Walton, *Struct. Bonding (Berlin)*, **62**, 1 (1985).
3. K. A. Conner and R. A. Walton, *Comprehensive Coordination Chemistry*, Vol. 4, Pergamon, Oxford, U.K., pp 125–213 (1987).
4. F. A. Cotton, N. F. Curtis, and W. R. Robinson, *Inorg. Chem.*, **4**, 1696 (1965).
5. F. A. Cotton, N. F. Curtis, B. F. G. Johnson, and W. R. Robinson, *Inorg. Chem.*, **4**, 326 (1965).
6. A. S. Kotel'nikova, M. I. Glinkina, T. V. Misatlova, and V. G. Lebedev, *Russ. J. Inorg. Chem.*, **21**, 547 (1976).
7. R. A. Bailey and J. A. McIntyre, *Inorg. Chem.*, **5**, 1940 (1966).
8. A. B. Brignole and F. A. Cotton, *Inorg. Synth.*, **13**, 82 (1972).
9. T. J. Barder and R. A. Walton, *Inorg. Chem.*, **21**, 2510 (1982).
10. F. A. Cotton, C. Oldham, and W. R. Robinson, *Inorg. Chem.*, **5**, 1798 (1966).
11. J. R. Ebner and R. A. Walton, *Inorg. Chem.*, **14**, 1987 (1975).

84. DI-µ-CHLORO-BIS[TRICARBONYLCHLORO-RUTHENIUM(II)]

$$Ru_3(CO)_{12} \xrightarrow{CHCl_3} [Ru(CO)_3Cl_2]_2$$

Submitted by A. MANTOVANI* and S. CENINI†
Checked by R. M. HEINTZ‡ and D. E. MORRIS‡

There are several literature reports[1-6] on the synthesis of $[Ru(CO)_3Cl_2]_2$. Direct chlorination of $Ru_3(CO)_{12}$[1] gives a mixture of compounds that are

---

*Cattedra di Chimica, Facoltà d'Ingegneria, Via Marzolo 9, 35100 Padova, Italy.
†Dipartimento di Chimica Inorganica e Metallorganica, Centro C.N.R., Via G. Venezian 21, 20133 Milano, Italy.
‡Corporate Research Department, Monsanto Co., St. Louis, MO 63166.

difficult to separate. Also, $H_2Ru(CO)_4$ reacts with carbon tetrachloride to give impure $[Ru(CO)_3Cl_2]_2$.[2] Carbonylation of $RuCl_3 \cdot 3H_2O$ under pressure leads to the desired product;[3] however, some difficulties were experienced in repeating this preparation. The reaction of $RuCl_3 \cdot 3H_2O$ with formic acid in the presence of hydrochloric acid is a rather complicated method for the synthesis of $[Ru(CO)_3Cl_2]_2$.[4] The displacement of the diene ligand from $[(diene)RuCl_2]_n$ (diene = benzene, norbornadiene) by carbon monoxide to give $[Ru(CO)_3Cl_2]_2$ has also been reported,[5] but the yields of the product were not satisfactory. The best method of synthesis so far reported seems to be the reaction of $Ru_3(CO)_{12}$ with chloroform under a low pressure of nitrogen at an elevated temperature.[6] The details of this synthesis are given here.

Procedure

A suspension of $Ru_3(CO)_{12}$* (0.5 g) in chloroform (20 mL), stabilized with ethanol, which favors the reaction,[6] is placed in a glass liner constructed to fit a 100-mL autoclave. A glass liner must be used to prevent the formation of decomposition products, presumably formed because of the presence of HCl.

The autoclave is charged with nitrogen (5 atm) and then heated at 110° in an oil bath for 6 h.†

After venting and cooling the autoclave, the glass liner and its contents are removed, and the suspension of the white $[Ru(CO)_3Cl_2]_2$ is collected on a filter. The solid is washed with a little chloroform and dried under vacuum. This first portion (0.32–0.37 g; 53–60% yield) gives an analytically pure product.

Anal. Calcd. for $[Ru(CO)_3Cl_2]_2$: C, 14.1; Ru, 39.5; Cl, 27.7. Found: C, 14.6; Ru, 38.7; Cl, 28.1. The pale-yellow mother liquor, to which the chloroform washings from the previous filtration are added, is evaporated under vacuum to a small volume (~ 5 mL) and cooled to $\sim -10°$. The pale-yellow precipitate is collected on a filter. This second portion of the product is recrystallized from a solution of hot 1,2-dichloroethane by adding *n*-hexane, and this yields an additional 0.05 to 0.1 g of product.

Properties

The white $[Ru(CO)_3Cl_2]_2$ turns to orange-brown at 215° and decomposes above 315°. It is slightly soluble in chloroform and 1,2-dichloroethane. It is

---

*See synthesis in section 55 in this volume.

†The checkers report that the pressure in the reactor increases from 5 to 10 atm on heating to 110°, and they find it desirable to carry out the reaction for 8 h.

readily soluble in methanol and tetrahydrofuran (THF), but this solvent also acts as a ligand to give $Ru(CO)_3(THF)Cl_2$.[3] In carbon tetrachloride its infrared spectrum shows $v(CO)$ = 2140 (s), 2081 (s), and 2076 (s) cm^{-1}. An X-ray determination has shown that the molecular structure is of C_{2h} symmetry:[7] The chlorine bridges can be easily broken by THF or other ligands, e.g., pyridine, triphenylphosphine, and nitriles.[8]

The THF adduct is a good starting material for the synthesis of cis- or trans-$Ru(CO)_2L_2Cl_2$ complexes [L = PPh_3, $AsPh_3$, py; L_2 = bipy, $C_2H_4(PPh_3)_2$].[3]

References

1. B. F. G. Johnson, R. D. Johnston, and J. Lewis, *J. Chem. Soc. A*, **1969**, 792.
2. J. D. Cotton, M. I. Bruce, and F. G. A. Stone, *J. Chem. Soc. A*, **1968**, 2162.
3. M. I. Bruce and F. G. A. Stone, *J. Chem. Soc. A*, **1967**, 1238.
4. R. Cotton and R. H. Farthing, *Aust. J. Chem.*, **24**, 903 (1971).
5. R. B. King and P. N. Kapoor, *Inorg. Chem.*, **11**, 336 (1972).
6. G. Braca, G. Sbrana, E. Benedetti, and P. Pino, *L. Chim. Ind.* (Milan), **51**, 1245 (1969).
7. S. Merlino and G. Montagnoli, *Acta Cryst.*, **B24**, 424 (1965).
8. E. Benedetti, G. Braca, G. Sbrana, F. Salvetti, and B. Grassi, *J. Organomet. Chem.*, **37**, 361 (1972).

85. DIHYDRIDOTETRAKIS(TRIPHENYLPHOSPHINE)-RUTHENIUM(II)

$$RuCl_2[P(C_6H_5)_3]_3 + Na[BH_4] + P(C_6H_5)_3 \rightarrow RuH_2[P(C_6H_5)_3]_4$$
$$+ \text{ other products}$$

Submitted by R. YOUNG* and G. WILKINSON*
Checked by S. KOMIYA† and A. YAMAMOTO†

The complex $RuH_2[P(C_6H_5)_3]_4$ can be formed in several ways, but the interaction with sodium tetrahydroborate(1−) of $RuCl_2[P(C_6H_5)_3]_3$[1] is the most convenient. This dihydride, the hydridochlorido $RuHCl[P(C_6H_5)_3]_4$, as well as the dichloride have an extensive chemistry of stoichiometric as well as catalytic reactions.[2,3]

Procedure

■ **Caution.** *Hydrogen is an explosion and fire hazard. The reaction should therefore be carried out in an efficient fume hood in the absence of flame or spark sources. The isolated product is somewhat air and light sensitive and should be stored under argon in the dark.*

A mixture of benzene (60 mL) and methanol (100 mL) containing triphenylphosphine (6 g, 22.9 mmol) is placed in a 250-mL, round-bottomed, three-necked flask fitted with a reflux condenser,‡ a gas inlet tube that will protrude beneath the surface of the solvent, and a magnetic stirrer. After purging the apparatus with hydrogen for ~5 min $RuCl_2[P(C_6H_5)_3]_3$ (1.0 g, 0.87 mmol) is added through a side neck. Dry, finely ground sodium tetrahydroborate(1−) (1.5 g, 0.04 mol) is added in approximately five 0.3-g portions, over a period of 20 min, with rapid stirring. The solution is stirred at room temperature for 1 h, during which time the initial red solution changes to brown and then yellow and the complex is precipitated as bright-yellow microcrystals. Degassed methanol (100 mL) is added to the solution, and the product is collected on a sintered filter under argon, washed with argon-purged methanol, and dried *in vacuo*.§ Typical yields based on ruthenium are 1.02 to 1.08 g (85–90%).

---

*Inorganic Chemistry Laboratories, Imperial College of Science and Technology, London, SW7 2AY.
†Research Laboratory of Resources Utilization, Tokyo Institute of Technology, Ookayama, Meguro, Tokyo 152, Japan.
‡The reflux condenser is required to prevent stripping off of solvent during purging.
§ The checkers report that the complex can be recrystallized from benzene or toluene to give a sample that decomposes in a sealed tube at a temperature of 220° (uncorrected).

Anal. Calcd. for $C_{72}H_{62}P_4Ru$: C, 75.0; H, 5.4; P, 10.7. Found: C, 75.1; H, 5.7; P, 10.4.

Properties

The complex is sparingly soluble in benzene, toluene, tetrahydrofuran, and acetone. The complex is moderately stable to air in the solid state but is readily oxidized in solution. In solution the complex dissociates to give free triphenylphosphine and $RuH_2[P(C_6H_5)_3]_3$, which may then reversibly add nitrogen or hydrogen to produce $Ru(N_2)H_2[P(C_6H_5)_3]_3$ and RuH_4-$[P(C_6H_5)_3]_3$, respectively. In addition to phosphine bands the infrared spectrum shows a single, medium-strong Ru—H stretching frequency at 2080 cm^{-1}.

References

1. P. S. Hallman, T. A. Stephenson, and G. Wilkinson, *Inorg. Synth.*, **12**, 238 (1970).
2. F. H. Jardine, *Progr. Inorg. Chem.*, **31**, 265 (1984).
3. G. G. Hlatky and R. H. Crabtree, *Coord. Chem. Rev.*, **65**, 1 (1985) (an extensive review on $RuCl_2[P(C_6H_5)_3]_3$ and its derivatives, with 357 references).

86. TRIS(2,2'-BIPYRIDINE)RUTHENIUM(II) DICHLORIDE HEXAHYDRATE

Submitted by JOHN A. BROOMHEAD* and CHARLES G. YOUNG*
Checked by PAM HOOD†

The tris(2,2'-bipyridine)ruthenium(II) complex cation, along with its substituted derivatives, is an important species in the study of electron transfer and likely solar energy conversion reactions. It was first prepared in low yield by pyrolysis of ruthenium trichloride with the ligand.[1] An improved yield was obtained by reflux of these reagents in ethanol,[2] but the method suffers from the long reaction time (72 h). Another synthesis[3] uses $K_4[Ru_2Cl_{10}O]\cdot H_2O$, which itself requires a separate preparation. The method described here uses commercial ruthenium trichloride, is of high yield, and takes about 1 h. It may also be used to prepare analogous 1,10-phenanthroline and related diimine ligand complexes.

---

*Department of Chemistry, Faculty of Science, Australian National University, Canberra, A.C.T. 2600, Australia.
†Department of Chemistry, Oklahoma State University, Stillwater, OK 74048.

The Nature of Commercial Hydrated Ruthenium Trichloride

Samples of $RuCl_3 \cdot xH_2O$ may contain Ru(IV), various oxo- and hydroxy-chlorocomplexes, and nitrosyl species. Also, considerable variation in reaction times and product distribution may accompany the use of samples from different sources.[4] In this synthesis, consistently high yields of $[Ru(bipy)_3]Cl_2$ can be obtained if the $RuCl_3 \cdot xH_2O$ is oven treated prior to use.

Drying Procedure

Commercial $RuCl_3 \cdot xH_2O$ is dried in an oven at 120° for 3 h. It is then finely ground in a mortar and returned to the oven for a further 1 h prior to use. It is convenient to store the "dried" $RuCl_3$ at this temperature.

Procedure

$$RuCl_3 + 3C_{10}H_8N_2 \xrightarrow{NaH_2PO_2/H_2O} [Ru(C_{10}H_8N_2)_3]Cl_2 \cdot 6H_2O$$

"Dried" $RuCl_3$ (0.4 g, 1.93 mmol), 2,2'-bipyridine (0.9 g, 5.76 mmol), and water (40 mL) are placed in a 100-mL flask fitted with a reflux condenser. Freshly prepared sodium phosphinate (sodium hypophosphite) solution (2 mL) is added and the mixture heated at the boil for 30 min. [The sodium phosphinate solution is prepared by the careful addition of sodium hydroxide pellets to about 2 mL of 31% phosphinic acid (hypophosphorous acid) until a slight cloudy precipitate is obtained. Phosphinic acid is then added dropwise until the precipitate just redissolves.]

During reflux, the initial green solution changes to brown and finally orange. It is filtered to remove traces of undissolved material and potassium chloride (12.6 g) added to the filtrate to precipitate the crude product. The solution and solid are then heated at the boil to give a deep-red solution, which on cooling to room temperature yields beautiful, red plate-like crystals. These are filtered off, washed with ice-cold 10% aqueous acetone (2 × 5 mL) and acetone (30 mL), and air-dried. The yield is 1.15 g (80%). The product may be recrystallized from boiling water (~ 2.8 mL g^{-1}) and then air-dried.

Anal. Calcd. for $C_{30}Cl_2H_{36}N_6O_6Ru$. C, 48.13; Cl, 9.47; H, 4.85; N, 11.22. Found: C, 48.30; Cl, 9.62; H, 4.85; N, 11.27.

Properties

Aqueous solutions of $[Ru(bipy)_3]Cl_2 \cdot 6H_2O$ have two characteristic absorption maxima at 428 nm (shoulder $\varepsilon = 11,700$) and 454 nm ($\varepsilon = 14,000$), which have been assigned to metal ligand charge-transfer transitions.[5] The charge-transfer excited state is relatively long-lived in solution (lifetime ~600 ns),[6] and the luminescence spectrum (λ_{max} 600 nm) has been assigned to heavy-metal perturbed triplet–singlet phosphorescence of the excited state.[7,8] The complex has been resolved into its optical enantiomers using the iodide antimony(+)tartrate salt. Also the redox potentials of various derivatives have been measured.[9]

References

1. F. H. Burstall, *J. Chem. Soc.*, **1936**, 173.
2. R. A. Palmer and T. S. Piper, *Inorg. Chem.*, **5**, 864 (1966).
3. F. P. Dwyer, *J. Proc. Roy. Soc. NSW*, **83**, 134 (1949).
4. J. M. Fletcher, W. E. Gardner, E. W. Hooper, K. R. Hyde, F. M. Moore, and J. L. Woodhead, *Nature*, **199**, 1089 (1963).
5. J. P. Paris and W. W. Brandt, *J. Am. Chem. Soc.*, **81**, 5001 (1959).
6. C.-T. Lin, W. Böttcher, M. Chou, C. Creutz, and N. Sutin, *J. Am. Chem. Soc.*, **98**, 6536 (1976).
7. J. N. Demas and G. A. Crosby, *J. Mol. Spectrosc.*, **26**, 72 (1968).
8. F. E. Lytle and D. M. Hercules, *J. Am. Chem. Soc.*, **91**, 253 (1969).
9. W. W. Brandt, F. P. Dwyer, and E. C. Gyarfas, *Chem. Rev.*, **54**, 959 (1954).

87. DICHLOROBIS[μ-METHYLENEBIS(DIPHENYL-PHOSPHINE)]DIPALLADIUM(I) (Pd–Pd)

Submitted by ALAN L. BALCH* and LINDA S. BENNER*
Checked by JOHN D. BASIL†

■ **Caution.** *Phosphines such as those used here should be handled with extreme care. Use only in a well-ventilated hood*!

Methylene bis(diphenylphosphine) [bis(diphenylphosphino)methane] is a useful ligand for constructing binuclear metal complexes with novel chemical properties. The flexibility of this ligand allows the metal–metal separations in trans-bridged, binuclear species to vary from 2.1 to 3.5 Å.[1] Mononuclear complexes in which bis(diphenylphosphino)methane acts as a chelating ligand are also known but are less common. The synthesis of $Pd_2(Ph_2PCH_2PPh_2)_2Cl_2$ reported here is more reliable than the original synthesis,[2] which starts with [Pd(CO)Cl]$_n$ and gives highly variable yields. The present procedure also gives $Pd_2(Ph_2PCH_2PPh_2)_2Cl_2$ free of contami-

---
*Department of Chemistry, University of California, Davis, CA 95616.
†Department of Chemistry, Case Western Reserve University, Cleveland, OH 44106.

nation by $Pd_2(Ph_2PCH_2PPh_2)_2(\mu\text{-}CO)Cl_2$.[3]

$$4Ph_2PCH_2PPh_2 + 2(PhCN)_2PdCl_2 + Pd_2(PhHC=CHC(O)CH$$
$$=CHPh)_3 \cdot CHCl_3$$
$$\rightarrow 2Pd_2(Ph_2PCH_2PPh_2)_2Cl_2 + 4PhCN + 3PhCH=CHC(O)CH=CHPh$$
$$+ CHCl_3$$

Procedure

Under a nitrogen atmosphere, a 100-mL Schlenk flask containing a magnetic stirring bar is charged with 50 mL oxygen-free dichloromethane, bis(benzonitrile)dichloropalladium(II)[4] (0.412 g, 1.08 mmol), tris(1,5-diphenyl-1,4-pentadien-3-one)dipalladium chloroform solvate[5] (0.545 g, 0.527 mmol), and bis(diphenylphosphino)methane (0.819 g, 2.13 mmol). A reflux condenser is fitted to the flask, and the stirred solution is heated under reflux in a nitrogen atmosphere for 30 min. After cooling, the red solution, which at this stage is no longer oxygen sensitive, is filtered. The filtrate is condensed to a volume of 10 mL by the use of a rotary evaporator. Methanol (100 mL) is added to the dichloromethane solution to precipitate the product. The red-brown solid is collected by filtration and washed with methanol (3 × 10 mL) and diethyl ether (10 mL). The product is purified by dissolution in dichloromethane, filtration of the solution, and reprecipitation of the product by the addition of methanol. The orange-red, crystalline product is dried under vacuum.

Anal. Calcd. for $C_{25}H_{22}ClP_2Pd$: C, 57.06; H, 4.21; Cl, 6.74. Found: C, 57.40; H, 4.01; Cl, 7.00.

Properties

The compound $Pd_2(Ph_2PCH_2PPh_2)_2Cl_2$ is an orange-red solid that is moderately soluble in dichloromethane and chloroform and slightly soluble in benzene and toluene. It is air stable both as a solid and in solution. The ^1H NMR spectrum in CD_2Cl_2 solution exhibits the methylene resonance at 4.17 ppm. It is a 1:4:6:4:1 quintet ($J_{P-H} = 4.0$ Hz) as a result of virtual coupling of the protons to the four phosphorus atoms. The ^{31}P NMR spectrum in $CDCl_3$ solution consists of a singlet at -2.5 ppm with respect to external 85% H_3PO_4.[3] In solution, $Pd_2(Ph_2PCH_2PPh_2)_2Cl_2$ undergoes addition of a number of small molecules, including carbon monoxide,[3] isocyanides,[3] sulfur dioxide,[6] atomic sulfur,[6] and activated acetylenes.[7] Addition of these small molecules involves their insertion into the Pd–Pd bond. As a consequence, the Pd–Pd bond breaks and the Pd–Pd separation increases by ~0.5 Å.

References

1. M. M. Olmstead, C. H. Lindsay, L. S. Benner, and A. L. Balch, *J. Organometal. Chem.*, **179**, 289 (1979).
2. R. Colton, R. H. Farthing, and M. J. McCormick, *Aust. J. Chem.*, **26**, 2607 (1973).
3. L. S. Benner and A. L. Balch, *J. Am. Chem. Soc.*, **100**, 6099 (1978).
4. M. S. Kharasch, R. C. Seyler, and F. R. Mayo, *J. Am. Chem. Soc.*, **60**, 882 (1938).
5. T. Ukai, H. Kawazura, Y. Ishii, J. J. Bonnet, and J. A. Ibers, *J. Organometal. Chem.*, **65**, 253 (1974).
6. A. L. Balch, L. S. Benner, M. M. Olmstead, *Inorg. Chem.*, **18**, 2996 (1979).
7. A. L. Balch, C.-H. Lee, C. H. Lindsay, and M. M. Olmstead, *J. Organometal. Chem.*, **177**, C22 (1979).

88. (η^3-ALLYL)PALLADIUM(II) COMPLEXES

Submitted by YOSHITAKA TATSUNO,* TOSHIKATSU YOSHIDA,* and SEI OTSUKA*
Checked by NAJEEB AL-SALEM† and BERNARD L. SHAW†

(η^3-Allyl)(η^5-cyclopentadienyl)palladium(II), first prepared by B. L. Shaw,[1] is a labile organopalladium compound useful for preparations of various Pd(0) complexes. The present preparation from bis(η^3-allyl)di-μ-chloro-dipalladium(II) follows the method of Shaw.

A. BIS(η^3-ALLYL)DI-μ-CHLORO-DIPALLADIUM(II)[2]

$$2Na_2PdCl_4 + 2CH_2=CHCH_2Cl + 2CO + 2H_2O \rightarrow$$
$$(\eta^3\text{-}C_3H_5)_2Pd_2Cl_2 + 4NaCl + 2CO_2 + 4HCl$$

Procedure

■ **Caution.** *The preparation should be performed in a well-ventilated hood.*

A 200-mL, two-necked, round-bottomed flask equipped with a magnetic stirring bar, a gas inlet tube, and a condenser topped with a bubbler is

---
*Department of Chemistry, Faculty of Engineering Science, Osaka University, Toyonaka, Osaka, Japan 560.
†School of Chemistry, University of Leeds, Leeds LS2 9JT, United Kingdom.

charged with an aqueous solution of palladium(II) chloride (4.44 g, 25 mmol) and sodium chloride (2.95 g, 50 mmol) in 10 mL of H_2O, followed by methanol (60 mL) and allyl chloride (3-chloro-1-propene) (6.0 g, 67 mmol). Carbon monoxide is passed slowly (2–2.5 L h^{-1}) under stirring through the reddish-brown solution by way of a gas-inlet tube for 1 h. The bright yellow suspension thus obtained is poured into water (300 mL) and extracted with chloroform (2 × 100 mL). The extract is washed with water (two 150-mL portions) and dried over calcium chloride. Evaporation under reduced pressure (20 torr) gives yellow crystals. Yield: 4.3 g (93%). The crude product can be used without further purification. The analytically pure sample can be obtained by recrystallization from a mixture of dichloromethane and hexane. The compound decomposes at 155–156°.

Anal. Calcd. for $C_6H_{10}Cl_2Pd_2$: C, 19.49; H, 2.73. Found: C, 19.60; H, 2.75.

Properties

The air-stable, yellow, crystalline compound is soluble in benzene, chloroform, acetone, and methanol. The ^1H NMR spectrum ($CDCl_3$) shows two doublets at δ 3.03 (anti CH_2, J = 12.0 Hz) and 4.10 ppm (syn CH_2, J = 7.1 Hz), and a triplet at δ 5.48 ppm (CH) in a relative ratio of 2:2:1. A variety of reactions of this compound are summarized in several reviews.[3]

B. (η^3-ALLYL)(η^5-CYCLOPENTADIENYL)PALLADIUM(II)

(η^3-C_3H_5)$_2$Pd$_2$Cl$_2$ + 2NaC$_5$H$_5$ → 2Pd(η^3-C_3H_5)(η^5-C_5H_5) + 2NaCl

Procedure

■ **Caution.** *(η^3-Allyl)(η^5-cyclopentadienyl)palladium is volatile and has an unpleasant odor. As the toxicity is unknown, all manipulations should be carried out in an efficient hood. All solvents are dried over sodium metal and distilled under nitrogen.*

A tetrahydrofuran (THF) solution of cyclopentadienide[4] is prepared by adding freshly distilled cyclopentadiene[5] to a sodium suspension in THF. The concentration of the resulting pale-pink solution can be determined by titration with acid.

In a 300-mL three-necked flask equipped with a three-way stopcock, a pressure-equalizing dropping funnel, and a Teflon-coated magnetic stirring bar, is placed bis(η^3-allyl)di-μ-chloro-dipalladium(II) (9.9 g, 27 mmol). The flask is evacuated and filled with nitrogen three times. THF (100 mL) and benzene (100 mL) are added through the three-way stopcock under nitrogen, with a syringe, to give clear yellow solution. The flask is then cooled with an ice–sodium chloride mixture to $-20°$. A THF solution of sodium cyclopentadienide (54 mmol in 28 mL of THF) is transferred by syringe to a nitrogen-flushed dropping funnel and is then added dropwise to the cooled solution with stirring at $-20°$. The solution changes slowly from yellow to a dark red. After 1 h the ice bath is removed and the temperature of the reaction mixture is allowed to reach room temperature with stirring. The stirring is continued for an additional 30 min. The solvents are removed by distillation *in vacuo* (30–60 torr) to give a dark-red solid. If the pressure is lower than 30 torr, a considerable amount of the palladium complex sublimes at 25°. The solid residue is extracted with hexane (80 mL) and the extract is filtered through a dried filter paper under a nitrogen atmosphere in a filtration funnel as shown

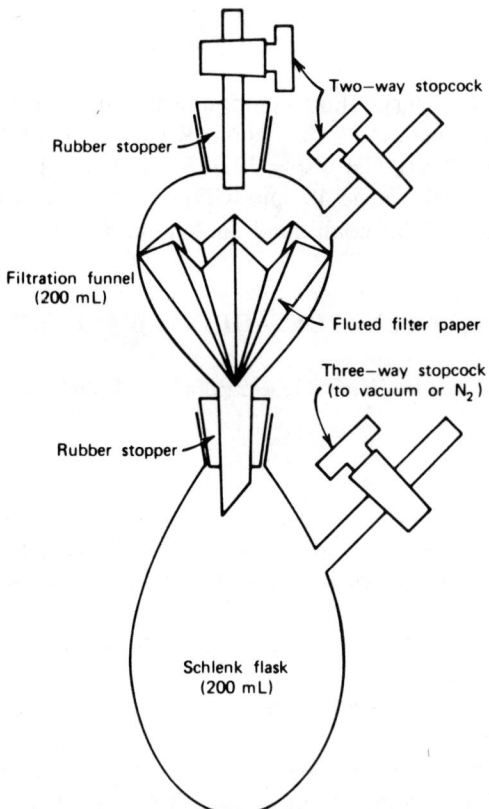

Fig. 1. Simple apparatus for filtering under inert atmosphere.

in Fig. 1.* The red filtrate is evaporated *in vacuo* (30–60 torr), affording red needles of (η^3-allyl)(η^5-cyclopentadienyl)palladium(II). The yield is about 9.2 g (80%).†

The product can be used for most preparations of palladium(0) complexes. This compound can be further purified by sublimation at 40° under 30 torr, with slightly decreased yield.

Anal. Calcd. for $C_8H_{10}Pd$: C, 45.20; H, 4.74. Found: C, 45.20; H, 4.77.

Properties

(η^3-Allyl)(η^5-cyclopentadienyl)palladium(II) is an easily sublimed compound with an unpleasant odor. It forms red, needle-like crystals that decompose at 61°. In the solid state it is fairly stable, although it decomposes gradually at room temperature to give a black solid that is insoluble in hexane. It is therefore recommended that the complex be stored below −20° under nitrogen. The ^{1}H NMR spectrum (C_6D_6) shows signals at δ 2.14 (2H, doublet, $J = 11$ Hz), 3.11 (2H, doublet, $J = 6$ Hz), and 4.63 ppm (1H, complex) for the η^3-allyl protons, and a signal at δ 8.1 ppm (5H, singlet) for the cyclopentadienyl ring protons.

This compound reacts readily with alkyl isocyanides to give a cluster "Pd(CNR)$_2$",[6] and with bulky alkyl phosphines to give two coordinated palladium(0) complexes.[7]

References

1. B. L. Shaw, *Proc. Chem. Soc.*, **1960**, 247.
2. W. T. Dent, R. Long, and A. J. Wilkinson, *J. Chem. Soc.*, **1964**, 1585.
3. P. M. Maitlis, *The Organic Chemistry of Palladium*, Vol. I, Academic Press, New York, 1971; F. A. Hartley, *The Chemistry of Platinum and Palladium*, Applied Science Publishers, London, 1973.
4. F. G. A. Stone and R. West (eds.), *Advances in Organometallic Chemistry*, Vol. 2, Academic Press, New York, 1964; p. 365; R. B. King and F. G. A. Stone, *Inorg. Synth.* **7**, 99 (1963).
5. R. B. Moffett, *Org. Synth., Coll.* Vol. 4, 238 (1963).
6. E. O. Fischer and H. Werner, *Chem. Ber.*, **95**, 703 (1962); S. Otsuka, A. Nakamura, and Y. Tatsuno, *J. Am. Chem. Soc.*, **91**, 6994 (1969).
7. S. Otsuka, T. Yoshida, M. Matsumoto, and K. Nakatsu, *J. Am. Chem. Soc.*, **98**, 5850 (1976).

---

*As a regular glass filter will become clogged, it is not suitable for this filtration.
†The checker used a mechanical stirrer and obtained a 98% yield.

89. CYCLIC DIOLEFIN COMPLEXES OF PLATINUM AND PALLADIUM

Submitted by D. DREW* and J. R. DOYLE*
Checked by ALAN G. SHAVER†

Recently compounds containing cyclic polyolefins coordinated to platinum or palladium have received considerable attention as a result of the unique bonding found in these compounds and their possible use as intermediates in a variety of reactions. Several methods have been reported for the synthesis of these compounds, and among these procedures the displacement of ethene[1] from di-μ-chloro-dichlorobis(ethene)diplatinum(II) and benzonitrile[2] from dichlorobis(benzonitrile)palladium(II) are the most generally applied procedures. Both of these methods involve the preparation of intermediates before the isolation of the product, and in addition these intermediates tend to decompose on storage.

Platinum Compounds

The procedure for the preparation of the platinum compounds is an extension of the method described by Kharasch and Ashford.[3] A glacial acetic acid solution of chloroplatinic acid is mixed with the appropriate olefin, and in the ensuing reaction the platinum is reduced from the 4 + oxidation state to the 2 + state. The overall stoichiometry of these reactions is not known; however, the reduction of the platinum is accompanied by the partial oxidation of the olefin.

A. DICHLORO(η^4-1,5-CYCLOOCTADIENE)-PLATINUM(II)

$$C_8H_{12} + H_2PtCl_6(H_2O)_x \rightarrow C_8H_{12}PtCl_2 \ddagger$$

Procedure

In a 125-mL Erlenmeyer flask 5.0 g (8.41 mmol) of hydrated chloroplatinic acid is dissolved in 15 mL of glacial acetic acid and the solution heated to 75°.

---

*The University of Iowa, Iowa City, IA 52242.
†Massachusetts Institute of Technology, Cambridge, MA 02139.
‡*Note*: Several recently purchased samples of the commercially available hydrated chloroplatinic acid, labeled to contain 40% platinum by weight, were actually analyzed as 32.8% by weight, and the yields were computed on the basis of the latter percentage. Caveat emptor.

Then 6 mL of 1,5-cyclooctadiene* is added to the warm solution and the mixture swirled gently, cooled to room temperature, and diluted with 50 mL of water. The black suspension is stored for one hour at room temperature, and the crude product is collected on a Büchner funnel, washed with 50 mL of water, and finally 100 mL of diethyl ether. The crude product is suspended in 400 mL of dichloromethane and the mixture heated to the boiling point and kept at this temperature for 5 min. The solution is cooled, mixed with 5.0 g of chromatographic-grade silica gel, and allowed to settle. The supernatant liquid should be colorless; if not, add additional silica gel in 1-g portions until the solution is clear. The mixture is filtered and the residue washed with two 50-mL portions of dichloromethane. The dichloromethane solution, approximately 500 mL, is evaporated until the product commences to crystallize, about 75 mL. The hot solution is poured into 200 mL of petroleum ether (bp 60–70°), yielding a finely divided white product. The precipitate is washed with 50 mL of petroleum ether and dried. Yield is 2.55 g (80%).

Anal. Calcd. for $C_8H_{12}PtCl_2$: C, 25.68; H, 3.23. Found: C, 25.73; H, 3.41.

A small amount of product can be recovered by evaporation of the dichloromethane–petroleum ether filtrate. The product can be recrystallized to yield white macroscopic crystals by dissolving the white powder in 150 mL of boiling dichloromethane and evaporating the solution until crystallization commences.

The dibromo(1,5-cyclooctadiene)platinum(II) and the diiodo(1,5-cyclooctadiene)platinum(II) can be prepared by a procedure similar to that described for the preparation of the dichloro derivative.[6]

Properties

Dichloro(1,5-cyclooctadiene)platinum(II) is a white, air-stable solid. The compound is slightly soluble in solvents such as chloroform, acetic acid, sulfolane (tetrahydrothiophene 1,1-dioxide), and nitromethane. It decomposes slowly on dissolution in dimethyl sulfoxide. The PMR spectrum of the compound in chloroform shows resonances at $\tau\, 4.38$, $J_{Pt-H} = 65$ Hz for the olefinic protons and $\tau\, 7.29$ for the methylene protons. The infrared spectrum in Nujol has strong absorption maxima at 1334, 1179, 1009, 871, 834, and 782 cm^{-1}.

---

*The following hydrocarbons have been successfully substituted for 1,5-cyclooctadiene in this procedure to yield the corresponding dichloro(olefin)platinum(II) derivatives: 1,3,5,7-cyclooctatetraene, dicyclopentadiene (3a,4,7,7a-tetrahydro-4,7-methanoindene), and bicyclo[2.2.1]hepta-2,5-diene (2,5-norbornadiene).

Palladium Compounds

B. DICHLORO(η^4-1,5-CYCLOOCTADIENE)-PALLADIUM(II)

$$2HCl + PdCl_2 \rightarrow H_2PdCl_4$$
$$H_2PdCl_4 + C_8H_{12} \rightarrow C_8H_{12}PdCl_2 + 2HCl$$

This compound has been prepared by the reaction of sodium tetrachloropalladate and the olefin[4] or by the displacement of carbon monoxide from $[COPdCl_2]_2$[5] or benzonitrile from $(C_6H_5C\equiv N)_2PdCl_2$. Bicyclo[2.2.1]hepta-2,5-diene (2,5-norbornadiene) may be substituted in the following procedure for 1,5-cyclooctadiene to yield (bicyclo[2.2.1]hepta-2,5-diene)dichloropalladium(II).

Procedure

Palladium(II) chloride (2.0 g, 11.3 mmol) is dissolved in 5 mL of concentrated hydrochloric acid by warming the mixture. The cool solution is diluted with 150 mL of absolute ethanol, filtered, and the residue and filter paper washed with 20 mL of ethanol. To the combined filtrate and washings is added 3.0 mL of 1,5-cyclooctadiene with stirring. The yellow product precipitates immediately, and after a 10-min storage is separated and washed with three 30-mL portions of diethyl ether. Yield is 3.10 g (96%, based on $PdCl_2$).

Anal. Calcd. for $C_8H_{12}PdCl_2$: C, 33.66; H, 4.24. Found: C, 34.26; H, 4.39.

The product can be obtained as macroscopic yellow crystals by dissolving the yellow powder in 200 mL of boiling methylene chloride and evaporating the hot solution until crystallization commences.

The dibromo(1,5-cyclooctadiene)palladium(II) can be prepared by a procedure similar to that described for the preparation of the dichloro derivative.[6]

Properties

Dichloro(1,5-cyclooctadiene)palladium(II) is an air-stable yellow solid. The compound is slightly soluble in solvents such as chloroform, sulfolane (tetrahydrothiophene 1,1-dioxide), and nitrobenzene and reacts with dimethyl sulfoxide to yield dichlorobis(dimethyl sulfoxide)palladium(II). The PMR spectrum of the compound in chloroform shows resonances at $\tau\,3.68$ for the olefinic protons and at $\tau\,7.31$ for the methylene protons. The infrared spectrum in Nujol has strong absorption maxima at 1489, 1419, 1337, 1088, 999, 867, 825, 794, 768, 325, and 295 cm^{-1}.

References

1. J. Anderson, *J. Chem. Soc.*, **1936**, 1042.
2. M. S. Kharasch, R. C. Seyler, and F. R. Mayo, *J. Am. Chem. Soc.*, **60**, 822 (1938).
3. M. S. Kharasch and T. Ashford, *J. Am. Chem. Soc.*, **58**, 1733 (1936).
4. J. Chatt, L. M. Vallarino, and L. M. Venanzi, *J. Chem. Soc.*, **1957**, 3413.
5. E. O. Fischer and H. Werner, *Chem. Ber.*, **93**, 2075 (1960).
6. D. Drew and J. R. Doyle, *Inorg. Synth.*, **13**, 47 (1972).

90. POTASSIUM TRICHLORO(ETHENE)PLATINATE(II) (ZEISE'S SALT)

$$K_2PtCl_4 + C_2H_4 + H_2O \xrightarrow{SnCl_2} K[PtCl_3(C_2H_4)] \cdot H_2O + KCl$$

Submitted by P. B. CHOCK,* J. HALPERN,* and F. E. PAULIK*
Checked by SAUL I. SHUPACK† and THOMAS P. DeANGELIS†

The original method[1] of preparation of Zeise's salt, $K[Pt(C_2H_4)Cl_3] \cdot H_2O$, and subsequent modifications thereof[2-4] all require either prolonged reaction times (7–14 days) or the use of high pressures. Furthermore, these procedures tend to yield products contaminated with potassium chloride and unreacted potassium tetrachloroplatinate(II). The improved procedure described below, which utilizes tin(II) chloride to catalyze the reaction between ethene and the tetrachloroplatinate(II),[5,6] results in the formation of Zeise's salt of high purity and in high yield within a few hours at atmospheric pressure.

Procedure

To 45 mL of 5 *M* aqueous hydrochloric acid in a 125-mL Erlenmeyer flask is added 4.5 g of potassium tetrachloroplatinate(II) (0.00108 mol). The flask is sealed with a rubber serum cap and deoxygenated immediately by flushing for 30 min with nitrogen or ethene through a polyethylene tube extending into the solution and attached to a needle inlet, with another needle as gas outlet. [Some undissolved potassium tetrachloroplatinate(II) may remain at

---

*University of Chicago, Chicago, IL 60637.
†Villanova University, Villanova, PA 19085.

this stage.] Then 40 mg of hydrated tin(II) chloride, $SnCl_2 \cdot 2H_2O$ (0.0002 mol)* is placed in a 5-mL flask, which is sealed with a serum cap and deoxygenated by flushing with pure nitrogen with needles as gas inlet and outlet. With a hypodermic syringe, 5 mL of deoxygenated distilled water is added to the tin(II) chloride, and the resulting suspension is transferred, also by means of a hypodermic syringe, to the flask containing the chloroplatinate(II). A stream of ethene is bubbled slowly through the resulting reaction mixture, which is shaken periodically. During the course of 2-4 h, the initially red-brown suspension turns yellow, and most of the solid dissolves as reaction proceeds. The reaction mixture is warmed to 40–45°C and clarified by filtering through a sintered-glass filter. (Do not use paper.) Cooling the filtrate in an ice bath yields a yellow precipitate of neelde-shaped crystals of Zeise's salt, $K[PtCl_3(C_2H_4)] \cdot H_2O$, which is separated by filtration, washed with a small amount of ice water, and air-dried at room temperature. The yield is 3.6 g (86%).† (Prolonged refrigeration of the mother liquor yields some additional product.) The infrared and visible–ultraviolet (Vis–UV) spectra (λ_{max}, 333 nm; ε_{max}, 230) of this product (which is unaffected by further recrystallizations from 5 M HCl) are in excellent accord with literature data.[7,8]

Pumping *in vacuo* for 16 h results in removal of the water of hydration, which yields $K[PtCl_3(C_2H_4)]$.

Anal. Calcd. for $C_2H_4Cl_3KPt$: C, 6.55; H, 1.09; Cl, 29.0. Found: C, 6.65; H, 1.09; Cl, 28.52.‡

Properties

Zeise's salt is obtained as well-formed, yellow, needle-shaped crystals. The compound is stable in the solid state at room temperature and decomposes with loss of ethene at about 180°C. The chemical, physical, and structural properties have been characterized thoroughly and are described in the literature.[9-13]

---

* Although the amount of tin(II) chloride used, and thereby the rate of the subsequent reaction, can be increased, the use of higher levels may be detrimental to the purity of the product and is not recommended.

† The checkers report that initial cooling in Dry Ice resulted in the precipitation of only 2 g of Zeise's salt. Evaporating the filtrate and adding just enough methanol (~10–15 mL) to dissolve the solids, filtering off the KCl and other impurities such as tin(II) chloride and unreacted K_2PtCl_4, followed by rapid evaporation of the methanol, yielded a further 1.5 g of Zeise's salt. Total yield: 85%.

‡ The checkers report that, using essentially the same method, they were able to prepare the analogous *cis*-2-butene platinum complex in 70% yield.

References

1. W. C. Zeise, *Mag. Pharm.*, **35,** 105 (1830).
2. I. I. Chernyaev and A. D. Hel'man, *Ann. Secteur Platine, Inst. Chim. gen.* (U.S.S.R.), **14,** 77 (1937).
3. J. Chatt and M. L. Searle, *Inorg. Synth.*, **5,** 210 (1957).
4. W. MacNevin, A. Giddings, and A. Foris, *Chem. Ind.* (London), **1958,** 577.
5. R. Cramer, E. L. Jenner, R. V. Lindsey, and U. G. Stolbert, *J. Am. Chem. Soc.*, **85,** 1691 (1963).
6. R. Pietropaolo, M. Graziani, and U. Belluco, *Inorg. Chem.*, **8,** 1506 (1969).
7. M. J. Grogan and K. Nakamoto, *J. Am. Chem. Soc.*, **88,** 5454 (1966).
8. J. R. Joy and M. Orchin, *J. Am. Chem. Soc.*, **81,** 305 (1959).
9. D. B. Powell and N. Sheppard, *Spectrochim. Acta*, **13,** 69 (1958).
10. J. W. Moore, *Acta Chem. Scand.*, **20,** 1154 (1966).
11. J. A. J. Jarvis, B. T. Kilbourn, and P. G. Owston, *Acta Cryst.*, **B27,** 366 (1971).
12. B. B. Bokii and G. A. Kukina, *Krystallografiya*, **2,** 3 (1967).
13. S. Maricic, C. R. Redpath, and J. A. Smith, *J. Chem. Soc.*, **1963,** 4905.

INDEX OF CONTRIBUTORS
Prepared by THOMAS E. SLOAN*

Adams, M. L., 27:314
Adams, R. D., 26:295, 303, 27:209
Aftandilian, V. D., 28:321
Ahmad, N., 28:81
Ahmed, K. J., 27:283
Aizpurua, J. M., 26:4
Albers, M. O., 26:52, 59, 68, 249, 28:140, 28:168, 179
Alexander, B. D., 27:214
Alonso, A. S., 26:393
Al-Salem, N., 28:342
Ammari, N., 27:128
Amundsen, A. R., 27:283
Andersen, R. A., 26:144, 27:146, 28:323
Anderson, G. K., 28:60
Angelici, R. J., 26:31, 77, 113, 27:294, 28:29, 92, 126, 148, 150, 155–158, 165, 28:186, 199, 230, 238, 278
Arliguie, T., 26:262, 271, 28:219, 221
Arndt, L. W., 26:335
Arnold, M. B., 27:322
Arnold, T., 28:196
Ash, C. E., 26:59, 28:168
Ashworth, T. V., 26:68
Attard, J. P., 26:303
Ayers, L. J., 26:200

Babin, J. E., 27:209
Backes-Dahmann, G., 27:39
Baker, R., 28:94
Bakir, M., 27:306
Balch, A. L., 27:222, 28:340
Ball, G. A., 26:388, 393
Bansemer, R. L., 27:14
Barder, T. J., 28:332
Barefield, E. K., 28:263
Barkhau, R. A., 26:388
Barthelmes, J. K.-H., 27:22, 214
Basil, J. D., 26:7, 28:305, 340
Basolo, F. 28:104, 160

Bassner, S., 28:236
Baudler, M., 27:227
Bauer, H., 26:117, 28:22
Beck, W., 26:92, 96, 106, 117, 126, 128, 26:231, 28:1, 5, 15, 22, 27, 63
Becker, G., 27:235, 240, 243, 249
Belefant, H., 26:12, 28:310
Bender, R., 26:341
Benner, L. S., 28:340
Bennett, M. A., 26:161, 171, 180, 200
Bercaw, J. E., 28:248, 317
Berg, D. J., 27:146
Bergman, R. G., 27:19
Bergounhou, C., 26:303, 360
Berke, H., 28:98
Bessel, C. A., 27:261
Bhattacharjee, M. N., 27:310, 312
Bianchini, C., 27:287
Binger, P., 26:204, 28:113, 119
Birdwhistell, R., 28:148
Bischoff, C. J., 26:335
Bishop, P. T., 28:37
Bisnette, M. B., 28:45
Blackman, G. S., 26:388
Blake, P. C., 26:150, 27:173
Blanski, R., 26:52, 28:179
Blohm, M. L., 26:286
Boncella, J. M., 27:146
Bonnet, J. J., 26:303, 360
Bosamle, A., 26:189
Bottomley, F., 27:59
Braam, J. M., 26:388
Braddock, J., 26:4
Bradley, J. S., 27:182
Braunstein, P., 26:312, 316, 319, 341, 26:356, 372, 27:188, 218
Bray, J., 28:86
Briat, B., 26:377
Briggs, D. A., 26:85
Brinkmann, A., 28:113, 119

*Chemical Abstracts Service, Columbus, OH.

Broomhead, J. A., 28:145, 338
Bruce, M. I., 26:171, 259, 262, 271, 324, 28:216, 219, 221, 270
Brunner, H., 27:69
Bruno, J. W., 27:173
Bryan, J. C., 28:326
Bryndza, H. E., 26:134
Buch, H. M., 26:204
Budge, J., 28:145
Bulls, A. R., 28:248
Burgess, K., 27:209, 218, 28:236
Burns, C. J., 27:146
Busby, D. C., 26:356
Busch, D. H., 27:261
Butler, I. S., 26:31, 28:140, 186

Cairns, C. J., 27:261
Calderazzo, F., 26:293, 28:192, 240
Cameron, C. J., 27:14
Canich, J. A., 27:329
Cariati, F., 28:123
Carr, C. J., 26:319
Carris, M. W., 28:263
Carty, A. J., 26:264, 28:248
Casalnuovo, J. A., 27:22
Casey, C. P., 26:231, 28:189
Cassidy, J., 27:182
Caulton, K. G., 27:14, 26
Cenini, S., 28:334
Ceriotti, A., 26:312, 316
Chakravorti, M. C., 27:294
Chaloupka, S., 26:126, 27:30, 28:27
Chaudhuri, M. K., 27:310, 312
Chebolu, V., 26:122, 28:56
Chen, A. Y-J., 26:52, 28:179
Chen, G. J.-J., 28:33, 38
Chen, Y-Y., 26:269
Chi, Y., 27:196
Chin, T. T., 28:196
Chini, P., 28:242
Chock, P. B., 28:349
Choi, D. J., 27:115
Choi, M. G., 27:294, 28:150, 186
Choudhury, D., 28:321
Clark, R. J., 26:12, 28:310
Clarke, M. J., 26:65
Clifford, D. A., 26:117, 28:22
Colbran, S. B., 26:264, 309
Collman, J. P., 28:92
Colsman, M. R., 27:329

Compton, S. J., 26:391
Contant, R., 27:104, 128
Corey, J. Y., 26:4
Cotton, F. A., 26:219, 27:306, 28:45, 28:110, 306, 332
Cotton, R., 26:81
Coucouvanis, D., 27:39, 47
Coulson, D. R., 28:107
Coville, N. J., 26:52, 59, 28:140, 168, 28:179
Cowan, R. L., 27:30
Cowie, A. G., 26:286
Cowley, A. H., 27:235, 240, 253
Cox, D. D., 26:388, 391, 393
Cozak, D., 28:186
Crabtree, R. H., 26:117, 122, 27:8, 22, 27:317, 28:22, 56, 88
Cramer, R., 28:86
Cramer, R. E., 27:177
Crascall, L. E., 28:126
Crawford, S. S., 26:155
Crumbliss, A. L., 28:68
Curtis, M. D., 27:69, 28:150
Cushing, M. A., 27:290, 28:94, 129
Cutler, A. R., 26:96, 231, 28:5

D'Aniello, Jr., M. J., 26:319
Darensbourg, D. J., 27:295, 28:236
Darensbourg, M. Y., 26:59, 169, 335, 28:168, 173
Darkwa, J., 27:51
Darst, K. P., 28:201
Dase, V., 26:259, 28:216
Davison, A., 28:92
Deacon, G. B., 26:17, 27:136, 142, 158, 27:161, 28:286, 291
DeAngelis, T. P., 28:349
Deeming, A. J., 26:289, 28:232
Degremont, A., 26:316, 319, 27:188
Dehnicke, K., 26:36
Derringer, D. R., 27:306
Diel, B. N., 28:315
Dietrich, H., 28:98
Dilworth, J. R., 28:33, 36–38
Dixon, N. E., 28:70
Doherty, N. M., 28:189
Dolan, P. J., 26:1
Domaille, P. J., 27:96
Dombek, B. D., 28:186, 199
Domingos, A. J. P., 28:52
Dominguez, R., 26:17, 28:291

Doran, S. L., 27:283
Doyle, J. R., 28:346
Doyle, R. A., 28:29, 126, 230
Drake, S. R., 28:230
Drew, D., 28:346
Drezdzon, M. A., 26:246

Edelman, M. A., 27:146
Edwin, J., 28:273
Elliott, G. P., 26:184
Ellis, J. E., 26:335, 28:192, 211
Enemark, J. H., 27:39
Engler, E. M., 26:391, 393
Evans, D., 28:79
Evans, W. J., 26:17, 27:155, 28:291, 28:297

Fackler, Jr., J. P., 26:7, 85, 27:177, 28:305
Fagan, P. J., 28:207, 317
Feng, S., 28:326
Ferris, E. C., 27:59
Feser, R., 28:280
Finke, R. G., 27:85, 128, 28:203
Fischer, E. O., 26:40
Floriani, C., 28:263
Forschner, T. C., 26:96, 231, 28:5
Forster, D., 28:88
Forsyth, C. M., 27:142
Fournier, M., 27:77–81, 83
Freeman, W. A., 27:314
Frith, S. A., 28:273
Fronzaglia, A. F., 28:45
Fuchita, Y., 26:208
Fumagalli, A., 26:372
Furuya, F. R., 26:246

Gaines, D. F., 26:1
Gangopadhyay, T., 27:294
Gard, G. L., 27:329
Garlaschelli, L., 28:211, 245
Geiger, Jr., W. E., 28:273
Genthe, W., 27:161
Geoffroy, G. L., 26:341, 28:236
George, T. A., 28:36
Giering, W. P., 28:207
Gilbert, R. J., 28:94
Gilbert, T. M., 27:19
Gilje, J. W., 27:177
Ginsberg, A. P., 27:222
Giolando, D. M., 27:51
Giordano, G., 28:88, 242

Gladfelter, W. L., 26:246, 286, 27:295, 28:245
Gladysz, J. A., 26:169, 27:59, 28:201
Glavee, G. N., 26:31, 28:238
Glinka, K., 27:227
Goel, A. B., 26:211
Goodson, P. A., 26:231
Gray, R. L., 28:236
Green, C. A., 27:128
Grelbig, T., 27:332
Gretz, E., 28:60
Grevels, F. W., 28:47
Grice, N., 28:52
Grim, S. O., 28:107
Grubbs, R. H., 26:44
Grumley, W., 28:145
Gudel, H. U., 26:377

Hackett, P., 28:148
Hadjikyriacou, A. I., 27:39, 47
Hall, S. W., 27:339
Halpern, J., 28:349
Halstead, G. W., 28:300
Hameister, C., 28:270
Hammershoi, A., 26:24
Hampton, M. D., 27:339
Hankey, D. R., 26:150
Hansongnern, K., 26:65
Hansson, E., 26:24
Harder, V., 28:280
Harter, P., 28:98
Hawkins, C. J., 26:24
Hay, M. S., 28:150
Haymore, B. L., 27:300, 327, 28:43
Hazin, P. N., 27:173
Head, R. A., 28:132
Heaton, D. E., 26:7, 28:305
Hedden, D., 27:322, 28:132
Heinrich, D. D., 27:177
Heintz, R. M., 28:88, 334
Heitmiller, R. F., 28:321
Henly, T., 26:324
Herrmann, G., 26:259, 28:216
Herve, G., 27:85, 96, 118, 120
Hill, E. W., 27:182
Hiraki, K., 26:208
Hlatky, G. G., 27:8
Hoff, C., 27:1
Hofmann, W., 28:280
Hollis, L. S., 27:283
Holmes, R. R., 27:258

Hood, P., 28:338
Horvath, I. T., 26:303
Howell, J. A. S., 28:52
Hoyano, J. K., 28:196
Hriljac, J., 26:280
Huckett, S. I., 26:113
Hur, N. H., 27:77–79
Hwang, L.-S., 27:196

Iggo, J. A., 26:225
Ismail, A. A., 26:31
Ittel, S. D., 27:290, 28:94, 98, 129, 28:270

Jaouen, G., 26:31
Jastrzebski, J. T. B. H., 26:150
Jeannin, Y., 27:111, 115
Jensen, C. M., 26:259, 28:216
Jeong, J. H., 27:177
Jeske, G., 27:158
Johnson, B. F. G., 28:47, 52
Jones, D. G., 28:113
Jones, N. L., 26:259, 28:216
Jones, R. A., 26:7, 27:227, 28:305
Jones, W. D., 26:180, 28:280
Jud, J., 26:341
Jung, C., 28:257

Kabir, S. E., 26:289, 28:232
Kacmarcik, R. T., 26:68
Kaesz, H. D., 26:52, 243, 269, 27:191, 28:84, 179, 238
Kafsz, H. D., 26:351
Kamil, W. A., 27:258
Kass, V., 26:161
Kauffman, G. B., 27:314
Keenan, S., 28:203
Keister, J. B., 27:196
Kim, C. C., 26:59, 28:168
King, R. B., 28:45
Kirchner, E., 27:235, 240
Klemperer, W. G., 27:71, 74, 77–81, 27:83, 104
Knox, S. A. R., 28:189
Koczon, L., 27:65
Kolis, J. W., 26:246
Komiya, S., 28:337
Krause, R. A., 26:65
Krickemeyer, E., 27:47
Kroll, W. R., 28:104
Krusic, P. J., 28:203

Kubas, G. J., 27:1, 8, 28:29, 68
Kubat-Martin, K. A., 27:8
Kubota, M., 28:92
Kuksuk, R. M., 27:314
Kwak, W. S., 27:123

Labinger, J. A., 28:267
Laguna, A., 26:85
Laguna, M., 26:85
Lambrecht, J., 28:98
La Monica, G., 28:123
Lappert, M. F., 26:144, 150, 27:146, 164, 27:168, 173
Latten, J. L., 26:180, 200
Lauher, J. W., 27:188
Lavigne, G., 26:262, 271, 360, 28:219, 28:221
Lawless, G. A., 27:146
Lawrance, G. A., 28:70
Lawrence, S. H., 26:1
Lay, P. A., 28:70
Layh, M., 27:235, 240
Lebreton, P. R., 27:314
Lee, W., 28:136
Legzdins, P., 28:196
Lehner, H., 27:218
Leidl, R., 26:128, 28:63
Leuenberger, B., 26:377
Levan, K. R., 26:17, 28:291
Levison, J. J., 28:81
Lewis, J., 28:52
Li, T., 27:310, 312
Libby, E., 27:306
Liddell, M. J., 26:171
Lin, M., 28:60
Lindner, E., 26:142, 161, 189
Lindsell, W. E., 26:328, 27:222, 224
Lippard, S. J., 27:283
Lipson, L. G., 28:84
Liu, J., 27:111, 118, 120
Lochschmidt, S., 27:253
Longoni, G., 26:312, 316
Loveday, P. A., 28:230
Low, A. A., 27:188
Lucas, C. R., 28:267
Luetkens, Jr., M. L., 26:7, 28:305
Lukehart, C. M., 28:199, 201
Lundquist, E. G., 27:26
Luo, X. L., 27:317
Lyon, D. K., 27:85

Index of Contributors

Mcauley, N. M., 26:184
McCall, J. M., 27:59, 65
McCarley, R. E., 28:37
McCleverty, J. A., 28:45, 84, 86
McCormick, F. B., 26:249
McDermott, A., 26:40
McDonàld, J. W., 28:33, 38, 145
McGrady, N., 28:317
McGrath, M. P., 27:22
Mckinney, R. J., 26:155
MacKinnon, P. I., 27:158
Maclaughlin, S. A., 26:264
Madden, D. P., 28:94
Maddox, M. L., 28:84
Mahaffy, C. A. L., 28:136
Maitlis, P. M., 28:110
Malito, J. T., 28:196
Mange, V., 28:263
Manning, A. R., 28:148
Manriquez, J. M., 28:317
Mantell, D. R., 27:295
Mantovani, A., 28:334
Manzer, L. E., 28:260
Marchionna, M., 26:316
Marko, L., 26:243
Marks, T. J., 27:136, 28:260, 286, 300, 28:317
Marmolejo, G., 27:59, 65
Martin, T. R., 26:144
Martinengo, S., 26:369, 372, 28:211, 242, 28:245
Martin-Frere, J., 27:111, 115
Matrana, B. A., 26:243
Matsuda, T., 28:119, 122
Matt, D., 27:218
Maurya, M. R., 26:36
Mayer, J. M., 28:326
Maynard, R. B., 27:177
Mayr, A., 26:40
Mays, M. J., 26:225
Meier, M., 28:104
Meier, W., 27:69
Meli, A., 27:287
Mellea, M. F., 26:122, 28:56
Merola, J. S., 26:68, 27:19
Mihelcic, J. M., 26:122, 28:56
Miller, D. C., 28:148
Mills, J. L., 27:339
Mintz, E. A., 28:260
Moehring, G. A., 27:14
Monzyk, B., 28:68

Moody, D. C., 26:1
Moriarty, K. J., 28:248
Morris, D. E., 28:88, 334
Morris, M. J., 28:189
Morse, S. A., 27:295
Moy, D., 28:104
Mrowca, J. J., 28:77, 123
Mueting, A. M., 27:22, 214
Muller, A., 27:47, 51
Muniyappan, T., 28:321
Murchie, M. P., 27:332
Murillo, C. A., 27:306
Murphy, C. J., 27:51
Murray, H. H., 26:7, 85, 28:305

Natarajan, K., 26:295
Nelson, C. K., 26:1
Neukomm, H., 28:280
Neumann, H., 26:161, 171
Nguyen, S., 26:134
Nicholls, J. C., 28:278
Nicholls, J. N., 26:280, 289, 295, 28:232
Nicholson, B. K., 26:271, 324, 28:221
Nickel, S., 27:155, 161, 28:297
Niedenzu, P., 27:339
Nielson, A. J., 27:300, 327, 28:315, 28:323
Nitschke, J., 26:113
Noda, H., 26:369
Norman, N. C., 27:235, 240
Novak, B., 26:328, 351
Nucciarone, D., 26:264

Oberdorf, K., 26:155
Ohgomori, Y., 27:287, 290
Ohlmeyer, M. J., 27:218
Olgemoller, B., 26:117, 126, 28:22, 27
Olgemoller, L., 26:126, 28:27
Onderdelinden, A. L., 28:90
Oosthuizen, H. E., 26:68
Opitz, J., 28:98
Osborn, J. A., 28:77, 79
Otsuka, S., 28:113, 119, 122, 342
Ovalles, C., 27:295

Pain, G. N., 26:17, 171, 28:291
Paine, Jr., R., 28:79
Pakulski, M., 27:235, 240, 253
Palamidis, E., 27:161
Palomo, C., 26:4
Pampaloni, G., 28:192

Pamploni, G., 26:293, 28:240
Pangagiotidou, P., 26:81
Papageorgiou, F., 26:360
Park, Y. K., 26:169
Parry, R. W., 28:79
Parshall, G. W., 28:81, 119, 122, 242
Passmore, J., 27:332
Patel, V. D., 28:248
Paulik, F. E., 28:349
Pauson, P. L., 28:136
Pederson, S. F., 26:44
Peet, W. G., 28:119, 122
Pell, S. D., 26:65
Pergola, R. D., 28:211, 245
Pettit, R., 28:52
Pfeffer, M., 26:208, 211
Pickner, H. C., 27:295
Pignolet, L. H., 27:22, 214
Piva, G., 26:312
Pombeiro, A. J. L., 28:43
Pope, M. T., 27:111, 115, 118, 120, 27:123
Powell, G. L., 28:332
Pray, A. R., 28:321

Quick, M. H., 28:155–158

Raab, K., 26:106, 28:15
Rapko, B., 27:128
Raston, C. L., 26:144
Rau, A., 26:161
Rauchfuss, T. B., 27:51, 65
Rausch, M. D., 28:136, 248
Raymond, K. N., 28:300
Reed, P. E., 26:388, 391
Rees, Jr., W. S., 27:339
Regitz, M., 27:243, 249
Reimer, K. J., 28:155–158
Reimer, S., 27:222
Rettig, M. F., 28:110
Reuvers, J. G. A., 28:47
Riaz, U., 27:69
Richards, R. L., 28:33, 38, 43
Rickard, C. E. F., 28:315
Rigsbee, J. T., 27:150
Robinson, B. H., 26:309
Robinson, S. D., 28:81
Roe, J., 27:188
Roesky, H. W., 27:332
Romanelli, M. G., 28:104
Roper, W. R., 26:184

Rosch, W., 27:243, 249
Rose, J., 26:312, 316, 319, 356
Rosenberg, E., 26:328, 351
Rosenblum, M., 28:207
Roundhill, D. M., 27:322, 28:132
Rowley, S., 27:222

Sailor, M., 26:280
Samuels, S.-B., 28:207
Sancho, J., 26:44
San Filippo, Jr., J., 28:203
Santure, D. J., 26:219
Sappa, E., 26:365
Sargeson, A. M., 28:70
Satek, L. C., 28:107
Sattelberger, A. P., 26:7, 219, 28:305
Sauer, N. N., 26:77, 165
Schaeffer, R., 26:1
Scherer, O. J., 27:224
Schertz, L. D., 28:317
Schloter, K., 28:5
Schmid, G., 27:214
Schmidpeter, A., 27:253
Schmidt, H., 27:243, 249
Schmidt, M., 27:235, 240
Schmidt, S. P., 26:113, 28:160
Schmock, F., 26:36
Scholter, K., 26:96
Schrock, R. R., 26:44
Schumann, H., 27:155, 158, 161, 28:297
Schunn, R. A., 28:90, 94
Schwalb, J., 27:224
Schwartz, J., 28:257, 267
Sears, Jr., C. T., 28:92
Selover, J. C., 28:201
Sen, A., 26:122, 128, 27:164, 28:56, 60, 28:63
Seppelt, K., 27:332
Sethuraman, P. R., 27:123
Seyam, A. M., 28:300
Shapley, J. R., 26:77, 106, 215, 309, 26:324, 27:196, 28:15, 160, 165, 184
Sharp, P. R., 28:326
Shaver, A. G., 27:59, 65, 28:155–158, 346
Shaw, B. L., 28:342
Shawkataly, O. B., 26:324
Shawl, E. T., 28:92
Sherban, M. M., 26:65
Sherlock, S., 28:245
Shore, S. G., 26:1, 360, 365, 27:339
Shreeve, J. M., 27:258

Shriver, D. F., 26:246
Shupack, S. I., 28:349
Shyu, S.-G., 26:225
Siedle, A. R., 27:317
Sikora, D. J., 28:248
Silavwe, N., 27:80, 81
Silva, R., 28:173
Simerly, S., 27:65
Simpson, J., 26:309
Singh, A., 27:164, 168
Singleton, E., 26:52, 59, 68, 249, 28:168, 179
Siriwardane, U., 26:360, 365
Sitzmann, H., 27:224
Smith, J., 28:36
Smith, J. M., 28:315
Smith, R. G., 27:164
Snow, M. R., 26:81
Sorato, C., 26:134
Sowa, Jr., J. R., 28:92, 278
Sowa, L. M., 26:388
Soye, P., 28:148
Spencer, J. L., 26:155, 27:26, 28:113, 28:126, 129, 273, 278
Springborg, J., 26:24
Springer, W., 28:98
Stebler, A., 26:377
Stecher, H. A., 27:164
Stern, E. W., 27:283
Stone, F. G. A., 27:191
Storhoff, B. N., 27:322
Strauss, S. H., 27:329
Streitwieser, Jr., A., 27:150
Strickland, D. A., 27:196
Strong, H., 28:203
Strycker, S., 28:321
Sun, X., 27:123
Sunkel, K., 26:96, 231, 28:5
Suss-Fink, G., 26:259, 269, 28:216
Swamy, K., 27:258
Swanson, B. I., 28:196
Swinger, A. G., 28:270
Syamal, A., 26:36

Takacs, J., 26:243, 27:168, 28:47
Takeuchi, K. J., 27:261
Tamres, M., 28:321
Tatsuno, Y., 28:342
Taube, H., 28:70
Taylor, R. G., 27:173
Templeton, J. L., 28:326

Terjeson, R. J., 27:329
Teze, A., 27:85, 96, 118, 120
Therien, M. J., 28:173
Thomas, R. R., 26:128, 28:63
Thorn, D. L., 26:200
Thouvenot, R., 27:128
Tilley, T. D., 27:146
Todaro, A., 26:96, 28:5
Torrence, G. P., 28:199
Tramontano, V., 26:65
Trogler, W. C., 26:113, 27:30, 28:160, 28:173
Tuong, T. D., 26:17, 27:136, 28:286, 291

Ugo, R., 28:123
Uhl, G., 27:243, 249
Uhl, W., 27:243, 249
Uhm, H. L., 28:140
Ulibarri, T. A., 27:155, 28:297
Underhill, A. E., 26:388
Urban, G., 26:96, 28:5
Urbancic, M. A., 26:77, 28:160, 165
Uremovich, M. A., 27:329
Uson, R., 26:85
Uttley, M. F., 28:81

Vahrenkamp, H., 26:341, 351, 27:191
Valle, G., 27:306
Valle, M., 26:365
van der Ent, A., 28:90
van der Sluys, L. S., 28:29
Van Koten, G., 26:150
Vargas, M. D., 26:280, 289, 295, 28:232
Vaughn, G. D., 26:169
Venanzi, L. M., 26:126, 134, 27:30, 28:27
Virgil, S. C., 26:44
Vogel, G. C., 26:341
Vogelbacher, U.-J., 27:243, 249

Wachter, J., 27:69
Wachter, W. A., 28:300
Wailes, P. C., 28:257
Wakatsuki, Y., 26:189
Wallis, R. C., 28:270
Walton, R. A., 27:14, 306, 28:332
Wang, H. H., 26:388, 27:22
Wang, R.-C., 27:77–79
Wang, W., 26:219
Warfield, L. T., 28:201
Watanabe, Y., 27:290
Watt, D., 26:372

Wayda, A. L., 27:142, 150
Webb, T. R., 28:110
Wedemann, P., 28:113, 119
Weigold, H., 28:257
Werner, H., 28:280
White, J. E., 28:45
White, R. P., 27:209, 28:189
Whiteker, G. T., 28:189
Whitmire, K. H., 26:243, 27:186
Whittlesey, B. R., 26:106, 28:15
Wilkinson, D. L., 27:136, 142, 28:286
Wilkinson, G., 28:77, 79, 84, 337
Williams, J. M., 26:386, 388, 391, 393
Williams, M. L., 26:262, 271, 27:191, 28:219, 221
Williamson, S. M., 26:12, 28:310
Wimmer, F. L., 26:81
Witt, M., 27:332
Wittmann, D., 26:40
Wojcicki, A., 26:225
Wonchoba, E. R., 28:81, 242
Wong, C.-M., 27:332
Worrell, J. H., 27:310, 312

Xu, D. Q., 26:351, 27:191

Yaghi, O. M., 27:83, 104
Yamamoto, A., 28:337
Yamamoto, T., 26:204
Yamazaki, H., 26:189, 369
Yan, S., 27:22
Yasufuku, K., 26:369
Yates, J. E., 26:249
Yeh, W. Y., 26:309
Yoshida, T., 28:113, 119, 122, 342
Young, C. G., 28:338
Young, R., 28:337
Youngdahl, K., 26:169

Zammit, M. G., 26:319
Zeile, J. V., 28:199, 201
Zheng, X., 28:37
Zubieta, J., 27:123, 28:36
Zuccaro, C., 26:293, 28:240

SUBJECT INDEX
Prepared by THOMAS E. SLOAN*

Names used in this Subject Index for Volumes 26–30 are based upon IUPAC *Nomenclature of Inorganic Chemistry*, Second Edition (1970), Butterworths, London; IUPAC *Nomenclature of Organic Chemistry*, Sections A, B, C, D, E, F, and H (1979), Pergamon Press, Oxford, U.K.; and the Chemical Abstracts Service *Chemical Substance Name Selection Manual* (1978), Columbus, Ohio. For compounds whose nomenclature is not adequately treated in the above references, American Chemical Society journal editorial practices are followed as applicable.

Inverted forms of the chemical names (parent index headings) are used for most entries in the alphabetically ordered index. Organic names are listed at the "parent" based on Rule C-10, *Nomenclature of Organic Chemistry*, 1979 Edition. Coordination compounds, salts and ions are listed once at each metal or central atom "parent" index heading. Simple salts and binary compounds are entered in the usual uninverted way, e.g., *Sulfur oxide* (S_8O), *Uranium (IV) chloride* (UCl_4).

All ligands receive a separate subject entry, e.g., *2,4-Pentanedione*, iron complex. The headings *Ammines, Carbonyl complexes, Hydride complexes*, and *Nitrosyl complexes* are used for the NH_3, CO, H, and NO ligands.

Acetaldehyde, iron complex, 26:235
  manganese and rhenium complexes, 28:199, 201
Acetic acid, chromium, molybdenum, and tungsten complexes, 27:297
  palladium complex, 26:208
  rhodium complex, 27:292
  tungsten complex, 26:224
—, chloro-, ruthenium complex, 26:256
—, trichloro-, ruthenium complex, 26:256
—, trifluoro-, ruthenium complex, 26:254
  tungsten complexes, 26:222
Acetone, iridium complex, 26:123
  molybdenum and tungsten complexes, 26:105, 28:14
—, dibenzylidene-, *see* 1,4-pentadien-3-one, 1,5-diphenyl-, 28:110
Acetonitrile, molybdenum and tungsten complexes, 26:122, 133
  molybdenum complex, 28:34, 37

osmium complex, 26:290
palladium complex, 26:128
ruthenium complex, 26:356
ruthenium (II) complex, 26:69
transition metal complexes, 28:63–66
Acetylene, diphenyl-, molybdenum complex, 26:102
Actinide complexes, 28:286
Acyl isocyanide, chromium complexes, 26:31
Ammines, ruthenium, 26:66
Ammoniodicobaltotetracontatungstotetraarsenate (23–),
  $[(NH_4)As_4W_{40}O_{140}[Co(H_2O)]_2]^{23-}$, tricosaammonium, nonadecahydrate, 27:119
Ammonium ammoniodicobaltotetracontatungstotetraarsenate (23–),
  $((NH_4)_{23}[(NH_4)As_4W_{40}O_{140}-[Co(H_2O)]_2])$, nonadecahydrate, 27:119

*Chemical Abstracts Service, Columbus, OH.

Ammonium dihydrogen pentamolybdobis
  [(4-aminobenzyl)phosphonate](4−),
  [(NH$_4$)$_2$H$_2$[Mo$_5$O$_{15}$-
  (NH$_2$C$_6$H$_4$CH$_2$PO$_3$)$_2$]],
  pentahydrate, 27:126
Ammonium pentamolybdobis (ethylphosphonate (4−), [(NH$_4$)$_4$-
  [Mo$_5$O$_{15}$(C$_2$H$_5$PO$_3$)$_2$]], 27:125
Ammonium pentamolybdobis (methylphosphonate) (4−), [(NH$_4$)$_4$-
  [Mo$_5$O$_{15}$(CH$_3$PO$_3$)$_2$]), dihydrate,
  27:124
Ammonium pentamolybdobis-
  (phenylphosphate)(4−), [(NH$_4$)$_4$-
  [Mo$_5$O$_{15}$(C$_6$H$_5$PO$_3$)$_2$]], pentahydrate,
  27:125
Ammonium pentamolybdobis (phosphonate)-
  (4−), [(NH$_4$)$_4$[(HPO$_3$)$_2$Mo$_5$O$_{15}$]),
  tetrahydrate, 27:123
Ammonium [(1R) (endo, anti)]-3-bromo-1,7-
  dimethyl-2-oxobicyclo [2.2.1] heptane-7-
  methanesulfonate, 26:24
Ammonium sodiohenicosatungstononaantimonate-
  (18−), [(NH$_4$)$_{18}$[NaSb$_9$W$_{21}$O$_{86}$],
  tetracosahydrate, 27:120
Ammonium sodiotricontatungstopentaphosphate (14−),
  [(NH$_4$)$_{14}$[NaP$_5$W$_{30}$O$_{110}$], hentricontahydrate, 27:115
Anisole, chromium complex, 28:137
Aromatic ketones, metallation of, 26:155
Arsine, 1,2-diphenylbis(dimethyl)-, gold complex, 26:89
  nickel complex, 28:103
—, triphenyl-, iron complex, 28:171
  nickel complex, 28:103
Aryl oxide complexes of lanthanide metals,
  27:164
Ascorbic acid, platinum complex, 27:283
Aurate(I), bis(pentafluorophenyl)-, bis[1,2-phenylenebis(dimethylarsine)]gold(I),
  26:89
—, chloro(pentafluorophenyl)-,
  (benzyl)triphenylphosphonium, 26:88
Azide, μ-nitrido-bis(triphenylphosphorus)-
  (1+), 26:286
Azobenzene, manganese complex, 26:173
  palladium complex, 26:175

Benzamide, N-[2-
  (diphenylphosphino)phenyl]-, 27:323
Benzenamine, tungsten complex, 27:301
—, N,N-dimethyl-, chromium complex,
  28:139
Benzene, chromium complex, 28:139
—, 1,2-bis[(trimethylsilyl)methyl]-, 26:148
  lithium complex, 26:148
—, 1-bromo-2,4,6-tri-$tert$-butyl-, 27:236
—, 1,3-butadiene-1,4-diylbis-, cobalt complex, 26:195
—, chloro-, chromium complex, 28:139
—, 1,2-diiodo-, iridium complex, 26:125
—, 1,2-ethenediylbis-, platinum complex,
  26:140
—, 1,1'-(1,2-ethynediyl)bis-, cobalt complex,
  26:192
—, ethynyl-ytterbium complex, 27:143
—, fluoro-, chromium complex, 28:139
—, hexamethyl-, ruthenium complex, 26:181
—, 2-isocyano-1,3-dimethyl-, iron complexes,
  26:53–57, 28:180–184
Benzene, methyl-, cobalt complex, 26:309
  lutetium complex, 27:162
  manganese complexes, 26:172
Benzene, pentafluoro-, gold complexes,
  26:86–90
Benzenemethanamine, N,N-dimethyl-, lutetium complex, 27:153
  lithium complex, 26:152
  palladium complex, 26:212
—, N,N,4-trimethyl-, lithium complex,
  26:152
Benzenemethanol, 2-phenylphosphino-, manganese complex, 26:169
Benzenesulfonic acid, 4-methyl-, rhodium
  complex, 27:292
Benzenethiol, osmium complex, 26:304
1,2-Benzisothiazol-3(2H)-one, 1,1-dioxide,
  chromium and vanadium complexes,
  27:307, 309
Benzo[h]quinoline, ruthenium complex,
  26:177
Benzoic acid, rhodium complex, 27:292
—, 3-fluoro-, rhodium complex, 27:292
Benzonitrile, palladium and platinum complexes, 28:60–62
  platinum complex, 26:345
  ruthenium(II) complex, 26:70

Subject Index

Benzoyl isocyanide, chromium, 26:34, 35
chromium complex, 26:32
Benzylideneacetone, *see* 3-butene-2-one, 4-phenyl-, 28:52
Bicyclo[2.2.1]hepta-2,5-diene, ruthenium complex, 26:250
Bicyclo[2.2.1]heptane-7-methanesulfonate, 3-bromo-1,7-dimethyl-2-oxo-, [(1R)-(*endo, anti*)]-, ammonium, 26:24
Bicyclo[2.2.1]hept-2-ene, platinum complex, 28:127
2,2'-Bi-1,3-dithiolo[4,5-*b*]-[1,4]dithiinylidene, 26:386
2,2'-Bi-1,3-dithiolo[4,5-*b*][1,4]dithiinylidene fluorosulfate, 26:393
2,2'-Bipyridine, nickel complex, 28:103
palladium complex, 27:319
rhenium complex, 26:82
ruthenium complex, 28:338
tungsten complex, 27:303
Bis(2,2'-bi-1,3-dithiolo[4,5-*b*]-[1,4]dithiinylidene) perrhenate, superconducting, 26:391
Borane, trimethyl-, 27:339
Borate(1−), tetradecahydronona-, potassium, 26:1
—, tetrafluoro-, iridium(III) complexes, 26:117
—, tetrafluoro-, molybdenum and tungsten complexes, 26:96
—, tetrafluoro-, rhenium complex, 26:108
(+)-α-Bromocamphor-π-sulfonate, *see* bicyclo[2.2.1]heptane-7-methanesulfonate, 3-bromo-1,7-dimethyl-2-oxo-, [(1R)-(*endo, anti*)]-, 26:24
Butanoic acid, 3-oxo-, methyl ester, rhodium complex, 27:292
3-Butenamide, nickel complex, 26:206
2-Butenedioic acid, 2-(dimethylphosphinothioyl)-, dimethyl ester, manganese complex, 26:163
3-Butene-2-one, 4-phenyl-, iron complex, 28:52
tert-Butyl isocyanide, chromium, molybdenum, and tungsten complexes, 28:143
—, nickel complex, 28:99
tert-Butyl isocyanide, *see* Propane, 2-isocyano-2-methyl-, 28:224

1-Butyne, 3,3-dimethyl-, mercury-molybdenum-ruthenium complexes, 26:329
2-Butyne, cobalt-molybdenum-ruthenium, 27:194

Cadmium dichloride, 28:322
Carbide, iron complex, 26:246
Carbido carbonyl ruthenium clusters, 26:280
Carbon, ruthenium cluster complexes, 26:281
Carbon dioxide, rhenium complex, 26:111
rhenium complexes, 28:20
Carbonyl complexes, chromium, 26:32, 28:137–139
chromium, molybdenum, and tungsten, 26:343, 27:297
cobalt, 28:273, 275
cobalt, copper, and ruthenium, 26:358
cobalt, iron, and ruthenium, 26:352
cobalt-gold-iron, 27:188
cobalt-gold-ruthenium, 26:327
cobalt-platinum, 26:370
cobalt-molybdenum-nickel, 27:192
cobalt-molybdenum-ruthenium, 27:194
cobalt and ruthenium, 26:176, 177
gold-osmium, 27:209, 211
hafnium, titanium, and zirconium, 28:248–255
iridium, 26:117, 28:23–26, 92
iridium, osmium, rhodium, and ruthenium, 28:216–247
iridium and rhodium, 28:213, 214
iron, 26:53–57, 27:183
iron-tungsten, 26:336
iron with Group 15 ligands, 26:59–63
manganese, 26:114, 28:155–158
manganese and rhenium, 28:15, 17, 199, 28:201
mercury-molybdenum-ruthenium, 26:329
molybdenum, 26:84, 27:3
molybdenum and platinum, 26:345
molybdenum and tungsten, 26:96, 28:5
nickel, 26:312
nickel, osmium, and ruthenium, 26:362
niodium, 28:192
osmium, 26:187
osmium and ruthenium, 27:196–207
platinum, 26:316
platinum-rhodium, 26:373

Carbonyl complexes (*Continued*)
  rhenium, 26:77, 28:19
  rhodium, 27:291
  ruthenium, 26:259, 28:47, 54
  tungsten, 26:40, 27:4
Carbonyl substituted metal complexes, 28:136
Carboxy, rhenium complexes, 28:21
Cerium, bis[η^5-1,3-bis(trimethylsilyl)cyclopentadienyl]di-μ-chloro-bis(tetrahydrofuran)lanthanum-, 27:170
—, tetrakis [η^5-1,3-bis-(trimethylsilyl)cyclopentadienyl]di-μ-chloro-di-, 27:171
Cesium decatungstophosphate(7−), ($Cs_7[PW_{10}O_{36}]$), 27:101
Cesium β-divanadodecatungstophosphate-(5−), ($Cs_5[\beta\text{-}PV_2W_{10}O_{40}]$), 27:103
Cesium pentatungstodiphosphate(6−), ($Cs_6[P_2W_5O_{23}]$), 27:101
Cesium α-1,2,3-trivanadononatungstophosphate(6−), ($Cs_6[\alpha\text{-}PV_3W_9O_{40}]$), 27:100
Cesium vanadodecatungstophosphate(5−), ($Cs_5([\gamma\text{-}PV_2W_{10}O_{40}]$), 27:102
Chalcogenide transition metal complexes, 27:39
Chiral compounds, trinuclearmetal clusters, 27:191
Chromate(1−), (acetato)pentacarbonyl-, μ-nitrido-bis(triphenylphosphorus)(1+), 27:297
—, hydridononacarbonyliron-, μ-nitrido-bis(triphenylphosphorus)(1+), 26:340
—, tricarbonyl (η^5-cyclopentadienyl)-, sodium, compound with 1,2-dimethoxyethane (1:2), 26:343
Chromate(2−), nonacarbonyliron-, bis[μ-nitrido-bis(triphenylphosphorus)(1+)], 26:339
Chromate(3−), nonabromodi-, tricesium, 26:379
—, nonabromodi-, trirubidium, 26:379
—, nonachlorodi-, tricesium, 26:379
Chromate(VI), fluorotrioxo-, pyridinium, 27:310
Chromium, (η^6-anisole)tricarbonyl-, 28:137
—, (η^6-benzene)tricarbonyl-, 28:139

—, (benzoyl isocyanide) dicarbonyl (η^6-methyl benzoate)-, 26:32
—, (benzoyl isocyanide) pentacarbonyl-, 26:34, 35
—, bis (*tert*-butyl isocyanide)tetracarbonyl-, *cis*-, 28:143
—, (*tert*-butyl isocyanide)pentacarbonyl-, 28:143
—, dicarbonyl(η^5-cyclopentadienyl)nitrosyl-, 28:196
—, (μ-disulfido-S:S) (μ-η^2:η^2-disulfido)bis(η^5-pentamethylcyclopentadienyl)-μ-thio-di-, (*Cr*–*Cr*), 27:69
—, hexacarbonylbis (η^5-cyclopentadienyl)di-, 28:148
—, tricarbonyl (η^6-chlorobenzene)-, 28:139
—, tricarbonyl (η^6-*N,N*-dimethylbenzenamine)-, 28:139
—, tricarbonyl (η^6-fluorobenzene)-, 28:139
—, tricarbonyl (η^6-methylbenzoate)-, 28:139
—, tricarbonyltris(propionitrile)-, 28:32
—, tris(*tert*-butyl isocyanide)-tricarbonyl-, *fac*-, 28:143
Chromium(II), tetraaquabis(1,2-benzisothiazol-3(2*H*)-one 1,1-dioxidato)-, dihydrate, 27:309
Chromium(III), *cis*-dichlorobis(1,2-ethanediamine)-, chloride, resolution of, 26:24, 27
—, dichlorobis(1,2-ethanediamine)-, Λ-*cis*-, chloride, monohydrate, resolution of, 26:28
Chromium carbonyl acyl isocyanides, 26:31
Chromium trichloride, 28:322
Cluster compounds, chiral, 27:191
  iridium, osmium, rhodium, and ruthenium, 28:216–247
  molybdenum–sulfur, 27:47
  transition metal, 27:182
Cobalt, (acetonitrile)dodecacarbonylcopper-rutheniumtri-, 26:359
—, (μ_3-2-butyne)nonacarbonylrutheniumdi-, 27:194
—, *cyclo*-[μ_3-1(η^2):2(η^2):3(η^2)-2-butyne]-octacarbonyl-1$\kappa^2 C$, 2$\kappa^3 C$, 3$\kappa^3 C$-[1(η^5)-cyclopentadienyl]molybdenumruthenium-, (*Co*–*Mo*)(*Co*–*Ru*) (*Mo*–*Ru*), 27:194
—, chlorotris(triphenylphosphine)-, 26:190

—, (η^5-cyclopentadienyl) (1,4-bis(methoxycarbonyl)-2-methyl-3-phenyl-1,3-butadiene-1,4-diyl](triphenylphosphine)-, 26:197
—, (η^5-cyclopentadienyl)bis(triphenylphosphine)-, 26:191
—, (η^5-cyclopentadienyl) (2,3-dimethyl-1,4-diphenyl-1,3-butadiene-1,4-diyl)(triphenylphosphine)-, 26:195
—, (η^5-cyclopentadienyl) [η^2-1,1'-(1,2-ethynediyl)bis(benzene)] (triphenylphosphine)-, 26:192
—, (η^5-cyclopentadienyl) (methyl 3-phenyl-η^2-2-propynoate) (triphenylphosphine)-, 26:192
—, dodecacarbonyltris(triphenylphosphine)trigoldtriruthenium-, 26:327
—, cyclo-μ_3-ethylidyne-1:2:3-κ^3C-pentacarbonyl-1κ^2C, 2κ^3C-bis[1,3(η^5)-cyclopentadienyl]molybdenumnickel-, (Co–Mo)(Co–Ni) (Mo–Ni), 27:192
—, heptacarbonyl[1,2-ethanediylbis(diphenylphosphine)]-platinumdi-, 26:370
—, nonacarbonyl (μ_3-phenylphosphinidene)irondi-, 26:353
—, nonacarbonyl-μ_3-thio-irondi-, 26:245
—, nonacarbonyl-μ_3-thio-rutheniumdi-, 26:352
—, octacarbonyl(η^5-cyclopentadienyl)-μ_3-ethylidynemolybdenumdi-, 27:193
—, tricarbonyl[2-(phenylazo)phenyl-C^1,N^2]-, 26:176
—, tris(η^5-cyclopentadienyl)bis(μ_3-phenylmethylidyne)tri-, 26:309
—, undecacarbonylrutheniumdi-, 26:354
Cobalt (I), bis(η^2-ethene)(η^5-pentamethylcyclopentadienyl)-, 28:278
—, (η^5-cyclopentadienyl)bis(trimethylphosphine)-, 28:281
—, (η^5-cyclopentadienyl)bis(trimethyl phosphite)-, 28:283
—, dicarbonyl(η^5-pentamethylcyclopentadienyl)-, 28:273
Cobalt (II), (2,3,10,11,13,19-hexamethyl-3,10,14,18,21,25-hexaazabicyclo[10.7.7]-hexacosa-1,11,13,18,20,25-hexaene-$\kappa^4N^{14,18,21,25}$)-, bis[hexafluorophosphate(1 −)], 27:270

Cobalt(III), carbonyldiiodo (η^5-pentamethylcyclopentadienyl)-, 28:275
—, di-μ-iodo-bis[iodo(η^5-pentamethylcyclopentadienyl)-, 28:276
Cobaltate(1 −), dodecacarbonylirontri-, tetraethylammonium, 27:188
—, dodecacarbonylirontri-, (triphenylphosphine)gold(1 +), 27:188
—, dodecacarbonylrutheniumtri-, tetraethylammonium, 26:358
Cobalt dichloride, 28:322
Cobalt dioxygen carriers, 27:261
Complexes with weakly bonded anions, 28:1
Copper(1 +), tetrakis(acetonitrile)-, hexafluorophosphate(1 −), 28:68
Copper, (acetonitrile)dodecacarbonyl-tricobalt ruthenium-, 26:359
Copper dichloride, 28:322
Crystal growth, 26:377
Cyanate, tungsten complex, 26:42
Cycloheptatriene, molybdenum complex, 27:4
1,2-Cyclohexanediamine, cis-, trans-(R,R)-, and trans-(S,S)-, platinum complex, 27:283
Cyclohexyl isocyanide, nickel complex, 28:101
Cyclometallapolyselanes, 27:59
Cyclometallation reactions, 26:171
1,5-Cyclooctadiene, iridium complex, 26:122
  nickel complex, 28:94
  osmium–rhodium complex, 27:29
  palladium and platinum complexes, 28:346–348
  platinum complex, 28:126, 128
  rhodium complex, 28:88
  ruthenium complex, 26:69, 253
1,3,5,7-Cyclooctatetraene, lithium complex, 28:127
  lutetium complex, 27:152
Cyclooctene, iridium and rhodium complexes, 28:90, 91
  platinum complex, 26:139
1,3-Cyclopentadiene, chromium, molybdenum, and tungsten, 28:196, 197
  chromium, molybdenum and tungsten complexes, 26:343
  cobalt complex, 26:191

1,3-Cyclopentadiene (*Continued*)
  cobalt-molybdenum-nickel complexes, 27:192
  cobalt-molybdenum-ruthenium, 27:194
  cobalt and rhodium complexes, 28:280, 281
  hafnium, titanium, and zirconium complexes, 28:248-260
  iron complexes, 26:232, 28:208, 210
  lanthanide complexes, 28:293-295
  lutetium complex, 27:161
  mercury-molybdenum-molybdenum and tungsten complexes, 26:96
  molybdenum complexes, 27:63, 28:11
  molybdenum-neodymium complex, 27:158
  nickel, osmium, and niobium complex, 28:267
  osmium-zirconium complex, 27:27
  palladium complex, 28:343
  platinum complexes, 26:345
  ruthenium complex, 26:178, 333, 362
  thallium complex, 28:315
  thorium and uranium complexes, 28:301, 302
  titanium complex, 27:60
  titanium and vanadium complexes, 28:261-266
  tungsten complexes, 27:67, 28:153
  uranium complex, 27:177
—, 1,3-bis(trimethylsilyl)-, lanthanide-lithium complexes, 27:169
—, 1-methyl-, chromium, molybdenum and tungsten hexacarbonyl complexes 28:148
  preparation and purification of, 27:52
  titanium complex, 27:52
  vanadium complex, 27:54
—, 1,2,3,4,5-pentamethyl-, 28:317
—, 1,2,3,4,5-pentamethylchromium complex, 27:69
  cobalt complexes, 28:273, 275
  hafnium, titanium, and zirconium complexes, 28:253-255
  iridium complex, 27:19
  samarium complex, 27:155
  titanium complex, 27:62
  ytterbium complex, 27:148
η^5-Cyclopentadienyl lanthanide complexes, 26:17

Decacyclo[9.9.1.0^{2,10}·O^{3,7}·O^{4,9}·O^{6,8}·O^{12,20}·O^{13,17}·O^{14,19}·O^{16,18}]-henicosaphosphide(3 −), 27:228

1,3,2,4-Diazadiphosphetidine, 1,3-di-*tert*-butyl-2,4-dichloro-, *cis*-, 27:258
Diborane(6), 27:215
Dicobalt iron and dicobalt ruthenium cluster complexes, 26:351
Diethyl ether, ytterbium complex, 27:148
  ruthenium complex, 27:198
Dinitrogen complexes, of molybdenum and tungsten, 28:33
Diphosphene, bis(2,4,6-tri-*tert*-butylphenyl)-, 27:241
—, bis[tris(trimethylsilyl)methyl]-, 27:241
Dithiocarbamic acid, diethyl-, molybdenum complex, 28:45
Dithiocarbonic acid, rhodium complex, 27:287
Dithioformic acid, iron complex, 28:186
1,3-Dithiolo[4,5-*b*][1,4]dithiin-2-thione, 26:389
Dodecatungstosilicic acid, (H$_4$[α-SiW$_{12}$O$_{40}$]), hydrate, 27:93
Dysprosium, tetrakis[η^5-1,3-bis(trimethylsilyl)cyclopentadienyl]di-μ-chloro-di-, 27:171
—, tris(2,6-di-*tert*-butyl-4-methylphenoxo)-, 27:167

Erbium, hexachloroheptakis(tetrahydrofuran)di-, 27:140
—, tetrakis[η^5-1,3-bis(trimethylsilyl)cyclopentadienyl]di-μ-chloro-di-, 27:171
—, trichloride-tetrahydrofuran(2:7), 27:140
—, tris(2,6-di-*tert*-butyl-4-methylphenoxo)-, 27:167
Ethanamine, 1,1-dimethyl-*N*-(trimethylsilyl)-, 27:327
—, 1,1-dimethyl-, tungsten complex, 27:301
—, *N*-ethyl-*N*-methyl-, tungsten, 26:40
Ethane, cobalt-molybdenum-nickel complexes, 27:192
—, 1,2-dimethoxy-, compound with cyclopentadienylsodium(1:1), 26:341
—, 1,2-dimethoxy-, tungsten complex, 26:343
  ytterbium complex, 28:295
1,2-Ethanediamine, chromium complex, resolution of, 26:24
  platinum complex, 27:314
—, monohydrochloride, platinum complex, 27:315

—, N,N,N',N'-tetramethyl-, lithium complex, 26:148
Ethanone, 1-phenyl-, manganese complex, 26:156
Ethene, cobalt complex, 28:278
molybdenum complexes, 26:102, 28:11
platinum complex, 28:129
rhenium complex, 26:110
rhodium complex, 28:86
Europium, tetrakis [η^5-1,3-bis(trimethylsilyl)cyclopentadienyl]di-μ-chloro-di-, 27:171

Ferrate(1 −), carbidododecacarbonyl-hydridotetra-, tetraethylammonium, 27:186
—, dodecacarbonyl[μ_4-(methoxycarbonyl)methylidyne]tetra-, tetraethylammonium, 27:184
—, dodecacarbonyltricobalt-, tetraethylammonium, 27:188
—, dodecacarbonyltricobalt-, (triphenylphosphine)gold(1 +), 27:188
—, hydridononacarbonylchromium-, μ-nitrido-bis(triphenylphosphorus)(1 +), 26:340
—, hydridononacarbonylmolybdenum-, μ-nitrido-bis(triphenylphosphorus)(1 +), 26:340
—, hydridononacarbonyltungsten-, μ-nitrido-bis(triphenylphosphorus)(1 +), 26:336
—, hydridotetracarbonyl-, μ-nitrido-bis(triphenylphosphorus)(1 +), 26:336
Ferrate(2 −), μ_4-carbido-dodecacarbonyltetra-, bis[μ-nitrido-bis(triphenylphosphorus)(1 +)], 26:246
—, carbidododecacarbonyltetra-, bis(tetraethylammonium), 27:187
—, carbidohexadecacarbonylhexa-, bis(tetraethylammonium), 27:183
—, nonacarbonylchromium-, bis[μ-nitrido-bis(triphenylphosphorus)(1 +)], 26:339
—, nonacarbonylmolybdenum-, bis[μ-nitrido-bis(triphenylphosphorus)(1 +)], 26:339
—, nonacarbonyltungsten-, bis[μ-nitrido-bis(triphenylphosphorus)(1 +)], 26:339
—, octacarbonyldi-, disodium, 28:203
—, tetracarbonyl-, disodium, 28:203
—, undecacarbonyltri-, bis[μ-nitrido-bis(triphenylphosphorus)(1 +)], 28:203
—, undecacarbonyltri-, disodium, 28:203

Fluorosulfate, tetrabutylammonium, 26:393
Formic acid, rhenium complex, 26:112
Furan, tetrahydro-, lanthanide complexes, 28:293–295
actinide and lanthanide complexes, 28:289, 290
iron complex, 26:232
lanthanide–lithium complexes, 27:169
lutetium complex, 27:152
magnesium complex, 26:147
molybdenum complex, 28:35, 36
neodymium complex, 27:158
neodymium and samarium complexes, 26:20
samarium complex, 27:155
ytterbium complex, 27:139

Gadolinium, tetrakis[η^5-1,3-bis(trimethylsilyl)cyclopentadienyl]di-μ-chloro-di-, 27:171
Gold, chloro(triphenylphosphine)-, 26:325
—, decacarbonylbis(triethylphosphine)triosmiumdi-, 27:211
—, decacarbonylbis(triphenylphosphine)triosmiumdi-, 27:211
—, decacarbonyl-μ-hydrido(triethylphosphine)triosmium-, 27:210
—, decacarbonyl-μ-hydrido(triphenylphosphine)triosmium-, 27:209
—, dodecacarbonyltris(triphenylphosphine)cobalttriruthenumtri-, 26:327
—, hexachlorododecakis(triphenylphosphine)pentapentaconta-, 27:214
—, octacarbonyl-1κ^4C,2κ^4C-μ-(diphenylphosphino)-1:2κP-(triphenylphosphine)-3κP-triangulo-dimanganese-, 26:229
Gold(1 +), chloro-1κCl-bis(triethylphosphine-κP)bis(triphenylphosphine)-2κP, 3κP-triangulo-platinumdi-, trifluoromethanesulfonate, 27:218
—, μ_3-oxo-tris[(triphenylphosphine)-, tetrafluoroborate(1 −), 26:326
—, (triphenylphosphine)-, dodecacarbonyltricobaltferrate(1 −), 27:188
Gold(I), bis[1,2-phenylenebis(dimethylarsine)]-, bis(pentafluorophenyl)aurate(I), 26:89
—, chloro(tetrahydrothiophene)-, 26:86

—, (pentafluorophenyl)(tetrahydrothiophene)-, 26:86
—, (pentafluorophenyl)-μ-thiocyanato(triphenylphosphine)di-, 26:90
Gold(III), tris(pentafluorophenyl)(tetrahydrothiophene)-, 26:87
Gold mixed-metal clusters, 26:324
Group 6 pentacarbonyl acetates, 27:295
Group 15 iron carbonyl complexes, 26:59–63

Hafnium, dicarbonylbis(η^5-cyclopentadienyl)-, 28:252
—, dicarbonylbis(η^5-penta-methylcyclopentadienyl)-, 28:255
cyclo-Heptasulfur(1+), bromo-, hexafluoroantimonate(1−), 27:336
—, bromo-, hexafluoroarsenate(1−), 27:336
—, bromo-, tetrasulfur(2+) hexafluoroarsenate(1−) (4:1:6), 27:338
—, iodo-, hexafluoroantimonate(1−), 27:333
—, iodo-, hexafluoroarsenate(1−), 27:333
—, iodo-, tetrasulfur(2+) hexafluoroarsenate(1−) (4:1:6), 27:337
cyclo-Heptasulfur(3+), μ-iodo-bis(4-iodo-, tris[hexafluoroantimonate(1−)], -2 arsenic trifluoride, 27:335
Heterobimetallic hydrido complexes, 27:26
3,10,14,18,21,25-Hexaazabicyclo[10.7.7]hexacosa-1,11,13,18,20,25-hexaene, 2,3,10,11,13,19-hexamethyl-, cobalt complex, 27:270
3,10,14,18,21,25-Hexaazabicyclo[10.7.7]hexacosa-1,11,13,18,20,25-hexaene, 2,3,10,13,19-hexamethyl-, nickel complex, 27:268
3,10,14,18,21,25-Hexaazabicyclo[10.7.7]hexacosa-1,11,13,18,20,25-hexaene, 2,3,10,11,13,19-hexamethyl-, tris[hexafluorophosphate(1−)], 27:269
3,11,15,19,22,26-Hexaazatricyclo[11.7.7.1^{5,9}]octacosa-1,5,7,9(28),12,14,19,21,26-nonaene, 3,11-dibenzyl-14,20-dimethyl-2,12-diphenyl-, iron complex, 27:280
3,11,15,19,22,26-Hexaazatricyclo[11.7.7.1^{5,9}]octacosa-1,5,7,9(28),12,14,19,21,26-nonaene, 3,11-dibenzyl-14,20-dimethyl-2,12-diphenyl-, nickel complex, 27:277

3,11,15,19,22,26-Hexaazatricyclo[11.7.7.1^{5,9}]octacosa-1,5,7,9(28),12,14,19,21,26-nonaene, 3,11-dibenzyl-14,20-dimethyl-2,12-diphenyl-, tris[hexafluorophosphate(1−)], 27:278
2,4-Hexadienedioic acid, 3-methyl-4-phenyl, dimethyl ester, cobalt complex, 26:197
Holmium, tetrakis[η^5-1,3-bis(trimethylsilyl)cyclopentadienyl]di-μ-chloro-di-, 27:171
—, tris(2,6-di-*tert*-butyl-4-methylphenoxo)-, 27:167
Hydrazine, ruthenium(II) complexes, 26:73
—, methyl-, ruthenium(II) complexes, 26:74
Hydrido complexes, gold-osmium, 27:209
iridium, 26:117, 27:19
iron, 26:244, 27:186
iron-tungsten, 26:336
manganese, 26:226
mercury-molybdenum-ruthenium, 26:329
molybdenum, 27:9
molybdenum and tungsten, 28:7
nickel, osmium, and ruthenium, 26:362
osmium, 26:186, 293, 28:236
osmium and ruthenium, 27:196–207
platinum, 26:135, 136, 27:32
rhenium, 26:77, 27:15
rhodium, 28:81, 82
ruthenium, 26:181, 28:219
tungsten, 27:10
zirconium, 28:257, 259
Hydrogen, molybdenum complex, 27:3
tungsten complex, 27:6
Hydrogen sulfide, titanium complex, 27:66
tungsten complex, 27:67

Iridate(1−), tetracarbonyl-, μ-nitrido-bis(triphenylphosphorus)(1+), 28:214
Iridium, carbonylchlorobis(triphenylphosphine)-, *trans*-, 28:92
—, di-μ-chloro-bis[bis(cyclooctene)-, 28:91
—, dodecacarbonyltetra-, 28:245
—, tetrahydrido(η^5-pentamethylcyclopentadienyl)-, 27:19
Iridium(1+), bis(acetone)dihydridobis(triphenylphosphine)-, tetrafluoroborate-(1−), 28:57
—, [1,4-butanediylbis(diphenylphosphine)]-pentahydridodi-, tetrafluoroborate-(1−), 27:26

Subject Index

—, (η^4-1,5-cyclooctadiene)bis(triphenylphosphine)-, tetrafluoroborate(1 —), 28:56
—, (η^4-1,5-cyclooctadiene)[1,3-propanediylbis(diphenylphosphine)]-, tetrafluoroborate(1 —), 27:23
—, diaquadihydridobis(triphenylphosphine)-, tetrafluoroborate(1 —), 28:58
—, (1,2-diiodobenzene)dihydridobis(triphenylphosphine)-, tetrafluoroborate-(1 —), 28:59
—, pentahydridobis[(1,3-propanediylbis(diphenylphosphine)]di-, tetrafluoroborate(1 —), 27:22
Iridium(2+), heptahydrido[1,3-propanediylbis(diphenylphosphine)]tri-, bis[tetrafluoroborate(1 —)], 27:22
—, tris[1,2-ethanediylbis(diphenylphosphine)]heptahydridotri-, bis[tetrafluoroborate(1 —)], 27:25
Iridium(I), chlorotris(triphenylphosphine)-, 26:201
—, (η^4-1,5-cyclooctadiene)bis(triphenylphosphine)-, tetrafluoroborate(1 —), 26:122
Iridium(III), bis(acetone)dihydridobis(triphenylphosphine)-tetrafluoroborate-(1 —), 26:123
—, carbonylchlorohydrido[tetrafluoroborato(1 —)]bis(triphenylphosphine)-, 26:117
—, carbonylchlorohydrido[tetrafluoroborato (1 —)]bis(triphenylphosphine)-, 28:23
—, carbonylchloromethyl[tetrafluoroborato-(1 —)]bis(triphenylphosphine)-, 26:118
—, carbonylhydridobis(trifluoromethanesulfonato)bis(triphenylphosphine)-, 26:120
—, carbonylhydridobis(trifluoromethanesulfonato)bis(triphenylphosphine)-, 28:26
—, chloro(dinitrogen)hydrido-[tetrafluoroborato(1 —)bis(triphenylphosphine)-, 26:119
—, chloro[2-(diphenylphosphino)phenyl-C^1,P]hydridobis(triphenylphosphine)-, (OC-6-53)-, 26:202
—, diaquadihydridobis(triphenylphosphine)-, tetrafluoroborate(1 —), 26:124
—, (1,2-diiodobenzene)dihydridobis(triphenylphosphine)-, tetrafluoroborate(1 —), 26:125

Iron(0), tetracarbonyl(tributylphosphine)-, 28:171
—, tetracarbonyl(triphenylarsine)-, 28:171
—, tetracarbonyl(triphenylstibine)-, 28:171
—, tricarbonylbis(tributylphosphine)-, 28:177
—, tricarbonylbis(tricyclohexylphosphine)-, 28:176
—, tricarbonylbis(trimethylphosphine)-, 28:177
—, tricarbonylbis(triphenylphosphine)-, 28:176
—, tricarbonyl(4-phenyl-3-butene-2-one)-, 28:52
Iron, acetyldicarbonyl(η^5-cyclopentadienyl)-, 26:239
—, carbidotridecacarbonyltetra-, 27:185
—, carbonyltetrakis(2-isocyano-1,3-dimethylbenzene)-, 26:57
—, dicarbonyl(η^5-cyclopentadienyl)(2-methyl-1-propenyl-κC^1)-, 28:208
—, dicarbonyl(η^5-cyclopentadienyl)[(methylthio)thiocarbonyl]-, 28:186
—, dicarbonyltris(2-isocyano-1,3-dimethylbenzene)-, 26:56
—, nonacarbonyldihydrido-μ_3-thio-tri-, 26:244
—, nonacarbonyl(μ_3-phenylphosphinidene)dicobalt-, 26:353
—, nonacarbonyl-μ_3-thio-dicobalt-, 26:245
—, pentakis(2-isocyano-1,3-dimethylbenzene)-, 26:57
—, tetracarbonyl complexes containing group 15 donor ligands, 28:168
—, tetracarbonyl(dimethylphenylphosphine)-, 26:61
—, tetracarbonyl(2-isocyano-1,3-dimethylbenzene)-, 26:53
—, tetracarbonyl(methyldiphenylphosphine)-, 26:61
—, tetracarbonyl(tributylphosphine)-, 26:61
—, tetracarbonyl(tricyclohexylphosphine)-, 26:61
—, tetracarbonyl(triethyl phosphite)-, 26:61
—, tetracarbonyl(trimethyl phosphite)-, 26:61
—, tetracarbonyl(triphenylarsine)-, 26:61
—, tetracarbonyl(triphenylphosphine)-, 26:61
—, tetracarbonyl(triphenyl phosphite)-, 26:61
—, tetracarbonyl(triphenylstibine)-, 26:61
—, tricarbonylbis(2-isocyano-1,3-dimethylbenzene)-, 26:54

Iron(1 +), μ-acetyl-C:O-bis[dicarbonyl-(η^5-cyclopentadienyl)-, hexafluorophosphate(1 −), 26:235
—, μ-acetyl-2κC^1: 1κO-pentacarbonyl-1$\kappa^3 C$,2$\kappa^2 C$-bis[1,2-(η^5-cyclopentadienyl)]molybdenum-, hexafluorophosphate(1 −), 26:239
—, μ-acetyl-2κC^1:1κO-tetracarbonyl-1$\kappa^3 C$,2$\kappa^2 C$-bis[1,2-(η^5-cyclopentadienyl)](triphenylphosphine-1κP)-molybdenum-, hexafluorophosphate(1 −), 26:241
—, μ-acetyl-2κC^1:1κO-tricarbonyl-1$\kappa^2 C$,2κC-bis[1,2-(η^5-cyclopentadienyl)](triphenylphosphine-2κP)di-, hexafluorophosphate(1 −), 26:237
—, (η^5-cyclopentadienyl)dicarbonyl-(tetrahydrofuran)-, hexafluorophosphate-(1 −), 26:232
—, dicarbonyl(η^5-cyclopentadienyl)(η^2-2-methyl-1-propene)-, tetrafluoroborate-(1 −), 28:210
—, dicarbonyl(η^5-cyclopentadienyl)(thiocarbonyl)-, trifluoromethanesulfonate, 28:186
Iron(II), [3,11-dibenzyl-14,20-dimethyl-2,12-diphenyl-3,11,15,19,22,26-hexaazatricyclo[11.7.7.1^{5,9}]octacosa-1,5,7,9(28),12,14,19,21,26-nonaene-$\kappa^4 N^{15,19,22,26}$]-,bis[hexafluorophosphate(1 −)], 27:280
Iron dioxygen carriers, 27:261
Iron isocyanide complexes, 28:179
Iron trichloride, 28:322
Isocyanide, iron complexes, 26:52
Isocyanide complexes of chromium molybdenum and tungsten, 28:140
Isopolyoxo metalates, 27:74
Isopropyl phosphite, nickel complex, 28:101

Lanthanide aryloxy complexes, 27:164
Lanthanide complexes, 28:286
Lanthanide η^5-cyclopentadienyl complexes, 26:17
Lanthanide trichlorides, 27:136
Lanthanum, bis[η^5-1,3-bis(trimethylsilyl)cyclopentadienyl]di-μ-chloro-bis(terahydrofuran)lithium-, 27:170
—, tetrakis[η^5-1,3-bis(trimethylsilyl)cyclopentadienyl]di-μ-chloro-di-, 27:171
—, tris(2,6-di-*tert*-butyl-4-methylphenoxo)-, 27:166
—, tris(2,6-di-*tert*-butyl-phenoxo)-, 27:167
Leaving groups, 28:2
Lithioheptadecatungstodiphosphate(9 −), [α_1-LiP$_2$O$_{17}$W$_{61}$]$^{9-}$, nonapotassium, eicosahydrate, 27:109
Lithium, bis[η^5-1,3-bis(trimethylsilyl)cyclopentadienyl]di-μ-chloro-bis(tetrahydrofuran)cerium-, 27:170
—, bis[η^5-1,3-bis(trimethylsilyl)cyclopentadienyl]di-μ-chloro-bis(tetrahydrofuran)lithium-, 27:170
—, bis[η^5-1,3-bis(trimethylsilyl)cyclopentadienyl]di-μ-chloro-bis(tetrahydrofuran)-neodymium-, 27:170
—, bis[η^5-1,3-bis(trimethylsilyl)cyclopentadienyl]di-μ-chloro-bis(tetrahydrofuran)-praseodymium-, 27:170
—, bis[η^5-1,3-bis(trimethylsilyl)cyclopentadienyl]di-μ-chloro-bis(tetrahydrofuran)-scandium-, 27:170
—, bis[η^5-1,3-bis(trimethylsilyl)cyclopentadienyl]di-μ-chloro-bis(tetrahydrofuran)-ytterbium-, 27:170
—, bis[η^5-1,3-bis(trimethylsilyl)cyclopentadienyl]di-μ-chloro-bis(tetrahydrofuran)yttrium-, 27:169
—, [η^5-1,3-bis(trimethylsilyl)cyclopentadienyl]-, 27:170
—, bis(trimethylsilyl)phosphide, (Li[P[Si(CH$_3$)$_3$]$_2$]), -tetrahydrofuran(1:2), 27:243
—, (1,3,5,7-cyclooctatetraene)di-, 28:127
—, (diethyl ether)[8-(dimethylamino)-1-naphthyl]-, 26:154
—, [2-[(dimethylamino)methyl]-5-methylphenyl]-, 26:152
—, [2-[(dimethylamino)methyl]phenyl]-, 26:152
—, [2-[(dimethylamino)phenyl]methyl]-, 26:153
—, [2-(methylphenylphosphino)ethyl]-, 27:178
—, μ-[($\alpha_1,\alpha',$1,2-η:$\alpha,\alpha',$1,2-η)-1,2-phenylenebis[(trimethylsilyl)methlene]]-bis (N, N, N', N'-tetramethyl-1, 2-ethanediamine)di-, 26:148

Lithium chloride, 28:322
Lithium dihydrogen phosphide, (LiH$_2$P), 27:228
Lithium heptaphosphide (Li$_3$P$_7$), 27:227
Lithium hexadecaphosphide, (Li$_2$P$_{16}$), 27:227
Lithium organic compounds, cyclometallated, 26:150
Lithium potassium hydrogen octatetracontatungstooctaphosphate(40−), (Li$_5$K$_{28}$H$_7$[P$_8$H$_{48}$O$_{184}$], dononacontahydrate, 27:110
Lutetium, bis(η^5-cyclopentadienyl)(tetrahydrofuran)-p-tolyl-, 27:162
—, bis(η^5-cyclopentadienyl)(tetrahydrofuran)[(trimethylsilyl)methyl]-, 27:161
—, chloro(η^8-1,3,5,7-cyclooctatetraene)(tetrahydrofuran)-, 27:152
—, [2-[(dimethylamino)methyl]-phenyl-C^1,N](η^8-1,3,5,7-cyclooctatetraene)(tetrahydrofuran)-, 27:153
—, tetrakis[η^5-1,3-bis(trimethylsilyl)cyclopentadienyl]di-μ-chloro-di-, 27:171

Magnesium, chloro(2,2-dimethylpropyl)-, 26:46
—, cyclotri[μ-1,2-phenylenebis(methylene)]hexakis(tetrahydrofuran)tri-, 26:147
Main-group compounds, 27:322
Manganate(1−), (μ-diphenylphosphino)-bis(tetracarbonyl-, (Mn–Mn), μ-nitridobis(triphenylphosphorus)(1+), 26:228
Manganate(III), trifluorosulfato-, dipotassium, 27:312
Manganese, acetylpentacarbonyl-, 28:199
—, (2-acetylphenyl-C,O)tetracarbonyl-, 26:156
—, μ-(azodi-2,1-phenylene-C^1,N^2:$C^{1'}$,N^1) octacarbonyldi-, 26:173
—, benzylpentacarbonyl-, 26:172
—, bromopentacarbonyl-, 28:156
—, (μ-chloromercurio)-(μ-diphenyl-phosphino)-bis(tetracarbonyl-, (Mn–Mn), 26:230
—, μ-(diphenylphosphino)-μ-hydridobis(tetracarbonyl-, (Mn–Mn), 26:226
—, octacarbonylbis(μ-dimethylphosphinothioyl-P:S)di-, 26:162
—, octacarbonyl-1$\kappa^4 C$,2$\kappa^4 C$-μ-[carbonyl-2κC:1κO-6-(diphenylphosphino-2κP)-1,2-phenylene-2κC^1:1κC^2]di-, 26:158
—, octacarbonyl-1$\kappa^4 C$,2$\kappa^4 C$-(μ-diphenylphosphino)-1:2κP-(triphenylphosphine)-3κP-$triangulo$-golddi-, 26:229
—, pentacarbonylchloro-, 28:155
—, pentacarbonyliodo-, 28:157, 158
—, pentacarbonyl[tetrafluoroborato-(1−)]-, 28:15
—, tetracarbonyl[2-(dimethylphosphinothioyl)-1,2-bis(methoxycarbonyl)ethenyl-C,S]-, 26:163
—, tetracarbonyl{[2-(diphenylphosphino)phenyl]hydroxymethyl-C,P}-, 26:169
—, tetracarbonyl[2-(phenylazo)phenyl-C^1,N^2]-, 26:173
—, tricarbonyl[η^2-2,3,4,5-tetrakis(methoxycarbonyl)-2,2-dimethyl-1H-phospholium]-, 26:167
—, tricarbonyl[η^2-3,4,5,6-tetrakis(methoxycarbonyl)-2,2-dimethyl-2H-1,2-thiaphosphorin-2-ium]-, 26:165
Manganese(I), pentacarbonylmethyl-, 26:156
—, pentacarbonyl(trifluoromethanesulfonato)-, 26:114
Mercury-bridged transition metal clusters, 26:328
Metal chlorides, anhydrous, 28:321
Metallacyclic complexes, 26:142
Metallation, of aromatic ketones, 26:155
Methane, bromo-, ruthenium and osmium complexes, 27:201, 205
—, chloro-, osmium complex, 27:205
—, isocyano-, tungsten complex, 28:43
—, tris(trimethylsilyl)-, 27:238
Methanesulfonic acid, trifluoro-, from chloride salts, 28:72
—, trifluoro-, from chloro complexes, 28:74
—, trifluoro-, from solid state reactions, 28:75
—, trifluoro-, from sulfate salts, 28:73
—, trifluoro-, iridium and platinum complexes, 28:26, 27
iridium(III) complexes, 26:177
manganese and rhenium complexes, 26:113
platinum complex, 26:126
transition metal complexes, 28:70–76

—, trifluoro-, regeneration of complexes of, 28:76
—, trifluoro-, using silver trifluoromethanesulfonate, 28:73
Methanol, platinum complex, 26:135
  tungsten complex, 26:45
Methyl, iridium complex, 26:118
  manganese complex, 26:156
  osmium complex, 27:206
  rhenium complexes, 26:107, 28:16
Methyl acetate, iron complex, 27:184
  osmium complex, 27:204
Methyl acrylate, ruthenium complex, 28:47
Methyl benzoate, chromium complex, 26:32
Methylene, osmium complex, 27:206
Molybdate(1−), (acetato)pentacarbonyl-, μ-nitrido-bis(triphenylphosphorus)(1+), 27:297
—, hydridononacarbonyliron-, μ-nitrido-bis(triphenylphosphorus)(1+), 26:340
—, tricarbonyl(η^5-cyclopentadienyl)-, sodium, compound with 1,2-dimethoxyethane(1:2), 26:343
Molybdate(2−), nonacarbonyliron-, bis[μ-nitrido-bis(triphenylphosphorus)(1+)], 26:339
—, thio-, $(Mo_2S_{10.56})^{2-}$, bis(tetraphenylphosphonium), 27:42
Molybdate(IV), tris(μ-sulfido)tris(disulfido)-μ_3-thio-*triangulo*-tri-, diammonium, hydrate, 27:48, 49
Molybdate(IV,VI), η^2-disulfido)di-μ-thio-trithiodi-, bis(tetraphenylphosphonium), 27:44
Molybdate(V), bis(η^2-disulfido)di-μ-thio-dithiodi-, bis(tetraphenylphosphonium), 27:45
—, bis(μ-disulfido)tetrakis(sulfido)di-, diammonium, dihydrate, 27:48, 49
—, di-μ-thio-tetrathiodi-, bis-(tetraphenylphosphonium), 27:43
—, pentachlorooxo-, diammonium, 26:36
Molybdate(VI), $[Mo_8O_{26}]^{4-}$, tetrakis(tetrabutylammonium), 27:78
—, $[Mo_6O_{19}]^{2-}$, bis(tetrabutylammonium), 27:77
—, tetrathio-, bis(tetraphenylphosphonium), 27:41

Molybdenum, bis(acetonitrile)tetrachloro-, 28:34
—, [bis(benzonitrile)platinum]-hexacarbonylbis(η^5-cyclopentadienyl)di-, (2*Mo*–*Pt*), 26:345
—, bis(*tert*-butyl isocyanide)tetracarbonyl-, *cis*-, 28:143
—, bis(diethyldithiocarbamato)dinitrosyl-, *cis*-, 28:145
—, bis(dinitrogen)bis[1,2-ethane-diylbis(diphenylphosphine)]-, *trans*-, 28:38
—, (*tert*-butyl isocyanide)pentacarbonyl-, 28:143
—, *cyclo*-[μ_3-1(η^2):2(η^2):3(η^2)-2-butyne]octacarbonyl-1κ^2C,2κ^3C,3κ^3C-[1(η^5)-cyclopentadienyl] cobalt-ruthenium-, (*Co*–*Mo*) (*Co*–*Ru*) (*Mo*–*Ru*), 27:194
—, carbonyl(η^5-cyclopentadienyl) [tetrafluoroborato(1−)]-, 28:5
—, dicarbonyl(η^5-cyclopentadienyl)-hydrido(triphenylphosphine)-, 26:98
—, dicarbonyl(η^5-cyclopentadienyl)nitrosyl-, 28:196
—, dicarbonyl(η^5-cyclopentadienyl)-[tetrafluoroborato(1−)]-(triphenylphosphine)-, 26:98
—, *cyclo*-μ_3-ethylidyne-1:2:3-κ^3C-Pentacarbonyl-1κ^2C, 2κ^3C-bis[1,3(η^5)-cyclopentadienyl]cobaltnickel-, (*Co*–*Mo*)(*Co*–*Ni*) (*Mo*–*Ni*), 27:192
—, hexacarbonylbis(η^5-cyclopentadienyl)bis(triphenylphosphine)diplatinumdi-, 26:347, 348
—, hexacarbonylbis(η^5-cyclopentadienyl)di-, 28:148
—, octacarbonyl(η^5-cyclopentadienyl)-μ_3-ethylidynedicobalt-, 27:193
—, tetracarbonylbis(η^5-cyclopentadienyl)di-, (*Mo*–*Mo*), 28:152
—, tetrachlorobis(tetrahydro-, 28:35–37
—, tricarbonyl(η^5-cyclopentadienyl) [tetrafluoroborato(1−)]-, 26:96
—, tricarbonyl(dihydrogen)bis(tricyclohexylphosphine)-, 27:3
—, tricarbonyltris(propionitrile)-, 28:31
—, tris(*tert*-butyl isocyanide)-tricarbonyl-, *fac*-, 28:143

Molybdenum(0), tricarbonyl(cycloheptatriene)-, 28:45
Molybdenum(1+), (acetone)tricarbonyl(η^5-cyclopentadienyl)-, tetrafluoroborate(1−), 26:105
—, (acetone)tricarbonyl(η^5-cyclopentadienyl)-, tetrafluoroborate-(1−), 28:14
—, carbonyl(η^5-cyclopentadienyl)bis(diphenylacetylene)-, tetrafluoroborate(1−), 26:102
—, carbonyl(η^5-cyclopentadienyl) (diphenylacetylene) (triphenylphosphine)-, tetrafluoroborate(1−), 26:104
—, tricarbonyl(η^5-cyclopentadienyl) (η^2-ethene)-, tetrafluoroborate(1−), 26:102
Molybdenum(2+), tetrakis(acetonitrile)dinitrosyl-, cis-, bis[tetrafluoroborate(1−)], 28:65
Molybdenum(I), dicarbonyl(η^5-cyclopentadienyl) (η^3-cyclotriphosphorus)-, 27:224
—, tetracarbonylbis(η^5-cyclopentadienyl) (μ-η^2:η^2-diphosphorus)di-, 27:224
Molybdenum(II), dibromotetracarbonyl-, 28:145
—, dicarbonylbis[1,2,-ethanediylbis(diphenylphosphine)]fluoro-, hexafluorophosphate(1−), 26:84
—, tetrakis(acetonitrile)dinitrosyl-, cis-, bis[tetrafluoroborate(1−)], 26:132
—, tricarbonylbis(diethyl-dithiocarbamato)-, 28:145
Molybdenum(III), trichlorotris(tetrahydrofuran)-, 28:36
—, tris(acetonitrile)trichloro-, 28:37
Molybdenum(IV), bis(η^5-cyclopentadienyl) [tetrasulfido(2−)]-, 27:63
—, tetrahydridotetrakis(methyldiphenylphosphine)-, 27:9
Molybdenum(VI), hexahydridotris(tricyclohexylphosphine)-, 27:13
Molybdenum sulfur clusters, 27:47
Molybdenum tetrachloride oxide, 28:325
Molybdobis[(4-aminobenzyl)phosphonate] (4−), [Mo$_5$O$_{15}$(NH$_2$C$_6$H$_4$CH$_2$PO$_3$)$_2$]$^{4-}$, bis(tetramethylammonium) dihydrogen, tetrahydrate and diammonium dihydrogen, pentahydrate, 27:126
Molybdobis[(2-aminoethyl)phosphonate] (4−), [Mo$_5$O$_{15}$(NH$_2$C$_2$H$_4$PO$_3$)$_2$]$^{4-}$, sodium tetramethylammonium dihydrogen, pentahydrate, 27:126
Molybdobis(ethylphosphonate) (4−), [Mo$_5$O$_{15}$(C$_2$H$_5$PO$_3$)$_2$]$^{4-}$, tetraammonium, 27:125
Molybdobis(methylphosphonate) (4−), [Mo$_5$O$_{15}$(CH$_3$PO$_3$)$_2$]$^{4-}$, tetraammonium, dihydrate, 27:124
Molybdobis(phenylphosphate) (4−) [Mo$_5$O$_{15}$(C$_6$H$_5$PO$_3$)$_2$]$^{4-}$, tetraammonium, pentahydrate, 27:125
Molybdobis(phosphonate) (4−), [(HPO$_3$)$_2$Mo$_5$O$_{15}$]$^{4-}$, tetraammonium, tetrahydrate, 27:123

1-Naphthalenamine, N,N-dimethyl-, lithium complex, 26:154
2,3-Naphthalenediol, in preparation of cis-tetraamminedihaloruthenium(III) complexes, 26:66
Neodymium, bis[η^5-1,3-bis(trimethylsilyl)cyclopentadienyl]di-μ-chloro-bis(tetrahydrofuran)lithium-, 27:170
—, tert-butylbis(η^5-cyclopentadienyl) (tetrahydrofuran)-, 27:158
—, hexachlorotris(tetrahydrofuran)di-, 27:140
—, tetrakis[η^5-1,3-bis(trimethylsilyl)cyclopentadienyl]di-μ-chloro-di-, 27:171
—, tris(2,6-di-tert-butyl-4-methylphenoxo)-, 27:167
Neodymium(III), tris(η^5-cyclopentadienyl) (tetrahydrofuran)-, 26:20
Neodymium trichloride–tetrahydrofuran (2:3), 27:140
Nickel, cyclo-μ_3-ethylidyne-1:2:3-κ^3C-pentacarbonyl-1κ^2C,2κ^3C-bis[1,3(η^5)-cyclopentadienyl]cobaltmolybdenum-, (Co−Mo) (Co−Ni) (Mo−Ni), 27:192
—, nonacarbonyl(η^5-cyclopentadienyl)tri-μ-hydrido-triosmium-, 26:362
—, nonacarbonyl(η^5-cyclopentadienyl)tri-μ-hydrido-triruthenium-, 26:363
—, nonacarbonyltris(η^5-cyclopentadienyl)triosmiumtri-, 26:365
Nickel(0), bis(2,2'-bipyridine)-, 28:103

—, bis(1,5-cyclooctadiene)-, 28:94
—, bis[1,2-ethanediylbis(dimethylphosphine)]-, 28:101
—, bis[1,2-ethanediylbis(diphenylphosphine)]-, 28:103
—, bis(1,10-phenanthroline)-, 28:103
—, bis[1,2-phenylenebis(dimethylarsine)]-, 28:103
—, tetrakis(*tert*-butyl isocyanide)-, 28:99
—, tetrakis(cyclohexyl isocyanide)-, 28:101
—, tetrakis(diethylphenylphosphine)-, 28:101
—, tetrakis(dimethyl phenylphosphonite)-, 28:101
—, tetrakis(isopropyl phosphite)-, 28:101
—, tetrakis(methyldiphenylphosphine)-, 28:101
—, tetrakis(tributylphosphine)-, 28:101
—, tetrakis(triethylphosphine)-, 28:101
—, tetrakis(triethyl phosphite)-, 28:101
—, tetrakis(trimethylphosphine)-, 28:101
—, tetrakis(trimethyl phosphite)-, 28:101
—, tetrakis(triphenylarsine)-, 28:103
—, tetrakis(triphenylphosphine)-, 28:102
—, tetrakis(triphenyl phosphite)-, 28:101
—, tetrakis(triphenylstibine)-, 28:103
Nickel(II), [3,11-bis(benzoyl)-2,12-dimethyl-1,5,9,13-tetraazacyclohexadeca-1,3,9,11-tetraenato(2−)-κ^4-$N^{1,5,9,13}$]-, 27:273
—, [3,11-bis[α-(benzylamino)benzylidene]-2,12-dimethyl-1,5,9,13-tetraazacyclohexadeca-1,4,9,12-tetraene-$\kappa^4 N^{1,5,9,13}$]-, bis[hexafluorophosphate(1−)], 27:276
—, [3,11-bis(α-methoxybenzylidene)-2,12-dimethyl-1,5,9,13-tetraazacyclohexadeca-1,4,9,12-tetraene-$\kappa^4 N^{1,5,9,13}$]-, bis[hexafluorophosphate(1−)], 27:275
—, [butanamidato(2−)-C^4,N](tricyclohexylphosphine)-, 26:206
—, [3,11-dibenzyl-14,20-dimethyl-2,12-diphenyl-3,11,15,19,22,26-hexaazatricyclo[11.7.7.1^{5,9}]octacosa-1,5,7,9(28),12,14,19,21,26-nonaene-$\kappa^4 N^{15,19,22,26}$]-, bis[hexafluorophosphate(1−)], 27:277
—, [2,12-dimethyl-3,11-bis(1-methoxyethylidene)-1,5,9,13-tetraazacyclohexadeca-1,4,9,12-tetraene-$\kappa^4 N^{1,5,9,13}$]-, bis[hexafluorophosphate(1−)], 27:264
—, [2,12-dimethyl-3,11-bis[1-(methylamino)ethylidene]-1,5,9,13-tetraazacyclohexadeca-1,4,9,12-tetraene-$\kappa^4 N^{1,5,9,13}$]-, bis[hexafluorophosphate(1−)], 27:266
—, [4,10-dimethyl-1,5,9,13-tetraazacyclohexadeca-1,3,9,11-tetraenato(2−)-$\kappa^4 N^{1,5,9,13}$]-, 27:272
—, (2,3,10,13,19-hexamethyl-3,10,14,18,21,25-hexaazabicyclo[10.7.7]hexacosa-1,11,13,18,20,25-hexaene-$\kappa^4 N^{14,18,21,25}$)-, bis[hexafluorophosphate(1−)], 27:268
—, [2-methylpropanamidato(2−)-C^3,N](tricyclohexylphosphine)-, 26:205
Nickelate(2−), hexa-μ-carbonyl-hexacarbonylhexa-, bis(tetramethylammonium), 26:312
Nickel clusters with ruthenium or osmium, 26:360
Nickel dichloride, 28:322
Niobate(1−), hexacarbonyl-, sodium, 28:192
Niobium(IV), dichlorobis(η^5-cyclopentadienyl)-, 28:267
Nitrido ruthenium clusters, 26:286, 287
Nitrile complexes, of chromium, molybdenum, and tungsten, 28:29
Nitrogen, iridium complex, 28:25
iridium(III) complex, 26:119
molybdenum complex, 28:38
tungsten complex, 28:41
Nitrosyls complexes, chromium, molybdenum, and tungsten, 28:196, 197
molybdenum, 26:132, 28:145
molybdenum and tungsten, 28:65, 66
tungsten, 26:133

Octacyclo[7.7.0.0^{2,6}·0^{3,8}·O^{5,7}·O^{10,14}·O^{11,16}·O^{13,15}]hexadecaphosphide(2−), 27:228
Olefin complexes of platinum, 28:126
Organolithium compounds, cyclometallated, 26:150
Organometallic fluoro complexes, 26:81
Organophosphonatomolybdates and tungstates, 27:123
Osmate(1−), μ-carbonyl-decacarbonyl-μ-hydrido-tri-, μ-nitrido-bis-(triphenylphosphorus) (1+), 28:236

Osmate(2−), octadecacarbonylhexa-, bis[μ-nitrido-bis(triphenylphosphorus)(1+)], 26:300
—, pentadecacarbonylpenta-, bis[μ-nitrido-bis(triphenylphosphorus)(1+)], 26:299
Osmium, (acetonitrile)undecacarbonyltri-, 26:290
—, (μ-benzenethiolato)decacarbonyl-μ-hydrido-tri-, 26:304
—, bis(acetonitrile)decacarbonyltri-, 26:292, 28:234
—, bis[1,1(η^5)-cyclopentadienyl]tris(dimethylphenylphosphine-2κP)-tri-μ-hydrido-hydrido-1κH-zirconium-, 27:27
—, (μ_3-bromomethylidyne)-nonacarbonyl-tri-μ-hydrido-*triangulo*-tri-, 27:205
—, μ_3-carbonylnonacarbonyl-μ_3-thio-tri-, 26:305
—, carbonyl(thiocarbonyl)tris(triphenylphosphine)-, 26:187
—, carbonyl(5-thioxo-1,3-pentadiene-1,5-diyl-C^1,C^5,S)bis(triphenylphosphine)-, 26:188
—, [2(η^4)-1,5-cyclooctadiene]tris(dimethylphenylphosphine-1κP)-tri-μ-hydrido-rhodium-, 27:29
—, decacarbonylbis[μ-(triethylphosphine)gold]tri-, 27:211
—, decacarbonylbis[μ-(triphenylphosphine)gold]tri-, 27:211
—, decacarbonyl-di-μ-hydrido-μ-methylene-*triangulo*-tri-, 27:206
—, decacarbonyldihydridotri-, 26:367
—, decacarbonyl-μ-hydrido(μ-methoxymethylidyne)-*triangulo*-tri-, 27:202
—, decacarbonylhydridomethyltri-, 27:206
—, decacarbonyl-μ-hydrido[μ-(triethylphosphine)gold]tri-, 27:210
—, decacarbonyl-μ-hydrido[μ-(triphenylphosphine)gold]tri-, 27:209
—, dodecacarbonyldi-μ_3-thio-tetra-, 26:307
—, dodecacarbonyltetra-μ-hydrido-*tetrahedro*-tetra-, 28:240
—, dodecacarbonyltetra-μ-hydrido-*tetrahedro*-tetra-, 26:293
—, dodecacarbonyltri-, 28:230
—, nonacarbonyl(μ_3-chloromethylidyne)tri-μ-hydrido-*triangulo*-tri-, 27:205

—, nonacarbonyl(η^5-cyclopentadienyl)tri-μ-hydrido-nickeltri-, 26:362
—, nonacarbonyldi-μ_3-thio-tri-, 26:306
—, nonacarbonyl-tri-μ-hydrido[μ_3-(methoxycarbonyl)methylidyne]-*triangulo*-tri-, 27:204
—, nonacarbonyl-tri-μ-hydrido(μ_3-methoxymethylidyne)-*triangulo*-tri-, 27:203
—, nonacarbonyltris(η^5-cyclopentadienyl)trinickeltri-, 26:365
—, octadecacarbonyldihydridohexa-, 26:301
—, octadecacarbonylhexa-, 26:295
—, tridecacarbonyldi-μ_3-thio-tetra-, 26:307
—, undecacarbonyl(pyridine)-, 26:291
—, undecacarbonyl(pyridine)tri-, 28:234
Osmium(II),
   dichloro(thiocarbonyl)tris(triphenylphosphine)-, 26:185
—, dichlorotris(triphenylphosphine)-, 26:184
—, dihydrido(thiocarbonyl)tris(triphenylphosphine)-, 26:186
Osmium(III),
   trichlorotris(dimethylphenylphosphine)-, *mer*-, 27:27
Osmium selenido clusters, 26:308
Oxygen carriers, cobalt and iron, 27:261

Palladium, bis(benzonitrile)dichloro-, 28:61
—, bis(1,1,1,5,5,5-hexafluoro-2,4-pentanedionato)-, 27:318
—, di-μ-chloro-bis[2-(phenylazo)phenyl-C^1,N^2]di-, 26:175
Palladium(0), bis(1,5-diphenyl-1,4-pentadien-3-one)-, 28:110
—, tetrakis(triethyl phosphite)-, 28:105
—, tetrakis(triphenylphosphine)-, 28:107
—, bis(di-*tert*-butylphenylphosphine)-, 28:114
—, bis(tri-*tert*-butylphosphine)-, 28:115
—, bis(tricyclohexylphosphine)-, 28:114
Palladium(2+), tetrakis(acetonitrile)-, bis[tetrafluoroborate(1−)], 28:63
Palladium(I), dichlorobis[μ-methylene-bis(diphenylphosphine)]di-(*Pd–Pd*), 28:340
—, tetrakis(*tert*-butyl isocyanide)di-μ-chloro-di-, 28:110
—, *tert*-butyl isocyanide, palladium complex, 28:110

Palladium(II), (η^3-allyl) (η^5-cyclopentadienyl)-, 28:343
—, (2,2'-bipyridine) (1,1,1,5,5,5-hexafluoro-2,4-pentanedionato)-, 1,1,1,5,5,5-hexafluoro-2,4-dioxo-3-pentanide, 27:319
—, bis(η^3-allyl)di-μ-chloro-di-, 28:342
—, [bis[2-(diphenylphosphino)ethyl]phenylphosphine] (1,1,1,5,5,5-hexafluoro-2,4-pentanedionato)-, 1,1,1,5,5,5-hexafluoro-2,4-dioxo-3-pentanide, 27:320
—, chloro[2-(2-pyridinylmethyl)phenyl-C^1,N] (3,5-dimethylpyridine)-, 26:210
—, di-μ-acetato-bis[2-(2-pyridinylmethyl)phenyl-C^1,N]di-, 26:208
—, di-μ-chloro-bis[2-[(dimethylamino)methyl]phenyl-C^1,N]di-, 26:212
—, di-μ-chloro-bis[2-(2-pyridinylmethyl)phenyl-C^1,N]di-, 26:209
—, di-μ-chloro-bis(8-quinolyl methyl-C,N)di-, 26:213
—, dichloro(η^4-1,5-cyclooctadiene)-, 28:348
—, tetrakis(acetonitrile)-, bis[tetrafluoroborate(1−)], 26:128
Palladium hexafluoro-2,4-pentanedionato complexes, 27:317
Palladium and platinum di-coordinate phosphine complexes, 28:113
1,4-Pentadien-3-one, 1,5-diphenyl-, palladium complex, 28:110
2,4-Pentadienthial, osmium complex, 26:188
2,4-Pentanedione, 1,1,1,5,5,5-hexafluoro-, palladium complexes, 27:318–320
2-Pentenedioic acid, 3-methyl-2-(phenylmethyl)-, dimethyl ester, cobalt complex, 26:197
Perrhenate, bis(2,2'-bi-1,3-dithiolo[4,5-b][1,4]dithiinylidene), superconducting, 26:391
—, tetrabutylammonium, 26:391
1,10-Phenanthroline, nickel complex, 28:103
Phenol, 2,6-di-*tert*-butyl-, actinide and lanthanide complexes, 27:166
—, 2,6-di-*tert*-butyl-4-methyl-, actinide and lanthanide complexes, 27:166
Phenol, rhodium complex, 27:292
Phenyl, platinum complex, 26:136
Phenyl isocyanide, 2,6-dimethyl-, *see* benzene, 2-isocyano-1,3-dimethyl-, 28:180

Phosphide, bis(trimethylsilyl)-, lithium, -tetrahydrofuran(1:2), 27:243
Phosphide, dilithium, (Li_2P_{16}), 27:227
Phosphide, lithium dihydrogen, (LiH_2P), 27:228
Phosphide, trilithium, (Li_3P_7), 27:227
Phosphide, trisodium, (Na_3P_{21}), 27:227
Phosphine, bis[2-(diphenylphosphino)ethyl]phenyl-, palladium complex, 27:320
—, 1,4-butanediylbis(diphenyl-, iridium complex, 27:26
—, di-*tert*-butylphenyl-, palladium and platinum complexes, 28:114, 116
—, diethylphenyl-, nickel complex, 28:101
platinum complex, 28:135
—, dimethylphenyl-, iron complex, 28:171
osmium complex, 27:27
osmium-rhodium complex, 27:29
osmium-zirconium complex, 27:27
ruthenium complex, 26:273
tungsten complex, 27:11
—, (2,2-dimethylpropylidyne)-, 27:249, 251
—, [2,2-dimethyl-1-(trimethylsiloxy)propylidene] (trimethylsilyl)-, 27:250
—, diphenyl-, manganese complex, 26:158
ruthenium complex, 26:264
—, [2-[(diphenylphosphino)methyl]-2-methyl-1,3-propanediyl]bis(diphenyl-, rhodium complex, 27:287
—, 1,2-ethanediylbis(dimethyl-, nickel complex, 28:101
—, 1,2-ethanediylbis(diphenyl-, iridium complex, 27:25
molybdenum complex, 26:84, 28:38
nickel complex, 28:103
platinum complexes, 26:370, 28:135
tungsten complex, 28:41
—, ethylmethylphenyl-, lithium complex, 27:178
—, ethynediylbis(diphenyl-, ruthenium complex, 26:277
—, methyldiphenyl-, iron complex, 28:171
molybdenum complex, 27:9
nickel complex, 28:101
tungsten complex, 27:10
—, methylenebis(diphenyl-, palladium complex, 28:340
ruthenium complex, 26:276
—, phenyl-, cobalt-iron complex, 26:353

—, 1,3-propanediylbis(diphenyl-, iridium complex, 27:22
—, tributyl-, iron complex, 28:171
  nickel complex, 28:101
—, tri-*tert*-butyl-, palladium complex, 28:115
—, (2,4,6-tri-*tert*-butyl-phenyl)-, 27:237
—, (2,4,6-tri-*tert*-butyl-phenyl)(trimethylsilyl)-, 27:238
—, tricyclohexyl-, iron complex, 28:171
  molybdenum complexes, 27:3, 13
  nickel complex, 26:205
  palladium and platinum complex, 28:114, 116
  platinum complex, 28:130
  rhodium complex, 27:291
  tungsten complex, 27:6
—, triethyl-, gold-osmium complex, 27:210, 211
  nickel complex, 28:101
  platinum complexes, 26:126, 135–140
—, triisopropyl-, platinum complex, 28:120
  rhodium complex, 27:292
  tungsten complex, 27:7
—, trimethyl-, 26:7
  preparation of, 28:305
  cobalt and rhodium complexes, 28:280, 281
  iron complex, 28:177
  nickel complex, 28:101
  tungsten complexes, 27:304, 28:327, 28:329
—, triphenyl-, 28:57
  cobalt complex, 26:190
  cobalt-gold-iron complex, 27:188
  cobalt-gold-ruthenium complex, 26:327
  gold complexes, 26:90, 325
  gold-manganese complex, 26:229
  gold-osmium complexes, 27:209, 211
  gold-platinum complex, 27:218
  iridium complexes, 26:122, 117
  iron complexes, 26:61, 28:170
  molybdenum complex, 28:13
  molybdenum-platinum complexes, 26:347
  molybdenum and tungsten complexes, 26:98, 28:7
  nickel complex, 28:102
  osmium complex, 26:184
  palladium complex, 28:107
  platinum complexes, 27:36, 28:135
  rhenium complexes, 27:15, 16
  rhodium complexes, 27:292, 27:222, 28:77–83
  ruthenium complex, 26:181
  tungsten complex, 28:40
—, tris(trimethylsilyl)-, 27:243
Phosphine sulfide, dimethyl-, 26:162
  manganese complex, 26:162
Phosphinothioyl cyclo-cotrimerization, 26:161
1*H*-Phospholium, 2,3,4,5-tetrakis(methoxycarbonyl)-2,2-dimethyl-, manganese complex, 26:167
Phosphonium, (benzyl)triphenyl-, chloro(pentafluorophenyl)aurate(I), 26:88
—, (dithiocarboxy)triethyl-, rhodium complex, 27:288
Phosphonous acid, phenyl-, dimethyl ester, nickel complex, 28:101
Phosphonous dichloride, (2,4,6-tri-*tert*-butylphenyl)-, 27:236
—, [tris(trimethylsilyl)methyl]-, 27:239
Phosphorane, dimethylmethylenephenyl-, uranium complex, 27:177
Phosphorodifluoridic acid, rhenium complex, 26:83
Phosphorous acid, tris(4-methylphenyl)ester, ruthenium complex, 28:227
Phosphorus(1+), μ-nitrido-bis(triphenyl-, acetate, 27:296
—, μ-nitrido-bis(triphenyl)-, azide, 26:286
Phosphorus, di-, and *cyclo*-tri, molybdenum complexes, 27:224
Phosphorus, *tetrahedro*-tetra-, rhodium complex, 27:222
Phosphorus trifluoride, 26:12
  preparation of, 28:301
Phthalic acid, rhodium complex, 27:291
Platinate(2−), dodecacarbonyltetrarhodium-, bis[μ-nitrido-bis(triphenylphosphorus)(1+)], 26:375
—, hexa-μ-carbonyl-hexacarbonylhexa-, bis(tetrabutylammonium), 26:316
—, pentakis[tri-μ-carbonyltricarbonyltri-, bis(tetraethylammonium), 26:320
—, tetradecacarbonyltetrarhodium-, bis[μ-nitrido-bis(triphenylphosphorus)(1+)], 26:373
—, tetrakis[tri-μ-carbonyl-tricarbonyltri-, bis(tetraethylammonium), 26:321

—, tris[tri-μ-carbonyltricarbonyltri-, bis(tetraethylammonium)], 26:322
Platinate(II), trichloro(ethene), potassium, 28:349
Platinum, bis(benzonitrile)dichloro-, 26:345
—, dichloro[1,2-ethanediylbis(diphenylphosphine)]-, 26:370
—, heptacarbonyl[1,2-ethanediylbis(diphenylphosphine)]dicobalt-, 26:370
—, hexacarbonylbis(η^5-cyclopentadienyl)bis(triphenylphosphine)dimolybdenum-di-, 26:347, 348
Platinum(0), bis(1,5-cyclooctadiene)-, 28:126, 128
—, bis(di-*tert*-butylphenylphosphine)-, 28:116
—, bis(diethylphenylphosphine)-(ethene)-, 28:135
—, bis(ethene)(tricyclohexylphosphine)-, 28:130
—, bis(tricyclohexylphosphine)-, 28:116
—, [1,2-ethanediylbis(diphenylphosphine)](ethene)-, 28:135
—, (ethene)bis(triethylphosphine)-, 28:133
—, (ethene)bis(triisopropylphosphine)-, 28:135
—, (ethene)bis(triphenylphosphine)-, 28:135
—, tetrakis(triethylphosphine)-, 28:122
—, tetrakis(triethyl phosphite)-, 28:106
—, tetrakis(triphenylphosphine)-, 28:124
—, tris(bicyclo[2.2.1]hept-2-ene)-, 28:127
—, tris(ethene)-, 28:129
—, tris(triethylphosphine)-, 28:120
—, tris(triisopropylphosphine)-, 28:120
—, tris(triphenylphosphine)-, 28:125
Platinum(II), [ascorbato(2−)-C^2,O^5]-(*cis*-1,2-cyclohexanediamine)-, 27:283
—, chlorobis(triethylphosphine)(trifluoromethanesulfonato)-, *cis*-, 26:126, 28:27
—, chloro(*cis*-1,2-diphenylethenyl)bis(triethylphosphine)-, *trans*-, 26:140
—, [*trans*-(R,R)-1,2-cyclohexanediamine]-diiodo-, 27:284
—, (η^3-cyclooctenyl)bis(triethylphosphine)-, tetraphenylborate(1−), 26:139

—, dichlorobis(1,2-ethanediamine monohydrochloride)-, *trans*-, 27:315
—, dichloro(η^4-1,5-cyclooctadiene-, 28:346
—, di-μ-hydrido-hydridotetrakis(triethylphosphine)di-, tetraphenylborate(1−), 27:34
—, di-μ-hydrido-hydridotetrakis(triphenylphosphine)di-, tetraphenylborate(1−), 27:36
—, μ-hydrido-dihydridotetrakis(triethylphosphine)di-, tetraphenylborate(1−), 27:32
—. μ-hydrido-hydridophenyltetrakis(triethylphosphine)di-, tetraphenylborate(1−), 26:136
—, hydrido(methanol)bis(triethylphosphine)-, *trans*-, trifluoromethanesulfonate), 26:135
—, (3-methoxy-3-oxo-κO-propyl-κC^1)bis(triethylphosphine)-, *cis*-, tetraphenylborate(1−), 26:138
Platinum(IV), dichlorobis(1,2-ethanediamine)-, chloride, *cis*-, 27:314
Platinum(1+), chloro-1κCl-bis(triethylphosphine-1κP)bis(triphenyl-phosphine)-2κP,3κP-*triangulo*-digold-, trifluoromethanesulfonate, 27:218
Platinum hydrido tertiary phosphine cations, 27:30
Platinum tri-coordinate phosphine complexes, 28:119
Polynuclear transition metal complexes, 26:215
Polyoxo anions of transition metals, 27:71
Potassium α-dodecatungstosilicate(4−), (K$_4$[α-SiW$_{12}$O$_{40}$]), heptadecahydrate, 27:93
Potassium β-dodecatungstosilicate(4−), (K$_4$[β-SiW$_{12}$O$_{40}$]), nonahydrate, 27:94
Potassium γ-decatungstosilicate(8−), (K$_8$[γ-SiW$_{10}$O$_{36}$]), dodecahydrate, 27:88
Potassium α_2-heptadecatungstodiphosphate(10−), (K$_{10}$[α_2-P$_2$O$_{17}$O$_{61}$]), eicosahydrate, 27:107
Potassium hexaiodorhenate(VI), 27:294
Potassium hydrogen A-β-1,2,3-trivanadononatungstosilicate(7−), (K$_6$H[A-β-SiW$_9$V$_3$O$_{40}$]), trihydrate, 27:129

Potassium α_1-lithioheptadecatungstodiphosphate(9—), ($K_9[\alpha_1$-$LiP_2W_{17}O_{62}]$), eicosahydrate, 27:109
Potassium α-octadecatungstodiphosphate-(6—), ($K_6[P_2W_{18}O_{62}]$), tetradecahydrate, 27:105
Potassium β-octadecatungstodiphosphate-(6—), ($K_6[\beta$-$P_2W_{18}O_{62}]$, nonadecahydrate, 27:105
Potassium trifluorosulfatomanganate(III), ($K_2[MnF_3(SO_4)]$), 27:312
Potassium α-undecatungstosilicate(8—), ($K_8[\alpha$-$SiW_{11}O_{39}]$), tridecahydrate, 27:89
Potassium β_2-undecatungstosilicate(8—), ($K_8[\beta_2$-$SiW_{11}O_{39}]$), tetradecahydrate, 27:91
Potassium α-vanadoundecatungstophosphate(4—), ($K_4[\alpha$-$PVW_{11}O_{40}]$), hydrate, 27:99
Praseodymium, bis[η^5-1,3-bis(trimethylsilyl)cyclopentadienyl]di-μ-chloro-bis(tetrahydrofuran)lithium-, 27:170
—, tetrakis[η^5-1,3-bis(trimethylsilyl) cyclopentadienyl]di-μ-chloro-di-, 27:171
—, tris(2,6-di-*tert*-butyl-4-methylphenoxo)-, 27:167
Propane, 2,2-dimethyl-, magnesium complex, 26:46
  tungsten complexes, 26:47, 50
—, 2-isocyano-2-methyl-, ruthenium complex, 26:275
—, 2-methyl-, neodymium complex, 27:158
Propanenitrile, tungsten complex, 27:4
Propanoic acid, 2,2-dimethyl-, tungsten complex, 26:223
2-Propenamide, 2-methyl-, nickel complex, 26:205
1-Propene, palladium complex, 28:342
  ruthenium complex, 26:251
—, 2-methyl-, iron complexes, 28:208, 210
2-Propenoic acid, methylester, platinum complex, 26:138
Propionitrile, chromium, molybdenum and tungsten complexes, 28:30–32
Propylidyne, 2,2-dimethyl-, tungsten(VI), complexes, 26:44

2-Propynoic acid, 3-phenyl-, methyl ester, cobalt complex, 26:192
Pyridine, osmium complex, 26:291
  vanadium complex, 27:308
—, 3,5-dimethyl-, palladium complex, 26:210
—, 2-(phenylmethyl)-, palladium complex, 26:208
4-Pyridinecarboxylic acid, rhodium complex, 27:292
Pyridinium fluorotrioxochromate(VI), 27:310

Quinoline, 8-methyl-, palladium complex, 26:213

Resolution, of [ascorbato (2—)-C^2, O^5](1,2-cyclohexanediamine)platinum(II), 27:283
  of *cis*-dichlorobis(1,2-ethanediamine) chromium (III) chloride, 26:24
Rhenate(III), octachlorodi-, bis (tetrabutylammonium), 28:332
Rhenate(IV), hexaiodo-, dipotassium, 27:294
Rhenium, acetylpentacarbonyl-, 28:201
—, (2,2'-bipyridine)tricarbonylfluoro-, 26:82
—, (2,2'-bipyridine)tricarbonyl(phosphorodifluorido)-, 26:83
—, bromopentacarbonyl-, 28:162
—, dodecacarbonyltetrafluorotetra-, tetrahydrate, 26:82
—, octacarbonylbis(μ_3-carbon dioxide)tetra-, 28:20
—, octadecacarbonylbis(μ_3-carbon dioxide)tetra-, 26:111
—, pentacarbonylchloro-, 28:161
—, pentacarbonylhydrido-, 26:77
—, pentacarbonyliodo-, 28:163
—, pentacarbonylmethyl-, 26:107
—, pentacarbonyl [tetrafluoroborato(1—)]-, 28:15, 17
—, tetracarbonylcarboxy-, 26:112
Rhenium(I), pentacarbonyl[tetrafluoroborate(1—)]-, 26:108
—, pentacarbonyl (trifluoromethanesulfonato)-, 26:115
Rhenium(IV), octahydridotetrakis-(triphenylphosphine)di-, 27:16
Rhenium(VII), heptahydridobis-(triphenylphosphine)-, 27:15
Rhenium(1+), pentacarbonyl (η^2-ethene)-, tetrafluoroborate(1—), 26:110

Rhodate(1−), tetracarbonyl-, μ-nitrido-bis(triphenylphosphorus)(1+), 28:213
Rhodate(2−), dodecacarbonylplatinum-tetra-, bis [μ-nitrido-bis(triphenylphosphorus) (1+)], 26:375
—, tetradecacarbonylplatinum-tetra-, bis[μ-nitrido-bis(triphenylphosphorus)(1+)], 26:373
Rhodium, carbonylchlorobis-(triphenylphosphine)-, *trans*-, 28:79
—, chloro[[2-[(diphenylphosphino)methyl]-2-methyl-1,3-propanediyl]bis(diphenylphosphine)] (dithiocarbonato)-, 27:289
—, chloro[[2-[(diphenylphosphino)methyl]-2-methyl-1,3-propanediyl]bis-(diphenylphosphine)][(dithiocarboxy)-triethylphosphoniumato]-, 27:288
—, [2(η^4)-1,5-cyclooctadiene]tris-(dimethylphenylphosphine-1κP)-tri-μ-hydrido-osmium-, 27:29
—, tetracarbonyldichlorodi-, 28:84
—, tri-μ-carbonyl-nonacarbonyltetra-, 28:242
Rhodium(I), (acetato)carbonylbis (triisopropylphosphine)-, 27:292
—, (benzoato) carbonylbis (tricyclohexylphosphine)-, 27:292
—, carbonyl (3-fluorobenzoato)bis-(triphenylphosphine)-, 27:292
—, carbonylhydridotris(triphenylphosphine)-, 28:82
—, carbonyl (hydrogen phthalato)bis-(tricyclohexylphosphine)-, 27:291
—, carbonyliodobis(tricyclo-hexylphosphine)-, 27:292
—, carbonyl (4-methylbenzenesulfonato)bis(tricyclohexylphosphine)-, 27:292
—, carbonyl (methyl 3-oxobutanoato-O)bis-(triisopropylphosphine)-, 27:292
—, carbonylphenoxobis(triisopropylphosphine)-, 27:292
—, carbonyl (4-pyridinecarboxylato)bis(triisopropylphosphine)-, 27:292
—, chlorobis (cyclooctene)-, 28:90
—, chloro(η^2-*tetrahedro*-tetraphosphorus)-bis(triphenylphosphine)-, 27:222
—, chlorotris(triphenylphosphine)-, 28:77
—, (η^5-cyclopentadienyl)bis(trimethylphosphine)-, 28:280

—, (η^5-cyclopentadienyl)bis(trimethylphosphite)-, 28:284
—, di-μ-chlorobis(η^4-1,5-cyclooctadiene)di-, 28:88
—, di-μ-chloro-tetrakis-(ethene)di-, 28:86
—, hydridotetrakis(triphenylphosphine)-, 28:81
Rhodium(III), [[2-[(diphenylphosphino)methyl]-2-methyl-1,3-propanediyl]bis(diphenylphosphine)] (dithiocarbonato)-, tetraphenylborate(1−), 27:287
Ruthenate(1−), decacarbonyl-1κ^3C, 2κ^3C, 3κ^4C-μ-hydrido-1:2κ^2H-bis(triethylsilyl)-1κSi,2κSi-*triangulo*-tri-, μ-nitrido-bis(triphenylphosphorus)(1+), 26:269
—, dodecacarbonyltricobalt-, tetraethylammonium, 26:358
—, hexadecacarbonylnitridohexa-, μ-nitrido-bis(triphenylphosphorus)(1+), 26:287
—, tetrachlorobis(acetonitrile)-, tetraethylammonium, 26:356
—, tetradecacarbonylnitridopenta-, μ-nitrido-bis(triphenylphosphorus)(1+), 26:288
Ruthenate(2−), μ_5-carbido-tetradecacarbonylpenta-, bis[μ-nitrido-bis(triphenylphosphorus)(1+)], 26:284
—, μ_5-carbido-tetradecacarbonylpenta-, disodium, 26:284
Ruthenium, (acetonitrile) dodecacarbonyltricobaltcopper-, 26:359
—, (μ_3-bromomethylidyne)-nonacarbonyltri-μ-hydrido-*triangulo*-tri-, 27:201
—, bromononacarbonyl (3,3-dimethyl-1-butynyl)mercurytri-, 26:332
—, (μ_3-2-butyne)nonacarbonyl-dicobalt-, 27:194
—, *cyclo*-[μ_3-1(η^2):2(η^2):3(η^2)-2-butyne]octacarbonyl-1κ^2C,2κ^3C,3κ^3C-[1(η^5)-cyclopentadienyl] cobaltmolybdenum-, (*Co–Mo*) (*Co–Ru*) (*Mo–Ru*), 27:194
—, μ_6-carbido-heptadecacarbonylhexa-, 26:281
—, μ_5-carbido-pentadecacarbonylpenta-, 26:283
—, decacarbonyl(dimethylphenylphosphine)-(2-isocyano-2-methylpropane)tri-, 26:275

—, decacarbonyl(dimethylphenylphosphine) tetrahydrido[tris(4-methylphenyl) phosphite]tetra-, 26:278
—, decacarbonyl-μ-hydrido(μ-methoxymethylidyne)-*triangulo*-tri-, 27:198
—, decacarbonyl[methylenebis-(diphenylphosphine)]tri-, 26:276
—, dodecacarbonyltetra-μ-hydrido-tetra-, 28:219
—, dodecacarbonyltri-, 26:259
—, dodecacarbonyltris (triphenylphosphine)-cobalttrigoldtri-, 26:327
—, dodecacarbonyltetra-μ-hydrido-tetra-, 26:262
—, [μ-ethynediylbis(diphenylphosphine)]bis-[undecacarbonyltri-, 26:277
—, (μ_4-mercury)bis[nonacarbonyl-(μ_3-3,3-dimethyl-1-butynyl)-*triangulo*-tri-, 26:333
Ruthenium, nonacarbonyl (η^5-cyclopentadienyl)tri-μ-hydrido-nickeltri-, 26:363
—, nonacarbonyl (μ_3-3,3-dimethyl-1-butynyl)-μ-hydrido-*triangulo*-tri-, 26:329
—, nonacarbonyl(3,3-dimethyl-1-butynyl)-(iodomercury)-*triangulo*-tri-, 26:330
—, nonacarbonyl(μ_3-3,3-dimethyl-1-butynyl-{μ-[tricarbonyl(η^5-cyclopentadienyl)-molybdenum]mercury}-*triangulo*-tri-, 26:333
—, nonacarbonyl-μ-hydrido-(μ-diphenylphosphido)-tri-, 26:264
—, nonacarbonyl-μ_3-thio-dicobalt-, 26:352
—, nonacarbonyl-tri-μ-hydrido-(μ_3-methoxymethylidyne)-*triangulo*-tri-, 27:200
—, tetracarbonylbis(η^5-cyclopentadienyl)di-, 28:189
—, tetracarbonyl(η^2-methyl acrylate)-, 28:47
—, undecacarbonyldicobalt-, 26:354
—, undecacarbonyl(dimethylphenylphosphine)tri-, 26:273
—, undecacarbonyltetrahydrido[tris (4-methylphenyl) phosphite] tetra-, 26:277
Ruthenium(0), tricarbonyl (1,5-cyclooctadiene)-, 28:54
Ruthenium(II), μ-aqua-bis(μ-chloroacetato)bis[(chloroacetato) (η^4-cycloocta-1,5-diene)-, 26:256

—, μ-aqua-bis (μ-trichloroacetato)bis[(η^4-bicyclo[2.2.1]hepta-2,5-diene)-(trichloroacetato)-, 26:256
—, μ-aqua-bis (μ-trifluoroacetato)bis[(η^4-cycloocta-1,5-diene) (trifluoroacetato)-, 26:254
—, (η^4-bicyclo[2.2.1]hepta-2,5-diene)bis(η^3-2-propenyl)-, 26:251
—, (η^4-bicyclo[2.2.1]hepta-2,5-diene) dichloro-, 26:250
—, bis (acetonitrile)dichloro(η^4-1,5-cyclooctadiene)-, 26:69
—, bis(benzo[h]quinolin-10-yl-C^{10},N) dicarbonyl-, *cis*-, 26:177
—, bis(benzonitrile) dibromo(η^4-1,5-cyclooctadiene)-, 26:71
—, bis(benzonitrile)dichloro (η^4-1,5-cyclooctadiene)-, 26:70
—, chloro(η^5-cyclopentadienyl)bis-(triphenylphosphine)-, 28:270
—, chloro(η^6-hexamethylbenzene)hydrido-(triphenylphosphine)-, 26:181
—, (η^4-cycloocta-1,5-diene)bis(η^3-2-propenyl)-, 26:254
—, (η^4-1,5-cyclooctadiene)tetrakis-(hydrazine)-, bis[tetraphenylborate-(1−)], 26:73
—, (η^4-1,5-cyclooctadiene)tetrakis-(methylhydrazine)-, bis[hexafluorophosphate(1−)], 26:74
—, (η^4-1,5-cyclooctadiene)tetrakis (methylhydrazine)-, bis[tetraphenylborate-(1−)], 26:74
—, (η^5-cyclopentadienyl) [2-[(diphenoxyphosphino)oxy]phenyl-C,P](triphenyl phosphite-P)-, 26:178
—, di-μ-chloro-bis (tricarbonylchloro-, 28:334
—, dichloro(η^4-cycloocta-1,5-diene)-, 26:253
—, di-μ-chloro-(η^4-1,5-cyclooctadiene)-, polymer, 26:69
—, dihydridotetrakis (triphenylphosphine)-, 28:337
—, [2-(diphenylphosphino)-phenyl-C^1, P]-(η^6-hexamethylbenzene)-hydrido-, 26:182
—, tetrakis (acetonitrile) (η^4-1,5-cyclooctadiene)-, bis[hexafluorophosphate(1−)], 26:72

—, tris (acetonitrile) bromo(η^4-1,5-cyclooctadiene)-, hexafluorophosphate-(1—), 26:72
—, tris(acetonitrile)chloro(η^4-1,5-cyclooctadiene)-, hexafluorophosphate-(1—), 26:71
—, tris(2,2′-bipyridine)-, dichloride, hexahydrate, 28:338
Ruthenium(III), tetraamminedibromo-, cis-, bromide, 26:67
—, tetraamminedichloro-, cis-, chloride, 26:66
Ruthenium dinuclear carboxylate complexes, 26:249
Ruthenium mixed-metal clusters, 26:356

Saccharinates, of chromium and vanadium, 27:306
Samarium, tetrakis [η^5-1, 3-bis(trimethylsilyl)cyclopentadienyl]di-μ-chloro-di-, 27:171
—, trichlorobis(tetrahydrofuran)-, 27:140
—, tris(2, 6-di-tert-butyl-phenoxo)-, 27:166
Samarium(II), bis(η^5-pentamethylcyclopentadienyl)bis(tetrahydrofuran)-, 27:155
Samarium(III), tris(η^5-cyclopentadienyl)(tetrahydrofuran)-, 26:21
Samarium trichloride-2-tetrahydrofuran-, 27:140
Scandium, bis[η^5-1, 3-bis(trimethylsilyl)cyclopentadienyl]di-μ-chloro-bis(tetrahydrofuran)lithium-, 27:170
—, tetrakis [η^5-1,3-bis(trimethylsilyl)cyclopentadienyl]di-μ-chloro-di-, 27:171
—, tris(2,6-di-tert-butyl-4-methylphenoxo)-, 27:167
—, tris(2,6-di-tert-butyl-phenoxo)-, 27:167
Selenium osmium carbonyl clusters, 26:308
Selenium titanium complex, 27:61
Silanamine, 1,1,1-trimethyl-N-(trimethylsilyl)-, ytterbium complex, 27:148
Silane, bromotrimethyl-, 26:4
—, methoxytrimethyl-, 26:44
—, tetramethyl-, lutetium complex, 27:161
—, triethyl-, ruthenium complex, 26:269
Sodiohenicosatungstononaantimonate(18—), [$NaSb_9W_{21}O_{86}$]$^{18-}$, octadecaammonium, tetracosahydrate, 27:120

Sodiotetracontatungstotetraarsenate(27—), [$NaAs_4W_{40}O_{140}$]$^{27-}$, heptacosasodium, hexacontahydrate, 27:118
Sodiotricontatungstopentaphosphate(14-), [$NaP_5W_{30}O_{110}$]$^{14-}$, tetradecaammonium, hentricontahydrate, 27:115
Sodium, cyclopentadienyl-, compound with 1,2-dimethoxyethane (1:1), 26:341
Sodium henicosaphosphide, (Na_3P_{21}), 27:227
Sodium nonatungstophosphate (9—), (Na_9 [A-PW_9O_{34}], hydrate, 27:100
Sodium α-nonatungstosilicate(10—), (Na_{10}-[α-SiW_9O_{34}]), 27:87
Sodium β-nonatungstosilicate (10—), (Na_9H [β-SiW_9O_{34}]), tricosahydrate, 27:88
Sodium α-pentadecatungstodiphosphate-(12—), (Na_{12} [α-$P_2W_{15}O_{56}$], tetracosahydrate, 27:108
Sodium sodiotetracontatungstotetraarsenate-(27—), (Na_{27}[$NaAs_4W_{40}O_{140}$], hexacontahydrate, 27:118
Sodium tetramethylammonium dihydrogen pentamolybdobis[(2-aminoethyl) phosphonate] (4—), $Na[(CH_3)_4N]$ H_2-[$Mo_5O_{15}(NH_2C_2H_4PO_3)_2$], pentahydrate, 27:126
Sodium β_1-undecatungstosilicate (8—), ($Na_8[\beta_1$-$SiW_{11}O_{39}$]), 27:90
Solid state synthesis, 26:377
Stibine, triphenyl-, iron complex, 28:171 nickel complex, 28:103
λ^6-Sulfane, (2-bromoethenyl)-pentafluoro-, 27:330
—, ethynylpentafluoro-, 27:329
Sulfur, chromium complex, 27:69
cobalt, iron, and ruthenium complexes, 26:352
iron complex, 26:244
molybdenum cluster complex, 27:48, 49
molybdenum complexes, 27:41, 27:42
osmium complex, 26:305
titanium complex, 27:52
vanadium complex, 27:54
Sulfur molybdenum clusters, 27:47
Sulfur oligomers, 27:332
Superconductors from synthetic metals, 26:386
Synthetic metals, 26:386

Terbium, tetrakis[η^5-1, 3-bis(trimethylsilyl)-
cyclopentadienyl]di-μ-chloro-di-, 27:171
1,5,9,13-Tetraazacyclohexadeca-1,3,9,11-
tetraene, 3,11-bis(benzoyl)-2,12-
dimethyl-, nickel complex, 27:273
—, 4,10-dimethyl-, nickel complex, 27:272
1,5,9,13-Tetraazacyclohexadeca-1,4,9,12-
tetraene, 3,11-bis [α-
(benzylamino)benzylidene]-2,12-
dimethyl-, nickel complex, 27:276
—, 3,11-bis(α-methoxybenzylidene)-2,12-dimethyl-, nickel complex, 27:275
—, 2,12-dimethyl-3,11-bis (1-methoxyethylidene)-, nickel complex, 27:264
—, 2,12-dimethyl-3,11-bis[1-(methylamino)ethylidene]-, nickel complex, 27:266
Tetrabutylammonium μ_3-[(η^5-cyclopentadienyl) trioxotitanate (IV)]-A-β-1,2,3-trivanadononatungstosilicate (4—), [[(C$_4$H$_9$)$_4$N]$_4$[A-β-(η^5-C$_5$H$_5$)TiSiW$_9$V$_3$O$_{40}$]], 27:132
Tetrabutylammonium decatungstate(VI), [(C$_4$H$_9$)$_4$N]$_4$ [W$_{10}$O$_{32}$], 27:81
Tetrabutylammonium dimolybdate(VI), [(C$_4$H$_9$)$_4$N]$_4$[Mo$_2$O$_7$], 27:79
Tetrabutylammonium γ-dodecatungstosilicate(4—), ([(C$_4$H$_9$)$_4$N]$_4$[γ-SiW$_{12}$O$_{40}$]), 27:95
Tetrabutylammonium hexamolybdate(VI), [(C$_4$H$_9$)$_4$N]$_2$ [Mo$_6$O$_{19}$], 27:77
Tetrabutylammonium hexatungstate(VI), [(C$_4$H$_9$)$_4$N]$_2$ [W$_6$O$_{19}$], 27:80
Tetrabutylammonium octamolybdate(VI), [(C$_4$H$_9$)$_4$N]$_4$ [Mo$_8$O$_{26}$], 27:78
Tetrabutylammonium trihydrogen decavanadate(V), [(C$_4$H$_9$)$_4$N]$_3$H$_3$[V$_{10}$O$_{28}$]), 27:83
Tetrabutylammonium trihydrogen A-β-1,2,3-trivanadononatungstosilicate(7—), [[(C$_4$H$_9$)$_4$N]$_4$H$_3$ [A-β-SiW$_9$V$_3$O$_{40}$]], 27:131
Tetramethylammonium dihydrogen pentamolybdobis[(4-aminobenzyl)-phosphonate]-(4—), [(CH$_3$)$_4$N]$_2$H$_2$[Mo$_5$O$_{15}$(NH$_2$C$_6$H$_4$CH$_2$PO$_3$)$_2$]], tetrahydrate, 27:127
Tetrasulfur (2+), bromo-$cyclo$-heptasulfur-(1+) hexafluoroarsenate(1—) (1:4:6), 27:338
—, iodo-$cyclo$-heptasulfur(1+) hexafluoroarsenate (1—) (1:4:6), 27:337
Thallium, cyclopentadienyl-, 28:315
2H-1,2-Thiaphosphorin-2-ium, tetrakis-(methoxycarbonyl)-2,2-dimethyl-, manganese complex, 26:165
Thiocarbonyl complexes, iron, 28:186
osmium, 26:185
Thiocyanate, gold complex, 26:90
Thio osmium clusters, 26:301
Thiophenetetracarboxylate acid, tetramethylester, 26:166
Thiophene, tetrahydro-, gold complexes, 26:85–87
Thorium(IV), bis[η^5-1,3-bis(trimethylsilyl)-cyclopentadienyl]dichloro-, 27:173
—, chlorotris(η^5-cyclopentadienyl)-, 28:302
Thorium tetrachloride, 28:322
Thulium, tetrakis[η^5-1,3-bis(trimethylsilyl)cyclopentadienyl] di-μ-chloro-di-, 27:171
Titanate(3—), nonabromodi-, tricesium, 26:379
—, nonabromodi-, trirubidium, 26:379
—, nonachlorodi-, tricesium, 26:379
—, nonachlorotri-, trirubidium, 26:379
Titanium, bis(η^5-methylcyclopentadienyl)-(pentasulfido-S^1:S^5)- 27:52
—, dicarbonylbis(η^5-cyclopentadienyl)-, 28:250
—, dicarbonylbis(η^5-pentamethylcyclopentadienyl)-, 28:253
Titanium(III), chlorobis(η^5-cyclopentadienyl)-, 28:261
Titanium(IV), bis(η^5-cyclopentadienyl)-bis(hydrogen sulfido)-, 27:66
—, bis(η^5-cyclopentadienyl)[pentasulfido-(2—)]-, 27:60
—, bis(η^5-pentamethylcyclopentadienyl)-[trisulfido(2—)]-, 27:62
Titanium (V), bis(η^5-cyclopentadienyl)-[pentaselenido(2—)]-, 27:61
Titanium bromide, (TiBr$_3$), 26:382
Toluene, α-2,4-cyclopentadiene-1-yl-, chromium molybdenum and tungsten hexacarbonyl complexes, 28:148
Transition metal chalcogenide complexes, 27:39

Transition metal complexes with weakly bonded anions, 26:92
Transition metal polyoxo anions, 27:71
Tributylammonium pentatungstobis(phenylphosphonate)(4−), [[(C$_4$H$_9$)$_3$NH]$_4$[W$_5$O$_{15}$-(C$_6$H$_5$PO$_3$)$_2$]]], 27:127
Tricyclo[2.2.1.0^{2,6}]heptaphosphide (3−), 27:228
Triethyl phosphite, iron complex, 28:171
 nickel, palladium, and platinum complexes, 28:104
 nickel complex, 28:101
Trimethyl phosphite, cobalt and rhodium complexes, 28:283, 284
 iron complex, 28:171
 nickel complex, 28:101
Trinuclear complexes, 26:243
Triphenyl phosphite, iron complex, 28:171
 nickel complex, 28:101
 ruthenium complex, 26:178
Triphosphenium, 1,1,1,3,3,3-hexakis(dimethylamino)-, tetraphenylborate(1−), 27:256
—, 1,1,1,3,3,3-hexaphenyl-, tetrachloroaluminate(1−), 27:254
Triphosphenium salts, 27:253
1H-1,2,3-Tripholium, 3,3,4,5-tetrahydro-1,1,3,3-tetraphenyl-, hexachlorostannate(2−) (2:1), 27:255
Tris(4-methylphenyl) phosphite, ruthenium complex, 26:277
Tungstate (1−), (acetato) pentacarbonyl-, μ-nitrido-bis(triphenylphosphorus)(1+), 27:297
—, hydridononacarbonyliron-, μ-nitrido-bis(triphenylphosphorus)(1+), 26:336
—, tricarbonyl (η^5-cyclopentadienyl)-, sodium, compound with 1,2-dimethoxyethane (1:2), 26:343
Tungstate(2−), nonacarbonyliron-, bis[μ-nitrido-bis(triphenylphosphorus)(1+)], 26:339
Tungstate(4−), aquadihydroxohenhexacontaoxobis[trioxoarsenato(III)]henicosa-, tetrarubidium, tetratricontahydrate, 27:113
Tungstate(VI), [W$_6$O$_{19}$]$^{2-}$, bis(tetrabutylammonium), 27:80

—, [W$_{10}$O$_{32}$]$^{4-}$, tetrakis(tetrabutylammonium), 27:81
Tungsten, aquahexahydroxoheptapentacontaoxobis[trioxoarsenato(III)]henicosa-, hydrate, 27:112
—, bis (*tert*-butyl isocyanide)-tetracarbonyl-, *cis*-, 28:143
—, bis (dinitrogen)bis[1,2-ethane-diylbis(diphenylphosphine)]-, *trans*-, 28:41
—, (*tert*-butyl isocyanide)pentacarbonyl-, 28:143
—, carbonyl (η^5-cyclopentadienyl) [tetrafluoroborato (1−)]-, 28:5
—, dicarbonyl (η^5-cyclopentadienyl)-hydrido-(triphenylphosphine)-, 26:98
—, dicarbonyl (η^5-cyclopentadienyl)nitrosyl-, 28:196
—, dicarbonyl (η^5-cyclopentadienyl) [tetrafluoroborato(1−)] (tetraphenylphosphine)-, 28:7
—, dicarbonyl (η^5-cyclopentadienyl) [tetrafluoroborato(1−)] (triphenylphosphine)-, 26:98
—, hexacarbonylbis(η^5-cyclopentadienyl)di-, 28:148
—, tetracarbonylbis(η^5-cyclopentadienyl)di-, (*W—W*), 28:153
—, tetracarbonyl[(diethylamino)methylidyne](isocyanato)-, *trans*-, 26:42
—, tetrachlorobis(triphenylphosphine)-, 28:40
—, tetrachloro[1,2-ethanediylbis(diphenylphosphine)]-, 28:41
—, tricarbonyl(η^6-cycloheptatriene)-, 27:4
—, tricarbonyl(η^5-cyclopentadienyl) [tetrafluoroborato (1−)]-, 26:96
—, tricarbonyl(dihydrogen)bis-(tricyclohexylphosphine)-, 27:6
—, tricarbonyl(dihydrogen)bis-(triisopropylphosphine)-, 27:7
—, tricarbonyltris(propanenitrile)-, 27:4
—, tricarbonyltris(propionitrile)-, 28:30
—, trichlorotrimethoxy-, 26:45
—, trimethoxy-, trichloride, *see* tungsten, trichlorotrimethoxy-, 26:45
—, tris(*tert*-butyl isocyanide)tricarbonyl-, *fac*-, 28:143
Tungsten(0), bis[1,2-ethanediylbis-(diphenylphosphine)]bis-(isocyanomethane)-, *trans*-, 28:43

Tungsten(II), dichlorotetrakis(dimethylphenylphosphine)-, 28:330
—, dichlorotetrakis(methyldiphenylphosphine)-, 28:331
—, dichlorotetrakis(trimethylphosphine)-, 28:329
—, tetrakis(acetato)di-, (W-^{4}-W), 26:224
—, tetrakis (acetonitrile)dinitrosyl-, cis-, bis[tetrafluoroborate(1−)], 26:133
—, tetrakis (2, 2-dimethylpropanoato)di-, (W-^{4}-W), 26:223
—, tetrakis(trifluoroacetato)di-, (W-^{4}-W), 26:222
Tungsten(IV), bis(η^5-cyclopentadienyl)bis(hydrogen sulfido)-, 27:67
—, tetrachlorobis(methyldiphenylphosphine)-, 28:328
—, tetrachlorotris(trimethylphosphine)-, 28:327
—, tetrahydridotetrakis(methyldiphenylphosphine)-, 27:10
Tungsten(VI), (2,2′-bipyridine)dichloro-[(1,1-dimethylethyl)imido] (phenylimido)-, 27:303
—, dichloro[(1,1-dimethylethyl)imido]-(phenylimido)bis(trimethylphosphine)-, 27:304
—, hexahydridotris (dimethylphenylphospine)-, 27:11
—, tetrachlorobis(1,1-dimethylethanamine)bis[(1,1-dimethylethyl)imido]bis(μ-phenylimido)di-, 27:301
—, trichloro(1,2-dimethoxyethane)(2,2-dimethylpropylidyne)-, 26:50
—, tris(2,2-dimethylpropyl) (2,2-dimethylpropylidyne)-, 26:47
Tungsten(1+), (acetone)tricarbonyl (η^5-cyclopentadienyl)-, tetrafluoroborate-(1−), 26:105
—, (acetone) tricarbonyl (η^5-cyclopentadienyl)-, tetrafluoroborate-(1−), 28:14
—, pentacarbonyl[(diethylamino)-methylidyne]-, tetrafluoroborate (1−), 26:40
Tungsten(2+), tetrakis(acetonitrile)dinitrosyl-, cis-, bis[tetrafluoroborate(1−)], 28:66
Tungsten carboxylate dinuclear complexes, 26:219

Tungsten tetrachloride, 26:221
Tungsten tetrachloride oxide, 28:324
Tungstobis(phenylphosphonate) (4−), [W_5O_{15} ($C_6H_5PO_3$)$_2$]$^{4-}$, tetrakis (tributylammonium), 27:127
Tungstodiphosphate(6−), [$P_2W_5O_{23}$]$^{6-}$, hexacesium, 27:101
Tungstodiphosphate(6−), [β-$P_2W_{18}O_{62}$]$^{6-}$, hexapotassium, nonadecahydrate, 27:105
Tungstodiphosphate(6−), [$P_2W_{18}O_{62}$]$^{6-}$, hexapotassium, tetradecahydrate, 27:104
Tungstodiphosphate(10−), [α_2-$P_2W_{17}O_{61}$]$^{10-}$), decapotassium, eicosahydrate, 27:107
Tungstodiphosphate(12−), [α-$P_2W_{15}O_{56}$]$^{12-}$ dodecasodium, tetracosahydrate, 27:108
Tungstooctaphosphate(40−), [$P_8W_{48}O_{184}$]$^{40-}$, pentalithium octacosapotassium heptahydrogen, dononacontahydrate, 27:110
Tungstophosphate(7−), [$PW_{10}O_{36}$]$^{7-}$, heptacesium, 27:101
Tungstophosphate(9−), [A-PW_9O_{34}]$^{9-}$, nonasodium, hydrate, 27:100
Tungstophosphates, vanadium(V)-substituted, 27:96
Tungstosilicate, [α-SiW_9O_{34}]$^{10-}$, decasodium, 27:87
Tungstosilicate(4−), [α-$SiW_{12}O_{40}$]$^{4-}$, tetrapotassium, heptadecahydrate, 27:93
Tungstosilicate(4−), [β-$SiW_{12}O_{40}$]$^{4-}$, tetrapotassium, nonahydrate, 27:94
Tungstosilicate(4−), [γ-$SiW_{12}O_{40}$]$^{4-}$, tetrakis (tetrabutylammonium), 27:95
Tungstosilicate(8−), [α-$SiW_{11}O_{39}$]$^{8-}$, octapotassium, tridecahydrate, 27:89
Tungstosilicate(8−), [β_1-$SiW_{11}O_{39}$]$^{8-}$, octasodium, 27:90
Tungstosilicate(8−), [β_2-$SiW_{11}O_{39}$]$^{8-}$, octapotassium, tetradecahydrate, 27:91
Tungstosilicate(8−), [γ-$SiW_{10}O_{36}$]$^{8-}$, octapotassium, dodecahydrate, 27:88
Tungstosilicate(10−), [β-SiW_9O_{34}]$^{10-}$, nonasodium hydrogen, tricosahydrate, 27:88
Tungstosilicic acid, (H_4[α-$SiW_{12}O_{40}$]), hydrate, 27:93
Tungstosilicic acids, α-, β-, γ-, 27:85

Uranium(IV), bis(η^5-1,3-bis(trimethylsilyl)-cyclopentadienyl]dibromo-, 27:174
—, bis (η^5-1,3-bis(trimethylsilyl)cyclopentadienyl]dichloro-, 27:174
—, bis(η^5-1,3-bis (trimethylsilyl)cyclopentadienyl]diiodo-, 27:176
—, chlorotris(η^5-cyclopentadienyl)-, 28:301
—, tris(η^5-cyclopentadienyl)[(dimethylphenylphosphoranylidene)methyl]-, 27:177

Vanadate(3—), nonabromodi-, tricesium, 26:379
—, nonabromodi-, trirubidium, 26:379
—, nonachlorodi-, tricesium, 26:379
—, nonachlorodi-, trirubidium, 26:379
Vanadate(V), [$V_{10}O_{28}$]$^{6-}$, tris (tetrabutylammonium) trihydrogen, 27:83
Vanadium, bis(η^5-methylcyclopentadienyl)-(μ-disulfido-S:S') (μ-η^2:η^2-disulfido)-μ-thio-di-, 27:54
—, (μ-disulfido-S:S')bis(η^5-methyl cyclopentadienyl)di-μ-thio-di-, 3:55
Vanadium(II), bis(1,2-benzisothiazol-3(2H)-one 1,1-dioxidato)tetrakis(pyridine)-, —-2-pyridine, 27:308
—, tetraaquabis(1,2-benzisothiazol-3(2H)-one 1,1-dioxidato)-, dihydrate, 27:307
Vanadium(III), chlorobis(η^5-cyclopentadienyl)-, 28:262
Vanadium(V)-substituted dodecatungstophosphates, 27:96
Vanadocene, 28:263
Vanadononatungstosilicate (4—), μ_3-[(η^5-cyclopentadienyl)trioxotitanate(IV)]-A-β-1,2,3-tri-, [A-β-(η^5-C_5H_5)-TiSiW$_9$V$_3$O$_{40}$]$^{4-}$, tetrakis (tetrabutylammonium), 27:132
Vanadodecatungstophosphate(5—), [γ-PV$_2$W$_{10}$O$_{40}$], pentacesium, 27:102
Vanadononatungstophosphate(6—), [α-1,2,3,-PV$_3$W$_9$O$_{40}$]$^{6-}$, hexacesium, 27:100
Vanadononatungstosilicate(7—), (A-β-1,2,3-SiW$_9$V$_3$O$_{40}$]$^{7-}$, hexapotassium hydrogen, trihydrate, 27:129
Vanadononatungstosilicate(7—), [A-β-1,2,3-SiW$_9$V$_3$O$_{40}$]$^{7-}$, tetrakis(tetrabutylammonium) trihydrogen, 27:131
Vanadoundecatungstophosphate(4—), [α-PVW$_{11}$O$_{40}$]$^{4-}$, tetrapotassium, hydrate, 27:99

Vaska-type rhodium complexes, 27:290

Water, iridium complex, 26:124
ruthenium complex, 26:254
vanadium and chromium complexes, 27:307, 309

1,2-Xylylene, magnesium complex, 26:147
1,2-Xylylene transfer reagents, 26:144

Ytterbium, bis[bis(trimethylsilyl)amido]bis(diethyl ether)-, 27:148
—, bis[η^5-1,3-bis(trimethylsilyl)cyclopentadienyl]di-μ-chloro-bis(tetrahydrofuran)lithium-, 27:170
—, (diethyl ether)bis(η^5-pentamethylcyclopentadienyl)-, 27:148
—, tetrakis [η^5-1,3-bis-(trimethylsilyl)cyclopentadienyl]di-μ-chloro-di-, 27:171
—, trichlorotris(tetrahydrofuran)-, 27:139, 28:289
—, tris(2,6-di-$tert$-butyl-4-methylphenoxo)-, 27:167
Ytterbium(II), bis(η^5-cyclopentadienyl) (1,2-dimethoxyethane)-, 26:22
—, bis(phenylethnyl)-, 27:143
Ytterbium trichloride-3-tetrahydrofuran, 27:139
Ytterbium diiodide, 27:147
Yttrium, bis[η^5-1,3-bis-(trimethylsilyl)cyclopentadienyl]di-μ-chloro-bis(tetrahydrofuran)lithium-, 27:169
—, tetrakis[η^5-1,3-bis(trimethylsilyl) cyclopentadienyl] di-μ-chloro-di-, 27:171
—, tris(2,6-di-$tert$-butyl-4-methylphenoxo)-, 27:167
—, tris(2,6-di-$tert$-butyl-phenoxo)-, 27:167

Zeise's salt, 28:349
Zinc dichloride, 28:322
Zirconium, bis(η^5-cyclopentadienyl) dihydrido-, 28:257
—, bis[1,1(η^5)-cyclopentadienyl]tris-(dimethylphenylphosphine-2κP)-tri-μ-hydrido-1-hydrido-1κH-osmium-, 27:27
—, chlorobis (η^5-cyclopentadienyl)hydrido-, 28:259
—, dicarbonylbis(η^5-cyclopentadienyl)-, 28:251
—, dicarbonylbis(η^5-pentamethylcyclopentadienyl)-, 28:254

FORMULA INDEX
Prepared by THOMAS E. SLOAN*

The Formula Index, as well as the Subject Index, is a Cumulative Index for Volumes 26–30. The Index is organized to allow the most efficient location of specific compounds and groups of compounds related by central metal ion or ligand grouping.

The formulas entered in the Formula Index are for the total composition of the entered compound, e.g., F_6NaU for sodium hexafluorouranate(V). The formulas consist solely of atomic symbols (abbreviations for atomic groupings are not used) and arranged in alphabetical order with carbon and hydrogen always given last, e.g., $Br_3CoN_4C_4H_{16}$. To enhance the utility of the Formula Index, all formulas are permuted on the symbols for all metal atoms, e.g., $FeO_{13}Ru_3C_{13}H_{13}$ is also listed at $Ru_3FeO_{13}C_{13}H_{13}$. Ligand groupings are also listed separately in the same order, e.g., $N_2C_2H_8$, 1,2-Ethanediamine, cobalt complexes. Thus individual compounds are found at their total formula in the alphabetical listing; compounds of any metal may be scanned at the alphabetical position of the metal symbol; and compounds of a specific ligand are listed at the formula of the ligand, e.g., NC for Cyano complexes.

Water of hydration, when so identified, is not added into the formulas of the reported compounds, e.g., $Cl_{0.30}N_4PtRb_2C_4 \cdot 3H_2O$.

$AlCl_4P_3C_{36}H_{30}$, Aluminate(1−), tetrachloro-, 1,1,1,3,3,3-hexaphenyltriphosphenium, 27:254

$AsBrF_6S_7$, Arsenate(1−)hexafluoro-, cyclo-heptasulfur(1+), bromo-, 27:336

$AsC_{18}H_{15}$, Arsine, triphenyl-, iron complex, 28:171
  nickel complex, 28:103

AsF_6IS_7, Arsenate(1−), hexafluoro-, iodo-cyclo-heptasulfur(1+), 27:333

$AsFeO_4C_{22}H_{15}$, Iron, tetracarbonyl-(triphenylarsine)-, 26:61

$As_2C_{10}H_{16}$, Arsine, 1,2-phenylenebis(dimethyl-, gold complex, 26:89
  nickel complex, 28:103

$As_2H_4O_{70}Rb_4W_{21} \cdot 34H_2O$, Tungstate(4−), aquadihydroxohenhexacontaoxobis[trioxoarsenato(III)]henicosa-, tetrarubidium, tetratricontahydrate, 27:113

$As_2H_8O_{70}W_{21} \cdot xH_2O$, Tungsten, aquahexahydroxoheptapentacontaoxobis[trioxoarsenato(III)]henicosa-, hydrate, 27:112

$As_4Au_2F_{10}C_{32}H_{32}$, Gold(I), bis[1,2-phenylenebis(dimethylarsine)]-, bis(pentafluorophenyl)aurate(I), 26:89

$As_4Co_2H_{100}N_{24}O_{142}W_{40} \cdot 19H_2O$, Ammoniodicobaltotetracontatungstotetraarsenate(23−), tricosaammonium, nonadecahydrate, 27:119

$As_4Na_{28}O_{140}W_{40} \cdot 60H_2O$, Sodiotetracontatungstotetraarsenate(27−), heptacosasodium, hexacontahydrate, 27:118

$As_4NiC_{20}H_{32}$, Nickel(0), bis[1,2-phenylenebis(dimethylarsine)]-, 28:103

$As_4NiC_{72}H_{60}$, Nickel(0), tetrakis(triphenylarsine)-, 28:103

$As_6Br_4F_{36}S_{32}$, Arsenate(1−), hexafluoro-, bromo-cyclo-heptasulfur(1+) tetrasulfur(2+) (6:4:1), 27:338

*Chemical Abstracts Service, Columbus, OH.

$As_6F_{36}I_4S_{32}$, Arsenate(1−), hexafluoro-, iodo-*cyclo*-heptasulfur(1+) tetrasulfur-(2+) (6:4:1), 27:337

$AuClF_5PC_{31}H_{22}$, Aurate(I), chloro(pentafluorophenyl)-, (benzyl)triphenylphosphonium, 26:88

$AuClPC_{18}H_{15}$, Gold, chloro(triphenylphosphine)-, 26:325

$AuClSC_4H_8$, Gold(I), chloro(tetrahydrothiophene)-, 26:86

$AuCo_3FeO_{12}PC_{30}H_{15}$, Gold(1+), (triphenylphosphine)-, dodecacarbonyltricobaltferrate(1−), 27:188

$AuF_5SC_{10}H_8$, Gold(I), (pentafluorophenyl)-(tetrahydrothiophene)-, 26:86

$AuF_{15}SC_{22}H_8$, Gold(III), tris(pentafluorophenyl)-(tetrahydrothiophene)-, 26:87

$AuHO_{10}Os_3PC_{16}H_{15}$, Osmium, decacarbonyl-$\mu$-hydrido[$\mu$-(triethylphosphine)gold]tri-, 27:210

$AuHO_{10}Os_3PC_{28}H_{15}$, Osmium, decacarbonyl-$\mu$-hydrido[$\mu$-(triphenylphosphine)gold]tri-, 27:209

$AuMn_2O_8P_2C_{38}H_{25}$, Gold, octacarbonyl-1κ^4C, 2κ^4C-μ-(diphenylphosphino)-1:2κP-(triphenylphosphine)-3κP-*triangulo*-dimanganese-, 26:229

$Au_2As_2F_{10}C_{32}H_{32}$, Gold(I), bis[1,2-phenylenebis(dimethylarsine)]-, bis(pentafluorophenyl)aurate(I), 26:89

$Au_2ClF_3O_3P_4PtSC_{49}H_{60}$, Gold(1+), chloro-1$\kappa Cl$-bis(triethylphosphine-1$\kappa P$)bis(triphenylphosphine-)-2κP, 3κP-*triangulo*-platinumdi-, trifluoromethanesulfonate, 27:218

$Au_2F_5NPSC_{25}H_{15}$, Gold(I), (pentafluorophenyl)-μ-thiocyanato-(triphenylphosphine)di-, 26:90

$Au_2O_{10}Os_3P_2C_{22}H_{30}$, Gold, decacarbonylbis(triethylphosphine)triosmiumdi-, 27:211

$Au_2O_{10}Os_3P_2C_{46}H_{30}$, Osmium, decacarbonylbis[μ-(triphenylphosphine)gold]tri-, 27:211

$Au_3BF_4OP_3C_{54}H_{45}$, Gold(1+), μ_3-oxotris[(triphenylphosphine)-, tetrafluoroborate(1−), 26:326

$Au_3CoO_{12}P_3Ru_3C_{66}H_{45}$, Ruthenium, dodecacarbonyltris(triphenylphosphine)-cobalt-trigoldtri-, 26:327

$Au_{55}Cl_6P_{12}C_{216}H_{180}$, Gold, hexachlorododecakis(triphenylphosphine)pentapentaconta-, 27:214

$BAu_3F_4OP_3C_{54}H_{45}$, Gold(1+), μ_3-oxotris[(triphenylphosphine)-, tetrafluoroborate(1−), 26:326

BC_3H_9, Borane, trimethyl-, 27:339

$BClF_4IrN_2N_2P_2C_{36}H_{31}$, Iridium(III), chloro(dinitrogen)hydrido[tetrafluoroborato(1−)]bis(triphenylphosphine)-, 26:119

$BClF_4IrN_2P_2C_{36}H_{31}$, Iridium(III), chloro-(dinitrogen)hydrido[tetrafluoroborato-(1−)]bis(triphenylphosphine)-, 28:25

$BClF_4IrOP_2C_{37}H_{31}$, Iridium(III), carbonylchlorohydrido[tetrafluoroborato(1−)]-bis(triphenylphosphine)-, 26:117, 28:23

$BClF_4IrOP_2C_{38}H_{33}$, Iridium(III), carbonylchloromethyl[tetrafluoroborato-(1−)]bis(triphenylphosphine)-, 26:118

BF_4, Borate(1−), tetrafluoro-, molybdenum and tungsten complexes, 26:96–105

$BF_4FeO_2C_{11}H_{13}$, Iron(1+), dicarbonyl(η^5-cyclopentadienyl) (η^2-2-methyl-1-propene)-, tetrafluoroborate(1−), 28:210

$BF_4H_5Ir_2C_{56}H_{56}$, Iridium(1+), [1,4-butanedibis(diphenylphosphine)]di-, tetrafluoroborate(1−), 27:26

$BF_4H_5Ir_2P_4C_{54}H_{52}$, Iridium(1+), pentahydridobis[1,3-propanediylbis(diphenylphosphine)]di-, tetrafluoroborate(1−), 27:22

$BF_4I_2IrC_{42}H_{36}$, Iridium(III), (1,2-diiodobenzene)dihydridobis(triphenylphosphine)-, tetrafluoroborate(1−), 26:125, 28:59

$BF_4IrO_2P_2C_{38}H_{36}$, Iridium(1+), diaquadihydridobis(triphenylphosphine)-, tetrafluoroborate(1−), 26:124, 28:58

$BF_4IrO_2P_2C_{42}H_{44}$, Iridium(1+), bis(acetone)-dihydridobis(triphenylphosphine)-, tetrafluoroborate(1−), 26:123, 28:57

$BF_4IrP_2C_{35}H_{38}$, Iridium(1+), (η^4-1,5-cyclooctadiene)[1,3-

propanediylbis(diphenylphosphine)]-, tetrafluoroborate(1 −), 27:23

$BF_4IrP_2C_{44}H_{42}$, Iridium(1 +), (η^4-1,5-cyclooctadiene)bis(triphenylphosphine)-, tetrafluoroborate(1 −), 26:122, 28:56

$BF_4MoOC_{34}H_{25}$, Molybdenum(1 +), carbonyl-(η^5-cyclopentadienyl)bis(diphenylacetylene)-, tetrafluoroborate(1 −), 26:102

$BF_4MoOPC_{38}H_{30}$, Molybdenum(1 +), carbonyl(η^5-cyclopentadienyl)(diphenylacetylene)-(triphenylphosphine)-, tetrafluoroborate(1 −), 26:104

$BF_4MoO_2PC_{25}H_{20}$, Molybdenum, dicarbonyl(η^5-cyclopentadienyl) [tetrafluoroborato(1 −)] (triphenylphosphine)-, 26:98

$BF_4MoO_3C_8H_5$, Molybdenum, tricarbonyl-(η^5-cyclopentadienyl) [tetrafluoroborato(1 −)]-, 26:96, 28:5

$BF_4MoO_3C_{10}H_9$, Molybdenum(1 +), tricarbonyl(η^5-cyclopentadienyl) (η^2-ethene)-, tetrafluoroborate(1 −), 26:102

$BF_4MoO_4C_{11}H_{11}$, Molybdenum(1 +), (acetone)tricarbonyl(η^5-cyclopentadienyl)-, tetrafluoroborate(1 −), 26:105

$BF_4NO_5WC_{10}H_{10}$, Tungsten(1 +), pentacarbonyl[(diethylamino)methylidyne]-, tetrafluoroborate(1 −), 26:40

$BF_4O_2PWC_{25}H_{20}$, Tungsten, dicarbonyl-(η^5-cyclopentadienyl) [tetrafluoroborato(1 −)]triphenylphosphine)-, 26:98

$BF_4O_3WC_8H_5$, Tungsten, tricarbonyl(η^5-cyclopentadienyl) [tetrafluoroborato-(1 −)]-, 26:96, 28:5

$BF_4O_4WC_{11}H_{11}$, Tungsten(1 +), (acetone)tricarbonyl(η^5-cyclopentadienyl)-, tetrafluoroborate-(1 −), 26:105

$BF_4O_5ReC_5$, Rhenium, pentacarbonyl[tetrafluoroborato(1 −)]-, 26:108

$BF_4O_5ReC_7H_4$, Rhenium(1 +), pentacarbonyl(η^2-ethene)-, tetrafluoroborate(1 −), 26:110

$BH_3P_4Pt_2C_{48}H_{80}$, Platinum(II), di-μ-hydrido-hydridotetrakis(triethylphosphine)ditetraphenylborate(1 −), 27:34

Platinum(II), μ-hydrido-dihydridotetrakis(triethylphosphine)di-, tetraphenylborate(1 −), 27:32

$BH_3P_4Pt_2C_{96}H_{80}$, Platinum(II), di-μ-hydrido-hydridotetrakis(triphenylphosphine)di-, tetraphenylborate(1 −), 27:36

$BN_6P_3C_{36}H_{56}$, Borate(1 −), tetraphenyl-, 1,1,1,3,3,3-hexakis(dimethylamino)triphosphenium, 27:256

$BOP_3RhS_2C_{66}H_{59}$, Borate(1 −), tetraphenyl-, [[2-[(diphenylphosphino)methyl]-2-methyl-1,3-propanediyl]bis(diphenylphosphine)](dithiocarbonato)rhodium(III), 27:287

$BO_2P_2PtC_{40}H_{57}$, Platinum(II), (3-methoxy-3-oxo-κO-propyl-κC^1)bis(triethylphosphine)-, tetraphenylborate(1 −), 26:138

$BP_2PtC_{44}H_{63}$, Platinum(II), (η^3-cyclooctenyl)bis(triethylphosphine)-, tetraphenylborate(1 −), 26:139

$BP_4PtC_{54}H_{33}$, Platinum(II), μ-hydrido-hydridophenyltetrakis(triethylphosphine)-di-, tetraphenylborate(1 −), 26:136

$B_2F_8H_7Ir_3P_6C_{78}H_{72}$, Iridium(2 +), tris[1,2-ethanediylbis(diphenylphosphine)]heptahydridotri-, bis[tetrafluoroborate(1 −)], 27:25

$B_2F_8H_7Ir_3P_6C_{81}H_{78}$, Iridium(2 +), heptahydridotris[1,3-propanediylbis(diphenylphosphine)]tri-, bis[tetrafluoroborate(1 −)], 27:22

$B_2F_8MoN_6O_2C_8H_{12}$, Molybdenum(2 +), tetrakis(acetonitrile)dinitrosyl-, cis-, bis-[tetrafluoroborate(1 −)], 26:132, 28:65

$B_2F_8N_4PdC_8H_{12}$, Palladium(2 +), tetrakis-(acetonitrile)-, bis[tetrafluoroborate(1 −)], 26:128, 28:63

$B_2F_8N_6O_2WC_8H_{12}$, Tungsten(2 +), tetrakis(acetonitrile)dinitrosyl-, cis-, bis-[tetrafluoroborate(1 −)], 26:133, 28:66

B_2H_6, Diborane(6), 27:215

$B_2N_8RuC_{56}H_{68}$, Ruthenium(II), (η^4-1,5-cyclooctadiene)tetrakis(hydrazine)-, bis-[tetraphenylborate(1 −)], 26:73

$B_2N_8RuC_{60}H_{76}$, Ruthenium(II), (η^4-1,5-cyclooctadiene)tetrakis(methylhydrazine)-, bis[tetraphenylborate(1 −)], 26:74

B$_9$H$_{14}$K, Borate(1−), tetradecahydronona-potassium, 26:1

BrAsF$_6$S$_7$, cyclo-Heptasulfur(1+), bromo-, hexafluoroarsenate(1−), 27:336

BrCH$_3$, Methane, bromo-, osmium and ruthenium complexes, 27:201, 205

BrC$_{18}$H$_{29}$, Benzene, 1-bromo-2,4,6-tri-*tert*-butyl-, 27:236

BrF$_5$SC$_2$H$_2$, λ^6-Sulfane, (2-bromoethenyl)pentafluoro-, 27:330

BrF$_6$N$_3$PRuC$_{14}$H$_{21}$, Ruthenium(II), tris-(acetonitrile)bromo(η^4-1,5-cyclooctadiene)-, hexafluorophosphate(1−), 26:72

BrF$_6$S$_7$Sb, cyclo-Heptasulfur(1+), bromo-, hexafluoroantimonate(1−), 27:336

BrH$_3$O$_9$Os$_3$C$_{10}$, Osmium, (μ_3-bromomethylidyne)nonacarbonyltri-μ-hydrido-*triangulo*-tri-, 27:205

BrH$_3$O$_9$Ru$_3$C$_{10}$, Ruthenium, (μ_3-bromomethylidyne)nonacarbonyl-tri-μ-hydrido-*triangulo*-tri-, 27:201

BrHgO$_9$Ru$_3$C$_{15}$H$_9$, Ruthenium, (bromomercury)nonacarbonyl(3,3-dimethyl-1-butynyl)-*triangulo*-tri-, 26:332

BrMnO$_5$C$_5$, Manganese, bromopentacarbonyl-, 28:156

BrNO$_4$SC$_{10}$H$_{18}$, Bicyclo[2.2.1]heptane-7-methanesulfonate, 3-bromo-1,7-dimethyl-2-oxo-, [(1R)-(*ENDO, ANTI*)]-, ammonium, 26:24

BrO$_5$ReC$_5$, Rhenium, bromopentacarbonyl-, 28:162

BrSiC$_3$H$_9$, Silane, bromotrimethyl, 26:4

Br$_2$MoO$_4$C$_4$, Molybdenum(II), dibromotetracarbonyl-, 28:145

Br$_2$N$_2$RuC$_{22}$H$_{22}$, Ruthenium(II), bis(benzonitrile)dibromo(η^4-1,5-cyclooctadiene)-, 26:71

Br$_2$Si$_2$UC$_{22}$H$_{42}$, Uranium(IV), bis[η^5-1,3-bis(trimethylsilyl)cyclopentadienyl]dibromo-, 27:174

Br$_3$H$_{12}$N$_4$Ru, Ruthenium(III), tetramminedibromo-, *cis*-, bromide, 26:67

Br$_3$Ti, Titanium tribromide, 26:382

Br$_4$As$_6$F$_{36}$S$_{32}$, cyclo-Heptasulfur(1+), bromo-, tetrasulfur(2+) hexafluoroarsenate(1−) (4:1:6), 27:338

Br$_9$Cr$_2$Cs$_3$, Chromate(3−), nonabromodi-, tricesium, 26:379

Br$_9$Cr$_2$Rb$_3$, Chromate(3−), nonabromodi-, trirubidium, 26:379

Br$_9$Cs$_3$Ti$_2$, Titanate(3−), nonabromodi-, tricesium, 26:379

Br$_9$Cs$_3$V$_2$, Vanadate(3−), nonabromodi-, tricesium, 26:379

Br$_9$Rb$_3$Ti$_2$, Titanate(3−), nonabromodi-, trirubidium, 26:379

Br$_9$Rb$_3$V$_2$, Vanadate(3−), nonabromodi-, trirubidium, 26:379

C, Carbide, iron complex, 26:246
 ruthenium cluster complexes, 26:281–284

CHF$_3$O$_3$S, Methanesulfonic acid, trifluoro-, iridium, manganese, and rhenium complexes, 26:114, 115, 120
 platinum complex, 26:126

CHOS$_2$, Dithiocarbonic acid, 27:287

CH$_2$, Methylene, osmium complex, 27:206

CH$_2$O$_2$, Formic acid, rhenium complex, 26:112

CH$_3$, Methyl, iridium complex, 26:118
 manganese complex, 26:156
 osmium complex, 27:206
 rhenium complexes, 26:107, 28:16

CH$_3$Br, Methane, bromo-, osmium and ruthenium complexes, 27:201, 205

CH$_3$Cl, Methane, chloro-, osmium complex, 27:205

CH$_4$O, Methanol, platinum complexes, 26:135
 tungsten complex, 26:45

CNa$_2$O$_{14}$Ru$_5$C$_{14}$, Ruthenate(2−)μ_5-carbido-tetradecacarbonyl-, disodium, 26:284

CO, Carbon monoxide, chromium complexes, 26:32, 34, 35, 28:137–139
 chromium, molybdenum, and tungsten complexes, 26:343, 27:297, 28:48, 196, 197
 cobalt complexes, 28:273, 275
 cobalt, copper, and ruthenium complexes, 26:358, 359
 cobalt-gold-ruthenium complexes, 26:327
 cobalt-iron complex, 26:244, 27:188
 cobalt, iron, and ruthenium complexes, 26:352–354
 cobalt-molybdenum-nickel complexes, 27:192

cobalt-molybdenum-ruthenium complex, 27:194
cobalt, platinum, and rhodium complex, 26:370
cobalt and ruthenium complexes, 26:176, 177
gold-osmium complexes, 27:209, 211
iridium complexes, 26:117–120, 28:23, 24
iridium, osmium, rhodium, and ruthenium, 28:216–247
iridium and rhodium complexes, 28:213
iron complex, 26:53–57, 232–241, 246, 27:183–185
iron-tungsten complex, 26:336–340
iron with Group 15 ligands, 26:59–63
manganese complexes, 26:156–158, 162–173, 226–230, 28:155–158
manganese and rhenium complexes, 26:114, 115, 28:199, 201
mercury-molybdenum-ruthenium complexes, 26:329–335
molybdenum, palladium, and platinum complexes, 26:345
molybdenum complexes, 26:84, 27:3, 28:151, 152
molybdenum and tungsten complexes, 26:96–105, 28:5
nickel complex, 26:312
nickel, osmium, and ruthenium complexes, 26:362–367
niobium complex, 28:192
osmium complexes, 26:187, 188, 290–293, 295–301, 304–307
platinum complexes, 26:316–322
rhenium complexes, 26:77, 82, 83, 107, 28:15–21
rhodium complexes, 27:291, 28:79
ruthenium complex, 26:259, 262, 264, 269, 273, 275–278, 281–284, 287, 288, 28:47
ruthenium and osmium complexes, 27:198–207
tungsten complexes, 26:40, 42, 27:4, 6
CO_2, Carbon dioxide, rhenium complex, 26:111
$CO_{14}P_2Ru_5C_{50}H_{15}$, Ruthenate(2−), μ_5-carbido-tetradecacarbonylpenta-, bis[μ-nitrido-bis(triphenylphosphorous)-(1+)], 26:284
$CO_{15}Ru_5C_{15}$, Ruthenium, μ_5-carbido-pentadecacarbonylpenta-, 26:283
$CO_{17}Ru_6C_{17}$, Ruthenium, μ_6-carbido-heptadecacarbonylhexa-, 26:281
CS, Thiocarbonyls, iron, 28:186
osmium, 26:185–187
$C_2HCl_3O_2$, Acetic acid, trichloro-, ruthenium complex, 26:256
$C_2HF_3O_2$, Acetic acid, trifluoro-, ruthenium complex, 26:254
tungsten complex, 26:222
C_2HF_5S, λ^6-Sulfane, ethynylpentafluoro-, 27:329
$C_2H_2BrF_5S$, λ^6-Sulfane, (2-bromoethenyl)pentafluoro-, 27:330
$C_2H_3ClO_2$, Acetic acid, chloro-, ruthenium complex, 26:256
C_2H_3N, Acetonitrile, cobalt, copper, and ruthenium complexes, 26:356, 359
molybdenum complex, 28:34
molybdenum, palladium, and tungsten complexes, 26:128–133
osmium complex, 26:290, 292
ruthenium(II) complexes, 26:69–72
transition metal complexes, 28:63–66
C_2H_3N, Methane, isocyano-, tungsten complex, 28:43
C_2H_4O, Acetaldehyde, iron complex, 26:235–241
manganese and rhenium complexes, 28:199, 201
$C_2H_4O_2$, Acetic acid, chromium, molybdenum, and tungsten complexes, 27:297
palladium complex, 26:208
rhodium complex, 27:292
tungsten complex, 26:224
C_2H_4, Ethene, cobalt complex, 28:278
molybdenum complex, 26:102–105
platinum complexes, 28:129, 130–134
rhenium complexes, 26:110, 28:19
rhodium complex, 28:86
$C_2H_4S_2$, Methyl dithioformate, 28:186
C_2H_6, Ethane, cobalt-molybdenum-nickel complex, 27:192
C_2H_6O, Dimethyl ether, ruthenium complex, 27:198
C_2H_7PS, Phosphine sulfide dimethyl-, 26:162
manganese complex, 26:162
$C_2H_8N_2$, 1,2-Ethanediamine, chromium complex, resolution of, 26:24–27
platinum complexes, 27:314, 315

C_3H_5N, Propanenitrile, chromium, molybdenum, and tungsten complexes, 28:30–32
tungsten complex, 27:4
C_3H_6, Propene, palladium complex, 28:342
ruthenium complex, 26:251
C_3H_6O, Acetone, iridium complex, 26:123
molybdenum and tungsten complexes, 26:105
$C_3H_6O_2$, Methyl acetate, iron complex, 27:184
osmium complex, 27:204
C_3H_9B, Borane, trimethyl-, 27:339
$C_3H_9O_3P$, Trimethyl phosphite, cobalt and rhodium complexes, 28:283, 284
$C_3H_9O_3P$, Trimethyl phosphite, iron complex, 28:171
nickel complex, 28:101
iron complex, 26:61
C_3H_9P, Phosphine, methyl-, iron complex, 28:177
C_3H_9P, Phosphine, trimethyl-, cobalt and rhodium complexes, 28:280, 281
complexes, 28:327, 329
nickel complex, 28:101
preparation of, 28:305
tungsten complexes, 27:304, 28:327, 329
C_4H_6, 2-Butyne, cobalt-molybdenum-ruthenium complex, 27:194
$C_4H_6O_2$, Methyl acrylate, platinum complex, 26:138
ruthenium complex, 28:47
C_4H_7NO, 3-Butenamide, nickel complex, 26:206
C_4H_8O, Furan, tetrahydro-, actinide and lanthanide complexes, 28:289, 290
iron complex, 26:232
lanthanide complexes, 28:293–295
lanthanide lithium complexes, 27:170
lutetium complex, 27:152
magnesium complex, 26:147
molybdenum complex, 28:35
neodymium complex, 27:158
neodymium and samarium complexes, 26:20
samarium complex, 27:155
C_4H_8, Propene, 2-methyl-, iron complexes, 28:208, 210
C_4H_8S, Thiophene, tetrahydro-, gold complexes, 26:85–87

C_4H_9NO, 2-Propenamide, 2-methyl-, nickel complex, 26:205
C_4H_{10}, Propane, 2-methyl-, neodymium complex, 27:158
$C_4H_{10}O$, Diethyl ether, ytterbium complex, 27:148
$C_4H_{10}O_2$, Ethane, 1,2-dimethoxy-, solvates of chromium, molybdenum, and tungsten carbonyl cyclopentadienyl complexes, 26:343
Ethane, 1,2-dimethoxy-, tungsten complex, 26:50
ytterbium complex, 26:22
$C_4H_{10}O_2 \cdot NaC_5H_5$, Ethane, 1,2-dimethoxy-, compd. with cyclopentadienylsodium, 26:341
$C_4H_{11}N$, Ethanamine, 1,1-dimethyl-, tungsten complex, 27:301
$C_4H_{12}OSi$, Silane, methoxytrimethyl-, 26:44
$C_4H_{12}Si$, Silane, tetramethyl-, lutetium complex, 27:161
$C_5F_6O_2H_2$, 2,4-Pentanedione, 1,1,1,5,5,5-hexafluoro-, palladium complexes, 27:318–320
$C_5H_4S_5$, 1,3-Dithiolo[4,5-b] [1,4]-dithiin-2-thione, 26:389
C_5H_5N, Pyridine, osmium complex, 26:291, 28:234
vanadium complex, 27:308
C_5H_6, 1,3-Cyclopentadien, cobalt complex, 26:309
chromium, molybdenum, and tungsten complexes, 26:343
cobalt complex 26:191–197
cobalt-molybdenum-nickel complex, 27:193
cobalt and rhodium complexes, 28:280, 281
iron complex, 26:232–241, 28:186
lanthanide complexes, 28:293–295
lutetium complex, 27:161
mercury-molybdenum-ruthenium complex, 26:333
molybdenum complexes, 27:63, 28:151, 152
molybdenum, palladium, and platinum complexes, 26:345
molybdenum and tungsten complexes, 26:96–105
neodymium complex, 27:158

neodymium, samarium, and ytterbium complexes, 26:20
nickel, osmium, and ruthenium complexes, 26:362-367
niobium complex, 28:267
osmium-zirconium complex, 27:27
palladium complex, 28:343
ruthenium complex, 26:178
thallium complex, 28:315
thorium and uranium complexes, 28:301, 302
titanium complex, 27:60
titanium and vanadium complexes, 28:261-266
tungsten complex, 27:67
uranium complex, 27:177

C_5H_6S, 2,4-Pentadienthial, osmium complex, 26:188

$C_5H_8O_3$, Butanoic acid, 3-oxo-, methyl ester, rhodium complex, 27:292

C_5H_9N, tert-Butyl isocyanide, chromium, molybdenum, and tungsten complexes, 28:143
nickel complex, 28:99
palladium complex, 28:110
ruthenium complex, 26:275, 28:224

C_5H_9P, Phosphine, (2,2-dimethylpropylidyne)-, 27:249, 251

$C_5H_{10}O_2$, Propanoic acid, 2,2-dimethyl-, tungsten complex, 26:223

$C_5H_{11}NS_2$, Dithiocarbamic acid, diethyl-, molybdenum complex, 28:145

C_5H_{12}, Propane, 2,2-dimethyl-, magnesium and tungsten complexes, 26:46, 47, 50

$C_5H_{13}N$, Ethanamine, N-ethyl-N-methyl-, tungsten complex, 26:40, 42

C_6HF_5, Benzene, pentafluoro-, gold complexes, 26:86-90

$C_6H_4I_2$, Benzene, 1,2-diiodo-, iridium complex, 26:125

C_6H_5, Phenyl, platinum complex, 26:136

C_6H_5Cl, Benzene, chloro-, chromium complex, 28:139

C_6H_5F, Benzene, fluoro-, chromium complex, 28:139

$C_6H_5NO_2$, 4-Pyridinecarboxylic acid, rhodium complex, 27:292

C_6H_6, Benzene, chromium complex, 28:139

C_6H_6O, Phenol, rhodium complex, 27:292

C_6H_6S, Benzenethiol, osmium complex, 26:304

C_6H_7N, Benzenamine, tungsten complex, 27:301

C_6H_7P, Phosphine, phenyl-, cobalt-iron complex, 26:353

C_6H_8, 1,3-Cyclopentadiene, 1-methyl-, chromium, molybdenum, and tungsten hexacarbonyl complexes, 28:148
preparation and purification of, 27:52
titanium complex, 27:52
vanadium complex, 27:54

$C_6H_8O_6$, Ascorbic acid, platinum complex, 27:283

C_6H_{10}, 1-Butyne, 3,3-dimethyl-, mercury-molybdenum-ruthenium complex, 26:329-335

$C_6H_{14}N_2$, 1,2-Cyclohexanediamine, cis-, trans-(R,R)-, and trans-(S,S)-, platinum complex, 27:283

$C_6H_{15}O_3P$, Triethyl phosphite, iron complex, 26:61, 28:171
nickel complex, 28:101, 104-106

$C_6H_{15}P$, Phosphine, triethyl-, gold-osmium complexes, 27:209, 211
gold-platinum complex, 27:218
nickel complex, 28:101
platinum complexes, 26:126, 135-140

$C_6H_{16}N_4$, 1,2-Ethanediamine, N,N,N',N'-tetramethyl-, lithium complex, 26:148

$C_6H_{16}P_2$, Phosphine, 1,2-ethanediylbis(dimethyl-, nickel complex, 28:101

$C_6H_{16}Si$, Silane, triethyl-, ruthenium complex, 26:269

$C_6H_{18}LiPSi_2 \cdot 2OC_4H_8$, Phosphide, bis(trimethylsilyl)-, lithium, -2-tetrahydrofuran, 27:243, 248

$C_7H_5FO_2$, Benzoic acid, 3-fluoro-, rhodium complex, 27:292

C_7H_5N, Benzonitrile, palladium complex, 28:61, 62
platinum complex, 26:345
ruthenium complexes, 26:70-72

$C_7H_5NO_3S$, 1,2-Benzisothiazol-3(2H)-one, 1,1-dioxide, chromium and vanadium complex, 27:307, 309

$C_7H_6O_2$, Benzoic acid, rhodium complex, 27:292

C_7H_8, Benzene, methyl-, cobalt complex, 26:309
  lutetium complex, 27:162
  manganese complex, 26:172
C_7H_8, Bicyclo[2.2.1]hepta-2,5-diene, ruthenium complex, 26:250, 251, 256
—, Cycloheptatriene, molybdenum complex, 28:45
C_7H_8O, Anisole, chromium complex, 28:137
$C_7H_8O_2S$, Benzenesulfonic acid, 4-methyl-, rhodium complex, 27:292
C_7H_9, Cycloheptatriene, tungsten complex, 27:4
C_7H_9N, Pyridine, 3,5-dimethyl-, palladium complex, 26:210
C_7H_{10}, Bicyclo[2.2.1]hept-2-ene, platinum complex, 28:127
$C_7H_{11}N$, Cyclohexyl isocyanide, nickel complex, 28:101
$C_7H_{16}PS_2$, Phosphonium, (dithiocarboxy)triethyl-, rhodium complex, 27:288
$C_7H_{19}NSi$, Ethanamine, 1,1-dimethyl-N-(trimethylsilyl)-, 27:327
C_8H_5NO, Benzoyl isocyanide, chromium complex, 26:32, 34, 35
C_8H_6, Benzene, ethynyl-, ytterbium complex, 27:143
$C_8H_6O_4$, Phthalic acid, rhodium complex, 27:291
C_8H_8, 1,3,5,7-Cyclooctatetraene, lithium complex, 28:127
  lutetium complex, 27:152
C_8H_8O, Ethanone, 1-phenyl-, manganese complex, 26:156-158
$C_8H_8O_2$, Methyl benzoate, chromium complex, 26:32
C_8H_{10}, o-Xylylene, magnesium complex, 26:147
$C_8H_{11}N$, Benzenamine, N,N-dimethyl-, chromium complex, 28:139
$C_8H_{11}O_2P$, Phosphonous acid, phenyl-, dimethyl ester, nickel complex, 28:101
$C_8H_{11}P$, Phosphine dimethylphenyl-, iron complex, 26:61, 28:171
  molybdenum complex, 27:11
  osmium complex, 27:27
  osmium-rhodium complex, 27:29
  osmium-zirconium complex, 27:27
  ruthenium complex, 26:273
  tungsten complex, 28:330

C_8H_{12}, 1,5-Cyclooctadiene, iridium complex, 26:122
  nickel complex, 28:94
  osmium-rhodium complex, 27:29
  palladium and platinum complexes, 28:346-348
  platinum complex, 28:126
  rhodium complex, 28:88
  ruthenium complexes, 26:69-72, 253-256, 28:54
$C_8H_{12}O_4PS$, 2-Butenedioic acid, 2-(dimethylphosphinothioyl)-, dimethyl ester, manganese complex, 26:163
C_8H_{14}, Cyclooctene, iridium and rhodium complexes, 28:90, 91
  platinum complex, 26:139
C_9H_9N, Benzene, 2-isocyano-1,3-dimethyl-, iron complexes, 26:53:57
$C_9H_{13}N$, Benzenamine, N,N,2-trimethyl-, lithium complex, 26:153
—, Benzenemethanamine, N,N-dimethyl-, lithium complex, 26:152
  lutetium complex, 27:153
  palladium complex, 26:212
$C_9H_{13}P$, Phosphine, ethylmethylphenyl-, lithium complex, 27:178
—, Phosphorane, dimethylmethylenediphenyl-, uranium complex, 27:177
$C_9H_{21}O_3P$, Isopropyl phosphite, nickel complex, 28:101
$C_9H_{21}P$, Phosphine, triisopropyl-, platinum complex, 28:120
  rhodium complex, 27:292
  tungsten complex, 27:7
$C_9H_{27}PSi_3$, Phosphine, tris(trimethylsilyl)-, 27:243
$C_{10}H_8N_2$, 2,2'-Bipyridine, nickel complex, 28:103
  palladium complex, 27:319
  rhenium complexes, 26:82, 83
  ruthenium complex, 28:338
  tungsten complex, 27:303
$C_{10}H_8O_2$, 2,3-Naphthalenediol, in prepn. of cis-tetraamminedihaloruthenium(III) complexes, 26:66, 67
—, 2-Propynoic acid, 3-phenyl-, methyl ester, cobalt complex, 26:192
$C_{10}H_8S_8$, 2,2'-Bi-1,3-dithiolo[4,5-b] [1,4]-dithiinylidene, 26:386
$C_{10}H_9N$, Quinoline, 8-methyl-, palladium complex, 26:213

$C_{10}H_{10}O$, 3-Butene-2-one, 4-phenyl-, iron complex, 28:52
$C_{10}H_{15}N$, Benzenemethanamine, N,N,4-trimethyl-, lithium complex, 26:152
$C_{10}H_{15}P$, Phosphine, diethylphenyl-, nickel complex, 28:101
platinum complex, 28:135
$C_{10}H_{16}As_2$, Arsine, 1,2-phenylenebis(dimethyl-, gold complex, 26:89
nickel complex, 28:103
$C_{10}H_{16}$, 1,3-Cyclopentadiene, 1,2,3,4,5-pentamethyl-, 28:317
chromium complex, 27:69
cobalt complexes, 28:273, 275
iridium complex, 27:19
samarium complex, 27:155
titanium complex, 27:62
ytterbium complex, 27:148
$C_{10}H_{18}BrNO_4S$, Bicyclo[2.2.1]heptane-7-methanesulfonate, 3-bromo-1,7-dimethyl-2-oxo-, [(1R)-(ENDO, ANTI)]-, ammonium, 26:24
$C_{10}H_{27}Cl_2PSi_3$, Phosphonous dichloride, [tris(trimethylsilyl)methyl]-, 27:239
$C_{10}H_{28}Si_3$, Methane, tris(trimethylsilyl)-, 27:238
$C_{11}H_{22}Si_2$, 1,3-Cyclopentadiene, 1,3-bis(trimethylsilyl)-, lanthanide-lithium complexes, 27:170
$C_{11}H_{27}OPSi_2$, Phosphine, [2,2-dimethyl-1-(trimethylsiloxy)propylidene](trimethylsilyl)-, 27:250
$C_{12}H_8N_2$, 1,10-Phenanthroline, nickel complex, 28:103
$C_{12}H_{10}N_2$, Azobenzene, cobalt and palladium complexes, 26:175, 176
manganese complex, 26:173
$C_{12}H_{11}N$, Pyridine, 2-(phenylmethyl)-, palladium complex, 26:208–210
$C_{12}H_{11}P$, Phosphine, diphenyl-, manganese complex, 26:158, 226–230
ruthenium complex, 26:264
$C_{12}H_{12}O_8S$, Thiophenetetracarboxylic acid, tetramethyl ester, 26:166
$C_{12}H_{12}$, Toluene, α-2,4-cyclopentadien-1-yl-, chromium, molybdenum and tungsten hexacarbonyl complexes, 28:148
$C_{12}H_{13}N$, Naphthalenamine, N,N-dimethyl-, lithium complex, 26:154

$C_{12}H_{18}$, Benzene, hexamethyl-, ruthenium complex, 26:181, 182
$C_{12}H_{27}P$, Phosphine, tributyl-, iron complex, 26:61
nickel complex, 28:101
—, Phosphine, tri-tert-butyl-, palladium complex, 28:115
$C_{13}H_9N$, Benzo[b]quinoline, ruthenium complex, 26:177
$C_{13}H_{13}P$, Phosphine, methyldiphenyl-, iron complex, 26:61
molybdenum complex, 27:9
nickel complex, 28:101
tungsten complex, 27:10
$C_{14}H_{10}$, Acetylene, diphenyl-, molybdenum complex, 26:102–105
molybdenum complex, 28:11
$C_{14}H_{12}$, Benzene, 1,2-ethendiylbis-, 26:192
platinum complex, 26:140
$C_{14}H_{18}O_8P$, 1H-Phospholium, 2,3,4,5-tetrakis(methoxycarbonyl)-2,2-dimethyl-, manganese complex, 26:167
$C_{14}H_{20}O_8PS$, 2H-1,2-Thiaphosphorin-2-ium, 3,4,5,6-tetrakis(methoxycarbonyl)-2,2-dimethyl, manganese complex, 26:165
$C_{14}H_{22}O$, Phenol, 2,6-di-tert-butyl-, actinide and lanthanide complexes, 27:166
$C_{14}H_{23}P$, Phosphine, di-tert-butylphenyl-, palladium and platinum complexes, 28:114, 116
$C_{14}H_{24}N_4$, 1,5,9,13-Tetraazacyclohexadeca-1,3,9,11-tetraene, 4,10-dimethyl-, nickel complex, 27:272
$C_{14}H_{26}Si_2$, Benzene, 1,2-bis[(trimethylsilyl)methyl]-, 26:148
lithium complex, 26:148
$C_{15}H_{16}O_4$, 2,4-Hexadienedioic acid, 3-methyl-4-phenyl-, dimethyl ester, cobalt complex, 26:197
—, 2-Pentenedioic acid, 3-methyl-2-(phenylmethyl)-, dimethyl ester, cobalt complex, 26:197
$C_{15}H_{24}O$, Phenol, 2,6-di-tert-butyl-4-methyl-, actinide and lanthanide complexes, 27:166
$C_{17}H_{14}O$, 1,4-pentadien-3-one, 1,5-diphenyl-, palladium complex, 28:110
$C_{18}H_{15}As$, Arsine, triphenyl-, iron complex, 26:61
nickel complex, 28:103
$C_{18}H_{15}O_3P$, Triphenyl phosphite, iron com-

$C_{18}H_{15}O_3P$ (Continued)
   plex, 26:16, 28:171
   nickel complex, 28:101
   ruthenium complex, 26:178
$C_{18}H_{15}P$, Phosphine, triphenyl-, cobalt complex, 26:190-197
   cobalt-gold-iron complex, 27:188
   cobalt-gold-ruthenium complex, 26:327
   gold complex, 26:90, 325, 326
   gold-manganese complex, 26:229
   gold-osmium complexes, 27:209, 211
   gold and platinum complex, 27:218
   iridium complexes, 26:117-120, 122-125, 201, 202, 28:23-26, 58, 59
   iron complexes, 26:61, 237, 241, 28:170
   molybdenum complex, 28:13
   molybdenum, palladium, and platinum complexes, 26:347
   molybdenum and tungsten complexes, 26:98-105
   nickel complex, 28:102
   osmium complex, 26:184-188
   palladium complex, 28:107
   platinum complex, 27:36, 28:124, 125
   rhenium complex, 27:15
   rhodium complex, 27:222, 28:77-83
   ruthenium complex, 26:181, 182
   tungsten complex, 28:40
$C_{18}H_{15}Sb$, Stibine, triphenyl-, iron complex, 26:61
   nickel complex, 28:103
$C_{18}H_{18}$, Benzene, 1,3-butadiene-1,4-diylbis-, cobalt complex, 26:195
$C_{18}H_{29}Br$, Benzene, 1-bromo-2,4,6-tri-*tert*-butyl-, 27:236
$C_{18}H_{31}P$, Phosphine, (2,4,6-tri-*tert*-butylphenyl)-, 27:237
$C_{18}H_{33}P$, Phosphine, tricyclohexyl-, iron complex, 26:61
   molybdenum complex, 27:3
   nickel complexes, 26:205, 206
   palladium and platinum complexes, 28:114, 116
   platinum complex, 28:130
   rhodium complex, 27:291
   tungsten complex, 27:6
$C_{19}H_{17}OP$, Benzenemethanol, 2-(diphenylphosphino)-, manganese complex, 26:169

$C_{20}H_{32}N_4O_2$, 1,5,9,13-Tetraazacyclohexadeca-1,4,9,12-tetraene, 2,12-dimethyl-3,11-bis(1-methoxyethylidene)-, nickel complex, 27:264
$C_{20}H_{34}N_6$, 1,5,9,13-Tetraazacyclohexadeca-1,4,9,12-tetraene, 2,12-dimethyl-3,11-bis[1-(methylamino)ethylidene]-, nickel complex, 27:266
$C_{20}H_{54}P_2Si_6$, Diphosphene, bis[tris(trimethylsilyl)methyl]-, 27:241, 242
$C_{21}H_{21}O_3P$, Phosphorous acid, tris(4-methylphenyl) ester, ruthenium complex, 26:277, 278, 28:227
$C_{21}H_{39}PSi$, Phosphine, (2,4,6-tri-*tert*-butylphenyl) (trimethylsilyl)-, 27:238
$C_{25}H_{20}NOP$, Benzamide, 2-(diphenylphosphino)-*N*-phenyl-, 27:324
—, Benzamide, *N*-[2-(diphenylphosphino)phenyl]-, 27:323
$C_{25}H_{22}P_2$, Phosphine, methylenebis(diphenyl-, palladium complex, 28:340
   ruthenium complex, 26:276
$C_{26}H_{20}P_2$, Phosphine, ethynediylbis(diphenyl-, ruthenium complex, 26:277
$C_{26}H_{24}P_2$, Phosphine, 1,2-ethanediylbis(diphenyl-, iridium complex, 27:25
   molybdenum complex, 26:84
   nickel complex, 28:103
   platinum complexes, 26:370, 28:135
   tungsten complex, 28:43
$C_{26}H_{44}N_6$, 3,10,14,18,21,25-Hexaazabicyclo[10.7.7]hexacosa-1,11,13,18,20,25-hexaene, 2,3,10,11,13,19-hexamethyl-, cobalt complex, 27:270
—, 3,10,11,14,18,21,25-Hexaazabicyclo[10.7.7]hexacosa-1,11,13,18,20,25-hexaene, 2,3,10,13,19-hexamethyl-, nickel complex, 27:268
$C_{26}H_{47}F_{18}N_6P_3$, 3,10,14,18,21,25-Hexaazabicyclo[10.7.7]hexacosa-1,11,13,18,20,25-hexaene, 2,3,10,11,13,19-hexamethyl-, tris[hexafluorophosphate-(1−)], 27:269
$C_{27}H_{26}P_2$, Phosphine, 1,3-propanediylbis(diphenyl-, iridium complex, 27:22

$C_{28}H_{28}P_2$, Phosphine, 1,4-butanediylbis(diphenyl-, iridium complex, 27:26

$C_{28}H_{32}N_4O_2$, 1,5,9,13-Tetraazacyclohexadeca-1,3,9,11-tetraene, 3,11-dibenzoyl-2,12-dimethyl-, nickel complex, 27:273

$C_{30}H_{36}N_4O_2$, 1,5,9,13-Tetraazacyclohexadeca-1,4,9,12-tetraene, 3,11-bis(α-methoxybenzylidene)-2,12-dimethyl-, nickel complex, 27:275

$C_{34}H_{33}P_3$, Phosphine, bis[2-(diphenylphosphino)ethyl]phenyl-, palladium complex, 27:320

$C_{36}H_{30}N_4P_2$, Phosphorus(1+), μ-nitrido-bis(triphenyl)-, azide, 26:286

$C_{36}H_{58}P_2$, Diphosphene, bis(2,4,6-tri-*tert*-butylphenyl)-, 27:241

$C_{38}H_{33}NO_2P_2$, Phosphorus(1+), μ-nitrido-bis(triphenyl-, acetate, 27:296

$C_{41}H_{39}P_3$, Phosphine, [2-[(diphenylphosphino)methyl]-2-methyl-1,3-propanediyl]bis(diphenyl-, rhodium complex, 27:287

$C_{42}H_{46}N_6$, 1,5,9,13-Tetraazacyclohexadeca-1,4,9,12-tetraene, 3,11-bis[α-(benzylamino)benzylidene]-2,12-dimethyl-, nickel complex, 27:276

$C_{50}H_{52}N_6$, 3,11,15,19,22,26-Hexaazatricyclo[11.7.7.1^{5,9}]octacosa-1,5,7,9(28),12,14,19,21,26-nonaene, 3,11-dibenzyl-14,20-dimethyl-2,12-diphenyl-, iron complex, 27:280

$C_{50}H_{52}N_6$, 3,11,15,19,22,26-Hexaazatricyclo[11.7.7.1^{5,9}]octacosa-1,5,7,9(28),12,14,19,21,26-nonaene, 3,11-dibenzyl-14,20-dimethyl-2,12-diphenyl-, nickel complex, 27:277

$C_{50}H_{55}F_{18}N_6P_3$, 3,11,15,19,22,26-Hexaazatricyclo[11.7.7.1^{5,9}]octacosa-1,5,7,9(28),12,14,19,21,26-nonaene, 3,11-dibenzyl-14,20-dimethyl-2,12-diphenyl-, tris[hexafluorophosphate(1−)], 27:278

$CdCl_2$, Cadmium dichloride, 28:322

$CeCl_2LiO_2Si_4C_{30}H_{58}$, Cerium, bis[η^5-1,3-bis(trimethylsilyl)cyclopentadienyl]di-μ-chloro-bis(tetrahydrofuran)lithium-, 27:170

$Ce_2Cl_2Si_4C_{44}H_{84}$, Cerium, tetrakis[η^5-1,3-bis(trimethylsilyl)cyclopentadienyl]di-μ-chloro-di, 27:171

$ClAuF_5PC_{31}H_{22}$, Aurate (I), chloro(pentafluorophenyl)-, (benzyl)triphenylphosphonium, 26:88

$ClAuPC_{18}H_{15}$, Gold, chloro(triphenylphosphine)-, 26:325

$ClAuSC_4H_8$, Gold (I), chloro(tetrahydrothiophene)-, 26:86

$ClAu_2F_3O_3P_4PtSC_{49}H_{60}$, Platinum(1+), chloro-1-κ*Cl*bis(triethylphosphine-1κ*P*)bis(triphenylphosphine)-2κ*P*,3κ*P-triangulo*-digold-, trifluoromethanesulfonate, 27:218

$ClBF_4IrN_2P_2C_{31}$, Iridium(III), chloro(dinitrogen)hydrido[tetrafluoroborate(1−)]bis(triphenylphosphine)-, 26:119

$ClBF_4IrN_2P_2C_{36}H_{31}$, Iridium(III), chloro(dinitrogen)hydrido[tetrafluoroborato(1−)]bis(triphenylphosphine)-, 28:25

$ClBF_4IrOP_2C_{37}H_{31}$, Iridium(III), carbonylchlorohydrido[tetrafluoroborato(1−)]bis(triphenylphosphine)-, 26:117, 28:23

$ClBF_4IrOP_2C_{38}H_{33}$, Iridium(III), carbonylchloromethyl[tetrafluoroborato(1−)]bis(triphenylphosphine)-, 26:118

$ClCH_3$, Methane, chloro-, osmium complex, 27:205

ClC_6H_5, Benzene, chloro-, chromium complex, 28:139

$ClCoP_3C_{54}H_{45}$, Cobalt, chlorotris(triphenylphosphine)-, 26:190

$ClCrO_3C_9H_5$, Chromium, tricarbonyl(η^6-chlorobenzene)-, 28:139

$ClF_3O_3PtSC_{13}H_{30}$, Platinum(II), chlorobis(triethylphosphine)(trifluoromethanesulfonato)-, *cis*-, 26:126

$ClF_6N_3PRuC_{14}H_{21}$, Ruthenium(II), tris(acetonitrile)chloro(η^4-1,5-cyclooctadiene)-, hexafluorophosphate(1−), 26:71

$ClH_3O_9Os_3C_{10}$, Osmium, nonacarbonyl(μ$_3$-chloromethylidyne)tri-μ-hydrido-*triangulo*-tri-, 27:205

$ClHgMn_2O_8PC_{20}H_{10}$, Manganese, μ-(chloromercurio)-μ-(diphenylphosphino)-

ClHgMn$_2$O$_8$PC$_{20}$H$_{10}$ (Continued)
  bis(tetracarbonyl-, (Mn–Mn), 26:230
ClIrOP$_2$C$_{37}$H$_{30}$, Iridium, carbonylchlorobis-
  (triphenylphosphine)-, trans-, 28:92
ClIrP$_3$C$_{54}$H$_{45}$, Iridium(I),
  chlorotris(triphenylphosphine)-, 26:201
—, Iridium(III), chloro-[2-
  (diphenylphosphino)phenyl-
  C^1,P]hydridobis-
  (triphenylphosphine)-, (OC-6-53)-,
  26:202
ClLi, Lithium chloride, 28:322
ClLuOC$_{12}$H$_{16}$, Lutetium, chloro(η^8-1,3,5,7-
  cyclooctatetraene) (tetrahydrofuran)-,
  27:152
ClMgC$_5$H$_{11}$, Magnesium, chloro(2,2-
  dimethylpropyl)-, 26:46
ClMnO$_5$C$_5$, Manganese, pentacarbonyl-
  chloro-, 28:155
ClN$_2$PdC$_{19}$H$_{19}$, Palladium(II), chloro[2-(2-
  pyridinylmethyl)phenyl-C^1,N] (3,5-
  dimethylpyridine)-, 26:210
ClOP$_2$RhC$_{36}$H$_{30}$, Rhodium,
  carbonylchlorobis(triphenylphosphine)-,
  trans-, 28:79
ClOP$_3$RhS$_2$C$_{42}$H$_{39}$, Rhodium, chloro[[2-
  [(diphenylphosphino)methyl]-2-methyl-
  1,3-propanediyl]bis(diphenylphosphine)](dithiocarbonato)-, 27:289
ClO$_2$C$_2$H$_3$, Acetic acid, chloro-, ruthenium
  complex, 26:256
ClO$_5$ReC$_5$, Rhenium, pentacarbonylchloro-,
  28:161
ClPRuC$_{30}$H$_{34}$, Ruthenium(II), chloro(η^6-
  hexamethylbenzene)hydrido(triphenylphosphine)-, 26:181
ClP$_2$PtC$_{26}$H$_{41}$, Platinum(II), chloro-(cis-1,2-
  diphenylethenylbis(triethylphosphine)-,
  trans-, 26:140
ClP$_2$RuC$_{41}$H$_{35}$, Ruthenium(II), chloro(η^5-
  cyclopentadienyl)bis(triphenyl-
  phosphine)-, 28:270
ClP$_3$RhC$_{54}$H$_{45}$, Rhodium(I),
  chlorotris(triphenylphosphine)-, 28:77
ClP$_4$RhS$_2$C$_{48}$H$_{54}$, Rhodium, chloro[[2-
  [(diphenylphosphino)methyl]-2-methyl-
  1,3-propanediyl]bis(diphenylphosphine)]
  [(dithiocarboxy)triethylphospho-
  niumato]-, 27:288
ClP$_6$RhC$_{36}$H$_{30}$, Rhodium(I), chloro(η^2-
  tetrahedro-tetraphosphorus)bis(triphenylphosphine)-, 27:222
ClRhC$_{16}$H$_{28}$, Rhodium(I),
  chlorobis(cyclooctene)-, 28:90
ClThC$_{15}$H$_{15}$, Thorium(IV), chlorotris(η^5-
  cyclopentadienyl)-, 28:302
ClTiC$_{10}$H$_{10}$, Titanium(III), chlorobis(η^5-
  cyclopentadienyl)-, 28:261
ClUC$_{15}$H$_{15}$, Uranium(IV), chlorotris(η^5-
  cyclopentadienyl)-, 28:301
ClVC$_{10}$H$_{10}$, Vanadium(III), chlorobis(η^5-
  cyclopentadienyl)-, 28:262
ClZrC$_{10}$H$_{11}$, Zirconium, chlorobis(η^5-
  cyclopentadienyl)hydrido-, 28:259
Cl$_2$Cd, Cadmium dichloride, 28:322
Cl$_2$CeLiO$_2$Si$_4$C$_{30}$H$_{58}$, Cerium, bis[η^5-1,3-
  bis(trimethylsilyl)cyclopentadienyl]di-μ-
  chloro-bis(tetrahydrofuran)lithium-,
  27:170
Cl$_2$Ce$_2$Si$_4$C$_{44}$H$_{84}$, Cerium, tetrakis[η^5-1,3-
  bis(trimethylsilyl)cyclopentadienyl]di-μ-
  chloro-di-, 27:171
Cl$_2$Co, Cobalt dichloride, 28:322
Cl$_2$Cu, Copper dichloride, 28:322
Cl$_2$Dy$_2$Si$_4$C$_{44}$H$_{84}$, Dysprosium,
  tetrakis[η^5-1,3-
  bis(trimethylsilyl)cyclopentadienyl]di-
  μ-chloro-di-, 27:171
Cl$_2$Er$_2$Si$_4$C$_{44}$H$_{84}$, Erbium,
  tetrakis[η^5-1,3-bis(trimethylsilyl)-
  cyclopentadienyl]di-μ-chloro-di-, 27:171
Cl$_2$Eu$_2$Si$_4$C$_{44}$H$_{84}$, Europium,
  tetrakis[η^5-1,3-bis(trimethylsilyl)-
  cyclopentadienyl]di-μ-chloro-di-, 27:171
Cl$_2$Gd$_2$Si$_4$C$_{44}$H$_{84}$, Gadolinium, tetrakis[η^5-
  1,3-bis(trimethylsilyl)-
  cyclopentadienyl]di-μ-chloro-di-, 27:171
Cl$_2$Ho$_2$Si$_4$C$_{44}$H$_{84}$, Holmium, tetrakis[η^5-
  1,3-bis(trimethylsilyl)-
  cyclopentadienyl]di-μ-chloro-di-, 27:171
Cl$_2$Ir$_2$C$_{32}$H$_{56}$, Iridium, di-μ-chloro-
  bis[bis(cyclooctene)-, 28:91
Cl$_2$LaLiO$_2$Si$_4$C$_{30}$H$_{58}$, Lanthanum, bis[η^5-
  1,3-bis(trimethylsilyl)-
  cyclopentadienyl]di-μ-chloro-
  bis(tetrahydrofuran)lithium-, 27: 170
Cl$_2$La$_2$Si$_4$C$_{44}$H$_{84}$, Lanthanum, tetrakis[η^5-
  1,3-bis(trimethylsilyl)cyclopentadienyl]-
  di-μ-chloro-di-, 27:171
Cl$_2$LiNdO$_2$Si$_4$C$_{30}$H$_{58}$, Neodymium,

bis[η^5-1,3-bis(trimethylsilyl)-cyclopentadienyl]di-μ-chloro-bis(tetrahydrofuran)lithium-, 27:170

$Cl_2LiO_2PrSi_4C_{30}H_{58}$, Praseodymium, bis[$\eta^5$-1,3-bis(trimethylsilyl)-cyclopentadienyl]di-μ-chloro-bis-(tetrahydrofuran)lithium-, 27:170

$Cl_2LiO_2ScSi_4C_{30}H_{58}$, Scandium, bis[$\eta^5$-1,3-bis(trimethylsilyl)-cyclopentadienyl]di-μ-chloro-bis(tetrahydrofuran)lithium-, 27:170

$Cl_2LiO_2Si_4YC_{30}H_{58}$, Yttrium, bis[$\eta^5$-1,3-bis(trimethylsilyl)cyclopentadienyl]di-μ-chloro-bis(tetrahydrofuran)lithium-, 27:170

$Cl_2LiO_2Si_4YbC_{30}H_{58}$, Ytterbium, bis[$\eta^5$-1,3-bis(trimethylsilyl)cyclopentadienyl]di-μ-chloro-bis(tetrahydrofuran)lithium-, 27:170

$Cl_2Lu_2Si_4C_{44}H_{84}$, Lutetium, tetrakis[η^5-1,3-bis(trimethylsilyl)-cyclopentadienyl]di-μ-chloro-di-, 27:171

$Cl_2N_2P_2C_8H_{18}$, 1,3,2,4,-Diazadiphosphetidine, 1,3-di-*tert*-butyl-2,4-dichloro-, *cis*-, 27:258

$Cl_2N_2P_2WC_{16}H_{32}$, Tungsten(VI), dichloro[(1,1-dimethylethyl)imido](phenylimido)-bis(trimethylphosphine)-, 27:304

$Cl_2N_2PdC_{14}H_{10}$, Palladium, bis(benzonitrile)dichloro-, 28:61

$Cl_2N_2Pd_2C_{18}H_{24}$, Palladium(II), di-μ-chloro-bis[2-[(dimethylamino)methyl]phenyl-C^1,N]di-, 26:212

$Cl_2N_2Pd_2C_{20}H_{16}$, Palladium(II), di-μ-chlorobis(8-quinolymethyl-C,N)di-, 26:213

$Cl_2N_2Pd_2C_{24}H_{20}$, Palladium(II), di-μ-chlorobis[2-(2-pyridimylmethyl)phenyl-C^1,N]di-, 26:209

$Cl_2N_2PtC_{14}H_{10}$, Platinum, bis(benzonitrile)dichloro-, 26:345

$Cl_2N_2RuC_{12}H_{18}$, Ruthenium(II), bis(acetonitrile) dichloro(η^4-1,5-cyclooctadiene)-, 26:69

$Cl_2N_2RuC_{22}H_{22}$, Ruthenium(II), bis(benzonitrile)dichloro(η^4-1,5-cyclooctadiene)-, 26:70

$Cl_2N_4Pd_2C_{20}H_{36}$, Palladium(I), tetrakis(*tert*-butyl isocyanide)di-μ-chlorodi-, 28:110

$Cl_2N_4Pd_2C_{24}H_{18}$, Palladium, di-μ-chloro-bis[2-(phenylazo)phenyl-C^1,N^2]di-, 26:175

$Cl_2N_4WC_{20}H_{22}$, Tungsten(VI), (2,2'-bipyridine)dichloro[(1,1-dimethylethyl)imido](phenylimido)-, 27:303

$Cl_2N_6O_6RuC_{30}H_{36} \cdot 6H_2O$, Ruthenium(II), tris(2,2'-bipyridine)-, dichloride, hexahydrate, 28:338

$Cl_2NbC_{10}H_{10}$, Niobium(IV), dichlorobis(η^5-cyclopentadienyl)-, 28:267

$Cl_2Nd_2Si_4C_{44}H_{84}$, Neodymium, tetrakis-[η^5-1,3-bis(trimethylsilyl)-cyclopentadienyl]di-μ-chloro-di-, 27:171

Cl_2Ni, Nickel dichloride, 28:322

$Cl_2O_4Rh_2C_4$, Rhodium, tetracarbonyldichlorodi-, 28:84

$Cl_2OsP_3C_{54}H_{45}$, Osmium(II), dichlorotris-(triphenylphosphine)-, 26:184

$Cl_2OsP_3SC_{55}H_{45}$, Osmium(II), dichloro(thiocarbonyl)tris(triphenylphosphine)-, 26:185

$Cl_2PC_{18}H_{29}$, Phosphonous dichloride, (2,4,6-tri-*tert*-butylphenyl)-, 27:236

$Cl_2PSi_3C_{10}H_{27}$, Phosphonous dichloride, [tris(trimethylsilyl)methyl]-, 27:239

$Cl_2P_2PtC_{26}H_{24}$, Platinum, dichloro[1,2-ethanediylbis(diphenylphosphine)]-, 26:370

$Cl_2P_4Pd_2C_{50}H_{44}$, Palladium(I), dichlorobis-[μ-methylenebis(diphenylphosphine)]di-, (*Pd–Pd*), 28:340

$Cl_2P_4WC_{12}H_{36}$, Tungsten(II), dichlorotetrakis(trimethylphosphine)-, 28:329

$Cl_2P_4WC_{32}H_{44}$, Tungsten(II), dichlorotetrakis(dimethylphenylphosphine)-, 28:330

$Cl_2P_4WC_{52}H_{52}$, Tungsten(II), dichlorotetrakis(methyldiphenylphosphine)-, 28:331

$Cl_2PdC_8H_{12}$, Palladium(II), dichloro(η^4-1,5-cyclooctadiene)-, 28:348

$Cl_2Pd_2C_6H_{10}$, Palladium(II), bis(η^3-allyl)di-μ-chloro-di-, 28:342

$Cl_2Pr_2Si_4C_{44}H_{84}$, Praseodymium, tetrakis-[η^5-1,3-bis(trimethylsilyl)-cyclopentadienyl]di-μ-chloro-di-, 27:171

$Cl_2PtC_8H_{12}$, Platinum(II), dichloro(η^4-1,5-cyclooctadiene)-, 28:346

$Cl_2Rh_2C_8H_{16}$, Rhodium(I), di-μ-chloro-tetrakis(ethene)di-, 28:86

$Cl_2Rh_2C_{16}H_{24}$, Rhodium(I), di-μ-chloro-bis-(η^4-1,5-cyclooctadiene)di-, 28:88

$Cl_2RuC_7H_8$, Ruthenium(II), (η^4-bicyclo-[2.2.1]hepta-2,5-diene)dichloro-, 26:250

$Cl_2RuC_8H_{12}$, Ruthenium(II), dichloro(η^4-cycloocta-1,5-diene)-, 26:253

—, Ruthenium(II), di-μ-chloro-(η^4-1,5-cyclooctadiene)-, polymer, 26:69

$Cl_2Sc_2Si_4C_{44}H_{84}$, Scandium, tetrakis[η^5-1,3-bis(trimethylsilyl)cyclopentadienyl]di-μ-chloro-di-, 27:171

$Cl_2Si_2ThC_{22}H_{42}$, Thorium(IV), bis(η^5-1,3-bis(trimethylsilyl)cyclopentadienyl]dichloro-, 27:173

$Cl_2Si_2UC_{22}H_{42}$, Uranium(IV), bis(η^5-1,3-bis(trimethylsilyl)-cyclopentadienyl]dichloro-, 27:174

$Cl_2Si_4Sm_2C_{44}H_{84}$, Samarium, tetrakis-[η^5-1,3-bis(trimethylsilyl)-cyclopentadienyl]di-μ-chloro-di-, 27:171

$Cl_2Si_4Tb_2C_{44}H_{84}$, Terbium, tetrakis[η^5-1,3-bis(trimethylsilyl)-cyclopentadienyl]di-μ-chloro-di-, 27:171

$Cl_2Si_4Tm_2C_{44}H_{84}$, Thulium, tetrakis[η^5-1,3-bis(trimethylsilyl)cyclopentadienyl]di-μ-chloro-di-, 27:171

$Cl_2Si_4Y_2C_{44}H_{84}$, Yttrium, tetrakis[η^5-1,3-bis(trimethylsilyl)-cyclopentadienyl]di-μ-chloro-di-, 27:171

$Cl_2Si_4Yb_2C_{44}H_{84}$, Ytterbium, tetrakis[η^5-1,3-bis(trimethylsilyl)cyclopentadienyl]di-μ-chloro-di-, 27:171

Cl_2Zn, Zinc dichloride, 28:322

Cl_3Cr, Chromium trichloride, 28:322

$Cl_3CrN_4C_4H_{16}$, Chromium(III), dichlorobis(1,2-ethanediamine)-, Λ-cis-, chloride, and monohydrate, resolution of, 26:24, 27, 28

$Cl_3Er \cdot 7/2C_4H_8O$, Erbium trichloride–tetrahydrofuran(2:7), 27:140, 28:290

Cl_3Fe, Iron trichloride, 28:322

$Cl_3H_{12}N_4Ru$, Ruthenium(III), tetraamminedichloro-, cis-, chloride, 26:66

$Cl_3KPtC_2H_4$, Platinate(II), trichloro-(ethene)-, potassium, 28:349

$Cl_3MoN_3C_6H_9$, Molybdenum(III), tris(acetonitrile)trichloro-, 28:37

$Cl_3MoO_3C_{12}H_{24}$, Molybdenum(III), trichlorotris(tetrahydrofuran)-, 28:36

$Cl_3Nd \cdot 3/2C_4H_8O$, Neodymium trichloride–tetrahydrofuran(2:3), 27:140, 28:290

$Cl_3O_2C_2H$, Acetic acid, trichloro-, ruthenium complex, 26:256

$Cl_3O_2SmC_8H_{16}$, Samarium, trichlorobis(tetrahydrofuran)-, 27:140

$Cl_3O_2WC_9H_{19}$, Tungsten(VI), trichloro(1,2-dimethoxyethane)(2,2-dimethylpropylidyne)-, 26:50

$Cl_3O_3WC_3H_9$, Tungsten, trichlorotrimethoxy-, 26:45

$Cl_3O_3YbC_{12}H_{24}$, Ytterbium, trichlorotri(tetrahydrofuran)-, 26:139, 28:289

$Cl_3OsP_3C_{24}H_{33}$, Osmium(III), trichlorotris(dimethylphenylphosphine)-, mer-, 27:27

$Cl_3Sm \cdot 2C_4H_8O$, Samarium trichloride–2tetrahydrofuran, 27:140, 28:290

$Cl_3Yb \cdot 3C_4H_8O$, Ytterbium trichloride–3tetrahydrofuran, 27:139, 28:289

$Cl_4AlP_3C_{36}H_{30}$, Aluminate(1−), tetrachloro-, 1,1,1,3,3,3-hexaphenyltriphosphenium, 27:254

$Cl_4MoN_2C_4H_6$, Molybdenum, bis(acetonitrile)tetrachloro-, 28:34

Cl_4MoO, Molybdenum tetrachloride oxide, 28:325

$Cl_4MoO_2C_8H_{16}$, Molybdenum, tetrachlorobis(tetrahydrofuran)-, 28:35

$Cl_4N_3RuC_{12}H_{26}$, Ruthenate(1−), tetrachlorobis(acetonitrile)tetraethylammonium, 26:356

$Cl_4N_4PtC_4H_{16}$, Platinum(IV), dichlorobis(1,2-ethanediamine)-, dichloride, cis, 27:314

$Cl_4N_4PtC_4H_{18}$, Platinum(II), dichlorobis(1,2-ethanediamine monohydrochloride)-, $trans$-, 27:315

$Cl_4N_6W_2C_{28}H_{25}$, Tungsten(VI), tetrachlorobis(1,1-dimethylethanamine)bis[(1,1-dimethylethyl)imido]bis(μ-phenylimido)di-, 27:301

Cl$_4$OW, Tungsten tetrachloride oxide, 28:324
Cl$_4$O$_6$Ru$_2$C$_6$, Ruthenium(II), di-μ-chloro-bis(tricarbonylchloro-, 28:334
Cl$_4$O$_9$Ru$_2$C$_{24}$H$_{34}$, Ruthenium(II)μ-aquabis(μ-chloroacetato)bis-[(chloroacetato)-η^4-cycloocta-1,5-diene)-, 26:256
Cl$_4$P$_2$WC$_{26}$H$_{24}$, Tungsten, tetrachloro[1,2-ethanediylbis(diphenylphosphine)]-, 28:41
Cl$_4$P$_2$WC$_{26}$H$_{26}$, Tungsten(IV), tetrachloro-bis(methyldiphenylphosphine)-, 28:328
Cl$_4$P$_2$WC$_{36}$H$_{30}$, Tungsten, tetrachlorobis(triphenylphosphine)-, 28:40
Cl$_4$P$_3$WC$_9$H$_{27}$, Tungsten(IV), tetrachloro-tris(trimethylphosphine)-, 28:327
Cl$_4$Th, Thorium tetrachloride, 28:322
Cl$_4$W, Tungsten tetrachloride, 26:221
Cl$_5$H$_8$MoN$_2$O, Molybdate(V), pentachloro-oxo-, diammonium, 26:36
Cl$_6$Au$_{55}$P$_{12}$C$_{216}$H$_{180}$, Gold, hexachlorodo-decakis(triphenylphosphine)penta-pentaconta-, 27:214
Cl$_6$Er$_2$O$_7$C$_{28}$H$_{56}$, Erbium, hexachlorohepta-kis(tetrahydrofuran)di-, 27:140
Cl$_6$Nd$_2$O$_3$C$_{12}$H$_{24}$, Neodymium, hexachloro-tris(tetrahydrofuran)di-, 27:140
Cl$_6$P$_6$SnC$_{52}$H$_{48}$, Stannate(1 —), hexachloro-, 3,3,4,5-tetrahydro-1,1,3,3-tetraphenyl-1H-1,2,3-triphospholium, 27:255
Cl$_8$N$_2$Re$_2$C$_{32}$H$_{72}$, Rhenate(III), octachlor-odi-, bis(tetrabutylammonium), 28:332
Cl$_9$Cr$_2$Cs$_3$, Chromate(3—), nonachlorodi-, tricesium, 26:379
Cl$_9$Cs$_3$Ti$_2$, Titanate(3—), nonachlorodi-, tri-cesium, 26:379
Cl$_9$Cs$_3$V$_2$, Vanadate(3—), nonachlorodi-, tri-cesium, 26:379
Cl$_9$Rb$_3$Ti$_2$, Titanate(3—), nonachlorodi-, trirubidium, 26:379
Cl$_9$Rb$_3$V$_2$, Vanadate(3—), nonachlorodi-, trirubidium, 26:379
Cl$_{12}$O$_9$Ru$_2$C$_{22}$H$_{18}$, Ruthenium(II), μ-aqua-bis(μ-trichloroacetato)bis(η^4-bicyclo-[2.2.1]heptadiene) (trichloroacetato)-, 26:256
CoAu$_3$O$_{12}$P$_3$Ru$_3$C$_{66}$H$_{45}$, Ruthenium, dodecacarbonyltris(triphenylphosphine)-cobalt trigoldtri-, 26:327

CoC$_{14}$H$_{23}$, Cobalt(I), bis(η^2-ethene)(η^5-pentamethylcyclopentadienyl)-, 28:278
CoClP$_3$C$_{54}$H$_{45}$, Cobalt, chlorotris(triphenylphosphine)-, 26:190
CoCl$_2$, Cobalt dichloride, 28:322
CoF$_{12}$N$_6$P$_2$C$_{26}$H$_{44}$, Cobalt(II), (2,3,10,11,13,19-hexamethyl-3,10,14,18,21,25-hexaazabicyclo[10.7.7]hexacosa-1,11,13,18,20,25-hexaene-$\kappa^4 N^{14,18,21,25}$)-, bis[hexafluorophosphate(1—)], 27:270
CoI$_2$OC$_{11}$H$_{15}$, Cobalt(III), carbonyldiiodo-(η^5-pentamethylcyclopentadienyl)-, 28:275
CoMoNiO$_5$C$_{17}$H$_{13}$, Nickel, cyclo-μ_3-ethylidyne-1:2:3-κ^3C-pentacarbonyl-1κ^2C-, 2κ^3C-bis[1,3(η^5)-cyclopentadienyl]cobaltmolybdenum-, (Co-Mo) (Co-Ni) (Mo-Ni), 27:192
CoMoO$_8$RuC$_{17}$H$_{11}$, Cobalt, cyclo-[μ_3-1(η^2):2(η^2):3(η^2)-2-butyne]-octacarbonyl-1κ^2C, 2κ^3C, 3κ^3C-[1(η^5)-cyclopentadienyl]molybdenumruth-enium-, (Co-Mo) (Co-Ru) (Mo-Ru), 27:194
CoN$_2$O$_3$C$_{15}$H$_9$, Cobalt, tricarbonyl[2-(phenylazo)phenyl-C^1,N^2]-, 26:176
CoO$_2$C$_{12}$H$_{15}$, Cobalt(I), dicarbonyl(η^5-pentamethylcyclopentadienyl)-, 28:273
CoO$_2$PC$_{33}$H$_{28}$, Cobalt, (η^2-2-propynoate)(triphenylphosphine)-, 26:192
CoO$_4$PC$_{38}$H$_{34}$, Cobalt, (η^5-cyclopentadienyl)[1,3-bis(methoxycarbonyl)-2-methyl-4-phenyl-1,3-butadiene-1,4-diyl](triphenylphosphine)-, 26:197
CoO$_6$P$_2$C$_{11}$H$_{23}$, Cobalt(I), (η^5-cyclopentadienyl)bis(trimethyl phosphite)-, 28:283
CoPC$_{37}$H$_{30}$, Cobalt, (η^5-cyclopentadienyl)[η^2-1,1'-(1,2-ethynediyl)bisbenzene](triphenylphos-phine)-, 26:192
CoPC$_{41}$H$_{38}$, Cobalt, (η^5-cyclopentadienyl)-(2,3-dimethyl-1,4-diphenyl-1,3-butadiene-1,4-diyl)(triphenylphosphine)-, 26:195
CoP$_2$C$_{11}$H$_{23}$, Cobalt(I), (η^5-cyclopenta-dienyl)bis(trimethylphosphine)-, 28:281

$CoP_2C_{44}H_{42}$, Cobalt, (η^5-cyclopentadienyl)bis(triphenylphosphine)-, 26:191

$Co_2As_4H_{100}N_{24}O_{142}W_{40} \cdot 19H_2O$, Ammoniodicobaltotetracontatungstotetraarsenate(23−), tricosaammonium, nonadecahydrate, 27:119

$Co_2FeO_9PC_{15}H_5$, Iron, nonacarbonyl(μ_3-phenylphosphimidene)dicobalt-, 26:353

$Co_2FeO_9SC_9$, Iron, nonacarbonyl-μ_3-thiodicobalt-, 26:245

$Co_2I_4C_{20}H_{30}$, Cobalt(III), di-μ-iodobis[iodo(η^5-pentamethylcyclopentadienyl)-, 28:276

$Co_2MoO_8C_{15}H_8$, Cobalt, octacarbonyl(η^5-cyclopentadienyl-μ_3-ethylidynemolybdenumdi, 27:193

$Co_2O_7P_2PtC_{33}H_{24}$, Cobalt, heptacarbonyl[1,2-ethanediylbis(diphenylphosphine)]-platinumdi-, 26:370

$Co_2O_9RuC_{13}H_6$, Cobalt, μ_3-2-butynenonacarbonylrutheniumdi-, 27:194

CoO_9RuSC_9, Ruthenium, nonacarbonyl-μ_3-thio-dicobalt-, 26:352

$Co_2O_{11}RuC_{11}$, Ruthenium, undecacarbonyldicobalt-, 26:354

$Co_3AuFeO_{12}PC_{30}H_{15}$, Cobalt(1−), dodecacarbonylirontri-, (triphenylphosphine)gold(1+), 27:188

$Co_3C_{29}H_{25}$, Cobalt, tris(η^5-cyclopentadienyl)bis(μ-phenylmethylidyne)tri-, 26:309

$Co_3CuNO_{12}C_{14}H_3$, Ruthenium, (acetonitrile)dodecacarbonyltricobaltcopper-, 26:359

$Co_3FeNO_{12}C_{20}H_{20}$, Ferrate(1−), dodecacarbonyltricobalt-, tetraethylammonium, 27:188

$Co_3NO_{12}RuC_{12}H_{20}$, Ruthenate(1−), dodecacarbonyltricobalt-, tetraethylammonium, 26:358

$CrClO_3C_9H_5$, Chromium, tricarbonyl(η^6-chlorobenzene)-, 28:139

$CrFNO_3C_5H_6$, Chromate(VI), fluorotrioxo-, pyridinium, 27:310

$CrCl_3N_4C_4H_{16}$, Chromium(III), dichlorobis(1,2-ethanediamine)-, Λ-cis-, chloride, and monohydrate resolution of, 26:24, 27

$CrCl_3$, Chromium trichloride, 28:322

$CrFO_3C_9H_5$, Chromium, tricarbonyl(η^6-fluorobenzene)-, 28:139

$CrFeNO_9P_2C_{45}H_{31}$, Chromate(1−)hydridononacarbonyliron-, μ-nitridobis(triphenylphosphorus)(1+), 26:338

$CrFeN_2O_9P_4C_{81}H_{60}$, Chromate(2−), nonacarbonyliron-, bis[μ-nitridobis(triphenylphosphorus)(1+)], 26:339

$CrNO_3C_7H_5$, Chromium, dicarbonyl(η^5-cyclopentadienyl)nitrosyl-, 28:196

$CrNO_3C_{11}H_{11}$, Chromium, tricarbonyl(η^6-N,N-dimethylbenzenamine)-, 28:139

$CrNO_5C_{10}H_9$, Chromium, (tert-butyl isocyanide)pentacarbonyl-, 28:143

$CrNO_5C_{18}H_{13}$, Chromium, (benzoyl isocyanide)dicarbonyl(η^6-methyl benzoate)-, 26:32

$CrNO_6C_{13}H_5$, Chromium, (benzoyl isocyanide)pentacarbonyl-, 26:34, 35

$CrNO_7P_2C_{43}H_{33}$, Chromate(1−), (acetato)pentacarbonyl-, μ-nitridobis(triphenylphosphorus)(1+), 27:297

$CrN_2O_4C_{14}H_{18}$, Chromium, bis(tert-butyl isocyanide)tetracarbonyl-, cis-, 28:143

$CrN_2O_{10}S_2C_{14}H_{16} \cdot 2H_2O$, Chromium(II), tetraaquabis(1,2-benzisothiazol-3(2H)-one 1,1-dioxidato)-, dihydrate, 27:309

$CrN_3O_3C_{12}H_{15}$, Chromium, tricarbonyltris(propionitrile)-, 28:32

$CrN_3O_3C_{18}H_{27}$, Chromium, tris(tert-butyl isocyanide)tricarbonyl-, fac-, 28:143

$CrNaO_3C_8H_5 \cdot 2C_4H_{10}O_2$, Chronate(1−), tricarbonyl(η^6-cyclopentadienyl)-, sodium, compd. with 1,2-dimethoxyethane(1:2), 26:343

$CrO_3C_9H_6$, Chromium, (η^6-benzene)tricarbonyl-, 28:139

$CrO_4C_{10}H_8$, Chromium, (η^6-anisole)tricarbonyl-, 28:137

$CrO_5C_{11}H_8$, Chromium, tricarbonyl(η^6-methyl benzoate)-, 28:139

$Cr_2Br_9Cs_3$, Chromate(3−), nonabromoditricesium, 26:379

$Cr_2Br_9Rb_3$, Chromate(3−), nonabromoditrirubidium, 26:379

$Cr_2Cl_9Cs_3$, Chromate(3−), nonachloroditricesium, 26:379

$Cr_2O_6C_{16}H_{10}$, Chromium, hexacarbonylbis(η^5-cyclopentadienyl)di-, 28:148

$Cr_2S_5C_{20}H_{30}$, Chromium, (μ-disulfido-

S:S)(μ-η^2: η^2-disulfido)bis(η^5-pentamethylcyclopentadienyl)-μ-thio-di-, (Cr–Cr), 27:69

Cs$_3$Br$_9$Cr$_2$, Chromate(3–), nonabromodi-, tricesium, 26:379

Cs$_3$Br$_9$Ti$_2$, Titanate(3–), nonabromodi-, tricesium, 26:379

Cs$_3$Br$_9$V$_2$, Vanadate(3–), nonachlorodi-, tricesium, 26:379

Cs$_3$Cl$_9$Cr$_2$, Chromate(3–), nonachlorodi-, tricesium, 26:379

Cs$_3$Cl$_9$Ti$_2$, Titanate(3–), nonachlorodi-, tricesium, 26:379

Cs$_3$Cl$_9$V$_2$, Vanadate(3–), nonachlorodi-, tricesium, 26:379

Cs$_5$O$_{40}$PV$_2$W$_{10}$, Divanadodecatungstophosphate(5–), γ-, pentacesium, 27:102

Cs$_6$O$_{23}$P$_2$W$_5$, Pentatungstodiphosphate(6–), hexacesium, 27:101

Cs$_6$O$_{40}$PV$_3$W$_9$, Trivanadononatungstophosphate(6–), α-1,2,3-, hexacesium, 27:100

Cs$_7$O$_{36}$PW$_{10}$, Decatungstophosphate(7–), hexacesium, 27:101

CuCl$_2$, Copper dichloride, 28:322

CuCo$_3$NO$_{12}$C$_{14}$H$_3$, Ruthenium, (acetonitrile)dodecacarbonyltricobaltcopper-, 26:359

CuF$_6$N$_4$PC$_8$H$_{12}$, Copper(1+), tetrakis(acetonitrile)-, hexafluorophosphate(1–), 28:68

DyO$_3$C$_{45}$H$_{69}$, Dysprosium, tris(2,6-di-tert-butyl-4-methylphenoxo)-, 27:167

Dy$_2$Cl$_2$Si$_4$C$_{44}$H$_{84}$, Dysprosium, tetrakis[η^5-1,3-bis(trimethylsilyl)cyclopentadienyl]-di-μ-chloro-di-, 27:171

ErCl$_3$ · 7/2C$_4$H$_8$O, Erbium trichloride-tetrahydrofuran(2:7), 27:140, 28:290

ErO$_3$C$_{45}$H$_{69}$, Erbium, tris(2,6-di-tert-butyl-4-methylphenoxo)-, 27:167

Er$_2$Cl$_2$Si$_4$C$_{44}$H$_{84}$, Erbium, tetrakis[η^5-1,3-bis(trimethylsilyl)cyclopentadienyl]di-μ-chloro-di-, 27:171

Er$_2$Cl$_6$O$_7$C$_{28}$H$_{56}$, Erbium, hexachloroheptakis(tetrahydrofuran)di-, 27:140

Eu$_2$Cl$_2$Si$_4$C$_{44}$H$_{84}$, Europium, tetrakis[η^5-1,3-bis(trimethylsilyl)cyclopentadienyl]di-μ-chloro-di-, 27:171

FC$_6$H$_5$, Benzene, fluoro-, chromium complex, 28:139

FCrNO$_3$C$_5$H$_6$, Chromate(VI), fluorotrioxo-, pyridinium, 27:310

FCrO$_3$C$_9$H$_5$, Chromium, tricarbonyl(η^6-fluorobenzene)-, 28:139

FNO$_3$SC$_{16}$H$_{36}$, Fluorosulfate, tetrabutylammonium, 26:393

FN$_2$O$_3$ReC$_{13}$H$_8$, Rhenium, (2,2'-bipyridine)tricarbonylfluoro-, 26:82

FO$_2$C$_7$H$_5$, Benzoic acid, 3-fluoro-, rhodium complex, 27:292

FO$_3$P$_2$RhC$_{44}$H$_{34}$, Rhodium(I), carbonyl(3-fluorobenzoato)bis(triphenylphosphine)-, 27:292

FO$_3$S$_9$C$_{10}$H$_8$, 2,2'-Bi-1,3-dithiolo[4,5-b]-[1,4]dithiinylidene fluorosulfate, 26:393

F$_2$HO$_2$P, Phosphorodifluoridic acid, rhenium complex, 26:83

F$_2$N$_2$O$_5$PReC$_{13}$H$_8$, Rhenium, (2,2'-bipyridine)tricarbonyl(phosphorodifluoridato)-, 26:83

F$_3$Au$_2$ClO$_3$P$_4$PtSC$_{49}$H$_{60}$, Platinum(1+), chloro-1-κCl-bis(triethylphosphine-1κP)bis(triphenylphosphine)-2κP,3κP-triangulo-digold-, trifluoromethanesulfonate, 27:218

F$_3$ClO$_3$PtSC$_{13}$H$_{30}$, Platinum(II), chlorobis(triethylphosphine)(trifluoromethanesulfonato)-, cis-, 26:126

F$_3$ClO$_3$PtSC$_{37}$H$_{30}$, Platinum(II), chlorobis(triphenylphosphine)(trifluoromethanesulfonato)-, cis-, 28:27

F$_3$FeO$_5$S$_2$C$_9$H$_5$, Iron(1+), dicarbonyl(η^5-cyclopentadienyl)(thiocarbonyl)-, trifluoromethanesulfonate, 28:186

F$_3$K$_2$MnO$_4$S, Manganate(III), trifluorosulfato-, dipotassium, 27:312

F$_3$MnO$_8$SC$_6$, Manganese(I), pentacarbonyl(trifluoromethanesulfonato)-, 26:114

F$_3$O$_2$CH, Acetic acid, trifluoro-, ruthenium complex, 26:254
tungsten complex, 26:222

F$_3$O$_3$SCH, Methanesulfonic acid, trifluoro-, 28:70

F_3O_3SCH (Continued)
  iridium, manganese and rhenium complexes, 26:114, 115, 120
  iridium and platinum complexes, 28:26, 27
  platinum complex, 26:126
$F_3O_4P_2PtSC_{14}H_{35}$, Platinum(II), hydrido(methanol)bis(triethylphosphine)-, trans-, trifluoromethanesulfonate, 26:135
$F_3O_8ReSC_6$, Rhenium(I), pentacarbonyl(trifluoromethanesulfonato)-, 26:115
F_3P, Phosphorus trifluoride, preparation of, 26:12, 28:310
$F_4Au_3BOP_3C_{54}H_{45}$, Gold(1+), μ_3-oxo[tris[(triphenylphosphine)-, tetrafluoroborate(1−), 26:326
F_4B, Borate(1−), tetrafluoro-, molybdenum and tungsten complexes, 26:96–105
$F_4BClIrN_2P_2C_{36}H_{31}$, Iridium(III), chloro(dinitrogen)hydrido[tetrafluoroborato(1−)]bis(triphenylphosphine)-, 26:119
$F_4BClIrOP_2C_{37}H_{31}$, Iridium(III), carbonylchlorohydrido[tetrafluoroborato(1−)]bis(triphenylphosphine)-, 26:117, 28:23
$F_4BClIrOP_2C_{38}H_{33}$, Iridium(III), carbonylchloromethyl[tetrafluoroborato(1−)bis(triphenylphosphine)-, 26:118
$F_4BFeO_2C_{11}H_{13}$, Iron(1+), dicarbonyl(η^5-cyclopentadienyl)(η^2-2-methyl-1-propene)-, tetrafluoroborate(1−), 28:210
$F_4BH_5Ir_2C_{56}H_{56}$, Iridium(1+), [1,4-butanedibis(diphenylphosphine)di-, tetrafluoroborate(1−), 27:26
$F_4BH_5Ir_2P_4C_{54}H_{52}$, Iridium(1+), pentahydridobis[1,3-propanediylbis(diphenylphosphine)]di-, tetrafluoroborate(1−), 27:22
$F_4BI_2IrC_{42}H_{36}$, Iridium(1+), (1,2-diiodobenzene)dihydridobis(triphenylphosphine)-, tetrafluoroborate(1−), 26:125, 28:59
$F_4BIrO_2P_2C_{36}H_{36}$, Iridium(1+), diaquadihydridobis(triphenylphosphine)-, tetrafluoroborate(1−), 26:124, 28:58
$F_4BIrO_2P_2C_{42}H_{44}$, Iridium(1+), bis(acetone)dihydridobis(triphenylphosphine)-, tetrafluoroborate(1−), 26:123, 28:57
$F_4BIrP_2C_{35}H_{38}$, Iridium(1+), (η^4-1,5-cyclooctadiene)[1,3-propanediylbis(diphenylphosphine)]-, tetrafluoroborate(1−), 27:23
$F_4BIrP_2C_{44}H_{42}$, Iridium(1+), (η^4-1,5-cyclooctadiene)bis(triphenylphosphine)-, tetrafluoroborate(1−), 26:122, 28:56
$F_4BMoOC_{34}H_{25}$, Molybdenum(1+), carbonyl(η^5-cyclopentadienyl)bis(diphenylacetylene)-, tetrafluoroborate(1−), 26:102
$F_4BMoOPC_{38}H_{30}$, Molybdenum(1+), carbonyl(η^5-cyclopentadienyl)(diphenylacetylene)(triphenylphosphine)-, tetrafluoroborate(1−), 26:104
$F_4BMoO_2PC_{25}H_{20}$, Molybdenum, dicarbonyl(η^5-cyclopentadienyl)[tetrafluoroborato(1−)](triphenylphosphine)-, 26:98
$F_4BMoO_3C_8H_5$, Molybdenum, tricarbonyl(η^5-cyclopentadienyl)[tetrafluoroborato(1−)]-, 26:96, 28:5
$F_4BMoO_3C_{10}H_9$, Molybdenum(1+), tricarbonyl(η^5-cyclopentadienyl)(η^2-ethene)-, tetrafluoroborate(1−), 26:102
$F_4BMoO_4C_{11}H_{11}$, Molybdenum(1+), (acetone)tricarbonyl(η^5-cyclopentadienyl)-, tetrafluoroborate(1−), 26:105
$F_4BNO_5WC_{10}H_{10}$, Tungsten(1+), pentacarbonyl[(diethylamino)methylidyne]-, tetrafluoroborate(1−), 26:40
$F_4BO_2PWC_{25}H_{20}$, Tungsten, dicarbonyl(η^5-cyclopentadienyl)[tetrafluoroborato(1−)](triphenylphosphine)-, 26:98, 28:7
$F_4BO_3WC_8H_5$, Tungsten, tricarbonyl(η^5-cyclopentadienyl)[tetrafluoroborato(1−)]-, 26:96, 28:5
$F_4BO_4WC_{11}H_{11}$, Tungsten(1+), (acetone)tricarbonyl(η^5-cyclopentadienyl)-, tetrafluoroborate(1−), 26:105, 28:14
$F_4BO_5ReC_5$, Rhenium, pentacarbonyl[tetrafluoroborato(1−)]-, 26:108
$F_4BO_5ReC_7H_4$, Rhenium(1+), pentacarbonyl(η^2-ethene)-, tetrafluoroborate(1−), 26:110

$F_4O_{12}Re_4C_{12} \cdot 4H_2O$, Rhenium, dodecarbonyltetrafluorotetra-, tetrahydrate, 26:82

$F_5AuClPC_{31}H_{22}$, Aurate(I), chloro(pentafluorophenyl)-, (benzyl)triphenylphosphenium, 26:88

$F_5AuSC_{10}H_8$, Gold(I), (pentafluorophenyl)tetrahydrothiophene)-, 26:86

$F_5Au_2NPSC_{25}H_{15}$, Gold(I), (pentafluorophenyl)-μ-thiocyanato(triphenylphosphine)di-, 26:90

$F_5BrSC_2H_2$, λ^6-Sulfane, (2-bromoethenyl)pentafluoro-, 27:330

F_5CH, Benzene, pentafluoro-, gold complexes, 26:86–90

F_5SC_2H, λ^6-Sulfane, ethynylpentafluoro-, 27:329

F_6AsBrS_7, Arsenate(1−)hexafluoro-, cyclo-heptasulfur(1+), bromo-, 27:336

F_6AsIS_7, Arsenate(1−), hexafluoro-, iodo-cyclo-heptasulfur(1+), 27:333

$F_6BrN_3PRuC_{14}H_{21}$, Ruthenium(II), tris(acetonitrile)bromo(η^4-1,5-cyclooctadiene)-, hexafluorophosphate-(1−), 26:72

F_6BrS_7Sb, Antimonate(1−), hexafluoro-, bromo-cyclo-heptasulfur(1+), 27:336

$F_6ClN_3PRuC_{14}H_{21}$, Ruthenium(II), tris(acetonitrile)chlor(η^4-1,5-cyclooctadiene)-, hexafluorophosphate-(1−), 26:71

$F_6CuN_4PC_8H_{12}$, Copper(1+), tetrakis(acetonitrile)-, hexafluorophosphate(1−), 28:68

$F_6FeMoO_5P_2C_{34}H_{28}$, Iron(1+), μ-acetyl-$2\kappa^5C^1:1\kappa O$-tetracarbonyl-$1\kappa^3C,2\kappa^2C$-bis[1,2-(η^5-cyclopentadienyl)](triphenylphosphine-$1\kappa P$)molybdenum-, hexafluorophosphate(1−), 26:241

$F_6FeMoO_6PC_{17}H_{13}$, Iron(1+), μ-acetyl-$2\kappa C^1\kappa O$-pentacarbonyl-$1\kappa^3C,2\kappa^2C$-bis[1,2-(η^5-cyclopentadienyl)]-molybdenum-, hexafluorophosphate-(1−), 26:239

$F_6FeO_2PC_{11}H_{13}$, Iron(1+), (η^5-cyclopentadienyl)dicarbonyl(tetrahydrofuran)-, hexafluorophosphate(1−), 26:232

$F_6Fe_2O_4P_2C_{33}H_{28}$, Iron(1+), μ-acetyl-$2\kappa C^1:1\kappa O$-tricarbonyl-$1\kappa^2C,2\kappa C$-bis[1,2-(η^5-cyclopentadienyl)](triphenylphosphine-$2\kappa P$)di-, hexafluorophosphate-(1−), 26:237

$F_6Fe_2O_5PC_{10}H_{13}$, Iron(1+), μ-acetyl-$C:O$-bis[dicarbonyl(η-cyclopentadienyl)-, hexafluorophosphate(1−), 26:235

F_6IS_7Sb, Antimonate(1−), hexafluoro-, iodo-cyclo-heptasulfur(1+), 27:333

$F_6IrO_7P_2S_2C_{39}H_{31}$, Iridium(III), carbonylhydridobis(trifluoromethanesulfonato)bis(triphenylphosphine)-, 28:26

$F_6IrO_7P_2S_2C_{39}H_{31}$, Iridium(III), carbonylhydridobis(trifluoromethanesulfonato)bis(triphenylphosphine)-, 26:120

$F_6N_6NiP_2C_{42}H_{46}$, Nickel(II), [3,11-bis[α-(benzylamino)benzylidene]-2,12-dimethyl-1,5,9,13-tetraazacyclohexadeca-1,4,9,12-tetraene-$\kappa^4N^{1,5,9,13}$]-, bis[hexafluorophosphate(1−)], 27:276

$F_6O_2C_5H_2$, 2,4-Pentanedione, 1,1,1,5,5,5-hexafluoro-, palladium complexes, 27:318–320

$F_7MoO_2P_5C_{54}H_{48}$, Molybdenum(II), dicarbonylbis[1,2-ethanediylbis(diphenylphosphine)]fluoro-, hexafluorophosphate(1−), 26:84

$F_8B_2H_7Ir_3P_6C_{78}H_{72}$, Iridium(2+), tris[1,2-ethanediylbis(diphenylphosphine)]heptahydridotri-, bis[tetrafluoroborate(1−)], 27:25

$F_8B_2H_7Ir_3P_6C_{81}H_{78}$, Iridium(2+), heptahydridotris[1,3-propanediylbis(diphenylphosphine)]tri-, bis[tetrafluoroborate(1−)], 27:22

$F_8B_2MoN_6O_2C_8H_{12}$, Molybdenum(2+), tetrakis(acetonitrile)dinitrosyl-, cis-, bis[tetrafluoroborate(1−)], 26:123, 28:65

$F_8B_2N_4PdC_8H_{12}$, Palladium(2+), tetrakis(acetonitrile)-, bis[tetrafluoroborate(1−)], 26:128, 28:63

$F_8B_2N_6O_2WC_8H_{12}$, Tungsten(2+), tetrakis(acetonitrile)dinitrosyl-, cis-, bis[tetrafluoroborate(1−)], 26:133, 28:66

$F_{10}As_4AuC_{32}H_{32}$, Gold(I), bis[1,2-phenylenebis(dimethylarsine)]-, bis(pentafluorophenyl)aurate(I), 26:89

$F_{12}CoN_6P_2C_{26}H_{44}$, Cobalt(II),

$F_{12}CoN_6P_2C_{26}H_{44}$ (Continued) (2,3,10,11,13,19-hexamethyl-3,10,14,18,21,25-hexaazabicyclo[10.7.7]hexacosa-1,11,13,18,20,25-hexaene-$\kappa^4 N^{14,18,21,25}$)-, bis[hexafluorophosphate(1−)], 27:270

$F_{12}FeN_6P_2C_{50}H_{52}$, Iron(II), [3,11-dibenzyl-14,20-dimethyl-2,12-diphenyl-3,11,15,19,22,26-hexaazatricyclo[11.7.7.1^{5,9}]octacosa-1,5,7,9-(28),12,14,19,21,26-nonaene-$\kappa^4 N^{15,19,22,26}$]-, bis[hexafluorophosphate(1−)], 27:280

$F_{12}N_2O_4PdC_{20}H_{10}$, Palladium(II), (2,2'-bipyridine)(1,1,1,5,5,5-hexafluoro-2,4-pentanedionato)-, 1,1,1,5,5,5-hexafluoro-2,4-dioxo-3-pentanide, 27:319

$F_{12}N_4NiO_2P_2C_{20}H_{32}$, Nickel(II), [2,12-dimethyl-3,11-bis(1-methoxyethylidene)-1,5,9,13-tetraazacyclohexadeca-1,4,9,12-tetraene-$\kappa^4 N^{1,5,9,13}$]-, bis[hexafluorophosphate(1−)], 27:264

$F_{12}N_4NiO_2P_2C_{30}H_{36}$, Nickel(II), [3,11-bis(α-methoxybenzylidene)-2,12-dimethyl-1,5,9,13-tetraazacyclohexadeca-1,4,9,12-tetraene-$\kappa^4 N^{1,5,9,13}$]-, bis[hexafluorophosphate(1−)], 27:275

$F_{12}N_4P_2RuC_{16}H_{24}$, Ruthenium(II), tetrakis(acetonitrile)(η^4-1,5-cyclooctadiene)-, bis[hexafluorophosphate(1−)], 26:72

$F_{12}N_6NiP_2C_{20}H_{34}$, Nickel(II), [2,12-dimethyl-3,11-bis[1-(methylamino)ethylidene]-1,5,9,13-tetraazacyclohexadeca-1,4,9,12-tetraene-$\kappa^4 N^{1,5,9,13}$]-, bis[hexafluorophosphate(1−)], 27:266

$F_{12}N_6NiP_2C_{26}H_{44}$, Nickel(II), (2,3,10,11,13,19-hexamethyl-3,10,14,18,21,25-hexaazabicyclo[10.7.7]hexacosa-1,11,13,18,20,25-hexaene-$\kappa^4 N^{14,18,21,25}$)-, bis[hexafluorophosphate(1−)], 27:268

$F_{12}N_6NiP_2C_{50}H_{52}$, Nickel(II), [3,11-dibenzyl-14,20-dimethyl-2,12-diphenyl-3,11,15,19,22,26-hexaazatricyclo[11.7.7.1^{5,9}]octacosa-1,5,7,9(28),12,14,19,21,26-nonaene-$\kappa^4 N^{15,19,22,26}$]-,

bis[hexafluorophosphate(1−)], 27:277

$F_{12}N_8P_2RuC_{12}H_{36}$, Ruthenium(II), ($\eta^4$-cyclooctadiene)tetrakis(methylhydrazine)-, bis[hexafluorophosphate-(1−), 26:74

$F_{12}O_4P_3PdC_{44}H_{35}$, Palladium(II), [bis[2-(diphenylphosphino)ethyl]phenylphosphine](1,1,1,5,5,5-hexafluoro-2,4-pentanedionato)-, 1,1,1,5,5,5-hexafluoro-2,4-dioxo-3-pentanide, 27:320

$F_{12}O_4PdC_{10}H_2$, Palladium, bis(1,1,1,5,5,5-hexafluoro-2,4-pentanedionato)-, 27:318

$F_{12}O_8W_2C_8$, Tungsten(II), tetrakis(trifluoroacetato)di-, (W−4−W), 26:222

$F_{12}O_9Ru_2C_{24}H_{26}$, Ruthenium(II), μ-aqua-bis(μ-trifluoroacetato)bis[(η^4-cycloocta-1,5-diene)(trifluoroacetato)-, 26:254

$F_{15}AuSC_{22}H_8$, Gold(III), tris(pentafluorophenyl)(tetrahydrothiophene)-, 26:87

$F_{18}I_3S_{14}Sb_3 \cdot 2AsF_3$, Antimonate(1−), hexafluoro-, μ-iodo-bis(4-iodo-cyclo-heptasulfur)(3+)(3:1), −2(arsenic trifluoride), 27:335

$F_{18}N_6P_3C_{26}H_{47}$, 3,10,14,18,21,25-Hexaazabicyclo[10.7.7]hexacosa-1,11,13,18,20,25-hexaene, 2,3,10,11,13,19-hexamethyl-, tris[hexafluorophosphate-(1−)], 27:269

$F_{18}N_6P_3C_{50}H_{55}$, 3,11,15,19,22,26-Hexaazatricyclo[11.7.7.1^{5,9}]-octacosa-1,5,7,9(28),12,14,19,21,26-nonaene, 3,11-dibenzyl-14,20-dimethyl-2,12-diphenyl-, tris[hexafluorophosphate(1−)], 27:278

$F_{36}As_6Br_4S_{32}$, Arsenate(1−), hexafluoro-, bromo-cyclo-heptasulfur(1+) tetrasulfur(2+)(6:4:1), 27:338

$F_{36}As_6I_4S_{32}$, Arsenate(1−), hexafluoro-, iodo-cyclo-heptasulfur(1+) tetrasulfur(2+)(6:4:1), 27:337

$FeAsO_4C_{22}H_{15}$, Iron, tetracarbonyl-(triphenylarsine)-, 26:61

$FeAuCo_3O_{12}PC_{30}H_{15}$, Ferrate(1−), dodecacarbonyltricobalt-, (triphenylphosphine)gold(1+), 27:188

$FeBF_4O_2C_{11}H_{13}$, Iron(1+), dicarbonyl(η^5-cyclopentadienyl)(η^2-2-methyl-1-

propene)-, tetrafluoroborate(1−), 28:210
FeCl$_3$, Iron trichloride, 28:322
FeCo$_2$O$_9$PC$_{15}$H$_5$, Iron, nonacarbonyl(μ_3-phenylphosphinidene)dicobalt-, 26:353
FeCo$_2$O$_9$SC$_9$, Iron, nonacarbonyl-μ_3-thiodicobalt-, 26:245, 352
FeCo$_3$NO$_{12}$C$_{20}$H$_{20}$, Ferrate(1−), dodecacarbonyltricobalt-, tetraethylammonium, 27:188
FeCrNO$_9$P$_2$C$_{45}$H$_{31}$, Chromate(1−)hydridononacarbonyliron-μ-nitrido-bis(triphenylphosphorus)(1+), 26:338
FeCrN$_2$O$_9$P$_4$C$_{81}$H$_{60}$, Chromate(2−), nonacarbonyliron-, bis[μ-nitrido-bis(triphenylphosphorus)(1+)], 26:339
FeF$_3$O$_5$S$_2$C$_9$H$_5$, Iron(1+), dicarbonyl(η^5-cyclopentadienyl)(thiocarbonyl)-, trifluoromethanesulfonate, 28:186
FeF$_6$MoO$_5$P$_2$C$_{34}$H$_{28}$, Iron(1+), μ-acetyl-2κC^1:1κO-tetracarbonyl- 1$\kappa^3 C$,2$\kappa^2 C$-bis[1,2-(η^5-cyclopentadienyl)](triphenylphosphine-1κP)-molybdenum-, hexafluorophosphate(1−), 26:241
FeF$_6$MoO$_6$PC$_{17}$H$_{13}$, Iron(1+), μ-acetyl-2κC^1:1κO-pentacarbonyl-1$\kappa^3 C$,2$\kappa^2 C$-bis[1,2-η^5-cyclopentadienyl]]molybdenum-, hexafluorophosphate(1−), 26:239
FeF$_6$O$_2$PC$_{11}$H$_{13}$, Iron(1+), (η^5-cyclopentadienyl)dicarbonyl(tetrahydrofuran)-, hexafluorophosphate(1−), 26:232
FeF$_{12}$N$_6$P$_2$C$_{50}$H$_{52}$, Iron(II), [3,11-dibenzyl-14,20-dimethyl-2,12-diphenyl-3,11,15,19,22,26-hexaazatricyclo-[11.7.7.1^{5,9}]octacosa-1,5,7,9(28), 12,14,19,21,26-nonaene-$\kappa^4 N^{15,19,22,26}$]-, bis[hexafluorophosphate(1−)], 27:280
FeMoNO$_9$P$_2$C$_{45}$H$_{31}$, Molybdate(1−), hydridononacarbonyliron-, μ-nitrido-bis(triphenylphosphorus)(1+), 26:338
FeMoN$_2$O$_9$P$_4$C$_{81}$H$_{60}$, Molybdate(2−), nonacarbonyliron-bis[μ-nitrido-bis(triphenylphosphorus)(1+)], 26:339
FeNO$_4$C$_{13}$H$_9$, Iron, tetracarbonyl(2-isocyano-1,3-dimethylbenzene)-, 26:53
FeNO$_4$P$_2$C$_{40}$H$_{31}$, Ferrate(1−), hydridotetracarbonyl-, μ-nitrido-bis(triphenylphosphorus)(1+), 26:336

FeNO$_9$WC$_{45}$H$_{31}$, Tungstate(1−), hydridononacarbonyliron-, μ-nitrido-bis(triphenylphosphorus)(1+), 26:336
FeN$_2$O$_3$C$_{21}$H$_{18}$, Iron, tricarbonylbis(2-isocyano-1,3-dimethylbenzene)-, 26:54
FeN$_2$O$_9$P$_4$WC$_{81}$H$_{60}$, Tungstate(2−), nonacarbonyliron-, bis[μ-nitrido-bis-(triphenylphosphorus)(1+)], 26:339
FeN$_3$O$_2$C$_{29}$H$_{27}$, Iron, dicarbonyltris(2-isocyano-1,3-dimethylbenzene)-, 26:56
FeN$_4$OC$_{37}$H$_{36}$, Iron, carbonyltetrakis(2-isocyano-1,3-dimethylbenzene)-, 26:57
FeN$_5$C$_{45}$H$_{45}$, Iron, pentakis(2-isocyano-1,3-dimethylbenzene)-, 26:57
FeNa$_2$O$_4$C$_4$, Ferrate(2−), tetracarbonyl-, disodium, 28:203
FeO$_2$C$_{11}$H$_{12}$, Iron, dicarbonyl(η^5-cyclopentadienyl)(2-methyl-1-propenyl-κ-C^1)-, 28:208
FeO$_2$S$_2$C$_9$H$_8$, Iron, dicarbonyl(η^5-cyclopentadienyl)[(methylthio)thiocarbonyl]-, 28:186
FeO$_3$C$_9$H$_8$, Iron, acetyldicarbonyl(η^5-cyclopentadienyl)-, 26:239
FeO$_3$P$_2$C$_9$H$_{18}$, Iron(0), tricarbonylbis(trimethylphosphine)-, 28:177
FeO$_3$P$_2$C$_{25}$H$_{54}$, Iron(0), tricarbonylbis(tributylphosphine)-, 28:177
FeO$_3$P$_2$C$_{39}$H$_{30}$, Iron(0), tricarbonylbis(triphenylphosphine)-, 28:176
FeO$_3$P$_2$C$_{39}$H$_{66}$, Iron(0), tricarbonylbis(tricyclohexylphosphine)-, 28:176
FeO$_4$C$_{13}$H$_{10}$, Iron(0), tricarbonyl(4-phenyl-3-butene-2-one)-, 28:52
FeO$_4$PC$_{12}$H$_{11}$, Iron, tetracarbonyl-(dimethylphenylphosphine)-, 26:61, 28:171
FeO$_4$PC$_{16}$H$_{27}$, Iron, tetracarbonyl(tributylphosphine)-, 26:61
FeO$_4$PC$_{17}$H$_{13}$, Iron, tetracarbonyl-(methyldiphenylphosphine)-, 26:61
FeO$_4$PC$_{22}$H$_{15}$, Iron, tetracarbonyl(triphenylphosphine)-, 26:61
FeO$_4$PC$_{22}$H$_{33}$, Iron, tetracarbonyl-(tricyclohexylphosphine)-, 26:61
FeO$_4$SbC$_{22}$H$_{15}$, Iron, tetracarbonyl-(triphenylstibine)-, 26:61

FeO$_7$PC$_3$H$_9$, Iron, tetracarbonyl(trimethyl phosphite)-, 26:61, 28:171

FeO$_7$PC$_{10}$H$_{15}$, Iron, tetracarbonyl(triethyl phosphite)-, 26:61

FeO$_7$PC$_{22}$H$_{15}$, Iron, tetracarbonyl(triphenyl phosphite)-, 26:61, 28:171

Fe$_2$F$_6$O$_4$P$_2$C$_{33}$H$_{28}$, Iron(1+), μ-acetyl-2κC^1:1κO-tricarbonyl-1$\kappa^2 C$,2κC-bis[1,2-(η^5-cyclopentadienyl)](triphenylphosphine-2κP)di-, hexafluorophosphate(1−), 26:237

Fe$_2$F$_6$O$_5$PC$_{16}$H$_{13}$, Iron(1+), μ-acetyl-C:O-bis[dicarbonyl(η-cyclopentadienyl)-, hexafluorophosphate(1−), 26:235

Fe$_2$Na$_2$O$_8$C$_8$, Ferrate(2−), octacarbonyldi-, disodium, 28:203

Fe$_3$N$_2$O$_{11}$P$_4$C$_{83}$H$_{60}$, Ferrate(2−), undecacarbonyltri-, bis[μ-nitrido-bis(triphenylphosphorus)(1+)], 28:203

Fe$_3$Na$_2$O$_{11}$C$_{11}$, Ferrate(2−), undecacarbonyltri-, disodium, 28:203

Fe$_3$O$_9$SC$_9$H$_2$, Iron, nonacarbonyldihydrido-μ_3-, 26:244

Fe$_4$HNO$_{12}$C$_{21}$H$_{20}$, Ferrate(1−), carbidododecacarbonylhydridotetra-, tetraethylammonium, 27:186

Fe$_4$NO$_{12}$P$_2$C$_{49}$H$_{30}$, Ferrate(2−), μ_4-carbidododecacarbonyltetra-, bis[μ-nitridobis(triphenylphosphorus)(1+)], 26:246

Fe$_4$NO$_{14}$C$_{23}$H$_{23}$, Ferrate(1−), dodecacarbonyl[μ_4-(methoxycarbonyl)methylidyne]tetra-, tetraethylammonium, 27:184

Fe$_4$N$_2$O$_{12}$C$_{29}$H$_{40}$, Ferrate(2−), carbidododecacarbonyltetra-, bis(tetraethylammonium), 27:187

Fe$_4$O$_{13}$C$_{14}$, Iron, carbidotridecacarbonyltetra-, 27:185

Fe$_6$N$_2$O$_{16}$C$_{33}$H$_{40}$, Ferrate(2−), carbidohexadecacarbonylhexa-, bis(tetraethylammonium), 27:183

Gd$_2$Cl$_2$Si$_4$C$_{44}$H$_{84}$, Gadolinium, tetrakis[η^5-1,3-bis(trimethylsilyl)cyclopentadienyl]di-μ-chloro-di-, 27:171

H, Hydride, iridium complex, 26:117, 119, 120, 123–125, 202

iron complex, 26:244
iron-tungsten, 26:336–340
manganese complex, 26:226
molybdenum and tungsten complexes, 26:98–105
nickel, osmium, and ruthenium complexes, 26:362, 363, 367
osmium complex, 26:186, 293, 301, 304
platinum complexes, 26:135, 136
rhenium, 28:165
ruthenium complex, 26:181, 182, 262, 264, 269, 277, 278, 329
zirconium complex, 28:257, 259

HAuO$_{10}$Os$_3$PC$_{16}$H$_{15}$, Osmium, decacarbonyl-μ-hydrido[μ-(triethylphosphine)gold]tri-, 27:210

HAuO$_{10}$Os$_3$PC$_{28}$H$_{15}$, Osmium, decacarbonyl-μ-hydrido[μ-(triphenylphosphine)gold]tri-, 27:209

HBClF$_4$IrN$_2$P$_2$C$_{36}$H$_{30}$, Iridium(III), chloro(dinitrogen)hydrido[tetrafluoroborato(1−)]bis(triphenylphosphine)-, 26:119

HBClF$_4$IrOP$_2$C$_{37}$H$_{30}$, Iridium(III), carbonylchlorohydrido[tetrafluoroborato(1−)]bis(triphenylphosphine)-, 26:117

HClIrP$_3$C$_{54}$H$_{44}$, Iridium(III), chloro[2-(diphenylphosphino)phenyl-C^1,P]hydridobis(triphenylphosphine, (OC-6-53)-, 26:202

HClPRuC$_{30}$H$_{33}$, Ruthenium(II), chloro(η^6-hexamethylbenzene)hydrido(triphenylphosphine)-, 26:181

HClZrC$_{10}$H$_{10}$, Zirconium, chlorobis(η^5-cyclopentadienyl)hydrido-, 28:259

HF$_2$O$_2$P, Phosphorodifluoridic acid, rhenium complex, 26:83

HF$_3$O$_4$P$_2$PtSC$_{14}$H$_{34}$, Platinum(II), hydrido(methanol)bis(triethylphosphine)-, $trans$-, trifluoromethanesulfonate, 26:135

HF$_6$I$_2$O$_7$P$_2$S$_2$C$_{39}$H$_{30}$, Iridium(III), carbonylhydridobis(trifluoromethanesulfonato)bis(triphenylphosphine)-, 26:120

HFeNO$_4$P$_2$C$_{40}$H$_{31}$, Ferrate(1−), hydridotetracarbonyl-, μ-nitrido-bis(triphenylphosphorus) (1+), 26:336

HFeNO$_9$WC$_{45}$H$_{30}$, Tungstate(1−), hydridononacarbonyliron-, μ-nitrido-bis-

(triphenylphosphorus)(1+), 26:336
HFe$_4$NO$_{12}$C$_{21}$H$_{20}$, Ferrate(1−), carbidododecacarbonylhydridotetra-, tetraethylammonium, 27:186
HK$_6$O$_{40}$SiV$_3$W$_{9.3}$H$_2$O, 1,2,3-Trivanadononatungstosilicate(7−), A-β-, hexapotassium hydrogen, trihydrate, 27:129
HMn$_2$O$_8$PC$_{20}$H$_{10}$, Manganese, μ-(diphenylphosphino)-μ-hydridobis(tetracarbonyl)-, (Mn−Mn), 26:226
HMoO$_2$PC$_{25}$H$_{20}$, Molybdenum, dicarbonyl(η^5-cyclopentadienyl) hydrido(triphenylphosphine)-, 26:98
HNO$_{10}$P$_2$Ru$_3$Si$_2$C$_{58}$H$_{60}$, Ruthenate(1−), decacarbonyl-1κ^3C,2κ^3C,3κ^4C-μ-hydrido-1:2κ^2H-bis (triethylsilyl)-1κSi, 2κSi-triangulo-tri-, μ-nitridobis(triphenylphosphorus) (1+), 26:269
HNO$_{11}$P$_2$Os$_3$C$_{47}$H$_{30}$, Osmate(1−), μ-carbonyldecacarbonyl-μ-hydrido-tri-, μ-nitrido-bis(triphenylphosphorus)(1+), 28:236
HNa$_9$O$_{34}$SiW$_{9.23}$H$_2$O, Nonatungstosilicate-(10−), β-, nonasodium hydrogen, tricosahydrate, 27:88
HOP$_3$RhC$_{55}$H$_{45}$, Rhodium(I), carbonylhydridotris(triphenylphosphine)-, 28:82
HO$_2$PWC$_{25}$H$_{20}$, Tungsten, dicarbonyl(η^5-cyclopentadienyl)hydrido(triphenylphosphine)-, 26:98
HO$_5$ReC$_5$, Rhenium, pentacarbonylhydrido-, 26:77
HO$_9$PRu$_3$C$_{21}$H$_{10}$, Ruthenium, nonacarbonyl-μ-hydrido-(μ-diphenylphosphido)tri-, 26:264
HO$_9$Ru$_3$C$_{15}$H$_{10}$, Ruthenium, nonacarbonyl (μ_3-3,3-dimethyl-1-butynyl)-μ-hydridotriangulo-, tri-, 26:329
HO$_{10}$Os$_3$C$_{11}$H$_3$, Osmium, decacarbonylhydridomethyltri-, 27:206
HO$_{10}$Os$_3$SC$_{16}$H$_5$, Osmium, (μ-benzenethiolato)decacarbonyl-μ-hydrido-tri-, 26:304
HO$_{11}$Os$_3$C$_{12}$H$_3$, Osmium, decacarbonyl-μ-hydrido(μ-methoxymethylidyne)-triangulo-tri-, 27:202
HPRuC$_{30}$H$_{32}$, Ruthenium(II), [2-(diphenylphosphino)phenyl-C^1,P](η^6-hexamethylbenzene)hydrido-, 36:182

HP$_4$RhC$_{72}$H$_{60}$, Rhodium(I), hydridotetrakis(triphenylphosphine)-, 28:81
HRu$_3$O$_{11}$C$_{12}$H$_3$, Ruthenium, decacarbonyl-μ-hydrido(μ-methoxymethylidyne)-triangulo-tri-, 27:198
H$_2$, Hydrogen, molybdenum complex, 27:3
tungsten complex, 27:6
H$_2$BF$_4$I$_2$IrC$_{42}$H$_{34}$, Iridium(1+), (1,2-diiodobenzene) dihydridobis(triphenylphosphine)-, tetrafluoroborate(1−), 26:125, 28:59
H$_2$BF$_4$IrO$_2$P$_2$C$_{36}$H$_{34}$, Iridium(1+), diaquadihydridobis(triphenylphosphine)-, tetrafluoroborate(1−), 26:124, 28:58
H$_2$BF$_4$IrO$_2$P$_2$C$_{42}$H$_{42}$, Iridium(1+), bis(acetone)dihydridobis(triphenylphosphine)-, tetrafluoroborate(1−), 26:123, 28:57
H$_2$BP$_4$PtC$_{54}$H$_{31}$, Platinum(II), μ-hydrido-hydridophenyltetrakis(triethylphosphine)ditetraphenylborate(1−), 26:136
H$_2$Fe$_3$O$_9$SC$_9$, Iron, nonacarbonyldihydrido-μ_3-thiotri-, 26:244
H$_2$LiP, Lithium dihydrogen phosphide, (LiH$_2$P), 27:228
H$_2$O, Water, chromium and vanadium complexes, 27:307, 309
iridium complex, 26:123
ruthenium complex, 26:254−256
H$_2$O$_{10}$Os$_3$C$_{10}$, Osmium, decacarbonyldihydridotri-, 26:367
H$_2$O$_{10}$Os$_3$C$_{11}$H$_2$, Osmium, decacarbonyl-di-μ-hydrido-μ-methylene-triangulo-tri-, 27:206
H$_2$O$_{18}$Os$_6$C$_{18}$, Osmium, octadecacarbonyldihydridohexa-, 26:301
H$_2$OsP$_3$SC$_{55}$H$_{45}$, Osmium(II), dihydrido(thiocarbonyl)tris(triphenylphosphine)-, 26:186
H$_2$P$_4$RuC$_{72}$H$_{60}$, Ruthenium(II), dihydridotetrakis(triphenylphosphine)-, 28:337
H$_2$S, Hydrogen sulfide, titanium complex, 27:66
tungsten complex, 27:67
H$_2$ZrC$_{10}$H$_{10}$, Zirconium, bis(η^5-cyclopentadienyl)dihydrido-, 28:257
H$_3$BP$_4$Pt$_2$C$_{48}$H$_{80}$, Platinum(II), di-μ-hydrido-hydridotetrakis(triethylphos-

$H_3BP_4Pt_2C_{48}H_{80}$ (Continued)
phine)di-, tetraphenylborate(1−), 27:34
—, Platinum(II), μ-hydrido-dihydridotetrakis(triethylphosphine)di-, tetraphenylborate(1−), 27:32

$H_3BP_4Pt_2C_{96}H_{80}$, Platinum(II), di-μ-hydrido-hydridotetrakis(triphenylphosphine)-di-, tetraphenylborate(1−), 27:36

$H_3BrO_9Os_3C_{10}$, Osmium, (μ_3-bromomethylidyne)nonacarbonyltri-μ-hydrido-triangulo-tri-, 27:205

$H_3BrO_9Ru_3C_{10}$, Ruthenium, (μ_3-bromomethylidyne)nonacarbonyl-tri-μ-hydrido-triangulo-tri-, 27:201

$H_3ClO_9Os_3C_{10}$, Osmium, nonacarbonyl (μ_3-chloromethylidyne)tri-μ-hydrido-triangulo-tri-, 27:205

$H_3NiO_9Os_3C_{14}H_5$, Osmium, nonacarbonyl-(η^5-cyclopentadienyl)tri-μ-hydrido-nickeltri-, 26:362

$H_3NiO_9Ru_3C_{14}H_5$, Ruthenium, nonacarbonyl-(η^5-cyclopentadienyl)tri-μ-hydrido-nickel-tri-, 26:363

$H_3O_{10}Os_3C_{11}H_3$, Osmium, nonacarbonyl-tri-μ-hydrido(μ_3-methoxymethylidyne)-triangulo-tri-, 27:203

$H_3O_{10}Ru_3C_{11}H_3$, Ruthenium, nonacarbonyltri-μ-hydrido(μ_3-methoxymethylidyne)-triangulo-tri-, 27:200

$H_3O_{11}Os_3C_{12}H_3$, Osmium, nonacarbonyl-tri-μ-hydrido[μ_3-(methoxycarbonyl)-methylidyne]-triangulo-tri-, 27:204

$H_3OsRhP_3C_{32}H_{45}$, Rhodium, [2(η^4)-1,5-cyclooctadiene]tris(dimethylphenylphosphine-1κP)-tri-μ-hydrido-osmium-, 27:29

$H_4As_2O_{70}Rb_4W_{21} \cdot 34H_2O$, Tungstate(4−), aquadihydroxohenhexacontaoxobis-[trioxoarsenato(III)]henicosa-, tetra-rubidium, tetratricontahydrate, 27:113

$H_4IrC_{10}H_{15}$, Iridium, tetrahydrido(η^5-pentamethylcyclopentadienyl)-, 27:19

$H_4MoP_4C_{52}H_{52}$, Molybdenum(IV), tetrahydridotetrakis(methyldiphenylphosphine)-, 27:9

H_4N, Hydrazine, ruthenium(II), complexes, 26:72

$H_4O_{12}Os_4C_{12}$, Osmium, dodecacarbonyl-tetra-μ-hydrido-tetrahedro-tetra-, 26:293

$H_4O_{12}Ru_4C_{12}$, Ruthenium, dodecacarbonyl-tetra-μ-hydrido-tetra-, 26:262

$H_4O_{13}P_2Ru_4C_{39}H_{32}$, Ruthenium, decacarbonyl-(dimethylphenylphosphine)-tetrahydrido[tris-(4-methylphenyl) phosphite]tetra-, 26:278

$H_4O_{14}PRu_4C_{32}H_{21}$, Ruthenium, undecacarbonyltetrahydrido[tris(4-methylphenyl) phosphite]tetra-, 26:277, 28:227

$H_4O_{40}SiW_{12} \cdot xH_2O$, Dodecatungstosilicic acid, α-, hydrate, 27:93

$H_4OsP_3ZrC_{34}H_{43}$, Zirconium, bis[1,1(η^5)-cyclopentadienyl]tris(dimethylphenylphosphine-2κP)-tri-μ-hydrido-hydrido-1κH-osmium-, 27:27

$H_4P_4WC_{52}H_{52}$, Tungsten(IV), tetrahydridotetrakis(methyldiphenylphosphine)-, 27:10

$H_5BF_4Ir_2C_{56}H_{56}$, Iridium(1+), [1,4-butanedibis(diphenylphosphine)]di-, tetrafluoroborate(1−), 27:26

$H_5BF_4Ir_2P_4C_{54}H_{52}$, Iridium(1+), pentahydridobis[1,3-propanediylbis-(diphenylphosphine)]di-, tetrafluoroborate(1−), 27:22

H_5N_2C, Hydrazine, methyl-, ruthenium(II), complexes, 26:72

H_6B_2, Diborane(6), 27:215

$H_6MoP_3C_{54}H_{99}$, Molybdenum(IV), hexahydridotris(tricyclohexylphosphine)-, 27:13

$H_6P_3WC_{24}H_{33}$, Tungsten(IV), hexahydrido-tris(dimethylphenylphosphine)-, 27:11

$H_7B_2F_8Ir_3P_6C_{78}H_{72}$, Iridium(2+), tris[1,2-ethanediylbis(diphenylphosphine)]heptahydridotri-, bis[tetrafluoroborate(1−)], 27:25

$H_7B_2F_8Ir_3P_6C_{81}H_{78}$, Iridium(2+), heptahydridotris[1,3-propanediylbis-(diphenylphosphine)]tri-, bis[tetrafluoroborate(1−)], 27:22

$H_7K_{28}Li_5O_{184}P_8W_{48} \cdot 92H_2O$, Octatetracontatungstooctaphosphate(40−), pentalithium octacosapotassium heptahydrogen, dononacontahydrate, 27:110

$H_7P_2ReC_{36}H_{30}$, Rhenium(VII), heptahydridobis(triphenylphosphine)-, 27:15

$H_8As_2O_{70}W_{21} \cdot xH_2O$, Tungsten, aquahexahydroxoheptapentacontaoxobis[trioxoarsenato(III)]henicosa-, hydrate, 27:112

$H_8Cl_5MoN_2O$, Molybdate(V), pentachlorooxo-, diammonium, 26:36

$H_8P_4Re_2C_{72}H_{60}$, Rhenium(IV), octahydridotetrakis(triphenylphosphine)di-, 27:16

$H_{12}Br_3N_4Ru$, Ruthenium(III), tetraamminedibromo-, cis-, bromide, 26:67

$H_{12}Cl_3N_4Ru$, Ruthenium(III), tetraamminedichloro-, cis-, chloride, 26:66

$H_{14}B_9K$, Borate(1−), tetradecahydronona-, potassium, 26:1

$H_{18}Mo_5N_4O_{21}P_2 \cdot 4H_2O$, Pentamolybdobis(phosphonate) (4−), tetraammonium, tetrahydrate, 27:123

$H_{64}N_{14}NaO_{110}P_5W_{30} \cdot 31H_2O$, Sodiotricontatungstopentaphosphate(14−), tetradecaammonium, hentricontahydrate, 27:115

$H_{72}N_{18}NaO_{86}Sb_9W_{21} \cdot 24H_2O$, Sodiohenicosatungstononaantimonate(18−), octadecaammonium, tetracosahydrate, 27:120

$H_{100}As_4Co_2N_{24}O_{142}W_{40} \cdot 19H_2O$, Ammoniodicobaltotetracontatungstotetraarsenate(23−), tricosaammonium, nonadecahydrate, 27:119

$HfO_2C_{12}H_{10}$, Hafnium, dicarbonylbis(η^5-cyclopentadienyl)-, 28:252

$HfO_2C_{22}H_{30}$, Hafnium, dicarbonylbis(η^5-pentamethylcyclopentadienyl)-, 28:255

$HgBrO_9Ru_3C_{15}H_9$, Ruthenium, (bromomercury)nonacarbonyl(3,3-dimethyl-1-butynyl)-triangulo-tri-, 26:332

$HgClMn_2O_8PC_{20}H_{10}$, Manganese, μ-(chloromercurio)-μ-(diphenylphosphino)bis(tetracarbonyl-, (Mn–Mn), 26:230

$HgIO_9Ru_3C_{15}H_9$, Ruthenium, nonacarbonyl-(3,3-dimethyl-1-butynyl) (iodomercury)-triangulo-tri-, 26:330

$HgMoO_{12}Ru_3C_{23}H_{14}$, Ruthenium, nonacarbonyl-(μ_3-3,3-dimethyl-1-butynyl) {μ-[tricarbonyl(η^5-cyclopentadienyl)molybdenum]-mercury}-triangulo-tri-, 26:333

$HgO_{18}Ru_6C_{30}H_{18}$, Ruthenium, (μ_4-mercury)bis[nonacarbonyl (μ_3-3,3-dimethyl-1-butynyl)-, triangulo-tri-, 26:333

$HoO_3C_{45}H_{69}$, Holmium, tris(2,6-di-tert-butyl-4-methylphenoxo)-, 27:167

$Ho_2Cl_2Si_4C_{44}H_{84}$, Holmium, tetrakis[η^5-1,3-bis(trimethylsilyl)cyclopentadienyl]di-μ-chloro-di-, 27:171

$IAsF_6S_7$, cyclo-Heptasulfur(1+), iodo-, hexafluoroarsenate(1−), 27:333

IF_6S_7Sb, cyclo-Heptasulfur(1+), iodo-, hexafluoroantimonate(1−), 27:333

$IHgO_9Ru_3C_{15}H_9$, Ruthenium, nonacarbonyl-(3,3-dimethyl-1-butynyl) (iodomercury)-triangulo-tri-, 26:330

$IMnO_5C_5$, Manganese, pentacarbonyliodo-, 28:157, 158

$IOP_2RhC_{37}H_{66}$, Rhodium(I), carbonyliodobis(tricyclohexylphosphine)-, 27:292

IO_5ReC_5, Rhenium, pentacarbonyliodo-, 28:163

$I_2BF_4IrC_{42}H_{36}$, Iridium(1+), (1,2-diiodobenzene)dihydridobis(triphenylphosphine)-, tetrafluoroborate(1−), 28:59

$I_2BF_4IrC_{42}H_{36}$, Iridium(III), (1,2-diiodobenzene)dihydridobis(triphenylphosphine)-, tetrafluoroborate(1−), 26:125

$I_2C_6H_4$, Benzene, 1,2-diiodo-, iridium complex, 26:125

$I_2CoOC_{11}H_{15}$, Cobalt(III), carbonyldiiodo(η^5-pentamethylcyclopentadienyl)-, 28:275

$I_2N_2PtC_6H_{14}$, Platinum(II), [trans-(R,R)-1,2-cyclohexanediamine]-diiodo-, 27:284

$I_2Si_2UC_{22}H_{42}$, Uranium(IV), bis(η^5-1,3-bis(trimethylsilyl)cyclopentadienyl]-diiodo-, 27:176

I_2Yb, Ytterbium diiodide, 27:147

$I_3F_{18}S_{14}Sb_3 \cdot 2AsF_3$, cyclo-Heptasulfur(3+), μ-iodo-bis(4-iodo-, tris[hexafluoroantimonate(1−)] −2(arsenic trifluoride), 27:335

$I_4As_6F_{36}S_{32}$, cyclo-Heptasulfur(1+), iodo-, tetrasulfur(2+) hexafluoroarsenate(1−)-(4:1:6), 27:337

$I_4Co_2C_{20}H_{30}$, Cobalt(III), di-μ-iodobis[iodo(η^5-pentamethylcyclopentadienyl)-, 28:276

I_6K_2Re, Rhenate(IV), hexaiodo-, dipotassium, 27:294

IrBClF$_4$N$_2$P$_2$C$_{36}$H$_{31}$, Iridium(III), chloro-(dinitrogen)hydrido[tetrafluoroborato(1−)]bis(triphenylphosphine)-, 26:119

IrBClF$_4$OP$_2$C$_{37}$H$_{31}$, Iridium(III), carbonylchlorohydrido[tetrafluoroborato-(1−)]bis(triphenylphosphine)-, 26:117, 28:23

IrBClF$_4$OP$_2$C$_{38}$H$_{33}$, Iridium(III), carbonylchloromethyl [tetrafluoroborato-(1−)]bis(triphenylphosphine)-, 26:118

IrBF$_4$I$_2$C$_{42}$H$_{36}$, Iridium(1+), (1,2-diiodobenzene)dihydridobis(triphenylphosphine)-, tetrafluoroborate(1−), 26:125, 28:59

IrBF$_4$O$_2$P$_2$C$_{38}$H$_{36}$, Iridium(1+), diaquadihydridobis(triphenylphosphine)-, tetrafluoroborate(1−), 26:124, 28:58

IrBF$_4$O$_2$P$_2$C$_{42}$H$_{44}$, Iridium(1+), bis(acetone)dihydridobis(triphenylphosphine)-, tetrafluoroborate(1−), 26:123, 28:57

IrBF$_4$P$_2$C$_{35}$H$_{38}$, Iridium(1+), (η^4-1,5-cyclooctadiene) [1,3-propanediylbis(diphenylphosphine)]-, tetrafluoroborate(1−), 27:23

IrBF$_4$P$_2$C$_{44}$H$_{42}$, Iridium(1+), (η^4-1,5-cyclooctadiene)bis(triphenylphosphine)-, tetrafluoroborate(1−), 26:122, 28:56

IrClOP$_2$C$_{37}$H$_{30}$, Iridium, carbonylchlorobis(triphenylphosphine)-, *trans*-, 28:92

IrClP$_3$C$_{54}$H$_{45}$, Iridium(I), chlorotris(triphenylphosphine)-, 26:201

—, Iridium(III), chloro[2-diphenylphosphino)phenyl-C^1, P]hydridobis(triphenylphosphine)-, (*OC*-6-53)-, 26:202

IrF$_6$O$_7$P$_2$S$_2$C$_{39}$H$_{31}$, Iridium(III), carbonylhydridobis(trifluoromethanesulfonato)bis(triphenylphosphine)-, 26:120, 28:26

IrH$_4$C$_{10}$H$_{15}$, Iridium, tetrahydrido(η^5-pentamethylcyclopentadienyl)-, 27:19

IrNO$_4$P$_2$C$_{40}$H$_{30}$, Iridate(1−), tetracarbonyl-, μ-nitrido-bis(triphenylphosphorus)(1+), 28:214

Ir$_2$BF$_4$H$_5$C$_{56}$H$_{56}$, Iridium(1+), [1,4-butanedibis(diphenylphosphine)]di-, tetrafluoroborate(1−), 27:26

Ir$_2$BF$_4$H$_5$P$_4$C$_{54}$H$_{52}$, Iridium(1+), pentahydridobis [1,3-propanediylbis(diphenylphosphine)]di-, tetrafluoroborate (1−), 27:22

Ir$_2$Cl$_2$C$_{32}$H$_{56}$, Iridium, di-μ-chlorobis[bis(cyclooctene)-, 28:91

Ir$_3$B$_2$F$_8$H$_7$P$_6$C$_{78}$H$_{72}$, Iridium(2+), tris[1,2-ethanediylbis(diphenylphosphine)]heptahydridotri-, bis[tetrafluoroborate(1−)], 27:25

Ir$_3$B$_2$F$_8$H$_7$P$_6$C$_{81}$H$_{78}$, Iridium(2+), heptahydridotris[1,3-propanediylbis(diphenylphosphine)]tri-, bis[tetrafluoroborate(1−)], 27:22

Ir$_4$O$_{12}$C$_{12}$, Iridium, dodecacarbonyltetra-, 28:245

KB$_9$H$_{14}$, Borate(1−), tetradecahydronona-, potassium, 26:1

KCl$_3$PtC$_2$H$_4$, Platinate(II), trichloro(ethene)-, potassium, 28:349

K$_2$F$_3$MnO$_4$S, Manganate(III), trifluorosulfato-, dipotassium, 27:312

K$_2$I$_6$Re, Rhenate(IV), hexaiodo-, dipotassium, 27:294

K$_4$O$_{40}$PVW$_{11}$·xH$_2$O, Vanadoundecatungstophosphate(4−), α-, tetrapotassium, hydrate, 27:99

K$_4$O$_{40}$SiW$_{12}$·9H$_2$O, Dodecatungstosilicate(4−), β-, tetrapotassium, nonahydrate, 27:94

K$_4$O$_{40}$SiW$_{12}$·17H$_2$O, Dodecatungstosilicate(4−), α-, tetrapotassium, heptadecahydrate, 27:93

K$_6$HO$_{40}$SiV$_3$W$_9$·3H$_2$O, 1,2,3-Trivanadononatungstosilicate(7−), A-β-, hexapotassium hydrogen, trihydrate, 27:129

K$_6$O$_{62}$P$_2$W$_{18}$·14H$_2$O, Octadecatungstodiphosphate(6−), α-, hexapotassium, tetradecahydrate, 27:105

K$_6$O$_{62}$P$_2$W$_{18}$·19H$_2$O, Octadecatungstodiphosphate(6−), β-, hexapotassium, nonadecahydrate, 27:105

K$_8$O$_{36}$SiW$_{10}$·12H$_2$O, Decatungstosilicate(8−), γ-, octapotassium, dodecahydrate, 27:88

K$_8$O$_{39}$SiW$_{11}$·13H$_2$O, Undecatungstosilicate(8−), α-, octapotassium, tridecahydrate, 27:89

K$_8$O$_{39}$SiW$_{11}$·14H$_2$O, Undecatungstosilicate(8−), β_2-, octapotassium, tetradecahydrate, 27:91

$K_9LiO_{61}P_2W_{17}\cdot 20H_2O$, Lithioheptadecatungstodiphosphate(9−), α_1-, nonapotassium, eicosahydrate, 27:109

$K_{10}O_{61}P_2W_{17}\cdot 20H_2O$, Heptadecatungstodiphosphate(10−), α_2-, decapotassium, eicosahydrate, 27:107

$K_{28}H_7Li_5O_{184}P_8W_{48}\cdot 92H_2O$, Octatetracontatungstooctaphosphate(40−), pentalithium octacosapotassium heptahydrogen, dononacontahydrate, 27:110

$LaCl_2LiO_2Si_4C_{30}H_{58}$, Lanthanum, bis[$\eta^5$-1,3-bis(trimethylsilyl)cyclopentadienyl]-di-μ-chloro-bis(tetrahydrofuran)-lithium-, 27:170

$LaO_3C_{43}H_{63}$, Lanthanum, tris(2,6-di-*tert*-butylphenoxo)-, 27:167

$LaO_3C_{45}H_{69}$, Lanthanum, tris(2,6-di-*tert*-butyl-4-methylphenoxo)-, 27:166

$La_2Cl_2Si_4C_{44}H_{84}$, Lanthanum, tetrakis[η^5-1,3-bis(trimethylsilyl)cyclopentadienyl]-di-μ-chloro-di-, 27:171

LiCl, Lithium chloride, 28:322

$LiCl_2CeO_2Si_4C_{30}H_{58}$, Cerium, bis[$\eta^5$-1,3-bis(trimethylsilyl)cyclopentadienyl]di-μ-chloro-bis(tetrahydrofuran) lithium-, 27:170

$LiCl_2LaO_2Si_4C_{30}H_{58}$, Lanthanum, bis[$\eta^5$-1,3-bis(trimethylsilyl)cyclopentadienyl]-di-μ-chloro-bis(tetrahydrofuran)-lithium-, 27:170

$LiCl_2NdO_2Si_4C_{30}H_{58}$, Neodymium, bis[$\eta^5$-1,3-bis(trimethylsilyl)cyclopentadienyl]-di-μ-chloro-bis(tetrahydrofuran)-lithium-, 27:170

$LiCl_2O_2PrSi_4C_{30}H_{58}$, Praseodymium, bis-[$\eta^5$-1,3-bis(trimethylsilyl)-cyclopentadienyl]di-μ-chloro-bis(tetrahydrofuran)lithium-, 27:170

$LiCl_2O_2ScSi_4C_{30}H_{58}$, Scandium, bis[$\eta^5$-1,3-bis(trimethylsilyl) cyclopentadienyl]di-μ-chloro-bis(tetrahydrofuran)-lithium-, 27:170

$LiCl_2O_2Si_4YC_{30}H_{58}$, Yttrium, bis[$\eta^5$-1,3-bis(trimethylsilyl)cyclopentadienyl]di-μ-chloro-bis(tetrahydrofuran)-lithium-, 27:170

$LiCl_2O_2Si_4YbC_{30}H_{58}$, Ytterbium, bis[$\eta^5$-1,3-bis(trimethylsilyl)cyclopentadienyl]di-μ-chloro-bis(tetrahydrofuran)-lithium-, 27:170

$LiK_9O_{61}P_2W_{17}\cdot 20H_2O$, Lithioheptadecatungstodiphosphate(9−), α_1-, nonapotassium, eicosahydrate, 27:109

$LiNC_9H_{12}$, Lithium, [2-[(dimethylamino)methyl]phenyl]-, 26:152

—, Lithium, [2-[(dimethylamino)phenyl]methyl]-, 26:153

$LiNC_{10}H_{14}$, Lithium, [2-[(dimethylamino)methyl]-5-methylphenyl]-, 26:152

$LiNOC_{16}H_{22}$, Lithium, (diethyl ether) [8-(dimethylamino)-1-naphthyl]-, 26:154

$LiPC_9H_{12}$, Lithium, [2-(methylphenylphosphino)ethyl]-, 27:178

$LiPSi_2C_6H_{18}\cdot 20C_4H_8$, Phosphide, bis-(trimethylsilyl)-, lithium, −2tetrahydrofuran, 27:243, 248

$LiSi_2C_{11}H_{21}$, Lithium, [η^5-1,3-bis(trimethylsilyl) cyclopentadienyl]-, 27:170

$Li_2C_8H_8$, Lithium, (1,3,5,7-cyclooctatetraene)di-, 28:127

$Li_2N_4Si_2C_{26}H_{56}$, Lithium, μ-[(α,α', 1,2-η:α,α', 1,2-η)-1,2-phenylenebis[(trimethylsilyl)methylene]] bis(N,N,N',N'-tetramethyl-1,2-ethanediamine)di-, 26:148

Li_2P_{16}, Lithium hexadecaphosphide, (Li_2P_{16}), 27:227

Li_3P_7, Lithium heptaphosphide, (Li_3P_7), 27:227

$Li_5H_7K_{28}O_{184}P_8W_{48}\cdot 92H_2O$, Octatetracontatungstooctaphosphate (40−), pentalithium octacosapotassium heptahydrogen, dononacontahydrate, 27:110

$LuClOC_{12}H_{16}$, Lutetium, chloro(η^8-1,3,5,7-cyclooctatetraene) (tetrahydrofuran)-, 27:152

$LuNOC_{21}H_{28}$, Lutetium, [(2-[(dimethylamino)methyl]phenyl-C^1, N](η^8-1,3,5,7-cyclooctatetraene) (tetrahydrofuran)-, 27:153

$LuOC_{21}H_{25}$, Lutetium, bis(η^5-cyclopentadienyl) (tetrahydrofuran)-*p*-tolyl-, 27:162

LuOSiC$_{18}$H$_{29}$, Lutetium, bis(η^5-cyclopentadienyl) (tetrahydrofuran) [(trimethylsilyl)methyl]-, 27:161

Lu$_2$Cl$_2$Si$_4$C$_{44}$H$_{84}$, Lutetium, tetrakis[η^5-1,3-bis(trimethylsilyl)cyclopentadienyl]-di-μ-chloro-di-, 27:171

MgClC$_5$H$_{11}$, Magnesium, chloro(2,2-dimethylpropyl)-, 26:46

Mg$_3$O$_4$C$_{48}$H$_{72}$, Magnesium, cyclotri[μ-1,2-phenylenebis(methylene)]hexakis (tetrahydrofuran)tri-, 26:147

MnBrO$_5$C$_5$, Manganese, bromopentacarbonyl-, 28:156

MnClO$_5$C$_5$, Manganese, pentacarbonylchloro-, 28:155

MnF$_3$K$_2$O$_4$S, Manganate(III), trifluorosulfato-, dipotassium, 27:312

MnF$_3$O$_8$SC$_6$, Manganese(I), pentacarbonyl(trifluoromethanesulfonato)-, 26:114

MnIO$_5$C$_5$, Manganese, pentacarbonyliodo-, 28:157, 158

MnN$_2$O$_4$C$_{16}$H$_9$, Manganese, tetracarbonyl[2-(phenylazo)phenyl-C^1,N^2]-, 26:173

MnO$_4$PC$_{22}$H$_{14}$, Manganese, octacarbonyl-1κ^4C, 2κ^4C-μ-[carbonyl-2κC:1κO-6-(diphenylphosphino-2κP)-o-phenylene-2κC^1:1κC^2]di-, 26:158

MnO$_5$C$_6$H$_3$, Manganese, pentacarbonylmethyl-, 26:156

MnO$_5$C$_{12}$H$_7$, Manganese, (2-acetylphenyl-C,O)tetracarbonyl-, 26:156

—, Manganese, benzylpentacarbonyl-, 26:172

MnO$_5$PC$_{23}$H$_{16}$, Manganese, tetracarbonyl{[2-(diphenylphosphino)phenyl]-hydroxymethyl-C,P}-, 26:169

MnO$_6$C$_7$H$_3$, Manganese, acetylpentacarbonyl-, 29:199

MnO$_8$PSC$_{12}$H$_{12}$, Manganese, tetracarbonyl[2-(dimethylphosphinothioyl)-1,2-bis(methoxycarbonyl)-ethenyl-C,S]-, 26:163

MnO$_{11}$PC$_{17}$H$_{18}$, Manganese, tricarbonyl-[η^2-2,3,4,5-tetrakis(methoxycarbonyl)-2,2-dimethyl-1H-phospholium]-, 26:167

MnO$_{11}$PSC$_{17}$H$_{18}$, Manganese, tricarbonyl-[η^2-3,4,5,6-tetrakis(methoxycarbonyl)-2,2-dimethyl-2H-1,2-thiaphosphorin-2-ium]-, 26:165

Mn$_2$AuO$_8$P$_2$C$_{38}$H$_{25}$, Gold, octacarbonyl-1κ^4C,2κ^4C-μ-(diphenylphosphino)-1:2κP-(triphenylphosphine)-3κP-triangulo-dimanganese-, 26:229

Mn$_2$ClHgO$_8$PC$_{20}$H$_{10}$, Manganese, μ-(chloromercurio)-μ-(diphenylphosphino)bis(tetracarbonyl-, (Mn–Mn), 26:230

Mn$_2$NO$_8$P$_3$C$_{56}$H$_{40}$, Manganate(1−), μ-(diphenylphosphino)bis(tetracarbonyl-, (Mn–Mn), μ-nitrido-bis(triphenylphosphorus)(1+), 26:228

Mn$_2$N$_2$O$_8$C$_{20}$H$_8$, Manganese, μ-(azodi-2,1-phenylene-C^1,N^2:C^1,N^1)octacarbonyldi-, 26:173

Mn$_2$O$_8$PC$_{20}$H$_{11}$, Manganese, μ-(diphenylphosphino)-μ-hydrido-bis(tetracarbonyl-, (Mn–Mn), 26:226

Mn$_2$O$_8$P$_2$S$_2$C$_{12}$H$_{12}$, Manganese, octacarbonyl-bis(μ-dimethylphosphinothioyl-P,S)di-, 26:162

MoBF$_4$OC$_{34}$H$_{25}$, Molybdenum(1+), carbonyl(η^5-cyclopentadienyl)-bis(diphenylacetylene)-, tetrafluoroborate(1−), 26:102, 28:11

MoBF$_4$OPC$_{38}$H$_{30}$, Molybdenum(1+), carbonyl(η^5-cyclopentadienyl)(diphenylacetylene)(triphenylphosphine)-, tetrafluoroborate(1−), 26:104

MoBF$_4$O$_2$PC$_{25}$H$_{20}$, Molybdenum, dicarbonyl(η^5-cyclopentadienyl) [tetrafluoroborato(1−)](triphenylphosphine)-, 26:98

MoBF$_4$O$_3$C$_8$H$_5$, Molybdenum, tricarbonyl (η^5-cyclopentadienyl) [tetrafluoroborato(1−)]-, 26:96, 28:5

MoBF$_4$O$_3$C$_{10}$H$_9$, Molybdenum(1+), tricarbonyl(η^5-cyclopentadienyl)(η^2-ethene)-, tetrafluoroborate(1−), 26:102

MoBF$_4$O$_4$C$_{11}$H$_{11}$, Molybdenum(1+), (acetone)tricarbonyl(η^5-cyclopentadienyl)-, tetrafluoroborate(1−), 26:105

MoB$_2$F$_8$N$_6$O$_2$C$_8$H$_{12}$, Molybdenum(II), tetrakis(acetonitrile)dinitrosyl-, cis-, bis[tetrafluoroborate(1−)], 26:132, 28:65

MoBr$_2$O$_4$C$_4$, Molybdenum(II), dibromotetracarbonyl-, 28:145

MoCl$_3$N$_3$C$_6$H$_9$, Molybdenum(III), tris-

(acetonitrile)trichloro-, 28:37
$MoCl_3O_3C_{12}H_{24}$, Molybdenum(III), trichlorotris(tetrahydrofuran)-, 28:36
$MoCl_4N_2C_4H_6$, Molybdenum, bis(acetonitrile)-tetrachloro-, 28:34
$MoCl_4O_2C_8H_{16}$, Molybdenum, tetrachlorobis(tetrahydrofuran)-, 28:35
$MoCl_5H_8N_2O$, Molybdate(V), pentachlorooxo-, diammonium, 26:36
$MoCoNiO_5C_{17}H_{13}$, Nickel, cyclo-μ_3-ethylidyne-1:2:3-κ^3C-pentacarbonyl-1κ^2C,2κ^3C-bis[1,3(η^5)-cyclopentadienyl]cobaltmolybdenum-, (Co–Mo) (Co–Ni) (Mo–Ni), 27:192
$MoCoO_8RuC_{17}H_{11}$, Molybdenum, cyclo-[μ_3-1(η^2):-2(η^2):3(η^2)-2-butyne]octacarbonyl-1κ^2C, 2κ^3C, 3κ^3C-[1(η^5)cyclopentadienyl]cobaltruthenium-, (Co–Mo)(Co–Ru) (Mo–Ru), 27:194
$MoCo_2O_8C_{15}H_8$, Molybdenum, octacarbonyl(η^5-cyclopentadienyl)-μ_3-ethylidynedicobalt-, 27:193
$MoF_6FeO_5P_2C_{34}H_{28}$, Iron(1+), μ-acetyl-2κC^1: 1κO-tetracarbonyl-1κ^3C, 2κ^2C-bis[1,2-(η^5-cyclopentadienyl)] (triphenylphosphine-1κP)molybdenum-, hexafluorophosphate(1–), 26:241
$MoF_6FeO_6PC_{17}H_{13}$, Iron(1+), μ-acetyl-2κC^1:-1κO-pentacarbonyl-1κ^3C, 2κ^2C-bis[1,2-(η^5-cyclopentadienyl)]molybdenum-, hexafluorophosphate(1–), 26:239
$MoF_7O_2P_5C_{54}H_{48}$, Molybdenum(II), dicarbonylbis[1,2-ethanediylbis(diphenylphosphine)]fluoro-, hexafluorophosphate(1–), 26:84
$MoFeNO_9P_2C_{45}H_{31}$, Molybdate(1–), hydridononacarbonyliron-, μ-nitridobis(triphenylphosphorus)(1+), 26:338
$MoFeN_2O_9P_4C_{81}H_{60}$, Molybdate(2–), nonacarbonyliron-, bis[μ-nitridobis(triphenylphosphorus) (1+)], 26:339
$MoH_4P_4C_{52}H_{52}$, Molybdenum(IV), tetrahydridotetrakis(methyldiphenylphosphine)-, 27:9
$MoH_6P_3C_{54}H_{99}$, Molybdenum(IV), hexahydridotris(tricyclohexylphosphine)-, 27:13
$MoHgO_{12}Ru_3C_{23}H_{14}$, Ruthenium, nonacarbonyl-(μ_3-3,3-dimethyl-1-butynyl) {μ-[tricarbonyl (η^5-cyclopentadienyl)molybdenum]-mercury}-triangulo-tri-, 26:333
$MoNO_3C_7H_5$, Molybdenum, dicarbonyl(η^5-cyclopentadienyl)nitrosyl-, 28:196
$MoNO_5C_{10}H_9$, Molybdenum, (tert-butyl isocyanide)pentacarbonyl-, 28:143
$MoNO_7P_2C_{43}H_{33}$, Molybdate(1–), (acetato)pentacarbonyl-, μ-nitridobis(triphenylphosphorus)(1+), 27:297
$MoN_2N_4C_{14}H_{18}$, Molybdenum, bis(tert-butyl isocyanide) tetracarbonyl-, cis-, 28:143
$MoN_2O_3S_4C_{13}H_{20}$, Molybdenum(II), tricarbonylbis(diethyldithiocarbamato)-, 28:145
$MoN_3O_3C_{12}H_{15}$, Molybdenum, tricarbonyltris(propionitrile)-, 28:31
$MoN_3O_3C_{18}H_{27}$, Molybdenum, tris(tert-butyl isocyanide)tricarbonyl-, fac-, 28:143
$MoN_4O_2S_4C_{10}H_{20}$, Molybdenum, bis(diethyldithiocarbamato)dinitrosyl-, cis-, 28:145
$MoN_4P_4C_{48}H_{52}$, Molybdenum, bis(dinitrogen)bis[1,2-ethanediylbis(diphenylphosphine)]-, trans-, 28:38
$MoNaO_3C_8H_5 \cdot 2C_4H_{10}O_2$, Molybdate(1–), tricarbonyl(η^5-cyclopentadienyl)-, sodium, compd. with 1,2-dimethoxyethane (1:2), 26:343
$MoOCl_4$, Molybdenum tetrachloride oxide, 28:325
$MoO_2PC_{25}H_{21}$, Molybdenum, dicarbonyl(η^5-cyclopentadienyl)hydrido-(triphenylphosphine)-, 26:98
$MoO_2P_3C_7H_5$, Molybdenum(I), dicarbonyl-(η^5-cyclopentadienyl) (η^3-cyclotriphosphorus)-, 27:224
$MoO_3C_{10}H_8$, Molybdenum(0), tricarbonyl (cycloheptatriene)-, 28:45
$MoO_3P_2C_{39}H_{68}$, Molybdenum, tricarbonyl(dihydrogen)bis(tricyclohexylphosphine)-, 27:3
$MoP_2S_4C_{48}H_{40}$, Molybdate(VI), tetrathio-, bis(tetraphenylphosphonium), 27:41
$MoS_4C_{10}H_{10}$, Molybdenum(IV), bis(η^5-cyclopentadienyl) [tetrasulfido(2–)]-, 27:63
$Mo_2N_2O_6PtC_{30}H_{20}$, Molybdenum,

$Mo_2N_2O_6PtC_{30}H_{20}$ (Continued)
[bis(benzonitrile)platinum]-
hexacarbonylbis(η^5-cyclopentadienyl)di-,
(2Mo-Pt), 26:345

$Mo_2N_2O_7C_{32}H_{72}$, Dimolybdate(VI),
bis(tetrabutylammonium), 27:79

$Mo_2N_2S_{12}H_8 \cdot 2H_2O$, Molybdate(V), bis($\mu$-sulfido)tetrakis (disulfido)di-, diammonium, dihydrate, 27:48, 49

$Mo_2O_4C_{14}H_{10}$, Molybdenum,
tetracarbonylbis(η^5-cyclopentadienyl)di-,
(Mo-Mo), 28:152

$Mo_2O_4P_2C_{14}H_{10}$, Molybdenum(I), tetra-carbonylbis(η^5-cyclopentadienyl) (μ-η^2:η^2-diphosphorus)di-, 27:224

$Mo_2O_6C_{16}H_{10}$, Molybdenum,
hexacarbonylbis(η^5-cyclopentadienyl)di-,
28:148

$Mo_2O_6P_2Pd_2C_{52}H_{40}$, Molybdenum,
hexacarbonylbis(η^5-cyclopentadienyl)bis(triphenylphosphine)dipalladiumdi-, 26:348

$Mo_2O_6P_2Pt_2C_{52}H_{40}$, Molybdenum,
hexacarbonylbis(η^5-cyclopentadienyl)bis(triphenylphosphine)diplatinumdi-, 26:347

$Mo_2P_2S_6C_{48}H_{40}$, Molybdate(V), di-μ-thio-tetrathiodi-, bis(tetraphenylphosphonium), 27:43

$Mo_2P_2S_7C_{48}H_{40}$, Molybdate(IV,VI), (η^2-disulfido)di-μ-thio-trithiodi-, bis(tetraphenylphosphonium), 27:44

$Mo_2P_2S_8C_{48}H_{40}$, Molybdate(V), bis(η^2-disulfido)di-μ-thio-dithiodi-, bis(tetraphenylphosphonium), 27:45

$Mo_2P_2S_{10.56}C_{48}H_{40}$, Molybdate(2−), thio-, $(Mo_2S_{10.56})^2$ bis(tetraphenylphosphonium), 27:42

$Mo_3N_2S_{13}H_8 \cdot XH_2O$, Molybdate(IV),
tris(μ-disulfido)tris(disulfido)-μ_3-thio-*triangulo*-tri-, diammonium, hydrate,
27:48, 49

$Mo_5H_{18}N_4O_{21}P_2 \cdot 4H_2O$,
Pentamolybdobis(phosphonate) (4−),
tetraammonium, tetrahydrate, 27:123

$Mo_5N_3O_{21}P_2C_8H_{26} \cdot 5H_2O$,
Pentamolybdobis[(2-aminoethyl)phosphonate] (4−), sodium tetramethylammonium dihydrogen, pentahydrate,
27:126

$Mo_5N_4O_{21}P_2C_2H_{22} \cdot 2H_2O$,
Pentamolybdobis(methylphosphonate)-
(4−), tetraammonium, dihydrate, 27:124

$Mo_5N_4O_{21}P_2C_4H_{26}$, Pentamolybdobis-(ethylphosphonate)(4−), tetraammonium, 27:125

$Mo_5N_4O_{21}P_2C_{12}H_{26} \cdot 5H_2O$,
Pentamolybdobis(phenylphosphate)-
(4−), tetraammonium, pentahydrate,
27:125

$Mo_5N_4O_{21}P_2C_{14}H_{26} \cdot 5H_2O$,
Pentamolybdobis[(4-aminobenzyl)-
phosphonate](4−), diammonium
dihydrogen, pentahydrate, 27:126

$Mo_5N_4O_{21}P_2C_{22}H_{42} \cdot 4H_2O$,
Pentamolybdobis[(4-aminobenzyl)-
phosphonate](4−), bis(tetramethylammonium)dihydrogen, tetrahydrate,
27:127

$Mo_6N_2O_{19}C_{32}H_{72}$, Hexamolybdate(VI),
bis(tetrabutylammonium), 27:77

$Mo_8N_4O_{26}C_{64}H_{144}$, Octamolybdate(VI),
tetrakis(butylammonium), 27:78

N, Nitride, ruthenium, cluster complexes,
26:287, 288

$NAu_2F_5PSC_{25}H_{15}$, Gold(I) (pentafluorophenyl)-μ-thiocyanato(triphenylphosphine)di-, 26:90

$NBF_4O_5WC_{10}H_{10}$, Tungsten(1+)
pentacarbonyl[(diethylamino)methylidyne]-, tetrafluoroborate(1−), 26:40

$NBrO_4SC_{10}H_{18}$, Bicyclo[2.2.1]heptane-7-methanesulfonate, 3-bromo-1,7-dimethyl-2-oxo-, [(1R)-($ENDO$,
$ANTI$)]-, ammonium, 26:24

NC_2H_3, Acetonitrile, cobalt, copper, and
ruthenium complexes, 26:356, 359
molybdenum complex, 28:34
molybdenum, palladium, and tungsten
complexes, 26:128–133
osmium complex, 26:290, 292
ruthenium(II) complexes, 26: 69–72
transition metal complexes, 28:63–66

NC_2H_3, Methane, isocyano-, tungsten complex, 28:43

NC_3H_5, Propanenitrile, chromium, molybdenum, and tungsten complexes,
28:30–32

tungsten complex, 27:4
NC$_4$H$_{11}$, Ethanamine, 1,1-dimethyl-, tungsten complex, 27:301
NC$_5$H$_5$, Pyridine, osmium complex, 26:291, 28:234
vanadium complex, 27:308
NC$_5$H$_9$, *tert*-Butyl isocyanide, chromium, molybdenum, and tungsten complexes, 28:143
nickel complex, 28:99
palladium complex, 28:110
ruthenium complex, 26:275, 28:224
NC$_5$H$_{13}$, Ethanamine, N-ethyl-methyl-, tungsten complex, 26:40, 42
NC$_6$H$_7$, Benzenamine, tungsten complex, 27:301
NC$_7$H$_5$, Benzonitrile, palladium complex, 28:61, 62
platinum complex, 26:345
ruthenium complexes, 26:70–72
NC$_7$H$_9$, Pyridine, 3,5-dimethyl-, palladium complex, 26:210
NC$_7$H$_{11}$, Cyclohexyl isocyanide, nickel complex, 28:101
NC$_8$H$_{11}$, Benzenamine, N,N-dimethyl-, chromium complex, 28:139
NC$_9$H$_9$, Benzene, 2-isocyano-1,3-dimethyl-, iron complexes, 26:53–57
NC$_9$H$_{13}$, Benzenemethanamine, N,N-dimethyl-, lithium complex, 26:152
lutetium complex, 27:153
palladium, 26:212
NC$_{10}$H$_9$, Quinoline, 8-methyl-, palladium complex, 26:213
NC$_{10}$H$_{15}$, Benzenemethanamine, N,N,2-trimethyl, lithium complex, 26:153
NC$_{12}$H$_{11}$, Pyridine, 2-(phenylmethyl)-palladium complex, 26:208–210
NC$_{12}$H$_{13}$, Naphthalenamine, N,N-dimethyl-, lithium complex, 26:154
NC$_{13}$H$_9$, Benzo[h]quinoline, ruthenium complex, 26:177
NCo$_3$CuO$_{12}$C$_{14}$H$_3$, Ruthenium, (acetonitrile)dodecacarbonyltricobaltcopper-, 26:359
NCo$_3$FeO$_{12}$C$_{20}$H$_{20}$, Ferrate(1−), dodecacarbonyltricobalt-, tetraethylammonium, 27:188
NCo$_3$O$_{12}$RuC$_{12}$H$_{20}$, Ruthenate(1−), dodecacarbonyltricobalt-, tetraethylammonium, 26:358
NCrFO$_3$C$_5$H$_6$, Chromate(VI), fluorotrioxo-, pyridinium, 27:310
NCrFeO$_9$P$_2$C$_{45}$H$_{31}$, Chromate(1−), hydridononacarbonyliron-, μ-nitrido-bis(triphenylphosphorus)(1+), 26:338
NCrO$_3$C$_7$H$_5$, Chromium, dicarbonyl(η^5-cyclopentadienyl)nitrosyl-, 28:196
NCrO$_3$C$_{11}$H$_{11}$, Chromium, tricarbonyl(η^6-N,N-dimethylbenzenamine)-, 28:139
NCrO$_5$C$_{10}$H$_9$, Chromium, (*tert*-butyl isocyanide)pentacarbonyl-, 28:143
NCrO$_5$C$_{18}$H$_{13}$, Chromium, (benzoyl isocyanide)dicarbonyl(η^6-methyl benzoate)-, 26:32
NCrO$_7$P$_2$C$_{43}$H$_{33}$, Chromate(1−), (acetato)pentacarbonyl-, μ-nitrido-bis(triphenylphosphorus)(1+), 27:297
NFO$_3$SC$_{16}$H$_{36}$, Fluorosulfate, tetrabutylammonium, 26:393
NFeMoNO$_9$P$_2$C$_{45}$H$_{31}$, Molybdate(1−), hydridononacarbonyliron-μ-nitrido-bis(triphenylphosphorus)(1+), 26:338
NFeO$_4$C$_{13}$H$_9$, Iron(0), tetracarbonyl(2-isocyano-1,3-dimethylbenzene)-, 28:180
NFeO$_4$P$_2$C$_{40}$H$_{31}$, Ferrate(1−), hydridotetracarbonyl-, μ-nitrido-bis(triphenylphosphorus)(1+), 26:336
NFeO$_9$WC$_{45}$H$_{31}$, Tungstate(1−), hydridononacarbonyliron-, μ-nitrido-bis(triphenylphosphorus)(1+), 26:336
NFe$_4$HO$_{12}$C$_{21}$H$_{20}$, Ferrate(1−), carbidododecacarbonylhydridotetra-, tetraethylammonium, 27:186
NFe$_4$O$_{12}$P$_2$C$_{49}$H$_{30}$, Ferrate(2−), μ$_4$-carbidododecacarbonyltetra-, bis[μ-nitrido-bis(triphenylphosphorus)(1+)], 26:246
NFe$_4$O$_{14}$C$_{23}$H$_{23}$, Ferrate(1−), dodecacarbonyl[μ^4-(methoxycarbonyl)-methylidyne]tetra-, tetraethylammonium, 27:184
NH$_3$, Ammines, ruthenium(III), 26:66, 67
NIrO$_4$P$_2$C$_{40}$H$_{30}$, Iridate(1−), tetracarbonyl-, μ-nitrido-bis(triphenylphosphorus)(1+), 28:214
NLiC$_9$H$_{12}$, Lithium, [2-[(dimethylamino)methyl]phenyl]-, 26:152
—, Lithium, [2-[(dimethylamino)phenyl]methyl]-, 26:153

NLiC$_{10}$H$_{14}$, Lithium, [2-[(dimethylamino)methyl]-5-methylphenyl]-, 26:152
NLiOC$_{16}$H$_{22}$, Lithium, (diethyl ether) [8-(dimethylamino)-1-naphthyl]-, 26:154
NLuOC$_{21}$H$_{28}$, Lutetium, [(2-[(dimethylamino)methyl]phenyl-C^1,N] (η^8-1,3,5,7-cyclooctatetraene) (tetrahydrofuran)-, 27:153
NMn$_2$O$_8$P$_3$C$_{56}$H$_{40}$, Manganate(1−), μ-(diphenylphosphino)-bis(tetracarbonyl)-, (Mn–Mn), μ-nitrido-bis(triphenylphosphorus)(1+), 26:228
NMoO$_3$C$_7$H$_5$, Molybdenum, dicarbonyl(η^5-cyclopentadienyl)nitrosyl-, 28:196
NMoO$_5$C$_{10}$H$_9$, Molybdenum, (*tert*-butyl isocyanide)pentacarbonyl-, 28:143
NMoO$_7$P$_2$C$_{43}$H$_{33}$, Molybdate(1−), (acetato)pentacarbonyl-, μ-nitrido-bis(triphenylphosphorus)(1+), 27:297
NNiOPC$_{22}$H$_{40}$, Nickel(II), [butanamidato(2−)-C^4,N] (tricyclohexylphosphine)-, 26:206
—, Nickel(II), [2-methylpropanamidato(2−)-C^3,N] (tricyclohexylphosphine)-, 26:205
NO, Nitrosyls, chromium, molybdenum, and tungsten, 28:196, 197
molybdenum and tungsten, 26:132, 133, 28:65, 66
NOC, Cyanate, tungsten complex, 26:42
NOC$_4$H$_7$, 3-Butenamide, nickel complex, 26:206
NOC$_4$H$_9$, 2-Propenamide, 2-methyl-, nickel complex, 26:205
NOC$_8$H$_5$, Benzoyl isocyanide, chromium complex, 26:32, 34, 35
NOPC$_{25}$H$_{20}$, Benzamide, 2-(diphenylphosphino)-N-phenyl-, 27:324
NOPC$_{25}$H$_{22}$, Benzamide, N-[2-(diphenylphosphino)phenyl]-, 27:323
NO$_2$C$_6$H$_5$, 4-Pyridinecarboxylic acid, rhodium complex, 27:292
NO$_2$P$_2$C$_{38}$H$_{33}$, Phosphorus(1+), μ-nitrido-bis(triphenyl)-, acetate, 27:296
NO$_3$P$_2$RhC$_{25}$H$_{46}$, Rhodium(I), carbonyl(4-pyridinecarboxylato)bis(triisopropylphosphine)-, 27:292
NO$_3$SC$_7$H$_5$, 1,2-Benzisothiazol-3(2H)-one, 1,1-dioxide, chromium and vanadium complex, 27:307, 309
NO$_3$WC$_7$H$_5$, Tungsten, dicarbonyl(η^5-cyclopentadienyl)nitrosyl-, 28:196
NO$_4$P$_2$RhC$_{40}$H$_{30}$, Rhodate(1−), tetracarbonyl-, μ-nitrido-bis(triphenylphosphorus)(1+), 28:213
NO$_4$ReC$_{16}$H$_{36}$, Perrenate, tetrabutylammonium, 26:391
NO$_5$WC$_{10}$H$_9$, Tungsten, (*tert*-butyl isocyanide)pentacarbonyl-, 28:143
NO$_7$P$_2$WC$_{43}$H$_{33}$, Tungstate(1−), (acetato)pentacarbonyl-, μ-nitrido-bis(triphenylphosphorus)(1+), 27:297
NO$_{10}$PRu$_3$C$_{23}$H$_{20}$, Ruthenium, decacarbonyl(dimethylphenylphosphine)(2-isocyano-2-methylpropane)tri-, 26:275, 28:224
NO$_{10}$P$_2$Ru$_3$Si$_2$C$_{58}$H$_{61}$, Ruthenate(1−), decacarbonyl-1$\kappa^3$$C$,2$\kappa^3$$C$,3$\kappa^4$$C$-$\mu$-hydrido-1:2-$\kappa^2$$H$-bis(triethylsilyl)-1$\kappa$$Si$,2$\kappa$$Si$-*triangulo*-tri-, μ-nitrido-bis(triphenylphosphorus)(1+), 26:269
NO$_{11}$Os$_3$C$_{13}$H$_3$, Osmium, (acetonitrile)undecacarbonyltri-, 26:290
NO$_{11}$Os$_3$C$_{16}$H$_5$, Osmium, undecacarbonyl(pyridine)tri-, 26:291
NO$_{11}$P$_2$Os$_3$C$_{47}$H$_{31}$, Osmate(1−), μ-carbonyldecacarbonyl-μ-hydrido-tri-, μ-nitrido-bis(triphenylphosphorus)(1+), 28:236
NSC, Thiocyanate, gold complex, 26:90
NS$_2$C$_5$H$_{11}$, Dithiocarbamic acid, diethyl-, molybdenum complex, 28:145
NSiC$_7$H$_{19}$, Ethanamine, 1,1-dimethyl-N-(trimethylsilyl)-, 27:327
NSi$_2$C$_6$H$_{19}$, Silanamine, 1,1,1-trimethyl-N-(trimethylsilyl)-, ytterbium complex, 27:148
N$_2$BClF$_4$IrP$_2$C$_{30}$H$_{31}$, Iridium(III), chloro(dinitrogen)hydrido[tetrafluoroborato(1−)]bis(triphenylphosphine)-, 26:119
N$_2$BClF$_4$IrP$_2$C$_{36}$H$_{31}$, Iridium(III), chloro(dinitrogen)hydrido[tetrafluoroborato(1−)]bis(triphenylphosphine)-, 28:25
N$_2$Br$_2$RuC$_{22}$H$_{22}$, Ruthenium(II), bis(benzonitrile)dibromo(η^4-1,5-cyclooctadiene)-, 26:71
N$_2$C$_2$H$_8$, 1,2-Ethanediamine, chromium complex, resolution of, 26:24, 27, 28
platinum complexes, 27:314, 315

Formula Index 419

$N_2C_6H_{14}$, 1,2-Cyclohexanediamine, cis-, trans-(R,R)-, and trans-(S,S)-, platinum complex, 27:283

$N_2C_6H_{16}$, 1,2-Ethanediamine, N,N,N',N'-tetramethyl-, lithium complex, 26:148

$N_2C_{10}H_8$, 2,2'-Bipyridine, nickel complex, 28:103
palladium complex, 27:319
rhenium complexes, 26:82, 83
ruthenium complex, 28:338
tungsten complex, 27:303

$N_2C_{12}H_8$, 1,10-Phenanthroline, nickel complex, 28:103

$N_2C_{12}H_{10}$, Azobenzene, cobalt and palladium complexes, 26:175, 176
manganese complex, 26:173

$N_2ClPdC_{19}H_{19}$, Palladium(II), chloro[2-(2-pyridinylmethyl)phenyl-C^1,N] (3,5-dimethylpyridine)-, 26:210

$N_2Cl_2P_2C_8H_{18}$, 1,3,2,4-Diazadiphosphetidine, 1,3-di-*tert*-butyl-2,4-dichloro-, cis-, 27:258

$N_2Cl_2P_2WC_{16}H_{32}$, Tungsten(VI), dichloro-[(1,1-dimethylethyl)imido](phenylimido)bis(trimethylphosphine)-, 27:304

$N_2Cl_2PdC_{14}H_{10}$, Palladium, bis(benzonitrile)dichloro-, 28:61

$N_2Cl_2Pd_2C_{18}H_{24}$, Palladium(II), di-μ-chloro-bis[2-[(dimethylamino)methyl]-phenyl-C^1,N]di-, 26:212

$N_2Cl_2Pd_2C_{20}H_{16}$, Palladium(II), di-μ-chloro-bis(8-quinolylmethyl-C,N)di-, 26:213

$N_2Cl_2Pd_2C_{24}H_{20}$, Palladium(II), di-μ-chloro-bis[2-(2-pyridinylmethyl)phenyl-C^1,N]di-, 26:209

$N_2Cl_2PtC_{14}H_{10}$, Platinum, bis(benzonitrile)dichloro-, 26:345

$N_2Cl_2RuC_{12}H_{18}$, Ruthenium(II), bis(acetonitrile)dichloro(η^4-1,5-cyclooctadiene)-, 26:69

$N_2Cl_2RuC_{22}H_{22}$, Ruthenium(II), bis(benzonitrile)dichloro(η^4-1,5-cyclooctadiene)-, 26:70

$N_2Cl_4MoC_4H_6$, Molybdenum, bis(acetonitrile)tetrachloro-, 28:34

$N_2Cl_5H_8MoO$, Molybdate(V), pentachlorooxo-, diammonium, 26:36

$N_2Cl_8Re_2C_{32}H_{72}$, Rhenate(III), octachlorodi-, bis(tetrabutylammonium), 28:332

$N_2CoO_3C_{15}H_9$, Cobalt, tricarbonyl[2-(phenylazo)phenyl-C^1,N^2]-, 26:176

$N_2CrFeO_9P_4C_{81}H_{60}$, Chromate(2−), nonacarbonyliron-, bis[μ-nitrido-bis(triphenylphosphorus)(1+)], 26:339

$N_2CrO_4C_{14}H_{18}$, Chromium, bis(*tert*-butyl isocyanide)tetracarbonyl-, cis-, 28:143

$N_2CrO_{10}S_2C_{14}H_{16}\cdot 2H_2O$, Chromium(II), tetraaquabis(1,2-benzisothiazol-3(2H)-one 1,1-dioxidato)-, dihydrate, 27:309

$N_2FO_3ReC_{13}H_8$, Rhenium, (2,2'-bipyridine)tricarbonylfluoro-, 26:82

$N_2F_2O_5PReC_{13}H_8$, Rhenium, (2,2'-bipyridine)tricarbonyl(phosphorodifluoridate)-, 26:83

$N_2F_{12}O_4PdC_{20}H_{10}$, Palladium(II), (2,2'-bipyridine) (1,1,1,5,5,5-hexafluoro-2,4-pentanedionato)-, 1,1,1,5,5,5-hexafluoro-2,4-dioxo-3-pentanide, 27:319

$N_2FeMoO_9P_4C_{81}H_{60}$, Molybdate(2−), nonacarbonyliron-, bis[μ-nitrido-bis(triphenylphosphorus)(1+)], 26:339

$N_2FeO_3C_{21}H_{18}$, Iron, tricarbonylbis(2-isocyano-1,3-dimethylbenzene)-, 26:54

$N_2FeO_9P_4WC_{81}H_{60}$, Tungstate(2−), nonacarbonyliron-, bis[μ-nitrido-bis(triphenylphosphorus)(1+)], 26:339

$N_2Fe_3O_{11}P_4C_{83}H_{60}$, Ferrate(2−), undecacarbonyltri-, bis[μ-nitrido-bis(triphenylphosphorus)(1+)], 28:203

$N_2Fe_4O_{12}C_{29}H_{40}$, Ferrate(2−), carbidododecacarbonyltetra-, bis(tetraethylammonium), 27:187

$N_2Fe_6O_{16}C_{33}H_{40}$, Ferrate(2−), carbidohexadecacarbonylhexa-, bis(tetraethylammonium), 27:183

N_2H_4, Hydrazine, ruthenium(II), complexes, 26:72

N_2H_6C, Hydriazine, methyl-, ruthenium(II) complexes, 26:72

$N_2I_2PtC_6H_{14}$, Platinum(II), [trans-(R,R)-1,2-cyclohexanediamine]diiodo-, 27:284

$N_2MnO_4C_{16}H_9$, Manganese, tetracarbonyl[2-(phenylazo)phenyl-C^1,N^2]-, 26:173

$N_2Mn_2O_8C_{20}H_8$, Manganese, μ-(azodi-2,1-

$N_2Mn_2O_8C_{20}H_8$ (Continued)
  phenylene-$C^1,N^2:C^{1'},N^1$)octacarbonyl-
  di-, 26:173
$N_2MoO_3S_4C_{13}H_{20}$, Molybdenum(II),
  tricarbonylbis(diethyldithiocarbamato)-,
  28:145
$N_2MoO_4C_{14}H_{18}$, Molybdenum, bis(tert-
  butyl isocyanide)tetracarbonyl-, cis-,
  28:143
$N_2Mo_2O_6PtC_{30}H_{20}$, Molybdenum, [bis-
  (benzonitrile)platinum]hexacarbonyl-
  bis(η^5-cyclopentadienyl)di-, (2Mo–Pt),
  26:345
$N_2Mo_2O_7C_{32}H_{72}$, Dimolybdate(VI),
  bis(tetrabutylammonium), 27:79
$N_2Mo_2S_{12}H_8 \cdot 2H_2O$, Molybdate(V), bis($\mu$-
  sulfido)tetrakis(disulfido)di-, diammo-
  nium, dihydrate, 27:48, 49
$N_2Mo_3S_{13}H_8 \cdot XH_2O$, Molybdate(IV), tris($\mu$-
  disulfido)tris(disulfido)-μ_3-thio-triangulo-
  tri-, diammonium, hydrate, 27:48, 49
$N_2Mo_6O_{19}C_{32}H_{72}$, Hexamolybdate(VI),
  bis(tetrabutylammonium), 27:77
$N_2Ni_6O_{12}C_{20}H_{24}$, Nickelate(2–), hexa-$\mu$-
  carbonyl-hexacarbonylhexa-,
  bis(tetramethylammonium), 26:312
$N_2O_2RuC_{28}H_{16}$, Ruthenium(II), bis(benzo-
  [h]quinolin-10-yl-C^{10},N]dicarbonyl-,
  cis-, 26:177
$N_2O_2Si_4YbC_{16}H_{46}$, Ytterbium,
  bis[bis(trimethylsilyl)amido]bis(diethyl
  ether)-, 27:148
$N_2O_4Pd_2C_{28}H_{26}$, Palladium(II), di-μ-
  acetato-bis[2-(2-pyridinylmethyl)phenyl-
  C^1,N]di-, 26:208
$N_2O_4WC_{14}H_{18}$, Tungsten, bis(tert-butyl
  isocyanide)tetracarbonyl-, cis-, 28:143
$N_2O_5WC_{10}H_{10}$, Tungsten, tetracarbonyl-
  [(diethylamino)methylidyne](isocyanato)-, trans-, 26:42
$N_2O_6PtC_{12}H_{20}$, Platinum(II), [ascorbato-
  (2–)] (cis-1,2-cyclohexanediamine)-,
  27:283
$N_2O_{10}Os_3C_{14}H_6$, Osmium, bis(acetonitrile)-
  decacarbonyltri-, 26:292
$N_2O_{10}S_2VC_{14}H_{16} \cdot 2H_2O$, Vanadium(II),
  tetraaquabis(1,2-benzisothiazol-3(2H)-
  one 1,1-dioxidato)-, dihydrate, 27:307
$N_2O_{12}P_4PtC_{84}H_{60}$, Rhodate(2–), dode-
  cacarbonylplatinumtetra-, bis[μ-nitrido-
  bis(triphenylphosphine)(1+)], 26:375
$N_2O_{12}Pt_6C_{44}H_{72}$, Platinate(2–), hexa-$\mu$-
  carbonyl-hexacarbonylhexa-,
  bis(tetrabutylammonium, 26:316
$N_2O_{14}P_2Ru_5C_{50}H_{30}$, Ruthenate(1–), tetra-
  decacarbonylnitridopenta-, μ-nitrido-
  bis(triphenylphosphorus)(1+), 26:288
$N_2O_{14}P_4PtRh_4C_{86}H_{60}$, Rhodate(2–), tetra-
  decacarbonylplatinumtetra-, bis[μ-
  nitridobis(triphenylphosphorus)(1+)],
  26:373
$N_2O_{15}Os_5P_4C_{87}H_{60}$, Osmate(2–)penta-
  decacarbonylpenta-, bis[μ-
  nitrido-bis(triphenylphosphorus)(1+)],
  26:299
$N_2O_{16}P_2Ru_6C_{52}H_{30}$, Ruthenate(1–), hexa-
  decacarbonylnitridohexa-, μ-nitrido-bis-
  (triphenylphosphorus)(1+), 26:287
$N_2O_{18}Os_6P_4C_{90}H_{60}$, Osmate(2–), octade-
  cacarbonylhexa-, bis[μ-nitrido-bis-
  (triphenylphosphorus)(1+)], 26:300
$N_2O_{18}Pt_9C_{34}H_{40}$, Platinate(2–), tris[tri-μ-
  carbonyl-tricarbonyltri-, bis(tetraethylammonium), 26:322
$N_2O_{24}Pt_{12}C_{40}H_{40}$, Platinate(2–), tetrakis-
  [tri-μ-carbonyl-tricarbonyltri-, bis-
  (tetraethylammonium), 26:321
$N_2O_{30}Pt_{15}C_{46}H_{40}$, Platinate(2–), pentakis-
  [tri-μ-carbonyl-tricarbonyltri-, bis-
  (tetraethylammonium), 26:320
$N_2P_4WC_{56}H_{54}$, Tungsten(0), bis[1,2-
  ethanediylbis(diphenylphosphine)]bis-
  (isocyanomethane)-, trans-, 28:43
$N_2W_6O_{19}C_{32}H_{72}$, Hexatungstate(VI),
  bis(tetrabutylammonium), 27:80
$N_3BrF_6PRuC_{14}H_{21}$, Ruthenium(II), tris-
  (acetonitrile)bromo(η^4-1,5-
  cyclooctadiene)-,
  hexafluorophosphate(1–), 26:72
$N_3ClF_6PRuC_{14}H_{21}$, Ruthenium(II),
  tris(acetonitrile)chloro(η^4-1,5-
  cyclooctadiene)-, hexafluoro-
  phosphate(1–), 26:71
$N_3Cl_3MoC_6H_9$, Molybdenum(III),
  tris(acetonitrile)trichloro-, 28:37
$N_3Cl_4RuC_{12}H_{26}$, Ruthenate(1–),
  tetrachlorobis(acetonitrile)-
  tetramethylammonium, 26:356
$N_3CrO_3C_{12}H_{15}$, Chromium, tricarbonyltris-
  (propionitrile)-, 28:32

$N_3CrO_3C_{18}H_{27}$, Chromium, tris(*tert*-butyl isocyanide)tricarbonyl-, *fac*-, 28:143

$N_3FeO_2C_{29}H_{27}$, Iron, dicarbonyltris(2-isocyano-1,3-dimethylbenzene)-, 26:56

$N_3MoO_3C_{18}H_{27}$, Molybdenum, tris(*tert*-butyl isocyanide)tricarbonyl-, *fac*-, 28:143

$N_3Mo_5O_{21}P_2C_8H_{26} \cdot 5H_2O$, Pentamolybdobis[(2-aminoethyl)phosphonate](4−), sodium tetramethylammonium dihydrogen, pentahydrate, 27:126

$N_3O_3MoC_{12}H_{15}$, Molybdenum, tricarbonyltris(propionitrile)-, 28:31

$N_3O_3WC_{12}H_{15}$, Tungsten, tricarbonyltris(propanenitrile)-, 27:4

$N_3O_3WC_{12}H_{15}$, Tungsten, tricarbonyltris(propionitrile)-, 28:30

$N_3O_3WC_{18}H_{27}$, Tungsten, tris(*tert*-butyl isocyanide)tricarbonyl-, *fac*-, 28:143

$N_3V_{10}O_{28}C_{48}H_{111}$, Decavanadate(V), tris(tetrabutylammonium)trihydrogen, 27:83

$N_4B_2F_8PdC_8H_{12}$, Palladium(2+), tetrakis(acetonitrile)-, bis[tetrafluoroborate(1−)], 28:63

$N_4B_2F_8PdC_8H_{12}$, Palladium(II), tetrakis(acetonitrile)-, bis[tetrafluoroborate(1−)], 26:128

$N_4Br_3H_{12}Ru$, Ruthenium(III), tetraamminedibromo-, *cis*-, bromide, 26:67

$N_4C_{14}H_{24}$, 1,5,9,13-Tetraazacyclohexadeca-1,3,9,11-tetraene, 4,10-dimethyl-, nickel complex, 27:272

$N_4Cl_2Pd_2C_{20}H_{36}$, Palladium(I), tetrakis(*tert*-butyl isocyanide)di-μ-chlorodi-, 28:110

$N_4Cl_2Pd_2C_{24}H_{18}$, Palladium, di-μ-chlorobis[2-(phenylazo)phenyl-C^1,N^2]-di-, 26:175

$N_4Cl_2WC_{20}H_{22}$, Tungsten(VI), (2,2'-bipyridine)dichloro[(1,1-dimethylethyl)imido](phenylimido)-, 27:303

$N_4Cl_3CrC_4H_{16}$, Chromium(III), dichlorobis(1,2-ethanediamine)-, Λ-*cis*-, chloride, and monohydrate, resolution of, 26:24, 27, 28

$N_4Cl_3H_{12}Ru$, Ruthenium(III), tetraamminedichloro-, *cis*-, chloride, 26:66

$N_4Cl_4PtC_4H_{16}$, Platinum(IV), dichlorobis(1,2-ethanediamine)-, dichloride, *cis*-, 27:314

$N_4Cl_4PtC_4H_{18}$, Platinum(II), dichlorobis(1,2-ethanediamine monohydrochloride)-, *trans*-, 27:315

$N_4CuF_6PC_8H_{12}$, Copper(1+), tetrakis(acetonitrile)-, hexafluorophosphate(1−), 28:68

$N_4F_{12}NiO_2P_2C_{20}H_{32}$, Nickel(II), [2,12-dimethyl-3,11-bis(1-methoxyethylidene)-1,5,9,13-tetraazacyclohexadeca-1,4,9,12-tetraene-$\kappa^4N^{1,5,9,13}$]-, bis[hexafluorophosphate(1−)], 27:264

$N_4F_{12}NiO_2P_2C_{30}H_{36}$, Nickel(II), [3,11-bis($\alpha$-methoxybenzylidene)-2,12-dimethyl-1,5,9,13-tetraazacyclohexadeca-1,4,9,12-tetraene-$\kappa^4N^{1,5,9,13}$]-, bis[hexafluorophosphate(1−)], 27:275

$N_4F_{12}P_2RuC_{16}H_{24}$, Ruthenium(II), tetrakis(acetonitrile)(η^4-, 1,5-cyclooctadiene)-, bis[hexafluorophosphate(1−)], 26:72

$N_4FeOC_{37}H_{36}$, Iron, carbonyltetrakis(2-isocyano-1,3-dimethylbenzene)-, 26:57

$N_4H_{18}Mo_5O_{21}P_2 \cdot 4H_2O$, Pentamolybdobis(phosphonate)(4−), tetraammonium, tetrahydrate, 27:123

$N_4Li_2Si_2C_{26}H_{56}$, Lithium, μ-[(α,α', 1,2-η:-α,α', 1,2-η-1,2-phenylenebis[(trimethylsilyl)methylene]]-bis(N,N,N',N'-tetramethyl-1,2-ethanediamine)di-, 26:148

$N_4MoO_2S_4C_{10}H_{20}$, Molybdenum, bis(diethyldithiocarbamato)dinitrosyl-, *cis*-, 28:145

$N_4MoP_4C_{48}H_{52}$, Molybdenum, bis(dinitrogen)bis[1,2-ethanediylbis(diphenylphosphine)]-, *trans*-, 28:38

$N_4Mo_5O_{21}P_2C_2H_{22} \cdot 2H_2O$, Pentamolybdobis(methylphosphonate)(4−), tetraammonium, dihydrate, 27:124

$N_4Mo_5O_{21}P_2C_4H_{26}$, Pentamolybdobis(ethylphosphonate)(4−), tetraammonium, 27:125

$N_4Mo_5O_{21}P_2C_{12}H_{26} \cdot 5H_2O$, Pentamolybdobis(phenylphosphate)(4−), tetraammonium, pentahydrate, 27:125

$N_4Mo_5O_{21}P_2C_{14}H_{26} \cdot 5H_2O$, Pentamol-

$N_4Mo_5O_{21}P_2C_{14}H_{26} \cdot 5H_2O$ (Continued) ybdobis[(4-aminobenzyl)phosphonate](4−), diammonium dihydrogen, pentahydrate, 27:126

$N_4Mo_5O_{21}P_2C_{22}H_{42} \cdot 4H_2O$, Pentamolybdobis-[(4-aminobenzyl)phosphonate]-(4−), bis-(tetramethylammonium) dihydrogen, tetrahydrate, 27:127

$N_4Mo_8O_{26}C_{64}H_{144}$, Octamolybdate(VI), tetrakis(butylammonium), 27:78

$N_4NiC_{14}H_{22}$, Nickel(II), [4,10-dimethyl-1,5,9,13-tetraazacyclohexadeca-1,3,9,11-tetraenato(2−)-$\kappa^4N^{1,5,9,13}$]-, 27:272

$N_4NiC_{20}H_{16}$, Nickel(0), bis(2,2′-bipyridine)-, 28:103

$N_4NiC_{20}H_{36}$, Nickel(0), tetrakis(*tert*-butyl isocyanide)-, 28:99

$N_4NiC_{24}H_{16}$, Nickel(0), bis(1,10-phenanthroline)-, 28:103

$N_4NiC_{28}H_{44}$, Nickel(0), tetrakis(cyclohexyl isocyanide)-, 28:101

$N_4NiO_2C_{28}H_{30}$, Nickel(II), [3,11-dibenxoyl-2,12-dimethyl-1,5,9,13-tetraazacyclohexadeca-1,3,9,11-tetraenato(2−)-$\kappa^4N^{1,5,9,13}$]-, 27:273

$N_4O_2C_{20}H_{32}$, 1,5,9,13-Tetraazacyclohexadeca-1,4,9,12-tetraene, 2,12-dimethyl-3,11-bis(1-methoxyethylidene)-, nickel complex, 27:264

$N_4O_2C_{28}H_{32}$, 1,5,9,13-Tetraazacyclohexadeca-1,3,9,11-tetraene, 3,11-dibenzoyl-2,12-dimethyl-, nickel complex, 27:273

$N_4O_2C_{30}H_{36}$, 1,5,9,13-Tetraazacyclohexadeca-1,4,9,12-tetraene, 3,11-bis-(α-methoxybenzylidene)-2,12-dimethyl-, nickel complex, 27:275

$N_4O_{21}P_2W_5C_{60}H_{122}$, Pentatungstobis(phenylphosphonate)(4−), tetrakis-(tributylammonium), 27:127

$N_4O_{40}SiTiV_3W_9C_{69}H_{149}$, 1,2,3-Trivanadononatungstosilicate(4−), μ_3-[(η^5-cyclopentadienyl)trioxotitanate(IV)]-, A-β-, tetrakis(tetrabutylammonium), 27:132

$N_4O_{40}SiV_3W_9C_{64}H_{147}$, 1,2,3-Trivanadononatungstosilicate(7−), A-β-, tetrakis(tetrabutylammonium)-trihydrogen, 27:131

$N_4O_{40}SiW_{12}C_{64}H_{144}$, Dodecatungstosilicate(4−), γ-, tetrakis(tetrabutylammonium), 27:95

$N_4P_2C_{36}H_{30}$, Phosphorus(1+), μ-nitridobis(triphenyl)-, azide, 26:286

$N_4P_4WC_{48}H_{52}$, Tungsten, bis(dinitrogen)bis-[1,2-ethanediylbis(diphenylphosphine)]-, *trans*-, 28:41

$N_4W_{10}O_{32}C_{64}H_{144}$, Decatungstate(VI), tetrakis(tetrabutylammonium), 27:81

$N_5FeC_{45}H_{45}$, Iron, pentakis(2-isocyano-1,3-dimethylbenzene)-, 26:57

$N_6BP_3C_{36}H_{56}$, Triphosphenium, 1,1,1,3,3,3-hexakis(dimethylamino)-, tetraphenylborate(1−), 27:256

$N_6B_2F_8MoO_2C_8H_{12}$, Molybdenum(2+), tetrakis(acetonitrile)dinitrosyl-, *cis*-, bis-[tetrafluoroborate(1−)], 26:132, 28:65

$N_6B_2F_8O_2WC_8H_{12}$, Tungsten(2+), tetrakis-(acetonitrile)dinitrosyl-, *cis*-, bis[tetrafluoroborate(1−)], 26:133, 28:66

$N_6C_{20}H_{34}$, 1,5,9,13-Tetraazacyclohexadeca-1,4,9,12-tetraene, 2,12-dimethyl-3,11-bis[1-(methylamino)ethylidene]-, nickel complex, 27:266

$N_6C_{26}H_{44}$, 3,10,14,18,21,25-Hexaazabicyclo-[10.7.7]hexacosa-1,11,13,18,20,25-hexaene, 2,3,10,11,13,19-hexamethyl-, cobalt complex, 27:270
nickel complex, 27:268

$N_6C_{42}H_{46}$, 1,5,9,13-Tetraazacyclohexadeca-1,4,9,12-tetraene, 3,11-bis[α-(benzylamino)benzylidene]-2,12-dimethyl-, nickel complex, 27:276

$N_6C_{50}H_{52}$, 3,11,15,19,22,26-Hexaazatricyclo[11.7.7.1^{5,9}]octacosa-1,5,7,9(28),12,14,19,21,26-nonaene, 3,11-dibenzyl-14,20-dimethyl-2,12-diphenyl-, iron complex, 27:280
nickel complex, 27:277

$N_6Cl_2O_6RuC_{30}H_{36} \cdot 6H_2O$, Ruthenium(II), tris-(2,2′-bipyridine)-, dichloride, hexahydrate, 28:338

$N_6Cl_4W_2C_{28}H_{25}$, Tungsten(VI), tetrachlorobis(1,1-dimethylethanamine)bis[(1,1-dimethylethyl)imido]-bis(μ-phenylimido)di-, 27:301

$N_6CoF_{12}P_2C_{26}H_{44}$, Cobalt(II), (2,3,10,11,13,19-hexamethyl-3,10,14,18,21,25-hexaazabicyclo-[10.7.7]hexacosa-1,11,13,18,20,25-hexaene-$\kappa^4N^{14,18,21,25}$)-, bis[hexafluorophosphate(1−)], 27:270

$N_6F_6NiP_2C_{42}H_{46}$, Nickel(II), [3,11-bis[α-(benzylamino)benzylidene]-2,12-dimethyl-1,5,9,13-tetraazacyclohexadeca-1,4,9,12-tetraene-$\kappa^4N^{1,5,9,13}$]-, bis[hexafluorophosphate(1−)], 27:276

$N_6F_{12}FeP_2C_{50}H_{52}$, Iron(II), [3,11-dibenzyl-14,20-dimethyl-2,12-diphenyl-3,11,15,19,22,26-hexaazatricyclo-[11.7.7.1^{5,9}]octacosa-1,5,7,9(28),12,14,19,21,26-nonaene-$\kappa^4N^{15,19,22,26}$]-, bis[hexafluorophosphate(1−)], 27:280

$N_6F_{12}NiP_2C_{26}H_{44}$, Nickel(II), (2,3,10,11,13,19-hexamethyl-3,10,14,18,21,25-hexaazabicyclo[10.7.7]hexacosa-1,11,13,18,20,25-hexaene-$\kappa^4N^{14,18,21,25}$)-, bis[hexafluorophosphate(1−)], 27:268

$N_6F_{12}NiP_2C_{50}H_{52}$, Nickel(II), [3,11-dibenzyl-14,20-dimethyl-2,12-diphenyl-3,11,15,19,22,26-hexaazatricyclo-[11.7.7.1^{5,9}] octacosa-1,5,7,9(28),12,14,19,21,26-nonaene-$\kappa^4N^{15,19,22,26}$]-, bis[hexafluorophosphate(1−)], 27:277

$N_6F_{18}P_3C_{26}H_{47}$, 3,10,14,18,21,25-Hexaazabicyclo[10.7.7]hexacosa-1,11,13,18,20,25-hexaene, 2,3,10,11,13,19-hexamethyl-, tris[hexafluorophosphate(1−)], 27:269

$N_6F_{18}P_3C_{50}H_{55}$, 3,11,15,19,22,26-Hexaazatricyclo[11.7.7.1^{5,9}]octacosa-1,5,7,9(28),12,14,19,21,26-nonaene, 3,11-dibenzyl-14,20-dimethyl-2,12-diphenyl-, tris[hexafluorophosphate(1−)], 27:278

$N_6NiP_2F_{12}C_{20}H_{34}$, Nickel(II), [2,12-dimethyl-3,11-bis[1-(methylamino)ethylidene]1,5,9,13-tetraazacyclohexadeca-1,4,9,12-tetraene-$\kappa^4N^{1,5,9,13}$]-, bis[hexafluorophosphate(1−)], 27:266

$N_6O_6S_2VC_{34}H_{28} \cdot 2NC_5H_5$, Vanadium(III), bis(1,2-benzisothiazol-3(2H)-one 1,1-dioxidato)tetrakis(pyridine)-, −2pyridine, 27:308

$N_8B_2RuC_{56}H_{68}$, Ruthenium(II), (η^4-1,5-cyclooctadiene)tetrakis(hydrazine)-, bis[tetraphenylborate(1−)], 26:73

$N_8B_2RuC_{60}H_{76}$, Ruthenium(II), (η^4-1,5-cyclooctadiene)tetrakis(methylhydrazine)-, bis[tetraphenylborate(1−)], 26:74

$N_8F_{12}P_2RuC_{12}H_{36}$, Ruthenium(II), ($\eta^4$-cyclooctadiene)tetrakis(methylhydrazine)-, bis[hexafluorophosphate(1−), 26:74

$N_{14}H_{64}NaO_{110}P_5W_{30} \cdot 31H_2O$, Sodiotricontatungstopentaphosphate(14−), tetradecaammonium, hentricontahydrate, 27:115

$N_{18}H_{72}NaO_{86}Sb_9W_{21} \cdot 24H_2O$, Sodiohenicosatungstononantimonate(18−), octadecaammonium, tetracosahydrate, 27:120

$N_{24}As_4Co_2H_{100}O_{142}W_{40} \cdot 19H_2O$, Ammoniodicobaltotetracontatungstotetraarsenate(23−), tricosaammonium, nonadecahydrate, 27:119

$NaC_5H_5 \cdot C_4H_{10}O_2$, Sodium, cyclopentadienyl-, compd. with 1,2-dimethoxyethane(1:1), 26:341

$NaCrO_3C_8H_5 \cdot 2C_4H_{10}O_2$, Chromate(1−), tricarbonyl(η^5-cyclopentadienyl)-, sodium, compd. with 1,2-dimethoxyethane(1:2), 26:343

$NaH_{64}N_{14}O_{110}P_5W_{30} \cdot 31H_2O$, Sodiotricontatungstopentaphosphate(14−), tetradecaammonium, hentricontahydrate, 27:115

$NaH_{72}N_{18}O_{86}Sb_9W_{21} \cdot 24H_2O$, Sodiohenicosatungstononantimonate(18−), octadecaammonium, tetracosahydrate, 27:120

$NaMoO_3C_8H_5 \cdot 2C_4H_{10}O_2$, Molybdate(1−), tricarbonyl(η^5-cyclopentadiene)-, sodium, compd. with 1,2-dimethoxyethane(1:2), 26:343

$NaNbO_6C_6$, Niobate(1−), hexacarbonyl-, sodium, 28:192

$NaO_3WCV_8H_5 \cdot 2C_4H_{10}O_2$, Tungstate(1−), tricarbonyl(η^5-cyclopentadienyl)-, sodium, compd. with 1,2-dimethoxyethane(1:2), 26:343

$Na_2FeO_4C_4$, Ferrate(2−), tetracarbonyl-, disodium, 28:203

$Na_2Fe_2O_8C_8$, Ferrate(2−), octacarbonyldi-, disodium, 28:203

Na$_2$Fe$_3$O$_{11}$C$_{11}$, Ferrate(2−), undecacarbonyltri-, disodium, 28:203

Na$_2$O$_{14}$Ru$_5$C$_{15}$, Ruthenate(2−), μ-carbidotetradecacarbonylpenta-, disodium, 26:284

Na$_3$P$_{21}$, Sodium henicosaphosphide, (Na$_3$P$_{21}$), 27:227

Na$_8$O$_{39}$SiW$_{11}$, Undecatungstosilicate(8−), β_1-, octasodium, 27:90

Na$_9$HO$_{34}$SiW$_9$ · 23H$_2$O, Nonatungstosilicate(10−), β-, nonasodium hydrogen, tricosahydrate, 27:88

Na$_9$O$_{34}$PW$_9$ · xH$_2$O, Nonatungstophosphate(9−), A-, nonasodium, hydrate, 27:100

Na$_{10}$O$_{34}$SiW$_9$, Nonatungstosilicate(10−), α-, decasodium, 27:87

Na$_{12}$O$_{56}$P$_2$W$_{15}$ · 24H$_2$O, Pentadecatungstodiphosphate(12−), α-, dodecasodium, tetracosahydrate, 27:108

Na$_{28}$As$_4$O$_{140}$W$_{40}$ · 60H$_2$O, Sodiotetracontatungstotetraarsenate(27−), heptacosasodium, hexacontahydrate, 27:118

NbCl$_2$C$_{10}$H$_{10}$, Niobium(IV), dichlorobis(η^5-cyclopentadienyl)-, 28:267

NbNaO$_6$C$_6$, Niobate(1−), hexacarbonyl-, sodium, 28:192

NdCl$_2$LiO$_2$Si$_4$C$_{30}$H$_{58}$, Neodymium, bis[η^5-1,3-bis(trimethylsilyl)cyclopentadienyl]di-μ-chloro-bis(tetrahydrofuran)lithium-, 27:170

NdCl$_3$ · 3/2C$_4$H$_8$O, Neodymium trichloride-tetrahydrofuran(2:3), 27:140, 28:290

NdOC$_{18}$H$_{27}$, Neodymium, tert-butylbis-(η^5-cyclopentadienyl)(tetrahydrofuran)-, 27:158

NdOC$_{19}$H$_{23}$, Neodymium(III), tris(η^5-cyclopentadienyl)(tetrahydrofuran)-, 26:20

NdO$_3$C$_{45}$H$_{69}$, Neodymium, tris(2,6-di-tert-butyl-4-methylphenoxo)-, 27:167

Nd$_2$Cl$_2$Si$_4$C$_{44}$H$_{84}$, Neodymium, tetrakis-[η^5-1,3-bis(trimethylsilyl)-cyclopentadienyl]di-μ-chloro-di-, 27:171

Nd$_2$Cl$_6$O$_3$C$_{12}$H$_{24}$, Neodymium, hexachlorotris(tetrahydrofuran)di-, 27:140

NiAs$_4$C$_{20}$H$_{32}$, Nickel(0), bis[1,2-phenylenebis(dimethylarsine)]-, 28:103

NiAs$_4$C$_{72}$H$_{60}$, Nickel(0), tetrakis(triphenylarsine)-, 28:103

NiC$_{16}$H$_{32}$, Nickel(0), bis(1,5-cyclooctadiene)-, 28:94

NiCl$_2$, Nickel dichloride, 28:322

NiCoMoO$_5$C$_{17}$H$_{13}$, Nickel, cyclo-μ$_3$-ethylidyne-1:2:3-κ^3C-pentacarbonyl-1κ^2C,2κ^3C-bis[1,3(η^5) cyclopentadienyl]cobaltmolybdenum-, (Co−Mo) (Co−Ni) (Mo−Ni), 27:192

NiF$_6$N$_6$P$_2$C$_{42}$H$_{46}$, Nickel(II), [3,11-bis[α-(benzylamino)benzylidene]-2,12-dimethyl-1,5,9,13-tetraazacyclohexadeca-1,4,9,12-tetraene-$\kappa^4N^{1,5,9,13}$]-, bis[hexafluorophosphate(1−)], 27:276

NiF$_{12}$N$_4$O$_2$P$_2$C$_{20}$H$_{32}$, Nickel(II), [2,12-dimethyl-3,11-bis(1-methoxyethylidene)-1,5,9,13-tetraazacyclohexadeca-1,4,9,12-tetraene-$\kappa^4N^{1,5,9,13}$]-, bis[hexafluorophosphate(1−)], 27:264

NiF$_{12}$N$_4$O$_2$P$_2$C$_{30}$H$_{36}$, Nickel(II), [3,11-bis-(α-methoxybenzylidene)-2,12-dimethyl-1,5,9,13-tetraazacyclohexadeca-1,4,9,12-tetraene-$\kappa^4N^{1,5,9,13}$]-, bis[hexafluorophosphate(1−)], 27:275

NiF$_{12}$N$_6$P$_2$C$_{26}$H$_{44}$, Nickel(II), (2,3,10,11,13,19-hexamethyl-3,10,14,18,21,25-hexaazabicyclo[10.7.7]hexacosa-1,11,13,18,20,25-hexaene-$\kappa^4N^{14,18,21,25}$)-, bis[hexafluorophosphate(1−)], 27:268

NiF$_{12}$N$_6$P$_2$C$_{50}$H$_{52}$, Nickel(II), [3,11-dibenzyl-14,20-dimethyl-2,12-diphenyl-3,11,15,19,22,26-hexaazatricyclo-[11.7.7.1^{5,9}]-octacosa-1,5,7,9(28),12,14,19,21,26-nonaene-κ^4-$N^{15,19,22,26}$]-, bis[hexafluorophosphate(1−)], 27:277

NiNOPC$_{22}$H$_{40}$, Nickel(II), [butanamidato-(2−)-C^4,N](tricyclohexylphosphine)-, 26:206

—, Nickel(II), [2-methylpropanamidato(2−)-C^3,N](tricyclohexylphosphine)-, 26:205

NiN$_4$C$_{14}$H$_{22}$, Nickel(II), [4,10-dimethyl-1,5,9,13-tetraazacyclohexadeca-1,3,9,11-tetraenato(2−)-$\kappa^4N^{1,5,9,13}$]-, 27:272

NiN$_4$C$_{20}$H$_{16}$, Nickel(0), bis(2,2′-bipyridine)-, 28:103

NiN$_4$C$_{20}$H$_{36}$, Nickel(0), tetrakis(*tert*-butyl isocyanide)-, 28:99

NiN$_4$C$_{24}$H$_{16}$, Nickel(0), bis(1,10-phenanthroline)-, 28:103

NiN$_4$C$_{28}$H$_{44}$, Nickel(0), tetrakis(cyclohexyl isocyanide)-, 28:101

NiN$_4$O$_2$C$_{28}$H$_{30}$, Nickel(II), [3,11-dibenzoyl-2,12-dimethyl-1,5,9,13-tetraazacyclohexadeca-1,3,9,11-tetraenato(2−)-$\kappa^4 N^{1,5,9,13}$]-, 27:273

NiN$_6$P$_2$F$_{12}$C$_{20}$H$_{34}$, Nickel(II), [2,12-dimethyl-3,11-bis[1-(methylamino)ethylidene]-1,5,9,13-tetraazacyclohexadeca-1,4,9,12-tetraene-$\kappa^4 N^{1,5,9,13}$]-, bis[hexafluorophosphate(1−)], 27:266

NiO$_8$P$_4$C$_{32}$H$_{44}$, Nickel(0), Tetrakis(dimethyl phenylphosphonite)-, 28:101

NiO$_9$Os$_3$C$_{14}$H$_8$, Osmium, nonacarbonyl(η^5-cyclopentadienyl)tri-μ-hydridonickeltri-, 26:362

NiO$_9$Ru$_3$C$_{14}$H$_8$, Ruthenium, nonacarbonyl-(η^5-cyclopentadienyl)tri-μ-hydridonickeltri-, 26:363

NiO$_{12}$P$_4$C$_{12}$H$_{36}$, Nickel(0), tetrakis-(trimethyl phosphite)-, 28:101

NiO$_{12}$P$_4$C$_{24}$H$_{60}$, Nickel(0), tetrakis(triethyl phosphite)-, 28:101

NiO$_{12}$P$_4$C$_{36}$H$_{84}$, Nickel(0), tetrakis(isopropyl phosphite)-, 28:101

NiO$_{12}$P$_4$C$_{72}$H$_{60}$, Nickel(0), tetrakis(triphenyl phosphite)-, 28:101

NiP$_4$C$_{12}$H$_{32}$, Nickel(0), bis[1,2-ethanediylbis(dimethylphosphine)]-, 28:101

NiP$_4$C$_{12}$H$_{36}$, Nickel(0), tetrakis(trimethylphosphine)-, 28:101

NiP$_4$C$_{24}$H$_{60}$, Nickel(0), tetrakis(triethylphosphine)-, 28:101

NiP$_4$C$_{40}$H$_{60}$, Nickel(0), tetrakis(diethylphenylphosphine)-, 28:101

NiP$_4$C$_{48}$H$_{108}$, Nickel(0), tetrakis(tributylphosphine)-, 28:101

NiP$_4$C$_{52}$H$_{48}$, Nickel(0), bis[1,2-ethanediylbis(diphenylphosphine)]-, 28:103

NiP$_4$C$_{52}$H$_{52}$, Nickel(0), tetrakis(methyldiphenylphosphine)-, 28:101

NiP$_4$C$_{72}$H$_{60}$, Nickel(0), tetrakis(triphenylphosphine)-, 28:102

NiSb$_4$C$_{72}$H$_{60}$, Nickel(0), tetrakis(triphenylstibine)-, 28:103

Ni$_3$O$_9$Os$_3$C$_{24}$H$_{15}$, Osmium, nonacarbonyltris(η^5-cyclopentadienyl)trinickeltri-, 26:365

Ni$_6$N$_2$O$_{12}$C$_{20}$H$_{24}$, Nickelate(2−), hexa-μ-carbonyl-hexacarbonylhexa-, bis(tetramethylammonium)-, 26:312

O, Oxide, gold complex, 26:326

OAu$_3$BF$_4$P$_3$C$_{54}$H$_{45}$, Gold(1+), μ_3-oxo-[tris[(triphenylphosphine)-, tetrafluoroborate(1−), 26:326

OBClF$_4$IrP$_2$C$_{37}$H$_{31}$, Iridium(III), carbonylchlorohydrido[tetrafluoroborato(1−)]bis(triphenylphosphine)-, 26:117, 28:23

OBClF$_4$IrP$_2$C$_{38}$H$_{33}$, Iridium(III), carbonylchloromethyl[tetrafluoroborato-(1−)]bis(triphenylphosphine)-, 26:118

OBF$_4$MoC$_{34}$H$_{25}$, Molybdenum(1+), carbonyl(η^5-cyclopentadienyl)bis(diphenylacetylene)-, tetrafluoroborate(1−), 26:102

OBF$_4$MoPC$_{38}$H$_{30}$, Molybdenum(1+), carbonyl(η^5-cyclopentadienyl)(diphenylacetylene)(triphenylphosphine)-, tetrafluoroborate(1−), 26:104

OBP$_3$RhS$_2$C$_{66}$H$_{59}$, Rhodium(III), [[2-[(diphenylphosphino)methyl]-2-methyl-1,3-propanediyl]bis(diphenylphosphine)](dithiocarbonato)-, tetraphenylborate(1−), 27:287

OCH$_4$, Methanol, platinum complexes, 26:123
  tungsten complex, 26:45

OC$_2$H$_4$, Acetaldehyde, iron complex, 26:235–241
  manganese and rhenium complexes, 28:199, 201

OC$_2$H$_6$, Dimethyl ether, ruthenium complex, 27:198

OC$_3$H$_6$, Acetone, iridium complex, 26:123
  molybdenum and tungsten complexes, 26:105

OC$_4$H$_8$, Furan, tetrahydro-, actinide and lanthanide complexes, 28:289, 290
  iron complex, 26:232
  lanthanide complexes, 28:293–295
  lanthanide-lithium complexes, 27:170

OC$_4$H$_8$ (Continued)
  lutetium complex, 27:152
  molybdenum complex, 28:35
  neodymium and samarium complexes, 26:20
  neodymium complex, 27:158
  samarium complex, 27:155
OC$_4$H$_{10}$, Diethyl ether, ytterbium complex, 27:148
OC$_6$H$_6$, Phenol, rhodium complex, 27:292
OC$_7$H$_8$, Anisole, chromium complex, 28:137
OC$_8$H$_8$, Ethanane, 1-phenyl-manganese complex, 26:156-158
OC$_{10}$H$_{10}$, 3-Butene-2-one, 4-phenyl-, iron complex, 28:52
OC$_{14}$H$_{22}$, Phenol, 2,6-di-tert-butyl-, actinide and lanthanide complexes, 27:166
OC$_{15}$H$_{24}$, Phenol, 2,6-di-tert-butyl-4-methyl-, actinide and lanthanide complexes, 27:166
OC$_{17}$H$_{14}$, 1,4-Pentadien-3-one, 1,5-diphenyl-, palladium complex, 28:110
OClIrP$_2$C$_{37}$H$_{30}$, Iridium, carbonylchlorobis(triphenylphosphine)-, trans-, 28:92
OClLuC$_{12}$H$_{16}$, Lutetium, chloro(η^8-1,3,5,7-cyclooctatetraene)(tetrahydrofuran)-, 27:152
OClP$_2$RhC$_{36}$H$_{30}$, Rhodium, carbonylchlorobis(triphenylphosphine)-, trans-, 28:79
OClP$_3$RhS$_2$C$_{42}$H$_{39}$, Rhodium, chloro[[2-[(diphenylphosphino)methyl]-2-methyl-1,3-propanediyl]bis(diphenylphosphine)](dithiocarbonato)-, 27:289
OCl$_4$Mo, Molybdenum tetrachloride oxide, 28:325
OCl$_4$W, Tungsten tetrachloride oxide, 28:324
OCl$_5$H$_8$MoN$_2$, Molybdate(V), pentachlorooxo-, diammonium, 26:36
OCoI$_2$C$_{11}$H$_{15}$, Cobalt(III), carbonyldiiodo(η^5-pentamethylcyclopentadienyl)-, 28:275
OFeN$_4$C$_{37}$H$_{36}$, Iron, carbonyltetrakis(2-isocyano-1,3-dimethylbenzene)-, 26:57
OH$_2$, Water, chromium and vanadium complexes, 27:307, 309
  iridium complex, 26:123
  ruthenium complex, 26:254-256
OIP$_2$RhC$_{37}$H$_{66}$, Rhodium(I), carbonyliodobis(tricyclohexylphosphine)-, 27:292
OLiNC$_{16}$H$_{22}$, Lithium, (diethyl ether)[8-(dimethylamino)-1-naphthyl]-, 26:154
OLuC$_{21}$H$_{25}$, Lutetium, bis(η^5-Cyclopentadienyl)(tetrahydrofuran)-p-tolyl-, 27:162
OLuNC$_{21}$H$_{28}$, Lutetium, [(2-[(dimethylamino)methyl]phenyl-C^1,N](η^8-1,3,5,7-cyclooctatetraene)(tetrahydrofuran)-, 27:153
OLuSiC$_{18}$H$_{29}$, Lutetium, bis(η^5-cyclopentadienyl)(tetrahydrofuran)[(trimethylsilyl)methyl]-, 27:161
ON, Nitrosyls, chromium, molybdenum, and tungsten, 28:196, 197
  molybdenum and tungsten, 26:132, 133
ONC, Cyanate, tungsten complex, 26:42
ONC$_4$H$_7$, 3-Butenamide, nickel complex, 26:206
ONC$_4$H$_9$, 2-Propenamide, 2-methyl-, nickel complex, 26:205
ONC$_8$H$_5$, Benzoyl isocyanide, chromium complex, 26:32, 34, 35
ONNiPC$_{22}$H$_{40}$, Nickel(II), [butanamidato(2−)-C^4,N](tricyclohexylphosphine)-, 26:206
—, Nickel(II), [2-methylpropanamidato(2−)-C^3,N](tricyclohexylphosphine)-, 26:205
ONPC$_{25}$H$_{20}$, Benzamide, 2-(diphenylphosphino)-N-phenyl-, 27:324
ONPC$_{25}$H$_{22}$, Benzamide, N-[2-(diphenylphosphino)phenyl]-, 27:323
ONdC$_{18}$H$_{27}$, Neodymium, tert-butylbis(η^5-cyclopentadienyl) (tetrahydrofuran)-, 27:158
ONdC$_{19}$H$_{23}$, Neodymium(III), tris(η^5-cyclopentadienyl) (tetrahydrofuran)-, 26:20
OOsP$_2$SC$_{42}$H$_{34}$, Osmium, carbonyl(5-thioxo-1,3-pentadiene-1,5-diyl-C^1,C^5,S)bis(triphenylphosphine)-, 26:188
OOsP$_3$SC$_{56}$H$_{45}$, Osmium, carbonyl(thiocarbonyl)tris(triphenylphosphine)-, 26:187
OPC$_3$H$_9$, Trimethyl phosphite, iron complex, 26:61
OPC$_{19}$H$_{17}$, Benzenemethanol, 2-(diphenylphosphine)-, manganese complex, 26:169
OPSi$_2$C$_{11}$H$_{27}$, Phosphine, [2,2-dimethyl-1-

(trimethylsiloxy)propylidene](trimethylsilyl)-, 27:250
$OP_3RhC_{55}H_{46}$, Rhodium(I), carbonylhydridotris(triphenylphosphine)-, 28:82
OS_2CH, Dithiocarbonic acid, 27:287
$OSiC_4H_{12}$, Silane, methoxytrimethyl-, 26:44
$OSmC_{19}H_{23}$, Samarium(III), tris(η^5-cyclopentadienyl) (tetrahydrofuran)-, 26:21
$OYbC_{24}H_{40}$, Ytterbium, (diethyl ether)bis-(η^5-pentamethylcyclopentadienyl)-, 27:148
$O_2BF_4FeC_{11}H_{13}$, Iron(1+), dicarbonyl(η^5-cyclopentadienyl) (η^2-2-methyl-1-propene)-, tetrafluoroborate(1−), 28:210
$O_2BF_4IrP_2C_{36}H_{36}$, Iridium(1+), diaquadihydridobis(triphenylphosphine)-, tetrafluoroborate(1−), 26:124, 28:58
$O_2BF_4IrP_2C_{42}H_{44}$, Iridium(1+), bis(acetone)dihydridobis(triphenylphosphine)-, tetrafluoroborate(1−), 26:123, 28:57
$O_2BF_4MoPC_{25}H_{20}$, Molybdenum, dicarbonyl(η^5-cyclopentadienyl)[tetrafluoroborato(1−)](triphenylphosphine)-, 26:98
$O_2BF_4PWC_{25}H_{20}$, Tungsten, dicarbonyl(η^5-cyclopentadienyl)[tetrafluoroborato(1−)](triphenylphosphine)-, 26:98
$O_2BP_2PtC_{40}H_{57}$, Platinum(II), (3-methoxy-3-oxo-κO-propyl-κC^1)bis(triethylphosphine)-, tetraphenylborate(1−), 26:138
$O_2B_2F_8MoN_6C_8H_{12}$, Molybdenum(2+), tetrakis(acetonitrile)dinitrosyl-, cis-, bis[tetrafluoroborate(1−)], 26:132, 28:65
$O_2B_2F_8N_6WC_8H_{12}$, Tungsten(2+), tetrakis(acetonitrile)dinitrosyl-, cis-, bis[tetrafluoroborate(1−)], 28:66
$O_2B_2F_8N_6WC_8H_{12}$, Tungsten(II), tetrakis(acetonitrile)dinitrosyl-, cis-, bis[tetrafluoroborate(1−)], 26:133
O_2C, Carbon dioxide, rhenium complex, 26:111
O_2CH_2, Formic acid, rhenium complex, 26:112
$O_2C_2H_4$, Acetic acid, chromium, molybdenum, and tungsten complexes, 27:297

palladium complex, 26:208
rhodium complex, 27:292
tungsten complex, 26:224
$O_2C_3H_6$, Methyl acetate, iron complex, 27:184
osmium complex, 27:204
$O_2C_4H_6$, Methyl acrylate, platinum ester, 26:138
ruthenium complex, 28:47
$O_2C_4H_{10} \cdot NaC_5H_5$, Ethane, 1,2-dimethoxy-, compd. with cyclopentadienylsodium(1:1), 26:341
$O_2C_4H_{10}$, Ethane, 1,2-dimethoxy-, solvates of chromium, molybdenum, and tungsten carbonyl cyclopentadienyl complexes, 26:343
tungsten complex, 26:50
ytterbium complex, 26:22, 28:295
$O_2C_5H_{10}$, Propanoic acid, 2,2-dimethyl-, tungsten complex, 26:223
$O_2C_7H_6$, Benzoic acid, rhodium complex, 27:292
$O_2C_8H_8$, Methyl benzoate, chromium complex, 26:32
$O_2C_{10}H_8$, 2,3-Naphthalenediol, in prepn. of cis-tetraamminedihaloruthenium(III) complexes, 26:66, 67
$O_2C_{10}H_8$, 2-Propynoic acid, 3-phenyl-, methyl ester, cobalt complex, 26:192
$O_2ClC_2H_3$, Acetic acid, chloro-, ruthenium complex, 26:256
$O_2Cl_2CeLiSi_4C_{30}H_{58}$, Cerium, bis[$\eta^5$-1,3-bis(trimethylsilyl)cyclopentadienyl]di-μ-chloro-bis(tetrahydrofuran)lithium-, 27:170
$O_2Cl_2LaLiSi_4C_{30}H_{58}$, Lanthanum, bis[$\eta^5$-1,3-bis(trimethylsilyl)cyclopentadienyl]di-μ-chloro-bis(tetrahydrofuran)lithium-, 27:170
$O_2Cl_2LiNdSi_4C_{30}H_{58}$, Neodymium, bis[$\eta^5$-1,3-bis(trimethylsilyl)cyclopentadienyl]di-μ-chloro-bis(tetrahydrofuran)lithium-, 27:170
$O_2Cl_2LiPrSi_4C_{30}H_{58}$, Praseodymium, bis[$\eta^5$-1,3-bis(trimethylsilyl)-cyclopentadienyl]di-μ-chloro-bis(tetrahydrofuran)lithium-, 27:170
$O_2Cl_2LiScSi_4C_{30}H_{58}$, Scandium, bis[$\eta^5$-1,3-bis(trimethylsilyl)-cyclopentadienyl]di-μ-chloro-bis(tetra-

$O_2Cl_2LiScSi_4C_{30}H_{58}$ (Continued) hydrofuran)lithium-, 27:170

$O_2Cl_2LiSi_4YC_{30}H_{58}$, Yttrium, bis[$\eta^5$-1,3-bis(trimethylsilyl)cyclopentadienyl]di-μ-chloro-bis(tetrahydrofuran)lithium-, 27:170

$O_2Cl_2LiSi_4YbC_{30}H_{58}$, Ytterbium, bis[$\eta^5$-1,3-bis(trimethylsilyl)cyclopentadienyl]di-μ-chloro-bis(tetrahydrofuran)lithium-, 27:170

$O_2Cl_3SmC_8H_{16}$, Samarium, trichlorobis(tetrahydrofuran)-, 27:140

$O_2Cl_3WC_9H_{19}$, Tungsten(VI), trichloro(1,2-dimethoxyethane)(2,2-dimethylpropylidyne)-, 26:50

$O_2Cl_4MoC_8H_{16}$, Molybdenum, tetrachlorobis(tetrahydrofuran)-, 28:35

$O_2CoC_{12}H_{15}$, Cobalt(I), dicarbonyl(η^5-pentamethylcyclopentadienyl)-, 28:273

$O_2CoPC_{33}H_{28}$, Cobalt, (η^5-cyclopentadienyl)(methyl 3-phenyl-η^2-2-propynoate)(triphenylphosphine)-, 26:192

$O_2FC_7H_5$, Benzoic acid, 3-fluoro-, rhodium complex, 27:292

O_2F_2HP, Phosphorodifluoridic acid, rhenium complex, 26:83

$O_2F_3C_2H$, Acetic acid, trifluoro-, ruthenium complex, 26:254
tungsten complex, 26:222

$O_2F_6C_5H_2$, 2,4-Pentanedione, 1,1,1,5,5,5-hexafluoro-, palladium complexes, 27:318–320

$O_2F_6FePC_{11}H_{13}$, Iron(1+), (η^5-cyclopentadienyl)dicarbonyl(tetrahydrofuran)-, hexafluorophosphate(1−), 26:232

$O_2F_7MoP_5C_{54}H_{48}$, Molybdenum(II), dicarbonylbis[1,2-ethanediylbis(diphenylphosphine)]-fluoro-, hexafluorophosphate(1−), 26:84

$O_2F_{12}N_4NiP_2C_{20}H_{32}$, Nickel(II), [2,12-dimethyl-3,11-bis(1-methoxyethylidene)1,5,9,13-tetraazacyclohexadeca-1,4,9,12-tetraene-$\kappa^4N^{1,5,9,13}$]-, bis[hexafluorophosphate(1−)], 27:264

$O_2F_{12}N_4NiP_2C_{30}H_{36}$, Nickel(II), [3,11-bis($\alpha$-methoxybenzylidene)-2,12-dimethyl-1,5,9,13-tetraazacyclohexadeca-1,4,9,12-tetraene-$\kappa^4N^{1,5,9,13}$]-, bis[hexafluorophosphate(1−)], 27:275

$O_2FeC_{11}H_{12}$, Iron, dicarbonyl(η^5-cyclopentadienyl)(2-methyl-1-propenyl-κ-C^1)-, 28:208

$O_2FeN_3C_{29}H_{27}$, Iron, dicarbonyltris(2-isocyano-1,3-dimethylbenzene)-, 26:56

$O_2FeS_2C_9H_8$, Iron, dicarbonyl(η^5-cyclopentadienyl)[(methylthio)thiocarbonyl]-, 28:186

$O_2HfC_{12}H_{10}$, Hafnium, dicarbonylbis(η^5-cyclopentadienyl)-, 28:252

$O_2HfC_{22}H_{30}$, Hafnium, dicarbonylbis(η^5-pentamethylcyclopentadienyl)-, 28:255

$O_2MoN_4S_4C_{10}H_{20}$, Molybdenum, bis(diethyldithiocarbamato)dinitrosyl-, cis-, 28:145

$O_2MoPC_{25}H_{21}$, Molybdenum, dicarbonyl(η^5-cyclopentadienyl)hydrido-(triphenylphosphine)-, 26:98

$O_2MoP_3C_7H_5$, Molybdenum(I), dicarbonyl-(η^5-cyclopentadienyl)(η^3-cyclotriphosphorus)-, 27:224

$O_2NC_6H_5$, 4-Pyridinecarboxylic acid, rhodium complex, 27:292

$O_2NP_2C_{38}H_{33}$, Phosphorus(1+), μ-nitridobis(triphenyl)-, acetate, 27:296

$O_2N_2RuC_{28}H_{16}$, Ruthenium(II)bis(benzo[h]-quinolin-10-yl-C^{10},N]dicarbonyl-, cis-, 26:177

$O_2N_2Si_4YbC_{16}H_{46}$, Ytterbium, bis[bis(trimethylsilyl)amido]bis(diethyl ether)-, 27:148

$O_2N_4C_{20}H_{32}$, 1,5,9,13-Tetraazacyclohexadeca-1,4,9,12-tetraene, 2,12-dimethyl-3,11-bis(1-methoxyethylidene)-, nickel complex, 27:264

$O_2N_4C_{28}H_{32}$, 1,5,9,13-Tetraazacyclohexadeca-1,3,9,11-tetraene, 3,11-dibenzoyl-2,12-dimethyl-, nickel complex, 27:273

$O_2N_4C_{30}H_{36}$, 1,5,9,13-Tetraazacyclohexadeca-1,4,9,12-tetraene, 3,11-bis(α-methoxybenzylidene)-2,12-dimethyl-, nickel complex, 27:275

$O_2N_4NiC_{28}H_{30}$, Nickel(II), [3,11-dibenzoyl-2,12-dimethyl-1,5,9,13-tetraazacyclohexadeca-1,3,9,11-

tetraenato(2—)-$\kappa^4 N^{1,5,9,13}$]-, 27:273
$O_2PC_8H_{11}$, Phosphonous acid, phenyl-, dimethyl ester, nickel complex, 28:101
$O_2PWC_{25}H_{21}$, Tungsten, dicarbonyl(η^5-cyclopentadienyl)hydrido(triphenylphosphine)-, 26:98
$O_2P_2RhC_{25}H_{47}$, Rhodium(I), carbonylphenoxobis(triisopropylphosphine)-, 27:292
$O_2PdC_{34}H_{28}$, Palladium(0), (1,5-diphenyl-1,4-pentadien-3-one)-, 28:110
$O_2SC_7H_8$, Benzenesulfonic acid, 4-methyl-, rhodium complex, 27:292
$O_2SmC_{28}H_{46}$, Samarium(II), bis(η^5-pentamethylcyclopentadienyl)bis(tetrahydrofuran)-, 27:155
$O_2TiC_{12}H_{10}$, Titanium, dicarbonylbis(η^5-cyclopentadienyl)-, 28:250
$O_2TiC_{22}H_{30}$, Titanium, dicarbonylbis(η^5-pentamethylcyclopentadienyl)-, 28:253
$O_2YbC_{14}H_{20}$, Ytterbium(II), bis(η^5-cyclopentadienyl)(1,2-dimethoxyethane)-, 26:22
$O_2ZrC_{12}H_{10}$, Zirconium, dicarbonylbis(η^5-cyclopentadienyl)-, 28:251
$O_2ZrC_{22}H_{30}$, Zirconium, dicarbonylbis(η^5-pentamethylcyclopentadienyl)-, 28:254
$O_3Au_2ClF_3P_4PtSC_{49}H_{60}$, Platinum(1+), chloro-1-κClbis(triethylphosphine-1κP)bis(triphenylphosphine)-2κP,3κP-triangulo-digold-, trifluoromethanesulfonate, 27:218
$O_3BF_4MoC_8H_5$, Molybdenum, tricarbonyl(η^5-cyclopentadienyl)[tetrafluoroborato(1—)]-, 26:96, 28:5
$O_3BF_4MoC_{10}H_9$, Molybdenum(1+)tricarbonyl(η^5-cyclopentadienyl)(η^2-ethene)-, tetrafluoroborate(1—), 26:102
$O_3BF_4WC_8H_5$, Tungsten, tricarbonyl(η^5-cyclopentadienyl)[tetrafluoroborato-(1—)]-, 26:96, 28:5
$O_3C_5H_8$, Butanoic acid, 3-oxo-, methyl ester, rhodium complex, 27:292
$O_3ClCrC_9H_5$, Chromium, tricarbonyl(η^6-chlorobenzene)-, 28:139
$O_3ClF_3PtSC_{13}H_{30}$, Platinum(II), chlorobis(triethylphosphine)(trifluoromethanesulfonato)-, cis-, 26:126, 28:27
$O_3Cl_3MoC_{12}H_{24}$, Molybdenum(III), trichlorotris(tetrahydrofuran)-, 28:36

$O_3Cl_3WC_3H_9$, Tungsten, trichlorotrimethoxy-, 26:45
$O_3Cl_3YbC_{12}H_{24}$, Ytterbium, trichlortris-(tetrahydrofuran)-, 27:139, 28:289
$O_3Cl_6Nd_2C_{12}H_{24}$, Neodymium, hexachlorotris(tetrahydrofuran)di-, 27:140
$O_3CoN_2C_{15}H_9$, Cobalt, tricarbonyl[2-(phenylazo)phenyl-C^1,N]-, 26:176
$O_3CrC_9H_6$, Chromium, (η^6-benzene)tricarbonyl-, 28:139
$O_3CrFNC_5H_6$, Chromate(VI), fluorotrioxo-, pyridinium, 27:310
$O_3CrNC_7H_5$, Chromium, dicarbonyl(η^5-cyclopentadienyl)nitrosyl-, 28:196
$O_3CrNC_{11}H_{11}$, Chromium, tricarbonyl(η^6-N,N-dimethylbenzenamine)-, 28:139
$O_3CrN_3C_{12}H_{15}$, Chromium, tricarbonyltris-(propionitrile)-, 28:32
$O_3CrN_3C_{18}H_{27}$, Chromium, tris(tert-butylisocyanide)tricarbonyl-, fac-, 28:143
$O_3CrNaC_8H_5 \cdot 2C_4H_{10}O_2$, Chromate(1—), tricarbonyl(η^5-cyclopentadienyl)-, sodium, compd. with 1,2-dimethoxyethane(1:2), 26:343
$O_3DyC_{45}H_{69}$, Dysprosium, tris(2,6-di-tert-butyl-4-methylphenoxo)-, 27:167
$O_3ErC_{45}H_{69}$, Erbium, tris(2,6-di-tert-butyl-4-methylphenoxo)-, 27:167
$O_3FCrC_9H_5$, Chromium, tricarbonyl(η^6-fluorobenzene)-, 28:139
$O_3FNSC_{16}H_{36}$, Fluorosulfate, tetrabutylammonium, 26:393
$O_3FN_2ReC_{13}H_8$, Rhenium, (2,2'-bipyridine)-tricarbonylfluoro-, 26:82
$O_3FP_2RhC_{44}H_{34}$, Rhodium(I), carbonyl(3-fluorobenzoato)bis(triphenylphosphine)-, 27:292
$O_3FS_9C_{10}H_8$, 2,2'-Bi-1,3-dithiolo[4,5-b]-[1,4]dithiinylidene fluorosulfate, 26:393
O_3F_3SCH, Methanesulfonic acid, trifluoro-, 28:70
iridium and platinum complexes, 28:26, 27
iridium, manganese and rhenium complexes, 26:114, 115, 120
platinum complex, 26:126
$O_3FeC_9H_8$, Iron, acetyldicarbonyl(η^5-cyclopentadienyl)-, 26:239
$O_3FeN_2C_{21}H_{18}$, Iron, tricarbonylbis(2-isocyano-1,3-dimethylbenzene)-, 26:54

$O_3FeP_2C_9H_{18}$, Iron(0), tricarbonylbis(trimethylphosphine)-, 28:177

$O_3FeP_2C_{25}H_{54}$, Iron(0), tricarbonylbis(tributylphosphine)-, 28:177

$O_3FeP_2C_{39}H_{30}$, Iron(0), tricarbonylbis(triphenylphosphine)-, 28:176

$O_3FeP_2C_{39}H_{66}$, Iron(0), tricarbonylbis(tricyclohexylphosphine)-, 28:176

$O_3HoC_{45}H_{69}$, Holmium, tris(2,6-di-*tert*-butyl-4-methylphenoxo)-, 27:167

$O_3LaC_{43}H_{63}$, Lanthanum, tris(2,6-di-*tert*-butylphenoxo)-, 27:167

$O_3LaC_{45}H_{69}$, Lanthanum, tris(2,6-di-*tert*-butyl-4-methylphenoxo)-, 27:166

$O_3MoC_{10}H_8$, Molybdenum(0), tricarbonyl(cycloheptatriene)-, 28:45

$O_3MoNC_7H_5$, Molybdenum, dicarbonyl(η^5-cyclopentadienyl)nitrosyl-, 28:196

$O_3MoN_2S_4C_{13}H_{20}$, Molybdenum(II), tricarbonylbis(diethyldithiocarbamato)-, 28:145

$O_3MoN_3C_{18}H_{27}$, Molybdenum, tris(*tert*-butylisocyanide)tricarbonyl-, *fac*-, 28:143

$O_3MoNaC_8H_{5.2}C_4H_{10}O_2$, Molybdate(1−), tricarbonyl(η^5-cyclopentadienyl)-, sodium, compd. with 1,2-dimethoxyethane(1:2), 26:343

$O_3MoP_2C_{39}H_{68}$, Molybdenum, tricarbonyl(dihydrogen)bis(tricyclohexylphosphine)-, 27:3

$O_3NP_2RhC_{25}H_{46}$, Rhodium(I), carbonyl(4-pyridinecarboxylato)bis(triisopropylphosphine)-, 27:292

$O_3NSC_7H_5$, 1,2-Benzisothiazol-3(2H)-one, 1,1-dioxide, chromium and vanadium complex, 27:307, 309

$O_3NWC_7H_5$, Tungsten, dicarbonyl(η^5-cyclopentadienyl)nitrosyl-, 28:196

$O_3N_3MoC_{12}H_{15}$, Molybdenum, tricarbonyltris(propionitrile)-, 28:31

$O_3N_3WC_{12}H_{15}$, Tungsten, tricarbonyltris-(propanenitrile)-, 27:4, 28:30

$O_3N_3WC_{18}H_{27}$, Tungsten, tris(*tert*-butyl isocyanide)tricarbonyl-, *fac*-, 28:143

$O_3NaWC_8H_5 \cdot 2C_4H_{10}O_2$, Tungstate(1−), tricarbonyl(η^5-cyclopentadienyl)-, sodium, compd. with 1,2-dimethoxyethane(1:2), 26:343

$O_3NdC_{45}H_{69}$, Neodymium, tris(2,6-di-*tert*-butyl-4-methylphenoxo)-, 27:167

$O_3PC_3H_9$, Trimethyl phosphite, cobalt and rhodium complexes, 28:283, 284
iron complex, 28:171
nickel complex, 28:101

$O_3PC_6H_{15}$, Triethyl phosphite, iron complex, 28:171
nickel complex, 28:101
nickel, palladium, and platinum complexes, 28:104–106

$O_3PC_9H_{21}$, Isopropyl phosphite, nickel complex, 28:101

$O_3PC_{18}H_{15}$, Triphenyl phosphite, iron complex, 28:171
nickel complex, 28:101
ruthenium complex, 26:178

$O_3PC_{21}H_{21}$, Tris(4-methylphenyl) phosphite, ruthenium complex, 26:277, 278, 28:227

$O_3P_2RhC_{21}H_{45}$, Rhodium(I), (acetato)carbonylbis(triisopropylphosphine)-, 27:292

$O_3P_2RhC_{44}H_{71}$, Rhodium(I), (benzoato)carbonylbis(tricyclohexylphosphine)-, 27:292

$O_3P_2RhC_{44}H_{73}$, Rhodium(I), carbonyl(4-methylbenzenesulfonato)bis(tricyclohexylphosphine)-, 27:292

$O_3P_2WC_{21}H_{44}$, Tungsten, tricarbonyl(dihydrogen)bis(triisopropylphosphine)-, 27:7

$O_3P_2WC_{39}H_{68}$, Tungsten, tricarbonyl(dihydrogen)bis(tricyclohexylphosphine)-, 27:6

$O_3PrC_{45}H_{69}$, Praseodymium, tris(2,6-di-*tert*-butyl-4-methylphenoxo)-, 27:167

$O_3RuC_{11}H_{12}$, Ruthenium(0), tricarbonyl(1,5-cyclooctadiene)-, 28:54

$O_3ScC_{42}H_{63}$, Scandium tris(2,6-di-*tert*-butylphenoxo)-, 27:167

$O_3ScC_{45}H_{69}$, Scandium, tris(2,6-di-*tert*-butyl-4-methylphenoxo)-, 27:167

$O_3SmC_{42}H_{63}$, Samarium, tris(2,6-di-*tert*-butylphenoxo)-, 27:166

$O_3WC_{10}H_9$, Tungsten, tricarbonyl(η^6-cycloheptatriene)-, 27:4

$O_3YC_{42}H_{63}$, Yttrium, tris(2,6-di-*tert*-butylphenoxo)-, 27:167

$O_3YC_{45}H_{69}$, Yttrium, tris(2,6-di-*tert*-butyl-4-methylphenoxo)-, 27:167

$O_3YbC_{45}H_{69}$, Ytterbium, tris(2,6-di-*tert*-butyl-4-methylphenoxo)-, 27:167

$O_4AsFeC_{22}H_{15}$, Iron, tetracarbonyl(triphenylarsine)-, 26:61

$O_4BF_4MoC_{11}H_{11}$, Molybdenum(1+), (acetone)tricarbonyl(η^5-cyclopentadienyl)-, tetrafluoroborate(1−), 26:105

$O_4BF_4WC_{11}H_{11}$, Tungsten(1+), (acetone)tricarbonyl(η^5-cyclopentadienyl)-, tetrafluoroborate(1−), 26:105

$O_4BrNSC_{10}H_{18}$, Bicyclo[2.2.1]heptane-7-methanesulfonate, 3-bromo-1,7-dimethyl-2-oxo-, [(1R)-(ENDO,ANTI)]-, ammonium, 26:24

$O_4Br_2MoC_4$, Molybdenum(II), dibromotetracarbonyl-, 28:145

$O_4C_8H_6$, Phthalic acid, rhodium complex, 27:291

$O_4C_{15}H_{16}$, 2,4-Hexadienedioic acid, 3-methyl-4-phenyl-, dimethyl ester, cobalt complex, 26:197

—, 2-Pentenedioic acid, 3-methyl-2-(phenylmethyl)-, dimethyl ester, cobalt complex, 26:197

$O_4Cl_2Rh_2C_4$, Rhodium, tetracarbonyldichlorodi-, 28:84

$O_4CoPC_{38}H_{34}$, Cobalt, (η^4-cyclopentadienyl)-[1,3-bis(methoxycarbonyl)-2-methyl-4-phenyl-1,3-butadiene-1,4-diyl](triphenylphosphine)-, 26:197

$O_4CrC_{10}H_8$, Chromium, (η^6-anisole)tricarbonyl-, 28:137

$O_4CrN_2C_{14}H_{18}$, Chromium, bis(*tert*-butyl isocyanide)tetracarbonyl-, *cis*-, 28:143

$O_4F_3K_2MnS$, Manganate(III), trifluorosulfato-, dipotassium, 27:312

$O_4F_3P_2PtSC_{14}H_{35}$, Platinum(II), hydrido(methanol)bis(triethylphosphine)-, *trans*-, trifluoromethanesulfonate, 26:135

$O_4F_6Fe_2P_2C_{33}H_{28}$, Iron(1+), μ-acetyl-$2\kappa C^1$:$1\kappa O$-tricarbonyl-$1\kappa^2 C, 2\kappa C$-bis[1,2-(η^5-cyclopentadienyl)] (triphenylphosphine-$2\kappa P$)di-, hexafluorophosphate(1−), 26:237

$O_4F_{12}N_2PdC_{20}H_{10}$, Palladium(II), (2,2'-bipyridine) (1,1,1,5,5,5-hexafluoro-2,4-pentanedionato)-, 1,1,1,5,5,5-hexafluoro-2,4-dioxo-3-pentanide, 27:319

$O_4F_{12}P_3PdC_{44}H_{35}$, Palladium(II), [bis[2-(diphenylphosphino)ethyl]phenylphosphine] (1,1,1,5,5,5-hexafluoro-2,4-pentanedionato)-, 1,1,1,5,5,5-hexafluoro-2,4-dioxo-3-pentanide, 27:320

$O_4F_{12}PdC_{10}H_2$, Palladium, bis(1,1,1,5,5,5-hexafluoro-2,4-pentanedionato)-, 27:318

$O_4FeC_{13}H_{10}$, Iron(0), tricarbonyl(4-phenyl-3-butene-2-one)-, 28:52

$O_4FeNC_{13}H_9$, Iron(0), tetracarbonyl(2-isocyano-1,3-dimethylbenzene)-, 28:180

$O_4FeNP_2C_{40}H_{31}$, Ferrate(1−), hydridotetracarbonyl-, μ-nitrido-bis(triphenylphosphorus)(1+), 26:336

$O_4FeNa_2C_4$, Ferrate(2−), tetracarbonyl-, disodium, 28:203

$O_4FePC_{12}H_{11}$, Iron, tetracarbonyl(dimethylphenylphosphine)-, 26:61

$O_4FePC_{12}H_{14}$, Iron(0), tetracarbonyl(dimethylphenylphosphine)-, 28:171

$O_4FePC_{16}H_{27}$, Iron, tetracarbonyl(tributylphosphine)-, 26:61

$O_4FePC_{17}H_{13}$, Iron, tetracarbonyl(methyldiphenylphosphine)-, 26:61

$O_4FePC_{22}H_{15}$, Iron, tetracarbonyl(triphenylphosphine)-, 26:61

$O_4FePC_{22}H_{33}$, Iron, tetracarbonyl(tricyclohexylphosphine)-, 26:61

$O_4FeSbC_{22}H_{15}$, Iron, tetracarbonyl(triphenylstibine)-, 26:61

$O_4IrNP_2C_{40}H_{30}$, Iridate(1−), tetracarbonyl-μ-nitrido-bis(triphenylphosphorus)(1+), 28:214

$O_4Mg_3C_{48}H_{72}$, Magnesium, cyclotri[μ-1,2-phenylenebis(methylene)]hexakis(tetrahydrofuran)tri-, 26:147

$O_4MnN_2C_{16}H_{19}$, Manganese, tetracarbonyl[2-(phenylazo)phenyl-C^1, N^2]-, 26:173

$O_4MnPC_{22}H_{14}$, Manganese, octacarbonyl-$1\kappa^4 C, 2\kappa^4 C$-μ-[carbonyl-$2\kappa C$:$1\kappa O$-6-(diphenylphosphino-$2\kappa P$)-o-phenylene-$2\kappa C^1$:$1\kappa C^2$]di-, 26:158

$O_4MoN_2C_{14}H_{18}$, Molybdenum, bis(*tert*-butyl isocyanide)tetracarbonyl-, *cis*-, 28:143

$O_4Mo_2C_{14}H_{10}$, Molybdenum, tetracar-

$O_4Mo_2C_{14}H_{10}$ (Continued)
bonylbis(η^5-cyclopentadienyl)di-, (Mo–Mo), 28:152

$O_4Mo_2P_2C_{14}H_{10}$, Molybdenum(I), tetracarbonylbis(η^5-cyclopentadienyl) (μ-η^2:η^2-diphosphorus)di-, 27:224

$O_4NP_2RhC_{40}H_{30}$, Rhodate(1−), tetracarbonyl-, μ-nitrido-bis(triphenylphosphorus) (1+), 28:213

$O_4NReC_{16}H_{36}$, Perrenate, tetrabutylammonium, 26:391

$O_4N_2Pd_2C_{28}H_{26}$, Palladium(II), di-μ-acetato-bis[2-(2-pyridinylmethyl)phenyl-C^1,N]di-, 26:208

$O_4N_2WC_{14}H_{18}$, Tungsten, bis(tert-butyl isocyanide)tetracarbonyl-, cis-, 28:143

$O_4PSC_8H_{12}$, 2-Butenedioic acid, 2-(dimethylphosphinothioyl)-, dimethyl ester, manganese complex, 26:163

$O_4P_2RhC_{24}H_{49}$, Rhodium(I), carbonyl(1-methoxy-1,3-butanedionato-o)bis(triisopropylphosphine)-, 27:292

$O_4Ru_2C_{14}H_{10}$, Ruthenium, tetracarbonylbis(η^5-cyclopentadienyl)di-, 28:189

$O_4S_{16}ReC_{20}H_{16}$, Bis(2,2'-bi-1,3-dithiolo-[4,5-b] [1,4]dithiinylidene) perrhenate, 26:391

$O_4W_2C_{14}H_{10}$, Tungsten, tetracarbonylbis-(η^5-cyclopentadienyl)di-, (W–W), 28:153

$O_5BF_4ReC_5$, Rhenium, pentacarbonyl[tetrafluoroborato(1−)]-, 26:108

$O_5BF_4ReC_7H_4$, Rhenium(1+), pentacarbonyl(η^2-ethene)-, tetrafluoroborate(1−), 26:110

$O_5BF_4WC_{10}H_{10}$, Tungsten(1+), pentacarbonyl[(diethylamino)methylidyne]-, tetrafluoroborate(1−), 26:40

O_5BrMnC_5, Manganese, bromopentacarbonyl-, 28:156

O_5BrReC_5, Rhenium, bromopentacarbonyl-, 28:162

O_5ClMnC_5, Manganese, pentacarbonylchloro-, 28:155

O_5ClReC_5, Rhenium, pentacarbonylchloro-, 28:161

$O_5CoMoNiC_{17}H_{13}$, Nickel, cyclo-μ_3-ethylidyne-1:2:3-κ^3C-pentacarbonyl-$1\kappa^2C$,-$2\kappa^3C$-bis[1,3(η^5)-cyclopentadienyl]cobaltmolybdenum-, (Co–Mo) (Co–Ni) (Mo–Ni), 27:192

$O_5CrC_{11}H_8$, Chromium, tricarbonyl(η^6-methyl benzoate)-, 28:139

$O_5CrNC_{10}H_9$, Chromium, (tert-butyl isocyanide)pentacarbonyl-, 28:143

$O_5CrNC_{18}H_{13}$, Chromium, (benzoyl isocyanide)dicarbonyl(η^6-methylbenzoate)-, 26:32

$O_5F_2N_2PReC_{13}H_8$, Rhenium, (2,2'-bipyridine)tricarbonyl(phosphorodifluoridato)-, 26:83

$O_5F_3FeS_2C_9H_5$, Iron(1+), dicarbonyl(η^5-cyclopentadienyl) (thiocarbonyl)-, trifluoromethanesulfonate, 28:186

$O_5F_6FeMoP_2C_{34}H_{28}$, Iron(1+), μ-acetyl-$2\kappa C^1$:$1\kappa O$-tetracarbonyl-$1\kappa^3C$,$2\kappa^2C$-bis[1,2-(η^5-cyclopentadienyl)] (triphenylphosphine-$1\kappa P$)molybdenum-, hexafluorophosphate(1−), 26:241

$O_5F_6Fe_2PC_{10}H_{13}$, Iron(1+), μ-acetyl-C:O-bis[dicarbonyl(η-cyclopentadienyl)]-, hexafluorophosphate(1−), 26:235

O_5HReC_5, Rhenium, pentacarbonylhydrido-, 26:77

O_5IMnC_5, Manganese, pentacarbonyliodo-, 28:157, 158

O_5IReC_5, Rhenium, pentacarbonyliodo-, 28:163

$O_5MnC_6H_3$, Manganese, pentacarbonylmethyl-, 26:156

$O_5MnC_{12}H_7$, Manganese, (2-acetylphenyl-C,O)tetracarbonyl-, 26:156

$O_5MnC_{12}H_7$, Manganese, benzylpentacarbonyl-, 26:172

$O_5MnPC_{23}H_{16}$, Manganese, tetracarbonyl{2-(diphenylphosphino)phenyl]hydroxymethyl-C,P}-, 26:169

$O_5MoNC_{10}H_9$, Molybdenum, (tert-butyl isocyanide)pentacarbonyl-, 28:143

$O_5NWC_{10}H_9$, Tungsten, (tert-butyl isocyanide)pentacarbonyl-, 28:143

$O_5N_2WC_{10}H_{10}$, Tungsten, tetracarbonyl-[(diethylamino)methylidyne] (isocyanato)-, trans-, 26:42

$O_5P_2RhC_{45}H_{71}$, Rhodium(I), carbonyl-(hydrogen phthalato)bis(tricyclohexylphosphine)-, 27:291

O_5ReC_5H, Rhenium, pentacarbonylhydrido-, 28:165

$O_5ReC_6H_3$, Rhenium, pentacarbonyl-

methyl-, 26:107
$O_6C_6H_8$, Ascorbic acid, platinum complex, 27:283
$O_6Cl_2N_6RuC_{30}H_{36} \cdot 6H_2O$, Ruthenium(II), tris(2,2'-bipyridine)-, dichloride, hexahydrate, 28:338
$O_6Cl_4Ru_2C_6$, Ruthenium(II), di-μ-chlorobis(tricarbonylchloro)-, 28:334
$O_6CoP_2C_{11}H_{23}$, Cobalt(I), (η^5-cyclopentadienyl)bis(trimethyl phosphite)-, 28:283
$O_6CrNC_{13}H_5$, Chromium, (benzoyl isocyanide)pentacarbonyl-, 26:34, 35
$O_6Cr_2C_{16}H_{10}$, Chromium, hexacarbonylbis(η^5-cyclopentadienyl)di-, 28:148
$O_6F_6FeMoPC_{17}H_{13}$, Iron(1+), μ-acetyl-2κC^1:1κO-pentacarbonyl-1$\kappa^3 C$,2$\kappa^2 C$-bis[1,2-(η^5-cyclopentadienyl)]molybdenum-, hexafluorophosphate(1−), 26:239
$O_6MnC_7H_3$, Manganese, acetylpentacarbonyl-, 29:199
$O_6Mo_2C_{16}H_{10}$, Molybdenum, hexacarbonylbis(η^5-cyclopentadienyl)di-, 28:148
$O_6Mo_2N_2PtC_{30}H_{20}$, Molybdenum, [bis(benzonitrile)platinum]-hexacarbonylbis(η^5-cyclopentadienyl)di-, (2$Mo-Pt$), 26:345
$O_6Mo_2P_2Pd_2C_{52}H_{40}$, Molybdenum, hexacarbonylbis(η^5-cyclopentadienyl)-bis(triphenylphosphine)dipalladiumdi-, 26:348
$O_6Mo_2P_2Pt_2C_{52}H_{40}$, Molybdenum, hexacarbonylbis(η^5-cyclopentadienyl)-bis(triphenylphosphine)diplatinumdi-, 26:347
$O_6N_2PtC_{12}H_{20}$, Platinum(II), [ascorbato(2−)] (cis-1,2-cyclohexanediamine)-, 27:283
$O_6N_6S_2VC_{34}H_{28} \cdot 2NC_5H_5$, Vanadium(III), bis(1,2-benzisothiazol-3(2H)-one 1,1-dioxidato)tetrakis(pyridine)-, −2pyridine, 27:308
O_6NbNaC_6, Niobate(1−), hexacarbonyl-, sodium, 28:192
$O_6P_2RhC_{11}H_{23}$, Rhodium(I), (η^5-cyclopentadienyl)bis(trimethyl phosphite)-, 28:284
$O_6P_2RuC_{41}H_{34}$, Ruthenium(II), (η^5-cyclopentadienyl)[2-[(diphenoxyphosphino)oxy]phenyl-C,P] (triphenyl phosphite-P)-, 26:178
O_6ReC_5H, Rhenium, tetracarbonylcarboxy-, 26:112
$O_6ReC_7H_3$, Rhenium, acetylpentacarbonyl-, 28:201
$O_6RuC_8H_6$, Ruthenium, tetracarbonyl(η^2-methyl acrylate)-, 28:47
$O_6W_2C_{16}H_{10}$, Tungsten, hexacarbonylbis-(η^5-cyclopentadienyl)di-, 28:148
$O_7Cl_6Er_2C_{28}H_{56}$, Erbium, hexachloroheptakis(tetrahydrofuran)di-, 27:140
$O_7Co_2P_2PtC_{33}H_{24}$, Cobalt, heptacarbonyl[1,2-ethanediylbis(diphenylphosphine)]-platinumdi-, 26:370
$O_7CrNP_2C_{43}H_{33}$, Chromate(1−), (acetato)pentacarbonyl-, μ-nitrido-bis(triphenylphosphorus) (1+), 27:297
$O_7F_6IrP_2S_2C_{39}H_{31}$, Iridium(III), carbonylhydridobis(trifluoromethanesulfonato)-bis(triphenylphosphine)-, 26:120, 28:26
$O_7FePC_3H_9$, Iron, tetracarbonyl(trimethyl phosphite)-, 26:61, 28:171
$O_7FePC_{10}H_{15}$, Iron, tetracarbonyl(triethyl phosphite)-, 26:61
$O_7FePC_{22}H_{15}$, Iron, tetracarbonyl(triphenyl phosphite)-, 26:61
$O_7MoNP_2C_{43}H_{33}$, Molybdate(1−), (acetato)pentacarbonyl-, μ-nitridobis(triphenylphosphorus) (1+), 27:297
$O_7Mo_2N_2C_{32}H_{72}$, Dimolybdate(VI), bis(tetrabutylammonium), 27:79
$O_7NP_2WC_{43}H_{33}$, Tungstate(1−), (acetato)pentacarbonyl-, μ-nitrido-bis(triphenylphosphorus) (1+), 27:297
$O_8AuMn_2P_2C_{38}H_{25}$, Gold, Octacarbonyl-1$\kappa^4 C$,2$\kappa^4 C$-μ-(diphenylphosphino)-1:2κP-(triphenylphosphine)-3κP-triangulo-dimanganese-, 26:229
$O_8ClHgMn_2PC_{20}H_{10}$, Manganese, μ-(chloromercurio)-μ-(diphenylphosphino)-bis(tetracarbonyl-, ($Mn-Mn$), 26:230
$O_8CoMoRuC_{17}H_{11}$, Ruthenium, cyclo-[μ_3-1(η^2):2(η^2):3(η^2)-2-butyne]octacarbonyl-1$\kappa^2 C$,2$\kappa^3 C$,3$\kappa^3 C$-[1(η^5)-cyclopentadienyl] cobaltmolybdenum-, ($Co-Mo$)($Co-Ru$)($Mo-Ru$), 27:194
$O_8Co_2MoC_{15}H_8$, Molybdenum, octacarbonyl(η^5-cyclopentadienyl)-μ_3-

$O_8Co_2MoC_{15}H_8$ (Continued)
ethylidynedicobalt-, 27:193
$O_8F_3MnSC_6$, Manganese(I), pentacarbonyl-(trifluoromethanesulfonato)-, 26:114
$O_8F_3ReSC_6$, Rhenium(I), pentacarbonyl-(trifluoromethanesulfonato)-, 26:115
$O_8F_{12}W_2C_8$, Tungsten(II), tetrakis(trifluoroacetato)di-, (W-4-W), 26:222
$O_8Fe_2Na_2C_8$, Ferrate(2−), octacarbonyldi-, disodium, 28:203
$O_8MnPSC_{12}H_{12}$, Manganese, tetracarbonyl-[2-(dimethylphosphinothioyl)-1,2-bis-(methoxycarbonyl)ethenyl-C,S]-, 26:163
$O_8Mn_2NP_3C_{56}H_{40}$, Manganate(1−), μ-(diphenylphosphino)bis(tetracarbonyl)-, (Mn–Mn), μ-nitrido-bis(triphenylphosphorus) (1+), 26:228
$O_8Mn_2N_2C_{20}H_8$, Manganese, μ-(azodi-2,1-phenylene-$C^1,N^2:C^{1'},N^1$)octacarbonyldi-, 26:173
$O_8Mn_2PC_{20}H_{11}$, Manganese, μ-(diphenylphosphino)-μ-hydridobis(tetracarbonyl-, (Mn–Mn), 26:226
$O_8Mn_2P_2S_2C_{12}H_{12}$, Manganese, octacarbonylbis(μ-dimethylphosphinothioyl-P,S)di-, 26:162
$O_8NiP_4C_{32}H_{44}$, Nickel(0), Tetrakis(dimethyl phenylphosphonite)-, 28:101
$O_8PC_{14}H_{18}$, 1H-Phospholium, 2,3,4,5-tetrakis(methoxycarbonyl)-2,2-dimethyl-, manganese complex, 26:167
$O_8PSC_{14}H_{20}$, 2H-1,2-Thiaphosphorin-2-ium, 3,4,5,6-tetrakis(methoxycarbonyl)-2,2-dimethyl-, manganese complex, 26:165
$O_8SC_{12}H_{12}$, Thiophenetetracarboxylic acid, tetramethyl ester, 26:166
$O_8W_2C_8H_{12}$, Tungsten(II), tetrakis-(acetato)di-, (W-4-W), 26:224
$O_8W_2C_{20}H_{36}$, Tungsten(II), tetrakis(2,2-dimethylpropanoato)di-, (W-4-W), 26:223
$O_9BrH_3Os_3C_{10}$, Osmium, (μ₃-bromomethylidyne)nonacarbonyltri-μ-hydrido-triangulo-tri-, 27:205
$O_9BrH_3Ru_3C_{10}$, Ruthenium, (μ₃-bromomethylidyne)nonacarbonyl-tri-μ-hydrido-triangulo-tri-, 27:201
$O_9BrHgRu_3C_{15}H_9$, Ruthenium, (bromomercury)nonacarbonyl(3,3-dimethyl-1-butynyl)-triangulo-tri-, 26:332
$O_9ClH_3Os_3C_{10}$, Osmium, nonacarbonyl(μ₃-chloromethylidyne)tri-μ-hydrido-triangulo-tri-, 27:205
$O_9Cl_4Ru_2C_{24}H_{34}$, Ruthenium(II), μ-aqua-bis(μ-chloroacetato)bis[(chloroacetato)-(η^4-cycloocta-1,5-diene)-, 26:256
$O_9Cl_{12}Ru_2C_{22}H_{18}$, Ruthenium(II), μ-aqua-bis(μ-trichloroacetato)bis[(η^4-bicyclo-[2.2.1]heptadiene) (trichloroacetato)-, 26:256
$O_9CoFeSC_9$, Iron, nonacarbonyl-μ-3-thio-dicobalt-, 26:245, 352
$O_9Co_2FePC_{15}H_5$, Iron, nonacarbonyl(μ-3-phenylphosphinidene)dicobalt-, 26:353
$O_9Co_2RuC_{13}H_6$, Ruthenium, μ₃-2-butyne-nonacarbonyldicobalt-, 27:194
$O_9Co_2RuSC_9$, Ruthenium, nonacarbonyl-μ₃-thio-dicobalt-, 26:352
$O_9CrFeN_2P_4C_{81}H_{40}$, Chromate(2−), nonacarbonyliron-bis[μ-nitrido-bis(triphenylphosphorus(1+)], 26:339
$O_9F_{12}Ru_2C_{24}H_{26}$, Ruthenium(II), μ-aqua-bis(μ-trifluoroacetato)bis[(η^4-cycloocta-1,5-diene) (trifluoroacetato)-, 26:254
$O_9FeMoNP_2C_{45}H_{31}$, Molybdate(1−), hydridononacarbonyliron-, μ-nitrido-bis(triphenylphosphorus) (1+), 26:338
$O_9FeMoN_2P_4C_{81}H_{60}$, Molybdate(2−), nonacarbonyliron-bis[μ-nitrido-bis-(triphenylphosphorus) (1+)], 26:339
$O_9FeNWC_{45}H_{31}$, Tungstate(1−), hydridononacarbonyliron-, μ-nitrido-bis(triphenylphosphorus) (1+), 26:336
$O_9FeN_2P_4WC_{81}H_{60}$, Tungstate(2−), nonacarbonyliron-, bis[μ-nitrido-bis(triphenylphosphorus) (1+)], 26:339
$O_9Fe_3SC_9H_2$, Iron, nonacarbonyldihydrido-μ₃-thiotri-, 26:244
$O_9HgIRu_3C_{15}H_9$, Ruthenium, nonacarbonyl(3,3-dimethyl-1-butynyl) (iodomercury)-triangulo-tri-, 26:330
$O_9NCrFeP_2C_{45}H_{31}$, Chromate(1−), hydridononacarbonyliron-, μ-nitrido-bis(triphenylphosphorus)(1+), 26:338
$O_9NiOs_3C_{14}H_8$, Osmium, nonacarbonyl(η^5-cyclopentadienyl)tri-

μ-hydrido-nickeltri-, 26:362
$O_9NiRu_3C_{14}H_8$, Ruthenium, nonacarbonyl(η^5-cyclopentadienyl)tri-μ-hydrido-nickeltri-, 26:363
$O_9Ni_3Os_3C_{24}H_{15}$, Osmium, nonacarbonyltris(η^5-cyclopentadienyl)trinickeltri-, 26:365
$O_9Os_3S_2C_9$, Osmium, nonacarbonyldi-μ_3-thio-tri-, 26:306
$O_9PRu_3C_{21}H_{11}$, Ruthenium, nonacarbonyl-μ-hydrido-(μ-diphenylphosphido)tri-, 26:264
$O_9Ru_3C_{15}H_{10}$, Ruthenium, nonacarbonyl(μ_3-3,3-dimethyl-1-butynyl)-μ-hydrido-*triangulo*-tri-, 26:329
$O_{10}AuHOs_3PC_{16}H_{15}$, Osmium, decacarbonyl-μ-hydrido[μ-triethylphosphine)gold]tri-, 27:210
$O_{10}AuHOs_3PC_{28}H_{15}$, Osmium, decacarbonyl-μ-hydrido[μ-(triphenylphosphine)gold]tri-, 27:209
$O_{10}Au_2Os_3P_2C_{22}H_{30}$, Osmium, decacarbonylbis[μ-(triethylphosphine)gold]tri-, 27:211
$O_{10}Au_2Os_3P_2C_{46}H_{30}$, Gold, decacarbonylbis(triphenylphosphine)triosmiumdi-, 27:211
$O_{10}CrN_2S_2C_{14}H_{16} \cdot 2H_2O$, Chromium(II), tetraaquabis(1,2-benzisothiazol-3(2H)-one 1,1-dioxidato)-, dihydrate, 27:309
$O_{10}HOs_3C_{11}H_3$, Osmium, decacarbonylhydridomethyltri-, 27:206
$O_{10}H_2Os_3C_{11}H_2$, Osmium, decacarbonyl-di-μ-hydrido-μ-methylene-*triangulo*-tri-, 27:206
$O_{10}H_3Os_3C_{11}H_3$, Osmium, nonacarbonyl-tri-μ-hydrido(μ_3-methoxymethylidyne)-*triangulo*-tri-, 27:203
$O_{10}H_3Ru_3C_{11}H_3$, Ruthenium, nonacarbonyl-tri-μ-hydrido(μ_3-methoxymethylidyne)-*triangulo*-tri-, 27:200
$O_{10}NPRu_3C_{23}H_{20}$, Ruthenium decacarbonyl(dimethylphenylphosphine)(2-isocyano-2-methylpropane)tri-, 26:275, 28:224
$O_{10}NP_2Ru_3Si_2C_{58}H_{61}$, Ruthenate(1−), decacarbonyl-1κ^3C,2κ^3C,3κ^4C-μ-hydrido-1:2κ^2H-bis(triethylsilyl)-1κSi,2κSi-*triangulo*-tri-, μ-nitrido-bis(triphenylphosphorus)(1+), 26:269
$O_{10}N_2Os_3C_{14}H_6$, Osmium, bis(acetonitrile)decacarbonyltri-, 26:292
$O_{10}N_2S_2VC_{14}H_{16} \cdot 2H_2O$, Vanadium(II), tetraaquabis(1,2-benzisothiazol-3(2H)-one 1,1-dioxidato)-, dihydrate, 27:307
$O_{10}Os_3C_{10}H_2$, Osmium, decacarbonyldihydridotri-, 26:367
$O_{10}Os_3SC_{10}$, Osmium, μ-3-carbonylnonacarbonyl-μ_3-thio-tri-, 26:305
$O_{10}Os_3SC_{16}H_6$, Osmium, (μ-benzenethiolato)decacarbonyl-μ-hydrido-tri-, 26:304
$O_{10}P_2Ru_3C_{35}H_{22}$, Ruthenium, decacarbonyl[methylenebis(diphenylphosphine)]tri-, 26:276
$O_{11}Co_2RuC_{11}$, Ruthenium, undecacarbonyldicobalt, 26:354
$O_{11}Fe_3N_2P_4C_{83}H_{60}$, Ferrate(2−), undecacarbonyltri-, bis[μ-nitrido-bis(triphenylphosphorous) (1+)], 28:203
$O_{11}Fe_3Na_2C_{11}$, Ferrate(2−), undecacarbonyltri-, disodium, 28:203
$O_{11}HOs_3C_{12}H_3$, Osmium, decacarbonyl-μ-hydrido(μ-methoxymethylidyne)-*triangulo*-tri-, 27:202
$O_{11}HRu_3C_{12}H_3$, Ruthenium, decacarbonyl-μ-hydrido(μ-methoxymethylidyne)-*triangulo*-tri-, 27:198
$O_{11}H_3Os_3C_{12}H_3$, Osmium, nonacarbonyl-tri-μ-hydrido[μ_3-(methoxycarbonyl)methylidyne]-*triangulo*-tri-, 27:204
$O_{11}MnPC_{17}H_{18}$, Manganese, tricarbonyl-[η^2-2,3,4,5-tetrakis(methoxycarbonyl)-2,2-dimethyl-1H-phospholium]-, 26:167
$O_{11}MnPSC_{17}H_{18}$, Manganese, tricarbonyl-[η^2-3,4,5,6-tetrakis(methoxycarbonyl)-2,2-dimethyl-2H-1,2-thiaphosphorin-2-ium]-, 26:165
$O_{11}NOs_3C_{13}H_3$, Osmium, (acetonitrile)-undecacarbonyltri-, 26:290
$O_{11}NOs_3C_{16}H_5$, Osmium, undecacarbonyl-(pyridine)tri-, 26:291
$O_{11}NP_2Os_3C_{47}H_{31}$, Osmate(1−), μ-carbonyldecacarbonyl-μ-hydrido-tri-, μ-nitrido-bis(triphenylphosphorus) (1+), 28:236
$O_{11}PRu_3C_{19}H_{11}$, Ruthenium, undecacarbonyl(dimethylphenylphosphine)tri-,

$O_{11}PRu_3C_{19}H_{11}$ (Continued) 26:273, 28:223

$O_{12}AuCo_3FePC_{30}H_{15}$, Ferrate(1−), dodecacarbonyltricobalt-, (triphenylphosphine)gold(1+), 27:188

$O_{12}Au_3CoP_3Ru_3C_{66}H_{45}$, Ruthenium, dodecacarbonyltris(triphenylphosphine)cobalttrigoldtri-, 26:327

$O_{12}Co_3CuNC_{14}H_3$, Ruthenium, (acetonitrile)dodecacarbonyltricobaltcopper-, 26:359

$O_{12}Co_3FeNC_{20}H_{20}$, Ferrate(1−), dodecacarbonyltricobalt-, tetraethylammonium, 27:188

$O_{12}Co_3NRuC_{12}H_{20}$, Ruthenate(1−), dodecacarbonyltricobalt-, tetraethylammonium, 26:358

$O_{12}F_4Re_4C_{12}\cdot 4H_2O$, Rhenium, dodecacarbonyltetrafluorotetra-, tetrahydrate, 26:82

$O_{12}Fe_4HNC_{21}H_{20}$, Ferrate(1−), carbidododecacarbonylhydridotetra-, tetraethylammonium, 27:186

$O_{12}Fe_4NP_2C_{49}H_{30}$, Ferrate(2−), μ_4-carbidododecacarbonyltetra-, bis[μ-nitrido-bis(triphenylphosphorus) (1+)], 26:246

$O_{12}Fe_4N_2C_{29}H_{40}$, Ferrate(2−), carbidododecacarbonyltetra-, bis(tetraethylammonium), 27:187

$O_{12}H_4Ru_4C_{12}$, Ruthenium, dedecacarbonyltetra-μ-hydrido-tetra-, 28:219

$O_{12}HgMoRu_3C_{23}H_{14}$, Ruthenium, nonacarbonyl-(μ-3,3-dimethyl-1-butynyl){μ-[tricarbonyl-(η^5-cyclopentadienyl)molybdenum]mercury}-triangulo-tri-, 26:333

$O_{12}Ir_4C_{12}$, Iridium, dodecacarbonyltetra-, 28:245

$O_{12}N_2Ni_6C_{20}H_{24}$, Nickelate(2−), hexa-$\mu$-carbonyl-hexacarbonylhexa-, bis(tetramethylammonium), 26:312

$O_{12}N_2P_4PtRh_4C_{84}H_{60}$, Rhodate(2−), dodecacarbonylplatinumtetra-, bis[μ-nitridobis(triphenylphosphine) (1+)], 26:375

$O_{12}NiP_4C_{12}H_{36}$, Nickel(0), tetrakis(trimethyl phosphite)-, 28:101

$O_{12}NiP_4C_{24}H_{60}$, Nickel(0), tetrakis(triethyl phosphite)-, 28:101

$O_{12}NiP_4C_{36}H_{84}$, Nickel(0), tetrakis(isopropyl phosphite)-, 28:101

$O_{12}NiP_4C_{72}H_{60}$, Nickel(0), tetrakis(triphenyl phosphite)-, 28:101

$O_{12}Os_3C_{12}H_4$, Osmium, dodecacarbonyltetra-μ-hydrido-tetrahedro-tetra-, 26:293

$O_{12}Os_3C_{12}$, Osmium, dodecacarbonyltri-, 28:230

$O_{12}Os_4C_{12}H_4$, Osmium, dodecacarbonyltetra-μ-hydrido-tetrahedro-tetra-, 28:240

$O_{12}Os_4S_2C_{12}$, Osmium, dodecacarbonyldi-μ_3-thio-tetra-, 26:307

$O_{12}P_4PdC_{24}H_{60}$, Palladium(0), tetrakis(triethyl phosphite)-, 28:105

$O_{12}P_4PtC_{24}H_{60}$, Platinum(0), tetrakis(triethyl phosphite)-, 28:106

$O_{12}Rh_4C_{12}$, Rhodium, tri-μ-carbonyl-nonacarbonyltetra-, 28:242

$O_{12}Ru_3C_{12}$, Ruthenium, dodecacarbonyltri-, 26:259

$O_{12}Ru_4C_{12}H_4$, Ruthenium, dodecacarbonyltetra-μ-hydrido-tetra-, 26:262

$O_{12}W_2Pt_6C_{44}H_{72}$, Platinate(2−), hexa-$\mu$-carbonyl-hexacarbonylhexa-, bis(tetrabutylammonium), 26:316

$O_{13}Fe_4C_{14}$, Iron, carbidotridecacarbonyltetra-, 27:185

$O_{13}Os_4S_2C_{13}$, Osmium, tridecacarbonyldi-μ_3-thio-tetra-, 26:307

$O_{13}P_2Ru_4C_{39}H_{36}$, Ruthenium, decacarbonyl(dimethylphenylphosphine)tetrahydrido[tris-(4-methylphenyl) phosphite]tetra-, 26:278

$O_{14}Fe_4NC_{23}H_{23}$, Ferrate(1−), dodecacarbonyl[μ^4-(methoxycarbonyl)-methylidyne]tetra-, tetraethylammonium, 27:184

$O_{14}N_2P_2Ru_5C_{50}H_{30}$, Ruthenate(1−), tetradecacarbonylnitridopenta-, μ-nitridobis(triphenylphosphorus)(1+), 26:288

$O_{14}N_2P_4PtRh_4C_{86}H_{60}$, Rhodate(2−), tetradecacarbonylplatinumtetra-, bis[μ-nitridobis(triphenylphosphorus)(1+)], 26:373

$O_{14}Na_2Ru_5C_{15}$, Ruthenate(2−), μ_5-carbidotetradecacarbidopenta-, disodium, 26:284

$O_{14}PRu_4C_{32}H_{25}$, Ruthenium, undecacarbonyltetrahydrido[tris(4-methylphenyl) phosphite]tetra-, 26:277

$O_{14}P_2Ru_5C_{51}H_{15}$, Ruthenate(2−), μ_5-carbidotetradecacarbonylpenta-, bis[μ-nitridobis(triphenylphosphorus)(1+)], 26:284

$O_{15}N_2Os_5P_4C_{87}H_{60}$, Osmate(2−), pentadecacarbonylpenta-, bis[μ-nitrido-bis-(triphenylphosphorus)(1+)], 26:299

$O_{15}Ru_5C_{16}$, Ruthenium, μ_5-carbidopentadecacarbonylpenta-, 26:283

$O_{16}Fe_6N_2C_{33}H_{40}$, Ferrate(2−), carbidohexadecacarbonylhexa-, bis(tetraethylammonium), 27:183

$O_{16}N_2P_2Ru_6C_{52}H_{30}$, Ruthenate(1−), hexadecacarbonylnitridohexa-, μ-nitrido-bis-(triphenylphosphorus)(1+)], 26:287

$O_{17}Ru_6C_{18}$, Ruthenium, μ_6-carbido-heptadecacarbonylhexa-, 26:281

$O_{18}HgRu_6C_{30}H_{18}$, Ruthenium, (μ_4-mercury)-bis[nonacarbonyl(μ_3-3,3-dimethyl-1-butynyl)-*triangulo*-tri-, 26:333

$O_{18}N_2Os_6P_4C_{90}H_{60}$, Osmate (2−), octadecacarbonylhexa-, bis[μ-nitrido-bis(triphenylphosphorus) (1+)], 26:300

$O_{18}N_2Pt_9C_{34}H_{40}$, Platinate(2−), tris[tri-μ-carbonyl-tricarbonyltri-, bis(tetraethylammonium), 26:322

$O_{18}Os_6C_{18}H_2$, Osmium, octadecacarbonyldihydridohexa-, 26:301

$O_{18}Os_6C_{18}$, Osmium, octadecacarbonylhexa-, 26:295

$O_{19}Mo_6N_2C_{32}H_{72}$, Hexamolybdate(VI), bis(tetrabutylammonium), 27:77

$O_{19}N_2W_6C_{32}H_{72}$, Hexatungstate(VI), bis(tetrabutylammonium), 27:80

$O_{21}H_{18}Mo_5N_4P_2 \cdot 4H_2O$, Pentamolybdobis(phosphonate) (4−), tetraammonium, tetrahydrate, 27:123

$O_{21}Mo_5N_3P_2C_8H_{26} \cdot 5H_2O$, Pentamolybdobis[(2-aminoethyl)phosphonate] (4−), sodium tetramethylammonium dihydrogen, pentahydrate, 27:126

$O_{21}Mo_5N_4P_2C_2H_{22} \cdot 2H_2O$, Pentamolybdobis(methylphosphonate)-(4−), tetraammonium, dihydrate, 27:124

$O_{21}Mo_5N_4P_2C_4H_{26}$, Pentamolybdobis(ethylphosphonate) (4−), tetraammonium, 27:125

$O_{21}Mo_5N_4P_2C_{12}H_{26} \cdot 5H_2O$, Pentamolybdobis(phenylphosphate)(4−), tetraammonium, pentahydrate, 27:125

$O_{21}Mo_5N_4P_2C_{14}H_{26} \cdot 5H_2O$, Pentamolybdobis[(4-aminobenzyl)phosphonate](4−), diammonium dihydrogen, pentahydrate, 27:126

$O_{21}Mo_5N_4P_2C_{22}H_{42} \cdot 4H_2O$, Pentamolybdobis[(4-aminobenzyl)phosphonate](4−), bis(tetramethylammonium) dihydrogen, tetrahydrate, 27:127

$O_{21}N_4P_2W_5C_{60}H_{122}$, Pentatungstobis(phenylphosphonate) (4−), tetrakis(tributylammonium), 27:127

$O_{22}P_2Ru_6C_{48}H_{20}$, Ruthenium, [$\mu$-ethynediylbis(diphenylphosphine)]bis-[undecacarbonyltri-, 26:277

$O_{22}Re_4C_{20}$, Rhenium, octadecacarbonylbis(μ_3-carbon dioxide)tetra-, 26:111

$O_{23}Cs_6P_2W_5$, Pentatungstodiphosphate-(6−), hexacesium, 27:101

$O_{24}N_2Pt_{12}C_{40}H_{40}$, Platinate(2−), tetrakis-[tri-μ-carbonyl-tricarbonyltri-, bis(tetraethylammonium), 26:321

$O_{26}Mo_8N_4C_{64}H_{144}$, Octamolybdate(VI), tetrakis(butylammonium), 27:78

$O_{28}N_3V_{10}C_{48}H_{111}$, Decavanadate(V), tris(tetrabutylammonium) trihydrogen, 27:83

$O_{30}N_2Pt_{15}C_{46}H_{40}$, Platinate(2−), pentakis[tri-μ-carbonyl-tricarbonyltri-, bis(tetraethylammonium), 26:320

$O_{32}N_4W_{10}C_{64}H_{144}$, Decatungstate(VI), tetrakis(tetrabutylammonium), 27:81

$O_{34}HNa_9SiW_9 \cdot 23H_2O$, Nonatungstosilicate(10−), β-, nonasodium hydrogen, tricosahydrate, 27:88

$O_{34}Na_9PW_9 \cdot xH_2O$, Nonatungstophosphate(9−), A-, nonasodium, hydrate, 27:100

$O_{34}Na_{10}SiW_9$, Nonatungstosilicate(10−), α-, decasodium, 27:87

$O_{36}Cs_7PW_{10}$, Decatungstophosphate(7−), hexacesium, 27:101

$O_{36}K_8SiW_{10} \cdot 12H_2O$, Decatungstosilicate(8

$O_{36}K_8SiW_{10} \cdot 12H_2O$ (Continued) −), γ-, octapotassium, dodecahydrate, 27:88

$O_{39}K_8SiW_{11} \cdot 14H_2O$, Undecatungstosilicate(8−), β_2-, octapotassium, tetradecahydrate, 27:91

$O_{39}K_8SiW_{11} \cdot 13H_2O$, Undecatungstosilicate(8−), α-, octapotassium, tridecahydrate, 27:89

$O_{39}Na_8SiW_{11}$, Undecatungstosilicate(8−), β_1-, octasodium, 27:90

$O_{40}Cs_5PV_2W_{10}$, Divanadodecatungstophosphate(5−), γ-, pentacesium, 27:102

$O_{40}Cs_6PV_3W_9$, Trivanadononatungstophosphate(6−), α-1,2,3-, hexacesium, 27:100

$O_{40}HK_6SiV_3W_9 \cdot 3H_2O$, 1,2,3-Trivanadononatungstosilicate(7−), A-β-, hexapotassium hydrogen, trihydrate, 27:129

$O_{40}H_4SiW_{12} \cdot xH_2O$, Dodecatungstosilicic acid, α-, hydrate, 27:93

$O_{40}K_4PVW_{11} \cdot xH_2O$, Vanadoundecatungstophosphate(4−), α-, tetrapotassium, hydrate, 27:99

$O_{40}K_4SiW_{12} \cdot 17H_2O$, Dodecatungstosilicate(4−), α-, tetrapotassium, heptadecahydrate, 27:93

$O_{40}K_4SiW_{12} \cdot 9H_2O$, Dodecatungstosilicate(4−), β-, tetrapotassium, nonahydrate, 27:94

$O_{40}N_4SiTiV_3W_9C_{69}H_{149}$, 1,2,3-Trivanadononatungstosilicate(4−), μ_3-[(η^5-cyclopentadienyl)trioxotitanate(IV)]-, A-β-, tetrakis(tetrabutylammonium), 27:132

$O_{40}N_4SiV_3W_9C_{64}H_{147}$, 1,2,3-Trivanadononatungstosilicate(7−), A-β-, tetrakis(tetrabutylammonium) trihydrogen, 27:131

$O_{40}N_4SiW_{12}C_{64}H_{144}$, Dodecatungstosilicate(4−), γ-, tetrakis(tetrabutylammonium), 27:95

$O_{56}Na_{12}P_2W_{15} \cdot 24H_2O$, Pentadecatungstodiphosphate(12−), α-, dodecasodium, tetracosahydrate, 27:108

$O_{61}K_{10}P_2W_{17} \cdot 20H_2O$, Heptadecatungstodiphosphate(10−), α_2-, decapotassium, eicosahydrate, 27:107

$O_{61}LiK_9P_2W_{17} \cdot 20H_2O$, Lithioheptadecatungstodiphosphate(9−), α_1-, nonapotassium, eicosahydrate, 27:109

$O_{62}K_6P_2W_{18} \cdot 19H_2O$, Octadecatungstodiphosphate(6−), β-, hexapotassium, nonadecahydrate, 27:105

$O_{62}K_6P_2W_{18} \cdot 14H_2O$, Octadecatungstodiphosphate(6−), α-, hexapotassium, tetradecahydrate, 27:105

$O_{70}As_2H_4Rb_4W_{21} \cdot 34H_2O$, Tungstate(4−), aquadihydroxohenhexacontaoxobis[trioxoarsenato(III)]henicosa-, tetrarubidium, tetratricontahydrate, 27:113

$O_{70}As_2H_8W_{21} \cdot xH_2O$, Tungsten, aquahexahydroxoheptapentacontaoxobis[trioxoarsenato(III)]henicosa-, hydrate, 27:112

$O_{86}H_{72}N_{18}NaSb_9W_{21} \cdot 24H_2O$, Sodiohenicosatungstononaantimonate-(18−), octadecaammonium, tetracosahydrate, 27:120

$O_{110}H_{64}N_{14}NaP_5W_{30} \cdot 31H_2O$, Sodiotricontatungstopentaphosphate(14−), tetradecaammonium, hentricontahydrate, 27:115

$O_{140}As_4Na_{28}W_{40} \cdot 60H_2O$, Sodiotetracontatungstotetraarsenate(27−), heptacosa-sodium, hexacontahydrate, 27:118

$O_{142}As_4Co_2H_{100}N_{24}W_{40} \cdot 19H_2O$, Ammoniodicobaltotetracontatungstotetraarsenate(23−), tricosaammonium, nonadecahydrate, 27:119

$O_{184}H_7K_{28}Li_5P_8W_{48} \cdot 92H_2O$, Octatetracontatungstooctaphosphate-(40−), pentalithium octacosapotassium heptahydrogen, dononacontahydrate, 27:110

$OsClP_3C_{54}H_{45}$, Osmium(II), dichlorotris(triphenylphosphine)-, 26:184

$OsCl_2P_3SC_{15}H_{45}$, Osmium(II), dichloro(thiocarbonyl)tris(triphenylphosphine)-, 26:185

$OsCl_3P_3C_{24}H_{33}$, Osmium(III), trichlorotris(dimethylphenylphosphine)-, mer-, 27:27

$OsH_3RhP_3C_{32}H_{45}$, Rhodium, [2(η^4)-1,5-cyclooctadiene]tris(dimethylphenylphosphine-1κP)-tri-μ-hydrido-osmium-, 27:29

$OsH_4P_3ZrC_{34}H_{43}$, Zirconium, bis[1,1(η^5)-

cyclopentadienyl]tris(dimethylphenyl-
phosphine-2κP)-tri-μ-hydrido-hydrido-
1κH-osmium-, 27:27
OsOP$_2$SC$_{42}$H$_{34}$, Osmium, carbonyl(5-
thioxo-1,3-pentadiene-1,5-diyl-
C^1,C^5,S)bis(triphenylphosphine)-, 26:188
OsOP$_3$SC$_{56}$H$_{45}$, Osmium, carbonyl-
(thiocarbonyl)tris(triphenylphosphine)-,
26:187
OsP$_3$SC$_{55}$H$_{47}$, Osmium(II), dihydrido-
(thiocarbonyl)tris(triphenylphosphine)-,
26:186
Os$_3$AuHO$_{10}$PC$_{16}$H$_{15}$, Osmium, decacarbonyl-μ-hydrido[μ-
(triethylphosphine)gold]tri-, 27:210
Os$_3$AuHO$_{10}$PC$_{28}$H$_{15}$, Osmium, decacarbonyl-μ-hydrido[μ-(triphenyl-
phosphine)gold]tri-, 27:209
Os$_3$Au$_2$O$_{10}$P$_2$C$_{22}$H$_{30}$, Osmium,
decacarbonylbis[μ-
(triethylphosphine)gold]tri-, 27:211
Os$_3$Au$_2$O$_{10}$P$_2$C$_{46}$H$_{30}$, Osmium,
decacarbonylbis[μ-(triphenyl-
phosphine)gold]tri-, 27:211
Os$_3$BrH$_3$O$_9$C$_{10}$, Osmium, (μ$_3$-
bromomethylidyne)nonacar-
bonyltri-μ-hydrido-triangulo-tri-,
27:205
Os$_3$ClH$_3$O$_9$C$_{10}$, Osmium, nonacarbonyl(μ$_3$-
chloromethylidyne)tri-μ-hydrido-
triangulo-tri-, 27:205
Os$_3$HO$_{10}$C$_{11}$H$_3$, Osmium, decacarbonyl-
hydridomethyltri-, 27:206
Os$_3$HO$_{11}$C$_{12}$H$_3$, Osmium, decacarbonyl-μ-
hydrido(μ-methoxymethylidyne)-
triangulo-tri-, 27:202
Os$_3$H$_2$O$_{10}$C$_{11}$H$_2$, Osmium, decacarbonyl-di-
μ-hydrido-μ-methylene-triangulo-tri-,
27:206
Os$_3$H$_3$O$_{10}$C$_{11}$H$_3$, Osmium, nonacarbonyl-
tri-μ-hydrido(μ$_3$-methoxymethylidyne)-
triangulo-tri-, 27:203
Os$_3$H$_3$O$_{11}$C$_{12}$H$_3$, Osmium, nonacarbonyl-
tri-μ-hydrido[μ$_3$-(methoxycarbonyl)-
methylidyne]-triangulo-tri-, 27:204
Os$_3$NO$_{11}$C$_{13}$H$_3$, Osmium, (acetonitrile)-
undecacarbonyltri-, 26:290
Os$_3$NO$_{11}$C$_{16}$H$_5$, Osmium, undecacarbonyl-
(pyridine)tri-, 26:291
Os$_3$NO$_{11}$P$_2$C$_{47}$H$_{31}$, Osmate(1−), μ-car-
bonyldecacarbonyl-μ-hydrido-tri-, μ-
nitrido-bis(triphenylphosphorus) (1+),
28:236
Os$_3$N$_2$O$_{10}$C$_{14}$H$_6$, Osmium, bis-
(acetonitrile)decacarbonyltri-, 26:292
Os$_3$NiO$_9$C$_{14}$H$_8$, Osmium, nonacarbonyl(η^5-
cyclopentadienyl)tri-μ-hydrido-
nickeltri-, 26:362
Os$_3$Ni$_3$O$_9$C$_{24}$H$_{15}$, Osmium,
nonacarbonyltris(η^5-
cyclopentadienyl)trinickeltri-, 26:365
Os$_3$O$_9$S$_2$C$_9$, Osmium, nonacarbonyldi-μ$_3$-
thio-tri-, 26:306
Os$_3$O$_{10}$C$_{10}$H$_2$, Osmium, decacarbonyl-
dihydridotri-, 26:367
Os$_3$O$_{10}$SC$_{10}$, Osmium, μ$_3$-carbonyl-
nonacarbonyl-μ$_3$-thio-tri-, 26:305
Os$_3$O$_{10}$SC$_{16}$H$_6$, Osmium, (μ-
benzenethiolato)decacarbonyl-μ-
hydrido-tri-, 26:304
Os$_3$O$_{12}$C$_{12}$, Osmium, dodecacarbonyltri-,
28:230
Os$_4$O$_{12}$C$_{12}$H$_4$, Osmium, dodecacarbonyl-
tetra-μ-hydrido-tetrahedro-tetra-, 26:293
Os$_4$O$_{12}$S$_2$C$_{12}$, Osmium, dodecacarbonyldi-
μ$_3$-thio-tetra-, 26:307
Os$_4$O$_{13}$S$_2$C$_{13}$, Osmium, tridecacarbonyldi-
μ$_3$-thio-tetra, 26:307
Os$_5$N$_2$O$_{15}$P$_4$C$_{87}$H$_{60}$, Osmate(2−), pentade-
cacarbonylpenta-, bis[μ-nitrido-
bis(triphenylphosphorus) (1+)], 26:299
Os$_6$N$_2$O$_{18}$P$_4$C$_{90}$H$_{60}$, Osmate(2−), octade-
cacarbonylhexa-, bis[μ-nitrido-
bis(triphenylphosphorus)(1+)], 26:300
Os$_6$O$_{18}$C$_{18}$, Osmium, octadecacarbonyl-
hexa-, 26:295
Os$_6$O$_{18}$C$_{18}$H$_2$, Osmium, octadecacarbonyl-
dihydridohexa-, 26:301

PAuClC$_{18}$H$_{15}$, Gold,
chloro(triphenylphosphine)-, 26:325
PAuClF$_5$C$_{31}$H$_{22}$, Aurate(I), chloro-
(pentafluorophenyl)-, (benzyl)triphenyl-
phosphorium, 26:88
PAuCo$_3$FeO$_{12}$C$_{30}$H$_{15}$, Ferrate(1−), dode-
cacarbonyltricobalt-, (triphenylphos-
phine)gold(1+), 27:188
PAuHO$_{10}$Os$_3$C$_{16}$H$_{15}$, Osmium, decacarbonyl-μ-hydrido[μ-
(triethylphosphine)gold]tri-, 27:210

PAuHO$_{10}$Os$_3$C$_{28}$H$_{15}$, Osmium, deca-
  carbonyl-μ-hydrido[μ-
  (triphenylphosphine)gold]tri-,
  27:209
PAu$_2$F$_5$NSC$_{25}$H$_{15}$, Gold(I),
  (pentafluorophenyl)-μ-thiocyanato-
  (triphenylphosphine)di-, 26:90
PBF$_4$MoOC$_{38}$H$_{30}$, Molybdenum(1+),
  carbonyl(η^5-cyclopentadienyl)
  (diphenylacetylene)(triphenylphosphine)-, tetrafluoroborate(1−),
  26:104
PBF$_4$MoO$_2$C$_{25}$H$_{20}$, Molybdenum,
  dicarbonyl(η^5-cyclopentadienyl) [tetra-
  fluoroborato(1−)] (triphenylphosphine)-, 26:98
PBF$_4$O$_2$WC$_{25}$H$_{20}$, Tungsten, dicarbonyl(η^5-
  cyclopentadienyl) [tetrafluoroborato-
  (1−)] (triphenylphosphine)-, 26:98
PBrF$_6$N$_3$RuC$_{14}$H$_{21}$, Ruthenium(II), tris-
  (acetonitrile)bromo(η^4-1,5-
  cyclooctadiene)-, hexafluoro-
  phosphate(1−), 26:72
PC$_3$H$_9$, Phosphine, methyl-, iron complex,
  28:177
PC$_3$H$_9$, Phosphine, trimethyl-, 26:7
  cobalt and rhodium complexes, 28:280,
    281, 28:280, 281
  nickel complex, 28:101
  preparation of, 28:305
  tungsten complex, 27:304, 28:327, 329
PC$_5$H$_9$, Phosphine, (2,2-dimethyl-
  propylidyne)-, 27:249, 251
PC$_6$H$_7$, Phosphine, phenyl-, cobalt-iron
  complex, 26:353
PC$_6$H$_{15}$, Phosphine, triethyl-, gold-osmium
  complexes, 27:209, 211
  gold-platinum complex, 27:218
  nickel complex, 28:101
  platinum complex, 26:126, 135–140
PC$_8$H$_{11}$, Phosphine, dimethylphenyl-, iron
  complex, 26:61, 28:171
  molybdenum complex, 27:11
  osmium complex, 27:27
  osmium-rhodium complex, 27:29
  osmium-zirconium complex, 27:27
  ruthenium complex, 26:273, 278
  tungsten complex, 28:330
PC$_9$H$_{13}$, Phosphine, ethylmethylphenyl-,
  lithium complex, 27:178

PC$_9$H$_{13}$, Phosphorane, dimethylmethylene-
  diphenyl-, uranium complex, 27:177
PC$_9$H$_{21}$, Phosphine, triisopropyl-, platinum
  complex, 28:120
  rhodium complex, 27:292
PC$_9$H$_{21}$, Phosphine, triisopropyl-, tungsten
  complex, 27:7
PC$_{10}$H$_{15}$, Phosphine, diethylphenyl-, nickel
  complex, 28:101
  platinum complex, 28:135
PC$_{12}$H$_{11}$, Phosphine, diphenyl-, manganese
  complex, 26:158, 226–230
  ruthenium complex, 26:264
PC$_{12}$H$_{27}$, Phosphine, tributyl-, iron complex, 26:61
  nickel complex, 28:101
—, Phosphine, tri-*tert*-butyl-, palladium
  complex, 28:115
PC$_{13}$H$_{13}$, Phosphine, methyldiphenyl-, iron
  complex, 26:61
  molybdenum complex, 27:9
  nickel complex, 28:101
  tungsten complex, 28:328
PC$_{14}$H$_{23}$, Phosphine, di-*tert*-butylphenyl-,
  palladium and platinum complexes,
  28:114, 116
PC$_{18}$H$_{15}$, Phosphine, triphenyl-, cobalt
  complex, 26:190–197
  cobalt-gold-iron complex, 27:188
  cobalt-gold-ruthenium complex, 26:327
  gold and platinum complex, 27:218
  gold complex, 26:325, 326, 27:214
  gold-manganese complex, 26:229
  gold-osmium complexes, 27:209, 211
  iridium complexes, 26:117–120, 122–125,
    201, 202, 28:23–26, 58, 59
  iron complexes, 26:237, 241, 28:170
  molybdenum and tungsten complexes,
    26:98–105
  molybdenum complex, 28:13
  molybdenum, palladium, and platinum
    complexes, 26:347
  nickel complex, 28:102
  osmium complex, 26:184–188
  palladium complex, 28:107
  platinum complexes, 28:124, 125, 27:37
  rhenium complex, 27:15
  rhodium complex, 27:222, 28:77–83
  ruthenium complex, 26:181, 182
  tungsten complex, 28:40

Formula Index   441

PC$_{18}$H$_{31}$, Phosphine, (2,4,6-tri-*tert*-butylphenyl)-, 27:237
PC$_{18}$H$_{33}$, Phosphine, tricyclohexyl-, iron complex, 26:61
  molybdenum complex, 27:3
  nickel complexes, 26:205, 206
  palladium and platinum complexes, 28:114, 116
  platinum complex, 28:130
  rhodium complex, 27:291
  tungsten complex, 27:6
PClF$_6$N$_3$RuC$_{14}$H$_{21}$, Ruthenium(II), tris(acetonitrile)chloro(η^4-1,5-cyclooctadiene)-, hexafluorophosphate(1 −)-, 26:71
PClRuC$_{30}$H$_{24}$, Ruthenium(II), chloro(η^6-hexamethylbenzene)hydrido(triphenylphosphine)-, 26:181
PCl$_2$C$_{18}$H$_{29}$, Phosphonous dichloride, (2,4,6-tri-*tert*-butylphenyl)-, 27:236
PCl$_2$Si$_3$C$_{10}$H$_{27}$, Phosphonous dichloride, [tris(trimethylsilyl)methyl]-, 27:239
PCoC$_{37}$H$_{30}$, Cobalt, (η^5-cyclopentadienyl)[η^2-1,1′-(1,2-ethynediyl)bisbenzene]-(triphenylphosphine)-, 26:192
PCoC$_{41}$H$_{38}$, Cobalt, (η^5-cyclopentadienyl)-(2,3-dimethyl-1,4-diphenyl-1,3-butadiene-1,4-diyl) (triphenylphosphine)-, 26:195
PCoO$_2$C$_{33}$H$_{28}$, Cobalt, (η^5-cyclopentadienyl) (methyl 3-phenyl-η^2-2-propynoate)(triphenylphosphine)-, 26:192
PCoO$_4$C$_{38}$H$_{34}$, Cobalt, (η^5-cyclopentadienyl) [1,4-bis(methoxycarbonyl)-2-methyl-3-phenyl-1,3-butadiene-1,4-diyl] (triphenylphosphine)-, 26:197
PCo$_2$FeO$_9$PC$_{15}$H$_5$, Iron, nonacarbonyl(μ^3-phenylphosphinedene)dicobalt-, 26:353
PCs$_5$O$_{40}$V$_2$W$_{10}$, Divanadodecatungstophosphate(5 −), γ-, pentacesium, 27:102
PCs$_6$O$_{40}$V$_3$W$_9$, Trivanadononatungstophosphate(6 −), α-1,2,3-, hexacesium, 27:100
PCs$_7$O$_{36}$W$_{10}$, Decatungstophosphate(7 −), hexacesium, 27:101
PCuF$_6$N$_4$C$_8$H$_{12}$, Copper(1 +), tetrakis(acetonitrile)-, hexafluorophosphate(1 −), 28:68
PF$_2$HO$_2$, Phosphorodifluoridic acid, rhenium complex, 26:83
PF$_2$N$_2$O$_5$ReC$_{13}$H$_8$, Rhenium, (2,2′-bipyridine)tricarbonyl(phosphorodifluoridato)-, 26:83
PF$_3$, Phosphorous trifluoride, preparation of, 26:12, 28:310
PF$_6$FeMoO$_6$C$_{17}$H$_{13}$, Iron(1 +), μ-acetyl-2κC^1:1κO-pentacarbonyl-1κ^3C,2κ^2C-bis[1,2-(η^5-cyclopentadienyl)] molybdenum-, hexafluorophosphate(1 −), 26:239
PF$_6$FeO$_2$C$_{11}$H$_{13}$, Iron(1 +), (η^5-cyclopentadienyl)dicarbonyl(tetrahydrofuran)-, hexafluorophosphate(1 −), 26:232
PF$_6$Fe$_2$O$_5$C$_{10}$H$_{13}$, Iron(1 +), μ-acetyl-C:O-bis[dicarbonyl(η-cyclopentadienyl)-, hexafluorophosphate(1 −), 26:235
PF$_7$MoO$_2$C$_{54}$H$_{48}$, Molybdenum(II), dicarbonylbis[1,2-ethanediylbis-(diphenylphosphine)]fluoro-, hexafluorophiosphate(1 −), 26:84
PFeO$_4$C$_{12}$H$_{11}$, Iron, tetracarbonyl(dimethylphenylphosphine)-, 26:61
PFeO$_4$C$_{12}$H$_{14}$, Iron(0), tetracarbonyl(dimethylphenylphosphine)-, 28:171
PFeO$_4$C$_{13}$H$_{11}$, Iron, tetracarbonyl-(dimethylphenylphosphine)-, 26:61
PFeO$_4$C$_{16}$H$_{27}$, Iron, tetracarbonyl-(tributylphosphine)-, 26:61
PFeO$_4$C$_{17}$H$_{13}$, Iron(0), tetracarbonyl-(methyldiphenylphosphine)-, 28:171
PFeO$_4$C$_{22}$H$_{15}$, Iron, tetracarbonyl-(triphenylphosphine)-, 26:61
PFeO$_4$C$_{22}$H$_{33}$, Iron, tetracarbonyl-(tricyclohexylphosphine)-, 26:61
PFeO$_7$C$_3$H$_9$, Iron, tetracarbonyl-(trimethyl phosphite)-, 26:61, 28:171
PFeO$_7$C$_{10}$H$_{15}$, Iron, tetracarbonyl-(triethyl phosphite)-, 26:61
PFeO$_7$C$_{22}$H$_{15}$, Iron, tetracarbonyl(triphenyl phosphite)-, 26:61
PH$_2$Li, Lithium dihydrogen phosphide, (LiH$_2$P), 27:228
PHgClMn$_2$O$_8$C$_{20}$H$_{10}$, Manganese, μ-(chloromercurio)-μ-(diphenylphosphino)-bis(tetracarbonyl-, (*Mn–Mn*), 26:230
PK$_4$O$_{40}$VW$_{11}$·xH$_2$O,

$PK_4O_{40}VW_{11} \cdot xH_2O$ *(Continued)*
Vanadoundecatungstophosphate(4 −), α-, tetrapotassium, hydrate, 27:99

$PLiC_9H_{12}$, Lithium, [2-(methylphenylphosphino)ethyl]-, 27:178

$PLiSi_2C_6H_{18} \cdot 2OC_4H_8$, Phosphide, bis(trimethylsilyl)-, lithium, −2tetrahydrofuran, 27:243, 248

$PMnO_4C_{22}H_{14}$, Manganese, octacarbonyl-$1\kappa^4C,2\kappa^4C$-μ-[carbonyl-$2\kappa C$:$1\kappa O$-6-(diphenylphosphino-$2\kappa P$)-o-phenylene-$2\kappa C^1,1\kappa C^2$]di-, 26:158

$PMnO_5C_{23}H_{16}$, Manganese, tetracarbonyl{2-(diphenylphosphino)phenyl]hydroxymethyl-C,P}-, 26:169

$PMnO_8SC_{12}H_{12}$, Manganese, tetracarbonyl-[2-(dimethylphosphinothioyl)-1,2-bis-(methoxycarbonyl)ethenyl-C,S]-, 26:163

$PMnO_{11}C_{17}H_{18}$, Manganese, tricarbonyl[η^2-2,3,4,5-tetrakis(methoxycarbonyl)-2,2-dimethyl-$1H$-phospholium]-, 26:167

$PMnO_{11}SC_{17}H_{18}$, Manganese, tricarbonyl[η^2-3,4,5,6-tetrakis(methoxycarbonyl)-2,2-dimethyl-$2H$-1,2-thiaphosphorin-2-ium]-, 26:165

$PMn_2O_8C_2H_{11}$, Manganese, μ-(diphenylphosphino)-μ-hydrido-bis(tetracarbonyl), (Mn—Mn), 26:226

$PMoO_2C_{25}H_{21}$, Molybdenum, dicarbonyl(η^5-cyclopentadienyl)-hydrido(triphenylphosphine)-, 26:98

$PNNiOC_{22}H_{40}$, Nickel(II), [butanamidato-$(2-)C^4,N$]-(tricyclohexylphosphine)-, 26:206

—, Nickel(II), [2-methylpropanamidato-$(2-)C^3,N$](tricyclohexylphosphine)-, 26:205

$PNOC_{25}H_{20}$, Benzamide, 2-(diphenylphosphino)-N-phenyl-, 27:324

$PNOC_{25}H_{22}$, Benzamide, N-[2-(diphenylphosphino)phenyl]-, 27:323

$PNO_{10}Ru_3C_{23}H_{20}$, Ruthenium, decacarbonyl(dimethylphenylphosphine)-(2-isocyano-2-methylpropane)tri-, 26:275, 28:224

$PNa_9O_{34}W_9 \cdot xH_2O$, Nonatungstophosphate(9−), A-, nonasodium, hydrate, 27:100

$POC_{19}H_{17}$, Benzenemethanamine, 2-(diphenylphosphino)-, manganese complex, 26:169

$POSi_2C_{11}H_{27}$, Phosphine, [2,2-dimethyl-1-(trimethylsiloxy)propylidene]-(trimethylsilyl)-, 27:250

$PO_2C_8H_{11}$, Phosphonous acid, phenyl-, dimethyl ester, nickel complex, 28:101

$PO_2WC_{25}H_{21}$, Tungsten, dicarbonyl(η^5-cyclopentadienyl)hydrido(triphenylphosphine)-, 26:98

$PO_3C_3H_9$, Trimethyl phosphite, cobalt and rhodium complexes, 28:283, 284
iron complexes, 26:61, 28:171
nickel complex, 28:101

$PO_3C_6H_{15}$, Triethyl phosphite, iron complexes, 26:61, 28:171
nickel complexes, 28:101, 104–106

$PO_3C_9H_{21}$, Isopropyl phosphite, nickel complex, 28:101

$PO_3C_{18}H_{15}$, Triphenyl phosphite, iron complex, 26:61, 28:171
nickel complex, 28:101
ruthenium complex, 26:178

$PO_3C_{21}H_{21}$, Tris(4-methylphenyl) phosphite, ruthenium complexes, 26:277, 278, 28:227

$PO_4SC_8H_{12}$, 2-Butenedioic acid, 2-(dimethylphosphinothioyl)-, dimethyl ester, manganese complex, 26:163

$PO_8C_{14}H_{18}$, 1H-Phospholium, 2,3,4,5-tetrakis(methoxycarbonyl)-2,2-dimethyl-, manganese complex, 26:167

$PO_8SC_{14}H_{20}$, 2H-1,2-Thiaphosphorin-2-ium, 3,4,5,6-tetrakis(methoxycarbonyl)-2,2-di-methyl-, manganese complex, 26:165

$PO_9Ru_3C_{21}H_{11}$, Ruthenium, nonacarbonyl-μ-hydrido-(μ-diphenylphosphido)tri-, 26:264

$PO_{11}Ru_3C_{19}H_{11}$, Ruthenium, undecacarbonyl(dimethylphenylphosphine)-tri-, 26:273, 28:223

$PO_{14}Ru_4C_{32}H_{25}$, Ruthenium, undecacarbonyltetrahydrido-[tris(4-methylphenyl) Phosphite]tetra-, 26:277

$POsSC_{55}H_{47}$, Osmium(II), dihydrido-(thiocarbonyl)tris(triphenylphosphine)-, 26:186

$PPtC_{22}H_{41}$, Platinum(0), bis(ethene)-

(tricyclohexylphosphine)-, 28:130
PRuC$_{30}$H$_{33}$, Ruthenium(II), [2-(diphenylphosphino)phenyl-C^1P] (η^6-hexamethylbenzene)hydrido-, 36:182
PSC$_2$H$_7$, Phosphine sulfide, dimethyl-, and manganese complex, 26:162
PS$_2$C$_7$H$_{16}$, Phosphonium, (dithiocarboxy)triethyl-, rhodium complex, 27:288
PSiC$_{21}$H$_{39}$, Phosphine, (2,4,6-tri-*tert*-butylphenyl) (trimethylsilyl)-, 27:238
PSi$_3$C$_9$H$_{27}$, Phosphine, tris(trimethylsilyl)-, 27:243
PUC$_{24}$H$_{27}$, Uranium(IV), tris(η^5-cyclopentadienyl) [(dimethylphenylphosphoranylidene)methyl]-, 27:177
P$_2$, Phosphorus, di-, molybdenum complex, 27:224
P$_2$AuMn$_2$O$_8$C$_{38}$H$_{25}$, Gold, octacarbonyl-1κ^4C, 2κ^4C-μ-(diphenylphosphino)-1:2κP-(triphenylphosphine)-3κP-*triangulo*-dimanganese-, 26:229
P$_2$Au$_2$O$_{10}$Os$_3$C$_{22}$H$_{30}$, Osmium, decacarbonylbis[μ-(triethylphosphine)gold]tri-, 27:211
P$_2$Au$_2$O$_{10}$Os$_3$C$_{46}$H$_{30}$, Osmium, decacarbonylbis[μ-(triphenylphosphine)gold]tri-27:211
P$_2$BClF$_4$IrN$_2$C$_{36}$H$_{31}$, Iridium(III), chloro(dinitrogen)hydrido[tetrafluoroborato(1−)]bis(triphenylphosphine)-, 26:119
P$_2$BClF$_4$IrOC$_{37}$H$_{31}$, Iridium(III), carbonylchlorohydrido[tetrafluoroborato(1−)bis(triphenylphosphine)-, 26:117, 28:23
P$_2$BClF$_4$IrOC$_{38}$H$_{33}$, Iridium(III), carbonylchloromethyl[tetrafluoroborato(1−)]bis[triphenylphosphine)-, 26:118
P$_2$BF$_4$IrC$_{35}$H$_{38}$, Iridium(1+), (η^4-1,5-cyclooctadiene) [1,3-propanediylbis(diphenylphosphine)]-, tetrafluoroborate(1−), 27:23
P$_2$BF$_4$IrC$_{44}$H$_{42}$, Iridium(1+), (η^4-1,5-cyclooctadiene)bis(triphenylphosphine)-, tetrafluoroborate(1−), 26:122, 28:56
P$_2$BF$_4$IrO$_2$C$_{38}$H$_{36}$, Iridium (1+), diaquadihydridobis(triphenylphosphine)-, tetrafluoroborate(1−), 26:124, 28:58
P$_2$BF$_4$IrO$_2$C$_{42}$H$_{44}$, Iridium(1+), bis(acetone)dihydridobis-(triphenylphosphine)-, tetrafluoroborate(1−), 26:123, 28:57
P$_2$BO$_2$PtC$_{40}$H$_{57}$, Platinum(II), (3-methoxy-3-oxo-κO-propyl-κC^1)bis(triethylphosphine)-, tetraphenylborate-(1−), 26:138
P$_2$BPtC$_{44}$H$_{63}$, Platinum(II), (η^3-cyclooctene)bis(triethylphosphine)-, tetraphenylborate(1−), 26:139
P$_2$C$_6$H$_{16}$, Phosphine, 1,2-ethanediylbis(dimethyl-, nickel complex, 28:101
P$_2$C$_{25}$H$_{22}$, Phosphine, methylenebis-(diphenylpalladium complex, 28:340
ruthenium complex, 26:276
P$_2$C$_{26}$H$_{20}$, Phosphine, ethynediylbis(diphenyl-, ruthenium complex, 26:276, 28:226
P$_2$C$_{26}$H$_{24}$, Phosphine, 1,2-ethanediylbis-(diphenyl-, iridium complex, 27:25
molybdenum complex, 26:84
nickel complex, 28:103
platinum complexes, 26:370, 28:135
tungsten complex, 28:41
P$_2$C$_{27}$H$_{26}$, Phosphine, 1,3-propanediylbis(diphenyl-, iridium complex, 27:22
P$_2$C$_{28}$H$_{28}$, Phosphine, 1,4-butanediylbis(diphenyl-, iridium complex, 27:26
P$_2$C$_{36}$H$_{58}$, Diphosphene, bis(2,4,6-tri-*tert*-butylphenyl)-, 27:241, 242
P$_2$ClIrOC$_{37}$H$_{30}$, Iridium, carbonylchlorobis(triphenylphosphine)-, *trans*-, 28:92
P$_2$ClORhC$_{36}$H$_{30}$, Rhodium, carbonylchlorobis(triphenylphosphine)-, *trans*-, 28:79
P$_2$ClPtC$_{26}$H$_{41}$, Platinum(II), chloro(*cis*-1,2-diphenylethenyl)bis(triethylphosphine)-, *trans*-, 26:140
P$_2$ClRuC$_{41}$H$_{35}$, Ruthenium(II), chloro(η^5-cyclopentadienyl)bis(triphenylphosphine)-, 28:270
P$_2$Cl$_2$N$_2$C$_8$H$_{18}$, 1,3,2,4,-Diazadiphosphetidine, 1,3-di-*tert*-butyl-2,4-dichloro-,

$P_2Cl_2N_2C_8H_{18}$ (Continued)
cis-, 27:258

$P_2Cl_2N_2WC_{16}H_{32}$, Tungsten(VI), dichloro[(1,1-dimethylethyl)imido]-(phenylimido)bis(trimethylphosphine)-, 27:304

$P_2Cl_2PtC_{26}H_{24}$, Platinum, dichloro[1,2-ethanediylbis(diphenylphosphine)]-, 26:370

$P_2Cl_4WC_{26}H_{24}$, Tungsten, tetrachloro[1,2-ethanediylbis(diphenylphosphine)]-, 28:41

$P_2Cl_4WC_{26}H_{26}$, Tungsten(IV), tetrachlorobis(methyldiphenylphosphine)-, 28:328

$P_2Cl_4WC_{36}H_{30}$, Tungsten, tetrachlorobis-(triphenylphosphine)-, 28:40

$P_2CoC_{11}H_{23}$, (Cobalt(I), (η^5-cyclopentadienyl)bis(trimethylphosphine)-, 28:281

$P_2CoC_{44}H_{42}$, Cobalt, (η^5-cyclopentadienyl)-bis(triphenylphosphine)-, 26:191

$P_2CoF_{12}N_6C_{26}H_{44}$, Cobalt(II), (2,3,10,11,13,19-hexamethyl-3,10,14,18,21,25-hexaazabicyclo[10.7.7]hexacosa-1,11,13,18,20,25-hexaene-$\kappa^4N^{14,18,21,25}$)-, bis[hexafluorophosphate(1−)], 27:270

$P_2CoO_6C_{11}H_{23}$, Cobalt(I), (η^5-cyclopentadienyl)bis(trimethyl phosphite)-, 28:283

$P_2Co_2O_7PtC_{33}H_{24}$, Cobalt, heptacarbonyl-[1,2-ethanediylbis(diphenylphosphine)]-platinumdi-, 26:370

$P_2CrFeNO_9C_{45}H_{31}$, Chromate(1−), hydridononacarbonyliron-, μ-nitrido-bis(triphenylphosphorus)(1+), 26:338

$P_2CrNO_7C_{43}H_{33}$, Chromate(1−), (acetato)-pentacarbonyl-, μ-nitrido-bis(triphenylphosphorus)(1+), 27:297

$P_2Cs_6O_{23}W_5$, Pentatungstodiphosphate-(6−), hexacesium, 27:101

$P_2FO_3RhC_{44}H_{34}$, Rhodium(I), carbonyl(3-fluorobenzoato)bis(triphenylphosphine)-, 27:292

$P_2F_3O_4PtSC_{14}H_{35}$, Platinum(II), hydrido-(methanol)bis(triethylphosphine)-, trans-, trifluoromethanesulfonate, 26:135

$P_2F_6FeMoO_5C_{34}H_{28}$, Iron(1+), μ-acetyl-2-κC^1:1κO-tetracarbonyl-1κ^3C,2κ^2C-bis[1,2-(η^3-cyclopentadienyl)]-(triphenylphosphine-1κP)molybdenum-hexafluorophosphate(1−), 26:241

$P_2F_6Fe_2O_4C_{33}H_{28}$, Iron(1+), μ-acetyl-2κC^1:1κO-tricarbonyl-1κ^2C,2κC-bis[1,2-(η^5-cyclopentadienyl)]-(triphenylphosphine-2κP)di-, hexafluorophosphate(1−), 26:237

$P_2F_6IrO_7S_2C_{39}H_{31}$, Iridium(III), carbonyl-hydridobis(trifluoromethanesulfonato)-bis(triphenylphosphine)-, 26:120, 28:26

$P_2F_6N_6NiC_{42}H_{46}$, Nickel(II), [3,11-bis-[α-(benzylamino)benzylidene]-2,12-dimethyl-1,5,9,13-tetraazacyclohexadeca-1,4,9,12-tetraene-$\kappa^4N^{1,5,9,13}$]-, bis[hexafluorophosphate(1−)], 27:276

$P_2F_{12}FeN_6C_{50}H_{52}$, Iron(II), [3,11-dibenzyl-14,20-dimethyl-2,12-diphenyl-3,11,15,19,22,26-hexaazatricyclo-[11.7.7.1^{5,9}]octacosa-1,5,7,9-(28),12,14,19,21,26-nonaene-$\kappa^4N^{15,19,22,26}$]-, bis[hexafluorophosphate(1−)], 27:280

$P_2F_{12}N_4NiO_2C_{20}H_{32}$, Nickel(II), [2,12-dimethyl-3,11-bis(1-methoxyethylidene)-1,5,9,13-tetraazacyclohexadeca-1,4,9,12-tetraene-$\kappa^4N^{1,5,9,13}$]-, bis[hexafluorophosphate(1−)], 27:264

$P_2F_{12}N_4NiO_2C_{30}H_{36}$, Nickel(II), [3,11-bis-(α-methoxybenzylidene)-2,12-dimethyl-1,5,9,13-tetraazacyclohexadeca-1,4,9,12-tetraene-$\kappa^4N^{1,5,9,13}$]-, bis[hexafluorophosphate(1−)], 27:275

$P_2F_{12}N_4RuC_{16}H_{24}$, Ruthenium(II), tetrakis-(acetonitrile) (η^4-1,5-cyclooctadiene)-, bis[hexafluorophosphate(1−)], 26:72

$P_2F_{12}N_6NiC_{26}H_{44}$, Nickel(II), (2,3,10,11,13,19-hexamethyl-3,10,14,18,21,25-hexaazabicyclo[10.7.7]hexacosa-1,11,13,18,20,25-hexaene-$\kappa^4N^{14,18,21,25}$)-, bis[hexafluorophosphate(1−)], 27:268

$P_2F_{12}N_6NiC_{50}H_{52}$, Nickel(II), [3,11-dibenzyl-14,20-dimethyl-2,12-diphenyl-3,11,15,19,22,26-hexaazatricyclo-[11.7.7.1^{5,9}]octacosa-1,5,7,9(28),12,14,19,21,26-nonaene-κ^4-$N^{15,19,22,26}$]-, bis[hexa-fluorophosphate(1−)], 27:277

$P_2F_{12}N_8RuC_{12}H_{36}$, Ruthenium(II), ($\eta^4$-cyclooctadiene)tetrakis(methyl-

hydrazine)-, bis[hexafluorophosphate(1−), 26:74
$P_2FeMoNO_9C_{45}H_{31}$, Molybdate(1−), hydridononacarbonyliron-μ-nitridobis(triphenylphosphorus) (1+), 26:338
$P_2FeNO_4C_{40}H_{31}$, Ferrate(1−), hydridotetracarbonyl-, μ-nitridobis(triphenylphosphorus) (1+), 26:336
$P_2FeO_3C_9H_{18}$, Iron(0), tricarbonylbis(trimethylphosphine)-, 28:177
$P_2FeO_3C_{25}H_{54}$, Iron(0), tricarbonylbis(tributylphosphine)-, 28:177
$P_2FeO_3C_{39}H_{30}$, Iron(0), tricarbonylbis(triphenylphosphine)-, 28:176
$P_2FeO_3C_{39}H_{66}$, Iron(0), tricarbonylbis(tricyclohexylphosphine)-, 28:176
$P_2Fe_4NO_{12}C_{49}H_{30}$, Ferrate(2−), μ_4-carbidododecacarbonyltetra-, bis[μ-nitrido-bis(triphenylphosphorus)-(1+)], 26:246
$P_2H_7ReC_{36}H_{30}$, Rhenium(VII), heptahydridobis(triphenylphosphine)-, 27:15
$P_2H_{18}Mo_5N_4O_{21} \cdot 4H_2O$, Pentamolybdobis(phosphonate) (4−), tetraammonium, tetrahydrate, 27:123
$P_2IORhC_{37}H_{66}$, Rhodium(I), carbonyliodobis(tricyclohexylphosphine)-, 27:292
$P_2IrNO_4C_{40}H_{30}$, Iridate(1−), tetracarbonyl-, μ-nitrido-bis(triphenylphosphorus) (1+), 28:214
$P_2K_6O_{62}W_{18} \cdot 19H_2O$, Octadecatungstodiphosphate(6−), β-, hexapotassium, nonadecahydrate, 27:105
$P_2K_6O_{62}W_{18} \cdot 14H_2O$, Octadecatungstodiphosphate(6−), α-, hexapotassium, tetradecahydrate, 27:105
$P_2K_{10}O_{61}W_{17} \cdot 20H_2O$, Heptadecatungstodiphosphate(10−), α_2-, decapotassium, eicosahydrate, 27:107
$P_2LiK_9O_{61}W_{17} \cdot 20H_2O$, Lithioheptadecatungstodiphosphate(9−), α_1-, nonapotassium, eicosahydrate, 27:109
$P_2Mn_2O_8S_2C_{12}H_{12}$, Manganese, octacarbonylbis(μ-dimethylphosphinothioyl-P,S)di-, 26:162
$P_2MoNO_7C_{43}H_{33}$, Molybdate(1−), acetato)pentacarbonyl-, μ-nitridobis(triphenylphosphorus) (1+), 27:297
$P_2MoO_3C_{39}H_{68}$, Molybdenum, tricarbonyl(dihydrogen)bis(tricyclohexylphosphine)-, 27:3
$P_2MoS_4C_{48}H_{40}$, Molybdate(VI), tetrathio-, bis(tetraphenylphosphonium), 27:41
$P_2Mo_2O_4C_{14}H_{10}$, Molybdenum(I), tetracarbonylbis(η^5-cyclopentadienyl) (μ-η^2:η^2-diphosphorus)di-, 27:224
$P_2Mo_2O_6Pd_2C_{52}H_{40}$, Molybdenum, hexacarbonylbis(η^5-cyclopentadienyl)bis(triphenylphosphine)-dipalladiumdi-, 26:348
$P_2Mo_2O_6Pt_2C_{52}H_{40}$, Molybdenum, hexacarbonylbis(η^5-cyclopentadienyl)bis(triphenylphosphine)-diplatinumdi-, 26:347
$P_2Mo_2S_6C_{48}H_{40}$, Molybdate(V), di-μ-thio-tetrathiodi-, bis(tetraphenylphosphonium), 27:43
$P_2Mo_2S_8C_{48}H_{40}$, Molybdate(V), bis(η^2-disulfido)di-μ-thio-dithiodi-, bis(tetraphenylphosphonium), 27:45
$P_2Mo_2S_{10.56}C_{48}H_{40}$, Molybdate(2−), thio-, $(Mo_2S_{10.56})^{2-}$-bis(tetraphenylphosphonium), 27:42
$P_2Mo_2S_7C_{48}H_{40}$, Molybdate(IV, VI), (η^2-disulfido)di-μ-thio-trithiodi-, bis(tetraphenylphosphonium), 27:44
$P_2Mo_5N_3O_{21}C_8H_{26} \cdot 5H_2O$, Pentamolybdobis[(2-aminoethyl)phosphonate] (4−), sodium tetramethylammonium dihydrogen, pentahydrate, 27:126
$P_2Mo_5N_4O_{21}C_2H_{22} \cdot 2H_2O$, Pentamolybdobis(methylphosphonate)-(4−), tetraammonium, dihydrate, 27:124
$P_2Mo_5N_4O_{21}C_4H_{26}$, Pentamolybdobis(ethylphosphonate)-(4−), tetraammonium, 27:125
$P_2Mo_5N_4O_{21}C_{12}H_{26} \cdot 5H_2O$, Pentamolybdobis(phenylphosphate)-(4−), tetraammonium, pentahydrate, 27:125
$P_2Mo_5N_4O_{21}C_{14}H_{26} \cdot 5H_2O$, Pentamolybdobis[(4-aminobenzyl)phosphonate](4−), diammonium dihydrogen, pentahydrate, 27:126
$P_2Mo_5N_4O_{21}C_{22}H_{42} \cdot 4H_2O$, Pentamolybdobis[(4-aminobenzyl)phosphonate]-(4−), bis(tetramethylammonium)dihydrogen, tetrahydrate, 27:127

$P_2NO_2C_{38}H_{33}$, Phosphorus(1+), μ-nitrido-bis(triphenyl)-, acetate, 27:296

$P_2NO_3RhC_{25}H_{46}$, Rhodium(I), carbonyl(4-pyridinecarboxylato)bis(triisopropylphosphine)-, 27:292

$P_2NO_4RhC_{40}H_{30}$, Rhodate(1−), tetracarbonyl-,μ-nitrido-bis(triphenylphosphorus) (1+), 28:213

$P_2NO_7WC_{43}H_{33}$, Tungstate(1−), (acetato)pentacarbonyl-, μ-nitrido-bis(triphenylphosphorus) (1+), 27:297

$P_2NO_{10}Ru_3Si_2C_{58}H_{61}$, Ruthenate(1−), decacarbonyl-1κ^3C,2κ^3C,3κ^4C-μ-hydrido-1:2κ^2H-bis(triethylsilyl)-1κSi,2κSi-*triangulo*-tri-, μ-nitrido-bis(triphenylphosphorus) (1+), 26:269

$P_2NO_{11}Os_3C_{47}H_{31}$, Osmate(1−), μ-carbonyldecacarbonyl-μ-hydrido-tri-, μ-nitrido-bis(triphenylphosphorus) (1+), 28:236

$P_2N_2O_{14}Ru_5C_{50}H_{30}$, Ruthenate(1−), tetradecacarbonylnitridopenta-, μ-nitrido-bis(triphenylphosphorus) (1+), 26:288

$P_2N_2O_{16}Ru_6C_{52}H_{30}$, Ruthenate(1−), hexadecacarbonylnitridohexa-, μ-nitrido-bis(triphenylphosphorus) (1+), 26:287

$P_2N_4C_{36}H_{30}$, Phosphorus(1+), μ-nitrido-bis(triphenyl)-, azide, 26:286

$P_2N_4O_{21}W_5C_{60}H_{122}$, Pentatungstobis(phenylphosphonate) (4−), tetrakis(tributylammonium), 27:127

$P_2N_6NiF_{12}C_{20}H_{34}$, Nickel(II), [2,12-dimethyl-3.11-bis[1-(methylamino)ethylidene]-1,5,9,13-tetraazacyclohexadeca-1,4,9,12-tetraene-$\kappa^4N^{1,5,9,13}$]-, bis[hexafluorophosphate(1−)], 27:266

$P_2Na_{12}O_{56}W_{15}\cdot 24H_2O$, Pentadecatungstodiphosphate(12−), α-, dodecasodium, tetracosahydrate, 27:108

$P_2OOsSC_{42}H_{34}$, Osmium, carbonyl(5-thioxo-1,3-pentadiene-1,5-diyl-C^1,C^5,S)bis(triphenylphosphine)-, 26:188

$P_2O_2RhC_{25}H_{47}$, Rhodium(I), carbonylphenoxybis(triisopropylphosphine)-, 27:292

$P_2O_3RhC_{21}H_{45}$, Rhodium(I), (acetato)carbonylbis(triisopropylphosphine)-, 27:292

$P_2O_3RhC_{44}H_{71}$, Rhodium(I), (benzoato)carbonylbis(tricyclohexylphosphine)-, 27:292

$P_2O_3RhC_{44}H_{73}$, Rhodium(I), carbonyl(4-methylbenzenesulfonato)bis(tricyclohexylphosphine)-, 27:292

$P_2O_3WC_{21}H_{44}$, Tungsten, tricarbonyl(dihydrogen)bis(triisopropylphosphine)-, 27:7

$P_2O_3WC_{39}H_{68}$, Tungsten, tricarbonyl(dihydrogen)bis(tricyclohexylphosphine)-, 27:6

$P_2O_4RhC_{24}H_{49}$, Rhodium(I), carbonyl(1-methoxy-1,3-butanedionato-o)bis(triisopropylphosphine)-, 27:292

$P_2O_5RhC_{45}H_{71}$, Rhodium(I), carbonyl(hydrogen phthalato)bis(tricyclohexylphosphine)-, 27:291

$P_2O_6RhC_{11}H_{23}$, Rhodium(I), (η^5-cyclopentadienyl)bis(trimethyl phosphite)-, 28:284

$P_2O_6RuC_{41}H_{34}$, Ruthenium(II), (η^5-cyclopentadienyl) [2-[(diphenoxyphosphino)oxy]phenyl-C,P] (triphenyl phosphite-P)-, 26:178

$P_2O_{10}Ru_3C_{35}H_{22}$, Ruthenium, decacarbonyl[methylenebis(diphenylphosphine)]tri-, 26:276

$P_2O_{13}Ru_4C_{39}H_{36}$, Ruthenium, decacarbonyl-(dimethylphenylphosphine)tetrahydrido-[tris(4-methylphenyl) phosphite]tetra-, 26:278, 28:229

$P_2O_{14}Ru_5C_{51}H_{15}$, Ruthenate(2−), μ_5-carbidotetradecacarbonylpenta-, bis[μ-nitrido-bis(triphenylphosphorus)-(1+)], 26:284

$P_2O_{22}Ru_6C_{48}H_{20}$, Ruthenium, [$\mu$-ethynediyl-bis(diphenylphosphine)]bis[undecacarbonyltri]-, 26:277

$P_2PdC_{24}H_{54}$, Palladium(0), bis(tri-*tert*-butylphosphine)-, 28:115

$P_2PdC_{28}H_{46}$, Palladium(0), bis(di-*tert*-butylphenylphosphine)-, 28:114

$P_2PdC_{36}H_{66}$, Palladium(0), bis(tricyclohexylphosphine)-, 28:116

$P_2PtC_{14}H_{34}$, Platinum(0), (ethene)bis(triethylphosphine)-, 28:133

$P_2PtC_{20}H_{46}$, Platinum(0), (ethene)bis(triisopropylphosphine) (ethene)-, 28:135
$P_2PtC_{22}H_{34}$, Platinum(0), bis(diethylphenylphosphine)-, 28:135
$P_2PtC_{28}H_{28}$, Platinum(0), [1,2-ethanediylbis(diphenylphosphine)] (ethene)-, 28:135
$P_2PtC_{28}H_{46}$, Platinum(0), bis(di-*tert*-butylphenylphosphine)-, 28:116
$P_2PtC_{36}H_{66}$, Platinum(0), bis(tricyclohexylphosphine)-, 28:116
$P_2PtC_{38}H_{34}$, Platinum(0), (ethene)bis(triphenylphosphine)-, 28:135
$P_2RhC_{11}H_{23}$, Rhodium(I), (η^5-cyclopentadienyl)bis(trimethylphosphine)-, 28:280
$P_2Si_6C_{20}H_{54}$, Diphosphene, bis[tris(trimethylsilyl)methyl]-, 27:241
P_3, Phosphorus, *cyclo*-tri-, molybdenum complex, 27:224
$P_3AlCl_4C_{36}H_{30}$, Triphosphenium, 1,1,1,3,3,3-hexaphenyl-, tetrachloroaluminate(1−), 27:254
$P_3Au_3BF_4OC_{54}H_{45}$, Gold(1+), μ-oxo-[tris-[(triphenylphosphine)-, tetrafluoroborate(1−), 26:326
$P_3Au_3CoO_{12}Ru_3C_{66}H_{45}$, Ruthenium, dodecacarbonyltris(triphenylphosphine)cobalttrigoldtri-, 26:327
$P_3BN_6C_{36}H_{56}$, Triphosphenium, 1,1,1,3,3,3-hexakis(dimethylamino)-, tetraphenylborate(1−), 27:256
$P_3BORhS_2C_{66}H_{59}$, Rhodium(III), [[2-[(diphenylphosphino)methyl]-2-methyl-1,3-propanediyl]bis(diphenylphosphine)](dithiocarbonato)-, tetraphenylborate(1−), 27:287
$P_3C_{34}H_{33}$, Phosphine, bis[2-(diphenylphosphino)ethyl]phenyl-, palladium complex, 27:320
$P_3C_{41}H_{39}$, Phosphine, [2-[(diphenylphosphino)methyl]-2-methyl-1,3-propanediyl]bis(diphenyl)-, rhodium complex, 27:287
$P_3ClCoC_{54}H_{45}$, Cobalt, chlorotris(triphenylphosphine)-, 26:190
$P_3ClIrC_{54}H_{45}$, Iridium(I), chlorotris(triphenylphosphine)-, 26:201
$P_3ClORhS_2C_{42}H_{39}$, Rhodium, chloro[[2-[(diphenylphosphino)methyl]-2-methyl-1,3-propanediyl]bis(diphenylphosphine)](dithiocarbonato)-, 27:289
$P_3ClRhC_{54}H_{45}$, Rhodium(I), chlorotris(triphenylphosphine)-, 28:77
$P_3Cl_2OsC_{54}H_{45}$, Osmium(II), dichlorotri-(triphenylphosphine)-, 26:184
$P_3Cl_2OsSC_{55}H_{45}$, Osmium(II), dichloro(thiocarbonyl)tris(triphenylphosphine)-, 26:185
$P_3Cl_3OsC_{24}H_{33}$, Osmium(III), trichlorotris-(dimethylphenylphosphine)-, *mer*-, 27:27
$P_3Cl_4WC_9H_{27}$, Tungsten(IV), tetrachlorotris(trimethylphosphine)-, 28:327
$P_3F_{12}O_4PdC_{44}H_{35}$, Palladium(II), [bis[2-(diphenylphosphino)ethyl]phenylphosphine] (1,1,1,5,5,5-hexafluoro-2,4-pentanedionato)-, 1,1,1,5,5,5-hexafluoro-2,4-dioxo-3-pentanide, 27:320
$P_3F_{18}N_6C_{26}H_{47}$, 3,10,14,18,21,25-Hexaazabicyclo[10.7.7]hexacosa-1,11,13,18,20,25-hexaene, 2,3,10,11,13,19-hexamethyl-, tris[hexafluorophosphate(1−)], 27:269
$P_3F_{18}N_6C_{50}H_{55}$, 3,11,15,19,22,26-Hexaazatricyclo[11.7.7.1^{5,9}]octacosa-1,5,7,9(28),12,14,19,21,26-nonaene, 3,11-dibenzyl-14,20-dimethyl-2,12-diphenyl-, tris[hexafluorophosphate(1−)], 27:278
$P_3H_3OsRhC_{32}H_{45}$, Rhodium, [2(η^4)-1,5-cyclooctadiene]tris(dimethylphenylphosphine-1κP)-tri-μ-hydrido-osmium-, 27:29
$P_3H_4OsZrC_{34}H_{43}$, Zirconium, bis[1,1(η^5)-cyclopentadienyl]tris(dimethylphenylphosphine-2κP)-tri-μ-hydrido-hydrido-1κH-osmium-, 27:27
$P_3H_6MoC_{54}H_{99}$, Molybdenum(IV), hexahydridotris(tricyclohexylphosphine)-, 27:13
$P_3H_6WC_{24}H_{33}$, Tungsten(IV), hexahydridotris(dimethylphenylphosphine)-, 27:11
$P_3Mn_2NO_8C_{56}H_{40}$, Manganate(1−), μ-(diphenylphosphino)-bis(tetracarbonyl-, (*Mn–Mn*), μ-nitrido-bis(triphenylphosphorus) (1+), 26:228
$P_3MoO_2C_7H_5$, Molybdenum(I), dicarbonyl-(η^5-cyclopentadienyl) (η^3-*cyclo*-triphosphorus)-, 27:224
$P_3OOsC_{56}H_{45}$, Osmium, carbonyl(thiocarbonyl)tris(triphenylphosphine)-, 26:187

$P_3ORhC_{55}H_{46}$, Rhodium(I), carbonylhydridotris(triphenylphosphine)-, 28:82

$P_3PtC_{18}H_{45}$, Platinum(0), tris(triethylphosphine)-, 28:120

$P_3PtC_{27}H_{63}$, Platinum(0), tri(triisopropylphosphine)-, 28:120

$P_3PtC_{54}H_{45}$, Platinum(0), tris(triphenylphosphine)-, 28:125

P_4, Phosphorus, *tetrahedro*-tetra-, rhodium complex, 27:222

$P_4Au_2ClF_3O_3PtSC_{49}H_{60}$, Platinum(1+), chloro-1-κ*Cl*bis(triethylphosphine-1κ*P*)bis(triphenylphosphine)-2κ*P*,3κ*P*-*triangulo*-digold-, trifluoromethanesulfonate, 27:218

$P_4BF_4H_5Ir_2C_{54}H_{52}$, Iridium(1+), pentahydridobis[1,3-propanediylbis(diphenylphosphine)]di-, tetrafluoroborate(1−), 27:22

$P_4BH_3Pt_2C_{48}H_{80}$, Platinum(II), di-μ-hydrido-hydridotetrakis(triethylphosphine)di-, tetraphenylborate(1−), 27:34

$P_4BH_3Pt_2C_{48}H_{80}$, Platinum(II), μ-hydrido-dihydridotetrakis(triethylphosphine)di-, tetraphenylborate(1−), 27:32

—, Platinum(II), di-μ-hydrido-hydridotetrakis(triphenylphosphine)di-, tetraphenylborate(1−), 27:36

$P_4BPtC_{54}H_{33}$, Platinum(II), μ-hydrido-hydridophenyltetrakis(triethylphosphine)di-, tetraphenylborate(1−), 26:136

$P_4ClRhS_2C_{48}H_{54}$, Rhodium, chloro-[[2-[(diphenylphosphino)methyl]-2-methyl-1,3-propanediyl]bis(diphenylphosphine)][(dithiocarboxy)triethylphosphoniumato]-, 27:288

$P_4Cl_2Pd_2C_{50}H_{44}$, Palladium(I), dichlorobis[μ-methylenebis(diphenylphosphine)]di-, (*Pd–Pd*), 28:340

$P_4Cl_2WC_{12}H_{36}$, Tungsten(II), dichlorotetrakis(trimethylphosphine)-, 28:329

$P_4Cl_2WC_{32}H_{44}$, Tungsten(II), dichlorotetrakis(dimethylphenylphosphine)-, 28:330

$P_4Cl_2WC_{52}H_{52}$, Tungsten(II), dichlorotetrakis(methyldiphenylphosphine)-, 28:331

$P_4CrFeN_2O_9C_{81}H_{60}$, Chromate(2−), nonacarbonyliron-, bis[μ-nitrido-bis(triphenylphosphorus) (1+)], 26:339

$P_4FeMoN_2O_9C_{81}H_{60}$, Molybdate(2−), nonacarbonyliron-, bis[μ-nitrido-bis(triphenylphosphorus) (1+)], 26:339

$P_4FeN_2O_9WC_{81}H_{60}$, Tungstate(2−), nonacarbonyliron-, bis[μ-nitrido-bis(triphenylphosphorus) (1+)], 26:339

$P_4Fe_3N_2O_{11}C_{83}H_{60}$, Ferrate(2−), undecacarbonyltri-, bis[μ-nitrido-bis(triphenylphosphorous) (1+)], 28:203

$P_4H_4MoC_{52}H_{52}$, Molybdenum(IV), tetrahydridotetrakis(methyldiphenylphosphine)-, 27:9

$P_4H_4WC_{52}H_{52}$, Tungsten(IV), tetrahydridotetrakis(methyldiphenylphosphine)-, 27:10

$P_4H_8Re_2C_{72}H_{60}$, Rhenium(IV), octahydridotetrakis(triphenylphosphine)di-, 27:16

$P_4MoN_4H_{48}H_{52}$, Molybdenum, bis(dinitrogen)bis[1,2-ethanediylbis(diphenylphosphine)]-, *trans*-, 28:38

$P_4N_2O_{12}PtRh_4C_{84}H_{60}$, Rhodate(2−), dodecacarbonylplatinumtetra-, bis[μ-nitrido-bis(triphenylphosphine) (1+)], 26:375

$P_4N_2O_{14}PtRh_4C_{86}H_{60}$, Rhodate(2−), tetradecacarbonylplatinumtetra-, bis[μ-nitrido-bis(triphenylphosphorus) (1+)], 26:373

$P_4N_2O_{15}Os_5C_{87}H_{60}$, Osmate(2−), pentadecacarbonylpenta-, bis[μ-nitrido-bis(triphenylphosphorus) (1+)], 26:299

$P_4N_2O_{18}Os_6C_{90}H_{60}$, Osmate(2−), octadecacarbonylhexa-, bis[μ-nitrido-bis(triphenylphosphorus) (1+)], 26:308

$P_4N_2WC_{56}H_{54}$, Tungsten(0), bis[1,2-ethanediylbis(diphenylphosphine)]bis-(isocyanomethane)-, *trans*-, 28:43

$P_4N_4WC_{48}H_{52}$, Tungsten, bis(dinitrogen)bis[1,2-ethanediylbis(diphenylphosphine)]-, *trans*-, 28:41

$P_4NiC_{12}H_{32}$, Nickel(0), bis[1,2-ethanediylbis(dimethylphosphine)]-, 28:101

$P_4NiC_{12}H_{36}$, Nickel(0), tetrakis(trimethylphosphine)-, 28:101

$P_4NiC_{24}H_{60}$, Nickel(0), tetrakis(triethylphosphine)-, 28:101

$P_4NiC_{40}H_{60}$, Nickel(0), tetrakis-(diethylphenylphosphine)-, 28:101

$P_4NiC_{48}H_{108}$, Nickel(0), tetrakis-(tributylphosphine)-, 28:101

$P_4NiC_{52}H_{48}$, Nickel(0), bis[1,2-ethanediylbis(diphenylphosphine)]-, 28:103

$P_4NiC_{52}H_{52}$, Nickel(0), tetrakis(methyldiphenylphosphine)-, 28:101

$P_4NiC_{72}H_{60}$, Nickel(0), tetrakis(triphenylphosphine)-, 28:102

$P_4NiO_8C_{32}H_{44}$, Nickel(0), Tetrakis(dimethyl phenylphosphonite)-, 28:101.

$P_4NiO_{12}C_{12}H_{36}$, Nickel(0), tetrakis(trimethyl phosphite)-, 28:101

$P_4NiO_{12}C_{24}H_{60}$, Nickel(0), tetrakis(triethyl phosphite)-, 28:101

$P_4NiO_{12}C_{36}H_{84}$, Nickel(0), tetrakis(isopropyl phosphite)-, 28:101

$P_4NiO_{12}C_{72}H_{60}$, Nickel(0), tetrakis(triphenyl phosphite)-, 28:101

$P_4O_{12}PdC_{24}H_{60}$, Palladium(0), tetrakis(triethyl phosphite)-, 28:105

$P_4O_{12}PtC_{24}H_{60}$, Platinum(0), tetrakis(triethyl phosphite)-, 28:106

$P_4PdC_{72}H_{60}$, Palladium(0), tetrakis(triphenylphosphine)-, 28:107

$P_4PtC_{24}H_{60}$, Platinum(0), tetrakis(triethylphosphine)-, 28:122

$P_4PtC_{72}H_{60}$, Platinum(0), tetrakis(triphenylphosphine)-, 28:124

$P_4RhC_{72}H_{61}$, Rhodium(I), hydridotetrakis(triphenylphosphine)-, 28:81

$P_4RuC_{72}H_{62}$, Ruthenium(II), dihydridotetrakis(triphenylphosphine)-, 28:337

$P_5H_{64}N_{14}NaO_{110}W_{30} \cdot 31H_2O$, Sodiotricontatungstopentaphosphate(14−), tetradecaammonium, hentricontahydrate, 27:115

$P_6B_2F_8H_7Ir_3C_{78}H_{72}$, Iridium(2+), tris[1,2-ethanediylbis(diphenylphosphine)]heptahydridotri-, bis[tetrafluoroborate(1−)], 27:25

$P_6B_2F_8H_7Ir_3C_{81}H_{78}$, Iridium(2+), heptahydridotris[1,3-propanediylbis(diphenylphosphine)]tri-, bis[tetrafluoroborate(1−)], 27:22

$P_6ClRhC_{36}H_{30}$, Rhodium(I), chloro(η^2-*tetrahedro*-tetraphosphorus)bis(triphenylphosphine)-, 27:222

$P_6Cl_6SnC_{52}H_{48}$, 1H-1,2,3-Triphospholium, 3,3,4,5-tetrahydro-1,1,3,3-tetraphenylhexachlorostannate(1−), 27:255

P_7, Tricyclo[2.2.1.0^{2,6}]heptaphosphide(3−), 27:228

P_7Li_3, Lithium heptaphosphide, (Li_3P_7), 27:227

$P_8H_7K_{28}Li_5O_{184}W_{48} \cdot 92H_2O$, Octatetracontatungstooctaphosphate(40−), pentalithium octacosapotassium heptahydrogen, dononacontahydrate, 27:110

$P_{12}Au_{55}Cl_6C_{216}H_{180}$, Gold, hexachlorododecakis(triphenylphosphine)pentapentaconta-, 27:214

P_{16}, Octacyclo[7.7.0.0$^{2,6} \cdot 0^{3,8} \cdot 0^{5,7} \cdot 0^{10,14} \cdot 0^{11,16} \cdot 0^{13,15}$]hexadecaphosphide(2−), 27:228

$P_{16}Li_2$, Lithium hexadecaphosphide, (Li_2P_{16}), 27:227

P_{21}, Decacyclo[9.9.1.0$^{2,10} \cdot 0^{3,7} \cdot 0^{4,9} \cdot 0^{6,8} \cdot 0^{12,20} \cdot 0^{13,17} \cdot 0^{14,19} \cdot 0^{16,18}$] henicosaphosphide(3−), 27:228

$P_{21}Na_3$, Sodium henicosaphosphide, (Na_3P_{21}), 27:227

$PdB_2F_8N_4C_8H_{12}$, Palladium(2+), tetrakis(acetonitrile)-, bis[tetrafluoroborate(1−)], 26:128, 28:63

PdC_8H_{10}, Palladium(II), (η^3-allyl)(η^5-cyclopentadienyl)-, 28:343

$PdClN_2C_{19}H_{19}$, Palladium(II), chloro[2-(2-pyridinylmethyl)phenyl-C^1,N] (3,5-dimethylpyridine)-, 26:210

$PdCl_2C_8H_{12}$, Palladium(II), dichloro(η^4-1,5-cyclooctadiene)-, 28:348

$PdCl_2N_2C_{14}H_{10}$, Palladium, bis(benzonitrile)dichloro-, 28:61

$PdF_{12}N_2O_4C_{20}H_{10}$, Palladium(II), (2,2'-bipyridine) (1,1,1,5,5,5-hexafluoro-2,4-pentanedionato)-, 1,1,1,5,5,5-hexafluoro-2,4-dioxo-3-pentanide, 27:319

$PdF_{12}O_4C_{10}H_2$, Palladium, bis(1,1,1,5,5,5-hexafluoro-2,4-pentanedionato)-, 27:318

$PdF_{12}O_4P_3C_{44}H_{35}$, Palladium(II), [bis[2-(diphenylphosphino)ethyl]phenylphosphine] (1,1,1,5,5,5-hexafluoro-2,4-pentanedionato)-, 1,1,1,5,5,5-hexafluoro-2,4-dioxo-3-pentanide, 27:320

$PdO_2C_{34}H_{28}$, Palladium(0), (1,5-diphenyl-

PdO$_2$C$_{34}$H$_{28}$ (*Continued*)
  1,4-pentadien-3-one)-, 28:110
PdO$_{12}$P$_4$C$_{24}$H$_{60}$, Palladium(0), tetrakis-
  (triethyl phosphite)-, 28:105
PdP$_2$C$_{24}$H$_{54}$, Palladium(0), bis(tri-*tert*-
  butylphosphine)-, 28:115
PdP$_2$C$_{28}$H$_{46}$, Palladium(0), bis(di-*tert*-
  butylphenylphosphine)-, 28:114
PdP$_2$C$_{36}$H$_{66}$, Palladium(0), bis(tricyclohexylphosphine)-, 28:116
PdP$_4$C$_{72}$H$_{60}$, Palladium(0), tetrakis-
  (triphenylphosphine)-, 28:107
Pd$_2$Cl$_2$C$_6$H$_{10}$, Palladium(II), bis(η^3-allyl)di-
  μ-chloro-di-, 28:342
Pd$_2$Cl$_2$N$_2$C$_{18}$H$_{24}$, Palladium(II), di-μ-
  chloro-bis[2-[(dimethylamino)methyl]-
  phenyl-C^1,N]di-, 26:212
Pd$_2$Cl$_2$N$_2$C$_{20}$H$_{16}$, Palladium(II), di-μ-
  chloro-bis(8-quinolylmethyl-C,N)di-,
  26:213
Pd$_2$Cl$_2$N$_2$C$_{24}$H$_{20}$, Palladium(II), di-μ-
  chloro-bis[2-(2-pyridinylmethyl)phenyl-
  C^1,N]di-, 26:209
Pd$_2$Cl$_2$N$_4$C$_{20}$H$_{36}$, Palladium(I), tetrakis-
  (*tert*-butyl isocyanide)di-μ-chloro-di-,
  28:110
Pd$_2$Cl$_2$N$_4$C$_{24}$H$_{18}$, Palladium, di-μ-chloro-
  bis[2-(phenylazo)phenyl-C^1,N^2]di-,
  26:175
Pd$_2$Cl$_2$P$_4$C$_{50}$H$_{44}$, Palladium(I), dichlorobis[μ-methylenebis(diphenylphos-
  phine)]di-, (*Pd–Pd*), 28:340
Pd$_2$Mo$_2$O$_6$P$_2$C$_{52}$H$_{40}$, Molybdenum, hexa-
  carbonylbis(η^5-cyclopentadienyl)bis(triphenylphosphine)dipalladiumdi-,
  26:348
Pd$_2$N$_2$O$_4$C$_{28}$H$_{26}$, Palladium(II), di-μ-
  acetato-bis[2-(2-pyridinylmethyl)phenyl-
  C^1,N]di-, 26:208
PrCl$_2$LiO$_2$Si$_4$C$_{30}$H$_{58}$, Praseodymium,
  bis[η^5-1,3-bis(trimethylsilyl)-
  cyclopentadienyl]di-μ-chloro-
  bis(tetrahydrofuran)lithium-, 27:170
PrO$_3$C$_{45}$H$_{69}$, Praseodymium, tris(2,6-di-*tert*-
  butyl-4-methylphenoxo)-, 22:167
Pr$_2$Cl$_2$Si$_4$C$_{44}$H$_{84}$, Praseodymium, tetrakis-
  [η^5-1,3-bis(trimethylsilyl)-
  cyclopentadienyl]di-μ-chloro-di-, 27:171
PtAu$_2$ClF$_3$O$_3$P$_4$SC$_{49}$H$_{60}$, Platinum(1+),
  chloro-1-κClbis(triethylphosphine-

1κP)bis(triphenylphosphine)-2κP,3κP-
  triangulo-digold-, trifluoromethanesulfonate, 27:218
PtBO$_2$P$_2$C$_{40}$H$_{57}$, Platinum(II), (3-methoxy-
  3-oxo-κO-propyl-κC^1)bis(triethylphosphine)-, tetraphenylborate(1–), 26:138
PtBP$_2$C$_{44}$H$_{63}$, Platinum(II), (η^3-
  cyclooctene)bis(triethylphosphine)-,
  tetraphenylborate(1–), 26:139
PtBP$_4$C$_{54}$H$_{33}$, Platinum(II), μ-hydrido-
  hydridophenyltetrakis(triethylphos-
  phine)di-, tetraphenylborate(1–), 26:136
PtC$_6$H$_{12}$, Platinum(0), tris(ethene)-, 28:129
PtC$_{16}$H$_{24}$, Platinum(0), bis(1,5-cycloocta-
  diene)-, 28:126
PtC$_{21}$H$_{30}$, Platinum(0), tris(bicyclo[2.2.1]-
  hept-2-ene)-, 28:127
PtClF$_3$O$_3$SC$_{13}$H$_3$, Platinum(II), chlorobis-
  (triethylphosphine)(trifluoromethanesulfonato)-, *cis*-, 26:126
PtClF$_3$O$_3$SC$_{13}$H$_{30}$, Platinum(II), chlorobis-
  (triphenylphosphine)(trifluoromethanesulfonato)-, *cis*-, 28:27
PtClP$_2$C$_{26}$H$_{41}$, Platinum(II), chloro-(*cis*-1,2-
  diphenylethenyl)bis(triethylphosphine)-,
  trans-, 26:140
PtCl$_2$C$_8$H$_{12}$, Platinum(II), dichloro(η^4-1,5-
  cyclooctadiene)-, 28:346
PtCl$_2$N$_2$C$_{14}$H$_{10}$, Platinum, bis(benzonitrile)dichloro-, 26:345
PtCl$_2$P$_2$C$_{26}$H$_{24}$, Platinum, dichloro[1,2-
  ethanediylbis(diphenylphosphine)]-,
  26:370
PtCl$_3$KC$_2$H$_4$, Platinate(II), trichloro-
  (ethene)-, potassium, 28:349
PtCl$_4$N$_4$C$_4$H$_{16}$, Platinum(IV),
  dichlorobis(1,2-ethanediamine)-, dichloride, *cis*-, 27:314
PtCl$_4$N$_4$C$_4$H$_{18}$, Platinum(II),
  dichlorobis(1,2-ethanediamine
  monohydrochloride)-, *trans*-, 27:315
PtCo$_2$O$_7$P$_2$C$_{33}$H$_{24}$, Cobalt, heptacarbonyl[1,2-ethanediylbis-
  (diphenylphosphine)]platinumdi-,
  26:370
PtF$_3$O$_4$P$_2$SC$_{14}$H$_{35}$, Platinum(II), hydrido-
  (methanol)bis(triethylphosphine)-, *trans*-,
  trifluoromethanesulfonate, 26:135
PtI$_2$N$_2$C$_6$H$_{14}$, Platinum(II), [*trans*-(R,R)-1,2-
  cyclohexanediamine]diiodo-, 27:284

PtMo$_2$N$_2$O$_6$C$_{30}$H$_{20}$, Molybdenum, [bis(benzonitrile)platinum]hexacarbonylbis(η^5-cyclopentadienyl)di-, (2Mo–Pt), 26:345

PtN$_2$O$_6$C$_{12}$H$_{20}$, Platinum(II), [ascorbato(2–)](cis-1,2-cyclohexanediamine)-, 27:283

PtN$_2$O$_{12}$P$_4$Rh$_4$C$_{84}$H$_{60}$, Rhodate(2–), dodecacarbonylplatinumtetra-, bis[μ-nitrido-bis(triphenylphosphine) (1+)], 26:375

PtN$_2$O$_{14}$P$_4$Rh$_4$C$_{86}$H$_{60}$, Rhodate(2–), tetradecacarbonylplatinumtetra-, bis[μ-nitrido-bis(triphenylphosphorus) (1+)], 26:373

PtO$_6$N$_2$C$_{12}$H$_{20}$, Platinum(II), [ascorbato(2–)][*trans*-(S,S)-1,2-cyclohexanediamine]-, 27:283

PtO$_{12}$P$_4$C$_{24}$H$_{60}$, Platinum(0), tetrakis(triethylphosphite)-, 28:106

PtPC$_{22}$H$_{41}$, Platinum(0), bis(ethene)(tricyclohexylphosphine)-, 28:130

PtP$_2$C$_{14}$H$_{34}$, Platinum(0), (ethene)bis(triethylphosphine)-, 28:133

PtP$_2$C$_{20}$H$_{46}$, Platinum(0), (ethene)bis(triisopropylphosphine)(ethene)-, 28:135

PtP$_2$C$_{22}$H$_{34}$, Platinum(0), bis(diethylphenylphosphine)-, 28:135

PtP$_2$C$_{28}$H$_{28}$, Platinum(0), [1,2-ethanediylbis(diphenylphosphine)](ethene)-, 28:135

PtP$_2$C$_{28}$H$_{46}$, Platinum(0), bis(di-*tert*-butylphenylphosphine)-, 28:116

PtP$_2$C$_{36}$H$_{66}$, Platinum(0), bis(tricyclohexylphosphine)-, 28:116

PtP$_2$C$_{38}$H$_{34}$, Platinum(0), (ethene)bis(triphenylphosphine)-, 28:135

PtP$_3$C$_{18}$H$_{45}$, Platinum(0), tris(triethylphosphine)-, 28:120

PtP$_3$C$_{27}$H$_{63}$, Platinum(0), tri(triisopropylphosphine)-, 28:120

PtP$_3$C$_{54}$H$_{45}$, Platinum(0), tris(triphenylphosphine)-, 28:125

PtP$_4$C$_{24}$H$_{60}$, Platinum(0), tetrakis(triethylphosphine)-, 28:122

PtP$_4$C$_{72}$H$_{60}$, Platinum(0), tetrakis(triphenylphosphine)-, 28:124

Pt$_2$BH$_3$P$_4$C$_{48}$H$_{80}$, Platinum(II), di-μ-hydrido-hydridotetrakis(triethylphosphine)di-, tetraphenylborate(1–), 27:34

Pt$_2$BH$_3$P$_4$C$_{48}$H$_{80}$, Platinum(II), μ-hydrido-dihydridotetrakis(triethylphosphine)di-, tetraphenylborate(1–), 27:32

Pt$_2$BH$_3$P$_4$C$_{96}$H$_{80}$, Platinum(II), di-μ-hydrido-hydridotetrakis(triphenylphosphine)- di-, tetraphenylborate(1–), 27:36

Pt$_2$Mo$_2$O$_6$P$_2$C$_{52}$H$_{40}$, Molybdenum, hexacarbonylbis(η^5-cyclopentadienyl)-bis(triphenylphosphine)diplatinumdi-, 26:347

Pt$_6$N$_2$O$_{12}$C$_{44}$H$_{72}$, Platinate(2–), hexa-μ-carbonyl-hexacarbonylhexa-, bis(tetrabutylammonium), 26:316

Pt$_9$N$_2$O$_{18}$C$_{34}$H$_{40}$, Platinate(2–), tris[tri-μ-carbonyl-tricarbonyltri-, bis(tetraethylammonium), 26:322

Pt$_{12}$N$_2$O$_{24}$C$_{40}$H$_{40}$, Platinate(2–), tetrakis-[tri-μ-carbonyl-tricarbonyltri-, bis-(tetraethylammonium), 26:321

Pt$_{15}$N$_2$O$_{30}$C$_{46}$H$_{40}$, Platinate(2–), pentakis-[tri-μ-carbonyl-tricarbonyltri-, bis-(tetraethylammonium), 26:320

Rb$_3$Br$_9$Cr$_2$, Chromate(3–), nonabromodi-, trirubidium, 26:379

Rb$_3$Br$_9$Ti$_2$, Titanate(3–), nonabromodi-, trirubidium, 26:379

Rb$_3$Br$_9$V$_2$, Vanadate(3–), nonabromodi-, trirubidium, 26:379

Rb$_3$Cl$_9$Ti$_2$, Titanate(3–), nonachlorodi-, tricesium, 26:379

Rb$_3$Cl$_9$V$_2$, Vanadate(3–), nonachlorodi-, trirubidium, 26:379

Rb$_4$As$_2$H$_4$O$_{70}$W$_{21}$·34H$_2$O, Tungstate(4–), aquadihydroxohenhexacontaoxobis-[trioxoarsenato(III)]henicosa-, tetrarubidium, tetratricontahydrate, 27:113

ReBF$_4$O$_5$C$_5$, Rhenium, pentacarbonyl[tetrafluoroborato(1–)]-, 26:108

ReBF$_4$O$_5$C$_7$H$_4$, Rhenium (1+), pentacarbonyl(η^2-ethene)-, tetrafluoroborate(1–), 26:110

ReBrO$_5$C$_5$, Rhenium, bromopentacarbonyl-, 28:162

ReClO$_5$C$_5$, Rhenium, pentacarbonylchloro-, 28:161

ReFN$_2$O$_3$C$_{13}$H$_8$, Rhenium, (2,2'-bipyridine)-tricarbonylfluoro-, 26:82, 83

ReF$_2$N$_2$O$_5$PC$_{13}$H$_8$, Rhenium, (2,2'-bipyridine)tricarbonyl(phosphorodifluoridato)-, 26:83
ReF$_3$O$_8$SC$_6$, Rhenium(I), pentacarbonyl-(trifluoromethanesulfonato)-, 26:115
ReF$_4$O$_{12}$C$_{12}$·4H$_2$O, Rhenium, dodecacarbonyltetrafluorotetra-, tetrahydrate, 26:82
ReHO$_5$C$_5$, Rhenium, pentacarbonylhydrido-, 26:77
ReH$_7$P$_2$C$_{36}$H$_{30}$, Rhenium(VII), heptahydridobis(triphenylphosphine)-, 27:15
ReIO$_5$C$_5$, Rhenium, pentacarbonyliodo-, 28:163
ReI$_6$K$_2$, Rhenate (IV), hexaiodo-, dipotassium, 27:294
ReNO$_4$C$_{16}$H$_{36}$, Perrenate, tetrabutylammonium, 26:391
ReO$_4$S$_{16}$C$_{20}$H$_{16}$, Bis(2,2'-bi-1,3-dithiolo-[4,5-b][1,4]dithiinylidene)perrhenate, 26:391
ReO$_5$C$_5$H, Rhenium, pentacarbonylhydrido-, 28:165
ReO$_5$C$_6$H$_3$, Rhenium, pentacarbonylmethyl-, 26:107
ReO$_6$C$_5$H, Rhenium, tetracarbonylcarboxy-, 26:112
ReO$_6$C$_7$H$_3$, Rhenium, acetylpentacarbonyl-, 28:201
Re$_2$Cl$_8$N$_2$C$_{32}$H$_{72}$, Rhenate(III), octachlorodi, bis(tetrabutylammonium), 28:332
Re$_2$H$_8$P$_4$C$_{72}$H$_{60}$, Rhenium(IV), octahydridotetrakis(triphenylphosphine)di-, 27:16
Re$_4$C$_{20}$O$_{22}$, Rhenium, octadecacarbonylbis-(μ_3-carbon dioxide)tetra-, 26:111
Re$_4$O$_{22}$C$_{20}$, Rhenium, octadecacarbonylbis(μ_3-carbon dioxide)tetra-, 28:20
RhBOP$_3$S$_2$C$_{66}$H$_{59}$, Rhodium(III), [[2-[(diphenylphosphino)methyl]-2-methyl-1,3-propanediyl]bis(diphenylphosphine)]-(dithiocarbonato)-, tetraphenylborate(1−), 27:287
RhClC$_{16}$H$_{28}$, Rhodium(I), chlorobis-(cyclooctene)-, 28:90
RhClOP$_2$C$_{36}$H$_{30}$, Rhodium, carbonylchlorobis(triphenylphosphine)-, trans-, 28:79
RhClOP$_3$S$_2$C$_{42}$H$_{39}$, Rhodium, chloro[[2-[(diphenylphosphino)methyl]-2-methyl-1,3-propanediyl]bis(diphenylphosphine)](dithiocarbonato)-, 27:289
RhClP$_3$C$_{54}$H$_{45}$, Rhodium(I), chlorotris(triphenylphosphine)-, 28:77
RhClP$_4$S$_2$C$_{48}$H$_{54}$, Rhodium, chloro[[2-[(diphenylphosphino)methyl]-2-methyl-1,3-propanediyl]bis(diphenylphosphine)][(dithiocarboxy)triethylphosphoniumato]-, 27:288
RhClP$_6$C$_{36}$H$_{30}$, Rhodium(I), chloro(η^2-tetrahedro-tetraphosphorus)bis(triphenylphosphine)-, 27:222
RhFO$_3$P$_2$C$_{44}$H$_{34}$, Rhodium(I), carbonyl(3-fluorobenzoato)bis(triphenylphosphine)-, 27:292
RhH$_3$OsP$_3$C$_{32}$H$_{45}$, Rhodium, [2(η^4)-1,5-cyclooctadiene]tris(dimethylphenylphosphine-1κP)-tri-μ-hydrido-osmium-, 27:29
RhIOP$_2$C$_{37}$H$_{66}$, Rhodium(I), carbonyliodobis(tricyclohexylphosphine)-, 27:292
RhNO$_3$P$_2$C$_{25}$H$_{46}$, Rhodium(I), carbonyl(4-pyridinecarboxylato)bis(triisopropylphosphine)-, 27:292
RhNO$_4$P$_2$C$_{40}$H$_{30}$, Rhodate(1−), tetracarbonyl-, μ-nitrido-bis(triphenylphosphorus)(1+), 28:213
RhOP$_3$C$_{55}$H$_{46}$, Rhodium(I), carbonylhydridotris(triphenylphosphine)-, 28:82
RhO$_2$P$_2$C$_{25}$H$_{47}$, Rhodium(I), carbonylphenoxybis(triisopropylphosphine)-, 27:292
RhO$_3$P$_2$C$_{21}$H$_{45}$, Rhodium(I), (acetato)carbonylbis(triisopropylphosphine)-, 27:292
RhO$_3$P$_2$C$_{44}$H$_{71}$, Rhodium(I), (benzoato)carbonylbis(tricyclohexylphosphine)-, 27:292
RhO$_3$P$_2$C$_{44}$H$_{73}$, Rhodium(I), carbonyl(4-methylbenzenesulfonato)bis(tricyclohexylphosphine)-, 27: 292
RhO$_4$P$_2$C$_{24}$H$_{49}$, Rhodium(I), carbonyl(1-methoxy-1,3-butanedionato-o)bis-(triisopropylphosphine)-, 27:292
RhO$_5$P$_2$C$_{45}$H$_{71}$, Rhodium(I), carbonyl(hydrogen phthalato)-bis(tricyclohexylphosphine)-, 27:291
RhO$_6$P$_2$C$_{11}$H$_{23}$, Rhodium(I), (η^5-cyclo-

pentadienyl)bis(trimethyl phosphite)-, 28:284

RhP$_2$C$_{11}$H$_{23}$, Rhodium(I), (η^5-cyclopentadienyl)bis(trimethylphosphine)-, 28:280

RhP$_4$C$_{72}$H$_{61}$, Rhodium(I), hydridotetrakis-(triphenylphosphine)-, 28:81

Rh$_2$Cl$_2$C$_8$H$_{16}$, Rhodium(I), di-μ-chlorotetrakis(ethene)di-, 28:86

Rh$_2$Cl$_2$C$_{16}$H$_{24}$, Rhodium(I), di-μ-chlorobis-(η^4-1,5-cyclooctadiene)di-, 28:88

Rh$_2$Cl$_2$O$_4$C$_4$, Rhodium, tetracarbonyldichlorodi-, 28:84

Rh$_4$N$_2$O$_{12}$P$_4$PtC$_{84}$H$_{60}$, Rhodate(2−), dodecacarbonylplatinumtetra-, bis[μ-nitrido-bis(triphenylphosphine)(1+)], 26:375

Rh$_4$N$_2$O$_{14}$P$_4$PtC$_{86}$H$_{60}$, Rhodate(2−), tetradecacarbonylplatinumtetra-, bis[μ-nitrido-bis(triphenylphosphorus) (1+)], 26:373

Rh$_4$O$_{12}$C$_{12}$, Rhodium, tri-μ-carbonyl-nonacarbonyltetra-, 28:242

RuB$_2$N$_8$C$_{60}$H$_{76}$, Ruthenium(II), (η^4-1,5-cyclooctadiene)tetrakis(methylhydrazine)-, bis[tetraphenylborate-(1−)], 26:74

RuB$_2$N$_8$RuC$_{56}$H$_{68}$, Ruthenium(II), (η^4-1,5-cyclooctadiene)tetrakis(hydrazine)-, bis-[tetraphenylborate(1−)], 26:73

RuBrF$_6$N$_3$PC$_{14}$H$_{21}$, Ruthenium(II), tris-(acetonitrile)bromo(η^4-1,5-cyclooctadiene)-, hexafluorophosphate(1−), 26:72

RuBr$_2$N$_2$C$_{22}$H$_{22}$, Ruthenium(II), bis(benzonitrile)dibromo(η^5-1,5-cyclooctadiene)-, 26:71

RuBr$_3$H$_{12}$N$_4$, Ruthenium(III), tetraamminedibromo-, cis-, bromide, 26:66

RuC$_{13}$H$_{18}$, Ruthenium(II), (η^4-bicyclo[2.2.1]-hepta-2,5-diene)bis(η^2-2-propenyl)-, 26:251

RuC$_{14}$H$_{22}$, Ruthenium(II), (η^4-cycloocta-1,5-diene)bis(η^3-2-propenyl)-, 26:254

RuClF$_6$N$_3$PC$_{14}$H$_{21}$, Ruthenium(II), tris-(acetonitrile)chloro(η^4-1,5-cyclooctadiene)-, hexafluorophosphate(1−), 26:71

RuClPC$_{30}$H$_{34}$, Ruthenium(II), chloro(η^6-hexamethylbenzene)hydrido-(triphenylphosphine)-, 26:181

RuClP$_2$C$_{41}$H$_{35}$, Ruthenium(II), chloro(η^5-cyclopentadienyl)bis(triphenylphosphine)-, 28:270

RuCl$_2$C$_7$H$_8$, Ruthenium(II), (η^4-bicyclo[2.2.1]hepta-2,5-diene)dichloro-, 26:250

RuCl$_2$C$_8$H$_{12}$, Ruthenium(II), dichloro(η^4-cycloocta-1,5-diene)-, 26:253

RuCl$_2$C$_8$H$_{12}$, Ruthenium(II), di-μ-chloro(η^4-1,5-cyclooctadiene)-, polymer, 26:69

RuCl$_2$N$_2$C$_{12}$H$_{18}$, Ruthenium(II), bis(acetonitrile)dichloro(η^4-1,5-cyclooctadiene)-, 26:69

RuCl$_2$N$_2$C$_{22}$H$_{22}$, Ruthenium(II), bis(benzonitrile)dichloro(η^5-1,5-cyclooctadiene)-, 26:70

RuCl$_2$N$_6$O$_6$C$_{30}$H$_{36}$·6H$_2$O, Ruthenium(II), tris-(2,2'-bipyridine)-, dichloride, hexahydrate, 28:338

RuCl$_3$H$_{12}$N$_4$, Ruthenium(III), tetraamminedichloro-, cis-, chloride, 26:66

RuCl$_4$N$_3$C$_{12}$H$_{26}$, Ruthenate(1−), tetrachlorobis(acetonitrile)-tetraethylammonium, 26:356

RuCoMoO$_8$C$_{17}$H$_{11}$, Ruthenium, cyclo-[μ_3-1(η^2):2(η^2):3(η^2)-2-butyne]-octacarbonyl-1κ^2C,2κ^3C,3κ^3C-[1(η^5)-cyclopentadienyl]cobaltmolybdenum-, (Co—Mo)(Co—Ru)(Mo—Ru), 27:194

RuCoO$_9$SC$_9$, Ruthenium, nonacarbonyl-μ_3-thio-dicobalt-, 26:352

RuCo$_2$O$_9$C$_{13}$H$_6$, Ruthenium, μ_3-2-butyne-nonacarbonyldicobalt-, 27:194

RuCo$_2$O$_{11}$C$_{11}$, Ruthenium, undecacarbonyldicobalt-, 26:354

RuCo$_3$NO$_{12}$C$_{12}$H$_{20}$, Ruthenate(1−), dodecacarbonyltricobalt-, tetraethylammonium, 26:358

RuF$_{12}$N$_4$P$_2$C$_{16}$H$_{24}$, Ruthenate(II), tetrakis-(acetonitrile)(η^4-1,5-cyclooctadiene)-, bis[hexafluorophosphate(1−)], 26:72

RuF$_{12}$N$_8$P$_2$C$_{12}$H$_{36}$, Ruthenium(II), (η^4-cyclooctadiene)tetrakis(methylhydrazine)-, bis[hexafluorophosphate-(1−)], 26:74

RuN$_2$O$_2$C$_{28}$H$_{16}$, Ruthenium(II), bis(benzo[h]quinolin-10-yl-C^{10},N]-

RuN$_2$O$_2$C$_{28}$H$_{16}$ *(Continued)*
  dicarbonyl-, *cis*, 26:177
RuO$_3$C$_{11}$H$_{12}$, Ruthenium(0), tricarbonyl(1,5-cyclooctadiene)-, 28:54
RuO$_6$C$_8$H$_6$, Ruthenium, tetracarbonyl(η^2-methyl acrylate)-, 28:47
RuO$_6$P$_2$C$_{41}$H$_{34}$, Ruthenium(II), (η^5-cyclopentadienyl)[2-[(diphenoxyphosphino)oxy]-phenyl-*C,P*](triphenyl phosphite-*P*)-, 26:178
RuPC$_{30}$H$_{33}$, Ruthenium(II), [2-(diphenyl-phosphino)phenyl-*C^1,P*](η^6-hexa-methylbenzene)hydrido-, 26:182
RuP$_4$C$_{72}$H$_{62}$, Ruthenium(II), dihydrido-tetrakis(triphenylphosphine)-, 28:337
Ru$_2$Cl$_4$O$_6$C$_6$, Ruthenium(II), di-μ-chloro-bis(tricarbonylchloro-, 28:334
Ru$_2$Cl$_4$O$_9$C$_{24}$H$_{34}$, Ruthenium(II), μ-aqua-bis(μ-chloroacetato)bis[(chloroacetato)(η^4-cycloocta-1,5-diene)-, 26:256
Ru$_2$Cl$_{12}$O$_9$C$_{22}$H$_{18}$, Ruthenium(II), μ-aqua-bis(μ-trichloroacetato)bis[(η^4-bicyclo-[2.2.1]heptadiene)(trichloroacetato)-, 26:256
Ru$_2$F$_{12}$O$_9$C$_{24}$H$_{26}$, Ruthenium(II), μ-aqua-bis(μ-trifluoroacetato)bis[(η^4-cycloocta-1,5-diene)(trifluoroacetato)-, 26:254
Ru$_2$O$_4$C$_{14}$H$_{10}$, Ruthenium, tetracarbonyl-bis(η^5-cyclopentadienyl)di-, 28:189
Ru$_3$Au$_3$CoO$_{12}$P$_3$C$_{66}$H$_{45}$, Ruthenium, dodecacarbonyltris(triphenylphosphine)-cobalttrigoldtri-, 26:327
Ru$_3$BrH$_3$O$_9$C$_{10}$, Ruthenium, (μ-bromomethylidyne)nonacarbonyl-tri-μ-hydrido-*triangulo*-tri-, 27:201
Ru$_3$BrHgO$_9$C$_{15}$H$_9$, Ruthenium, (bromomercury)nonacarbonyl(3,3-dimethyl-1-butynyl)-*triangulo*-tri-, 26:332
Ru$_3$HO$_{11}$C$_{12}$H$_3$, Ruthenium, decacarbonyl-μ-hydrido(μ-methoxymethylidyne)-*triangulo*-tri-, 27:198
Ru$_3$H$_3$O$_{10}$C$_{11}$H$_3$, Ruthenium, nonacarbonyl-tri-μ-hydrido(μ_3-methoxymethylidyne)-*triangulo*-tri-, 27:200
Ru$_3$HgIO$_9$C$_{15}$H$_9$, Ruthenium, nonacarbonyl(3,3-dimethyl-1-butynyl)-(iodomercury)-*triangulo*-tri-, 26:330
Ru$_3$MoHgO$_{12}$C$_{23}$H$_{14}$, Ruthenium, nonacarbonyl(μ_3-3,3-dimethyl-1-butynyl){μ-[tricarbonyl(η^5-cyclopentadienyl)molybdenum]-mercury}-*triangulo*-tri-, 26:333
Ru$_3$NO$_{10}$PC$_{23}$H$_{20}$, Ruthenium, decacarbonyl(dimethyl-phenylphosphine)(2-isocyano-2-methylpropane)tri-, 26:275, 28:224
Ru$_3$NO$_{10}$P$_2$Si$_2$C$_{58}$H$_{61}$, Ruthenate(1−), decacarbonyl-1κ^3C,2κ^3C,3κ^4C-μ-hydrido-1:2κH-bis(triethylsilyl)-1κSi,2κSi-*triangulo*-tri-, μ-nitrido-bis-(triphenylphosphorus)(1+), 26:269
Ru$_3$NiO$_9$C$_{14}$H$_8$, Ruthenium, nonacarbonyl-(η^5-cyclopentadienyl)tri-μ-hydrido-nickel-tri-, 26:363
Ru$_3$O$_9$C$_{15}$H$_{10}$, Ruthenium, nonacarbonyl(μ_3-3,3-dimethyl-1-butynyl)-μ-hydrido-*triangulo*-tri-, 26:329
Ru$_3$O$_9$PC$_{21}$H$_{11}$, Ruthenium, nonacarbonyl-μ-hydrido-(μ-diphenylphosphido)tri-, 26:264
Ru$_3$O$_{10}$P$_2$C$_{35}$H$_{22}$, Ruthenium, decacarbonyl[methylenebis(diphenylphos-phine)]tri-, 26:276
Ru$_3$O$_{11}$PC$_{19}$H$_{11}$, Ruthenium, undecacarbonyl(dimethyl-phenylphosphine)tri-, 26:273, 28:223
Ru$_3$O$_{12}$C$_{12}$, Ruthenium, dodecacarbonyl-tri-, 26:259
Ru$_4$H$_4$O$_{12}$C$_{12}$, Ruthenium, dodecacarbonyl-tetra-μ-hydrido-tetra-, 28:219
Ru$_4$O$_{12}$C$_{12}$H$_4$, Ruthenium, dodecacarbonyl-tetra-μ-hydrido-tetra-, 26:262
Ru$_4$O$_{13}$P$_2$C$_{39}$H$_{36}$, Ruthenium, decacar-bonyl(dimethylphenylphosphine)-tetrahydrido-[tris(4-methylphenyl) phosphite]tetra-, 26:278
Ru$_4$O$_{14}$PC$_{32}$H$_{25}$, Ruthenium, undecacarbonyltetrahydrido[tris(4-methylphenyl) phosphite]tetra-, 26:277
Ru$_5$N$_2$O$_{14}$P$_2$C$_{50}$H$_{30}$, Ruthenate(1−), tetradecacarbonylnitridopenta-, μ-nitrido-bis(triphenylphosphorus)(1+), 26:288
Ru$_5$Na$_2$O$_{14}$C$_{15}$, Ruthenate(2−), μ_5-carbido-tetradecacarbonylpenta-, disodium, 26:284

$Ru_5O_{15}C_{16}$, Ruthenium, μ_5-carbido-pentadecacarbonylpenta-, 26:283

$Ru_5P_2O_{14}C_{51}H_{15}$, Ruthenate(2−), μ_5-carbido-tetradecacarbonylpenta-, bis-[μ-nitrido-bis(triphenylphosphorus)-(1+)], 26:284

$Ru_6HgO_{18}C_{30}H_{18}$, Ruthenium, (μ_4-mercury)bis[nonacarbonyl(μ_3-3,3-dimethyl-1-butynyl)-*triangulo*-tri-, 26:333

$Ru_6N_2O_{16}P_2C_{52}H_{30}$, Ruthenate(1−), hexadecacarbonylnitridohexa-, μ-nitrido-bis-(triphenylphosphorus(1+), 26:287

$Ru_6O_{17}C_{18}$, Ruthenium, μ_6-carbido-heptadecacarbonylhexa-, 26:281

$Ru_6O_{22}P_2C_{48}H_{20}$, Ruthenium, [$\mu$-ethynediylbis(diphenylphosphine)]-bis[undecacarbonyltri]-, 26:277

S, Sulfur, cobalt and iron complexes, 26:244
  cobalt, iron, and ruthenium, complexes, 26:352
  osmium cluster complex, 26: 305–307
$SAuClC_4H_8$, Gold(I), chloro(tetrahydrothiophene)-, 26:86
$SAuF_5C_{10}H_8$, Gold(I), (pentafluorophenyl)(tetrahydrothiophene)-, 26:86
$SAuF_{15}C_{22}H_8$, Gold(III), tris(pentafluorophenyl)(tetrahydrothiophene)-, 26:87
$SAu_2ClF_3O_3P_4PtC_{49}H_{60}$, Platinum(1+), chloro-1-κClbis(triethylphosphine-1κP)-bis(triphenylphosphine)-2κP,3κP-*triangulo*-digold-, trifluoromethanesulfonate, 27:218
$SAu_2F_5NPSC_{25}H_{15}$, Gold(I), (pentafluorophenyl)-μ-thiocyanato-(triphenylphosphine)di-, 26:90
$SBrF_5C_2H_2$, λ^6-Sulfane, (2-bromoethenyl)pentafluoro-, 27:330
$SBrNO_4C_{10}H_{18}$, Bicyclo[2.2.1]heptane-7-methanesulfonate, 3-bromo-1,7-dimethyl-2-oxo-, [(1R)-(*ENDO, ANTI*)]-, ammonium, 26:24
SC, Thiocarbonyls, iron, 28:186
  osmium, 26:185–187
SC_4H_8, Thiophene, tetrahydro-, gold complexes, 26:85–87
SC_5H_6, 2,4-Pentadienthial, osmium complex, 26:188
SC_6H_6, Benzenethiol, osmium complex, 26:304

$SClF_3O_3PtC_{13}H_{30}$, Platinum(II), chlorobis-(triethylphosphine)(trifluoromethanesulfonato)-, *cis*-, 26:126, 28:27
$SCl_2OsP_3C_{55}H_{45}$, Osmium(II), dichloro(thiocarbonyl)tris(triphenylphosphine)-, 26:185
$SCo_2FeO_9C_9$, Iron, nonacarbonyl-μ_3-thiodicobalt-, 26:245, 352
$SCo_2O_9RuC_9$, Ruthenium, nonacarbonyl-μ_3-thio-dicobalt-, 26:352
$SFNO_3C_{16}H_{36}$, Fluorosulfate, tetrabutylammonium, 26:393
$SF_3K_2MnO_4$, Manganate(III), trifluorosulfato-, dipotassium, 27:312
$SF_3MnO_8C_6$, Manganese(I), pentacarbonyl-(trifluoromethanesulfonato)-, 26:114
SF_3O_3CH, Methanesulfonic acid, trifluoro-, 28:70
  iridium and platinum complexes, 28:26, 27
  iridium, manganese and rhenium complexes, 26:114, 115, 120
  platinum complex, 26:126
$SF_3O_4P_2PtC_{14}H_{35}$, Platinum(II), hydrido-(methanol)bis(triethylphosphine)-, *trans*-, trifluoromethanesulfonate, 26:135
$SF_3O_8ReC_6$, Rhenium(I), pentacarbonyl-(trifluoromethanesulfonato)-, 26:115
SF_5C_2H, λ^6-Sulfane, ethynylpentafluoro-, 27:329
$SFe_3O_9C_9H_2$, Iron, nonacarbonyldihydrido-μ_3-thiotri-, 26:244
SH_2, Hydrogen sulfide, titanium complex, 27:66
  tungsten complex, 27:67
$SMnO_8PC_{12}H_{12}$, Manganese, tetracarbonyl[2-(dimethylphosphinothioyl)-1,2-bis(methoxycarbonyl)ethenyl-*C,S*]-, 26:162
$SMnO_{11}PC_{17}H_{18}$, Manganese, tricarbonyl-[η^2-3,4,5,6-tetrakis(methoxycarbonyl)-2,2-dimethyl-2*H*-1,2-thiophosphorin-2-ium]-, 26:165
SNC, Thiocyanate, gold complex, 26:90
$SNO_3C_7H_5$, 1,2-Benzisothiazol-3(2*H*)-one 1,1-dioxide, chromium and vanadium complex, 27:307, 309
$SOOsPC_{56}H_{45}$, Osmium, carbonyl-(thiocarbonyl)tris(triphenylphosphine)-, 26:187

SOOsP$_2$C$_{45}$H$_{34}$, Osmium, carbonyl(5-thioxo-1,3-pentadiene-1,5-diyl-C^1,C^5,S) bis(triphenylphosphine)-, 26:188

SO$_2$C$_7$H$_8$, Benzenesulfonic acid, 4-methyl-, rhodium complex, 27:292

SO$_4$PC$_8$H$_{12}$, 2-Butenedioic acid, 2-(dimethylphosphinothioyl)-, dimethyl ester, manganese complex, 26:163

SO$_8$C$_{12}$H$_{12}$, Thiophenetetracarboxylic acid, tetramethyl ester, 26:166

SO$_{10}$Os$_3$C$_{10}$, Osmium, μ_3-carbonylnonacarbonyl-μ_3-thio-tri-, 26:305

SO$_{10}$Os$_3$C$_{16}$H$_6$, Osmium, (μ-benzenethiolato)-(decacarbonyl-μ-hydrido-tri-, 26:304

SOsP$_3$C$_{55}$H$_{47}$, Osmium(II), dihydrido(thiocarbonyl)tris(triphenylphosphine)-, 26:186

SPC$_2$H$_7$, Phosphine sulfide, dimethyl-, and manganese complex, 26:162

SPO$_8$C$_{14}$H$_{20}$, 2H-1,2-Thiaphosphorin-2-ium, 3,4,5,6-tetrakis(methoxycarbonyl)-2,2-dimethyl-, manganese complex, 26:165

S$_2$BOP$_3$RhC$_{66}$H$_{59}$, Rhodium(III), [[2-[(diphenylphosphino)methyl]-2-methyl-1,3-propanediyl]bis(diphenylphosphine)]-(dithiocarbonato)-, tetraphenylborate(1−), 27:287

S$_2$C$_2$H$_4$, Methyl dithioformate, 28:186

S$_2$ClOP$_3$RhC$_{42}$H$_{39}$, Rhodium, chloro[[2-[(diphenylphosphino)methyl]-2-methyl(1,3-propanediyl]bis(diphenylphosphine)] (dithiocarbonato)-, 27:289

S$_2$ClP$_4$RhC$_{48}$H$_{54}$, Rhodium, chloro[[2-[(diphenylphosphino)methyl]-2-methyl-1,3-propanediyl]bis(diphenylphosphine)] [(dithiocarboxy) triethylphosphoniumato]-, 27:288

S$_2$CrN$_2$O$_{10}$C$_{14}$H$_{16}$ · 2H$_2$O, Chromium(II), tetraaquabis(1,2-benzisothiazol-3(2H)-one 1,1-dioxidato)-, dihydrate, 27:309

S$_2$F$_3$FeO$_5$C$_9$H$_5$, Iron(1+), dicarbonyl(η^5-cyclopentadienyl) (thiocarbonyl)-, trifluoromethanesulfonate, 28:186

S$_2$F$_6$IrO$_7$P$_2$C$_{39}$H$_{31}$, Iridium(III), carbonylhydridobis(trifluoromethanesulfonato)bis(triphenylphosphine)-, 26:120, 28:26

S$_2$FeO$_2$C$_9$H$_8$, Iron, dicarbonyl(η^5-cyclopentadienyl)-[(methylthio)thiocarbonyl]-, 28:186

S$_2$Mn$_2$O$_8$P$_2$C$_{12}$H$_{12}$, Manganese, octacarbonylbis(μ-dimethylphosphinothioyl-P,S)di-, 26:162

S$_2$NC$_5$H$_{11}$, Dithiocarbamic acid, diethyl-, molybdenum complex, 28:145

S$_2$N$_2$O$_{10}$VC$_{14}$H$_{16}$ · 2H$_2$O, Vanadium(II), tetraaquabis(1,2-benzisothiazol-3(2H)-one 1,1-dioxidato)-, dihydrate, 27:307

S$_2$N$_6$O$_6$VC$_{34}$H$_{28}$ · 2NC$_5$H$_5$, Vanadium(III), bis(1,2-benzisothiazol-3(2H)-one 1,1-dioxidato)tetrakis(pyridine), -2pyridine, 27:308

S$_2$OCH, Dithiocarbonic acid, 27:287

S$_2$O$_9$Os$_3$C$_9$, Osmium, nonacarbonyl-μ_3-thio-tri-, 26:306

S$_2$O$_{12}$Os$_4$C$_{12}$, Osmium, dodecacarbonyldi-μ_3-thio-tetra-, 27:307

S$_2$O$_{13}$Os$_4$C$_{13}$, Osmium, tridecacarbonyldi-μ_3-thio-tetra-, 27:307

S$_2$PC$_7$H$_{16}$, Phosphonium, (dithiocarboxy)triethyl-, rhodium complex, 27:288

S$_2$TiC$_{10}$H$_{12}$, Titanium(IV), bis(η^5-cyclopentadienyl)bis(hydrogen sulfido)-, 27:66

S$_2$WC$_{10}$H$_{12}$, Tungsten(IV), bis(η^5-cyclopentadienyl)bis(hydrogen sulfido)-, 27:67

S$_3$TiC$_{20}$H$_{30}$, Titanium(IV), bis(η^5-pentamethylcyclopentadienyl)[trisulfido-(2−)]-, 27:62

S$_4$MoC$_{10}$H$_{10}$, Molybdenum(IV), bis(η^5-cyclopentadienyl)[tetrasulfido(2−)]-, 27:63

S$_4$MoN$_2$O$_3$C$_{13}$H$_{20}$, Molybdenum(II), tricarbonylbis(diethyldithiocarbamato)-, 28:145

S$_4$MoN$_4$O$_2$C$_{10}$H$_{20}$, Molybdenum, bis(diethyldithiocarbamato)dinitrosyl-, cis-, 28:145

S$_4$MoP$_2$C$_{48}$H$_{40}$, Molybdate(VI), tetrathio-, bis(tetraphenylphosphonium), 27:41

S$_4$V$_2$C$_{12}$H$_{14}$, Vanadium, (μ-disulfido-S:S')bis(η^5-methylcyclopentadienyl)-di-μ-thio-di-, 27:55

S$_5$C$_5$H$_4$, 1,3-Dithiolo[4,5,-b][1,4]dithiin-2-thione, 26:389

S$_5$Cr$_2$C$_{20}$H$_{30}$, Chromium, (μ-disulfido-S:S)(μ-η^2:η^2-disulfido)bis(η^5-pentamethylcyclopentadienyl)-μ-thio-di-, (Cr–Cr), 27:69

S$_5$TiC$_{10}$H$_{10}$, Titanium(IV), bis(η^5-cyclo-

pentadienyl)[pentasulfido(2−)]-, 27:60
$S_5TiC_{12}H_{14}$, Titanium, bis(η^5-methyl cyclopentadienyl)(pentasulfido-S^1:S^5)-, 27:52
$S_5V_2C_{12}H_{14}$, Vanadium, (μ-disulfido-S:S')-(μ-η^2:η^2-disulfido)bis(η-methyl-cyclopentadienyl)-μ-thio-di-, 27:54
$S_6Mo_2P_2C_{48}H_{40}$, Molybdate(V), di-μ-thio-tetrathiodi-, bis(tetraphenylphosphonium), 27:43
S_7AsBrF_6, cyclo-Heptasulfur(1+), bromo-, hexafluoroarsenate(1−), 27:336
S_7AsF_6I, cyclo-Heptasulfur(1+), iodo-, hexafluoroarsenate(1−), 27:333
S_7BrF_6Sb, cyclo-Heptasulfur(1+), bromo-, hexafluoroantimonate(1−), 27:336
S_7F_6ISb, cyclo-Heptasulfur(1+), iodo-, hexafluoroantimonate(1−), 27:333
$S_7Mo_2P_2C_{48}H_{40}$, Molybdate(IV,VI), (η^2-disulfido)-di-μ-thio-trithiodi-, bis(tetraphenylphosphonium), 27:44
$S_8C_{10}H_8$, 2,2'-Bi-1,3-dithiolo[4,5-b][1,4]-dithiinylidene, 26:386
$S_8Mo_2P_2C_{48}H_{40}$, Molybdate(V), bis(η^2-disulfido)di-μ-thio-dithiodi-, bis(tetraphenylphosphonium), 27:45
$S_9FO_3C_{10}H_8$, 2,2'-Bi-1,3-dithiolo[4,5-b]-[1,4]dithiinylidene fluorosulfate, 26:393
$S_{10.56}Mo_2P_2C_{48}H_{40}$, Molybdate(2−), thio-, $(Mo_2S_{10.56})^{2-}$ bis(tetraphenylphosphonium), 27:42
$S_{12}Mo_2N_2H_8 \cdot 2H_2O$, Molybdate(V), bis($\mu$-sulfido)tetrakis(disulfido)di-, diammonium, dihydrate, 27:48, 49
$S_{13}Mo_3N_2H_8 \cdot xH_2O$, Molybdate(IV), tris($\mu$-disulfido)tris(disulfido)-μ_3-thio-triangulo-tri-, diammonium, hydrate, 27:48, 49
$S_{14}F_{18}I_3Sb_3 \cdot 2AsF_3$, cyclo-Heptasulfur(3+), μ-iodo-bis(4-iodo-, tris[hexafluoroantimonate(1−)] - 2(arsenic trifluoride), 27:335
$S_{16}O_4ReC_{20}H_{16}$, Bis(2,2'-bi-1,3-dithiolo-[4,5-b][1,4]dithiinylidene) perrhenate, 26:391
$S_{32}As_6Br_4F_{36}$, cyclo-Heptasulfur(1+), bromo-, tetrasulfur(2+) hexafluoroarsenate(1−)(4:1:6), 27:338
$S_{32}As_6F_{36}I_4$, cyclo-Heptasulfur(1+), iodo-, tetrasulfur(2+) hexafluoroarsenate(1−)-(4:1:6), 27:337

$SbBrF_6S_7$, Antimonate(1−), hexafluoro-, bromo-cyclo-heptasulfur(1+), 27:336
$SbC_{18}H_{15}$, Stibine, triphenyl-, iron complex, 26:61
nickel complex, 28:103
SbF_6IS_7, Antimonate(1−), hexafluoro-, iodo-cyclo-heptasulfur(1+), 27:333
$SbFeO_4C_{22}H_{15}$, Iron(0), tetracarbonyl(triphenylstibine)-, 26:61, 28:171
$Sb_3F_{18}I_3S_{14} \cdot 2AsF_3$, Antimonate(1−), hexafluoro-, μ-iodo-bis(4-iodo-cyclo-heptasulfur)(3+)(3:1),—2(arsenic trifluoride), 27:335
$Sb_4NiC_{72}H_{60}$, Nickel(0), tetrakis(triphenylstibine)-, 28:103
$Sb_9H_{72}N_{18}NaO_{86}W_{21} \cdot 24H_2O$, Sodiohenicosa-tungstononaantimonate(18−), octadecaammonium, tetracosahydrate, 27:120
$ScCl_2LiO_2Si_4C_{30}H_{58}$, Scandium, bis[$\eta^5$-1,3-bis(trimethylsilyl)cyclopentadienyl]di-μ-chloro-bis(tetrahydrofuran)lithium-, 27:170
$ScO_3C_{43}H_{63}$, Scandium, tris(2,6-di-tert-butylphenoxo)-, 27:167
$ScO_3C_{45}H_{69}$, Scandium, tris(2,6-di-tert-butyl-4-methylphenoxo)-, 27:167
$Sc_2Cl_2Si_4C_{44}H_{84}$, Scandium, tetrakis[η^5-1,3-bis(trimethylsilyl)cyclopentadienyl]di-μ-chloro-di-, 27:171
Se, Silenium, Osmium, carbonyl clusters, 26:308
$Se_5TiC_{10}H_{10}$, Titanium(IV), bis(η^5-cyclo-pentadienyl)[pentaselenido(2−)]-, 27:61
$SiBrC_3H_9$, Silane, bromotrimethyl-, 26:4
SiC_4H_{12}, Silane, tetramethyl-, lutetium complex, 27:161
SiC_6H_{16}, Silane, triethyl-, ruthenium complex, 26:269
$SiC_{14}H_{26}$, Benzene, 1,2-bis[(trimethyl-silyl)methyl]-, lithium complex, 26:148
$SiHK_6O_{40}V_3W_9 \cdot 3H_2O$, 1,2,3-Trivanadononatungstosilicate(7−), A-β-, hexapotassium hydrogen, trihydrate, 27:129
$SiHNa_9O_{34}W_9 \cdot 23H_2O$, Nonatungsto-silicate(10−), β-, nonasodium

SiHNa$_9$O$_{34}$W$_9$·23H$_2$O (Continued)
  hydrogen, tricosahydrate, 27:88
SiH$_4$O$_{40}$W$_{12}$·xH$_2$O, Dodecatungstosilicic
  acid, α-, hydrate, 27:93
SiK$_4$O$_{40}$W$_{12}$·17H$_2$O,
  Dodecatungstosilicate(4−), α-,
  tetrapotassium, heptadecahydrate, 27:93
SiK$_4$O$_{40}$W$_{12}$·9H$_2$O,
  Dodecatungstosilicate(4−), β-,
  tetrapotassium, nonahydrate, 27:94
SiK$_8$O$_{36}$W$_{10}$·12H$_2$O, Decatungstosilicate-
  (8−), γ-, octapotassium, dodecahydrate,
  27:88
SiK$_8$O$_{39}$W$_{11}$·14H$_2$O,
  Undecatungstosilicate(8−), β$_2$-,
  octapotassium, tetradecahydrate, 27:91
SiK$_8$O$_{39}$W$_{11}$·13H$_2$O,
  Undecatungstosilicate(8−), α-,
  octapotassium, tridecahydrate, 27:89
SiLuOC$_{18}$H$_{29}$, Lutetium, bis(η^5-
  cyclopentadienyl)(tetrahydrofuran)[(tri-
  methylsilyl)methyl]-, 27:161
SiNC$_7$H$_{19}$, Ethanamine, 1,1-dimethyl-N-
  (trimethylsilyl)-, 27:327
SiN$_4$O$_{40}$TiV$_3$W$_9$C$_{69}$H$_{149}$, 1,2,3-
  Trivanadononatungstosilicate(4−), μ$_3$-
  [(η^5-cyclopentadienyl)trioxotitanate-
  (IV)]-, A-β-, tetrakis(tetrabutyl-
  ammonium), 27:132
SiN$_4$O$_{40}$V$_3$W$_9$C$_{64}$H$_{147}$, 1,2,3-
  Trivanadononatungstosilicate(7−), A-β-,
  tetrakis(tetrabutylammonium)
  trihydrogen, 27:131
SiN$_4$O$_{40}$W$_{12}$C$_{64}$H$_{144}$,
  Dodecatungstosilicate(4−), γ-,
  tetrakis(tetrabutylammonium), 27:95
SiNa$_8$O$_{39}$W$_{11}$, Undecatungstosilicate(8−),
  β$_1$-, octasodium, 27:90
SiNa$_{10}$O$_{34}$W$_9$, Nonatungstosilicate(10−), α-,
  decasodium, 27:87
SiOC$_4$H$_{12}$, Silane, methoxytrimethyl-, 26:44
SiPC$_{21}$H$_{39}$, Phosphine, (2,4,6-tri-*tert*-
  butylphenyl)(trimethylsilyl)-, 27:238
Si$_2$Br$_2$UC$_{22}$H$_{42}$, Uranium(IV), bis(η^5-1,3-
  bis(trimethylsilyl)cyclopentadienyl]di-
  bromo-, 27:174
Si$_2$C$_{11}$H$_{22}$, 1,3-Cyclopentadiene, 1,3-
  bis(trimethylsilyl)-, lanthanide-lithium
  complexes, 27:170
Si$_2$C$_{14}$H$_{26}$, Benzene, 1,2-
  bis[(trimethylsilyl)methyl]-, 26:148
Si$_2$Cl$_2$ThC$_{22}$H$_{42}$, Thorium(IV), bis(η^5-1,3-
  bis(trimethylsilyl)cyclopentadienyl]-
  dichloro-, 27:173
Si$_2$Cl$_2$UC$_{22}$H$_{42}$, Uranium(IV), bis(η^5-1,3-
  bis(trimethylsilyl)-
  cyclopentadienyl]dichloro-, 27:174
Si$_2$I$_2$UC$_{22}$H$_{42}$, Uranium(IV), bis(η^5-1,3-
  bis(trimethylsilyl)cyclopentadienyl]-
  diiodo-, 27:176
Si$_2$LiC$_{11}$H$_{21}$, Lithium, [η^5-1,3-
  bis(trimethylsilyl)cyclopentadienyl]-,
  27:170
Si$_2$Li$_2$N$_4$C$_{26}$H$_{56}$, Lithium, μ-[(α,α',1,2-η:-
  α,α',1,2-η)-1,2-phenylenebis[(trimethyl-
  silyl)methylene]]bis(N,N,N',N'-
  tetramethyl-1,2-ethane-diamine)di-,
  26:149
Si$_2$NC$_6$H$_{19}$, Silanamine, 1,1,1-trimethyl-N-
  (trimethylsilyl)-, ytterbium complex,
  27:148
Si$_2$NO$_{10}$P$_2$Ru$_3$C$_{58}$H$_{61}$, Ruthenate(1−),
  decacarbonyl-1κ^{3}C,2.kappa^{3}C,
  3.kappa^{4}C-μ-hydrido-1:2κ^{2}H-
  bis(triethylsilyl)-1κSi,2κSi-*triangulo*-tri-,
  μ-nitrido-bis(triphenylphosphorus)(1+),
  26:269
Si$_2$OPC$_{11}$H$_{27}$, Phosphine, [2,2-dimethyl-1-
  (trimethylsiloxy)propylidene](trimethylsilyl)-, 27:250
Si$_3$C$_{10}$H$_{28}$, Methane, tris(trimethylsilyl)-,
  27:238
Si$_3$Cl$_2$PC$_{10}$H$_{27}$, Phosphonous dichloride,
  [tris(trimethylsilyl)methyl]-, 27:239
Si$_3$PC$_9$H$_{27}$, Phosphine, tris(trimethylsilyl)-,
  27:243
Si$_4$Ce$_2$Cl$_2$C$_{44}$H$_{84}$, Cerium, tetrakis[η^5-1,3-
  bis(trimethylsilyl)cyclopentadienyl]di-μ-
  chloro-di-, 27:171
Si$_4$Cl$_2$CeLiO$_2$C$_{30}$H$_{58}$, Cerium, bis[η^5-1,3-
  bis(trimethylsilyl)cyclopentadienyl]di-μ-
  chloro-bis(tetrahydrofuran)lithium-,
  27:170
Si$_4$Cl$_2$Dy$_2$C$_{44}$H$_{84}$, Dysprosium, tetrakis[η^5-
  1,3-bis(trimethylsilyl)-
  cyclopentadienyl]di-μ-chloro-di-, 27:171
Si$_4$Cl$_2$Er$_2$C$_{44}$H$_{84}$, Erbium, tetrakis-
  [η^5-1,3-bis(trimethylsilyl)cyclopenta-
  dienyl]di-μ-chloro-di-, 27:171
Si$_4$Cl$_2$Eu$_2$C$_{44}$H$_{84}$, Europium,

tetrakis[η^5-1,3-bis(trimethylsilyl)cyclopentadienyl]di-μ-chloro-di-, 27:171

Si$_4$Cl$_2$Gd$_2$C$_{44}$H$_{84}$, Gadolinium, tetrakis[η^5-1,3-bis(trimethylsilyl)cyclopentadienyl]di-μ-chloro-di-, 27:171

Si$_4$Cl$_2$Ho$_2$C$_{44}$H$_{84}$, Holmium, tetrakis[η^5-1,3-bis(trimethylsilyl)cyclopentadienyl]di-μ-chloro-di-, 27:171

Si$_4$Cl$_2$LaLiO$_2$C$_{30}$H$_{58}$, Lanthanum, bis[η^5-1,3-bis(trimethylsilyl)cyclopentadienyl]di-μ-chloro-bis(tetrahydrofuran)lithium-, 27:170

Si$_4$Cl$_2$La$_2$C$_{44}$H$_{84}$, Lanthanum, tetrakis[η^5-1,3-bis(trimethylsilyl)cyclopentadienyl]di-μ-chloro-di-, 27:171

Si$_4$Cl$_2$LiNdO$_2$C$_{30}$H$_{58}$, Neodymium, bis[η^5-1,3-bis(trimethylsilyl)cyclopentadienyl]di-μ-chloro-bis(tetrahydrofuran)lithium-, 27:170

Si$_4$Cl$_2$LiO$_2$PrC$_{30}$H$_{58}$, Praseodymium, bis[η^5-1,3-bis(trimethylsilyl)cyclopentadienyl]di-μ-chloro-bis(tetrahydrofuran)lithium-, 27:170

Si$_4$Cl$_2$LiO$_2$ScC$_{30}$H$_{58}$, Scandium, bis[η^5-1,3-bis(trimethylsilyl)cyclopentadienyl]di-μ-chloro-bis(tetrahydrofuran)lithium-, 27:170

Si$_4$Cl$_2$LiO$_2$YC$_{30}$H$_{58}$, Yttrium, bis[η^5-1,3-bis(trimethylsilyl)cyclopentadienyl]di-μ-chloro-bis(tetrahydrofuran)lithium-, 27:170

Si$_4$Cl$_2$LiO$_2$YbC$_{30}$H$_{58}$, Ytterbium, bis[η^5-1,3-bis(trimethylsilyl)cyclopentadienyl]di-μ-chloro-bis(tetrahydrofuran)lithium-, 27:170

Si$_4$Cl$_2$Lu$_2$C$_{44}$H$_{84}$, Lutetium, tetrakis[η^5-1,3-bis(trimethylsilyl)cyclopentadienyl]di-μ-chloro-di-, 27:171

Si$_4$Cl$_2$Nd$_2$C$_{44}$H$_{84}$, Neodymium, tetrakis[η^5-1,3-bis(trimethylsilyl)cyclopentadienyl]di-μ-chloro-di-, 27:171

Si$_4$Cl$_2$Pr$_2$C$_{44}$H$_{84}$, Praseodymium, tetrakis[η^5-1,3-bis(trimethylsilyl)cyclopentadienyl]di-μ-chloro-di-, 27:171

Si$_4$Cl$_2$Sc$_2$C$_{44}$H$_{84}$, Scandium, tetrakis[η^5-1,3-bis(trimethylsilyl)cyclopentadienyl]di-μ-chloro-di-, 27:171

Si$_4$Cl$_2$Sm$_2$C$_{44}$H$_{84}$, Samarium, tetrakis[η^5-1,3-bis(trimethylsilyl)cyclopentadienyl]di-μ-chloro-di-, 27:171

Si$_4$Cl$_2$Tb$_2$C$_{44}$H$_{84}$, Terbium, tetrakis[η^5-1,3-bis(trimethylsilyl)cyclopentadienyl]di-μ-chloro-di-, 27:171

Si$_4$Cl$_2$Tm$_2$C$_{44}$H$_{84}$, Thulium, tetrakis[η^5-1,3-bis(trimethylsilyl)cyclopentadienyl]di-μ-chloro-di-, 27:171

Si$_4$Cl$_2$Y$_2$C$_{44}$H$_{84}$, Yttrium, tetrakis[η^5-1,3-bis(trimethylsilyl)cyclopentadienyl]di-μ-chloro-di-, 27:171

Si$_4$Cl$_2$Yb$_2$C$_{44}$H$_{84}$, Ytterbium, tetrakis[η^5-1,3-bis(trimethylsilyl)cyclopentadienyl]di-μ-chloro-di-, 27:171

Si$_4$N$_2$O$_2$YbC$_{16}$H$_{46}$, Ytterbium, bis[bis(trimethylsilyl)amido]bis(diethyl ether)-, 27:148

Si$_6$P$_2$C$_{20}$H$_{54}$, Diphosphene, bis[tris(trimethylsilyl)methyl]-, 27:241

SmCl$_3$O$_2$C$_8$H$_{16}$, Samarium, trichlorobis(tetrahydrofuran)-, 27:140

SmCl$_3$·2C$_4$H$_8$O, Samarium trichloride–2tetrahydrofuran, 27:140, 28:290

SmOC$_{19}$H$_{23}$, Samarium(III), tris(η^5-cyclopentadienyl)(tetrahydrofuran)-, 26:21

SmO$_2$C$_{28}$H$_{46}$, Samarium(II), bis(η^5-pentamethylcyclopentadienyl)bis(tetrahydrofuran)-, 27:155

SmO$_3$C$_{42}$H$_{63}$, Samarium, tris(2,6-di-*tert*-butylphenoxo)-, 27:166

Sm$_2$Cl$_2$Si$_4$C$_{44}$H$_{84}$, Samarium, tetrakis[η^5-1,3-bis(trimethylsilyl)cyclopentadienyl]di-μ-chloro-di-, 27:171

SnCl$_6$P$_6$C$_{52}$H$_{48}$, Stannate(1−), hexachloro-, 3,3,4,5-tetrahydro-1,1,3,3-tetraphenyl-1*H*-1,2,3-triphospholium, 27:255

Tb$_2$Cl$_2$Si$_4$C$_{44}$H$_{84}$, Terbium, tetrakis[η^5-1,3-bis(trimethylsilyl)cyclopentadienyl]di-μ-chloro-di-, 27:171

ThClC$_{15}$H$_{15}$, Thorium(IV), chlorotris(η^5-cyclopentadienyl)-, 28:302

ThCl$_2$Si$_2$C$_{22}$H$_{42}$, Thorium(IV), bis(η^5-1,3-bis(trimethylsilyl)cyclopentadienyl]dichloro-, 27:173

ThCl$_4$, Thorium tetrachloride, 28:322

TiClC$_{10}$H$_{10}$, Titanium(III), chlorobis(η^5-cyclopentadienyl)-, 28:261

TiN$_4$O$_{40}$SiV$_3$W$_9$C$_{69}$H$_{149}$, 1,2,3-Trivanadononatungstosilicate(4−), μ_3-[(η^5-cyclopentadienyl)trioxo-

$TiN_4O_{40}SiV_3W_9C_{69}H_{149}$ (Continued) titanate(IV)]-, A-β-, tetrakis(tetrabutylammonium), 27:132

$TiO_2C_{12}H_{10}$, Titanium, dicarbonylbis(η^5-cyclopentadienyl)-, 28:250

$TiO_2C_{22}H_{30}$, Titanium, dicarbonylbis(η^5-pentamethylcyclopentadienyl)-, 28:253

$TiS_2C_{10}H_{12}$, Titanium(IV), bis(η^5-cyclopentadienyl)bis(hydrogen sulfido)-, 27:66

$TiS_3C_{20}H_{30}$, Titanium(IV), bis(η^5-pentamethylcyclopentadienyl)[trisulfido(2−)]-, 27:62

$TiS_5C_{10}H_{10}$, Titanium(IV), bis(η^5-cyclopentadienyl)[pentasulfido(2−)]-, 27:60

$TiS_5C_{12}H_{14}$, Titanium, bis(η^5-methylcyclopentadienyl)(pentasulfido-S^1:S^5)-, 27:52

$TiSe_5C_{10}H_{10}$, Titanium(IV), bis(η^5-cyclopentadienyl)[pentaselenido(2−)]-, 27:61

$Ti_2Br_9Cs_3$, Titanate(3−), nonabromodi-, tricesium, 26:379

$Ti_2Br_9Rb_3$, Titanate(3−), nonabromodi-, trirubidium, 26:379

$Ti_2Cl_9Cs_3$, Titanate(3−), nonachlorodi-, tricesium, 26:379

$Ti_2Cl_9Rb_3$, Titanate(3−), nonachlorodi-, tricesium, 26:379

TlC_5H_5, Thallium, cyclopentadienyl-, 28:315

$Tm_2Cl_2Si_4C_{44}H_{84}$, Thulium, tetrakis[η^5-1,3-bis(trimethylsilyl)cyclopentadienyl]di-μ-chloro-di-, 27:171

$UBr_2Si_2C_{22}H_{42}$, Uranium(IV), bis(η^5-1,3-bis(trimethylsilyl)cyclopentadienyl]dibromo-, 27:174

$UClC_{15}H_{15}$, Uranium(IV), chlorotris(η^5-cyclopentadienyl)-, 28:301

$UCl_2Si_2C_{22}H_{42}$, Uranium(IV), bis(η^5-1,3-bis(trimethylsilyl)cyclopentadienyl]dichloro-, 27:174

$UI_2Si_2C_{22}H_{42}$, Uranium(IV), bis(η^5-1,3-bis(trimethylsilyl)cyclopentadienyl]diiodo-, 27:176

$UPC_{24}H_{27}$, Uranium(IV), tris(η^5-cyclopentadienyl)[(dimethylphenylphosphoranylidene)methyl]-, 27:177

$VC_{10}H_{10}$, Vanadocene, 28:263

$VClC_{10}H_{10}$, Vanadium(III), chlorobis(η^5-cyclopentadienyl)-, 28:262

$VK_4O_{40}PW_{11} \cdot xH_2O$, Vanadoundecatungstophosphate(4−), α-, tetrapotassium, hydrate, 27:99

$VN_2O_{10}S_2C_{14}H_{16} \cdot 2H_2O$, Vanadium(II), tetraaquabis(1,2-benzisothiazol-3(2H)-one 1,1-dioxidato)-, dihydrate, 27:307

$VN_6O_6S_2C_{34}H_{28} \cdot 2NC_5H_5$, Vanadium(III), bis(1,2-benzisothiazol-3(2H)-one 1,1-dioxidato)tetrakis(pyridine)-, −2pyridine, 27:308

$V_2Br_9Cs_3$, Vanadate(3−), nonabromodi-, tricesium, 26:379

$V_2Br_9Rb_3$, Vanadate(3−), nonabromodi-, trirubidium, 26:379

$V_2Cl_9Cs_3$, Vanadate(3−), nonachlorodi-, tricesium, 26:379

$V_2Cl_9Rb_3$, Vanadate(3−), nonachlorodi-, trirubidium, 26:379

$V_2Cs_5O_{40}PW_{10}$, Divanadodecatungstophosphate(5−), γ-, pentacesium, 27:102

$V_2S_4C_{12}H_{14}$, Vanadium, (μ-disulfido-S:S')-bis(η^5-methylcyclopentadienyl)-di-μ-thio-di-, 27:55

$V_2S_5C_{12}H_{14}$, Vanadium, (μ-disulfido-S:S')-(μ-η^2:η^2-disulfido)bis(η-methylcyclopentadienyl)-μ-thio-di-, 27:54

$V_3Cs_6O_{40}PW_9$, Trivanadononatungstophosphate(6−), α-1,2,3-, hexacesium, 27:100

$V_3HK_6O_{40}SiW_9 \cdot 3H_2O$, 1,2,3-Trivanadononatungstosilicate(7−), A-β-, hexapotassium hydrogen, trihydrate, 27:129

$V_3N_4O_{40}SiTiW_9C_{69}H_{149}$, 1,2,3-Trivanadononatungstosilicate(4−), μ_3-[(η^5-cyclopentadienyl)trioxotitanate-(IV)]-, A-β-, tetrakis(tetrabutylammonium), 27:132

$V_3N_4O_{40}SiW_9C_{64}H_{147}$, 1,2,3-Trivanadononatungstosilicate(7−), A-β-, tetrakis(tetrabutylammonium) trihydrogen, 27:131

$V_{10}N_3O_{28}C_{48}H_{111}$, Decavanadate(V), tris(tetrabutylammonium) trihydrogen, 27:83

WBF$_4$NO$_5$C$_{10}$H$_{10}$, Tungsten(1+), pentacarbonyl[(diethylamino)methylidyne]-, tetrafluoroborate(1−), 26:40

WBF$_4$O$_2$PC$_{25}$H$_{20}$, Tungsten, dicarbonyl(η^5-cyclopentadienyl)[tetrafluoroborato-(1−)]- (triphenylphosphine)-, 26:98

WBF$_4$O$_3$C$_8$H$_5$, Tungsten, tricarbonyl(η^5-cyclopentadienyl)[tetrafluoroborato-(1−)]-, 26:96, 28:5

WBF$_4$O$_4$C$_{11}$H$_{11}$, Tungsten(1+), (acetone)tricarbonyl(η^5-cyclopentadienyl)-, tetrafluoroborate(1−), 26:105

WB$_2$F$_8$N$_6$O$_2$C$_8$H$_{12}$, Tungsten(II), tetrakis(acetonitrile)dinitrosyl-, cis-, bis[tetrafluoroborate(1−)], 26:133, 28:66

WC$_{20}$H$_{42}$, Tungsten(VI), tris(2,2-dimethylpropyl) (2,2-dimethylpropylidyne)-, 26:47

WCl$_2$N$_2$P$_2$C$_{16}$H$_{32}$, Tungsten(VI), dichloro[(1,1-dimethylethyl)imido](phenylimido)bis(trimethylphosphine)-, 27:304

WCl$_2$N$_4$C$_{20}$H$_{22}$, Tungsten(VI), (2,2′-bipyridine)dichloro[(1,1-dimethylethyl)imido] (phenylimido)-, 27:303

WCl$_2$P$_4$C$_{12}$H$_{36}$, Tungsten(II), dichlorotetrakis(trimethylphosphine)-, 28:329

WCl$_2$P$_4$C$_{32}$H$_{44}$, Tungsten(II), dichlorotetrakis(dimethylphenylphosphine)-, 28:330

WCl$_2$P$_4$C$_{52}$H$_{52}$, Tungsten(II), dichlorotetrakis(methyldiphenylphosphine)-, 28:331

WCl$_3$O$_2$C$_9$H$_{19}$, Tungsten(VI), trichloro(1,2-dimethoxyethane)(2,2-dimethylpropylidyne)-, 26:50

WCl$_3$O$_3$C$_3$H$_9$, Tungsten, trichlorotrimethoxy-, 26:45

WCl$_4$O, Tungsten tetrachloride oxide, 28:324

WCl$_4$P$_2$C$_{26}$H$_{24}$, Tungsten, tetrachloro[1,2-ethanediylbis(diphenylphosphine)]-, 28:41

WCl$_4$P$_2$C$_{26}$H$_{26}$, Tungsten(IV), tetrachlorobis(methyldiphenylphosphine)-, 28:328

WCl$_4$P$_2$C$_{36}$H$_{30}$, Tungsten, tetrachlorobis(triphenylphosphine)-, 28:40

WCl$_4$P$_3$C$_9$H$_{27}$, Tungsten(IV), tetrachlorotris(trimethylphosphine)-, 28:327

WCl$_4$, Tungsten tetrachloride, 26:221

WFeNO$_9$C$_{45}$H$_{31}$, Tungstate(1−), hydridononacarbonyliron-, μ-nitrido-bis(triphenylphosphorus)(1+), 26:336

WFeN$_2$O$_9$P$_4$C$_{81}$H$_{60}$, Tungstate(2−), nonacarbonyliron-, bis[μ-nitrido-bis(triphenylphosphorus)(1+), 26:339

WH$_4$P$_4$C$_{52}$H$_{52}$, Tungsten(IV), tetrahydridotetrakis(methyldiphenylphosphine)-, 27:10

WH$_6$P$_3$C$_{24}$H$_{33}$, Tungsten(IV), hexahydridotris(dimethylphenylphosphine)-, 27:11

WNO$_3$C$_7$H$_5$, Tungsten, dicarbonyl(η^5-cyclopentadienyl)nitrosyl-, 28:196

WNO$_5$C$_{10}$H$_9$, Tungsten, (tert-butyl isocyanide)pentacarbonyl-, 28:143

WNO$_7$P$_2$C$_{43}$H$_{33}$, Tungstate(1−), (acetato)pentacarbonyl-, μ-nitrido-bis(triphenylphosphorus)(1+), 27:297

WN$_2$O$_4$C$_{14}$H$_{18}$, Tungsten, bis(tert-butyl isocyanide)tetracarbonyl-, cis-, 28:143

WN$_2$O$_5$C$_{10}$H$_{10}$, Tungsten, tetracarbonyl[(diethylamino)methylidyne](isocyanato)-, trans-, 26:42

WN$_2$P$_4$C$_{56}$H$_{54}$, Tungsten(0), bis[1,2-ethanediylbis(diphenylphosphine)]bis(isocyanomethane)-, trans-, 28:43

WN$_3$O$_3$C$_{12}$H$_{15}$, Tungsten, tricarbonyltris(propanenitrile)-, 27:4, 28:30

WN$_3$O$_3$C$_{18}$H$_{27}$, Tungsten, tris(tert-butyl isocyanide)tricarbonyl-, fac-, 28:143

WN$_4$P$_4$C$_{48}$H$_{52}$, Tungsten, bis(dinitrogen)bis[1,2-ethanediylbis(diphenylphosphine)]-, trans-, 28:41

WNaO$_3$C$_8$H$_5 \cdot$ 2C$_4$H$_{10}$O$_2$, Tungstate(1−), tricarbonyl(η^5-cyclopentadienyl)-, sodium, compd. with 1,2-dimethoxyethane(1:2), 26:343

WO$_2$PC$_{25}$H$_{21}$, Tungsten, dicarbonyl(η^5-cyclopentadienyl)hydrido(triphenylphosphine)-, 26:98

WO$_3$C$_{10}$H$_9$, Tungsten, tricarbonyl(η^6-

$WO_3C_{10}H_9$ (Continued) cycloheptatriene)-, 27:4

$WO_3P_2C_{21}H_{44}$, Tungsten, tricarbonyl(dihydrogen)bis(triisopropylphosphine)-, 27:7

$WO_3P_2C_{39}H_{68}$, Tungsten, tricarbonyl(dihydrogen)bis(tricyclohexylphosphine)-, 27:6

$WS_2C_{10}H_{12}$, Tungsten(IV), bis(η^5-cyclopentadienyl)bis(hydrogen sulfido)-, 27:67

$W_2Cl_4N_6C_{28}H_{25}$, Tungsten(VI), tetrachlorbis(1,1-dimethylethanamine)bis[(1,1-dimethylethyl)imido]bis(μ-phenylimido)di-, 27:301

$W_2F_{12}O_8C_8$, Tungsten(II), tetrakis(trifluoroacetato)di-, (W–4–W), 26:222

$W_2O_4C_{14}H_{10}$, Tungsten, tetracarbonylbis(η^5-cyclopentadienyl)di-, (W–W), 28:153

$W_2O_6C_{16}H_{10}$, Tungsten, hexacarbonylbis(η^5-cyclopentadienyl)di-, 28:148

$W_2O_8C_8H_{12}$, Tungsten(II), tetrakis(acetato)di-, (W–4–W), 26:224

$W_2O_8C_{20}H_{36}$, Tungsten(II), tetrakis(2,2-dimethylpropanoato)di-(W–4–W), 26:223

$W_5Cs_6O_{23}P_2$, Pentatungstodiphosphate(6–), hexacesium, 27:101

$W_5N_4O_{21}P_2C_{60}H_{122}$, Pentatungstobis(phenylphosphonate)(4–), tetrakis(tributylammonium), 27:127

$W_6N_2O_{19}C_{32}H_{72}$, Hexatungstate(VI), bis(tetrabutylammonium), 27:80

$W_9Cs_6O_{40}PV_3$, Trivanadononatungstophosphate(6–), α-1,2,3-, hexacesium, 27:100

$W_9HK_6O_{40}SiV_3 \cdot 3H_2O$, 1,2,3-Trivanadononatungstosilicate(7–), A-β-, hexapotassium hydrogen, trihydrate, 27:129

$W_9HNa_9O_{34}Si \cdot 23H_2O$, Nonatungstosilicate(10–), β-, nonasodium hydrogen, tricosahydrate, 27:88

$W_9N_4O_{40}SiTiV_3C_{69}H_{149}$, 1,2,3-Trivanadononatungstosilicate(4–), μ_3-[(η^5-cyclopentadienyl)trioxotitanate(IV)]-, A-β-, tetrakis(tetrabutylammonium), 27:132

$W_9N_4O_{40}SiV_3C_{64}H_{147}$, 1,2,3-Trivanadononatungstosilicate(7–), A-β-, tetrakis(tetrabutylammonium) trihydrogen, 27:131

$W_9Na_9O_{34}P \cdot xH_2O$, Nonatungstophosphate(9–), A-, nonasodium, hydrate, 27:100

$W_9Na_{10}O_{34}Si$, Nonatungstosilicate(10–), α-, decasodium, 27:87

$W_{10}Cs_5O_{40}PV_2$, Divanadodecatungstophosphate(5–), γ-, pentacesium, 27:102

$W_{10}Cs_7O_{36}P$, Decatungstophosphate(7–), hexacesium, 27:101

$W_{10}K_8O_{36}Si \cdot 12H_2O$, Decatungstosilicate(8–), γ-, octapotassium, dodecahydrate, 27:88

$W_{10}N_4O_{32}C_{64}H_{144}$, Decatungstate(VI), tetrakis(tetrabutylammonium), 27:81

$W_{11}K_4O_{40}PV \cdot xH_2O$, Vanadoundecatungstophosphate(4–), α-, tetrapotassium, hydrate, 27:99

$W_{11}K_8O_{39}Si \cdot 14H_2O$, Undecatungstosilicate(8–), β_2-, octapotassium, tetradecahydrate, 27:91

$W_{11}K_8O_{39}Si \cdot 13H_2O$, Undecatungstosilicate(8–), α-, octapotassium, tridecahydrate, 27:89

$W_{11}Na_8O_{39}Si$, Undecatungstosilicate(8–), β_1-, octasodium, 27:90

$W_{12}H_4O_{40}Si \cdot xH_2O$, Dodecatungstosilicic acid, α-, hydrate, 27:93

$W_{12}K_4O_{40}Si \cdot 17H_2O$, Dodecatungstosilicate(4–), α-, tetrapotassium, heptadecahydrate, 27:93

$W_{12}K_4O_{40}Si \cdot 9H_2O$, Dodecatungstosilicate(4–), β-, tetrapotassium, nonahydrate, 27:94

$W_{12}N_4O_{40}SiC_{64}H_{144}$, Dodecatungstosilicate(4–), γ-, tetrakis(tetrabutylammonium), 27:95

$W_{15}Na_{12}O_{56}P_2 \cdot 24H_2O$, Pentadecatungstodiphosphate(12–), α-, dodecasodium, tetracosahydrate, 27:108

$W_{17}K_{10}O_{61}P_2 \cdot 20H_2O$, Heptadecatungstodiphosphate(10–),

α_2-, decapotassium, eicosahydrate, 27:107

$W_{17}LiK_9O_{61}P_2 \cdot 20H_2O$, Lithioheptadecatungstodiphosphate-(9−), α_1-, nonapotassium, eicosahydrate, 27:109

$W_{18}K_6O_{62}P_2 \cdot 19H_2O$, Octadecatungstodiphosphate(6−), β-, hexapotassium, nonadecahydrate, 27:105

$W_{18}K_6O_{62}P_2 \cdot 14H_2O$, Octadecatungstodiphosphate(6−), α-, hexapotassium, tetradecahydrate, 27:105

$W_{21}As_2H_4O_{70}Rb_4 \cdot 34H_2O$, Tungstate(4−), aquadihydroxohenhexacontaoxobis[trioxoarsenato(III)]henicosa-, tetrarubidium, tetratricontahydrate, 27:113

$W_{21}As_2H_8O_{70} \cdot xH_2O$, Tungsten, aquahexahydroxoheptapentacontaoxobis-[trioxoarsenato(III)]henicosa-, hydrate, 27:112

$W_{21}H_{72}N_{18}NaO_{86}Sb_9 \cdot 24H_2O$, Sodiohenicosatungstononaantimonate (18−), octadecaammonium, tetracosahydrate, 27:120

$W_{30}H_{64}N_{14}NaO_{110}P_5 \cdot 31H_2O$, Sodiotricontatungstopentaphosphate-(14−), tetradecaammonium, hentricontahydrate, 27:115

$W_{40}As_4Co_2H_{100}N_{24}O_{142} \cdot 19H_2O$, Ammoniodicobaltotetracontatungstotetraarsenate(23−), tricosaammonium, nonadecahydrate, 27:119

$W_{40}As_4Na_{28}O_{140} \cdot 60H_2O$, Sodiotetracontatungstotetraarsenate-(27−), heptacosasodium, hexacontahydrate, 27:118

$W_{48}H_7K_{28}Li_5O_{184}P_8 \cdot 92H_2O$, Octatetracontatungstooctaphosphate(40−), pentalithium octacosapotassium heptahydrogen, dononacontahydrate, 27:110

$YCl_2LiO_2Si_4C_{30}H_{58}$, Yttrium, bis[$\eta^5$-1,3-bis(trimethylsilyl)cyclopentadienyl]di-μ-chloro-bis(tetrahydrofuran)lithium-, 27:170

$YO_3C_{43}H_{63}$, Yttrium, tris(2,6-di-*tert*-butylphenoxo)-, 27:167

$YO_3C_{45}H_{69}$, Yttrium, tris(2,6-di-*tert*-butyl-4-methylphenoxo)-, 27:167

$Y_2Cl_2Si_4C_{44}H_{84}$, Yttrium, tetrakis[η^5-1,3-bis(trimethylsilyl)cyclopentadienyl]di-μ-chloro-di-, 27:171

$YbC_{16}H_{10}$, Ytterbium(II), bis(phenylethynyl)-, 27:143

$YbCl_2LiO_2Si_4C_{30}H_{58}$, Ytterbium, bis[$\eta^5$-1,3-bis(trimethylsilyl)cyclopentadienyl]-di-μ-chloro-bis(tetrahydrofuran)lithium-, 27:170

$YbCl_3O_3C_{12}H_{24}$, Ytterbium, trichlorotris(tetrahydrofuran)-, 27:139, 28:289

$YbCl_3 \cdot 3C_4H_8O$, Ytterbium trichloride–3tetrahydrofuran, 27:139, 28:289

YbI_2, Ytterbium diiodide, 27:147

$YbN_2O_2Si_4C_{16}H_{46}$, Ytterbium, bis[bis(trimethylsilyl)amido]bis(diethyl ether)-, 27:148

$YbOC_{24}H_{40}$, Ytterbium, (diethyl ether)bis(η^5-pentamethylcyclopentadienyl)-, 27:148

$YbO_2C_{14}H_{20}$, Ytterbium(II), bis(η^5-cyclopentadienyl)(1,2-dimethoxyethane)-, 26:22

$YbO_3C_{45}H_{69}$, Ytterbium, tris(2,6-di-*tert*-butyl-4-methylphenoxo)-, 27:167

$Yb_2Cl_2Si_4C_{44}H_{84}$, Ytterbium, tetrakis[η^5-1,3-bis(trimethylsilyl)cyclopentadienyl]-di-μ-chloro-di-, 27:171

$ZnCl_2$, Zinc dichloride, 28:322

$ZrC_{10}H_{12}$, Zirconium, bis(η^5-cyclopentadienyl)dihydrido-, 28:257

$ZrClC_{10}H_{11}$, Zirconium, chlorobis(η^5-cyclopentadienyl)hydrido-, 28:259

$ZrH_4OsP_3C_{34}H_{43}$, Zirconium, bis[1,1(η^5)-cyclopentadienyl]tris(dimethylphenylphosphine-2κP)-tri-μ-hydrido-hydrido-1κH-osmium-, 27:27

$ZrO_2C_{12}H_{10}$, Zirconium, dicarbonylbis(η^5-cyclopentadienyl)-, 28:251

$ZrO_2C_{22}H_{30}$, Zirconium, dicarbonylbis(η^5-pentamethylcyclopentadienyl)-, 28:254